ANNUAL REVIEW OF
PLANT PHYSIOLOGY

ANNUAL REVIEW OF PLANT PHYSIOLOGY

WINSLOW R. BRIGGS, *Editor*
Harvard University

PAUL B. GREEN, *Associate Editor*
Stanford University

RUSSELL L. JONES, *Associate Editor*
University of California, Berkeley

VOLUME 24

1973

ANNUAL REVIEWS INC.   4139 EL CAMINO WAY   PALO ALTO, CALIFORNIA 94306

ANNUAL REVIEWS INC.
Palo Alto, California, USA

International Standard Book Number: 0-8243-0624-4
Library of Congress Catalog Card Number: A51-1660

Assistant Editor	Jean Heavener
Indexers	Mary Glass
	Leigh Dowling
Subject Indexers	Melissa Chapin
	Terry Chapin
Compositor	George Banta Co., Inc.

PRINTED AND BOUND IN THE UNITED STATES OF AMERICA
BY GEORGE BANTA COMPANY, INC.

PREFACE

Those of you who regularly read the preface to the *Annual Review of Plant Physiology* volumes are aware that during his many years of tenure as Editor of this *Review*, Leonard Machlis used the preface to thank various individuals—retiring members of the Editorial Committee, the Assistant Editor, and the authors who year after year accept the difficult task of preparing the reviews. I wish to continue this practice and in this case to begin by thanking Leonard Machlis himself. He was with the *Annual Review of Plant Physiology* from Volume 1, as Associate Editor from 1950 through 1958, and as Editor from 1958 until his resignation last year. His impact on plant physiology through this *Review* was enormous. I have worked with him as an Associate Editor since 1960 and shall sorely miss him. I only hope that I can maintain his high standards.

We shall also miss Roderic Park, who has resigned as Associate Editor. His contributions in this capacity have been substantial, his sound advice on topics and authors valuable, and his good editorial sense extremely useful. We welcome his replacement, Paul Green, and my replacement as Associate Editor, Russell Jones. Bruce Stowe is also leaving our advisory group, having completed his five-year term as a member of the Editorial Committee. We are very grateful for his careful consideration of topics and authors and his perceptive evaluation of people far outside his own field. We welcome his replacement, George Cheniae, and look forward to working with him for the coming five years.

Thus this year the Editor is doing his job for the first time, and the Associate Editors, Paul Green and Russell Jones, are doing their jobs for the first time. With the continued willingness of plant physiologists who are experts in the many diverse corners of the field to haunt the library and organize their reviews, we will continue to produce a volume of central importance to plant physiologists. We feel fortunate to have as Assistant Editor of this *Review* Jean Heavener, whose careful technical editing, persistence with tardy authors and editors, and guidance through the production process are invaluable in bringing the volume to eventual publication.

WINSLOW R. BRIGGS, *Editor*

CONTENTS

Paul J. Kramer

Ann. Rev. Plant Physiol. 1973. 24:1–24

SOME REFLECTIONS AFTER 40 YEARS ❖ 7539
IN PLANT PHYSIOLOGY

Paul J. Kramer

Department of Botany, Duke University, Durham, North Carolina

CONTENTS

INTRODUCTION

Leonard Machlis, former editor, informed me that "The prefatory chapter is to be a personal, philosophical, or historical essay . . . " and not a scientific review.

This is a difficult charge because to reminisce is to risk boring younger readers and being corrected by older ones, and a personal approach also bores some readers. Nevertheless, I shall risk the dangers of all three approaches. Some reference to the history of plant physiology seems desirable both because I find history interesting and because I believe it is impossible to understand the present situation of science without some knowledge of its past. Kierkegaarde, the Danish philosopher, spoke truly when he said that "Life can only be understood backward, but it can only be lived forward."

Nothing is Completely New

One of the clearest lessons from my experience and reading is that almost nothing is completely new. Too often people think they are making important new discoveries when they are merely rediscovering what has been known for decades or even centuries. Today's problems of population, poverty, and pollution seem new and unique, but they really are old problems in new forms. Scientists often are as bad as nonscientists in this respect. For example, criticism of science and scientists, particularly by the literary community, is not new. Writers, poets, and politicians of England during the 18th and early part of the 19th century were very critical of science, and Keats is said to have complained that Newton destroyed all the poetry of the rainbow by explaining how the colors are produced. After the battle between two ironclad warships in 1862, Henry Adams wrote: "Some day science may have the existence of mankind in its power and the human race will commit suicide by blowing up the world." Adams' prediction was correct in the sense that science has provided man with the knowledge and through technological development the power to blow himself up or to make himself healtheir and happier, as he chooses.

A Period of Rapid Change

The 40 years since I left graduate school probably have covered the period of most drastic change in the history of man. The changes have affected every area of life—social, economic, political, religious, and scientific—and they have affected everyone—men, women, and children, laborers and business men, priests and professors. Looking back, the dominant impression is the astonishing increase in size or amount of almost everything, except perhaps human happiness. Accompanying the increases in numbers have been increases in complexity of the problems of individual people, of cities, and of states. There have been notable advances in standards of living, housing, transportation, medical care, social services, and perhaps even in the concern of society for the welfare of the individual. However, it is doubtful if these advances have brought any significant increase in human happiness. Plumbing and central heating are more convenient than outdoor privies and coal stoves, autos are more convenient than horsedrawn vehicles, and television provides more alleged entertainment in a week than my parents saw in a year, but possession of these conveniences does not ensure happiness. Furthermore, the easy availability of so many material things seems to have decreased the initiative of some young people.

Unfortunately, many of the benefits provided by social and technological advances seem to have been cancelled by the problems produced by increase in population. Nowhere is this more evident than in the education of our children. I first attended one-room country schools, then a village school where the only teaching aids were a small library, a few maps, and a hand-cranked Victrola. However, my education was presided over by dedicated and fairly competent teachers who ruled their classrooms with a firm hand and permitted no disruption of the learning process by unruly students. As a result all of us learned at least the basic essentials, so far as our genotypes would permit. Today the lack of discipline in crowded public schools has largely cancelled out the possible benefits from the great improvements in curricula and textbooks and the increase in teaching aids which are provided. I am glad that I will not have to teach college students from these schools 10 years from now.

My only consolation, and a poor one, is that disorder in schools is not new either. At the country school attended by my father the first requisite of the teacher was enough brawn to be able to thrash the largest boys. Otherwise he would have been unable to teach effectively. Also, in many respects college boys were rougher and more unruly 50 or 75 years ago than they are today. Hazing, which sometimes endangered limbs and occasionally even life, and crude pranks that resulted in destruction of property were commonplace. There seems to be much less physical cruelty practiced today by college students, but whether there is less mental cruelty, I cannot say. However, it seems certain that physically hazing fellow students did less damage than enticing them into the use of drugs or persuading them to become "dropouts" from society.

The rate of change in the field of science and in my field of plant physiology during the past 40 years has been as rapid and as drastic as in other areas. To me the most notable changes have been the rapid increase in number of plant physiologists, the increase in amount, complexity, and cost of equipment, the increase in specialization, and the tremendous increase in volume of publication. Other important changes are the increased involvement of scientists in national and international affairs and the accompanying increase in amount of travel. Later we will examine some of these change in more detail and discuss how they have produced the situation in which plant physiology and science in general finds itself today.

Rate of Appearance of New Ideas

First, however, we might consider whether or not evaluation of change in terms of increases in numbers of workers and publications is too superficial. Although it is an easy test, many scientists are willing to admit that it is an unsatisfactory one. The important question concerns the rate at which really new and important ideas appear among the increasing numbers of publications. It seems probable that the number of important new ideas is not increasing nearly as rapidly as the output of scientific literature.

It is true that progress was very slow in the 18th and first half of the 19th century, but there were few workers and plant physiology was only an avocation for most of them. It was 50 to 75 years after Hales published his *Vegetable Staticks* in 1727 before Ingenhousz, Senebier, de Saussure, and others made important contributions to the understanding of photosynthesis and plant nutrition. In 1826 Dutrochet published his theory of osmosis, and in the middle of the 19th century Boussingault made important contributions to plant physiology. However, there was a rapid expansion, almost a burst of activity, in the second half of the 19th century, led by Sachs and his contemporaries. As a result, by 1900 there was a fairly good understanding of plant structure and of the major physiological processes of plants such as photosynthesis, respiration, translocation, mineral nutrition, and plant water relations. The cohesion theory of the ascent of sap had been developed and the occurrence of growth regulators was predicted.

Much important work was done in plant physiology during the first 30 years of the 20th century, but it was largely an extension of concepts and processes recognized and described in the 19th century. The work on permeability and salt accumulation was a continuation of work begun by Traube, Pfeffer, de Vries, and others in the second half of the 19th century. Ten elements already were regarded as essential to plants in 1900, and additions to this list resulted mostly from improvements in experimental technique. The idea of the permanent wilting point in soil was foreshadowed by Sachs' observation that sand contained less water than loam soil when plants growing in it wilted. Perhaps the idea advanced by Renner, Ursprung & Blum, and others, that water movement depends on suction force or water potential gradients rather than on osmotic pressure gradients can be regarded as a new concept. The work of Paal, Went, Thimann, and others on auxin in the first third of the 20th century was very important, but the existence of plant hormones was proposed by Sachs and Darwin in 1880. One of the few almost completely new ideas of this period was the role of photoperiod in controlling plant growth and flowering, described by Garner and Allard in 1920. However, even this was foreshadowed by the work of Bailey and Klebs.

Since 1930 many important contributions have been made, but most of them are additions to concepts already in existence. For example, the discovery of phytochrome supplied a means of explaining the action of certain wavelengths of light in connection with photoperiodism and various other reactions to light. Discovery of the Hill reaction clarified understanding of the photosynthetic process and opened up new areas of research. The electron microscope and cytochemistry have added much information about the fine structure of cells and the location of various cellular processes, but it has solved few problems and produced no new principles. Molecular biology created a better understanding of how genetic material is replicated, of how enzymes are produced, and of how they are turned on and off. However, the dogmas of molecular genetics really have contributed very little toward a better understanding of morphogenesis or the physiology of whole plants. Our increasing knowledge of plant processes results more from improvements in instrumentation and in experimental techniques than from development of new concepts.

Lest plant physiologists become too complacent over their many important achievements, let me remind them of a few of their failures. After a century of study we neither understand the structure of cell membranes nor exactly how ions are transported across them. In fact we do not even understand exactly how ions are transported across roots and into the xylem nor how organic compounds are translocated in the phloem. We still do not fully understand how auxin is transported or controls growth, how an embryo gives rise to a plant, or exactly how many enzymes perform their catalytic functions. After many decades of research we cannot explain why the protoplasm of plants of some species can survive desiccation or freezing while that of other species cannot. In fact not enough is known about any major plant process to satisfactorily explain how plants react to their environment. The deficiencies become especially obvious when we are asked to supply information to scientists attempting to construct models of plant growth. One of my colleagues claims this is because plant physiologists have given too much attention to causes and not enough to quantification of processes.

It is somewhat disturbing to find that although the number of plant physiologists has increased more than tenfold and publications in plant physiology even more during the past 40 years, it appears that few really new principles or processes have been described. Instead, our time and energy have been largely devoted to improving our understanding of processes already known, some for as long as 50 to 100 years. Perhaps Bentley Glass was correct in his suggestion that science has reached its "golden age" instead of attaining "endless horizons" (*Science*, Jan. 8, 1971). This situation supports the view of D. J. de Solla Price in his book, *Science Since Babylon*, that the "stature" of science increases much more slowly than the amount of information published. Glass suggested that the average originality of scientists is declining and that it is becoming more difficult to make new discoveries or break through the accepted views of the scientific establishment. This probably is correct, but experience indicates that it would be unduly pessimistic to suppose that nearly everything important has been done. In the past advances in scientific theory and instrumentation often have produced new bursts of progress after periods of stagnation. For example, in 1890 most physicists felt that they had made all possible important discoveries. Then in 1895 Röntgen announced the discovery of x rays, and within 2 years the dormant field of physics literally exploded into new activity that eventually affected chemistry and biology. Perhaps such a stimulus may appear in plant physiology, although there obviously is a limit to what can be discovered in a finite world. We already have mentioned many important problems which remain unsolved. The whole field of comparative plant physiology has been largely ignored, although recent work on C_3 versus C_4 plants, photorespiration, and crassulacean acid metabolism indicate its possibilities. Work in physiological genetics also should be very productive, both by explaining how existing adaptations came into existence and in producing new varieties of plants adapted to particular environments. The availability of phytotrons ought to stimulate research in both fields.

The Time Lag for New Ideas

This brings us to another interesting problem, the time lag between the proposal of new ideas and their acceptance by the scientific public. In these days of instant communication it would be expected that new ideas would spread more rapidly and win acceptance sooner than a century or more ago. Slow acceptance of new ideas was understandable in earlier days when communication was slower, but it did not always occur. It was only 20 years from the time Priestley announced that plants can improve air "injured by the burning of candles," until Ingenhousz described fairly clearly the role of photosynthesis in plant nutrition. In contrast Hales proposed in 1727 that transpiration in some way brings about the ascent of sap, but it was not until 1895 that an acceptable explanation of the ascent of sap, based on the high internal cohesive forces in water, was developed by Askenasy and Dixon. A half century or more later some writers still regarded the ascent of sap as an unsolved problem, and even today the cohesion theory is occasionally questioned. In 1928 Gradmann suggested that the flow of water through the soil-plant-atmosphere system might be treated as analogous to the flow of electricity in a conducting system, and this view was further developed by van den Honert in 1948. However, it was another 15 years before this convenient oversimplification came into common use. Enough data were available 30 years ago to justify the conclusion that significant amounts of water and salt are absorbed through suberized roots, and considerable new evidence has appeared in recent years. Nevertheless, even today few texts consider this possibility.

Perhaps new ideas which contradict well-established ones are generally accepted only after a new generation of scientists replaces the supporters of the old ideas. This is not altogether bad. Reasonable caution about accepting new ideas is desirable because many of them prove unsound after a period of testing, but the boundary between prudent caution and excessive conservatism is difficult to define, and excursions too far in either direction tend to hinder progress. Uncritical acceptance of every new idea results in cluttering the literature with useless or incorrect data, and excessive conservatism in accepting new ideas prevents needed changes.

THE RAPID GROWTH OF SCIENCE

We will now examine the recent rate of growth of science and scientists and its consequences in more detail.

Increase in Number of Scientists

The population explosion has been even more striking in science than in the country as a whole. From 1930 to 1970 the population of the United States increased from 123 million to 200 million, or 62 percent, but the population of scientists increased much more rapidly. The number of scientists has doubled every 10 years in recent decades, and so rapid was their increase in the late fifties and early sixties that observers estimated that if it were to continue at the same rate, by 2021 everyone in the United States would be working in science and technology. The rate of increase can be illustrated by the increase in membership of several plant science societies. The Botanical Society of America was founded in 1906 with 119 members; in 1930 it had 1200 members, and in 1970 the membership was about 3300. The American Society of Plant Physiologists was founded in 1924 with 76 charter members; in 1930 it had nearly 400 members and in 1970 over 2800. The great increase in numbers has resulted in an undesirable increase in anonymity as compared with smaller countries such as Australia and New Zealand where all established scientists know one another. Increasing numbers conceal both talent and mediocrity.

Increase in Publication

The rapid increase in numbers of plant physiologists is only a statistic until we look at the program of a national meeting. However, the increase in number of publications is a fact of which we are all uncomfortably aware because we must find some way to cope with it. Not only have many new journals appeared, but old ones have grown larger. *Plant Physiology* increased from one volume containing 47 papers and 636 pages in 1930 to two volumes containing 353 papers and 1673 pages in 1970. Even review journals such as the *Annual Review of Plant Physiology* are growing uncomfortably large. In 1930 I could at least skim through the important papers in the entire field of plant physiology, but today I cannot read, let alone evaluate, all of the papers in plant water relations. In addition there is a never-ending flood of books which ought to be at least examined.

Some Consequences of Increased Volume of Publication

The great increase in the number of plant physiologists and in the volume of published research can be treated as evidence of the vigorous condition of the science, but it also produces certain penalties. Two of these are the difficulty of reading the flood of literature and the pressure toward increased specialization. Many investigators are finding that even the search for abstracts takes too much time and are turning the task over to *Biological Abstracts* or some other computerized literature-searching service. These services can be very helpful, especially to investigators working in a narrow, well-defined field that can be characterized by a few key words. However, they are less successful for someone like the writer who is interested in several fields. Furthermore, there are serious dangers in relying too much on abstracts because one usually is unable to judge the quality of the research from an abstract. Also, important ideas often are buried in the body of a paper and are not mentioned in the abstract.

Possibly we are in more danger of being strangled intellectually by the mass of publications than we are of being strangled physiologically by air pollution. Some investigators expect computers to cope with the information problem, and they doubtless can provide the national storage and retrieval system first proposed by Vannevar Bush. Unfortunately, although computers can store and retrieve information they cannot evaluate it. Only scientists can do that, and if they attempt to keep up with the flood of literature, little time is left for research. The information problem was discussed recently by Brady and Branscomb (*Science*, March 3, 1972).

The Need for Generalists

It is regrettable that the increasing complexity of many explanations of physiological processes and the flood of literature are producing increased specialization just when we need more generalists. Specialization is necessary in research, but it sometimes results in physiologists who are familiar with only a single process or even a single enzyme system. Such investigators may be very productive, but they often tend to overemphasize the importance of the processes with which they are familiar and overlook the roles of other equally important processes. The increase in specialization has created a need for generalists who can assemble the results of specialists and produce a balanced, coherent picture of the role of various processes in plant growth. For example, there are specialists on the carbon pathway in photosynthesis, the energy transfer system, and the structure of chloroplasts, but the effectiveness of photosynthesis as a supply of energy for plant growth is limited more often by stomatal behavior, leaf structure, mineral nutrition, or water supply than by processes at the molecular level. An understanding of salt and water absorption by plants requires some knowledge of soil as a reservoir for salt and water, and of root structure, as well as of absorption mechanisms. In 1956 the late Professor E. N. Transeau wrote me that "progress in botany is being handicapped more by failure to organize and use existing knowledge than by the lack of knowledge," and I believe this is even more true today. There is a great need for scientists with training and interests broad enough to enable them to collect information from various areas, correlate and organize it, and use it to explain plant behavior.

Lost Information

Another result of the great amount of research being done is the large amount of information which remains unpublished or is lost in obscure publications. It sometimes is claimed that too much is published, and I agree that some papers should not have been published. On the other hand, there is no doubt that considerable important information which cost much effort and money remains unpublished. Sometimes this is because good investigators are poor writers or because they lack patience to deal with editorial criticism. More often, particularly in agricultural experiment stations and the federal government service, transfer of personnel terminates projects, leaving useful information which is filed and forgotten. Some of this wasted effort could be avoided by better organization of research so that more persons are familiar with each project and better continuity is assured. Better methods for cataloging and retrieving unpublished data also are needed. Too much time and money are wasted repeating research already done, and sometimes even published. Theoretically, when the abstracting services such as BIOSIS get most of the published material in a form where it can be searched by computers, the loss of published data can be greatly reduced, but it probably can never be entirely eliminated. The problem of lost unpublished data seems insoluble, yet it is costing much money and effort.

We hear from time to time that printed scientific journals will eventually disappear and will be replaced by abstracts and stores of tape from which investigators can have any desired article printed on request. This may represent an advance in efficiency, but I doubt if it will represent real progress. Loss of the opportunity to leaf through periodicals and browse in fields not directly related to the reader's interests would be unfortunate because it would make specialists even more narrow in their knowledge.

Increase in Amount and Cost of Equipment

Another important change which was just beginning 40 years ago is the great increase in amount and cost of instrumentation in our laboratories. In 1930 equipment was simple and inexpensive and a good laboratory could be equipped for a few thousand dollars. Beyond balances, ovens, and the usual glassware almost none of the equipment used in a plant physiology laboratory today was available in 1930. Spectrophotometers and pH meters were becoming available just before World War II, but infrared gas analyzers, magnetic and electrode-type oxygen analyzers, radioactive and stable isotopes and the instrumentation to measure them only became commercially available after the War. Most of the instrumentation containing electronic components has appeared in the past 20 years, and solid state circuits now have virtually replaced electron tubes. Although column chromatography was known long ago, most of its applications are fairly recent, and gas and thin layer chromatography are comparative newcomers to the laboratory, as is the refrigerated ultracentrifuge. Gaseous oxygen and sometimes CO_2 were measured with an Orsat apparatus, a collection of stopcocks and absorption bulbs guaranteed to test the patience of the investigator, and dissolved oxygen was measured by the classic Winkler method. Now we have instruments which measure gaseous or dissolved oxygen in 5 percent of the time formerly required. When I began measurement of CO_2 uptake by leaves, the CO_2 in the air stream was absorbed in alkali which was laboriously titrated. Later the amount of CO_2 absorbed was determined from measurement of the change in conductivity of the absorbing solution, a much more rapid method, and today we can make many measurements per hour with an infrared gas analyzer attached to a recorder or even to a computer terminal. Weighing was once a tedious process of adding weights and moving riders; but today we turn a knob and read the weight from a dial.

In the not so good old days we grew our plants in the open or in greenhouses where they were exposed to wide variations in temperature and light, and we were often forced to suspend research in the winter because our experimental material grew poorly with the short days and low light intensity of that season. Today we have plant growth chambers in which we can maintain controlled light intensity, temperature, and humidity in both summer and winter.

There was little direct recording of data in 1930, and that was done on clockwork-driven drums covered with smoked paper prepared by the investigator. Today we have a variety of instruments that can record data from many sensors, or our instruments can be connected directly to data logging equipment that stores the data in various ways until it can be processed. In the 1930s we worked up our data with a pencil and a slide rule or on noisy hand-operated calculators. Today we have quiet, rapid electronic calculators, or we can program our data into computers which calculate in seconds results which formerly required days or weeks. The availability of computers has stimulated the construction of models of plant processes and plant growth. To me the chief utility of model building is that it reveals important gaps in our knowledge of physiological processes.

In general, the history of plant physiology indicates that improvements in research techniques have always been followed by rapid advances in knowledge. This was true even in much earlier times. An understanding of the gas exchanges involved in photosynthesis was impossible until methods for separating oxygen and carbon dioxide and measuring the volumes of gases became available late in the 18th century. The availability of radioactive isotopes after World War II made it much easier to study uptake and translocation of ions and organic compounds, the carbon pathway in photosynthesis, and the steps in various other metabolic processes. The availability of infrared gas analyzers to measure CO_2 made possible numerous useful measurements of CO_2 uptake by many kinds of plants growing in a variety of environments. The introduction and improvement of thermocouple psychrometers stimulated much-needed quantitative research on the effects of water stress on plants, and the availability of diffusion porometers is stimulating measurements of stomatal behavior.

These improvements in instrumentation have greatly increased the kinds of processes we can measure and the speed with which we can measure them, thus increasing the scope and the quantity of research. They have also greatly increased the cost of doing research, with results that will be discussed later.

Change in Nature of Research

Another important change in plant physiology is in the nature of the research itself. There have been important changes over the years in what plant physiologists work on as well as how they work. The increasing interest in molecular biology was accompanied by a shift from whole plant physiology to research in biochemistry and biophysics and cellular physiology. This was very productive, but there now appears to be a healthy renewal of interest in the physiology of whole plants. Some of the scientists who rather arrogantly claimed in the late 1950s and early 1960s that the only worthwhile approach to biological programs was through molecular biology have now changed their views. I think it has become apparent to even the most devoted supporters of molecular biology that an organism is much more than a collection of molecules or even of cells and one cannot solve all the problems of organisms solely at the molecular level. To fully understand plant growth requires research at all levels, molecular, cellular, organismal, and community, and it is silly to claim that any one approach can do the job alone. Instead of being the master of whole organism biology, molecular biology really is its useful servant, helping to explain at the molecular level why organisms behave as they do.

At the same time that physiology was moving toward biochemistry and biophysics, quite a few ecologists moved toward physiological research at the whole plant level. This trend has largely broken down the boundary between autecology and plant physiology, a very desirable occurrence. An increasing number of agronomists, horticulturists, and foresters also are working on problems of whole plant physiology. It is very gratifying to see that workers in applied fields are finally accepting the idea that it is necessary to understand how plants grow in order to grow them most effectively. This idea was advanced by Samuel W. Johnson in his book, *How Crops Grow*, published in 1868, but it is only now being widely accepted.

THE PRESENT SITUATION OF PLANT PHYSIOLOGY

Today we find plant physiology a large, vigorous, productive branch of plant science. Along with other branches of science, however, it is faced with serious problems, some of them the result of its own success. Some consequences of the information explosion were discussed earlier.

Consequences of the High Cost of Research

Equally troublesome are the consequences of the increasing cost of carrying on research. Forty years ago a few hundred dollars accomplished more than several thousand will today. The increasing cost of research already was straining the finances of both private and public educational institutions before World War II. At that time the only large scale governmental support of research was in the field of agriculture, through the land grant colleges and agricultural experiment stations. Support from private sources such as the Rockefeller Foundation was very limited in amount and restricted in scope. Experience during the war showed the importance of science and technology and persuaded our political leaders of the need for continuing support of basic research in the universities. As a result, several government agencies began to support the basic research of university scientists by grants and contracts, and the National Science Foundation was brought into existence for this purpose. As the pressure created by World War II was relaxing, the success of the Russian space program gave another boost to the science support programs. These programs were largely successful and many of us have benefited from them. In fact most scientists not supported by agricultural funds became dependent on contracts and grants from governmental agencies to finance their research. The increasing dependence of science on government funds provoked various gloomy predictions, some of which proved correct, others incorrect.

The first concern was that investigators would become opportunists and work on whatever they could get funded, regardless of its importance. I believe this fear that science would be totally corrupted is largely untrue. This may be partly because scientists are usually rather individualistic and determined to work on the problems they consider important. Also many mission-oriented agencies were lenient in allowing scientists to work on what they preferred, even though it was not always relevant to the objectives of the granting agency. Fortunately much opportunistic research and research which seemed irrelevant to agencies supporting it has turned out to be worthwhile in its own right.

The second concern was that generous financial support through contracts and project grants to individual investigators would lure scientists away from teaching and destroy departmental and institutional loyalties. These fears were partly justified. Some university staff members were attracted from teaching to research in the 50s and 60s, and it is probable that some damage was done to teaching. Furthermore, the availability of research funds probably led to financing some scientists having little imagination and limited competence who should never have been encouraged to do independent research. It also is true that a few well-financed scientific entrepreneurs developed research empires without regard to the welfare of the departments which sheltered them. Such "operators" were relatively secure from departmental pressure because they would move elsewhere if their activities were curtailed. Fortunately, these abuses were limited in extent, and the damage done seems small compared to the good done by the large amount of money which became available for graduate and postgraduate training.

Another difficulty during the period of easy money was the temptation to let tenured staff members become dependent on "soft money," that is grant and contract funds, for part or all of their salaries. In some instances, salaries were even tied to the amount of grant money an investigator could obtain, a pernicious practice. This resulted in considerable embarrassment to some of our institutions when research funds began to shrink. Fortunately, some universities resolutely resisted this pressure, so they have found it much easier to live with reduced research support than those who succumbed to the lure of soft money.

The independence afforded to individual investigators by the project grant system used by some agencies has been a continued source of concern to university administrators, including departmental chairmen. They felt that programs were developing without regard to the needs of their departments and institutions. As a result there has been continued pressure for some type of institutional grants which could be used at the discretion of administrators to build better balanced programs, but this has always been resisted by individual scientists who preferred to take their chances at the national level rather than compete on the campus for their share of an institutional grant. The project grant system, therefore, continues to be the chief source of support for most plant scientists in spite of its disadvantages. I have occasionally suggested to administrators that they could control and guide development even with the project system if they really tried. It would require careful planning, tactful consultation with their scientists, and strong leadership to correlate grants requested by various individuals into a central plan, but it could be done. A beneficial effect of the high cost of research has been to force universities to cooperate in building and operating large and expensive laboratories. We doubtless will see more cooperation between departments and institutions as costs continue to rise.

Consequences of Dependence on Public Funds

Unfortunately, largely unpredicted difficulties have arisen which are restricting the supply of public funds for support of research. These restrictions arose partly because of the financial demands of the Vietnam war and our expanding social programs, partly because of inflation, and partly because the public lost confidence in science and technology and made them a scapegoat for some of its problems. To the extent that science is dependent on public funds, it quite properly is subject to scrutiny by the public and the public's representatives in Congress concerning the use it makes of those funds. A decade ago the public enthusiastically supported research in science and technology because it expected them to solve its problems. Today a considerable segment of the public suspects, incorrectly of course, that science and technology are creating more problems than they solve and question how much support they should receive. Fortunately, the attitude of the public seems to be improving.

The late president Kennedy said in a talk to the National Academy of Sciences, "Scientists alone can establish the objectives of their research, but society in extending support to science must take account of its own needs." In other words, through its representatives in Congress, society determines how much tax money will be allocated to science in competition with national defense, public welfare programs, transportation, the space program, etc. This is the inevitable result of the increased dependence of science on government funding since World War II. The only way to avoid pressure from society is to cease the use of tax funds, and this is impossible because modern science is too expensive to be supported by private funds.

Some of the complaints provoked by the leveling off of appropriations for the public support of scientific research suggest that many scientists feel they have some inherent right to expect support for their work. I see no reason why scientists should expect to be supported automatically by taxpayers simply because they regard their work as important. It sometimes is argued that all competent scientists are equally deserving of support, but I doubt if this is true in theory and it certainly is impractical, Lack of money to support everyone makes priorities necessary, and the demands of society make some kinds of problems more urgent than others. The public has rather limited appreciation of the benefits derived from scientific research, but it is very concerned about its health, welfare, and convenience. To maintain broad support for science requires better education of the public concerning the ways in which basic research contributes to its wellbeing and more attention to priorities by scientists (Bevan, *Science*, June 2, 1972).

Decision Making and Priorities

The dependence of modern science on the government for financial support raises troublesome questions concerning the methods used in allocating funds. Who should establish priorities and decide on the kinds and amounts of research to be supported? At present budgets for agencies such as the National Science Foundation are prepared by agency administrators and reviewed and adjusted by the Office of Management and Budget to fit administration policies. The amount of money actually appropriated by Congress is determined by conferences and compromises after hearings at which various interested parties can plead their case. Congress may appropriate more or less than the sum requested in the official budget, set limits on expenditures in particular areas, and designate funds for uses quite different from those desired by the majority of the scientific world. Basic research supported by the National Science Foundation has had to compete with such diverse activities as oceanography, teacher training, curriculum improvement, and international projects such as the International Biological Program. A recent addition to the competition is the program entitled Research Applied to National Needs (RANN), which is more the product of administrative and congressional desires to see increased applications of science to problem solving than of pressure from scientists. Only at the last stage, when applications are being evaluated by panels, do scientists have much influence over the kind of research supported.

Unfortunately, there is no entirely satisfactory alternative to the present system. There ought to be more input from scientists at an early stage in budget-making, but who is to supply it? It is questionable whether scientists are much more competent than congressmen to establish priorities. Most scientists and scientific organizations are as subject to bias from self interest as any other citizens, and each thinks generous support of his field not only is important to science but even essential to the national welfare. Since the amount of money available for the support of science is limited, choices have to be made among fields and within fields, and these ought to be made in some rational manner. To make rational choices requires sound criteria and scientific statesmen to apply them. Alvin Weinberg has dealt cogently with this problem in several articles, of which that in *Physics Today* (March, 1964) probably is the most accessible. He suggests that there are two kinds of criteria for choosing among competing areas of science, internal and external. The internal criteria center around two questions: (1) is the field ready for exploitation; and (2) are competent scientists available to exploit it? The external criteria are (1) technological merit; (2) scientific merit; and (3) social merit. Weinberg argues that the field which contributes most to other scientific fields has the most scientific merit, but social merit or relevance to human welfare is more difficult to evaluate. The average scientist lacks the broad background and perspective required to make such evaluations. Wise decisions concerning scientific priorities at the national level can be made only by persons broadly informed concerning science on a worldwide basis and sufficiently judicious to minimize personal scientific bias. Until we find some way of identifying and using more of these people in scientific decision making, we will doubtless continue to use the present system of pressures and counterpressures combined with political expediency. The need for a rational, coherent, long term, national policy for the support of science and technology was discussed by Bevan (*Science*, June 2, 1972).

LOOKING TO THE FUTURE

Looking toward the future, I wish to remind scientists that they are operating today in a very different situation from that which existed even in 1965. There are many reasons for the change, including a decrease of confidence in science, competition for money with the Vietnam war and social programs, increasing concern of the general public about the value of science, and the leveling off of college enrollments.

There is little doubt that science does not hold as high a place in public esteem as it did in 1960. This may be partly because some of its spokesmen promised the public too many immediate returns, especially in the medical field. There also is a tendency among young people to relate science to bombs and pollution, and many are distrustful of both science and technology, although this situation seems to be improving. Bevan suggested that refusal of scientists to establish priorities and their tendancy toward elitism has estranged them and contributed to loss of public confidence.

Probably the most interesting development of the last few years is the increasing power of public opinion. For the first time in the history of this country, technologists and engineers have been prevented by the force of public opinion from doing what they desired to do. Examples are the postponement of the Alaskan oil pipeline and several badly needed electric generating plants, cancellation of the SST, and stoppage of work on the Florida Barge Canal. Five years ago few people would have predicted that public opinion could operate so effectively to curb technology.

It is possible that public opinion may next turn to science and demand that it demonstrate that its activities are in the public interest, or at least what it thinks is the public interest. Biology and medical sciences are likely to feel the pressure first. Publicity about the possibilities of genetic management of humans already is producing suggestions that research in some areas ought to be forbidden. It is frightening to think of the effects on science if its activities were to fall under the censorship of some kind of government agency. The best defense is a strong effort to educate the public more effectively concerning the goals and the methods of science. Efforts to do this have not been very effective, but it must be done if science is to receive the public support essential for its continued progress.

The pressure of public opinion and the concern for human welfare is likely to produce more emphasis on teaching relative to research and on applied research relative to basic research. If not carried too far, this may be beneficial. There are instructors who neglect their students in order to carry on more research, and there are researchers who insist on working on trivial pet projects while pressing problems more useful to society are neglected. However, these situations are much less harmful to science and society than the damage that would be done by attempts to organize and direct all research toward what some government bureau imagines are socially desirable objectives. I believe that science has obligations to society and I believe that plant physiology has an obligation to assist in increasing the production of food. However, this can be accomplished in many ways, direct and indirect. I doubt if any governmental agency has the wisdom to predict what approaches will be most productive.

The only thing I am fairly certain about in the future is that there is likely to be as rapid change in the methods, the content, and the subjects of research in plant physiology in the next 40 years as occurred during the past 40 years. This emphasizes the desirability of adaptability and willingness to work on new and unfamiliar problems, and it increases the need for broad basic training in general principles to prevent early obsolescence. Perhaps formal retraining programs to update teachers and researchers will become routine. Plant physiologists who are well trained and willing to adapt to new conditions and new problems may find the next 40 years even more productive and exiciting than the past 40 years. However, the nature of the problems on which they work may be very different.

SOME PERSONAL OBSERVATIONS

I will now turn to a discussion of my personal relationship with plant physiology. This has been placed at the end of the chapter so those who do not enjoy reading reminiscences can skip it.

Education

My personal contact with plants began at a very early age because I was reared on a farm in southwestern Ohio. My parents were well informed and had an unusually large library in which I read omnivorously. I was equipped with pocket guides to birds and flowers at an early age and soon became reasonably well acquainted with the local flora and fauna. The first scientific paper that I can recall was an article by Garner on photoperiodism in the *United States Department of Agriculture Yearbook* for 1920. My professional contact with botany began in 1922, when I enrolled in a course in the botany of cultivated plants taught by Bruce Fink at Miami University. The choice of botany resulted from a conversation a few weeks earlier with Miami University President R. M. Hughes, in which he learned that I was interested in plants and was acquainted with Professor Fink. I enrolled in the course on the botany of cultivated plants because it sounded more interesting than the regular freshman course. This choice probably was fortunate because the introductory course of those days was exceedingly dry and formal and might have driven me away from botany. I found most of my college subjects interesting and considered majoring in areas so diverse as history and psychology but finally settled on botany chiefly because I enjoyed the laboratory work. My undergraduate training was adequate except in the physical sciences. Although this deficiency was partly repaired in graduate school, I never attained the level of proficiency in mathematics which we now require of our graduate students. I sometimes wonder if I would have gone into plant physiology if I had been forced to take all the courses we require today.

Professor Fink was a well-known lichenologist, an interesting teacher, and an energetic recruiter who sent many students to graduate school, mostly in mycology. He seemed able to teach almost anything from bacteriology or physiology to taxonomy with interest and authority. Although I disappointed him by my lack of interest in lichens, he obtained several offers of assistantships for me when I was ready to leave college. Among them was one from the University of Idaho which paid $750, a princely sum in those days. Furthermore, it gave me an opportunity to visit the Northwest, an interesting region which I had never seen. I spent the academic year 1926–27 at the University of Idaho with Floyd W. Gail, a very thorough teacher, but a rather temperamental man. I assisted in general botany and other courses and even gave a course in systematic botany for a small class. I also took frequent trips into the surrounding country and became fairly well acquainted with northern Idaho.

Professor Gail was very anxious for me to stay on at the University of Idaho and offered me an instructorship if I would spend the summer of 1927 taking course work at the University of Chicago. However, I decided not to do this. Instead I spent the summer working in the woods with a crew from the Department of Agriculture which was studying the germination of *Ribes* seedlings after forest fires and disturbance of the litter by logging. These seedlings were being studied because *Ribes* is the alternate host for *Cronartium ribicola*, which causes white pine blister rust, a serious disease of white pine. At the end of the summer I returned to Ohio by way of California and Arizona, getting my first view of the southwestern deserts. At the beginning of the winter term I enrolled as a graduate student in botany at the Ohio State University. As a student of Professor Fink, it was assumed that I would probably go into mycology or plant pathology, and I took several courses in these areas. However, I was not attracted by the taxonomic and morphological approaches used by the professor of mycology. Having taken courses in physiology and ecology with E. N. Transeau, I went to him for a problem. He sent me to read a paper by Burton E. Livingston which suggested that osmosis is not an important factor in water absorption. Transeau then proposed that I investigate the correctness of this idea. I was still woefully ignorant of the literature and of experimental methods, but with the advice of Transeau, Dr. B. S. Meyer, who was just beginning his career at Ohio State, and various fellow graduate students, I soon began making some progress in research. While a graduate student I spent one summer in northeastern Ohio with a barberry eradication team and another in southwestern Ohio as an assistant county agricultural agent, working with orchardists and truck crop producers. Later I acted as laboratory assistant to Professor O. F. Curtis of Cornell for the two summers which he spent as visiting professor at Ohio State University. This resulted in a lifelong friendship with Curtis.

The Early Days at Duke University

Positions were as scarce in 1931, when I received my doctoral degree, as they are today, but I finally obtained an instructorship at the then recently established Duke University. At this time I married Edith Vance, who had been a graduate student in botany at Ohio State and taught botany at Vassar for 2 years. She has played a very important part in my 40 years in plant physiology. In fact, it was on her advice that I applied for a position at Duke. My salary was about half what we pay a teaching assistant today, but the cost of living also was low and the shortage of openings made bargaining impossible. In the autumn of 1931, we moved to Duke University, where we were exposed to the stimulating and sometimes trying experience of participating in the conversion of undergraduate Trinity College into Duke University. At that time Duke had a biology department chaired by Professor A. S. Pearse, a vigorous animal ecologist who had come from Wisconsin a few years earlier. Pearse was an energetic promoter who helped start the Association of Southeastern Biologists and was the chief organizer of the Duke University Marine Laboratory at Beaufort, North Carolina. There were several good scientists and strong individuals in the group who set high standards for the younger people. Among these were F. G. Hall and G. T. Hargitt in Zoology, Ruth Addoms, H. L. Blomquist, and F. A. Wolf in Botany, and C. F. Korstian, Director of the Duke Forest. Hall gave me advice on scientific matters, Pearse pushed my advancement with the administration, and Korstian introduced me to the opportunities in tree physiology. Association with all these vigorous personalities doubtless contributed more to my development than I can now identify.

In 1932 we were joined by H. J. Oosting, an ecologist from Minnesota, and H. S. Perry, a geneticist from Cornell. Soon afterward we persuaded Dean Wannamaker to organize a separate botany department. This was a fortunate move because the separate botany and zoology groups developed to a degree that would never have occurred in a biology department. The botany department felt obliged to work hard in order to justify its independent status, and this pressure doubtless contributed significantly to the attainment of our present position among the leading botany departments in the country.

From the beginning the department gave a great deal of attention to teaching, and the development of a good introductory course in botany contributed significantly to its status on the campus. Nearly every member of the staff was publishing and most were already attending national scientific meetings, beginning the process of building a national and eventually an international reputation. In the period before and during World War II teaching loads were heavy, varying from 12 to 15 or occasionally 18 hours, yet we always found time for research. In view of that experience I have little sympathy for young people who claim that they are unable to do research while teaching 6 or 8 hours a week. As the years passed additional staff members were added and in line with the trend of the times our teaching loads decreased. However, the load of extracurricular work increased as the older staff members took on more committee assignments, national offices, editorial duties, and advisory tasks. As a result there actually was an increase in responsibilities and some of us grew busier as we grew older.

Three Simultaneous Careers

It sometimes seemed that I was carrying on three professional careers in addition to my responsibilities to my family. First was my teaching, both undergraudate and graduate. Second was my personal research program and the supervision of the programs of graduate students, which sometimes seemed equivalent to a full-time job. It resulted in the training of over 40 PhDs, more than 100 scientific papers, and three books. The third line of activity was service on a variety of committees both on and off the campus. During the late 1930s and 1940s, considerable time was spent on the executive committee of our then rather new graduate school and on the library council. Also, just at the end of World War II, the botany department took over management of the Sarah P. Duke Gardens, a large ornamental garden on the Duke University campus. As a result of my interest in ornamental horticulture, I became director and started a program of enlargement and improvement which continues today.

Teaching

It was emphasized by Dean Wannamaker that I was hired to teach, and I recall being warned not to get involved in research. This advice was carefully ignored as it was evident that a reasonable amount of research and publication was essential both for my advancement and for that of the department. Until the 1950s I taught both general botany and various courses in plant physiology. A. W. Naylor joined the department in 1952 and took over part of the teaching load in plant physiology. However, I continued to teach the introductory course until 1969. There may be a few good scientists who cannot teach effectively, but I have found the combination of teaching and research very satisfactory. It is stimulating because students raise many questions which would never occur to an investigator working alone in his laboratory. The teacher also is forced at every lecture to arrange his material and present his ideas in a reasonably orderly and logical manner. Scientists who do not teach lose a valuable experience, and there is no evidence that a modest teaching load necessarily reduces the research output. Perhaps most people have only a limited number of original ideas, regardless of the time available to them.

Research

My research program soon involved two areas, study of the absorption of water, started while a graduate student, and research on woody plants. Forty years ago absorption of water was generally treated somewhat vaguely as an osmotic process, although Otto Renner had stated about 1915 that osmosis is unimportant in absorption of water by transpiring plants. My experiments supported Renner's view that there are two absorption mechanisms, active absorption responsible for root pressure and guttation, and passive absorption by rapidly transpiring plants in which water is pulled in through the roots by the tension generated in the transpiring shoots. We also demonstrated in that period that factors such as low temperature and deficient aeration reduce absorption chiefly by increasing the resistance to water flow through the roots rather than by directly affecting the absorption mechanism. This led to research on the effects of flooding on herbaceous and woody plants and a study of cypress knees. My interest in woody plants led to a study of absorption through woody roots, and our laboratory was probably the first to measure uptake of radioactive elements by mycorrhizal roots of trees. From this developed a study of the absorbing zone of the young roots of various species and the observation that much salt uptake occurs some distance behind the root tip, where the xylem is well differentiated. One of our students also made careful comparisons of the salt accumulating capacity of the cortex and stele and demonstrated that contrary to some reports the stele can accumulate salt as effectively as the cortex. Other experiments led me to conclude that considerable salt is carried passively into roots in the transpiration stream, especially in mature plants, supporting the idea advanced by Hylmo in 1953. This view has not been generally accepted, but I think it is unavoidable unless we are to assume that roots are completely impermeable to ions, which certainly is not true.

Observations in our laboratory also support the idea that significant quantities of salt and water are absorbed through the suberized roots of perennial plants. In spite of much supporting evidence accumulated by various investigators during the past 40 years, this view seems not yet generally accepted. It appears to constitute another example of the difficulty of replacing a well-established concept by a new one.

By 1949 I felt prepared to bring out a book, *Plant and Soil Water Relationships*, which was used so widely that I meet plant scientists all over the world who studied it. Apparently it significantly stimulated interest in the field of plant water relations. It was replaced in 1969 by a completely new version which was much more difficult to write because of the increase in the amount of literature and in complexity of concepts. I had much valuable advice on the new version from R. O. Slatyer of Canberra, Australia, who was then writing his well-known book, *Plant Water Relationships*. I recall that an English reviewer complained that my first book on water relations was dangerous because it made complex problems appear simple. This amused and flattered me at the time because I was trying to make everything as simple as possible, but perhaps his criticism was justifiable. Probably I do tend to oversimplify complicated problems. However, I think it is better to have a student understand an oversimplified explanation than to drive him away by one which is too complex. If he is competent he can absorb the complete explanation later.

By the middle 1950s it was becoming clear that many of the contradictory results obtained from research on effects of water stress on plant growth resulted from the fact that the plant water stress had not been measured. The only reliable way to evaluate plant water stress is by direct measurement, but it was not until the 1950s that satisfactory methods were available. For over a decade some of the people working in my laboratory have studied various methods of measuring water potential, and I have spent considerable time publicizing the necessity for such measurements. The wide interest in and the progress made by various laboratories in measuring water potential was demonstrated at the symposium held at Utah State University in 1971. Unfortunately, even today there occasionally are workers who neglect to measure water potential, another example of how difficult it is to bring about changes in viewpoint. Workers in our laboratory are now making simultaneous measurements of soil, root, and leaf water potential which should improve our understanding of water movement in the soil-plant system.

Soon after arriving at Duke University I became acquainted with C. F. Korstian, who later became the first dean of the Duke University School of Forestry. Korstian felt that a good understanding of the physiology of trees is essential to productive research in forestry, and he spent considerable time educating me concerning problems in forestry. Our first experiments were simple observational studies of the relation of dormancy to photoperiod in various tree species. Many foresters regarded dormancy as an inherent characteristic of forest trees, and it was rather difficult to persuade them that the onset of dormancy in some species is related to the decreasing length of days in late summer and early autumn. Later research in the Earhart Laboratory at California Institute of Technology showed that thermoperiod also has important effects on the length of growing season and amount of growth of several tree species. These studies indicate that both photoperiod and thermoperiod may affect tree growth and distribution. We also became involved in an old question concerning the relative importance of light and water in the failure of pine seedlings to develop under hardwood stands. Studies by several graduate students of physiological differences between pines and hardwoods indicated that pine seedlings are unable to become established under forest stands where hardwood seedlings flourish because of their low rates of photosynthesis in the shade.

I was one of the first plant physiologists in the United States to work extensively with woody plants, but interest in this area increased rapidly and nearly half of my graduate students were foresters. Today there are many workers in this area, a number of whom obtained at least part of their training at Duke University. I am particularly gratified that I and some of my students have had a part in bringing about this increased appreciation of the role of plant physiology in forestry. In 1960 my former student T. T. Kozlowski and I brought out *Physiology of Trees*. Kozlowski has continued a career of writing and editing which is adding much to the literature of tree physiology and plant water relations.

It would be ungrateful to close the discussion of my research activities without recognizing the sources of funding which have supported it. In the earliest days when support was very limited, C. F. Korstian financed several graduate students working on tree physiology from School of Forestry funds. Others were supported by grants from the Duke University Research Council and by teaching assistantships in the Botany department. For about 20 years I received support from the Atomic Energy Commission for research assistants to work on various problems in salt absorption. Over the past 15 years the National Science Foundation has supported many research assistants and a number of postdoctoral research associates for work in the field of water relations. Without this support only a small fraction of the research which has been done in our laboratory could have been done. I particularly wish to acknowledge the contributions of these pre- and postdoctoral investigators who not only did research, but stimulated me by their questions and discussions.

Extracurricular Activities

I shall now return to an account of some of the extracurricular activities which occupied an increasing fraction of my time after 1950. In 1954 I attended my first international botanical congress in Paris, followed by those in Montreal (1959), Edinburgh (1964), and Seattle (1969). I was on the organizing committee for the Seattle congress and learned how much time and money is required to stage such affairs.

In 1955 a sabbatical semester was spent working on effects of photoperiod and thermoperiod on tree seedlings at the Earhart Laboratory of the California Institute of Technology. This experience was so profitable scientifically that I started a campaign to obtain funding for a phytotron in the eastern United States, preferably at Duke University. This campaign resulted in feasibility studies which culminated in construction of the biotron at the University of Wisconsin and a few years later the two-unit phytotron at Duke and North Carolina State universities, called the Southeastern Plant Environment Laboratories (SEPEL). This project occupied much time from 1956 to 1968. The history of SEPEL was told in *BioScience* for November 1970.

I spent the year 1960–61 at the National Science Foundation as Program Director for Regulatory Biology. This was followed by several years of service on panels and on the advisory committee of the Division of Biological and Medical Science. In addition, I have reviewed hundreds of proposals for various programs, a small return for the research support I have received over the past 15 years. I also was on the visiting committee for the biology department of Harvard University for 6 years and on a variety of committees to make site visits and inspect departments for various purposes. I also served terms on the Agricultural Board of the National Research Council, the Committee on Agricultural Science of the U. S. Department of Agriculture, the Board of Trustees of *Biological Abstracts*, the Sigma Xi National Lectureship Committee, and various nominating and organizing committees. Among the more interesting jobs was membership on the first editorial board of the *Annual Review of Plant Physiology*. These activities required considerable time and energy, but I learned much about biology in the United States and made many pleasant acquaintanceships.

Over the years I have served as an officer of several scientific societies, mostly in a rather routine manner. The only exception to this was my term as an officer of the American Institute of Biological Sciences which coincided with a severe crisis in its financial affairs. My connection with AIBS began with the organizing committee which met during 1946–47 and developed support for the organization. D. W. Bronk described the early history of AIBS in the July 1972 issue of *BioScience*. Experience during World War II had demonstrated the need for an organization which could speak for biologists as the American Physical Society and American Chemical Society spoke for physicists and chemists. It was very difficult to persuade botanists and zoologists of this need, but when the organization was brought into existence it soon began to fulfill numerous functions. Among the best known are early sponsorship of the Biological Sciences Curriculum Study and the Council of Biological Editors. AIBS sponsored so many projects for government agencies that if it had not been in existence in the 1950s, something similar would have had to be invented. Unfortunately, it was inadequately financed by biologists and as a result of overoptimistic management got into serious financial problems in 1962. It was the misfortune of James D. Ebert and myself to be officers of AIBS during this trying time. We and many other concerned biologists had a very difficult time before the problems of AIBS were solved. Unfortunately, many plant scientists continue to fail to see the need to support AIBS as a spokesman for biology.

Retrospect

Looking back, it seems that I have been involved in an unusually wide range of research activities, including the mechanism of and factors affecting water absorption, the absorbing zone of roots, the radial movement of salt and water into roots, the physiology of pine and hardwood competition, factors controlling the range of tree species, measurement of plant water stress, and effects of water stress on plant processes. Research in so many different fields would be almost impossible today because it no longer is possible to keep up with the literature in such diverse fields. However, it has been much more interesting than concentration in a narrow field.

In retrospect there is another interesting fact. None of the areas in which I worked were fashionable when I started. Little work was being done in plant water relations in the 1930s, but today the number of papers published exceeds my ability to read and digest them. Likewise, very little work was being done in forest tree physiology when I started, but now research is being done all over the world. When we started the campaign to obtain funds to build a phytotron I had no idea that it would go into operation just as a great surge of interest in environmental problems occurred.

Some people claim to have carefully planned their careers, but mine seems to have been a succession of accidents which were reasonably productive and satisfactory. My entry into botany was more or less an accident, my research on plant water relations started from a casual suggestion by a professor, and my work in tree physiology resulted from my contact with C. F. Korstian. I have generally operated on the philosophy that if I make the correct choices today, tomorrow will take care of itself. If the results have been generally satisfactory, it is because I was able to see what was important at various stages and was persistent in working on it. Also I was fortunate in having so many stimulating graduate students, postdoctoral research associates, and colleagues far and near, all of whom contributed to the success of our work. I will close with the suggestion that personal careers in science are really simply parts of a large cooperative and sometimes competitive enterprise, and their success probably depends as much on the scientific community as on the individual.

Ann. Rev. Plant Physiol. 1973. 24:25–46

ELECTROPOTENTIALS OF PLANT CELLS[1]

❖ 7540

N. Higinbotham

Department of Botany, Washington State University, Pullman, Washington

CONTENTS

[1] Abbreviations used: ATP (adenosine triphosphate); ATPase (adenosine triphosphatase); CCCP (carbonyl cyanide-*m*-chlorophenylhydrazone); DCMU (dichloro-phenyl-dimethylurea); DNP (2,4-dinitrophenol); ecp (electrochemical potential); E_j (Nernst potential of ion *j*); E_m (membrane potential); FCCP (carbonyl cyanide *p*-trifluoromethoxy-phenylhydrazone); P_j (permeability coefficient of ion *j*).

INTRODUCTION

The importance of cell electropotentials and ion transport in plants generally was forcefully brought to attention in this series by Jack Dainty in 1962 (20). In brief, the transfer of any ionizing substance during passive or active transport may result in charge separation and generation of electropotential gradients; these gradients affect the diffusion of any charged species, mineral or organic, provided the species is free to move independently. Even active transport via a carrier complex will be affected, providing the complex is not neutral. Nonpolar organic substances traversing the membrane may or may not be neutral during transport, e.g. whether sugars pass through membranes as sugar phosphates is an open question (10, 18). Many of the other organic substances which show mobility within cells and tissues carry charges, e.g. organic acids, amino acids, adenosine phosphates, etc. The significance of this for protoplasm, which is an intricate framework of membrane barrier systems, should be quite apparent; much of the biochemical regulatory processes must lie in controlling the access of substrates to the enzymes with which they react. The fact that virtually all of our knowledge of electrophysiology in plants rests upon the relationships in mineral movements should not obscure the vision of future pathways in the elucidation of regulatory systems.

In addition to Dainty's 1962 article, other recent discussions having an important bearing on this subject are those of Scott (114) on bioelectric fields, Sibaoka (116, 117) on rapid movements, MacRobbie (82), Slayman (120), Hope (63), and Anderson (1) on ion transport, and Gutknecht & Dainty (48) on ion transport in marine algae. More recent are the papers of the *Liverpool Workshop on Ion Transport in Plants* (2).

ELECTROPHYSIOLOGICAL METHODS

Cell Potentials

In general, the procedure used for measuring electropotentials in plant cells is to insert a microcapillary glass electrode with the aid of a micromanipulator into a cell under microscopic observation. The electrode is of the Ling-Gerard type (76) filled with concentrated KCl solution, usually 3 M but sometimes less (77), which provides a low resistance through the fine tip. In small cells, plant and animal, the tips are less than one micron, which seems necessary in order for the cytoplasm to make an electrically tight junction with the glass. Coarser electrodes up to 10 μ at the tip may be used with the giant-celled algae (145) such as the *Characeae*, *Valonia*, *Halicystis*, etc. The reference electrode is generally coarser but filled with 3 M KCl in agar. Coupling to a high input resistance ($10^{10}-10^{14}$ Ω) electrometer is by AgCl-Ag wires or calomel cells. The reference electrode is ordinarily placed in the solution directly bathing the cells, although in some cases where intact leaves have been used, there may be an appreciable distance between the electrodes with no apparent difficulties (151; R. L. Jefferies, private communication).

There are many theoretical as well as practical difficulties in bioelectrical work. Guggenheim (44) has said that electropotential differences between points in different media are not measurable and have not been physically defined. As expressed by Tasaki & Singer (137), the defining of resting potentials across membranes must be in operational terms. This by no means should be construed as indicating that resting potentials as measured are necessarily inaccurate, but only that in cells they represent complex functions of junction potentials, interactions with polyelectrolytes, and other factors too complicated to be evaluated in absolute terms on a firm theoretical basis. Changes induced in potentials by concentration gradients or action potentials are predictable and their magnitudes are as significant, for example, as those found with the pH electrode or other ion-selective electrodes. The theory and practice of electrophysiological methods and the nature of various electrodes have been dealt with extensively in recent years (26, 27, 33, 88, 89, 141–143).

Resistance Measurements

Most measurements of electrical resistance, or its reciprocal, conductance, in cells have utilized two loops, one to monitor the cell potential and the other to pass current of a known amount; thus two electrodes, one for each loop, are inserted into a cell (57, 146, 147). The change in PD induced by a small current of known magnitude permits calculation of the resistance. The resistance r in ohms cm^2 is given by:

$$r = E \cdot A / I \qquad\qquad 1.$$

where E is the change in cell electropotential; A is the cell membrane surface area; and I is the current passed. Insertion of two or more electrodes into a single cell of a giant-celled alga such as *Nitella* (34, 127, 148) or *Chara australis*[2] (34) may be accomplished with no apparent damage; however, Spanswick (127) has shown that 5–6 hours may be required after electrode insertion to obtain maximum values in PD and membrane resistance measurements.

Current passed to depolarize or hyperpolarize the cell may not have uniform density over the cell surface. To cope with this in long cylindrical cells, e.g. *Nitella* and *Chara*, as in nerves, the object is treated as a leaky cable (62, 146), and a simplified method has been developed for determining membrane resistance in such cells (61).

In higher plant tissues the problem is much more complicated, since in practice insertion of more than one microelectrode into a single cell appears to lower the potential as though the cell has been damaged, or the cytoplasm does not seal around the electrodes (43, 57). Furthermore, the cells are interconnected by plasmodesmata, consequently significant amounts of current pass from cell to cell rather than directly through the surface to the ground electrode (128).

A new method for measuring cell membrane resistance using a single electrode has been devised by Brennecke & Lindemann (13). This technique, which utilizes

---

[2] The species formerly referred to as *Chara australis* is now known as *C. corallina* (149). However, the epithet used in this review will be that of the original article.

a chopped current of brief pulses (about 50 μsec), permits measurement of both potentials and resistance; this new technique promises to be very useful in following membrane resistance changes when the external solution is modified.

Short-circuit Technique

One of the best criteria for active transport is the demonstration of net ion movement through a membrane separating two solutions of identical composition and having no electropotential gradient. For membranes generating a PD an electrical field is applied to clamp the potential at zero; then any net ion movement must be driven by a pump and can be measured by monitoring electrical current flow; tracer fluxes can also be measured. By this method, Ussing & Zerahn (140) demonstrated with frog skin the essential identity of active Na^+ transport, tracer labeled, with the flow of electrical current. Recent applications with plants also demonstrate active ion transports (46, 58, 135).

RESTING POTENTIALS

Diffusion Potential

Cells typically show a resting potential of about 100 mV, interior negative. Potentials may arise from a Donnan system, by diffusion where both cations and anions are mobile, or by an electrogenic pump. The source of the potential difference across cell membranes has commonly been thought to be diffusional and to result from the fact that the two ions of a salt usually have different mobilities. Thus in a two-phase system separated by a membrane across which there is a concentration gradient, at equilibrium, the following relation holds:

$$E = \frac{u^+ - u^-}{u^+ + u^-} \frac{RT}{zF} \ln \frac{[C]_1}{[C]_2} \qquad 2.$$

where E is the electropotential difference between phases 1 and 2; u^+ and u^- refer to the mobilities in the membrane of the cation and anion respectively; R is the gas constant (joules deg^{-1} $mole^{-1}$); T, the temperature; z, the ionic valence; F, the Faraday (coulomb $equivalent^{-1}$); and C, the concentration (more precisely the activity of the solute) in the phase indicated. If $u^- = 0$, that is, the membrane is impermeable to the anion, Equation 2 reduces to the Nernst equation:

$$E = \frac{RT}{zF} \ln \frac{[C^+]_1}{[C^+]_2} \qquad 3.$$

Thus, in this case E is determined solely by diffusion of the cation C^+. Plant cells under certain conditions may closely approach this relationship with the external cation being K^+ (54, 125) or H^+ (73, 125).

There is good reason to believe that in addition to a potential arising from diffusion there is active electrogenic transport, i.e. a transport resulting in movement of net charge and generation of electrical potential at the expense of metabolic energy (56, 119, 125, 126).

Some workers in ion transport believe that the membrane is impermeable to the independent movement of ions and that ion transfer takes place solely in combination with a carrier (28). Although the kinetics of ion uptake seem consistent with this view, there is ample evidence, in the discussion to follow, favoring the conclusion that a significant amount of independent ion movement occurs through membranes either inward, outward, or both.

K^+ Transport and Diffusion

The electrical properties of the large-celled algae *Chara* and *Nitella* have been studied extensively. With these cells it is possible to insert electrodes into the cytoplasm or vacuole and also to isolate samples from each compartment for chemical and tracer analysis. Thus compartment potentials, ion concentration gradients, and tracer fluxes have been well defined. With *Chara australis*,[2] in the absence of Ca^{++}, Hope & Walker (65) found a good agreement with the Goldman voltage equation (42, 60) describing passive diffusional relationships:

$$E = \frac{RT}{F} \ln \frac{P_K[K]_o + P_{Na}[Na]_o + P_{Cl}[Cl]_i}{P_K[K]_i + P_{Na}[Na]_i + P_{Cl}[Cl]_o} \qquad 4.$$

where P is the passive permeability coefficient of the ion indicated by the subscript, and bracketed ions represent activities. Hope & Walker, neglecting Cl^- since P_{Cl} is small, found that the changes in E induced by substituting K^+ for Na^+, and keeping the total concentration constant, was in accord with predictions of the equation. Since the Donnan potential of the wall was constant these results indicated that the ΔE values obtained quite certainly came about from ionic diffusion; the ratio P_{Na}/P_K was found to be about 0.04.

In the presence of Ca^{++}, increasing K^+ externally over a range from 0.1–10 mM has been found to have little effect on the resting potential of *Chara corallina*[2] (71). Similar results were found with *Nitella translucens* (130), and under these conditions the membrane potential could not be accounted for by diffusion (Equation 4); in the absence of Ca^{++} the results of Hope & Walker (65) with *Chara* were confirmed, namely that ΔE was related to diffusion of K^+ and Na^+.

In the cells of the fungus *Neurospora* and of seed plants, as in the algae, electropotential changes may be induced by changes in the diffusion gradient, particularly of K^+. In *Neurospora crassa*, Slayman (118) found that increasing K^+ externally depolarized the cell by about 45 mV/log unit concentration change; thus, between 0.1 to 1.0 mM KCl, E changed from about -230 mV to -185 mV. If K^+ is the only ion diffusing, the Nernst equation (Equation 3) predicts the maximum depolarization of 58 mV per log unit (at 20°C); if there is significant anion permeation, a lesser slope would be predicted [Slayman also reported a slope of 85 mV/log unit with $[K]_o$ in the range of 10–100 mM; this has not been explained but might be due to inhibition of an electrogenic pump (119)]. Addition of Ca^{++} to a medium containing K^+ and Na^+ caused hyperpolarization by about

40 mV; in the presence of Ca^{++} the depolarization induced by $[K]_o$ was reduced to 17 mV/log unit. The effect of Na^+ was similar but there was an appreciably lesser effect. In coleoptile cells of *Avena sativa*, which had been pretreated to remove Ca^{++}, the K^+ depolarization slope was about 50 mV/log unit $\Delta[K]_o$ (54). Na^+ had a lesser effect at low concentration but above 10 mM gave results similar to those with K^+. Increasing $[Ca]_o$ also caused hyperpolarization in *Avena* cells. However, it appears thus far that only *Avena* and *Neurospora* show this effect since in other tissues tested increasing $[Ca]_o$ causes depolarization (84). Barley roots are similar with respect to depolarization by increasing $[K]_o$ (96) as other tissues (78, 79), and it is presumed to be a general phenomenon.

With information on the concentration gradient across the cell membrane at flux equilibrium, the Nernst equation can be used to determine for each ion whether it is in an electrochemical potential balance. If an ion j is diffusing passively, its Nernst potential E_j is equal to the membrane potential E_m. If not, the following relation may be expected:

$$E_m - E_j = E_D{}^j \qquad\qquad 5.$$

where $E_D{}^j$ is the driving force on ion j and the sign of E_D indicates the direction. In more recent studies an effort has been made to distinguish between the major compartments of cells: the outside o, wall w, cytoplasm c, and the vacuole, v. In the large-celled algae, differences in electropotentials and ionic concentrations have been found across both the plasmalemma and tonoplast; however, the plasmalemma is the chief barrier in most cases, with the exception of certain marine algae, e.g. *Valonia ventricosa* (48). Using the Nernst equation as a criterion, the following generalizations can be made: (*a*) K^+ approaches an electrochemical potential equilibrium in a number of cases and appears to be moving in accord with the electrical gradient. In some of the exceptions the conclusion that K^+ is actively transported is based on relatively small discrepancies between E_m and E_K; this may not be justified in view of uncertainties in determination of absolute values of the PD and of K^+ activity in the cytoplasm (83). However, there are some cases in which the departure from electrochemical potential balance is so great that active transport of K^+ must be concluded. (*b*) Na^+ movement occurs against the electrochemical gradient from the cytoplasm to the outside and frequently in cells having a large central vacuole, also from the cytoplasm to the vacuole. This appears to be universal, or essentially so, for living cytoplasm of the various plant groups, including bacteria, as well as of animal species. (*c*) Anions (Cl^-, $NO_3{}^-$, $H_2PO_4{}^-$, $SO_4{}^{--}$) generally are accumulated against the electropotential gradient; thus active transport may be inferred. (*d*) The status of Ca^{++} and Mg^{++} is less certain largely because they have not been carefully investigated in this context. However, the information available suggests that the electrochemical potential gradient is from outside to inside; thus influx may be passive. There is a similar lack of evidence on transport of the trace metals, e.g. Fe^{++}, Mn^{++}, Zn^{++}, Co^{++}, etc; however, in these cases the small amounts involved could probably be bound to exchange sites.

In *Nitella flexilis* Kishimoto & Tazawa (72) found that K^+ gradients across both the plasmalemma and tonoplast closely approached the values predicted for a passive distribution. The electropotential across the plasmalemma (and the wall), E_{co}, was about -170 mV, while in the flowing cytoplasm there was 110–125 mM K$^+$ (with $[K]_o = 0.075$ mM). E_K was nearly equal to E_{co}. The electropotential across the tonoplast E_{vc} was about 8–12 mV, vacuole positive, where E_K was 8 mV. Similar results for K$^+$ have been found in *Chara australis* (141). On the basis of electrochemical data it has been inferred that K$^+$ is actively transported from outside to the cytoplasm in *Nitella translucens* (82), *Hydrodictyon africanum* (103), *Nitella clavata* (5, 6), and *Chaetomorpha darwinii* (25, 37). In *Chara australis* (64) and *Tolypella intricata* (82) K$^+$ may be passively distributed between the cytoplasm and vacuole, although in *Nitella translucens* and *Tolypella intricata* MacRobbie (80–82) found evidence suggesting transport to the vacuole via vesicles rather than by ordinary diffusion.

One of the better examples of active K$^+$ transport into the cytoplasm is that of *Nitella translucens* (131). The cytoplasm was found to be -138 mV relative to the outside and -18 mV with respect to the vacuole. The Nernst equation would predict a K$^+$ concentration of about 31 mM in the flowing cytoplasm (an activity of 24 mM), whereas chemical analysis showed the concentration to be 119 mM, far above the value expected with passive diffusion along the electropotential gradient. Thus K$^+$ is accumulated against an outward driving force of about 34 mV. In contrast with the active transport of K$^+$ at the plasmalemma, movement across the tonoplast appears to be passive.

In most cases the plasmalemma appears to be the major barrier with respect to ion movement and electrical resistance. However, in the marine species *Griffithsia monile* and *G. pulvinata*, Findlay, Hope & Williams (38) found an electrical resistance of 5000 Ωcm^2 for the tonoplast and only 200 Ωcm^2 for the plasmalemma. E_{co} was -85 mV and E_{vc} 25–35 mV; K$^+$ apparently moved passively into the cytoplasm but active transport into the vacuole was indicated. Other marine forms in which the electrochemical data afford good evidence of active K$^+$ transport from the cytoplasm to the vacuole include *Valonia ventricosa* (45) and *Chaetomorpha darwinii* (25, 37) among other examples (see 48). The high positive electropotentials of the vacuole relative to the cytoplasm in these forms—$+88$ mV in *Valonia* (45)—clearly indicates that the tonoplast is the major barrier in cells of some marine organisms.

In shoot and root tissue of etiolated pea and oat seedlings, none of the eight major nutrient ions (K$^+$, Na$^+$, Ca^{++}, Mg^{++}, Cl$^-$, NO$_3^-$, H$_2$PO$_4^-$, and SO$_4^{--}$) exactly conformed to the Nernst relationship (55). In a nutrient solution containing K$^+$ at 1 mM, E_K approached E_m but the Nernst potentials of other ions were far removed from their ecp balance; the data suggest inward passive movement of Na$^+$, Ca^{++}, and Mg^{++} but active inward transport for each of the anions. Etherton (29), in a study of the same tissues, found that at lower nutrient solution concentrations (including K$^+$ at 0.1 mM) the measured cell content of K$^+$ was much higher than that predicted from the cell potentials; at a high concentration (including K$^+$ at 10 mM) the reverse was true, i.e. the predicted K$^+$ content was much lower than that predicted from cell PD. Actually, the cell potential of each tissue was insensitive to changes in concentration of a complete nutrient solution ranging from a 0.1-fold to a 10-fold concentration; thus the Goldman voltage equation (Equation 4) does not apply, and it may be inferred that active ion transport rather than passive diffusion dominates the system. As in the algae and in *Neurospora* there is good evidence that the presence of Ca^{++} externally is an important factor in maintaining the insensitivity of PD to external concentration; in its absence the Nernst relation (and Goldman voltage equation) may be quite closely approached, at least with single salts in the concentration range above 1 mM K^+ (54). The explanation of this, as with the other cases, probably is that one or more ion pumps are electrogenic, and passive diffusion components of transport can dominate the electrophysiologic properties of the cell membrane only under restricted conditions. The major barrier in higher plant cells is the plasmalemma; however, the vacuole may be positive to the cytoplasm by a few millivolts (32, 93).

Components of K^+ Transport

Among the ions which can be studied by tracers K^+ has the greatest diffusive permeability. However, it has become apparent that K^+ movement frequently does not conform to predictions for passive diffusion. Considering the array of evidence as a whole, it seems likely that ion movement consists of several components among which may be: (*a*) independent ion diffusion in accord with electrical gradients; (*b*) ion exchange, e.g. K^+ for K^+, or other cations such as Na^+ or H^+, in an electrically neutral process; (*c*) active ion transport, e.g. by carriers or other pumps which may be either electrically neutral (facilitated diffusion, see 134) or electrogenic; and (*d*) movement simultaneously with a co-ion in an electrically neutral process. It seems likely that two or more of these processes are going on simultaneously, but unfortunately adequate estimates of the contribution of each are lacking.

Na^+ Transport

In view of the extensive evidence that Na^+ fails to be accumulated in the cytoplasm of either plant or animal cells, as would be predicted by the electrochemical potential gradient, an exclusion mechanism must be inferred. Since the permeability to Na^+ is relatively high in many cases, amounting to 0.68 P_K in oat coleoptile cells (95), an efflux pump at the plasmalemma is usually assumed (4, 36, 55, 82, 95, 97, 103, 120). Recently Poole (100) has reported that disks of red beet root become Na^+-selective following washing in distilled water at 10°C for several days; disks washed for 24 hours are K^+-selective. Poole has interpreted his results as indicating the development of a selective transport system for Na^+, but he has failed to show that the uptake is not in response to the electropotential gradient. According to his earlier work (98, 99), the cell PD would be about -165 mV when the external solution is 1 mM KCl. At this PD, and at equilibration, an internal concentration of 693 mM Na^+ would be predicted. Thus an active transport system for Na^+ need not be invoked, rather an impairment of a Na^+-efflux pump may be a preferable explanation. It should be noted that the Na^+-selective tissue also has greater permeability for K^+, in the absence of Na^+, than the K^+-selective tissue.

There is good evidence in several plants that Na^+ may be actively transported from the cytoplasm to the vacuole, against a small potential gradient, in some algae (82) and in some higher plants (95, 97). Thus Na^+ appears to be pumped out of the cytoplasm in both directions, the net effect being to maintain a lower concentration than that predicted from diffusion. The report of active accumulation of Na^+ by sunflower roots (11) might be explained by a pump at the tonoplast.

Many attempts have been made with plants to discover a coupled K^+/Na^+ pump such as those found in red blood cells (and other animal cells). K^+ moves in as Na^+ is pumped out; if there is a one-to-one exchange, the process is electrically neutral and Goldman's voltage equation should apply. Application of the cardiac glycoside, ouabain, or removal of external K^+ blocks the pump. In the last few years it has been found that the process is mediated by a membrane-bound (K^+-Na^+)—enhanced ATPase and that the pump operates in either direction (40). In the forward direction about three Na^+ ions are pumped out and about two K^+ ions are pumped in for each ATP molecule hydrolyzed; with steep concentration gradients and a low energy supply the pump proceeds backward and ATP is synthesized. Some partial reactions of this system have been described (41).

A number of attempts have been made to demonstrate the occurrence of a K^+/Na^+ pump in plant cells by the use of ouabain as a blocking agent (see 107). However, in only three cases has ouabain had the expected effect, namely in *Nitella translucens* (80), *Hydrodictyon africanum* (103, 104) and in *Allium cepa* (15). Raven (103) found that increasing external K^+ enhanced Na^+ efflux as well as K^+ influx; the two processes had the same Michaelis constant and were blocked by ouabain, more in the light than in the dark. However, in a later study, Raven (107) found some cultures of *Hydrodictyon africanum* which were insensitive to ouabain.

There is considerable evidence that ATPases are involved in K^+ and Na^+ transport in plants (39, 74, 82). Thus, the failure of ouabain to affect such transport, either of K^+ influx or of Na^+ efflux, and the lack of coupling between the two fluxes, suggests that ouabain may not affect the transport ATPases of plants. It is also quite likely that in most plants K^+ influx is driven by the potential gradient, thus ouabain should have no effect. The same cannot hold true for Na^+ extrusion. Cram (19), in a study of ion fluxes in carrot disks, found that ouabain depressed Na^+ efflux but not K^+ uptake.

Anion Transport

The major anions, NO_3^-, Cl^-, $H_2PO_4^-$, and SO_4^{--}, apparently are all accumulated against the ecp gradient. The evidence for this is fragmentary since most of the electrochemical studies of anion transport have been done with Cl^- for which there is a good tracer isotope. However, it has been shown that both root and shoot tissue of oat and pea seedlings accumulate each of these anions against a cell PD of -80 to -110 mV (55). Active uptake has been observed for $H_2PO_4^-$ in

Nitella translucens (123), for SO_4^{--} in several species of the *Characeae* (109, 110), and for Cl^- in many species (48, 82). An active inward transport for bicarbonate also has been found by Raven (105, 106) in *Hydrodictyon africanum* and by Smith (124) in several Characean species. A number of plants, including some angiosperms, have been reported to assimilate bicarbonate in photosynthesis, and thus active transport might be inferred.

In *Acetabularia mediterranea* there is a clear case for an electrogenic pump for Cl^-, since decreasing Cl^- in the external solution causes rapid depolarization (111). Hill (58), using short-circuit analysis, has found good evidence for electrogenic Cl^- transport across the salt glands on leaves of *Limonium vulgare*. Thus far there appears to be no good evidence in other plants for electrogenic anion pumps despite the general occurrence of active anion influx. In the Characeae (82) and in oat coleoptiles (54), substituting one anion for another, e.g. SO_4^{--} for Cl^-, results in quite small changes in cell PD. This would not be expected if there were specific anion pumps generating a potential.

Calcium

The importance of Ca^{++} for the structural integrity of cell membranes has been well documented; however, at the molecular level the role of Ca^{++} is not well understood. Previously mentioned is the fact that in the presence of Ca^{++} the cell PD is relatively insensitive to increases in the external concentration of K^+ (54, 130); increasing Ca^{++} may actually induce hyperpolarization in oat coleoptiles (54) and *Neurospora* (119). This "clamping effect" of Ca^{++} seems quite general, having been observed in many animal systems as well (115).

Recent work of A. Kylin & M. Kahr (private communication) has revealed in wheat and oat roots the presence of a Ca^{++}-enhanced ATPase; the activity was inhibited by Mg^{++} at higher concentration, but increasing Ca^{++} relieved the inhibition. If such an ATPase is plasmalemma-bound and active in ion transport, it may help explain the role of Ca^{++} with respect to enhancement of ion uptake and voltage effects.

The electrochemical relations of Ca^{++} are not well known. It appears to permeate the protoplast very slowly, apparently in a passive downhill direction (55, 132). The accumulated soluble Ca^{++} may be very low, particularly in the algae but also in higher plants, and by some criteria it might be regarded as a trace element (150).

H^+ and Electrogenic Pumps

There is increasing evidence that a large portion of the cell electropotential arises from one or more electrogenic ion pumps. An electrogenic pump may be defined as an active transport in which net charge is transferred across a membrane at the expense of metabolic energy; the result is that a potential is generated. The membrane potential then would represent the sum of the diffusion and the pump PDs, although complex interactions of the two systems are to be expected. Important criteria for such electrogenesis are as follows: (*a*) Since an electrogenic pump is dependent upon metabolic energy, a respiratory inhibitor should cause a rapid depolarization. (b) Removal of the ion pumped or lowering the temperature, thus blocking respiration, should also depolarize. (c) The occurrence of cell membrane potentials exceeding the Nernst potential of any of the diffusible ions is strong evidence for an electrogenic pump. Light induces such a hyperpolarization (68, 125, 126). (d) The occurrence of net ion transport is accompanied by electrical current dependent on a specific ion in short-circuit procedures (58).

Some years ago it was found that oat coleoptile cells were rapidly depolarized by DNP and that the inhibition was reversible (32). This was confirmed by later studies (50), and it has been shown that CN has a similar effect on cells of both oat coleoptiles and pea epicotyl (56). Reversible depolarization induced by DNP has also been found in roots of *Trianea* (*Limnobium*) *bogotensis* (79) and of wheat (122). The response in multicellular tissue to the addition and removal of inhibitors, while rapid, is relatively sluggish compared to Slayman's (119) results with *Neurospora crassa*, in which azide inhibits the PD (depolarizes) at a rate of 20 mV/sec; reversibility after removing the inhibitor is somewhat slower, requiring about 10 min in *Neurospora* and about 30 min in pea epicotyl for full recovery. These results, of course, constitute strong evidence for an electrogenic pump which contributes part of the measured PD, while ion diffusion contributes the remainder. At maximum inhibition by CN in pea epicotyl cells, a residual potential of about -64 mV remains of the -140 mV prior to inhibition (56). In *Neurospora* azide reduces the potential from about -200 mV to -30 mV (119).

Further evidence with pea epicotyl and oat coleoptiles that part of the membrane potential has its origin in a metabolically driven pump is the fact that at higher external concentration (including K^+ above 1 mM) the measured PD exceeds the value which could arise from diffusion, i.e. E_m exceeds the Nernst potential for any of the known diffusible ions (29, 56). An additional test of this hypothesis lies in the goodness-to-fit of the Goldman voltage equation (Equation 4) when the tissue is under maximum inhibition; that is, in the absence of any active transport the system should be governed by passive diffusion. Higinbotham, Graves & Davis (56) have shown that within the limits of our present knowledge of passive permeabilities a reasonable fit is obtained.

It could be argued that the effect of metabolic inhibitors on PD might be explained by hypotheses other than an electrogenic pump, e.g. by changes in the ionic permeabilities. This could not apply to the case above, and Slayman (119) showed clearly that azide had little or no effect on the electrical resistance of the cell membrane in *Neurospora* during the depolarization period. After several minutes there was an increase in resistance, possibly accounted for by reduced active transport.

H^+ Transport and Electrical Resistance

Which ion—or ions—may be subject to the electrogenic pump? Kitasato (73) found that increases in H^+ concentration externally caused marked depolarization in *Nitella clavata* in the pH range 4–7. When the membrane PD was clamped at the value of the K^+ equilibrium potential, the variation of the current necessary to maintain E_K was in accord with the expectations for large passive influxes of

H^+ ions. He postulated that there was an electrogenic H^+ efflux pump which balanced the passive H^+ influx. Furthermore he found that the sum of the conductances of K^+, Na^+, and Cl^- was far too small to account for total membrane conductance, but his measured current-flow values were consistent with the hypothesis that H^+ transfer, in the pH range 5–6, substantially accounted for the discrepancy. Kitasato's conclusions have been criticized as being subject on several grounds to other interpretations (147). However, Spanswick (125, 126) has carefully reviewed the evidence and concluded that explanations involving an electrogenic H^+ extrusion are more likely, but that Kitasato's explanation was not completely adequate.

In a study of *Nitella translucens* Spanswick (125) has reported data favoring an electrogenic H^+ pump and proposed a model which appears to adequately explain the presently known data. Spanswick noted that in artificial pond water including K^+ at 0.1 mM the PD was less negative than E_K and was relatively insensitive to changes in external K^+; thus the Goldman voltage equation did not apply. In view of Kitasato's (73) evidence for electrogenic H^+ extrusion, Spanswick sought the conditions under which E_m exceeded E_K. He found that in the dark at an external K^+ concentration of 0.5 mM, E_m closely approached E_K; however, in the light the cell became hyperpolarized by about 50 mV as expected if an electrogenic pump is present. In the dark the temperature coefficient for E_m was 0.97 mV deg^{-1}, but in the light, 2.5 mV deg^{-1}. In the light the current required to maintain the membrane potential at E_K was equivalent to a transfer of 20 pmoles cm^{-2} sec^{-1}; this constitutes a measure of the electrogenic flux. Since in *Characean* cells the flux of K^+, the most permeable of the major ions, is about 1 pmole cm^{-2} sec^{-1}, it is quite apparent that most conductance is not due to K^+, Na^+, Cl^-, etc. Spanswick's measurements are consistent with the finding of Spear et al (133) that H^+ effluxes were 5–20 pmole cm^{-2} sec^{-1}, and with the electrical current measurements of Kitasato (73) made while clamping E_m at E_K and changing the pH from 5 to 6. However, Kitasato attributed the conductance change to a decrease in passive influx of H^+, whereas Spanswick (125) argues that it is an increase in efflux through the electrogenic pump.

Spanswick's (125) model is for a voltage-dependent, light-stimulated, electrogenic pump for H^+. It is based in large part on theory and equations developed by Rapoport (102) to explain the quantitative relations of the Na^+/K^+ pump to metabolism in animal tissue. Important features of the model are that the electropotential has a regulatory action on the pump and that the membrane resistance is lowered when the pump is active. The conductance through the pump appears to account for the discrepancy which has previously existed between measured electrical resistances and those calculated from tracer fluxes of the major ions (20, 127). It can be anticipated that Spanswick's model will be the subject of considerable investigation in future work.

The cells of *Nitella* and *Chara* often show circular bands of $CaCO_3$ encrustation which alternate with bare green zones along the cell. In *Nitella clavata*, Spear et al (133) have shown that H^+ efflux and Cl^- influx are localized in the green zones; Cl^- efflux was not localized. A dependence of active anion transport on H^+ concentration or active efflux seems indicated.

In higher plants there is not such clear evidence for large passive influxes of H^+ as found in *Nitella* by Kitasato (73). A change of pH over the range 7.0 to 3.6 actually resulted in a small hyperpolarization of oat coleoptile cells (54). A 4-hour treatment at pH 3.5 did result in depolarization but also in deterioration of the tissue. A similar lack of sensitivity to pH in the range 4 to 7 was found in barley roots (96); a small depolarization was found in wheat roots (122). This by no means precludes the possibility that an electrogenic H^+ efflux pump exists in these tissues. If the internal H^+ concentration is limiting the pump, the increase in passive H^+ influx might be compensated for by increased active efflux. Certainly the electrical conductance through the membranes of oat coleoptile cells (57), even allowing for plasmodesmata, greatly exceeds the value predicted from tracer fluxes of K^+, Na^+, and Cl^- (95). Conductance of H^+ through an electrogenic pump, as in Spanswick's model, could explain the discrepancy. A test of this must lie in measuring membrane resistance under appropriate conditions; unfortunately this is greatly complicated in higher plant tissue since the measurement of interest is the resistance across the plasmalemma, not easily accomplished. And attempting to clamp the potential in small cells while measuring current during pH changes may yield results difficult to interpret.

Effect of Light

Exposure of green cells to light generally results in hyperpolarization and enhanced ion transport. Although light may induce transient changes in the cell PD (112, 144), Spanswick (125) has shown that in *Nitella translucens* steady resting potentials occur in light and dark. In the light the PD is about 50 mV more negative than in the dark, and, as noted above, this portion of the potential was identified with an electrogenic pump. In this species in the light K^+ and Cl^- influxes are 10- to 20-fold that in the dark (80); results with *Hydrodictyon africanum* are similar (103, 104), but in this case clear evidence for an electrogenic pump is lacking.

Green cells of higher plants also may be hyperpolarized by light. Again transient changes have been observed, but in time steady resting potentials occur in cells of the moss *Mnium* and in leaf mesophyll of *Atriplex spongiosa* and *Chenopodium album* (77). The PD exceeds that predicted from passive diffusion of the major ions, but at least in *Atriplex* it appears to be related to H^+ diffusion (94). Although the authors are inclined to explain their results on the basis of passive H^+ movement, the pH gradient seems unlikely to be great enough to account for the high cell PD and an electrogenic pump seems likely.

Jeschke (68) has found that leaf cells of *Elodea densa* in the light have a potential of about −180 mV, about 60 mV more negative than in the dark. In *Elodea canadensis* cells the PD is about −164 mV in the dark and as much as −296 mV in the light (126). Spanswick (private communication) believes this to be a world record and clearly beyond the range of passive diffusion potentials (126); the internal K^+ concentration is about 91 mM, giving an E_K value of -172 mV, close to the cell PD in the dark. The hyperpolarization induced by light is reversibly inhibited by azide, CN, CCCP, and DCMU, as expected with an electrogenic pump.

Although the electrogenic pumps described above appear to be light dependent, present data do not preclude the possibility that they may also operate in darkness. Such pumps do occur in nongreen cells, e.g. *Neurospora* (119), and in etiolated tissue of pea and oat seedlings (56).

The transient potential changes observed on switching from dark to light, or vice versa, are probably a result of ionic diffusion. In the light H^+ is taken up by chloroplasts (90) and K^+ is extruded (92); this would alter the ionic concentration gradients and, depending upon the rates of the fluxes, could lead to transient changes in diffusion potentials. Pallaghy & Lüttge (94) have advanced the argument that the transients in leaf mesophyll cells of *Atriplex spongiosa* arise in this manner from H^+ diffusion. Chloroplasts may constitute about half the volume of the cytoplasm and in *Tolypella intricata* may contain three to four times the K^+ concentration of the bulk cytoplasm (75); consequently K^+ efflux could lead to sharp changes in ionic gradients across both plasmalemma and the tonoplast.

Bulychev et al (16) recently have succeeded with *Peperomia metallica* leaves in measuring chloroplast potentials relative to those of the cytoplasm. They found two types of photoinduced signals, a fast response (seconds) making the PD positive by 10 to 30 mV relative to the cytoplasm but showing a decay, and a slower response (minutes) which appeared to be steady. Relative to the cytoplasm the PD could go negative up to 65 mV. The authors interpret their results as being consistent with Mitchell's (85) chemiosmotic hypothesis in which the fast positive response would result from H^+ inward transport; the magnitude of the fast response increases with light intensity and is greatly reduced by FCCP, an uncoupler of phosphorylation. Further studies of this system are likely to yield important information on ion transport in green cells.

If there is an electrogenic H^+ efflux pump at the plasmalemma and light-induced H^+ influx into chloroplasts, switching on the light should reduce H^+ concentration in the ground plasm and thus lead to cell depolarization. This is reported for *Nitella flexilis* (3). Since the pH in the vacuole shifts downward during depolarization, the authors attribute the effect to electrogenic H^+ transport to the vacuole; however, the influence of chloroplast transport is not excluded.

Source of Energy

Many studies using respiratory inhibitors and phosphorylation uncouplers indicate that ATP is a source of energy for ion transport [see reviews by MacRobbie (82) and Higinbotham (52)]. The strong depolarization induced by inhibitors in *Neurospora* (119), in etiolated seedling tissue (56), and in *Elodea* leaves (126) is consistent with the idea that ATP is the energy source. Probably the best evidence for this is found with *Neurospora*, in which it has been found that under CN inhibition the time course of ATP decay coincides with the curve for loss of membrane potential (121). Inhibition of electron transfer was appreciably faster as shown by the fluorescence of reduced pyridine nucleotides which accumulate under CN treatment.

In *Streptococcus faecalis*, Harold et al (49) found that during K^+ uptake there is a stoichiometric exchange for H^+ and Na^+. They argue that the primary event in this transport, which is driven by ATP hydrolysis, is the active electrogenic extrusion of H^+ or Na^+.

Potassium- and sodium-enhanced ATPases have been isolated from plant tissue (39, 74) and correlations have been found between K^+ uptake and ATPase activity (39). More recently ATPases having enhanced activity with Ca^{++} (A. Kylin & M. Kahr, private communication) and with Cl^- (59) have been isolated. A presumption is that these ATPases serve as transport sites, or part of the sites, in membranes as in the quite well-characterized system for the Na^+/K^+ pump of animal cells (41); however, both the function and location remain to be established in plants.

In view of the discovery of an electrogenic pump for H^+, there is now a tendency to identify this with the chemiosmotic theory of Mitchell (85, 86) developed for mitochondria and chloroplasts. According to this model, ATP hydrolysis (in the membrane) results in an electrogenic separation of H^+ and OH^- ions across the membrane with H^+ driven outward in mitochondria and inward with chloroplasts; the process is reversible in the presence of suitable pH gradients or electropotential gradients. In fact, ATP synthesis in chloroplasts has been achieved by increasing the external H^+ concentration (67). Also K^+ efflux in mitochondria may drive ATP synthesis (17). It will be recalled that the $Na^+ - K^+$ ATPase transport system of red blood cells is reversible (40). Such systems would have obvious biological advantages under appropriate conditions for conserving energy and permitting feedback controls in both electron flow processes and ion transport.

There is good evidence that some transport systems do not utilize ATP as the energy source. In *Nitella translucens*, Cl^- uptake has been shown to be inhibited under conditions in which ATP was available from photophosphorylation while cyclic electron flow was restricted (82); K^+ uptake was not affected. Further, Cl^- uptake was shown to be insensitive to low concentrations of CCCP which block K^+ uptake. Similar results have been found in *Hydrodictyon africanum* by Raven (104), but whether this is true for other plants is not clear. The evidence now suggests that the electrogenic Cl^- transport over the salt glands of *Limonium vulgare* (58) may require a Cl^--enhanced ATPase (59).

Although sources of energy other than ATP must occur (108), attempts to identify them have been unsuccessful generally. However, in isolated membrane vesicles of certain bacteria it has been shown that P-enolpyruvate only is required for uptake and phosphorylation of methyl-α-D-glucopyranoside (70). Spear et al (133) have suggested that triose phosphate may be an energy source in *Nitella clavata*. Since ion transport may require only a small portion of metabolic energy it may be difficult to identify the primary sources; isolated membrane vesicles seem to offer the best hope, providing their identity with membranes of intact cells can be demonstrated.

ACTION POTENTIALS

Algae

In nerve and muscle certain stimuli, e.g. electrical current, heat, etc, reaching a threshold value, induce a rapid depolarization of the membrane which is followed by a recovery to the resting PD. This is attributed to a sudden change in membrane permeability which permits rapid Na^+ influx, thus causing a change toward positivity; in fact, action potentials usually overshoot, i.e. exceed the resting PD, thus going slightly positive (23). Energy is not required for the depolarization but is needed for the recovery during which Na^+ is pumped out. Action potentials have also been studied in *Nitella* and *Chara*, although the duration of the spike is measured in seconds rather than milliseconds as in animal cells. In addition, in these vacuolated cells there are action PDs across both the plasmalemma and the tonoplast (34, 35); the action PD across the plasmalemma is much the larger.

Mullins (87) found in *Nitella* that during the action PD there was a rapid efflux of Cl^-, more than enough to account for the spike. Findlay & Hope (34, 35) have shown in *Chara australis* that an increase in Ca^{++} influx could also explain the action PD; they also found similar responses whether Cl^-, Br^- or NO_3 were used and thus the spike is not dependent upon a single anion. Associated with the depolarization is a rapid lowering of the resistance as revealed by current flow.

Higher Plants

Thus far no function for action PDs has been found in plants such as the *Characeae*, and ordinarly the impulse does not travel from cell to cell (63, 114). However, in certain tissues of plant parts showing mechanical movement, e.g. *Dionaea muscipula* and *Mimosa pudica* leaves, action potentials inducible by various stimuli are propagated to the sites of mechanical action and must therefore represent a communication system. Further details may be found in the fine reviews by Sibaoka (116, 117). Not all mechanical movements are associated with propagated action PDs, since they were not revealed by a careful study in *Utricularia* leaf bladder traps (136). The trap is reset by salt transport from the bladder lumen followed by passive water movement.

In cells of plants not showing rapid movements, action PDs have not been detected, although it seems likely that a careful search for them has not been made. The light-induced transients do not conform to action PDs, since they do not obey the "all-or-none" law according to which the maximum spike is obtained once a threshold stimulus is given; increasing the intensity of the stimulus does not increase the response. However, all transients in electropotentials are probably accompanied by rapid changes in ionic conductance and permeability of ions relative to one another; this implies a capacity of the membrane to change its structure, perhaps allosterically.

TISSUE AND ORGANS

Cell-to-Cell Communication

In multicellular plants cells are commonly interconnected by plasmodesmata, making a continuum of cytoplasm referred to as the symplast. Tyree (138), in an excellent treatment using irreversible thermodynamics, has developed a general theory of transport through the symplast. From a consideration of a range of data available in the literature he concludes that plasmodesmata constitute the pathway of least resistance in short-distance transport of all small solutes and that a concentration gradient of about 1 mN is sufficient for diffusive movement. The electrical resistance of plasmodesmata have been measured in *Nitella translucens* (129) and in higher plant tissues, *Elodea canadensis* leaves, oat coleoptiles, and corn roots (128). The specific resistance of walls bearing plasmodesmata is much less than elsewhere, by a factor of about 6 in *Elodea* and 50 in *Nitella*. In addition to playing a role in movement of solutes, hormones, etc, plasmodesmata may be necessary for propagation of action potentials in plants such as *Mimosa pudica*.

Electroosmosis and Streaming Potentials

The flow of an electrolyte solution through a pore when a potential is applied is called electroosmosis; a variant is to measure the pressure developed at zero potential. Streaming potentials are the converse, i.e. the potential developed when an electrolyte solution is forced through a pore. It has been tempting to invoke these electrokinetic processes in explanations of turgor pressure, action potentials, and translocation phenomena. However, under ordinary circumstances at least, the contributions of electroosmosis or streaming potentials are small (21) although measurable (7, 8). The results indicate the importance of such studies for ascertaining concentration profiles in the wall-plasmalemma, complex changes in relative permeabilities, and local osmotic effects. They are particularly relevant to current-induced transients, and Barry (7) suggests that these electrokinetic processes may be involved in rapid plant movements.

For maximum pressures or electropotentials in electroosmosis or solution flow, both water and ionic solutes must move through the same pathway; also there are restrictions on "pore" diameters. In fact, little is known about the dimensions of the pathways through plant cell membranes, but the selectivity is so great that solution flow must be considered to be relatively small. In *Valonia ventricosa* Gutknecht (47) has shown that the osmotic and diffusional permeability coefficients are identical; this indicates the absence of water-filled pores through which solution flow or electroosmosis could occur.

Tyree & Fensom (139) have considered the role of electrokinetic processes in flow through vascular bundles, and they conclude from their data that neither pressure flow nor electrokinetics adequately explain the movement. In the absence of more concrete evidence for electroosmotic flow, I am inclined to the view that pressure flow is more likely.

Root Exudate Potentials

The PD between the root exudate and the bathing medium is influenced by external concentration and inhibitors much like a cell PD (22). Recently evidence has been found in corn roots that intact cytoplasm lines the vessel over a distance of several centimeters after perforation plates become open (53); thus the exudate appears to be an extension of the central vacuoles of vessels and to represent a vacuole PD. Therefore it appears that the long-sought barrier to radial movement in roots may be the xylem vessel plasmalemma.

Morphogenetic Effects

Growth responses have often been found to be associated with electrical fields. Etherton (30) reported that indole-3-acetic acid hyperpolarized cells of oat coleoptiles. In a later study of corn coleoptile cells, it was found that the lower end of the cell was about 2 mV more negative than the upper end in either upright or inverted segments (31); this was believed to be related to gravity-induced increases in metabolic rate in lower portions as well as to the geoelectric effect of the coleoptile. In a coleoptile placed horizontally, more auxin moves in the lower half and this side of the organ becomes electrically positive, by about 70 mV, to the upper half. The displacement of auxin is an indirect result of gravity apparently, rather than being due to the electric field (24). Light induces a similar displacement of auxin in vertical organs (69). The positive potential is always associated with the higher auxin content, although auxin at 100 mM may reverse the sign (12). The effect of auxin is believed to be on selectivity of cell membrane diffusion; enhancement by auxin of ^{42}K uptake has been reported (51). Other interactions of auxin and electrical fields have been discussed by Scott (114). It is a curious fact that the application of an external field leads to transport of indole-3-acetic acid toward the electronegative side (113); bending is toward the electropositive side, contrary to the results with stimuli from gravity or light, a paradox thus far unexplained.

The functioning of phytochrome has also been associated with electrical phenomena, including changes in the properties of cell membranes. In the oat coleoptile, red light induces a rapid increase of as much as 15 mV in the potential at the surface of the organ near the tip relative to its base (91); a reverse effect of far-red light was found. Measurements of the effect of red and far-red light on cell PDs of Mung bean indicate that the response, if existent, must be small (101); as pointed out by Briggs & Rice (14), the changes in coleoptile surface PDs must represent the sum of effects on many cells.

It seems well established that imposed electrical fields affect the development and may control polarity in reproductive cells such as *Fucus* eggs, *Funaria* spores, and of *Equisetum* spores (9). For reviews of this subject and related electrical phenomena see Bentrup (9) and Jaffe (66).

CONCLUDING REMARKS

It seems apparent that with the concept of electrogenic pumps we have reached a new and important stage in predictability of electrochemical events at cell membranes. The discovery that H^+ may contribute significantly to electrical conductance appears to resolve the serious conflict previously existent between measured tracer fluxes and those predicted. At the molecular level serious problems remain with regard to the nature of selective ion receptor sites, e.g. carriers, and the immediate energy source. ATP is indicated for most but not all active ion transports, and we are still seeking evidence for the degree of passive movement.

With respect to intercellular relationships, electrophysiology is in an infantile stage of development, but a beginning has been made. The problems are exceedingly complex. The functioning of all cell membranes, including that of cell organelles, seems to be associated with electrical phenomena and ion transports as well as with biochemical reactions; there is a hierarchy from cell organelles to the whole plant. At each level cause and effect have yet to be sorted out, but it seems safe to predict that electrical effects will be found to play an important role.

ACKNOWLEDGMENTS

I am grateful to the National Science Foundation for Grant No. GB 19201-1. I should also like to express my sincere thanks to those who have kindly supplied me with reprints, preprints, and other communications; unfortunately not all matters of significance could be included. I am much indebted to D. L. Hendrix for aid in the literature search and to W. P. Anderson for helpful suggestions on the manuscript.

ADDENDUM

Since the literature search was concluded, two reports on isolated plasma membranes have appeared. Y. F. Lai and J. E. Thompson (*Plant Physiol.* 1972. 50: 451–57) isolated membrane material in vesicle form from cotyledons of *Phaseolus vulgaris* and showed that Na^+-K^+ stimulated ATPase activity was related to active cation extrusion; added ATP was the energy source. Although the author's interpretation is that the active extrusion is related to the function of cotyledons in exporting materials during germination, it is also possible that the membranes become inverted. T. K. Hodges, R. T. Leonard, C. E. Bracker and T. W. Keenan (*Proc. Nat. Acad. Sci. USA* 1972. 69:3307–11) isolated plasma membrane material from roots of *Avena sativa* and showed that the fraction having highest Na^+-K^+ stimulated ATPase was that most enriched with plasma membrane. By various criteria this work appears to have yielded the cleanest preparation of plasmalemma vesicles yet made.

Literature Cited

1. Anderson, W. P. 1972. *Ann. Rev. Plant Physiol.* 23:51–72
2. Anderson, W. P., Ed. *The Liverpool Workshop on Ion Transport in Plants.* London: Academic. In press
3. Andrianov, V. K., Bulychev, A. A., Kurella, G. A., Litvin, F. F. 1971. *Biophysics USSR* 16:1031–36
4. Barber, J., Shieh, Y. J. 1971. *First Eur. Biophys. Congr.* 3:107–10
5. Barr, C. E. 1965. *J. Gen. Physiol.* 49:181–97
6. Barr, C. E., Broyer, T. C. 1964. *Plant Physiol.* 39:48–52
7. Barry, P. H. 1970. *J. Membrane Biol.* 3:335–71
8. Barry, P. H., Hope, A. B. 1969. *Biophys. J.* 9:729–57
9. Bentrup, F.-W. 1968. *Z. Pflanzenphysiol.* 59:309–39
10. Bowen, J. E. 1972. *Plant Physiol.* 49:82–86
11. Bowling, D. J. F., Ansari, A. Q. 1971. *Planta* 98:323–29
12. Brauner, L., Diemer, R. 1967. *Planta* 77:1–31
13. Brennecke, R., Lindemann, B. 1971. *T.I.T. J. Life Sci.* 1:53–58
14. Briggs, W. R., Rice, H. V. 1972. *Ann. Rev. Plant Physiol.* 23:293–334
15. Brown, H. D., Jackson, R. T., Dupuy, H. J. 1964. *Nature* 202:722–23
16. Bulychev, A. A., Andrianov, V. K., Kurella, G. A., Litvin, F. F. 1972. *Nature* 236:175–76
17. Cockrell, R. S., Harris, E. J., Pressman, B. C. 1967. *Nature* 215:1487–88
18. Crafts, A. S., Crisp, C. E. 1971. *Phloem Transport in Plants.* San Francisco: Freeman. 481 pp.
19. Cram, W. J. 1968. *J. Exp. Bot.* 19: 611–16
20. Dainty, J. 1962. *Ann. Rev. Plant Physiol.* 13:379–402
21. Dainty, J., Croghan, P. C., Fensom, D. S. 1963. *Can. J. Bot.* 41:953–66
22. Davis, R. F., Higinbotham, N. 1969. *Plant Physiol.* 44:1383–92
23. Davson, H. 1959. *A Textbook of General Physiology.* Boston: Little, Brown. 846 pp. 2nd ed.
24. Dedolph, R. R., Breen, J. J., Gordon, S. A. 1965. *Science* 148:1100–1
25. Dodd, W. A., Pitman, M. G., West, K. R. 1966. *Aust. J. Biol. Sci.* 19:341–54
26. Durst, R. A. *Ion-Selective Electrodes.* Washington, D. C.: Nat. Bur. Stand. Spec. Publ. 314. 474 pp.
27. Eisenman, G., Bates, R., Mattock, G., Friedman, S. M. 1965. *The Glass Electrode.* New York: Interscience. 106 pp.
28. Epstein, E. 1972. *Mineral Nutrition of Plants.* New York: Wiley. 412 pp.
29. Etherton, B. 1963. *Plant Physiol.* 38: 581–85
30. Ibid 1970. 45:527–28
31. Etherton, B., Dedolph, R. R. 1972. *Plant Physiol.* 49:1019–20
32. Etherton, B., Higinbotham, N. 1960. *Science* 131:409–10
33. Feder, W., Ed. 1968. *Bioelectrodes. Ann. NY Acad. Sci.* 148:1–287
34. Findlay, G. P., Hope, A. B. 1964. *Aust. J. Biol. Sci.* 17:62–77
35. Ibid, 400–11
36. Findlay, G. P., Hope, A. B., Pitman, M. G., Smith, F. A., Walker, N. A. 1969. *Biochim. Biophys. Acta* 183: 565–76
37. Findlay, G. P., Hope, A. B., Pitman, M. G., Smith, F. A., Walker, N. A. 1971. *Aust. J. Biol. Sci.* 24:731–45
38. Findlay, G. P., Hope, A. B., Williams, E. J. 1970. *Aust. J. Biol. Sci.* 23:323–38
39. Fisher, J. D., Hansen, D., Hodges, T. K. 1970. *Plant Physiol.* 46:812–14
40. Garrahan, P. J., Glynn, I. M. 1967. *J. Physiol. London* 192:237–56
41. Glynn, I. M., Hoffman, J. F., Lew, Y. L. 1971. *Phil. Trans. Roy. Soc. London Ser. B* 262:91–102
42. Goldman, D. E. 1943. *J. Gen. Physiol.* 27:37–60
43. Greenham, C. G. 1966. *Planta* 69: 150–57
44. Guggenheim, E. A. 1957. *Thermodynamics.* New York: Interscience. 3rd ed. 476 pp.
45. Gutknecht, J. 1966. *Biol. Bull. Mar. Biol. Lab. Woods Hole* 130:331–44
46. Gutknecht, J. 1967. *J. Gen. Physiol.* 50:1821–34
47. Gutknecht, J. 1967. *Science* 158:787–88
48. Gutknecht, J., Dainty, J. 1968. *Oceanogr. Mar. Biol.* 6:163–200
49. Harold, F. M., Barda, J. R., Pavlosova, E. 1970. *J. Bacteriol.* 101:152–59
50. Higinbotham, N. 1964. *Abstr. 10th Int. Bot. Congr.,* 169–70
51. Higinbotham, N. 1968. *Abh. Deut. Akad. Wiss. Berlin* Nr. 4:167–77
52. Higinbotham, N. *Bot. Rev.* In press
53. Higinbotham, N., Davis, R. F.,

Mertz, S. M. Jr., Shumway, L. K. See Ref. 2
54. Higinbotham, N., Etherton, B., Foster, R. J. 1964. *Plant Physiol.* 39:196–203
55. Ibid 1967. 42:37–46
56. Higinbotham, N., Graves, J. S., Davis, R. F. 1970. *J. Membrane Biol.* 3:210–22
57. Higinbotham, N., Hope, A. B., Findlay, G. P. 1964. *Science* 143: 1448–49
58. Hill, A. E. 1967. *Biochim. Biophys. Acta* 135:461–65
59. Hill, B. S., Hill, A. E. See Ref. 2
60. Hodgkin, A. L., Katz, B. 1949. *J. Physiol.* 108:37–77
61. Hogg, J., Williams, E. J., Johnston, R. J. 1968. *Biochim. Biophys. Acta* 150:518–20
62. Hogg, J., Williams, E. J., Johnston, R. J. 1969. *J. Theor. Biol.* 24:317–24
63. Hope, A. B. 1971. *Ion Transport and Membranes: A Biophysical Outline.* London: Butterworths. 121 pp.
64. Hope, A. B., Walker, N. A. 1960. *Aust. J. Biol. Sci.* 13:277–91
65. Ibid 1961. 14:26–44
66. Jaffe, L. F. 1970. *Develop. Biol. Suppl.* 3:83–111
67. Jagendorf, A. T., Uribe, E. 1967. In *Energy Conversion by the Photosynthetic Apparatus*, 215–45. Brookhaven Symp. Biol. 19. Brookhaven Nat. Lab. 514 pp.
68. Jeschke, W. D. 1970. *Z. Pflanzenphysiol.* 62:158–72
69. Johnsson, A. 1965. *Physiol. Plant.* 18:574–76
70. Kaback, H. R. 1970. *Ann. Rev. Biochem.* 39:561–98
71. Kishimoto, U. 1959. *Ann. Rep. Sci. Works, Fac. Sci. Osaka Univ.* 7:115–46
72. Kishimoto, U., Tazawa, M. 1965. *Plant Cell Physiol.* 6:507–18
73. Kitasato, H. 1968. *J. Gen. Physiol.* 52:60–87
74. Kylin, A. See Ref. 2
75. Larkum, A. W. D. 1968. *Nature* 218:447–49
76. Ling, G. N., Gerard, R. W. 1949. *J. Cell Comp. Physiol.* 34:382–96
77. Lüttge, U., Pallaghy, C. K. 1969. *Z. Pflanzenphysiol.* 61:58–67
78. Lyalin, O. O. 1969. *Sb. Tr. Agron. Fiz.* (USSR) 24:144–60
79. Lyalin, O. O., Ktitorova, I. N. 1969. *Fiziol. Rast.* 16:261–71
80. MacRobbie, E. A. C. 1962. *J. Gen. Physiol.* 45:861–78

81. MacRobbie, E. A. C. 1969. *J. Exp. Bot.* 20:236–56
82. MacRobbie, E. A. C. 1970. *Quart. Rev. Biophys.* 3:251–94
83. Macklon, A. E. S., Higinbotham, N. 1970. *Plant Physiol.* 45:133–38
84. Macklon, A. E. S., MacDonald, I. R. 1966. *J. Exp. Bot.* 17:703–17
85. Mitchell, P. 1966. *Chemiosmotic Coupling in Oxidative and Photosynthetic Phosphorylation.* Bodmin, Cornwall, England: Glynn Res. 192 pp.
86. Mitchell, P. 1970. In *Membranes and Ion Transport*, ed. E. E. Bittar, 1:192–256. New York: Wiley-Interscience. 483 pp.
87. Mullins, L. J. 1962. *Nature* 196:986–87
88. Nastuk, W. L., Ed. 1963. *Physical Techniques in Biological Research*, Vol. 6. *Electrophysiological Methods Part A.* 425 pp.
89. Ibid 1964. Vol. 5. 460 pp.
90. Neumann, J., Jagendorf, A. T. 1964. *Arch. Biochem. Biophys.* 107:109–10
91. Newman, I. A., Briggs, W. R. 1971. *Plant Physiol.* 47s:1
92. Nobel, P. S. 1969. *Biochim. Biophys. Acta* 172:134–43
93. Nobel, P. S., Craig, R. L. 1971. *Plant Cell Physiol.* 12:653–56
94. Pallaghy, C. K., Lüttge, U. 1970. *Z. Pflanzenphysiol.* 62:417–25
95. Pierce, W. S., Higinbotham, N. 1970. *Plant Physiol.* 46:666–73
96. Pitman, M. G., Mertz, S. M. Jr., Graves, J. S., Pierce, W. S., Higinbotham, N. 1971. *Plant Physiol.* 47:76–80
97. Pitman, M. G., Saddler, H. D. W. 1967. *Proc. Nat. Acad. Sci.* 57:44–49
98. Poole, R. J. 1966. *J. Gen. Physiol.* 49:551–63
99. Poole, R. J. 1969. *Plant Physiol.* 44: 485–90
100. Ibid 1971. 47:735–39
101. Racusen, R. H. 1972. *Phytochrome-induced electrical interactions in Mung bean root tips.* MS thesis. Univ. Vermont, Burlington. 32 pp.
102. Rapoport, S. I. 1970. *Biophys. J.* 10:246–59
103. Raven, J. A. 1967. *J. Gen. Physiol.* 50:1607–25
104. Ibid, 1627–40
105. Raven, J. A. 1968. *J. Exp. Bot.* 19: 193–206
106. Raven, J. A. 1970. *Biol. Rev. Cambridge Phil. Soc.* 45:167–221
107. Raven, J. A. 1971. *Planta* 97:28–38

108. Robertson, R. N. 1968. *Protons, Electrons, Phosphorylation and Active Transport.* Cambridge Univ. Press. 96 pp.
109. Robinson, J. B. 1969. *J. Exp. Bot.* 20:201–11
110. Ibid, 212–20
111. Saddler, H. D. W. 1970. *J. Gen. Physiol.* 55:802–21
112. Schilde, C. 1968. *Z. Naturforsch.* 23:1369–76
113. Schrank, R. H. 1960. In *Encyclopedia of Science and Technology,* 363–67. New York: McGraw-Hill
114. Scott, B. I. H. 1967. *Ann. Rev. Plant Physiol.* 18:409–18
115. Shanes, A. M. 1958. *Pharmacol. Rev.* 10:59–164
116. Sibaoka, T. 1966. In *Nervous and Hormonal Mechanisms of Integration,* 49–73. *20th Symp. Soc. Exp. Biol.* Cambridge Univ. Press
117. Sibaoka, T. 1969. *Ann. Rev. Plant Physiol.* 20:165–84
118. Slayman, C. L. 1965. *J. Gen. Physiol.* 49:69–92
119. Ibid, 93–116
120. Slayman, C. L. 1970. *Am. Zool.* 10: 377–92
121. Slayman, C. L., Lu, C. Y.-H., Shane, L. 1970. *Nature* 226:274–76
122. Smirnova, I. N., Lyalin, O. O., Karmanov, V. G. 1968. *Fiziol. Rast. (CCCP)* 15:625–30
123. Smith, F. A. 1966. *Biochim. Biophys. Acta* 126:94–99
124. Smith, F. A. 1968. *J. Exp. Bot.* 19: 207–17
125. Spanswick, R. M. *Biochim. Biophys. Acta.* In press
126. Spanswick, R. M. See Ref. 2
127. Spanswick, R. M. 1970. *J. Exp. Bot.* 21:617–27
128. Spanswick, R. M. 1972. *Planta* 102: 215–27
129. Spanswick, R. M., Costerton, J. W. F. 1967. *J. Cell Sci.* 2:451–64
130. Spanswick, R. M., Stolarek, J., Williams, E. J. 1967. *J. Exp. Bot.* 18:1–16
131. Spanswick, R. M., Williams, E. J. 1964. *J. Exp. Bot.* 15:193–200
132. Ibid 1965. 16:463–73
133. Spear, D. G., Barr, J. K., Barr, C. E. 1969. *J. Gen. Physiol.* 54:397–414
134. Stein, W. D. 1967. *The Movement of Molecules Across Cell Membranes.* New York: Academic
135. Strunk, T. H. 1971. *J. Exp. Bot.* 23: 863–74
136. Sydenham, P. K., Findlay, G. P. See Ref. 2
137. Tasaki, I., Singer, I. 1968. *Ann. NY Acad. Sci.* 148:36–53
138. Tyree, M. T. 1970. *J. Theor. Biol.* 26:181–214
139. Tyree, M. T., Fensom, D. S. 1970. *J. Exp. Bot.* 21:304–24
140. Ussing, H. H., Zerahn, K. 1952. *Acta Physiol. Scand.* 23:110–27
141. Vorobiev, L. N. 1967. *Nature* 216: 1325–27
142. Vorobiev, L. N. 1968. *Abh. Deut. Akad. Wiss. Berlin Kl. Med.* Nr. 4:197–201
143. Vorobiev, L. N., Khitrov, Y. A. 1971. *Stud. Biophyzi. Berlin* 1:49–56
144. Vredenberg, W. J. 1970. *Biochim. Biophys. Acta* 223:230–39
145. Walker, N. A. 1955. *Aust. J. Biol. Sci.* 8:476–89
146. Ibid 1960. 13:468–78
147. Walker, N. A., Hope, A. B. 1969. *Aust. J. Biol. Sci.* 22:1179–95
148. Williams, E. J., Johnston, R. J., Dainty, J. 1964. *J. Exp. Bot.* 15:1–14
149. Wood, R. D., Imahori, K. 1964–65. *A Revision of the Characeae.* Weinheim: Cramer. Vol. I, 943 pp.; Vol. II, 789 pp.
150. Wyn Jones, R. G., Lunt, O. R. 1967. *Bot. Rev.* 33:407–26
151. Young, M., Jefferies, R. L., Sims, A. P. 1970. *Abh. Deut. Akad. Wiss. Berlin* Nr. 4 B and b:67–82

Ann. Rev. Plant Physiol. 1973. 24:47–68

PROTEIN BIOSYNTHESIS ❖ 7541

Saul Zalik and B. L. Jones[1]

Department of Plant Science, University of Alberta, Edmonton, Alberta, Canada

CONTENTS

INTRODUCTION

Several reviews have appeared in the last few years which are particularly relevant to the subject of protein synthesis. Among these are reviews of general protein biosynthesis (73, 91, 131), a review of ribosome structure and function relationships (76), and reviews specifically oriented toward protein biosynthesis in plants (19, 20). The subject of general plant protein biosynthesis has thus recently been covered in some detail.

In general, the pathways of protein biosynthesis seem to be similar in all the organisms that have been studied. Our current understanding of the role of ribosomes and the other components of protein synthesizing systems in plants,

[1] Present address: Department of Plant Science, University of Manitoba, Winnipeg, Manitoba.

as in animals, has depended considerably on information obtained from studies on prokaryotes, especially *Escherichia coli*. Protein synthesis in plants may occur in free or membrane-bound ribosomes of cytoplasmic, mitochondrial, or chloroplast origin. Depending upon their stage of development, plants contain differing amounts of ribosomes of these various classes. Therefore, despite the difficulties associated with the isolation of homogeneous ribosome preparations from chloroplasts, mitochondria, and the cytoplasm, plants afford a unique opportunity for comparative studies on protein synthesis during growth and development.

RIBOSOMES

Protein biosynthesis occurs on a very complicated system, and even in prokaryotic cells a good deal of uncertainty remains about the synthesis process. For example, although the ribosome is at the center of this process, much about its genesis and the role of its component parts is not known. Many physical and chemical characteristics of the ribosome have been determined, however. This review will begin with a brief description of some of these. The main emphasis, though, will be on the mechanism of protein synthesis and wherever possible will describe the steps that have been confirmed or elucidated in studies with plant materials.

Description of Ribosomes

Rather drastic treatment is required to obtain the different kinds of ribosomes from higher plants, so the different physical and biochemical properties ascribed to them may, to varying degrees, be a reflection of the methods employed in their isolation. Nevertheless, a fairly consistent picture of the chloroplast and cytoplasmic ribosomes of higher plants has emerged. The plant cytoplasmic ribosomes are characterized as having sedimentation values of 80S like mammalian cytoplasmic ribosomes, while chloroplast ribosomes are 70S like those of bacteria (9, 10, 14, 15, 39, 52, 64, 79, 88, 92, 118, 129, 135, 146, 150). Other features which distinguish these two classes of ribosomes are the molecular weights of their RNAs (87), the number and electrophoretic mobility of their proteins (69, 157), and their different sensitivities to some protein synthesis inhibitors (19, 41, 42, 123). An additional difference found recently by Payne & Dyer (122) is that 80S cytoplasmic ribosomes of plants, like those from animals, contain 5.8S RNA, while 70S chloroplast ones do not.

Although Wilson et al (162) reported that mitochondrial ribosomes from corn root sedimented as 66S particles, mitochondrial ribosomes from a number of higher plants have recently been reported by Leaver & Harmey (78) to have sedimentation values of 78–80S. If so, they would resemble plant and animal cytoplasmic ribosomes and, perhaps surprisingly, would be unlike the mitochondrial ribosomes of most other species that have been investigated (16–18), although it has been reported that mitochondrial ribosomes from *Tetrahymena* contain 80S ribosomes (2, 34). If the finding by Leaver & Harmey (78) is substantiated, it will compound the uncertainty of many earlier studies conducted with plant cytoplasmic ribosomes, since it is unlikely that the cytoplasmic preparations were devoid of mitochondrial ribosomes. In addition, it will be harder to resolve the problem of whether every chloroplast, cytoplasmic, or mitochondrial ribosome from a given plant contains exactly the same structural proteins or whether individual ribosomes may contain different structural protein complements. Clearly, a prime requirement in any such determinations must be that of obtaining a clean separation of each class of ribosomes.

In general, estimates of the dimensions of ribosomes in solution, obtained by sedimentation velocity and other methods, are somewhat larger than those determined by electron microscope studies. For the 70S ribosomes of prokaryotes which have been extensively studied, the dimensions are $200 \times 170 \times 170$ Å in the dry state and $290 \times 210 \times 120$ Å in solution (144). Chloroplast and cytoplasmic ribosomes of tobacco leaves were compared by Miller et al (108) using electron microscopy. The cytoplasmic ribosomes were described as being acorn-shaped and composed of two associated particles, one sphere-like and the other a cap-like unit. The dimensions of the 80S ribosome were 286 ± 28 Å $\times 222 \pm 25$ Å. The cytoplasmic ribosomes of peas, according to Bruskov & Odintsova (25), were 260 ± 10 Å $\times 190 \pm 10$ Å in size and those of bean 260 ± 10 Å $\times 200 \pm 10$ Å. Both groups of workers noted a cleft in the chloroplast ribosomes. Miller et al (108) recorded measurements of 268 ± 24 Å $\times 214 \pm 20$ Å for tobacco chloroplast ribosomes, whereas for peas and beans Bruskov & Odintsova (25) found dimensions of only 220 ± 10 Å $\times 170 \pm 10$ Å. According to the latter authors, the apparently larger dimensions of the tobacco chloroplast ribosomes may have resulted from cytoplasmic ribosome contamination and selective elimination of 70S ribosomes during the negative staining procedure employed by Miller et al (108).

Ribosome Subunits

All ribosomes are composed of subunits of unequal size. Ribosomes (70S) from prokaryotes and chloroplasts have subunits of 50S and 30S, while cytoplasmic ribosomes (80S) of eukaryotes have subunits of 60S and 40S. In his recent review, Kurland (76) points out that although Bretscher's (23) and Spirin's (143) models assign prominent roles to the subunit structure of ribosomes, there is still no demonstration that ribosome dissociation is obligatory following each round of protein synthesis in vivo.

Raacke (125) has questioned both of these models on two counts. First, she points to evidence against the donor molecule in peptide bond synthesis being peptidyl-tRNA (see *Elongation*). Secondly, she draws attention to the mechanical problem of translocating a molecule as large as a tRNA esterified to a protein, since this complex is bound to both the 50S and 30S subunits. Although she proposed a model for protein synthesis in which the donor for the peptide elongation reaction is peptidyl-5S RNA, it is not readily apparent how this would reduce the inherent complexities of translocation. Moreover, Fahnestock & Nomura (45), who tested this model by determining the activity of ribosomes containing 5S RNA with a chemically modified 3'-terminus, concluded that her model and probably any other model that involved formation of peptidyl-5S RNA is incorrect.

To understand the intricacies of the protein biosynthesis process, detailed information will be required on the dynamic three-dimensional relationship between the structural proteins and RNA in the ribosome, as well as details on the initiation, elongation, and release factors which are involved. Information on the protein components of ribosomes has been provided by the painstaking analysis of the proteins from *E. coli* ribosome subunits (76). The analysis of ribosomal RNA from *E. coli* has progressed to the stage where the nucleotide sequence of the 5S RNA has been described (24) and the sequencing of both the 16S and 23S RNAs are well in hand (46–48).

Reconstitution of Subunits

The elegant reconstitution experiments of Traub & Nomura (154), in which they were able to reconstitute active *E. coli* 30S ribosomal subunits from the purified structural proteins and RNA, provided a method for relating the function of the organelle with its molecular composition (113). Moreover, they demonstrated that the information for the correct assembly of this particle is contained in the structure of its molecular components and not in nonribosomal factors. Since then, Nomura & Erdmann (117) have succeeded in reconstituting *Bacillus stearothermophilus* 50S subunits from their dissociated molecular components, and Maruta et al (102) have reassembled functionally active *E. coli* 50S ribosomal particles from their component proteins and RNAs. Nomura's group (56, 114, 116) has also studied ribosome assembly in vivo and has concluded that the in vivo and in vitro assembly processes are somewhat different. Despite these studies and the information they have provided, the precise manner by which the ribosomal RNAs and proteins participate in, and direct, protein synthesis still eludes us.

PROCESS OF PROTEIN BIOSYNTHESIS

The mechanism of bacterial protein synthesis as it is presently understood is diagrammed in Figure 1. The figure and this text employ the uniform system of nomenclature recently adopted (31) for factors involved in protein biosynthesis.

Transfer RNAs

The first step of the protein biosynthesis process is the formation of aminoacyl-tRNAs. This is done by a two-stage reaction, both stages of which are catalyzed by a single enzyme, the aminoacyl-tRNA synthetase or ligase. This reaction can be represented as:

$$(a) \quad \text{Amino acid} + \text{ATP} \xrightarrow{\text{Enzyme}} \text{Enzyme-aminoacyl-AMP}$$

$$+ \text{ Pyrophosphate}$$

(b) Enzyme-aminoacyl-AMP + tRNA → Enzyme

+ aminoacyl-tRNA + AMP

The product of this reaction, aminoacyl-tRNA, contains an "activated" amino acid with enough energy in the amino acid-tRNA bond for formation of the lower energy peptide bond. In addition, binding of the tRNA adaptor molecule, with its anticodon, to the amino acid ensures that the amino acid will be placed in its correct position in the finished protein. The aminoacylation reaction can therefore control the amounts of free and aminoacylated tRNAs in the cell and might thus regulate the rate of protein biosynthesis.

Since proteins commonly contain 20 different amino acids, plants must possess at least an equal number of tRNAs. The existence of isoaccepting tRNA molecules (different tRNA molecules which are aminoacylated by the same amino acid), however, indicates that there must be more than 20 plant tRNAs. Isoaccepting tRNAs have now been shown for several of the amino acids. Soybeans, for example, contain at least six different leucine-accepting tRNA species (6) and peas contain two (33). In general, plants seem to contain multiple isoaccepting tRNAs for each amino acid (27, 58, 104, 156). Since there are probably at least three (cytoplasmic, chloroplast, and mitochondrial) protein synthesizing systems in plants, the question whether the presence of isoaccepting tRNAs reflects the presence of tRNAs specific to each of the three systems arises. Investigations of the isoaccepting tRNAs of cotton (40) have shown that of the four tRNAs which accept valine one is of chloroplast origin while two of the four isoleucine-accepting tRNAs are from the chloroplasts. Likewise, the chloroplasts contained one to three of the isoaccepting tRNAs for each of the other amino acids tested (104). These values represent the minimum number of both chloroplast and total iso-accepting species, and improvements in the separation methods may lead to the discovery of more of both total and chloroplast tRNAs. The amount of chloro-plast isoaccepting tRNAs is low in etioplasts, roots, and developing cotyledons (28), but their quantity increases on exposure of etioplasts to light or when germination begins. After germination there is apparently no change in the tRNAs during maturation of cotton plants (104). No differences were found in the levels of tRNAs of dividing and nondividing pea cells (156), although the amounts of some soybean cotyledon tRNAs did change with development and senescence (13). Cytoplasmic and chloroplast methionine-accepting tRNAs have also been found, and these are discussed in the section on protein initiation.

Aminoacyl-tRNA Synthetases

Aminoacyl-tRNA synthetase enzyme has been found in a variety of plants (11, 38, 82, 112, 124). The presence of isoaccepting tRNA species in higher plants raises the question of whether or not the isoaccepting tRNAs for each amino acid are acylated by a single aminoacyl-tRNA synthetase. There is evidence that a single E. coli leucyl-tRNA synthetase can acylate five different leucine-specific tRNAs (74). The situation appears to be quite different for the leucine isoaccept-

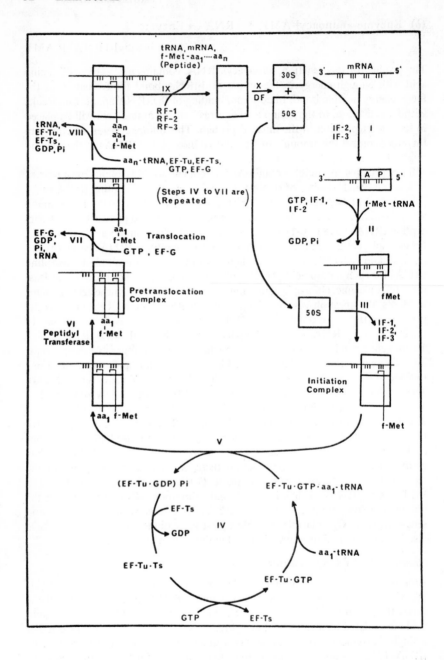

←‑《《《

Figure 1 Diagram of protein synthesis by bacterial ribosomes:

I. The binding of messenger RNA to the 30S ribosomal subunit. Binding is near the 5′ end of the mRNA cistron. Initiation factor IF-2 promotes the binding of synthetic nucleotides, but IF-3 is apparently also involved in the binding of "natural" messenger RNAs.

II. The binding of the initiating aminoacyl-tRNA to the 30S subunit-mRNA complex. This step requires GTP, IF-1, and IF-2. The bacterial initiating aminoacyl-tRNA is fMet-tRNA. Chloroplasts also use formylated methionyl-tRNA for initiation, while plant cytoplasmic ribosomes use nonformylated methionine-tRNA. The aminoacylated tRNA may bind directly to the P site or alternatively it may bind at some "pre-P" site and subsequently move to the P site. The 40S subunit·mRNA·aminoacyl-tRNA complex of wheat has been demonstrated.

III. The large (50S) ribosomal subunit binds to form the Initiation Complex.

IV. Through a series of replacement steps, GDP is removed from the elongation factor EF-Tu and replaced by GTP. Aminoacyl-tRNA is then added to form an EF-Tu·GTP ·aa-tRNA complex for use in V whenever the codon of the mRNA corresponding to the anticodon of the complex is at the A site of the ribosome.

V. Chain elongation begins. EF-Tu·GTP·aa_1-tRNA donates its aminoacyl-tRNA with a concomitant hydrolysis of GTP.

VI. The peptidyl transferase or peptide synthetase step. Peptidyl transferase enzyme (probably a part of the 50S ribosomal subunit) catalyzes peptide bond formation with the result that the nascent peptide chain is located at the A site after reaction.

VII. Translocation. In the presence of the elongation factor EF-G and GTP the mRNA and ribosome move with respect to each other to bring the next codon into position for translation. The nascent peptide is then located at the P site of the ribosome. The deacylated tRNA which was bound at the P site is released and GTP is hydrolyzed.

VIII. Amino acids are added to the carboxyl-terminal end of the growing peptide chain by repetitions of the steps V through VII.

IX. In the presence of one of the termination codons (UAG, UAA, or UGA) of the mRNA and the release factors appropriate to that codon, the finished peptide is released from the ribosome, together with the mRNA and the tRNA which carried in the car-boxyl-terminal amino acid. The ribosome may be released as a 70S monoribosome or as subunits which recombine in the absence of dissociation factor.

X. Ribosome dissociation. In the presence of dissociation factor (DF), which may be the initiation factor IF-3, the ribosomal subunits separate.

ing tRNAs of soybean seedlings (75). It was found (6) that a total leucyl-tRNA synthetase preparation from soybean hypocotyls could not acylate two (peaks five and six) of six leucine-accepting tRNA species of soybean even though a similar preparation from cotyledons could. This indicated that at least two different leucine aminoacyl-tRNA synthetases were involved, one of which was missing in the hypocotyl preparation. When the total cotyledon synthetase preparation was separated into three fractions on hydroxyapatite, it was found that one of the enzyme fractions could acylate the two (peaks five and six) tRNAs which were not acylated by the hypocotyl enzymes but could not charge the other four tRNAs. The other two synthetases, separated with hydroxyapatite, acylated only the remaining four tRNAs. The enzyme fraction which acylated the tRNA peaks five and six was absent from a total enzyme preparation obtained from hypocotyls. Soybeans thus contain at least three enzymes which attach the amino acid leucine onto leucine isoaccepting tRNAs. Since two of the enzymes seem to acylate the same four tRNAs, yet elute from hydroxyapatite differently, it may be that they contain common active sites but are altered at some point removed from the active site.

Just as there are cytoplasmic and chloroplast tRNAs in plants, there are also aminoacyl-tRNA synthetases which seem to be localized in the cytoplasm and in the chloroplasts of plants. *Euglena* (126), when grown in light, contains two aminoacyl-tRNA synthetases for both phenylalanine and isoleucine. In dark-grown *Euglena* or in bleached mutants which contain no chloroplasts, however, there is only one of each of the synthetases, indicating that the missing synthetases are probably localized in the chloroplasts. This conclusion is strengthened by the observation that light-induced isoleucyl and phenylalanyl-tRNAs, which are therefore presumably of chloroplast origin, can only be acylated by the chloroplast synthetases. Isoleucyl- but not phenylalanyl-tRNA synthetase is absent from the bleached *Euglena* mutant. The presence of the chloroplast phenylalanyl-tRNA synthetase in the bleached mutant, which contains no chloroplast DNA or structure, probably indicates that this chloroplast enzyme is synthesized in the cytoplasm and is coded by nuclear genes (126).

Among higher plants, chloroplast aminoacyl-tRNA synthetases have been demonstrated in *Phaseolus vulgaris* (27) and *Gossypium hirsutum* (cotton) (105). In bean chloroplasts it has been shown (27) that there are three leucine-specific tRNAs, which are only aminoacylated by the chloroplast enzymes. In addition, there is one chloroplast valine-specific tRNA which is recognized by only the chloroplast synthetase.

Merrick & Dure (105, 106) have investigated the charging by *E. coli* aminoacyl-tRNA synthetases of cotton cotyledon total tRNA and of tRNA isolated from partially purified cotton chloroplasts. They reported that total tRNA from cotyledons of young embryos or dry seeds, containing about 10% chloroplast tRNA, was only acylated 41% as much by *E. coli* synthetase as it was by cotton synthetase. At the same time, tRNA of green cotyledons, containing about 40% chloroplast tRNA, was charged 52%, and the tRNA from partially purified chloroplasts was charged 72% as much by *E. coli* synthetases as it was by cotton synthetases. This would indicate that the synthetases of *E. coli* can recognize and acylate chloroplast tRNA better than cytoplasmic tRNA. To test this hypothesis, the isoaccepting tRNAs for each of 19 amino acids were isolated from cotton and reacted with *E. coli* aminoacyl-tRNA synthetases. The result indicates that *E. coli* synthetases can aminoacylate all cotton tRNAs (both chloroplast and cytoplasmic) for some of the amino acids (arg, his, lys, met), they amino-acylate only chloroplast tRNAs in others (leu, ile, tyr), some chloroplast and some cytoplasmic tRNAs in a third group (asn, asp, cys, glu, gly, phe, ser, thr, and val), and would not charge any of the tRNAs of a fourth group (ala, gly, pro). When the converse heterologous aminoacylation was tried (i.e. tRNAs from *E. coli*, synthetases from cotton) all the isoaccepting tRNAs for three amino acids (ala, ile, val) were charged. Some, but not all, of the isoaccepting tRNAs of all the other amino acids were charged except that none of the isoaccepting glutamine tRNAs were charged. The finding that the cotton tRNAs completely charged by the *E. coli* synthetases are different from the *E. coli* ones completely charged by the cotton synthetases implies that the recognition mechanism governing the reaction is not stringent. However, no evidence of charging of a tRNA with an incorrect amino acid was noticed.

The specificity of leucyl-tRNA synthetase in plant species (peas and soybeans) has been tested by measuring the ability of aminoacyl-tRNA synthetases from the two species to charge separated leucine isoaccepting tRNAs from both peas and soybeans. While soybean synthetase could charge six tRNAs, the pea synthetase could only charge three, so differences exist even between plant species, and this fact should be considered whenever protein synthesis is being assayed in a system containing components from different species. Other cases of cross-reaction between tRNA and synthetases from different sources have been reviewed recently (20).

Different aminoacyl-tRNA synthetases from *Aesculus* (buckeye) (4, 5) and from bean leaves (151, 152) have been partially purified and characterized. A leucyl-tRNA synthetase from yeast has been crystallized and appears to be particularly amenable to X-ray structural analysis to atomic resolution, although heavy-atom isomorphous replacements have not yet been obtained (35).

The specificity of the synthetase enzymes for both amino acids and tRNAs must play a large part in determining that amino acids are inserted into protein in the correct order. It has recently been reported that *E. coli* phenylalanyl-tRNA synthetase enzyme can destroy leucyl-tRNA[Phe], in which a leucine has been incorrectly bound to a phenylalanine-specific tRNA (163). This process may serve as another means of preserving the fidelity of translation during protein biosynthesis.

Initiation

Over the last decade the steps involved in the synthesis of protein by bacteria have been elucidated, and the process was recently reviewed by Lucas-Lenard & Lipmann (91). It has been known for some time that the amino acid most commonly found at the N-terminal of *E. coli* proteins is methionine (158), indicating that methionine was probably the first amino acid incorporated into nascent protein. It is now apparent that *E. coli* cells begin protein synthesis by incorporating N-formylmethionine (f-Met), using a special methionyl transfer RNA, $tRNA_f^{Met}$ (initiator tRNA), whose methionine is enzymatically formylated after being bound to the tRNA. Nonterminally located methionine residues are inserted into the nascent peptide by a separate methionine-accepting tRNA, $tRNA_m^{Met}$. The methionine bound to this tRNA cannot be formylated, even though it is apparently placed into the tRNA by the same aminoacyl-tRNA synthetase that charges $tRNA_f^{Met}$.

It is now generally accepted that when proteins are synthesized on 80S cytoplasmic ribosomes of plants, the initiating tRNAs are methionine-accepting tRNAs charged with nonformylated methionine, whereas formylated methionine is the initiator for synthesis catalyzed by chloroplast (70S) ribosomes. The most thoroughly characterized protein synthesizing system from plants is that of wheat, which has been studied by several groups.

Chloroplasts (26, 133) and mitochondria (50, 141) of eukaryotes contain fMet-tRNA (N-formylmethionyl transfer RNA), while none is present in the cytoplasm. It has been demonstrated that fMet-tRNA is also the chain initiator of at least part of the protein synthesis in *Acetabularia* chloroplasts (12). From their studies on wheat germ, Leis & Keller (83) obtained three chromatographically distinct methionine tRNAs, and they postulated that protein chain initiation in the cytoplasm of wheat germ was brought about by the initiating methionyl-tRNA, which was present in major amount, without prior formylation. In an independent study, Tarrago et al (153) arrived at essentially the same conclusion. When Leis & Keller (84) extended their study to wheat leaves they found two chain-initiating methionine tRNAs. The one which was located in the cytoplasm was not formylated by wheat extracts and functioned in protein chain initiation in the cytoplasm without formylation. The other initiating tRNA was from the chloroplasts. It was formylated by transformylase present in wheat extracts and functioned in chain initiation in chloroplasts as formylmethionyl-tRNA. Further evidence for two methionine-tRNA systems in wheat—a chloroplast one which uses formylated methionine for initiation and a cytoplasmic one which initiates protein synthesis with unformylated methionine—was provided by Marcus et al (99) from studies which employed tobacco mosaic virus RNA (TMV-RNA) for messenger (94). In addition, they showed that the TMV proteins synthesized by cytoplasmic ribosomes contain unformylated methionine at their NH_2-terminal ends.

Protein synthesis initiation has also been studied in some detail in beans. *Phaseolus vulgaris* chloroplasts contain fMet-tRNA (26), while the cytoplasm contains only unformylated methionyl-tRNA. F-Met-tRNA and an active transformylase have also been shown in bean etioplasts (53). Work by Yarwood et al (164) has shown that protein synthesis initiation in *Vicia faba* is much like that of wheat, since the bean seeds contain two major $tRNA^{Met}$ species, presumably cytoplasmic, neither of which can be formylated. One of these apparently initiates synthesis with unformylated methionine while the other specifies internal methionine. A minor species of tRNA^{Met} was also found. After charging, its methionine group could be formylated with either *E. coli* or bean transformylase, and it is possibly the initiator tRNA of cellular organelles. A cell-free system from *Euglena* chloroplasts also translates viral RNA messengers and places fMet at the N-terminal end of synthesized viral coat protein (138). While virtually nothing is known about the initiation of protein synthesis in plant mitochondria (due to complications in obtaining pure plant mitochondrial ribosomes), it has been shown that bean (*Vicia faba*) mitochondria contain formylated methionyl-tRNA (53). Since mitochondrial ribosomes, like chloroplast ones, are generally similar to bacterial ribosomes, and since fMet-tRNA is the initiator of *Neurospora* mitochondrial protein synthesis (43), it seems probable that plant mitochondria also use fMet as the starting amino acid.

Initiation Factors

In bacterial protein synthesizing systems (73, 91) the initiation of protein synthesis (i.e. binding of mRNA, 30S ribosomal subunit, 50S ribosomal subunit, and fMet-tRNA) requires at least three protein factors: IF-1, IF-2, and IF-3 (31). Although the exact roles of these factors are not known, a number of their functions have been identified. The first step of the ribosomal cycle, binding of mRNA to the 30S ribosomal subunit, is stimulated by factor IF-2 (Figure 1, step I[2]). Factor IF-3 may also play a part in this binding when "natural" messenger RNA is used. IF-1, IF-2, and GTP participate in the binding of fMet-tRNA to the 30S subunit-mRNA complex. IF-3, or one of its subfractions, probably also acts as a dissociation factor, giving 30S and 50S subunits from 70S ribosomes. To further complicate matters, the three initiation factors can interact with one another and so modify their individual effects.

Ciferri (36) has stressed that a distinguishing feature for different classes of ribosomes is the nonexchangeability of their initiation factors. He showed that the ribosomal initiation factors of one species of bacteria could function in another and that they could also function for chloroplast and mitochondrial ribosomes but not cytoplasmic ones. This is in agreement with the findings of Richter et al (127).

The initiation factors of higher plants have been studied extensively by Marcus and his co-workers, who have published a series of papers outlining the steps occurring during initiation of protein biosynthesis in a wheat embryo system using TMV-RNA as a messenger RNA. When using a "natural" mRNA, where synthesis must begin with reading of the AUG initiation codon and the concomitant formation of a true initiation complex (in contrast to studies with poly U messenger), it was found (95) that protein synthesis began only after a lag period. Preincubation of ribosomes with a mix of ATP and two protein factors overcame the lag period and protein synthesis began at once, indicating that an initiation complex was probably formed before protein elongation began. In contrast to initiation systems of bacteria, the two protein factors from wheat

[2] Refers to the steps diagrammed in Figure 1 and used in the legend.

embryo needed to relieve the lag period were found in the cell cytoplasm not bound to ribosomes. In addition, ATP was specifically needed for wheat but not for *E. coli* protein synthesis initiation. The initiation complex was separated out by gradient centrifugation (96) and did not appear to contain either free or aminoacylated tRNA. Upon addition of aminoacylated tRNA the complex chased into polysomes. By showing that aurintricarboxylic acid (ATA) inhibited binding of the TMV-RNA messenger to wheat ribosomes (97), and using the inhibitor to prevent further binding of messenger, it was possible to show that two separable factors (factors C and D) were necessary before initiation could occur. The two factors have been completely separated (136), and protein chain initiation is dependent on both factors. The factors are apparently analogous to the bacterial IF-1 and IF-2.

The protein synthesis initiation inhibitor pactamycin has been used to show that initiation in the presence of factors C and D involves at least two steps (137). In the presence of pactamycin, Met-tRNA is bound to the ribosomes but cannot react with puromycin, while in the absence of pactamycin it reacts freely with puromycin. The same is found when factor C is replaced by a translocation factor from wheat, so it may be that factor C is responsible for a shift from some "pre-P" site into the P site (see *Elongation*). Marcus and his co-workers (161) have provided evidence that initiation in wheat embryo ribosomes proceeds through the formation of a 40S ribosomal subunit-mRNA-Met-tRNA complex which subsequently combines with a 60S ribosomal subunit to yield the 80S mono-ribosome which is functional in protein synthesis. Protein initiation in the wheat embryo system has been shown to be inhibited by potassium fluoride and 2-(4-methyl-2,6-dinitroanilino)-N-methyl-propionamide (MDMP), which inhibits the binding of the 60S ribosomal subunit to the 40S·mRNA·tRNA complex (160). MDMP may, therefore, prove useful in further studies of protein initiation in eukaryotes.

Dissociation of Ribosomes

The mechanism and significance of ribosomal dissociation is not fully under-stood, but it appears that within the bacterial cell 70S ribosomes exchange 30S and 50S subunits (71, 72). In one view the ribosomes are released as 70S mono-somes which then dissociate before initiation begins anew. The results obtained by Subramanian et al (148) support this. They isolated a protein factor (dissocia-tion factor, DF) from 30S subunits of *E. coli* ribosomes which promoted almost complete dissociation of 70S ribosomes from starved *E. coli* cells. Garcia-Patrone et al (51) found that antibiotics which affected the 30S ribosome subunit modified the ability of DF to dissociate the treated ribosomes, but antibiotics which affected the 50S subunit did not, lending support to the notion (149) that DF binds to the 30S subunit.

Subramanian & Davis (147) and Albrecht et al (1) tested each of the three *E. coli* initiation factors for dissociation activity. Only IF-3 acted as a DF in their studies. On the other hand, Miall et al (107) extensively purified a ribosome dis-sociation factor from *E. coli* and concluded, on the basis of purification procedure and molecular weight, that the active DF might be the same as the IF-1 described by Hershey et al (61). Dissociation factors associated with animal ribosomes have also been reported (89). In contrast to the above theory, where dissociation follows release of the ribosome from messenger, an alternative view (72) holds that, upon completion of polypeptide synthesis, *E. coli* ribosomes are released from polysomes directly in the form of ribosomal subunits. According to Kaempfer (72), the single ribosomes found in cells are not actively participating in the ribosome cycle, which uses only subunits.

Dissociation and reassociation studies on plant ribosomes have involved manipulation of the constitution of isolation and/or dissociation media (70, 101, 161) and the use of puromycin (101, 155) or N_2 gas pretreatment (85). Since the results obtained in dissociation studies have depended very much on the precise environmental conditions and the preparative and analytical methods used, it is not surprising that they have yielded variable results. For example, using the ultracentrifuge as a means of assessing dissociation may give misleading results since it has been shown that in the course of analytical ultracentrifugation ribosomes may dissociate (60, 65, 66).

Summary of Initiation

In summary, the initiation complex of prokaryotes is formed by the binding of mRNA and fMet-tRNA to the 30S ribosome subunit through the intervention of at least three initiation factors. The 50S subunit then joins to yield the 70S monoribosome. In plant cytoplasmic ribosomes a 40S ribosomal subunit·mRNA ·met-tRNA complex is formed, probably with the intervention of at least two factors. The 60S subunit then combines with this complex to form an 80S monoribosome. The monoribosomes formed, either 70S or 80S, are presumably ready at this stage to accept the next aminoacyl-tRNA and elongation of the peptide chain commences.

Elongation

Before discussing the elongation cycle, it may be helpful to explain the terminology used for the binding sites on the ribosome. The site to which the initiator tRNA, i.e. fMet-tRNA or Met-tRNA binds (Figure 1, step I) is referred to as the peptidyl (P) or donor site. The first step in elongation (V) is the binding of an aminoacyl-tRNA to a ribosomal site adjacent to the P site. This second site, where the incoming aminoacyl-tRNA attaches, is called the acceptor or A site. Peptide bond formation then occurs between the newly bound aminoacyl-tRNA at the A site and the peptidyl tRNA located at the P site. The P site is called the donor site since it donates its peptide chain to the amino group of the aminoacyl-tRNA located at the acceptor (A) site. The next step in the elongation process (VII) involves the translocation of the newly synthetized peptidyl tRNA from the A site to the P site with the concomitant advance of the ribosome along the messenger by the distance of one codon in the 5' to 3' direction. The acceptor (A) site is thus freed for the next elongation cycle.

Relatively complete study of the elongation factors of prokaryotes has defined a series of partial reactions in which they are involved. These have been detailed by Lucas-Lenard & Lipmann (91). Unlike the bacterial initiation factors, which are proteins loosely attached to the 30S subunit (128), the elongation factors are proteins found in the soluble fraction of the cell. There are at least three such elongation factors in the prokaryotic organisms: EF-Tu, EF-Ts, and EF-G (90, 138, 140). As for other eukaryotes in which two elongation factors, designated EF-1 and EF-2, have been isolated (31, 59), two factors have also been found in plants (37, 68, 81). However, it was recently reported that EF-1 from calf brain (111) and EF-1 from wheat germ (80) may exist in two different forms.

The series of reactions in which prokaryote elongation factors are involved may be described briefly as follows. Elongation factor EF-Tu binds with GTP and aminoacyl-tRNA in a stepwise manner (IV) to produce a ternary complex, EF-Tu·GTP·aa-tRNA, which reacts with ribosomes, transferring the aminoacyl-tRNA to the A site (V). An EF-Tu·GDP complex and inorganic phosphate are released (138, 159). The second elongation factor, EF-Ts, then displaces GDP from EF-Tu·GDP to form the complex EF-Ts·EF-Tu. Finally, GTP displaces EF-Ts to regenerate EF-Tu·GTP.

The peptide bond formed between the terminal carboxyl group of the peptidyl-tRNA at the P site and the α-amino group of the aminoacyl-tRNA at the A site is formed by the enzyme peptidyl transferase (VI) which is a component of the 50S ribosomal subunit (93, 110).

A third bacterial elongation factor, EF-G, is required, along with GTP, to translocate the newly formed peptidyl-tRNA to the P site and to move the ribosome along the mRNA (VII). Fairly direct evidence has recently been presented by Gupta et al (55) that for each amino acid incorporated into a polypeptide chain the ribosome moves three nucleotide units along the messenger RNA in the 5' to 3' direction. Hydrolysis of GTP is necessary for the translocation. The elongation cycle is repeated until a termination codon is read.

Two elongation factors have been isolated from wheat germ extract by Legocki & Marcus (81). They demonstrated that the wheat elongation factor EF-1 stimulates binding of aminoacyl-tRNA to ribosomes, while EF-2 acts as a translocase. It is assumed that the elongation factors from plants are like those from other eukaryotes and therefore function in a manner analogous to those of prokaryotes (38, 68, 81). Legocki (80) has carried out further extensive purification of the two factors from wheat germ and has used these to determine the partial reactions in which they participate. His results confirm the view that their function is similar to that of the elongation factors of bacteria and mammals.

The direction of chain elongation during *E. coli* protein biosynthesis has been well established as being from the amino terminal end toward the carboxyl end of the polypeptide chain. Indeed, it is generally accepted that all ribosomally synthesized proteins are made in this manner. However, Iwata & Kaji (67) have reported a heterodox result. They have found that when ribosome structural proteins were synthesized in the presence of a pulse of labeled amino acids, the amino acids at the NH_2-terminal end of the newly made proteins were more radioactive than those at the carboxyl end. Assuming that the synthesizing ribosomes contained incomplete nascent protein chains at the time of addition of the isotope, this would indicate that the proteins in question were being synthesized from the carboxyl end toward the NH_2-terminal end. At the same time, control experiments on the synthesis of soluble proteins indicated they were synthesized in the accepted NH_2-terminal to carboxyl-terminal direction. So far no explanation has been found for their anomalous results.

Chain Termination

A number of gene mutations of E. coli which cause premature chain termination were utilized in studies designed to specify which codons of mRNA control termination of protein synthesis (21, 22, 115, 130, 145). On the basis of these studies and of investigations utilizing synthetic polynucleotides (31, 77, 91), the triplet codons UAG, UAA, and UGA were found to serve as termination codons.

Two release factors, RF-1 which mediates release in response to termination codons UAG and UAA, and RF-2 which mediates release in response to UAA and UGA, have been isolated and purified (29, 134)). A third release factor, RF-3, has been isolated (30, 109) and it appears to stimulate the reactions of RF-1 and RF-2.

Summary of Termination

The process of termination may be depicted as follows: peptidyl-tRNA with the completed peptide chain is translocated from the A site to the P site. At this point, one of the three termination nucleotide codons, UAG, UAA, or UGA, is moved into the A site. In response to one of these codons, the appropriate termination factor binds to the ribosomal complex and the peptidyl tRNA is hydrolyzed, releasing the completed polypeptide (IX). Information concerning mammalian peptide termination is fragmentary and chain termination in plants has not been investigated, but both are assumed to use processes fundamentally similar to those of bacteria.

AMINO ACID INCORPORATING SYSTEMS

Incorporation System

It is apparent from the preceding sections, describing the presumed steps in amino acid activation and protein synthesis, that a large number of highly coordinated events are involved in protein synthesis. Simulating these conditions for in vitro studies is thus extremely complicated. In addition to the need for homogeneous ribosomes free of bacterial contamination, a complex series of specific reaction components are required in the reaction mixture. Although diverse reaction mixtures have been used to study protein synthesis, many constituents ($MgCl_2$, GTP, ATP, and an energy regenerating system) have been common to all.

One of the major differences in the synthesizing systems is whether a complete or a transfer system is employed. Where a complete system is used (e.g. 3, 98), the labeled amino acid(s) are added without previous attachment to tRNA. The reaction mixture therefore requires tRNA, the aminoacyl tRNA synthetase enzymes, and a mixture of radioactive and nonradioactive amino acids if "natural" messenger is used. If synthetic poly U is used as messenger, radioactive phenylalanine is the only amino acid added. When a transfer system is used (e.g. 8), the tRNA is charged with radioactive amino acid in a separate preliminary reaction so that the synthetase enzymes are not required in the incorporation reaction mixture. Another major difference in the reaction conditions of some studies has been in the source of messenger. The most commonly used messenger has been synthetic poly U referred to above. With poly U it is not possible to assay for initiation or termination factors. However, natural messenger RNA from various plant polyribosome sources has been used, and recently tobacco mosaic virus RNA has been utilized by Marcus and his co-workers (95, 96).

Amino Acid Incorporation by Plant Ribosomes

Studies on amino acid incorporation by wheat germ cytoplasmic ribosomes have been reported by Marcus & Feeley (98) and by Allende & Bravo (3). Since wheat germ ribosomes are apparently free of active messenger RNA, poly U was required. However, synthesis was activated when the seeds were allowed to imbibe water. Since then, Schultz et al (132) have presented evidence that endogenous messenger RNA is stored as a ribosome-mRNA complex in ungerminated wheat germ.

App & Gerosa (8) demonstrated that poly U, Mg^{2+}, K^+, GTP, and supernatant factors were required for amino acid incorporation by rice embryo ribosomes. Later, App (7) purified two factors from a rice embryo soluble fraction which he called transferases. The activity of washed ribosomes was stimulated by the addition of these transferases. Marcus & Feeley (98) and Allende & Bravo (3) also showed that poly U, Mg^{2+}, GTP, and a supernatant factor were required for activity in a wheat germ system. In addition, an energy regenerating system such as creatine phosphate and creatine phosphokinase was required. Reference to the factors isolated from wheat germ supernatant has been made under the discussion of elongation.

Incorporation of ^{14}C-phenylalanine by cytoplasmic ribosomes from wheat leaves was reported by Mehta et al (103). Parthier (120) described a poly U transfer system which he used in studying pea seedling cytoplasmic ribosomes. Washing the pea seedling ribosomes enhanced their ability to synthesize polyphenylalanine (54), but addition of pea seedling supernatant did not stimulate incorporation, and they (54) thought this was due to the presence of some inhibiting substance.

Recently Marei et al (100) isolated cytoplasmic ribosomes from fig fruit. In a complete system, ^{14}C-phenylalanine incorporation was dependent upon GTP, ATP, and magnesium, but was only slightly dependent on poly U. Examination of the ribosome preparation revealed many polysomes which were responsible for much of the incorporation.

Sissakian et al (139) obtained amino acid incorporation by ribosomes isolated from pea seedling chloroplasts. ATP did not stimulate activity. Boardman et al (15) have compared the amino acid incorporating ability of cytoplasmic and chloroplast ribosomes from tobacco leaves, while Hadziyev & Zalik (57) compared the rates of incorporation of wheat leaf chloroplast and cytoplasmic ribosomes. Both groups reported much higher rates of incorporation with the chloroplast than the cytoplasmic ribosomes. However, Tucker & Zalik (155), who compared purified 70S (chloroplast) and 80S (cytoplasmic) ribosomes from wheat leaves, found the rates of incorporation were the same. This seeming inconsistency may be explained by differences in methods of preparation and homogeneity of the ribosomes used.

Site of Protein Synthesis

As mentioned in the introduction, protein synthesis in plants may occur on free or membrane-bound ribosomes of cytoplasmic, mitochondrial, or chloroplast origin. The question then arises as to which synthesizing systems are responsible for synthesis of which proteins. Two main approaches to this problem have involved: (a) attempts at identifying the proteins synthesized by isolated organelles; and (b) the use of inhibitors which are presumed to block protein synthesis by one class of ribosomes but not by another.

In his comprehensive review of the inhibitors which have been used to elucidate ribosome function, Pestka (123) draws attention to the pleiotropic effects of many of these compounds. He indicates that antibiotics which predominantly affect the small ribosomal subunits may also influence the function of the large subunits or the converse may be true. Moreover, some of the inhibitors bind only to specific sites on the ribosome and the ribosome may be inhibited only when the ribosome is in one particular state. His concluding remark that "the effects of antibiotic binding to ribosomes is analogous to effects resulting from the toss of a jamming agent into complex machinery—almost anything can happen" is a point well taken. It should not be surprising then to find diverse and even contradictory results from experiments based on the use of inhibitors. For example, conflicting results based upon studies involving inhibitors are evident in the table presented by Boulter et al (20). The table shows that although it was inferred that the site of synthesis of NADP-triosephosphate dehydrogenase was the cytoplasmic ribosomes of peas and beans, it was thought the enzyme was synthesized by the chloroplast ribosomes of corn. Conflicting conclusions were likewise reached regarding the site of synthesis of this enzyme in *Euglena*.

Work by Hoober (62, 63) has shown that both chloroplast and cytoplasmic ribosomes of *Chlamydomonas reinhardii* are involved in the synthesis of chloroplast membrane polypeptides. Eytan & Ohad (44) also concluded that thylakoid membranes are assembled within the chloroplast from some proteins synthesized in the cytoplasm and others made in the chloroplast. From their studies on *Ochromonas danica*, Smith-Johannsen & Gibbs (142) postulated that one or more chloroplast ribosomal proteins are synthesized on chloroplast ribosomes, whereas mitochondrial ribosomal proteins are synthesized on cytoplasmic ribosomes.

Lizardi & Luck (86) have employed three complementary methods of studying the intracellular site of synthesis of mitochondrial proteins in *Neurospora crassa*. On the basis of these three types of studies, including mitochondrial protein synthesis in vitro and sensitivity of in vivo synthesis to two inhibitors, they concluded that most, if not all, of the 53 structural proteins of mitochondrial ribosomal subunits in *Neurospora crassa* are synthesized on cytoplasmic ribosomes.

Payne & Boulter (121) found that the amount of free and membrane-bound ribosomes of bean change during development of the bean cotyledons, and they proposed that the free and bound ribosomes each synthesize different groups of proteins. This is in agreement with Opik (119), who suggested that membrane-bound ribosomes in the cotyledons of developing seeds of *Phaseolus vulgaris* preferentially synthesize storage proteins.

Filippovich et al (49) separated pea chloroplast ribosomes into free and lamellar-bound ribosomes. They determined ^{14}C amino acid incorporation by each ribosome fraction and concluded that protein synthesis in the chloroplast was mainly by the lamellar-bound ribosome system. Similarly, Chen & Wildman (32) demonstrated the existence of both free and bound ribosomes in isolated tobacco chloroplasts, where more than half of the ribosomes were bound to thylakoid membranes. They concluded that the nascent peptides of bound ribosomes were different from those produced on the free ribosomes. In summary, it can only be said that, for higher plants, information as to which ribosomes synthesize any particular protein is fragmentary.

ACKNOWLEDGMENTS

We are grateful to the investigators who sent us manuscripts in preprint form. The preparation of this review was supported in part by a grant to Saul Zalik from the National Research Council of Canada.

Literature Cited

1. Albrecht, J., Stap, F., Voorma, H. O., van Knippenberg, P. H., Bosch, L. 1970. *FEBS Lett.* 6:297–301
2. Allen, N. E., Suyama, Y. 1972. *Biochim. Biophys. Acta* 259:369–77
3. Allende, J. E., Bravo, M. 1966. *J. Biol. Chem.* 241:5813–18
4. Anderson, J. W., Fowden, L. 1970. *Biochem. J.* 119:677–90
5. Ibid, 691–97
6. Anderson, M. B., Cherry, J. H. 1969. *Proc. Nat. Acad. Sci. USA* 62:202–9
7. App, A. A. 1969. *Plant Physiol.* 44:1132–38
8. App, A. A., Gerosa, M. M. 1966. *Plant Physiol.* 41:1420–24
9. Arglebe, C., Hall, T. C. 1969. *Plant Cell Physiol.* 10:171–82
10. Attardi, G., Amaldi, F. 1970. *Ann. Rev. Biochem.* 39:183–226
11. Attwood, M. M., Cocking, E. C. 1965. *Biochem. J.* 96:616–25
12. Bachmayer, H. 1970. *Biochim. Biophys. Acta* 209:584–86
13. Bick, M. D., Liebke, H., Cherry, J. H., Strehler, B. L. 1970. *Biochim. Biophys. Acta* 204:175–82
14. Boardman, N. K., Francki, R. I. B., Wildman, S. G. 1965. *Biochemistry* 4:872–76
15. Boardman, N. K., Francki, R. I. B., Wildman, S. G. 1966. *J. Mol. Biol.* 17:470–89
16. Borst, P. 1971. *Autonomy and Biogenesis of Mitochondria and Chloroplasts*, ed. N. K. Boardman, A. W. Linnane, R. M. Smillie, 260–66. Amsterdam: North-Holland
17. Borst, P. 1972. *Ann. Rev. Biochem.* 41:333–76
18. Borst, P., Grivell, L. A. 1971. *FEBS Lett.* 13:73–87
19. Boulter, D. 1970. *Ann. Rev. Plant Physiol.* 21:91–114
20. Boulter, D., Ellis, R. J., Yarwood, A. 1972. *Biol. Rev.* 47:113–75
21. Brenner, S., Barnett, L., Katz, E. R., Crick, F. H. C. 1967. *Nature* 213:449–50
22. Brenner, S., Beckwith, J. R. 1965. *J. Mol. Biol.* 13:629–37
23. Bretscher, M. S. 1968. *Nature* 218:675–77
24. Brownlee, G. G., Sanger, F., Barrell, B. G. 1967. *Nature* 215:735–36
25. Bruskov, V. I., Odintsova, M. S.

26. Burkard, G., Eclancher, B., Weil, J. H. 1969. *FEBS Lett.* 4:285–87
27. Burkard, G., Guillemaut, P., Weil, J. H. 1970. *Biochim. Biophys. Acta* 224:184–98
28. Burkard, G., Vaultier, J. P., Weil, J. H. 1972. *Phytochemistry* 11:1351–53
29. Capecchi, M. R. 1967. *Proc. Nat. Acad. Sci. USA* 58:1144–51
30. Capecchi, M. R., Klein, H. A. 1969. *Cold Spring Harbor Symp. Quant. Biol.* 34:469–77
31. Caskey, T., Leder, P., Moldave, K., Schlessinger, D. 1972. *Science* 176:195–97
32. Chen, J. L., Wildman, S. G. 1970. *Biochim. Biophys. Acta* 209:207–19
33. Cherry, J. H., Osborne, D. J. 1970. *Biochem. Biophys. Res. Commun.* 40:763–69
34. Chi, J. C. H., Suyama, Y. 1970. *J. Mol. Biol.* 53:531–56
35. Chirikjian, J. G., Wright, H. T., Fresco, J. R. 1972. *Proc. Nat. Acad. Sci. USA* 69:1638–41
36. Ciferri, O. In press
37. Ciferri, O., Parisi, B. 1970. *Progr. Nucl. Acid Res. Mol. Biol.* 10:121–44
38. Clark, J. M. Jr. 1958. *J. Biol. Chem.* 233:421–24
39. Clark, M. F. Matthews, R. E. F., Ralph, R. K. 1964. *Biochim. Biophys. Acta* 91:289–304
40. Dure, L. S. III, Merrick, W. C. 1971. See Ref. 16, 413–21
41. Ellis, R. J. 1969. *Science* 163:477–78
42. Ellis, R. J. 1970. *Planta* 91:329–35
43. Epler, J. L., Shugart, L. R., Barnett, W. E. 1970. *Biochemistry* 9:3575–79
44. Eytan, G., Ohad, I. 1970. *J. Biol. Chem.* 245:4297–4307
45. Fahnestock, S. R., Nomura, M. 1972. *Proc. Nat. Acad. Sci. USA* 69:363–65
46. Fellner, P. 1971. *Biochemie* 53:573–83
47. Fellner, P., Ehresmann, C. 1972. *Nature New Biol.* 239:1–5
48. Fellner, P., Ehresmann, C., Ebel, J. P. 1970. *Nature* 225:26–29
49. Filippovich, I. I., Tongur, A. M., Alina, B. A., Oparin, A. I. 1970. *Biokhimiya* 35:247–56
50. Galper, J. B., Darnell, J. E. Jr. 1969. *Biochem. Biophys. Res. Commun.* 34:205–13
51. Garcia-Patrone, M., Perazzolo,

C. A., Baralle, F., Gonzalez, N. S. 1971. *Biochim. Biophys. Acta* 246:291–99
52. Gualerzi, C., Cammarano, P. 1969. *Biochim. Biophys. Acta* 190:170–86
53. Guillemaut, P., Burkard, G., Weil, J. H. 1972. *Phytochemistry* 11:2217–19
54. Gulyas, A., Parthier, B. 1971. *Biochem. Physiol. Pflanz.* 162:60–74
55. Gupta, S. L., Waterson, J., Sopori, M. L., Weissman, S. M., Lengyel, P. 1971. *Biochemistry* 10:4410–21
56. Guthrie, C., Nashimoto, H., Nomura, M. 1969. *Cold Spring Harbor Symp. Quant. Biol.* 34:69–75
57. Hadziyev, D., Zalik, S. 1970. *Biochem. J.* 116:111–24
58. Hague, D. R., Kofoid, E. C. 1971. *Plant Physiol.* 48:305–11
59. Hardesty, B., Arlinghaus, R., Shaeffer, J., Schweet, R. 1963. *Cold Spring Harbor Symp. Quant. Biol.* 28:215–22
60. Hauge, J. G. 1971. *FEBS Lett.* 17:168–75
61. Hershey, J. W. B., Dewey, K. F., Thach, R. E. 1969. *Nature* 222:944–47
62. Hoober, J. K. 1970. *J. Biol. Chem.* 245:4327–34
63. Hoober, J. K. 1972. *J. Cell Biol.* 52:84–96
64. Hoober, J. K., Blobel, G. 1969. *J. Mol Biol.* 41:121–38
65. Infante, A. A., Baierlein, R. 1971. *Proc. Nat. Acad. Sci. USA* 68:1780–85
66. Infante, A. A., Krauss, M. 1971. *Biochim. Biophys. Acta* 246:81–99
67. Iwata, S., Kaji, H. 1971. *Proc. Nat. Acad. Sci. USA* 68:690–94
68. Jerez, C., Sandoval, A., Allende, J. E., Henes, C., Ofengand, J. 1969. *Biochemistry* 8:3006–14
69. Jones, B. L., Nagabhushan, N., Gulyas, A., Zalik, S. 1972. *FEBS Lett.* 23:167–70
70. Jones, B. L., Nagabhushan, N., Zalik, S. *Nucleic Acids and Proteins in Higher Plants.* Budapest: Academic. In press
71. Kaempfer, R. 1968. *Proc. Nat. Acad. Sci. USA* 61:106–13
72. Kaempfer, R. 1970. *Nature* 228:534–7
73. Kaji, H. 1970. *Int. Rev. Cytol.* 29:169–211
74. Kan, J., Sueoka, N. 1971. *J. Biol. Chem.* 246:2207–10
75. Kanabus, J., Cherry, J. H. 1971. *Proc. Nat. Acad. Sci. USA* 68:873–76

76. Kurland, C. G. 1972. *Ann. Rev. Biochem.* 41:377–408
77. Last, J. A. et al 1967. *Proc. Nat. Acad. Sci. USA* 57:1062–67
78. Leaver, C. J., Harmey, M. A. In press
79. Leaver, C. J., Ingle, J. 1971. *Biochem. J.* 123:235–43
80. Legocki, A. B. In press
81. Legocki, A. B., Marcus, A. 1970. *J. Biol. Chem.* 245:2814–18
82. Legocki, A. B., Pawelkiewicz, J. 1967. *Acta Biochem. Pol.* 14:313–22
83. Leis, J. P., Keller, E. B. 1970. *Biochem. Biophys. Res. Commun.* 40:416–21
84. Leis, J. P., Keller, E. B. 1970. *Proc. Nat. Acad. Sci. USA* 67:1593–99
85. Lin, C. Y., Key, J. L. 1971. *Plant Physiol.* 48:547–52
86. Lizardi, P. M., Luck, D. J. L. 1972. *J. Cell. Biol.* 54:56–74
87. Loening, U. E. 1968. *J. Mol. Biol.* 38:355–65
88. Loening, U. E., Ingle, J. 1967. *Nature* 215:363–67
89. Lubsen, N. H., Davis, B. D. 1972. *Proc. Nat. Acad. Sci. USA* 69:353–57
90. Lucas-Lenard, J., Lipmann, F. 1966. *Proc. Nat. Acad. Sci. USA* 55:1562–66
91. Lucas-Lenard, J., Lipmann, F. 1971. *Ann. Rev. Biochem.* 40:409–48
92. Lyttleton, J. W. 1962. *Exp. Cell Res.* 26:312–17
93. Maden, B. E. H., Traut, R. R., Monro, R. E. 1968. *J. Mol. Biol.* 35:333–45
94. Marcus, A. 1969. *Symp. Soc. Exp. Biol.* 23:143–60
95. Marcus, A. 1970. *J. Biol. Chem.* 245:955–61
96. Ibid, 962–69
97. Marcus, A., Bewley, J. D., Weeks, D. P. 1970. *Science* 167:1735–36
98. Marcus, A., Feeley, J. 1965. *J. Biol. Chem.* 240:1675–80
99. Marcus, A., Weeks, D. P., Leis, J. P., Keller, E. B. 1970. *Proc. Nat. Acad. Sci. USA* 67:1681–87
100. Marei, N., Gadallah, A. I., Kilgore, W. W. 1972. *Phytochemistry* 11:529–33
101. Martin, T. E., Wool, I. G., Castles, J. J. 1971. *Methods Enzymol.* 20:417–29
102. Maruta, H., Tsuchiya, T., Mizuno, D. 1971. *J. Mol. Biol.* 61:123–34
103. Mehta, S. L., Hadziyev, D., Zalik, S. 1969. *Biochim. Biophys. Acta* 195:515–22

104. Merrick, W. C., Dure, L. S. III *J. Biol. Chem.* In press
105. Merrick, W. C., Dure, L. S. III *J. Biol. Chem.* In press
106. Merrick, W. C., Dure, L. S. III 1971. *Proc. Nat. Acad. Sci. USA* 68:641–44
107. Miall, S. H., Kato, T., Tamaoki, T. 1970. *Nature* 226:1050–52
108. Miller, A., Karlsson, U., Boardman, N. K. 1966, *J. Mol. Biol.* 17: 487–89
109. Milman, G., Goldstein, J., Scolnick, E., Caskey, T. 1969. *Proc. Nat. Acad. Sci. USA* 63:183–90
110. Monro, R. E., Staehelin, T., Celma, M. L., Vazquez, D. 1969. *Cold Spring Harbor Symp. Quant. Biol.* 34:357–66
111. Moon, H.-M., Redfield, B., Weissbach, H. 1972. *Proc. Nat. Acad. Sci. USA* 69:1249–52
112. Moustafa, E., Lyttleton, J. W. 1963. *Biochim. Biophys. Acta* 68:45–55
113. Nashimoto, H., Held, W., Kaltschmidt, E., Nomura, M. 1971. *J. Mol. Biol.* 62:121–38
114. Nashimoto, H., Nomura, M. 1970. *Proc. Nat. Acad. Sci. USA* 67:1440–47
115. Nirenberg, M. et al 1965. *Proc. Nat. Acad. Sci. USA* 53:1161–68
116. Nomura, M. 1972. *Fed. Proc.* 31:18–20
117. Nomura, M., Erdmann, V. A. 1970. *Nature* 228:744–48
118. Odintsova, M. S., Bruskov, V. I., Golubeva, E. V. 1967. *Biokhimiya* 32:1047–59
119. Opik, H. 1968. *J. Exp. Bot.* 19:64–67
120. Parthier, B. 1971. *Biochem. Physiol. Pflanz.* 162:45–59
121. Payne, P. I., Boulter, D. 1969. *Planta* 84:263–71
122. Payne, P. I., Dyer, T. A. 1972. *Nature* 235:145–47
123. Pestka, S. 1971. *Ann. Rev. Microbiol.* 25:487–562
124. Peterson, P. J., Fowden, L. 1963. *Nature* 200:148–51
125. Raacke, I. 1971. *Proc. Nat. Acad. Sci. USA* 68:2357–60
126. Reger, B. J., Fairfield, S. A., Epler, J. L., Barnett, W. E. 1970. *Proc. Nat. Acad. Sci. USA* 67:1207–13
127. Richter, D., Lin, L., Bodley, J. W. 1971. *Arch. Biochem. Biophys.* 147: 186–91
128. Sabol, D., Sillero, M. A. G., Iwasaki, K., Ochoa, S. 1970. *Nature* 228: 1269–75
129. Sager, R., Hamilton, M. G. 1967.
130. Sarobhai, A. S., Stretton, A. O. W., Brenner, S., Bolle, A. 1964. *Nature* 201:13–17
131. Schreiber, G. 1971. *Angew. Chem. Int. Ed.* 10:638–51
132. Schultz, G. A., Chen, D., Katchalski, E. 1972. *J. Mol. Biol.* 66:379–90
133. Schwartz, J. H., Meyer, R., Eisenstadt, J. M., Brawerman, G. 1967. *J. Mol. Biol.* 25:571–74
134. Scolnick, E., Tompkins, R., Caskey, C. T., Nirenberg, M. W. 1968. *Proc. Nat. Acad. Sci. USA* 61:768–74
135. Scott, N. S., Munns, R., Graham, D., Smillie, R. M. 1971. See Ref. 16, 383–92
136. Seal, S. N., Bewley, J. D., Marcus, A. 1972. *J. Biol. Chem.* 247:2592–97
137. Seal, S. N., Marcus, A. 1972. *Biochem. Biophys. Res. Commun.* 46: 1895–1902
138. Shorey, R. L., Ravel, J. M., Garner, C. W., Shive, W. 1969. *J. Biol. Chem.* 244:2555–64
139. Sissakian, N. M., Filippovich, I. I., Svetailo, E. N., Aliyev, K. A. 1965. *Biochim. Biophys. Acta* 95:474–85
140. Skoultchi, A., Ono, Y., Moon, H.-M., Lengyel, P. 1968. *Proc. Nat. Acad. Sci. USA* 60:675–82
141. Smith, A. E., Marcker, K. 1968. *J. Mol. Biol.* 38:241–43
142. Smith-Johannsen, H., Gibbs, S. P. 1972. *J. Cell Biol.* 52:598–614
143. Spirin, A. S. 1969. *Cold Spring Harbor Symp. Quant. Biol.* 34:197–207
144. Spirin, A. S., Gavrilova, L. P. 1969. *The Ribosome.* New York: Springer-Verlag
145. Stretton, A. O. W., Brenner, S. 1965. *J. Mol. Biol.* 12:456–65
146. Stutz, E., Noll, H. 1967. *Proc. Nat. Acad. Sci. USA* 57:774–81
147. Subramanian, A. R., Davis, B. D. 1970. *Nature* 228:1273–75
148. Subramanian, A. R., Davis, B. D., Beller, R. J. 1969. *Cold Spring Harbor Symp. Quant. Biol.* 34:223–30
149. Subramanian, A. R., Ron, E. Z., Davis, B. D. 1968. *Proc. Nat. Acad. Sci. USA* 61:761–67
150. Svetailo, E. N., Filippovich, I. I., Sissakian, N. M. 1967. *J. Mol. Biol.* 24:405–15
151. Tao, K. L., Hall, T. C. 1971. *Biochem. J.* 121:495–501
152. Ibid, 975–81
153. Tarrago, A., Monasterio, O., Allende, J. E. 1970. *Biochem. Biophys. Res. Commun.* 41:765–73

154. Traub, P., Nomura, M. 1968. *Proc. Nat. Acad. Sci. USA* 59:777-84
155. Tucker, E., Zalik, S. In press
156. Vanderhoef, L. N., Key, J. L. 1970. *Plant Physiol.* 46:294-98
157. Vasconcelos, A. C. L., Bogorad, L. 1971. *Biochim. Biophys. Acta* 228: 492-502
158. Waller, J. P. 1963. *J. Mol. Biol.* 7:483-96
159. Waterson, J., Beaud, G., Lengyel, P. 1970. *Nature* 227:34-38
160. Weeks, D. P., Baxter, R. In press
161. Weeks, D. P., Verma, D. P. S., Seal, S. N., Marcus, A. 1972. *Nature* 236:167-68
162. Wilson, R. H., Hanson, J. B., Mollenhauer, H. H. 1968. *Plant Physiol.* 43:1874-77
163. Yarus, M. 1972. *Proc. Nat. Acad. Sci. USA* 69:1915-19
164. Yarwood, A., Boulter, D., Yarwood, J. N. 1971. *Biochem. Biophys. Res. Commun.* 44:353-61

Ann. Rev. Plant Physiol. 1973. 24:69-88

ROLE OF OXIMES IN NITROGEN METABOLISM IN PLANTS

❖ 7542

S. Mahadevan

Department of Biochemistry, Indian Institute of Science, Bangalore-12, India

CONTENTS

INTRODUCTION

Early work on the involvement of oximes in plant nitrogen metabolism has been a topic of controversy (5, 48, 63, 92, 136, 138, 145). Oximes were initially reported to be formed as condensation products of glyoxylic, oxalacetic, 2-ketoglutaric, or pyruvic acids and free hydroxylamine, the latter formed as an intermediate during either symbiotic nitrogen fixation or nitrate reduction via nitrite (63, 136, 138). Such oxime formation was believed to detoxify hydroxylamine otherwise harmful to plants, the oximino acids finally being reduced to amino acids. However, later studies with nitrogenase enzymes and nitrite reductases has clearly revealed that hydroxylamine is not a free intermediate during the reduction of either nitrogen or nitrite to ammonia (9, 48, 50). Formation of hydroxylamine in plants, if any, therefore had to be from some other source. The fairly widespread occurrence in plants and microorganisms of enzymes capable of reducing hydroxylamine to ammonia, and to a lesser extent oximes to amino acids (5, 63, 92), suggested that any free hydroxylamine could be thus metabolized and "detoxified." Besides oxime-reducing enzymes with properties similar to nitrate and hydroxylamine reductases (87), oxime transferring enzymes, the transoximases, have been demonstrated in plants (88). Their natural substrates and function are not clear.

The conversion of 3-indoleacetaldoxime to indoleacetonitrile by certain higher plants and fungi was reported in 1963 (71), and the reaction was shown to be enzymatic (64). This study was undertaken in order to find a precursor for the nitrile, whose formation from indolepyruvic acid oxime (116) and indoleacetaldoxime (70) had been speculated. An interesting feature of this conversion was that all the plants converting indoleacetonitrile to indoleacetic acid (71, 126) were also capable of the oxime to nitrile conversion, suggesting the possibility of their being successive steps in a metabolic sequence. Formation of indoleacetonitrile had been reported from the then recently discovered indolemethylglucosinolate (40) by the enzyme myrosinase acting at a lowered pH (137). The reaction appeared to be partly enzymic and partly nonenzymic. Indoleacetonitrile thus had two precursors, an aldoxime and a glucosinolate, but the relationship between them was not apparent at that time. Almost simultaneously, the conversion of 2,6-dichlorobenzaldoxime to the corresponding nitrile by bacteria grown on the oxime and their cell-free extract was reported (85). The growth regulatory ability of certain aldoximes was attributed to their conversion to the nitrile and subsequently to the acid (33).

In 1967, following an extensive search for precursors besides amino acids for the aglycone moiety of cyanogenic glycosides (21), Tapper et al (122) discovered that both isobutyraldoxime and 2-ketoisobutyric acid oxime were efficiently converted to linamarin, the β-glucoside of 2-hydroxyisobutyronitrile in flax plants. Similarly, phenylacetaldoxime and phenylpyruvic acid oxime were converted to prunasin, the β-glucoside of D-mandelonitrile in cherry laurel leaves (120). Almost simultaneously, the efficient conversion of phenylacetaldoxime to benzylglucosinolate in *Tropaeolum majus*, phenylpropionaldoxime to phenylethylglucosinolate in *Nasturtium officinale* (129), and of isobutyraldoxime to isopropyl glucosinolate in *Cochlearia officinalis* was demonstrated (119). Unlike their incorporation into cyanogenic glucosides, phenylpyruvic acid oxime and 2-ketoisovaleric acid oxime were not converted to the corresponding glucosinolates to any significant extent (119). Aldoximes were thus shown to be efficient precursors for at least two classes of widely distributed natural products, the nitriles such as cyanogenic glucosides and glucosinolates.

Earlier investigations had shown that amino acids were the precursors of the aglycone moiety of cyanogenic glucosides (20, 21, 35, 39) and of glucosinolates (28, 35, 65, 66, 133). The α-carbon and amino-nitrogen atoms of the amino acid were incorporated into the aglycone moieties as a unit without randomization as shown in Figure 1. These findings required that if aldoximes were intermediates in the conversion of amino acids to these two classes of compounds, then the carbon and nitrogen atoms in question be transferred via the aldoxime without randomization (Figure 1). This has indeed been shown to be the case (77, 120).

This review will concern itself with this new role of oximes in plant nitrogen

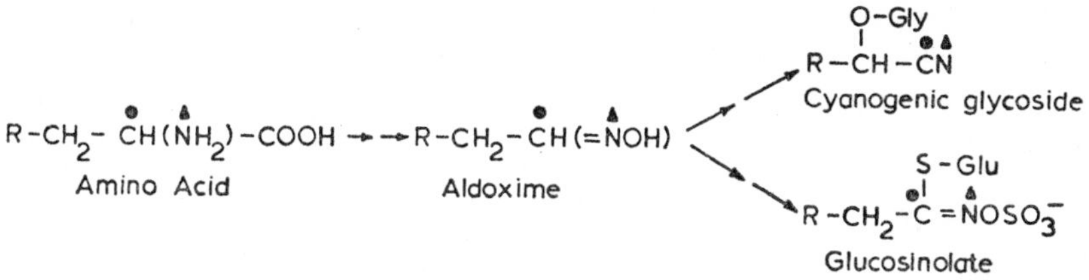

$$R-CH_2-\overset{\bullet}{\underset{Amino\ Acid}{CH}}(\overset{\blacktriangle}{NH_2})-COOH \rightarrow\rightarrow R-CH_2-\overset{\bullet}{\underset{Aldoxime}{CH}}(=\overset{\blacktriangle}{NOH})$$

$$\overset{O-Gly}{\underset{Cyanogenic\ glycoside}{R-CH-\overset{\bullet}{\overset{\blacktriangle}{CN}}}}$$

$$\overset{S-Glu}{\underset{Glucosinolate}{R-CH_2-\overset{\bullet}{C}=\overset{\blacktriangle}{NOSO_3^-}}}$$

Figure 1 Formation of cyanogenic glucosides and glucosinolates from amino acids.

metabolism wherein they serve as intermediates in the converison of amino acids to secondary plant products such as nitriles, cyanogenic glucosides, glucosinolates, and their further metabolic products. Research in this area up to 1968–69 has been reviewed by Conn (19), Conn & Butler (21), and Ettlinger & Kjaer (28).[1]

PATHWAYS OF OXIME METABOLISM

Pathways of oxime formation from amino acids and their further metabolism are depicted in Figure 2; the structures (roman numerals) represent classes of compounds with a variable R group which depends on the parent amino acid (I). The reactions (arabic numerals) are similarly type reactions. A specific R group need not be involved in all reactions. The fate of α-carbon and amino-nitrogen of the parent amino acid through the various steps are shown in boldface letters. Solid arrows indicate those reactions where at least one in vitro (enzymic) conversion has been reported. Broken arrows indicate in vivo conversions, demonstrated by isotope incorporation studies. Certain speculative conversions are represented similarly.

Aldoximes (IV) are the branch point for the formation of the following families of compounds: (i) cyanogenic glucosides (VIII), nitriles (V), 2-hydroxynitriles (VI) and hydrocyanic acid; (ii) glucosinolates (XIV), organic thio- and isothiocyanates (XVII, XV), inorganic thiocyanate, nitrile (V), amines (XVIII), and nitro compound (XX); (iii) non-nitrogenous organic acids (XXII), alcohols (XXV), and aldehydes (XXIV). Lower aldehyde (IX), acid (X), and alcohol (XVI) could also be formed via the oxime pathway. However, such non-nitrogenous acids, alcohols, and aldehydes are also formed from amino acids via the

---

[1] *Nomenclature:* Chemical names are used for specific compounds except common amino acids. Cyanogenic glucosides, being few in number, are referred to by their trivial names, the chemical name being used only the first time. The nomenclature for specific glucosinolates follows that suggested by Ettlinger & Kjaer (28), the "R" group chemically described and the functional group ($-C(S.Glu)-NOSO_4^-$) referred to as glucosinolate. The term "desthioglucosinolate" refers to the glucosinolate without the sulfate group. They are the S(β-D-glucopyranosyl) derivatives of the corresponding thiohydroximic acids.

keto acid or amine without involving oximes. Since cyanogenic glycosides and glucosinolates have never been shown to occur together in any plant, the branch point of aldoxime metabolism to these two classes of compounds appears to be a strict either/or process.

Current information on the conversion of specific aldoximes to various nitrogenous compounds in plants are summarized in Table 1.

Conversion of Amino Acids to Aldoximes

Steps involved in the conversion of amino acids to aldoximes are common for both cyanogenic glycoside and glucosinolate formation. Based on the structure of the 14 cyanogenic glycosides (19, 21, 104, 105) and the 55 glucosinolates (25, 28, 61, 62, 131, 132) isolated so far, the following protein and nonprotein amino acids

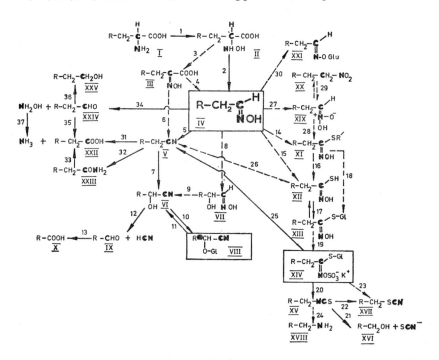

Figure 2   Pathways of oxime metabolism in plants.

I. Amino acid; II. N-hydroxy amino acid; III. 2-keto acid oxime; IV. aldoxime; V. nitrile; VI. 2-hydroxynitrile; VII. 2-hydroxyaldoxime; VIII. cyanogenic glycoside; IX. lower aldehyde; X. lower acid; XI. S-alkyl derivative of thiohydroximic acid; XII. thiohydroximic acid; XIII. desthioglucosinolate; XIV. glucosinolate; XV. organic isothiocyanate; XVI. lower alcohol; XVII. organic thiocyanate; XVIII. amine; XIX. *aci*-nitro tautomer of nitro compound; XX. nitro compound; XXI. β-O-glucoside of aldoxime; XXII. carboxylic acid; XXIII. carboxamide; XXIV. aldehyde; XXV. alcohol.

Table 1 Conversion of Aldoximes to Nitrogenous Compounds in Plants

Aldoxime	Product	Plant	Reference
Indoleacetaldoxime	Indoleacetonitrile	Several plants and fungi[a]	71, 72, 90
	Indolemethylgluco-sinolate	*Isatis tinctoria, Sinapis alba*	53, 73
	N-Sulfoindolemethyl glucosinolate	*I. tinctoria*	73
	Desthioindolemethyl glucosinolate[b]	*I. tinctoria, Sinapis alba*	53, 74
Isobutyraldoxime	Linamarin	*Linum usitatissimum*	120, 122
	Isopropyl glucosinolate	*Cochlearia officinalis*	119
	Isobutyraldoxime-O-β-glucoside[b]	*L. usitatissimum*	121
Phenylacetaldoxime	Benzyl glucosinolate	*Lepidium sativum, Tropaeolum majus*	119, 129
	Prunasin	*Prunus lauro-cerasus*	122
	1-Nitro-2-phenylethane	*T. majus*	79
	Phenylacetothio-hydroximate[c]	*T. majus*	80
Phenylpropion-aldoxime	Phenylethyl glucosinolate	*Nasturtium officinale*	129
p-Hydroxyphenyl-acetaldoxime	p-Hydroxybenzyl gluco-sinolate	*S. alba*	58
	Desthio-p-hydroxybenzyl glucosinolate[b]	*S. alba*	58
	p-Hydroxyphenylaceto-thiohydroximate	*S. alba*	59
	Triglochinin[d]	*Thalictrum aquilegifolium*	105
	Dhurrin[e]	*Sorghum vulgare*	32
	p-Hydroxyphenylace-tonitrile[f]	*Sorghum vulgare*	18
4-(Methylthio)-butyraldoxime	Allyl glucosinolate	*Armoracia lapathifolia*	77
5-(Methylthio)-valeraldoxime	2(S)-Hydroxy-3-butenyl-glucosinolate	*Brassica napobrassica*	69
2-Hydroxyisobutyr-aldoxime	Linamarin	*L. usitatissimum*	123
2-Hydroxyphenyl-acetaldoxime	Prunasin	*Prunus lauro-cerasus*	123

[a] *Musa* sp., *Cuscuta* sp., *Brassica* sp., *Helianthus annuus*, *Convolvulus* sp., *Avena sativa*, *Cucumis sativus*, *Gossypium hirsutum*, *Lycopersicon esculentum*.
Fungi: *Aspergillus, Penicillium, Fusaria, Gibberella fujikuroi*.
[b] Tentative identification.
[c] Identified as S-methyl derivative.

$$\underset{\underset{\displaystyle CH_2-COOCH_3}{|}}{\overset{\overset{\displaystyle O-Glu}{|}}{HOOC-CH=CH-C=C-CN}}$$

[d] (above structure)

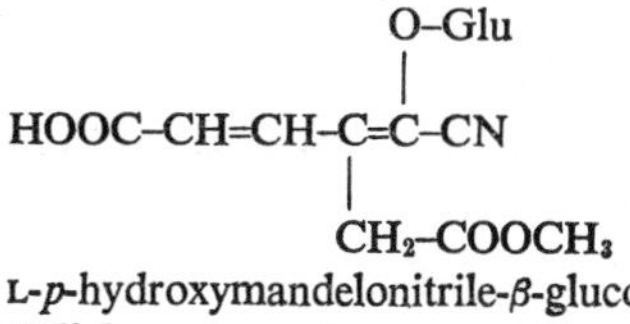

[e] L-p-hydroxymandelonitrile-β-glucoside.
[f] Cell-free conversion.

would appear to be metabolized via the oxime pathway. *Protein amino acids*: alanine, valine, leucine, isoleucine, phenylalanine, tyrosine, tryptophan, and possibly glutamic acid; *nonprotein amino acids*: 2-aminobutyric acid, homoleucine, homoisoleucine, homomethionine (and higher homologs up to 10 methylene groups), homophenylalanine, *m*-tyrosine, di- and tri-hydroxyphenylalanine, and certain other amino acids. As will be described later, extensive changes in the R group of amino acids after its entry into the oxime pathway may lead to end products having R groups with no resemblance to the R group of the parent amino acid.

A specific search for and identification of the amino acid γ-phenyl, α-amino butyric acid (homophenylalanine) in *Nasturtium officinale* as a precursor for phenylethylglucosinolate (35, 130, 146) is an indication that several nonprotein amino acids (e.g. higher homologs of leucine, isoleucine) whose natural occurrence has not been shown may yet be discovered (28).

A specific though limited search for the natural occurrence of aldoximes in general (52) or of a specific aldoxime (isobutyraldoxime) (121) has yielded negative results, indicating that their native pool size, like many true intermediates, may be very small. A brief report exists of the formation of labeled indoleacetaldoxime in cabbage plants grown in $^{14}CO_2$ atmosphere (56). The in vivo formation in plants of the following ^{14}C-aldoximes from the corresponding ^{14}C-amino acid has been demonstrated by trapping experiments or by the use of specific inhibitors (121): phenylacetaldoxime from phenylalanine in *T. majus* (129); indoleacetaldoxime from tryptophan in cabbage (56); isobutyraldoxime from valine in flax (121), *p*-OH phenylacetaldoxime from tyrosine in *Aubrietia graeca* (59) and sorghum (32); *p*-methoxyphenylacetaldoxime from *p*-methoxy phenylalanine in *Aubrietia hybrida* (59). The nonbactericidal oxime, 3-(oximinoacetamido) acrylamide, has been isolated along with the *aci*-nitro derivative antibiotic enteromycin in a *Streptomyces* species (144).

A N-hydroxylation step is a necessity if the α-carbon and amino-nitrogen of the amino acid has to be incorporated without randomization into the aglycone moiety of the glucosides via the aldoxime as shown in Figure 2, and both N-hydroxyamino acid (II) and 2-keto acid oxime (III) have been postulated as free intermediates (56, 60). N-Hydroxyamino acids are unstable compounds and N-hydroxyphenylalanine disproportionates nonenzymically to phenylalanine and phenylpyruvic acid oxime, the latter decomposing further to the nitrile (111). Though the natural occurrence of a free N-hydroxyamino acid has not been reported, several N-hydroxyamino acid analog and hydroxamic acids are known (127). The formation of the antibiotic hadacidin from glycine is apparently via N-hydroxyglycine (112). N-Hydroxylation of tryptophan by a particulate system from cabbage requiring NADPH has been reported (reaction 1), and a transformation of tryptophan to indoleacetaldoxime by a horseradish peroxidase/dihydroxyfumarate system has been claimed (56). Further pursuit of these reactions has apparently proved difficult (57).

The formation of benzylglucosinolate in *T. majus* plants from N-hydroxyphenylanine and the partial purification of an enzyme from *T. majus*, *N. officinale*, and *Sinapis alba* capable of converting N-hydroxyphenylanine to phenylacetal-doxime have been reported (reaction 2) (60). Some phenylacetonitrile was also formed. Significantly, phenylpyruvic acid oxime was not utilized by the enzyme. The reaction was optimally active at pH 8.5, showed oxygen uptake, was activated by FMN and phenazinemethosulfate but not FAD, and was inhibited by thiols but not mercury compounds or cyanide. However, attempts to demonstrate similar enzyme activity with N-hydroxy compounds in cyanogenic plants such as flax and sorghum have been consistently unsuccessful (18). The involvement of free keto acid oxime and of reactions 3 and 4 are uncertain in the conversion of amino acids to aldoximes (120).

Conversion of Aldoximes to Nitriles, Cyanogenic Glycosides, and Hydrocyanic Acid

Indoleacetonitrile (51), N-methoxyindoleacetonitrile and 4-methoxyindole-acetonitrile (117), phenylacetonitrile (37, 140), 1-cyano-2-hydroxy-3-butene and 1-cyano-2-hydroxy-3,4-epithiobutanes (24) have been isolated from plant materials also containing the corresponding glucosinolates, and are generally believed to be formed from the latter by the action of myrosinase during extraction (reaction 25) (100, 101, 128, 135, 137). The isolation of phenylacetonitrile from *Leptactinia senegambica* (28, 97) and the suspected occurrence of indoleaceto-nitrile in banana ovaries (103) and other plants (8) are examples of the presence of some of the above nitriles in nonglucosinolate containing plants. Labeled iso-butyronitrile was detected in flax plants administered ^{14}C-valine (121), and *p*-glucosyloxymandelonitrile has been identified as a cyanogenic compound in *Nandina domestica and Thalictrum aquilegifolium* (2, 106).

The enzyme indoleacetaldoxime hydrolyase (reaction 5) has been partially purified from the fungus *Gibberella fujikuroi* (107, 108) and shown to be rather specific for indoleacetaldoxime. The enzyme was optimally active at pH 7, and was activated by pyridoxal phosphate, ferrous ions, ascorbic and dehydroascorbic acids, and inhibited by thiols and thiol reagents. Other properties indicated that the enzyme was "regulatory" in nature, being formed optimally at low temperatures and in stationary cultures. An induced enzyme capable of dehydrating both *syn-* and *anti*-2,6-dichlorobenzaldoxime to the nitrile has been extracted from *Pseudomonas putrefaciens* grown on the same aldoxime (85). Cell-free preparations from cabbage (64) and sorghum (18) converted indoleacetaldoxime or *p*-hydroxyphenylacetaldoxime to the nitriles. Dehydration of aldoximes thus appears to be a mechanism in higher plants, fungi, and bacteria for the de novo formation of the nitrile group. The conversion of 2-keto acid oximes to cyano-genic compounds (120, 121) may not be via the aldoxime but by their direct non-enzymic dehydration and decarboxylation to the nitrile (reaction 6) (3, 116, 120). Isobutyronitrile was converted to linamarin in flax plants (44, 119, 120) and phenylacetonitrile to prunasin (D-mandelonitrile-β-glucoside) in cherry laurel leaves (44). Conversion of these nitriles to cyanogenic glycosides requires the α-oxidation of the nitrile (reaction 7), and $^{18}O_2$ was incorporated as the glucosidic linkage oxygen into linamarin and lotaustralin (2-hydroxy-2-methylbutyronitrile glucoside) in flax seedlings (148). The α-oxidation of nitriles has been shown in many plants (34). The α-oxidation of indoleacetonitrile to indolecarboxylic acid possibly via indolealdehyde (reactions 7, 12, 13) has been demonstrated by cell-free preparations of pea plants (124). The fate of the nitrile moiety, though not determined, may have been loss to hydrocyanic acid.

An alternative route for cyanogenic glycoside formation from aldoxime via 2-hydroxyaldoxime has been proposed (reactions 8 and 9) based on the incorporation of 2-hydroxyisobutyraldoxime and 2-hydroxyphenylacetaldoxime to linamarin and prunasin respectively (123). An α-oxidation of an aldoxime would then be required prior to dehydration. As Tapper and co-workers suggest, specificity studies on the enzymes involved would be required to clarify the point.

The glucosylation of a 2-hydroxynitrile would yield cyanogenic glycoside (reaction 10) and an enzyme named UDP-glucose: ketonecyanohydrin glucosyl transferase has been purified 120-fold from flax seedlings (42–44) capable of glucosylating both 2-hydroxyisobutyronitrile and 2-hydroxy-2-methylbutyronitrile to yield linamarin and lotaustralin. The activity towards both substrates remained constant during purification and hence only one enzyme appears to be involved for both reactions. The enzyme had a pH optimum between 8–9, glucosylated the cyanohydrins of aliphatic ketones but not aromatic aldehydes, and was specific for UDP-glucose as the glucosyl donor (42).

The accumulation of cyanogenic glycosides in plants suggest that they are end products of metabolism. However, a turnover of cyanogenic glycosides, such as of dhurrin in sorghum (11, 21), indicates that they may be only temporary end products. Enzymes capable of decomposing cyanogenic glucosides are always associated with these compounds, and the release of hydrocyanic acid (HCN) by "crushed plant" reactions has been known for over 170 years (21). Two enzymes, a β-glycosidase (reaction 11) and a 2-hydroxynitrile lyase (reaction 12) are involved in HCN production. β-Glycosidases acting specifically on cyanogenic glycosides have been purified from flax (12), sorghum (76), and almonds (46). Hydroxynitrile lyase from almonds containing FAD as a prosthetic group (4) and from sorghum not containing FAD (102) have been obtained as homogenous proteins and are stereospecific in their reactions.

The absence of hydrocyanic acid in intact plants has supported the idea that cyanogenesis is a "crushed plant" reaction where a sudden release of quantities of HCN result. HCN, however, is extensively metabolized by cyanogenic and noncyanogenic plants to asparagine and aspartic acid via β-cyanoalanine (21, 35) by reactions A and B indicated below:

A. HCN + Serine, Cysteine or o-methylserine $\xrightarrow{\text{(i)}}$ β-Cyanoalanine

B. β-Cyanoalanine $\xrightarrow{\text{(ii)}}$ asparagine $\rightarrow$ aspartic acid

C. β-Cyanoalanine $\rightarrow$ γ-glutamyl-β-cyanoalanine

$\rightarrow$ $\beta(\gamma$-glutamyl) aminopropionitrile

The enzymes β-cyanoalanine synthase (i) (47) and β-cyanoalanine hydrolase (ii) (31) have been purified from blue lupin. Any HCN formation in intact plants may therefore be immediately metabolized, as was shown in *Lotus sp.* where the amide group of asparagine became labeled when these plants were fed uniformly labeled valine (1). β-Cyanoalanine, γ-glutamyl-β-cyanoalanine, and β(γ-glutamyl) aminopropionitrile occur as natural constituents in certain legumes such as *Vicia* sp. and *Lathyrus* sp. (6, 7, 21, 94, 95, 99). The nitrile moiety in these compounds arise from preformed HCN (reactions A, B, and C). Whether or not aldoximes are indirectly involved in the formation of the nitrile moiety in these "noncyanogenic" plants, via reactions indicated in Figure 2 and the above reactions (A, B, and C), awaits investigation. The demonstration of the enzyme rhodanase, which converts cyanide to inorganic thiocyanate in the cyanogenic plant *Manihot utilissima*, may represent yet another way of cyanide metabolism (15).

Linamarin and lotaustralin have been identified as precursors of HCN in an unidentified cyanogenic psychrophilic basidiomycete (113). HCN was metabolized by this fungus to alanine and glutamic acids via α-aminopropionitrile and 4-amino-4-cyanobutyric acids by an enzymic analogy to the Strekker synthesis (114, 115). Enzymic hydrolysis of the nitriles by a nitrilase yielded the amino acids.

Conversion of Aldoximes to Glucosinolates and their Decomposition Products and Nitro Compounds

All experimental evidence indicates that thioglucosidation of aldoximes precedes sulfatation during the formation of glucosinolates (Figure 2). However, a large lacuna exists in our current knowledge of the initial step in glucosinolate formation, namely the incorporation of the divalent sulfur atom. Cysteine and methionine were more efficient divalent sulfur donors than thioglucose (78, 143) in the biosynthesis of benzyl, phenylethyl, and allyl glucosinolates. Cystine was another efficient sulfur donor in the formation of indolemethyl and N-sulfoindolemethyl glucosinolates from indoleacetaldoxime (74). Indications of an unidentified metabolite of cystine as the sulfur donor has been obtained. These results suggest the initial formation of an S-alkylthiohydroximate (XI) (reaction 14) where the alkyl group may be derived from cysteine, thiocysteine, or some yet unknown compound (58, 73).

The hypothesis that thiohydroximic acid (XII) may be an intermediate in glucosinolate biosynthesis (28, 77, 82) was experimentally verified by the efficient conversion of labeled phenylacetothiohydroximate to benzyl glucosinolate in *T. majus* (reactions 17 and 19) (134). ^{14}C-Phenylacetothiohydroximate was trapped as its S-methyl derivative in *T. majus* plants administered ^{14}C-phenylacetaldoxime (reactions 15 or 14-16) (80). A similar formation of *p*-hydroxyphenylacetothiohydroximate from *p*-hydroxyphenylacetaldoxime has been reported (59). These findings add thiohydroximates to the list of intermediates in glucosinolate biosynthesis, being either formed directly from the aldoxime (reaction 15) or more likely from a S-alkyl intermediate (reaction 16). Thiohydroximic acids are unstable compounds and decompose to nitriles (reaction 26) (29).

An enzyme, UDP-glucose: thiohydroximate glucosyltransferase, catalyzing the glucosylation of phenylacetothiohydroximate (reaction 17) by UDP-glucose to desthiobenzylglucosinolate, has been purified 20-fold from *T. majus* and similar activity demonstrated in cell free extracts of *S. alba*, *N. officinale*, and *A. lapathifolia* (80, 81). The *T. majus* enzyme glucosylated several thiohydroximates, though less efficiently than phenylacetothiohydroximate, but not phenylacetohydroximate, cysteine, or thiols. TDP-Glucose served to a certain extent as glucose donor but not ADP-, CDP-, or GDP-glucose. The enzyme was inhibited by sulfyhydryl reagents and stimulated by reducing agents such as thiols and ascorbic acid. The suggestion that desthioglucosinolate may be an intermediate in glucosinolate biosynthesis (28, 77, 83) was experimentally verified when desthiobenzylglucosinolate (aglycone labeled) was converted with about 60% efficiency to benzylglucosinolate in *T. majus* plants (reaction 19) (134). The same compound when glucose labeled showed lesser efficiency of incorporation, suggesting the occurrence of glucosyl transfer side reaction. A slow thioglucosidase cleavage of desthioindolemethyl glucosinolate was obtained with crude myrosinase (74) (reverse of reaction 17), and such an enzyme could possess glucosyl transfer activity. Desthio-3-butenyl glucosinolate was incorporated into both 3-butenyl- and 2-hydroxy-3-butenyl glucosinolates in two *Brassica napus* cultivars (52). Compounds tentatively identified as desthio-*p*-hydroxybenzyl glucosinolate (58) in *Sinapis alba* and as desthioindolemethyl glucosinolate in *Isatis tinctoria* (74) and *S. alba* (53) were formed following administration of labeled *p*-hydroxyphenyl-acetaldoxime or indoleacetaldoxime to these plants. A greater accumulation of the desthioindolemethyl glucosinolate occurred when potassium selenate was administered along with indoleacetaldoxime and cystine to *I. tinctoria* leaves (74); selenate apparently inhibited the sulfatation process. When this desthio compound was refed to *I. tinctoria* leaves, efficient conversion to both indolemethyl and N-sulfoindolemethyl glucosinolate occurred. The greater accumulation of the desthio compound following selenate inhibition is further proof that sulfatation is indeed the last step in glucosinolate biosynthesis. An in vitro demonstration of this reaction has not been made, but PAPS (134) or APS may be the sulfate donor.

The conversion of both 1-nitro-2-phenylethane and phenylacetaldoxime to benzyl glucosinolate in *T. majus* (79) lends support to a postulate based on chemical analogy, that the *aci*-nitro tautomer (XIX) may be a common intermediate in the conversion of these two compounds to the glucosinolate (reaction 27 and 29). The mercaptide attack on the *aci*-nitro tautomer would yield the S-alkyl thiohydroximate (XI) (reaction 28) already described (28). The percent conversion of the nitro compound was lower than the conversion of phenylalanine to the glucosinolate and was therefore even less efficient than phenylacetaldoxime as a precursor (129). The low incorporation has been attributed to the lower solubility, transport, or availability of the nitro compound at the site of glucosinolate biosynthesis (79). Interestingly, some conversion of phenylacetaldoxime to the nitro compound was observed in the same study, suggesting a precursor role for aldoximes in the formation of nitro compounds in plants. Both 1- nitro-2-phenylethane (41) and 3-nitro propionic acid (28) have been isolated from plants (Table 2) and phenylalanine (28, 41) and malonylmonohydroxamate (13) suggested as their respective precursors. Mention of the co-occurrence of an aldoxime and its *aci*-nitro tautomer derivative in a *Streptomyces* species (144) has been made.

As with cyanogenic glycosides, glucosinolates invariably occur in conjunction with enzymes capable of decomposing them. The "crushed-plant" reaction of glucosinolates yields a variety of products such as organic isothiocyanates (XV), organic thiocyanates (XVII), and nitriles (V) by reactions 20, 22 or 23, and 25 (28, 35, 137). Certain unstable organic isothiocyanates decompose to yield inorganic thiocyanate and the lower alcohol (XVI) (137). Enzymic isomerization of benzyl isothiocyanate to benzyl thiocyanate has been shown in *Lepidium* sp. (139). The origins of benzylamine in *Moringa pterygosperma* (14) and of *p*-hydroxybenzylamine in *S. alba* (67) are believed to be by the hydrolysis of the corresponding isothiocyanates (reaction 24). The detection of benzyl isothiocyanate vapors around green intact papaya fruits suggests that it may be a normal metabolite in this plant (118). Similarly the release of the same compound by roots of *Moringa pterygosperma* renders the soil in the immediate vicinity of the roots of this plant free of bacterial population (91).

The β-thioglucosidase (glucosinolase) responsible for the decomposition of glucosinolates has been purified from several sources (38, 110). The absolute dependence of the purified enzyme from *Sinapis alba* on ascorbic acid for its activity was explained as a coenzymic involvement of the vitamin in the reaction per se (27). Two isoenzymes with glucosinolase action have been obtained from *Sinapis alba*, of which only one isoenzyme had an absolute requirement for ascorbic acid for its activity while the other did not. Both isoenzymes had the same affinity for the substrate *p*-hydroxybenzyl glucosinolate, yielding the same products (141). These results go against the coenzymic role hypothesis. The ability of the glucosinolases to yield either the isothiocyanate or the nitrile as the product of reaction, depending on the reaction condition (128, 137), is intriguing and requires further study, since this enzyme (or complex of enzymes) probably decides the further metabolic fate of glucosinolates in plants.

Conversion of Aldoximes and Nitriles to Amides and Non-Nitrogenous Compounds

Though plants have been shown to be able to convert aldoximes and nitriles to acids (XXII), amides (XXIII), aldehydes (XXIV), and alcohols (XXV), a systematic study is wanting. As mentioned earlier, these products can be formed in routes not involving oximes, and therefore, unlike cyanogenic glucosides or glucosinolates, their presence in plants is not diagnostic of the involvement of the oxime pathway.

The conversion of indoleacetonitrile to indoleacetic acid (reaction 31) by

plants was initially investigated to explain why the nitrile acted as a growth hoɪ-
mone only in certain plants (125). The enzyme nitrilase purified from barley
leaves (75, 126) was rather nonspecific. Enzymes hydrolyzing nitriles exist in
bacteria (96) and fungi (72, 115) but appear to be more specific. Enzymic hydrol-
yses of nitriles do not proceed via a free amide (75, 96), which makes the reaction
mechanistically interesting. A study of the distribution of nitrilase activity in
higher plants using only one substrate, namely indoleacetonitrile, indicated it to
be rather restricted; of the 29 plants representing 22 families tested, only members of four families—Cruciferae, Graminae, Musaceae, and Strelitziaceae—
showed unambiguous conversion (126). The occurrence of both oxime dehydrating and nitrile hydrolyzing enzymes in plants containing neither cyanogenic glycosides nor glucosinolates, as *Musa*, *Avena*, and *Hordeum* and in fungi (cf Table 1)
(71, 90, 126), supports the hypothesis of a third pathway of oxime metabolism to
nonnitrogenous products operating in plants, though obviously further work is
needed to establish this.

Enzymes capable of hydrating nitriles to amides (XXIII) (reaction 32) appear
to be widespread in plants, and cyanamide hydration by cyanamidase (49), β-
cyanoalanine hydration to asparagine (31, 86), and indoleacetonitrile conversion
to the amide in red alga (98) are known. Deamidation of amides (reaction 33) in
plants has been reported (34, 142).

p-Hydroxyphenyl- and *p*-methoxyphenylacetaldoximes were converted to *p*-
hydroxy- and *p*-methoxyphenyl ethyl alcohols in *Sinapis alba* and *Aubrietia* sp.
respectively (reactions 34 and 36) (59). Similarly indoleacetaldoxime was converted to 3-indole-ethanol by yeast (72) and every plant tested and representing 17
families (90), including those capable of converting oxime to the nitrile (cf Table
2). Both *p*-hydroxyphenylethyl alcohol and indoleethanol are natural plant constituents (59, 93). Conversion of indoleacetaldoxime to indoleacetic acid via the
aldehyde (reactions 34, 35) (90) explains the growth promoting property of the
former compound in pea tissue (33).

Oxime formation and hydrolysis is a pH-dependent equilibrium process, and a
slow nonenzymic hydrolysis of indoleacetaldoxime to the aldehyde occurred at
acid pH values (90). Enzymatic conversion of the aldehyde to either alcohol or
acid or of hydroxylamine to ammonia (reaction 37) would shift the reaction towards further oxime hydrolysis. While this mechanism may explain the ubiquitous formation of the alcohol from the oxime in plants, some preliminary evidence
for an enzymic hydrolysis of indoleacetaldoxime in *Avena* has been obtained (90).
Further clarification of this reaction, which leads to yet another pathway of oxime
metabolism, is evidently needed.

The formation of a compound tentatively identified as isobutyraldoxime-*o*-β-
D-glucopyranoside (XXI) (reaction 30) in flax seedlings administered labeled
valine and inhibitors of linamarin synthesis such as DL-*o*-methylthreonine or DL-
2-methoxypropionaldoxime, has been explained as a block in the aldoxime to
nitrile conversion resulting in the induction of glucoside formation (121).

R Group Modifications of Amino Acids Metabolized via the Oxime Pathway

Chain elongation of amino acids via their 2-keto acids before their entry into the oxime pathway appears to be a common mechanism in the biosynthesis of several glucosinolates (35, 78, 133). The chain elongation process is analogous to 2-ketoglutaric acid formation from oxalacetic acid and acetyl CoA in the tricarboxylic acid cycle.

Modifications of the R groups of glucosinolates and cyanogenic glycosides during or after their formation lead to products having little resemblance to the parent amino acid. Characteristic modifications are methylthio group elimination and introduction of a double bond (77), hydroxylation (131), indole N-methoxylation and sulfonation (25, 28), ring cleavage (105), methylation (105), S-oxidation or reduction (17), and glucosylation (106). As seen in Table 1, all these modifications appear to occur after the aldoxime stage. Thus indoleacetaldoxime was converted to both indolemethyl and N-sulfoindolemethyl glucosinolates, 4-(methylthio) butyraldoxime to allylglucosinolate, 5-(methylthio) valeraldoxime to 2(S)-hydroxy-3-butenyl glucosinolate, and p-hydroxyphenylacetaldoxime to the branched chain cyanogenic glucoside triglochinin. N-Sulfonation occurred after desthioindolemethyl glucosinolate formation in *I. tinctoria* (74), and desthio 3-butenyl glucosinolate gave rise to both 3-butenyl and 2-hydroxy-3-butenyl glucosinolate in *B. napus* (52), suggesting methylthio elimination even at the desthioglucosinolate stage. In *Armoracia lapathifolia*, the thiohydroximate, desthio-3-methylthiopropyl, 3-methylthiopropyl, and 3-methylsulfinylpropyl glucosinolates were all converted to allyl glucosinolate. Methyl thio- elimination and double bond formation therefore occurred after glucosinolate formation (17). A reduction of 3-methyl sulfinyl propyl glucosinolate to 3-methylthiopropyl glucosinolate has been observed in *Cheiranthus kewensis* (16). The inability of *Sinapis alba* plants to convert administered sulfur-labeled benzyl glucosinolate to p-hydroxybenzyl glucosinolate shows that phenyl ring hydroxylation does not occur at the glucosinolate level (55).

Specificities in the Oxime Pathways

At least two types of genetically controlled specificities are encountered in the oxime pathways: (i) at the amino acid level (or even preamino acid level where chain elongation of a particular amino acid takes place); and (ii) at the aldoxime level which specifies the end product, cyanogenic glucoside, glucosinolate, or possibly even a non-nitrogenous compound. Obviously these specificities are qualitatively and quantitatively controlled by the enzymes involved. Paucity of current knowledge on the enzymes involved in these early reactions therefore shed little light on their specificities. The inhibitory action of 2-methoxypropionaldoxime in linamarin biosynthesis (121) and the competitive inhibition by phenylacetaldoxime during the enzymic dehydration of indoleacetaldoxime (107) are

Table 2 Distribution of the Oxime Pathways in Higher Plant Families

Class subclass	Glucosinolate[a]	Cyanogenic glycoside[b]	Oxime to nitrile[c] conversion	Miscellaneous
Dicots				
Magnoliidae	—	Berberidaceae[d] Ranunculaceae[d] Papaveraceae	—	Lauraceae (A) Anonaceae (A) Corynocarpaceae (B)
Hamameliidae	—	—	—	—
Caryophyllidae	Phytolaccaceae	—	—	Chenopodiaceae (C)
Dillenidae	Caricaceae Bixaceae[f] Capparaceae Cruciferae Moringaceae Resedaceae Tovariaceae	Flacourtiaceae Passifloraceae Sapotaceae Tiliaceae	Cruciferae[e] Cucurbitaceae Malvaceae	Violaceae (D)
Rosidae	Euphorbiaceae Limnanthaceae Salvadoraceae Tropaeolaceae	Euphorbiaceae Leguminasae Linaceae Melastomaceae Myrtaceae Olacaceae Rosaceae Rutaceae Umbelliferae	—	Malpighiaceae (D) Leguminosae (D) Euphorbiaceae (E) Sapindaceae (F)
Asteridae	—	Asclepediaceae Caprifoliaceae Compositae Convolvulaceae Myoporaceae Rubiaceae Scrophulariaceae	Compositae Convolvulaceae Cuscutaceae Solanaceae	Rubiaceae (G)
Monocots				
Alismatidae	—	Scheuchzeriaceae	—	—
Liliidae	—	—	—	—
Arecidae	—	—	—	—
Commeliniidae	—	Graminae	Musaceae[e] Graminae[e]	—

examples of specificity at the aldoxime dehydration step. The conversion of 2-hydroxyaldoximes to the cyanogenic glucosides could represent a lack of absolute specificity (reaction 9) (123). The specificity of the glucosylating enzymes during linamarin or lotaustralin formation (42) or benzyl glucosinolate formation (81) suggests partial control even at this level. The existence of close lines of cyanogenic and noncyanogenic *Trifolium repens* (22) and of high or low glucosinolate-producing cultivars of *Brassica napus* (54) reflect genetic variations in the levels of enzymes involved.

In *Sorghum*, rapid dhurrin formation occurs only during early stages of seedling growth, the glucoside level falling off thereafter (21). Indoleacetaldoxime hydrolyase activity in *Gibberella fujikuroi* was dependent on age and culture conditions, particularly temperature (107, 108). The levels of indolemethyl and N-sulfoindolemethyl glucosinolates were found to vary, but in a parallel manner, during the growth and development of *Isatis tinctoria* (26). During ontogenesis of *Sinapis alba*, a decrease in indolemethyl glucosinolate and an increase in *p*-hydroxybenzyl glucosinolate levels were observed (10), and has been interpreted as different enzyme systems operating at different stages of plant development. These examples illustrate the ontogenic control of the oxime pathways in plants resulting in variation in both the type and the amount of end products formed.

DISTRIBUTION OF OXIME PATHWAYS IN THE PLANT KINGDOM

A combination of the distribution of cyanogenic glycosides and glucosinolates in plants represents, at least partially, the distribution of oxime pathways in

—————————————————————————

←⦉⦉⦉

ᵃ (28).
ᵇ (2, 30, 45, 104).
ᶜ (71, 72, 90).
ᵈ Contains *p*-glucosyloxymandelonitrile (2, 106).
ᵉ Also nitrile to acid activity (126).
ᶠ Questioned (28).
A = 1-Nitro-2-phenylethane (28, 41).
B = Nitropropionic acid of karakin glucoside (89).
C = Girgensohnine (147).
D = 3-Nitropropionic acid (28, 89).
E = Ricinine.
F = Cyanogenic lipids (84, 109).

plants. Though cyanogenesis is widely distributed in higher plants comprising about 65 families in angiosperms, 2 in pteridophytes, and 3 in gymnosperms (45), the nature of the HCN producing compound in most of the plants is unknown. Present evidence indicates that 25 families (2, 30, 45, 104) contain cyanogenic glycosides. Likewise, glucosinolates so far have been shown in only 11 or 12 families (28). Plants belonging to 9 families possess the capacity to convert an aldoxime (indole or phenylacetaldoxime) to the nitrile (18, 71, 72, 90). The familywise distribution of these compounds or activity in each subclass of the angiosperms is given in Table 2, based on Cronquist's classification of plants (23). Table 2 also includes the distribution of nitro compounds and certain nitriles which may or may not be derived from oximes.

The oxime pathway leading to cyanogenic compounds is widely distributed in both dicots and monocots ranging from the primitive (Magnoliidae) to the advanced (Asteridae) subclasses, reflecting the spread of cyanogenesis in general in plants and having representation in every subclass of the angiosperms (45). The oxime pathway leading to glucosinolate formation is restricted to only 3 subclasses in the dicots, namely Caryophyllidae, Dilleniidae, and Rosidae. Their absence in Asteridae, which is described as arising from Rosidae (23), is interesting. Euphorbiaceae appears to be the sole exception where oxime pathways lead to both cynanogenic glycosides and glucosinolates, and this family has been described as polyphyletic in origin (28). Apart from Euphorbiaceae, Cruciferae, Resedaceae, and Capparaceae are glucosinolate-producing and cyanogenic, while the remaining 8 glucosinolate-forming families are not listed as cyanogenic (45).

Plants representing 9 families (Table 2) converted indoleacetaldoxime to the nitrile (71, 90). Four of these (Cucurbitaceae, Cuscutaceae, Malvaceae, and Musaceae) are neither cyanogenic (45) nor glucosinolate formers (28). Channeling of oximes to nondiagnostic products as described earlier may therefore be of widespread occurrence but could go undetected.

Cyanogenesis in fungi (68) and the formation of linamarin and lotaustralin in one psychrophilic basidiomycete (113) has been described. The ability to convert oxime to nitrile and nitrile to acid in several Moniliales and in an ascomycete (71, 126) shows the prevalence of the oxime pathways in fungi also.

CONCLUDING REMARKS

An intriguing and unanswered question in plant biochemistry is the raison d'etre of secondary plant products (36). A protective role has been offered to those compounds having toxic properties. If this were true, then the products derived by the oxime pathways are indeed admirably suited for this role, comprising as they do compounds such as HCN, nitriles, organic isothio- and thiocyanates. If a protective role must be given to oximes, then it is not by detoxifying hydroxylamine, as originally proposed, but by keeping away parasites and pests by the vast array of toxic compounds produced!

Several interesting and novel enzymic reactions have come to light during the study of the metabolism of oximes and nitriles. Investigating their mechanisms and control should be fruitful areas of work for the future.

ACKNOWLEDGMENT

Thanks are due to Professor B. B. Stowe, Yale University, for help in making available reprints of several papers included in this review.

Literature Cited

1. Abrol, Y. P., Conn, E. E. 1966. *Phytochemistry* 5:237–42
2. Abrol, Y. P., Conn, E. E., Stoker, J. R. 1966. *Phytochemistry* 5:1021–28
3. Ahmad, A., Spenser, I. D. 1961. *Can. J. Chem.* 39:1340–59
4. Becker, W., Benthin, U., Eschenhof, E., Pfeil, E. 1963. *Biochem. Z.* 337:156–66
5. Beevers, L., Hageman, R. H. 1969. *Ann. Rev. Plant Physiol.* 20:495–522
6. Bell, E. A. 1964. *Nature (London)* 203:378–80
7. Bell, E. A., Tirimanna, A. S. L. 1965. *Biochem. J.* 97:104–11
8. Bentley, J. A. 1958. *Ann. Rev. Plant Physiol.* 9:47–80
9. Bergersen, F. J. 1971. *Ann. Rev. Plant Physiol.* 22:121–40
10. Bergmann, F. 1970. *Z. Pflanzenphysiol.* 62:362–75
11. Bough, W. A., Gander, J. E. 1971. *Phytochemistry* 10:67–77
12. Butler, G. W., Bailey, R. W., Kennedy, L. D. 1965. *Phytochemistry* 4:369–81
13. Candlish, E., LaCroix, L. J., Unrau, A. M. 1969. *Biochemistry* 8:182–86
14. Chakravarti, R. N. 1955. *Bull. Calcutta Sch. Trop. Med.* 3:162–63
15. Chew, M. Y., Boey, C. G. 1972. *Phytochemistry* 11:167–69
16. Chisholm, M. D. 1972. *Phytochemistry* 11:197–202
17. Chisholm, M. D., Matsuo, M. 1972. *Phytochemistry* 11:203–7
18. Conn, E. E. Personal communication
19. Conn, E. E. 1969. *J. Agr. Food Chem.* 17:519–26
20. Conn, E. E., Akazawa, T. 1958. *Fed. Proc.* 17:205
21. Conn, E. E., Butler, G. W. 1969. In *Perspectives of Phytochemistry*, ed. J. B. Harborne, T. Swain, 47–74. London: Academic
22. Corkill, L. 1952. *N. Z. J. Sci. Technol.* A 34: 1–16
23. Cronquist, A. 1968 *The Evolution and Classification of Flowering Plants.* London: Nelson
24. Daxenbichler, M. E., VanEtten,C. H., Wolff, I. A. 1968. *Phytochemistry* 7: 989–96
25. Elliott, M. C., Stowe, B. B. 1970. *Phytochemistry* 9:1629–32
26. Elliott, M. C., Stowe, B. B. 1971. *Plant Physiol.* 48:498–503
27. Ettlinger, M. G., Dateo, G. P.,
   Harrison, B. H., Mabry, T. J., Thompson, C. P. 1961. *Proc. Nat. Acad. Sci. Wash.* 47:1875–80
28. Ettlinger, M. G., Kjaer, A. 1968. In *Recent Advances in Phytochemistry*, ed. T. J. Mabry, R. E. Alston, V. C. Runeckles, 1:59–144. New York: Appleton-Century-Crofts
29. Ettlinger, M. G., Lundeen, A. J. *J. Am. Chem. Soc.* 79:1764–65
30. Eyjolfsson, R. 1970. *Phytochemistry* 9:845–51
31. Farnden, K. J. F., Castric, P. A., Conn, E. E. 1971. *Plant Physiol. Suppl.* 47:17
32. Farnden, K. J. F., Rosen, M., Conn, E. E. 1972. *Plant Physiol. Suppl.* 49:38
33. Fawcett, C. H. 1964. *Nature (London)* 204:1200–1
34. Fawcett, C. H., Taylor, H. F., Wain, R. L., Wightman, F. 1958. *Proc. Roy. Soc.* B148:543–70
35. Fowden, L. 1967. *Ann. Rev. Plant Physiol.* 18:85–106
36. Fraenkel, G. S. 1959. *Science* 129:1466–70
37. Gadamer, J. 1899. *Chem. Ber.* 32: 2335–41
38. Gaines, R. D., Goering, K. J. 1962. *Arch. Biochem. Biophys.* 96:13–19
39. Gander, J. E. 1958. *Fed. Proc.* 17:226
40. Gmelin, R., Virtanen, A. I. 1961. *Suom. Kemistilehti* B34:15–18
41. Gottlieb, O. R. 1972. *Phytochemistry* 11:1537–70
42. Hahlbrock, K., Conn, E. E. 1970. *J. Biol. Chem.* 245: 917–22
43. Hahlbrock, K., Conn, E. E. 1971. *Phytochemistry* 10:1019–23
44. Hahlbrock, K., Tapper, B. A., Butler, G. W., Conn, E. E. 1968. *Arch. Biochem. Biophys.* 125:1013–16
45. Hegnauer, R. 1960. *Pharm. Zentralh.* 99:322–29
46. Helferich, B., Kleinschmidt, T. 1968. *Hoppe Seyler's Z. Physiol. Chem.* 349:25–28
47. Hendrickson, H. R., Conn, E. E. 1970. *J. Biol. Chem.* 244:2632–40
48. Hewitt, E. J., Hucklesby, D. P., Betts, G. F. 1968. In *Recent Aspects of Nitrogen Metabolism in Plants*, ed. E. J. Hewitts,C. V. Cutting, 47–81. First Long Ashton Symp. London: Academic
49. Hofmann, E., Latzko, E., Suss, A. 1954. *Z. Pflanzenernaehr. Dueng. Bodenk.* 66:193–202

50. Hucklesby, D. P., Hewitt, E. J. 1970. *Biochem. J.* 119:615–27
51. Jones, E. R. H., Henbest, H. B., Smith, G. F., Bentley, J. A. 1952. *Nature (London)* 169:485–87
52. Josefsson, E. 1971. *Physiol. Plant.* 24:161–75
53. Ibid 1972. In press
54. Josefsson, E., Appelqvist, L. A. 1968. *J. Sci. Food Agr.* 19:564–70
55. Kindl, H. 1965. *Monatsh. Chem.* 96:527–32
56. Kindl, H. 1968. *Hoppe Seyler's Z. Physiol. Chem.* 349:519–20
57. Kindl, H. Personal communication
58. Kindl, H., Schiefer, S. 1969. *Monatsh. Chem.* 100:1773–87
59. Kindl, H., Schiefer, S. 1971. *Phytochemistry* 10:1795–1802
60. Kindl, H., Underhill, E. W. 1968. *Phytochemistry* 7:745–56
61. Kjaer, A., Schuster, A. 1972. *Phytochemistry* 11:3045–48
62. Kjaer, A., Wagnieres, M. 1971. *Phytochemistry* 10:2195–98
63. Kretovich, W. L. 1965. *Ann. Rev. Plant Physiol.* 16:141–54
64. Kumar, S. A., Mahadevan, S. 1963. *Arch. Biochem. Biophys.* 103:516–18
65. Kutáček, M., Kralovà, M. 1972. *Biol. Plant.* 14:279–81
66. Kutáček, M., Procházka, Z., Veres, K. 1962. *Nature (London)* 194:393–94
67. Larsen, P. O. 1965. *Biochim. Biophys. Acta* 107:134–36
68. Lebeau, J. B., Dickson, J. G. 1953. *Phytopathology* 43:581–82
69. Lee, C. J., Serif, G. S. 1971. *Biochim. Biophys. Acta* 230:462–76
70. Linser, H., Mayer, H., Maschek, F. 1954. *Planta* 44:103–20
71. Mahadevan, S. 1963. *Arch. Biochem. Biophys.* 100:557–58
72. Mahadevan, S. Unpublished observations
73. Mahadevan, S., Stowe, B. B. 1972. In *Plant Growth Substances 1970*, ed. D. J. Carr, 117–26. Berlin: Springer Verlag
74. Mahadevan, S., Stowe, B. B. 1972. *Plant Physiol.* 50:43–50
75. Mahadevan, S., Thimann, K. V. 1964. *Arch. Biochem. Biophys.* 107:62–68
76. Mao, C. H., Anderson, L. 1967. *Phytochemistry* 6:473–83
77. Matsuo, M. 1968. *Tetrahedron Lett.*, 4101–4
78. Matsuo, M. 1968. In *Mechanisms of Reaction of Sulfur Compounds*, 3:107–13
79. Matsuo, M., Kirkland, D. F., Under-

hill, E. W. 1972. *Phytochemistry* 11:697–701
80. Matsuo, M., Underhill, E. W. 1969. *Biochem. Biophys. Res. Commun.* 36:18–23
81. Matsuo, M., Underhill, E. W. 1971. *Phytochemistry* 10:2279–86
82. Meakin, D. 1965. PhD thesis. Univ. Alberta, Calgary
83. Meakin, D. 1967 *Experientia* 23:174–75.
84. Mikolajczak, K. L., Smith, C. R. 1971. *Lipids* 6:349–50
85. Milborrow, B. V. 1963. *Biochem. J.* 87:255–58
86. Nartey, F. 1970 *Z. Pflanzenphysiol.* 62:398–400
87. Omura, H., Osajima, Y., Tsutsumi, M. 1967. *Enzymologia* 32:135–44
88. Omura, H., Tsutsumi, M. 1968. *Enzymologia* 34:187–97
89. Paris, B. 1963. In *Chemical Plant Taxonomy*, ed. T. Swain, 337–58. London: Academic
90. Rajagopal, R., Larsen, P. 1972. *Planta* 103:45–54
91. Rao, P. L. N., Ramanathan, S. 1960. *Antiseptic* 57:792–98
92. Rautanen, N. 1958. In *Encyclopedia of Plant Physiology*, ed. W. Ruhland, 8:212–23. Berlin: Springer-Verlag
93. Rayle, D. L., Purves, W. K. 1968. In *Biochemistry and Physiology of Plant Growth Substances*, ed. F. Wightman, G. Setterfield, 153–61. Ottawa: Runge
94. Ressler, C. 1962. *J. Biol. Chem.* 237:733–35
95. Ressler, C., Giza, Y. H., Nigam, S. N. 1963. *J. Am. Chem. Soc.* 85:2874–75
96. Robinson, W. G., Hook, R. H. 1964. *J. Biol. Chem.* 239:4257–62
97. Sabetay, S., Palfray, L., Trabaud, L. 1938. *C. R. Acad. Sci.* 207:540–42
98. Schiewer, G., Libbert, E. 1965. *Planta* 66:377–80
99. Schilling, E. D., Strong, F. M. 1954. *J. Am. Chem. Soc.* 76:2848
100. Schraudolf, H., Bergmann, F. 1965. *Planta* 67:75–95
101. Schraudolf, H., Weber, H. 1969. *Planta* 88:136–43
102. Seely, M. K., Criddle, R. S., Conn, E. E. 1966. *J. Biol. Chem.* 241:4457–62
103. Shanmugavelu, K. G., Rangaswami, G. 1962. *Nature (London)* 194:775–76
104. Sharples, D., Spring, M. S., Stoker, J. R. 1972. *Phytochemistry* 11:2999–3002
105. Ibid, 3069–71

106. Sharples, D., Stoker, J. R. 1969. *Phytochemistry* 8:597–601
107. Shukla, P. S., Mahadevan, S. 1968. *Arch. Biochem. Biophys.* 125: 873–83
108. Ibid 1970. 137:166–174
109. Siegler, D. S., Seaman, F., Mabry, T. J. 1971. *Phytochemistry* 10:485–87
110. Snowden, D. R., Gaines, R. D. 1969. *Phytochemistry* 8:1649–54
111. Spenser, I. D., Ahmad, A. 1961. *Proc. Chem. Soc.* 375
112. Stevens, R. L., Emery, T. F. 1966. *Biochemistry* 5:74–81
113. Stevens, D. L., Strobel, G. A. 1968. *J. Bacteriol.* 95:1094–1102
114. Strobel, G. A. 1966. *J. Biol. Chem.* 241:2618–21
115. Ibid 1967. 242:3265–69
116. Stowe, B. B. 1959. *Fortschr. Chem. Org. Naturst.* 14:248–97
117. Tamura, S., Nomoto, M., Nagao, M. 1972. See Ref. 73, 127–32
118. Tang, Chung-Shih 1971. *Phytochemistry* 10:117–21
119. Tapper, B. A., Butler, G. W. 1967. *Arch. Biochem. Biophys.* 120:719–21
120. Tapper, B. A., Butler, G. W. 1971. *Biochem. J.* 124:935–41
121. Tapper, B. A., Butler, G. W. 1972. *Phytochemistry* 11:1041–46
122. Tapper, B. A., Conn, E. E., Butler, G. W. 1967. *Arch. Biochem. Biophys.* 119:593–95
123. Tapper, B. A., Zilg, H., Conn, E. E. 1972. *Phytochemistry* 11:1047–53
124. Taylor, H. F., Wain, R. L. 1959. *Nature (London)* 184:1142
125. Thimann, K. V. 1953. *Arch. Biochem. Biophys.* 44:242–43
126. Thimann, K. V., Mahadevan, S. 1964. *Arch. Biochem. Biophys.* 105: 133–41
127. Thompson, J. F., Morris, C. J., Smith, I. K. 1969. *Ann. Rev. Biochem.* 38:137–58
128. Tookey, H. L., Wolff, I. A. 1970. *Can. J. Biochem.* 48:1024–28
129. Underhill, E. W. 1967. *Eur. J. Biochem.* 2:61–63
130. Underhill, E. W. 1968. *Can. J. Biochem.* 46:401–5
131. Underhill, E. W., Kirkland, D. F. 1972. *Phytochemistry* 11:1973–80
132. Ibid, 2085–88
133. Underhill, E. W., Wetter, L. R. 1966. In *Biosynthesis of Aromatic Compounds. Proc. 2nd Meet. Fed. Eur. Biochem. Soc.*, ed. T. Billek, 3:129–37. Oxford: Pergamon
134. Underhill, E. W., Wetter, L. R. 1969. *Plant Physiol.* 44: 584–90
135. VanEtten, C. H., Daxenbichler, M. E. 1971. *J. Agr. Food Chem.* 19:194–95
136. Virtanen, A. I. 1961. *Ann. Rev. Plant Physiol.* 12:1–12
137. Virtanen, A. I. 1962. *Arch. Biochem. Biophys. Suppl.* 1:200–8
138. Virtanen, A. I., Miettinen, J. K., 1963. In *Plant Physiology* ed. F. C. Steward, 3:539–668. New York: Academic
139. Virtanen, A. I., Saarivirta, M. 1962. *Suom. Kemistilehti* B35:102–4
140. Ibid, 248–49
141. Vose, J. R. 1972. *Phytochemistry* 11:1649–53
142. Webster, G. 1964. In *Modern Methods of Plant Analysis*, ed. H. F. Linskens, B. D. Sanwal, M. V. Tracey, 7:392–420. Berlin: Springer-Verlag
143. Wetter, L. R., Chisholm, M. D. 1968. *Can. J. Biochem.* 46:931–35
144. Wiley, P. F., Herr, R. R., MacKellar, F. A., Argondeles, A. D. 1965. *J. Org. Chem.* 30:2330–34
145. Wilson, P. W. 1958. See Ref. 92, 8:9–47
146. Yoshida, S. 1969. *Ann. Rev. Plant Physiol.* 20:41–62
147. Yurashevskii, N. K., Stepanova, N. L. 1946. *J. Gen. Chem. USSR* 16:141–44
148. Zilg, H., Tapper, B. A., Conn, E. E. 1972. *J. Biol. Chem.* 247:2384–86

Ann. Rev. Plant Physiol. 1973. 24:89–114

PHOTOPHOSPHORYLATION IN VIVO[1] ❖ 7543

W. Simonis and W. Urbach

Botanisches Institut I der Universität Würzburg, Germany

CONTENTS

INTRODUCTION

In spite of the increasing number of publications concerned with the problem of photophosphorylation (PP) in vivo, no comprehensive review on this subject

[1] The following abbreviations are used in this chapter: ADP (adenosine diphosphate); ATP (adenosine triphosphate); CCCP (carbonylcyanide *m*-chlorophenylhydrazone); CMU [3-(*p*-chlorophenyl)-1,1-dimethylurea]; DAD (2,3,5,6-tetramethyl-*p*-phenylenediamine); DBMIB (2,5-dibromo-3-methyl-6-isopropyl-*p*-benzoquinone); DCMU (3-(3,4-dichlorophenyl)-1,1-dimethylurea); DNP (2,4-dinitrophenol); DSPD (disalicylidenepropanediamine); FCCP (carbonyl cyanide *p*-trifluoromethoxyphenylhydrazone); HOQNO (*n*-heptyl-4-hydroxyquinoline-N-oxide); KCN (potassium cyanide); Pi (inorganic phosphate); PMS (phenazine methosulphate); Po-fraction (trichloracetic acid-soluble organic phosphate fraction); PP (photophosphorylation); PS-I (photosystem I); PS-II (photosystem II).

has appeared for a long time. The difficulties in measuring with intact cells the conversion of light energy into energy-rich phosphate bonds coupled with photosynthetic electron transport, the consistent lack of agreement on a concept of PP in vitro, and the much higher complexity in vivo may be the main reasons for this absence of recent reviews.

Most measurements of PP in vivo are dependent on or even questioned by supplementary processes like transport through membranes or by consumption of the phosphorylated products. It is also reasonable to consider if processes found in thylakoids with artificial systems have any relevance in living cells. Nevertheless, in the past years many interesting details on PP in vivo have been accumulated by developing various new methods, and these need to be summarized in order to provide a basis for further studies. After a brief summary on PP in vitro, this article will discuss the various reactions providing evidence for the occurrence of different processes of PP in living cells. Finally, we shall try to draw some conclusions on the significance and role of PP in vivo far away from a conclusive picture of this subject.

After the first attempts of Emerson, Stauffer & Umbreit (43) to show the participation of energy-rich phosphates in photosynthesis in vivo, many different methods have been introduced and many reactions linked to photosynthesis have been studied. The earlier papers were extensively reviewed by Simonis (200) and Kandler (108). In the meantime, several reviews have been published on only a few aspects of PP in vivo (4, 8, 51, 61, 75, 95, 128, 142). Papers on photosynthetic bacteria are not included in this article because their PP has already been dealt with in recent years (57, 260).

PHOTOPHOSPHORYLATION IN VITRO

For a discussion of the more complex findings of PP in vivo, it seems necessary to consider the current views on PP in vitro. In recent years several reviews, monographs, and symposia have appeared which cover different aspects of PP in vitro, including the photosynthetic electron transport systems, proton transport, formation of a membrane potential, and the relationship between structure and function of chloroplasts (8, 20, 51, 56, 61, 68, 74, 142, 161, 238, 281).

Today two schemes of the photosynthetic electron transport system are mainly in discussion: (a) the Z-scheme with the two photoreactions (PS-I and PS-II) in series, first proposed by Hill & Bendall (84); and (b) the parallel scheme with 3 light reactions (PS-I, PS-IIa, and PS-IIb) proposed by Arnon's group (116). Much experimental evidence favors the Z-scheme, but the parallel scheme tries to take into consideration also the localization of different photosystems in separate chloroplast membrane particles. Whereas according to the Z-scheme electrons from a cyclic system reenter the noncyclic chain between the two photosystems, in the parallel scheme a cyclic electron transport system acts separately with its own electron carriers and pigment system (PS-I). Recently Park & Sane (161) offered a modification of the discussed schemes with two PS-I reactions and one

PS-II reaction in which one PS-I reaction is located in the stroma lamellae and the other PS-I reaction in the grana lamellae together with PS-II. These electron transport systems are generally coupled with ATP synthesis. Therefore, two main types of photophosphorylating reactions must be distinguished in isolated chloroplasts: noncyclic and cyclic PP. If instead of the electron acceptors of the noncyclic electron transport (e.g. NADP, ferricyanide) oxygen reacts with the electron chain (a so-called "Mehler" reaction), this leads to a third type, the pseudocyclic PP (5, 141). In noncyclic as well as in pseudocyclic PP PS-I and PS-II are involved (or PS-IIa and PS-IIb in the parallel scheme). Cyclic PP, however, seems to depend on excitation of PS-I alone and proceeds in vitro only if cofactors like PMS, pyocyanine, or ferredoxin are added. The ferredoxin-catalyzed cyclic PP resembles more the cyclic PP in vivo (4), like the endogenous cyclic PP in intact chloroplasts which proceeds without addition of any cofactors and under anaerobic conditions (see below). In contrast to noncyclic and pseudocyclic PP, the cyclic type is only slightly sensitive to DCMU.

INVESTIGATIONS OF PHOTOPHOSPHORYLATION IN VIVO BY DIFFERENT REACTIONS

Methodical

The difficulties in directly measuring PP in vivo as is done in isolated chloroplasts have induced the investigation of a number of different reactions taken as indicator reactions, but recently attempts have been made to investigate PP in vivo also in a more direct way. Some of the reactions described below have been briefly summarized by Avron & Neumann (8) and more recently by Glagoleva & Zalensky (67).

The indirect test reactions widely used can be separated into several groups of processes, all correlated with the energy metabolism of photosynthesizing cells in the light. Energy-dependent uptake of organic substances (44, 106, 135, 167, 170, 232, 275) and ions (2, 100, 131, 134, 138, 176, 215, 244) in the light in different plant cells can be related to PP under certain conditions, if it is linked with ATP-consuming processes. Light-dependent incorporation of ^{32}P into soluble organic compounds (202, 205, 206, 212, 253, 257, 282) and insoluble phosphorylated compounds—mainly polyphosphates (179, 249, 264, 280)—can also be used for the investigation of PP in vivo. Recently, nitrogen fixation in blue-green algae and photosynthetic bacteria coupled with ATP consumption seems to provide a convenient approach for studies on PP in these cells (27, 49). The photoreduction of carbon dioxide with hydrogen in adapted algae is suitable to study PP reactions dependent on PS-I (14, 15, 111). The experiments of Nultsch and co-workers (150, 154, 156) support the view that photokinesis of blue-green algae and photosynthetic bacteria is linked with PP. Steady-state relaxation spectrophotometry permits measurement also of cyclic electron fluxes and thereby the coupled cyclic PP in vivo (89, 91, 189, 236, 237). Photoinhibition of respiration gives evidence for the functioning of cyclic PP in intact cells (79, 90, 181, 198, 237). Light-dependent conformational changes of chloroplasts in intact leaves and algae measured by changes in light scattering are likewise used as a convenient indicator reaction for PP in vivo (62, 80, 120).

A more direct approach to the light-dependent ATP synthesis in vivo is the measurement of the dark-light transition changes in the level of ATP, ADP, or Pi in intact cells. Also the decrease of the Pi level in photosynthesizing cells upon illumination by PP under anaerobic conditions can be used as a rather direct method in studying cyclic PP in vivo (105, 106, 115). Under specific conditions the increase of the ATP level after transition from dark to light even reflects the different processes of PP in living cells (23, 24, 53, 62, 109, 149, 190, 254, 255).

In addition to the different test reactions, inhibitors have also been applied to elucidate the processes of PP in vivo. Although inhibitors are important tools for experiments in vivo, they often influence additional processes. The inhibitors mainly used in the work with intact cells include: (a) compounds which inhibit noncyclic electron transport (CMU, DCMU); (b) compounds which in photosynthesis preferentially inhibit cyclic electron flow (antimycin A); (c) inhibitors of nonclyclic as well as of cyclic electron transport (DSPD, salicylaldoxime, DBMIB); (d) uncouplers of PP (CCCP or FCCP, DNP, desaspidin); and (e) energy-transfer inhibitors (Dio-9, phloridzin). The action of most of these inhibitors has been already extensively described (7, 8, 68, 174).

Photoassimilation of Organic Compounds

Light influences the incorporation of organic substances in photosynthetically active cells in several ways. Besides changes in permeability and viscosity, light may supply energy for an energy-dependent uptake into the cell and for metabolism (photoassimilation) of these organic substances (275). Of special interest for the elucidation of PP in vivo seems to be the uptake and photoassimilation of acetate and glucose.

Acetate assimilation can proceed in blue-green algae mainly as a light-dependent process (30). Light may be required for the synthesis of ATP used for the activation of acetate (87, 88). This photoassimilation of acetate is severely inhibited by DCMU (87, 88). Therefore, noncyclic electron flow coupled with ATP formation is involved. Photoassimilation of acetate has been reported also for flagellates (33) and unicellular green algae (167, 275). Several experiments provide evidence that acetate assimilation in green algae is also initiated by acetylphosphate or by the formation of acetyl-CoA (159, 276), for which ATP is required. Two different ways are known for acetate metabolism in Chlorella pyrenoidosa. In the dark under aerobic conditions, the tricarboxylic acid cycle proceeds and the enzymes of the glyoxylate pathway are functioning (224, 225). Organic acids, amino acids, and also carbohydrates and proteins are formed from the acetate (70). In the light, acetate is preferentially incorporated into lipids.

In the presence of O_2 and CO_2, the enzymes of the glyoxylate pathway are inactive and the activity of isocitrate lyase especially is low (34, 70, 224, 279). Under these conditions the acetate assimilation is severely inhibited by DCMU ($10^{-6}M$) (70, 272). In the presence of DCMU the incorporation pattern of acetate is changed to that in the dark. Therefore, in *Chlorella* cells the photoassimilation of acetate in the presence of O_2 and CO_2 is correlated with noncyclic electron flow. The ATP consumed in this process seems to be produced by noncyclic PP, and the reduction power may be needed for lipid formation.

On the other hand, cyclic PP can proceed also if the conditions are changed. In the light in air, but in the absence of CO_2, the enzymes of the glyoxylate cycle, preferentially isocitrate lyase, are active if acetate is present. The same is true if DCMU is added in the presence of CO_2 (224). Also, under N_2 in the light isocitrate lyase and malate synthetase are induced. The activation of these enzymes is saturated at low light intensities and occurs also in far-red light and in the presence of DCMU (224). Experiments with DNP in the dark indicate that the activation of isocitrate lyase depends on ATP. Therefore, in the light and in the presence of DCMU the activation of isocitrate lyase as well as the photoassimilation of acetate must be supported by cyclic PP.

Another type of acetate assimilation coupled to cyclic PP is found in *Chlamydobotrys stellata* (170, 171, 274, 275, 277, 278) and *Chlamydomonas mundana* (278). In these cases acetate photoassimilation is generally mediated by the active glyoxylate pathway (274, 278). In far-red light in the presence of O_2, incorporation of ^{14}C-acetate is observed in carbohydrates (272, 278). The reaction is then completely insensitive to DCMU. However, if oxygen is removed, DCMU suppresses the photoassimilation of acetate (44). The special significance of O_2 is not known. It is proposed (272, 275) that during acetate assimilation reduced intermediates may be formed which have to be reoxidized for the completion of the reaction. The aerobic photoassimilation of acetate is also inhibited by uncouplers like DNP and CCCP (274). It could be shown that the absorption spectrum of *Chlamydobotrys stellata* grown photoheterotrophically with acetate is shifted to long wavelengths together with an increase of the chlorophyll a/b ratio (274). This is correlated with structural changes of thylakoids (276). The participation of PS-I only in this aerobic photoassimilation of acetate is concluded also from the absence of any enhancement effect (278) and from the higher quantum yield in far-red light. With the assumption that polysaccharides are the main product and that ATP is generated exclusively by cyclic PP, Wiessner (273) estimated the quantum requirement for ATP in this reaction as 2 Einstein/mole ATP.

Photoassimilation of glucose is currently regarded as a reaction which permits measuring PP in vivo (31, 106, 140, 216, 232, 235, 283). It seems to be particularly convenient because theoretically the incorporation of glucose into carbohydrates (110, 233, 235) requires only ATP and should proceed without reducing power. According to our present knowledge, however, at least three partial reactions must be distinguished: (*a*) a light-stimulated adaptation of a transport system for glucose uptake (230, 232); (*b*) the energy-dependent membrane transport of sugar (47, 227, 235); and (*c*) the light-dependent accumulation by photoassimilation of carbohydrates as a sink (106, 135, 235).

The light-stimulated induction of a transport system for the uptake of glucose, indicated by a lag phase, could be found until now in only a few species of Chlorococcales (*Chlorella vulgaris, C. fusca*) (229, 230). The induction is suppressed by cycloheximide, a known inhibitor of protein synthesis. In the light the lag phase is extended by DCMU (230). The severe inhibition of the photoassimilation of glucose by DCMU (228) might therefore be due also to the sensitivity of processes during the lag phase. The membrane transport system is inducible by addition of glucose in the dark and in air and also in far-red light under anaerobic conditions. This suggests that cyclic PP operates as energy donor. The similarity to the induction of isocitrate lyase by acetate under comparable conditions in *Chlorella* is remarkable (224). In both cases cyclic PP may supply ATP for protein synthesis.

The energy-dependent membrane transport of sugar has been established by using substituted sugars, e.g. 3-*O*-methylglucose, which is not further metabolized. In the light its uptake proceeds under aerobic as well as under anaerobic conditions and is supported also by far-red light (227, 230). In the dark the uptake in *Chlorella* is inhibited by FCCP or by anaerobiosis (117, 118). Assuming a P/O ratio of 3 from measurements of increase of respiration and uptake of desoxyglucose, the "ATP" per mole of substrate required for initial uptake was estimated as 1.18 (39). In comparison with such experiments it may be suggested also that for anaerobic light-driven active sugar uptake 1 mole ATP is needed for 1 mole sugar taken up. In accordance with earlier experiments, Tanner (227) assumed that the anaerobic uptake of *O*-methylglucose in the light is driven by cyclic PP.

In most papers published so far, the different partial reactions have not been considered. The photoassimilation of glucose as a total process has been observed in several groups of plants: blue-green algae (48, 114, 214), unicellular green algae (106, 235), giant algal cells (216, 223), mosses (213), and in leaves of higher plants (12, 119, 135, 140). This total process is stimulated by light, depending on temperature (106, 216), and it is uncoupled by CCCP (216, 233) or DNP, indicating that glucose assimilation is an energy-dependent and ATP-mediated process. Including the energy used for the uptake in general, 3 moles of ATP are needed for the incorporation of 1 mole glucose into polysaccharides (39, 231, 232). The photoassimilation of glucose is saturated at low light intensities (106, 135, 224, 228, 233); it occurs also in the absence of CO_2 (12, 119, 216) under anaerobic conditions (110, 233). The ratio of glucose to $^{14}CO_2$ uptake increases from 0.55 (658 nm) to 16.4 (711 nm) (110). In experiments with green algae (195, 224, 233) and also with *Nitella* (216), concentrations of DCMU which severely reduce CO_2 assimilation are only slightly inhibitory to the anaerobic photoassimilation of glucose in white and far-red light. These results suggest dependence on a cyclic

PP driven by PS-I. In other experiments, however, a more or less strong inhibition by DCMU or by N_2 of glucose assimilation was observed (29, 119, 223, 228). These results can be explained in different ways. Besides the extension of a lag phase during the induction period of glucose uptake in *Chlorella* by DCMU (228, 229), a pseudocyclic PP which is DCMU-sensitive also may be involved (29, 119). Additionally it was shown that even in nitrogen a noncyclic electron transport can proceed. If reducible ions (NO_3^-, SO_4^{--}) are present in the medium, they can act as electron acceptors in vivo (111, 146, 162, 246). Finally we have to consider that cyclic PP itself may be sensitive to DCMU (103, 179, 180, 233). Higher concentrations of DCMU suppress the cyclic PP in vitro (3, 6). But it must be remembered that the cyclic electron transport is maintained only if an external reductant is present (8, 78).

For the differentiation between cyclic and noncyclic PP other inhibitors, e.g. antimycin A and salicylaldoxime, have been used (110, 228, 232, 233). Of these, antimycin A as an inhibitor of cyclic electron transport (151, 226, 257) inhibits the anaerobic photoassimilation of glucose, whereas CO_2 assimilation is much less affected. These results suggest that cytochrome b_6 may be involved only in cyclic PP and that cyclic PP is not closely related to CO_2 assimilation.

Uptake of Ions

Light-stimulated uptake of ions can be used as test reaction for PP in vivo if it has been shown that the ion uptake is an ATP-dependent and energy-linked process. Therefore, many attempts have been made to establish such a relationship, and in special cases correlations with the different types of PP could be shown (38, 45, 137, 138, 176). But light influences the influx of ions from the incubation medium into the cells of green plants in different ways. Changes of permeability or viscosity of the plasmalemma and cytoplasm may occur. Light affects also the electrical potential difference across the plasmalemma and the electrical conduction of this membrane (97, 133, 191). Likewise, the activity of enzymes is directly or indirectly changed by light which in this way may influence ion uptake.

Phosphate uptake is stimulated by light in unicellular algae as well as in cells of higher plants, as repeatedly observed (60, 100, 124, 139, 208, 215, 265, 270). In unicellular algae the light-stimulated phosphate uptake (46, 60, 124, 139, 208, 242, 244, 246, 264) is dependent on the temperature (17, 46), enhanced by Na^+ ions (211, 242–244), and inhibited by uncouplers like DNP (124, 208) or CCCP (244). It is saturated at quite low light intensities (46, 139) and inhibited by *o*-phenanthroline and DCMU under steady state conditions in white light and in the presence of air plus CO_2 (139, 207, 208). Therefore, the reactions which mediate phosphate uptake depend upon a noncyclic electron flow coupled with ATP formation. But light absorbed by PS-I can also support the light-dependent influx of phosphate. The action spectrum of phosphate uptake in *Anacystis* (46, 203, 204) in air plus CO_2 exhibits two maxima corresponding with the maxima of the absorption spectrum (430 nm, 629 nm, 680 nm), and the ratio of the relative quantum efficiency (PO_4/O_2) shows values of 1.0 (629 nm), 1.13 (650 nm), 2.8 (670 nm), 11.3 (680 nm), 29.5 (700 nm), 19.8 (710 nm) without regard to the photoinhibition of respiration (204). Therefore, in blue-green algae, PS-I including cyclic PP seems to be involved. Also in green algae under nitrogen and in far-red light, where DCMU exerts only a little inhibition (249), cyclic PP enables these algae to maintain phosphorus uptake.

Earlier experiments have shown that giant algal cells of Characeae and vacuolate cells of higher aquatic plants can accumulate phosphate against a chemical concentration gradient (32, 86) and this phosphate uptake is also enhanced by light (71, 100, 215, 270). Calculations based upon the internal concentration of phosphate and the electrical potential difference of the membrane of *Nitella translucens* (215) as well as *Elodea densa* (100) revealed that in addition to the strong dependence of the influx on temperature and light in a special range of the phosphate concentration in the medium, phosphate uptake is an energy-dependent active process. The stimulation by light of the phosphate influx and also the uptake in the dark are suppressed to a high extent if uncouplers like CCCP are added (71, 215). In contrast to the inhibitors of electron transfer Dio 9 or atebrine (98), which obviously influence also the uptake reaction at the plasmalemma, CCCP may act mainly on the ATP-generating process itself. Therefore it can be concluded that the active phosphate influx is an ATP-mediated process. The light-dependent phosphate uptake proceeds in nitrogen (215, 270) as well as in CO_2-free air (12, 140). CO_2 assimilation itself seems, therefore, not necessary for phosphate uptake. In experiments using *Nitella* and *Elodea*, no inhibition by DCMU at fairly high phosphate concentrations in air, white or far-red light, and in N_2 were found. Thus it was concluded that the light-dependent process is supported by cyclic PP (215, 270). But at low phosphate concentrations in the medium and in air plus CO_2, DCMU at higher concentrations can inhibit the phosphate uptake in *Elodea* (71), suggesting a participation of pseudocyclic PP.

The experiments so far reported indicate that the different phosphorylating processes—oxidative ph, cyclic, pseudocyclic, and noncyclic PP—can provide energy for the phosphate uptake into vacuolized cells of higher plants.

Sulfate uptake in relation to PP has often been investigated but the results are controversial and not easy to explain. Regarding the high dependence of sulfate influx on temperature (100, 187) and the great difference between the observed and the calculated electrochemical gradients (187) an active uptake of the sulfate ion must be supposed. In some organisms the influx of sulfate is stimulated by light, dependent on external sulfate concentration (100, 125, 242, 267) in others (e.g. *Chara australis* and *Ch. foetida*), however, the uptake in the dark is often higher than in the light (166, 187). The influx of sulfate is inhibited by DNP particularly in the dark but also in the light (123, 147, 188, 259). In a few experiments the influx of sulfate and at the same time the ATP level were measured. With increasing concentrations of mannitol in the incubation medium the ATP level and also the sulfate influx in *Elodea canadensis* leaves decrease (270). In *Limnophila* the inhibition of the sulfate as well as of the phosphate influx by CCCP is parallel to the decrease of the ATP level (166). The persistence of sulfate influx under nitrogen in the light has been used as evidence for the participation of cyclic PP (165, 269), but under these conditions sulfate uptake may be accompanied by sulfate reduction, the sulfate ion serving as electron acceptor for a noncyclic electron flow (92, 192, 241). Other experiments, however, using species of *Chara* show that CCCP is not effective in the light but only in the dark (166, 188). The lower sulfate influx in the light is increased by DCMU to the higher level of the uptake in the dark. This leads to the assumption that the respiratory source of ATP is sufficient for sulfate influx in the light (188). Only in those cases in nitrogen in which DCMU was ineffective, as has been shown in *Elodea canadensis* (270), and if uncouplers are effective, the sulfate influx may be supplied by cyclic PP. But so far a final proof has not been established.

Chloride uptake may be especially useful in studying the relationship of ion influxes to PP because this anion cannot be metabolized. Earlier considerations have shown that the chloride influx has to be an energy-dependent process (38, 137, 138, 177, 218). The uptake into algal cells or in leaves of higher plants is greatly enhanced by light (2, 9, 131, 136, 213). The action spectrum was found to be similar to that of photosynthesis (99, 131, 174). MacRobbie (136) has shown that the light-stimulated component of the chloride influx in cells of *Nitella translucens* was independent of CO_2 but severely inhibited by DCMU and also diminished by far-red light. In contrast, the potassium influx was insensitive to DCMU and far-red light was effective. By using CCCP as an uncoupler it could be shown that K^+ influx and CO_2 assimilation were greatly diminished, while Cl^- uptake was left almost unchanged (136, 172, 178, 217). Antimycin A, desaspidin, and DSPD, regarded as inhibitors of cyclic PP, affect the active K^+ influx under anaerobic conditions (174–176, 199). Therefore it was suggested that K^+ uptake is supported by cyclic PP, whereas the Cl^- transport may depend upon the noncyclic electron flow without requirement for ATP. But with some other species this concept could not be confirmed (35, 270). In nitrogen the light-dependent Cl^- influx was DCMU-insensitive (101, 270), and in several algae it was blocked by CCCP, sometimes even more than O_2 evolution (9, 130). Thus Cl^- uptake somehow must be correlated with ATP formation.

In *Elodea* recently a close similarity of the inhibition by CCCP of K^+ and Cl^- influxes and CO_2 fixation was found, indicating that both ions are translocated in a partially combined transport system dependent on ATP (96, 98, 99, 173). It could be shown that the relative quantum yield in far-red light of both fluxes drops to a similar degree as the quantum yield of O_2 evolution. Therefore it has to be considered that in the absence of CO_2 these ion influxes may be supported by pseudocyclic PP (99, 175). It may be supposed that besides the ATP dependence of these fluxes, noncyclic electron flow is required for the transport of ATP to the sites of ion transport. This is supported by the findings that ATP transport in the cell can be mediated by translocator systems provided by intermediates of the Calvin cycle (82, 83) or by other translocating substances. In these cases a differentiation between a requirement of ATP or NADPH for ion uptake seems nearly impossible (99).

Incorporation of Phosphate

In 1948 Gest & Kamen (60) used the incorporation of ^{32}P into organic phosphate compounds of intact cells during illumination to investigate PP in vivo. A few years later Simonis & Grube (206) found a light-stimulated incorporation of ^{32}P preferentially in the trichloroacetic acid-soluble organic phosphate fraction (Po-fraction) of *Elodea* leaves. This effect could be related to processes of PP in the leaf cells. Due to the limitation by phosphate uptake kinetics, however, experiments on ^{32}P incorporation during illumination of leaf or algal cells do not easily allow conclusions about the turnover of the phosphate compounds and about the rate of PP (222, 268). Nevertheless, they provide a convenient qualitative test reaction of PP in intact cells.

The first stages of these investigations of PP in vivo by ^{32}P experiments have been extensively reviewed by Simonis (200) and Kandler (108). It was shown in our laboratory that light stimulates the incorporation of ^{32}P into the Po-fraction of the green alga *Ankistrodesmus* in air by two to tenfold (210, 211, 243). The incorporation into other phosphate fractions is less enhanced by light in short time experiments up to several minutes. Experiments with various inhibitors support the assumption that this ^{32}P labeling is caused by processes of PP. Inhibitors of respiration, e.g. antimycin A, amytal, KCN, HOQNO, and salicylaldoxime, suppress ^{32}P incorporation much more efficiently in the dark than in the light (201, 210, 256, 257). Under anaerobic conditions the ^{32}P labeling of the Po-fraction in the dark is strongly inhibited in contrast to the incorporation in the light (201, 257), and CO_2 enhances this labeling (253). On the other hand, DCMU in low concentrations suppresses ^{32}P incorporation only in the light, as well as uncouplers like CCCP and desaspidin (201, 252, 253, 258).

The action spectrum of the light-dependent ^{32}P incorporation in green as well as in blue-green algae corresponds with the absorption spectrum of the photosynthetic pigments (46, 201, 209). Using this method of ^{32}P incorporation, it could be demonstrated that different types of PP proceed in intact cells (65, 202, 249, 257, 258). The light-stimulated incorporation of ^{32}P, especially in N_2 and in the presence of DCMU, which completely blocks photosynthesis, is strongly inhibited by antimycin A. This inhibitor, first used by Tagawa et al (226), prevents mainly the ferredoxin-catalyzed cyclic PP in isolated chloroplasts, but additional effects of this inhibitor have to be considered (see also 37, 42). Moreover, the incorporation in far-red light shows a stronger inhibition by antimycin A than in red light, whereas the effect of DCMU is inverse (249). In the light, ^{14}C fixation and ^{32}P incorporation in N_2+CO_2 are more sensitive to DCMU than the light-dependent incorporation of ^{32}P in N_2 (253). The latter reaction is also saturated at low light intensities in contrast to ^{32}P labeling in N_2+CO_2 and ^{14}C fixation.

Apart from special limitations discussed below, we concluded from these and other results that correspond to the findings of other authors (126, 157, 174, 179, 228) that a cyclic as well as a noncyclic PP is functioning in intact cells. In relation to the results of Tagawa et al (226) with antimycin A, we suggested that the cyclic PP in vivo may be comparable with the ferredoxin-catalyzed cyclic PP in chloroplasts (3). Investigations with DSPD, an inhibitor acting near or at the site of ferredoxin (11, 239), support this conclusion (65). From the experiments with DSPD and other inhibitors (252) which severely block cyclic PP without inhibiting total ^{14}C fixation it has been inferred that cyclic PP is not required for CO_2 assimilation as also suggested by other authors (110, 174, 175, 232).

Recently, however, Klob et al (115) pointed out that cyclic PP may be necessary for the start of CO_2 assimilation under anaerobic conditions, but not for steady state photosynthesis. The question of whether high-energy intermediates instead of ATP may be used directly for CO_2 assimilation in vivo, which has been concluded from experiments especially with phloridzin and Dio-9 in vitro (262) and in vivo (64, 173, 178, 252), is still unanswered. But it has to be considered that these inhibitors could show an additional effect possibly on the transport-ATPase in cellular membranes (98, 173, 178). Experiments with ^{32}P also gave some indications for the occurence of a pseudocyclic PP in vivo (249) as postulated by other authors (80, 175).

Polyphosphate formation in the light has been used also for the investigation of PP in vivo. Bacteria, blue-green and green algae, like other lower plants, are capable of accumulating large amounts of polyphosphate (40, 76, 121, 122, 145, 179, 248). In some of the experiments reported, the authors did not measure polyphosphate formation itself, but the values were estimated as "bound phosphate" or as TCA-insoluble phosphate, thus representing only a rough approximation. Illuminated algal cells convert inorganic and intermediary organic phosphates into polyphosphate as well in air without CO_2 (104, 247, 265, 280) as in nitrogen (126, 179, 247, 249). In air plus CO_2 and in white light the incorporation of ^{32}P into the insoluble phosphate of the blue-green alga *Anacystis*, like that of the green alga *Ankistrodesmus*, is inhibited to a large extent by DCMU. Also, the uncoupler CCCP decreases the incorporation considerably, indicating that noncyclic PP contributes to the formation of bound phosphate in the insoluble fraction under these conditions (139).

In nitrogen, light-dependent polyphosphate formation is higher than in nitrogen plus CO_2 (247, 280), and it increases even more with addition of oxygen (46, 126, 245, 247). Further experiments carried out under nitrogen show that polyphosphate formation is saturated at low intensities of white light (179, 245, 280). It occurs also in far-red light, particularly absorbed by the pigments of PS-I (179, 249). In nitrogen and also in far-red light polyphosphate formation is far less sensitive to DCMU than oxygen evolution in the presence of CO_2 (179, 202, 249). Correspondingly, antimycin A inhibits the labeling of polyphosphate in far-red light considerably more than in red light (249). These results together indicate that in the absence of CO_2 under nitrogen, polyphosphate formation is supported by cyclic PP. However, higher concentrations of DCMU inhibit the incorporation of phosphate into the polyphosphate fraction in nitrogen, but in red light more than in far-red light (249). This is an indication that even in nitrogen a noncyclic electron flow can proceed to some extent in red light (246) (see below). Inhibition of polyphosphate formation by DCMU at a concentration higher than $6.6 \times 10^{-6} M$ in far-red light of low intensity is discussed by van Rensen (179) as an indication that DCMU becomes effective by blocking a substance X between PS-II and I which may be likewise located in the cyclic and noncyclic electron transport chains.

Nitrogen Fixation in Blue-Green Algae

Many strains of blue-green algae are able to fix nitrogen in a light-dependent process if they contain nitrogenase. Photosynthetic electron transport may be expected to supply both ATP and reduced ferredoxin for nitrogen fixation, but the light stimulation of nitrogenase activity in unstarved cells of *Anabaena* was unaffected in the presence of CMU. From this it was assumed that photostimulation of nitrogen fixation is related to the consumption of ATP generated independently of noncyclic electron transport (36). The action spectrum of nitrogenase activity in *Anabaena* has a maximum at 675 nm, and therefore it resembles that of reactions mediated by PS-I (49). Stewart et al (219) introduced the use of an acetylene reduction technique as a simple method for detecting the potential for nitrogen fixing capacity. Heterocysts, probably sites of nitrogen fixation (50), have been isolated from *Anabaena*. They show an absorption spectrum which is similar to the action spectrum of C_2H_2 reduction (49) which indicates a very low content of phycocyanin and the presence of the long-wavelength form of chlorophyll *a*.

Direct evidence for the absence of PS-II and the presence of the active PS-I in isolated heterocysts of *Anabaena* was derived from the fact that illumination of heterocysts caused an oxidation of P_{700} and a cytochrome. On the other hand, no light-induced changes of fluorescence or an influence of DCMU could be observed (41). This together with an indication of the presence of nitrogenase activity in heterocysts (219) suggests that in heterocysts a functional PS-I may generate ATP by cyclic PP. But ATP formation in isolated heterocysts has not been proven as yet. It must be considered, however, that the ability to fix nitrogen is not invariably correlated only with heterocysts (186). Bothe (27) studied C_2H_2 reduction by *Anabaena* cells in vivo and found no effect by DCMU. But DSPD, a specific inhibitor of ferredoxin-dependent reactions, and also DBMIB affect CO_2 fixation and C_2H_2 reduction to a similar extent. These data are strong indications for cyclic PP in *Anacystis* cells in vivo. Therefore, C_2H_2 reduction in the presence of DCMU provides a very useful system for testing cyclic PP in vivo. Moreover, they indicate that ferredoxin catalyzes cyclic PP in vivo.

Photokinesis

Changes in the speed of the movement of motile microorganisms induced by light are called photokinesis. In comparison with photophobotaxis, i.e. the reversal of the direction of movement caused by sudden changes in light intensity, photokinesis appears as a useful indicator reaction for studies in PP. Nultsch and his group (150, 154) observed that the action spectra of photokinesis and photophobotaxis of blue-green algae belong to two different types. The action spectrum of photophobotaxis of *Phormidium uncinatum* is similar to that of CO_2 fixation and resembles the absorption spectrum of the phycobiline pigments. The action spectrum of photokinesis, however, is comparable to the absorption spectrum of chlorophyll *a*, indicating a reaction coupled with PS-I (150, 154, 156). Photokinesis is saturated at low light intensities in contrast to photophobotaxis. DCMU inhibits both reactions but photokinesis is more sensitive. Therefore, noncyclic electron flow seems to be involved (151, 157). Photokinesis in an argon atmosphere is not affected and proceeds even in argon in the presence of DCMU (152), whereas desaspidin in far-red light inhibits this reaction (153). HOQNO and antimycin A are also not effective, but they suppress the remaining mobility in the presence of DCMU (151) in a way similar to the ^{32}P incorporation in the Po fraction of *Ankistrodesmus* (257, 258). Therefore, it was concluded that noncyclic as well as cyclic PP contribute to the energy supply of photokinesis. Using *Anabaena*, the action spectrum of photokinesis and the responses to inhibitors are quite different (56). The results indicate that in this organism primarily noncyclic or pseudocyclic PP are involved. The study of photokinesis therefore seems to be helpful in elucidating the complex reactions of PP in vivo although in the case of photokinesis several additional effects must be considered (155, 156).

Photoreduction

Hydrogen-adapted algae (14, 58, 59) are able to carry out CO_2 photoreduction coupled to PP without O_2 evolution and are stabilized by DCMU (15, 19). Photoreduction proceeds in far-red light and low light intensities with a low quantum requirement (13, 15, 19, 59), and the enhancement effect is lacking (66). PP was measured by anaerobic glucose photoassimilation (110, 169, 234). Mutants of *Scenedesmus* (16, 19, 169, 266) lacking O_2 evolution but possessing an intact PS-I including cytochrome *f* and P_{700} can perform photoreduction (15, 19, 168, 169) and in close relationship also glucose assimilation (16, 19, 231, 232). These results indicate that cyclic PP is involved (16, 234). Measurements of fluorescence of Mn-deficient algae, however, show that to some extent H_2 activated by hydrogenase may enter also into PS-II during photoreduction (112, 113).

Steady-State Relaxation Spectrophotometry

Hoch (89) introduced a method that can be used to calculate the rates of electron flow through intermediates of the electron transport systems especially in intact cells. With this method (i.e. alternate oxidation and reduction of electron carriers in cells illuminated with modulated actinic light) an attempt has been made for the first time to estimate the magnitude of the cyclic electron flow and thereby of the coupled cyclic PP in vivo (91, 189, 236). Thus cyclic electron flow has been demonstrated in blue-green, green, and red algae by the discovery that the turnover of cytochrome f and P_{700} persists in the presence of a concentration of DCMU that completely blocks oxygen evolution (189, 236, 237). Similarly, certain mutant strains of *Chlamydomonas reinhardii* unable to evolve oxygen still possess an active cyclic electron transport system (237). The turnover rate of the investigated carriers increased in the presence of the uncoupler FCCP which indicates that PP was coupled to the cyclic electron transport system and that the rate of the PP step limits the magnitude of the electron flow in the absence of an uncoupler (189, 237).

The measured magnitude of this cyclic electron transport in algae argues against the assumption that it may be a minor process in photosynthesis (91, 236). Whereas the electron transport driven by PS-I through cytochrome or P_{700} appears essentially insensitive to DCMU, this inhibitor effects neither a change nor an increase in the estimated electron flux (91) if PS-II is additionally excited. These results indicate that the operation of PS-I in a cyclic electron flow is only indirectly influenced by the presence or absence of oxygen evolution in algal cells. Isolated chloroplasts behave quite differently and do not show a DCMU-insensitive cyclic electron flow (189). This result possibly could be explained by the use of not intact chloroplasts which can reduce ferricyanide and may have lost the ability of endogenous cyclic electron flow (see below). Experiments with mutant strains of *Chlamydomonas reinhardii* lacking different components of the photosynthetic electron transport chain indicate that cytochrome f and P_{700} but not cytochrome b_{559} are involved in the cyclic PP in vivo (237).

Photoinhibition of Respiration

Besides the photosynthetic evolution of oxygen, at least three processes determine the response of oxygen exchange to illumination: (*a*) the photosynthetic inhibition of respiration; (*b*) the photosynthetic stimulation of respiration; and (*c*) the photosynthetic uptake of oxygen (90, 94, 185, 263). Among these processes the photoinhibition of respiration seems to be closely related with PP. It is saturated at low light intensities and occurs also in the light in the presence of DCMU. Therefore PS-I seems to be involved. Ried (181, 182) measured in algal cells light-dark transitions of the oxygen exchange after short light pulses. The transients show two peaks. The first peak (T_1) is completely suppressed by DCMU. The second one (T_3), well separated from T_1, is unaffected by DCMU but is inhibited by antimycin A and glucose. All conditions which suppress T_3 eliminate the Kok effect. The wavelength dependence of the amplitude of T_1 and T_3 appears correlated with PS-II and PS-I respectively. Using light pulses and low light intensities, the action spectrum of T_1 in cells of *Chlorella fusca* (183) coincides rather closely with the absorption spectrum of PS-II. The action spectrum of T_3 with the maximum at 683 nm differs from the spectrum of PS-II by the absence of the chlorophyll b band but is similar to published action spectra of reactions driven by PS-I (102, 261).

Also, in *Anacystis*, transients after short light pulses have been observed but with more pronounced oscillations (184, 198). The fast component (T_1) is suppressed by DCMU, but the slow component together with the oscillations remains unaltered. If glycolysis is inhibited by NaF, no oscillations occur. The action spectrum of these oscillations shows all the characteristics of a PS-I spectrum. The inhibition of the oscillations by CCCP, desaspidin, or salicylaldoxime, together with the other findings, suggests that cyclic PP is the mediator of this photoinhibition (184).

Other observations using steady state relaxation spectrophotometry (89, 189, 236) support this conclusion. With mutant strains of *Chlamydomonas reinhardii*, which are capable of a cyclic electron flow, photoinhibition of respiration has been demonstrated. On the other hand, those mutant strains which do not show a cyclic turnover do not show photoinhibition of respiration either (237). Addition of FCCP increases the cyclic turnover and abolishes the oxygen exchange in all mutant strains which show a significant photoinhibition. All this indicates that the photoinhibition of respiration is supported by the cyclic electron transport coupled with PP. The interaction between photosynthesis and respiration may be regulated by means of the adenylate system (184, 185, 237). Healey & Myers (79), however, observed no effect of uncouplers on the Kok effect and therefore suggest that reductants from the respiratory pathway may supply electrons to PS-I. But the observation of an inhibition of glycolysis by far-red light is in accordance with the regulatory function of the adenylate system (85).

Light-Scattering Changes

Light-dependent conformational changes of chloroplasts in intact leaves and unicellular green algae measured by recording slow changes of transmission or light scattering at 535 nm have been used as a convenient indicator reaction for PP in vivo (62, 80, 120). This seems possible because a close relationship between the changes of light scattering and PP in vivo has been demonstrated, e.g. by uncouplers (62, 80). A critical investigation by Gimmler (62), in which he compared the light-induced scattering changes at 535 nm with dark-light transition changes of the ATP pool in intact cells of the green alga *Dunaliella*, supported the use of this method for in vivo measurements of PP.

In experiments with leaves in N_2 during illumination with far-red or low intensity red light, Heber (80) observed light-scattering changes indicating a cyclic electron flow. Excitation of PS-II with high intensity red light in N_2 suppressed scattering, and it is suggested that this effect is due to "over-reduction" of electron carriers within the cyclic flow. In far-red and low intensity red light, scattering was also strongly inhibited by addition of oxygen which appeared to interrupt the cyclic electron transport by oxidizing an electron acceptor at a site between PS-I and NADP. Finally, the scattering in air-CO_2 or in N_2 was suppressed by addition of CO_2. These results indicate that the cyclic PP in vivo proceeds only under anaerobic conditions and is controlled by PS-II and by the availability of electron acceptors (e.g. CO_2, O_2). In the presence of O_2 probably a "Mehler" reaction (141) occurs which may supply ATP by a pseudocyclic PP. Measurements of scattering changes of isolated intact chloroplasts in comparison to those in vivo show that in chloroplast preparations O_2 reacts with the electron transport chain less readily than in leaves. Thus in contrast to intact cells in isolated intact chloroplasts, a cyclic electron flow and in consequence a cyclic PP seems to proceed even in the presence of oxygen (120). The experiments of Gimmler (62), who measured scattering changes and dark-light transition changes of the ATP level of the green alga *Dunaliella*, support the findings that in the presence of O_2 cyclic PP plays a minor role in vivo.

Finally, the investigations of PP in vivo by scattering changes and measurements of endogenous ATP synthesis suggest the existence of two sites for phosphorylation in the noncyclic electron transport system (62). One is probably located before PS-II as postulated from earlier experiments of Böhme & Trebst (22) with chloroplasts. This assumption is based on the reactivation of the PP in vivo in the presence of the plastoquinone antagonist DBMIB by the Hill reagent benzoquinone and the sensitivity of this reaction to DCMU. The second site of ATP formation is resistent to DCMU and DBMIB in the presence of the electron donor DAD and should therefore be located between plastoquinone and PS-I. These two sites in a noncyclic PP in vivo should be able to maintain CO_2 assimilation if cyclic PP were indeed suppressed by an unappropriate redox level of electron carriers in the presence of oxygen.

Changes in Pi Level

One of the first attempts made to investigate PP in vivo was by measuring the dark-light transition change of inorganic phosphate in *Chlorella* by Kandler (105). He showed that the inorganic phosphate level in green algae decreases immediately upon illumination and increases again when the light is turned off. This Pi decrease can proceed under anaerobic conditions and shows an early light saturation (107). Changes in the level of Pi after transition from dark to light were also found in chloroplasts within leaf cells of *Elodea* and *Coleus* (190, 250).

Recent experiments using the light-induced Pi decrease in comparison with the anaerobic light-dependent glucose assimilation confirm the earlier results and indicate that the Pi decrease under anaerobic conditions is caused by cyclic PP (115). In the presence of oxygen the Pi decrease upon illumination is much smaller and slower, it is not inhibited by FCCP at concentrations which block the Pi decrease in N_2, and it most likely reflects not the noncyclic PP in vivo but the increase in phosphorylated intermediates of the carbon reduction cycle. Inhibition of cyclic PP under anaerobic conditions by FCCP within affecting steady state photosynthesis confirms earlier results of these authors according to which cyclic PP is not required for CO_2 assimilation. During the induction period of CO_2 assimilation, however, FCCP shows an inhibitory effect in N_2 but not in air (115). Therefore, it is concluded that cyclic PP is only necessary for the start of CO_2 assimilation under N_2, whereas in air another energy source, possibly oxidative phosphorylation, can replace it. This dark phosphorylation as well as the PP (possibly noncyclic) which is required during steady state photosynthesis must be less sensitive to FCCP than cyclic PP in vivo. Such a different sensitivity of steady-state photosynthesis, cyclic and noncyclic PP, to FCCP or CCCP has not been found by other authors in different organisms (98, 174, 252).

Changes at ATP Level

Besides the changes in the Pi level in dark-light and light-dark transitions, changes in the levels of other compounds involved in photosynthetic processes also have been investigated in leaf and algal cells (69, 82, 109, 163, 164, 190, 222). Comparison of the behavior of the levels of the different intermediates shows the close relationship of PP to other metabolic processes (carbon reduction cycle, glycolysis, respiration) and suggests also the proposed regulatory function of the adenylate system during photosynthesis in intact cells.

Special interest for the study of PP in vivo is due to the measurements of changes of the ATP levels in intact cells after transitions from dark to light (23, 53, 82, 149, 190, 222). Improving the analysis of ATP by the sensitive luciferin-luciferase method (221, 271) instead of measuring the cellular ATP pool after long time ^{32}P labeling or by the enzymatic assay has made it easier to determine small changes of the ATP level at dark-light transitions in short time intervals down to few seconds (24, 93, 129, 193, 220, 254, 255, 271). Also measurements of steady-state pool sizes of ATP under different conditions give some indications on the occurrence of PP, especially in correspondence to other processes (24, 25, 93, 126, 132, 255, 269). According to these experiments, the ATP level in intact cells remains fairly constant and must be rapidly regulated by various processes. Over a wide range, different temperatures, pH values, and Pi concentrations in the medium produce no distinct effect on the ATP level in *Anacystis* either in the light or in the dark (25). Also concentrations of DCMU up to $10^{-5}M$, which completely suppress ^{14}C fixation, do not lower but rather enhance the steady-state level of ATP in *Anacystis* (24, 25) and *Scenedesmus* (126). On the other hand, CCCP or FCCP generally diminish the ATP pool in most of the organisms investigated so far (25, 133, 166, 255). The same is observed with the energy transfer inhibitor Dio-9 in *Anacystis* (25).

These effects can be easily explained by a block of PP processes, whereas in the presence of DCMU cyclic PP and oxidative phosphorylation still proceed and the inhibition of CO_2 assimilation may result in an increase of the ATP pool by a lower consumption. After suddenly changing some environmental conditions (e.g. pH, CO_2 concentration), or after adding DCMU in *Anacystis* and *Anki-*

strodesmus in the light, no distinct effect on the ATP level could be observed (24, 25, 255), whereas addition of CCCP lowered the pool very rapidly (255). In Chlorella, oscillations for several minutes in the level of ATP upon changing the CO_2 concentrations or by light-dark and dark-light transition have been demonstrated (129), also indicating the existence of an efficient control system for the ATP pool in intact cells.

These observations agree with the findings that the steady-state level of ATP in cells is usually similar in the light and in the dark (23, 93, 193). Nevertheless, distinct changes of the ATP pool by the transition from dark to light occur but can only be measured in short time intervals. They show a rapid increase of the ATP level upon illumination and a decrease after turning off the light. This light-dependent increase of the ATP level in N_2 and far-red light in unicellular algae is insensitive to DCMU (24, 251, 254) but strongly inhibited by CCCP and also by DSPD (255), which preferentially blocks the ferredoxin-catalyzed cyclic PP in chloroplasts (239). This result supports earlier assumptions (26, 65, 252, 257) that ferredoxin is a cofactor of cyclic PP in vivo (4). The inhibition by DBMIB indicates a participation of plastoquinone in cyclic PP (255).

Experiments with Anacystis show a DCMU-insensitive ATP formation also under aerobic conditions (24). Relatively low light intensities are sufficient to reach the steady-state level of ATP. Gimmler (62), however, presented evidence that in Dunaliella under aerobic conditions a synthesis of ATP is not measurable either in far-red light or in the presence of DCMU in red light. These results support the conclusion of Heber (80) that in the presence of O_2, cyclic PP cannot proceed in vivo and instead a pseudocyclic PP is functioning. We measured light-induced changes of the ATP level in Ankistrodesmus under aerobic conditions and, under illumination with red light, we also found an inhibition of ATP formation by $10^{-5}M$ DCMU (251). But similar results could be obtained in far-red light of 721 nm, which preferentially excites PS-I and should not maintain a well-functioning pseudocyclic electron transport usually driven by PS-I and PS-II.

It has been shown by several authors that reactions occurring in far-red light also can be inhibited by DCMU (103, 179, 233). Thus the action of DCMU is still not completely understood. Besides the assumption that a pseudocyclic PP proceeds under these conditions, the inhibition by DCMU in far-red light in the presence of O_2 can also be explained by a change of the redox level of electron carriers by oxygen in a yet unknown way (compare 8). On the other hand, ATP formation by cyclic PP insensitive to DCMU could be demonstrated in leaves also under aerobic conditions (53, 158).

ATP synthesis by cyclic as well as by noncyclic PP has been observed in intact leaves during development (158). In greening leaves the cyclic PP appears well before noncyclic PP and reaches a maximal rate at about the time oxygen evolution commences. After this time cyclic PP remains constant while noncyclic PP increases.

Some attempts have been made to estimate the rates of PP in vivo by measuring either the initial rates of the increasing ATP level upon illumination or the decay of the ATP pool at steady-state conditions after turning off the light (24, 251, 255, 271). Using the green alga *Ankistrodesmus*, rates from 5 to 50 μmoles ATP per mg chlorophyll per hour for cyclic PP have been calculated (251, 255). The blue-green alga *Anacystis* achieves initial rates in white light and in air up to 180 μmoles per mg chlorophyll per hour (24). These are probably minimal rates that are limited by several factors, and the methods have to be improved to come closer to the real rates of PP in vivo. The method of following the dark-light transition changes of the ATP level in intact cells upon illumination is also suitable for the study of endogenous PP in intact chloroplasts (81, 103, 127, 255).

ENDOGENOUS PHOTOPHOSPHORYLATION IN CHLOROPLASTS

Because it is possible to isolate intact chloroplasts that are capable of fixing CO_2 at high rates similar to those measured in vivo, it seems reasonable to pay attention also to studies on endogenous PP of such chloroplasts for the interpretation of PP in vivo. But the number of investigations on endogenous PP in chloroplasts has been increased in recent years in such a way that this subject cannot be reviewed in this article in detail. The close relationship to PP in vivo, however, should at least be mentioned.

Improving the isolation procedure and incubation medium for chloroplasts leads to measurable rates of endogenous PP in presence of exogenous ADP and Pi (148). Endogenous pseudocyclic PP takes place in intact as well as in osmotically broken chloroplasts (52, 81, 103, 144, 148). Besides the pseudocyclic type, an endogenous cyclic PP also exists in isolated chloroplasts (1, 54, 55, 103, 143, 144, 194), but only in intact chloroplasts (Type A, 73) capable of fixing CO_2 at high rates without any addition of cofactors or enzymes. The endogenous cyclic type of PP generally behaves like the cyclic PP in vivo. It is relatively insensitive to inhibitors of PS-II such as CMU, DCMU, and *o*-phenanthroline and is strongly suppressed by antimycin A (54, 55, 103, 143, 144, 194) like the ferredoxin-catalyzed cyclic PP (226) of broken chloroplasts and the cyclic PP in algae (151, 228, 249, 257). The experiments of Forti & Rosa (54) with the plastoquinone antagonist DBMIB (21, 240) suggest that plastoquinone is not involved in endogenous cyclic PP.

CONCLUDING REMARKS

The results reported in this article are manifold and in several respects preliminary and controversial in detail. Nevertheless, they allow us to draw some summarizing conclusions on PP in vivo. In all reactions in vivo that are light dependent and depressed by uncouplers, PP seems to exist if the action spectrum follows the absorption spectra of the photosynthetic pigments. Secondary effects of the uncouplers have to be taken into account.

Noncyclic PP in intact cells seems to occur in the presence of CO_2, concomitant with O_2 evolution, if PS-II is sensitized especially by high light intensities. The reaction can be blocked by inhibitors of O_2 production. It is generally assumed that PP in an atmosphere without CO_2 or in N_2 has to be regarded as cyclic PP. But even under these conditions oxygen evolution can proceed at a low rate (10, 111, 162, 246) if reducible ions like NO_3^- or reducible intermediates in the cells are present (77, 146, 160, 246). These reactions seem also to be coupled with noncyclic PP (246, 249).

Pseudocyclic PP coupled to a pseudocyclic electron flow is commonly found in vitro (5, 28, 52, 81, 141, 148). It seems to occur in vivo under conditions similar to the noncyclic PP, only with the difference that instead of CO_2, oxygen acts as electron acceptor without measurable O_2 evolution. This reaction is not easy to establish yet in intact cells, but there exist several indications for the occurrence of this type of PP in vivo (29, 80, 90, 99, 119, 175, 202). Important evidence in favor of a pseudocyclic PP in vivo has been provided by Heber (80) by measuring light-scattering changes in leaves.

Cyclic PP in vivo has been found by many authors (53, 110, 151, 236, 257, 277) and is characterized by several criteria: dependence on PS-I, occurrence also under anaerobic conditions, insensitivity to inhibitors of O_2 evolution, inhibition by antimycin A and DSPD, and saturation at low light intensity (231, 237, 253). The significance of the latter criterion should be taken with some caution since some processes depending on open chain electron flow as well as the ATP level of the cell are also saturated at low light intensity. The ATP level is efficiently regulated, which is particularly shown by the fact that fast changes of external conditions that cause a switching to another type of PP do not change the ATP level except for rapid oscillations (129). The several reactions that have been used as indicator reactions for the cyclic PP demonstrate that many energy-dependent processes e.g. transport, movements, carbohydrate accumulation, and growth, may be supported by ATP synthesized in cyclic PP under proper conditions.

Whether the cyclic or even the pseudo-cyclic PP provides the "extra ATP" required for CO_2 assimilation is still an unsolved problem (81, 175, 232). Even the question of how many sites of ATP formation may be involved in PP in vivo is still unanswered. Some experiments indicate the presence of a second phosphorylation site (62, 179, 202, 231). The contribution of cyclic PP to the ATP generation of the cell seems to be different during various developmental stages. During the greening process of etiolated intact leaves the cyclic PP appears distinctly before noncyclic PP (72, 158). Experiments with young and old leaves indicate also that the changes in photosynthetic rates with aging of leaves are probably correlated with the change of the ratio between PS-I and PS-II activity during development (197). During the synchronous life cycle of green algae the activity of PS-I and cyclic PP remains more or less constant, whereas PS-II and noncyclic PP change activity (63, 195, 196, 203). Finally, it is worthwhile to consider that cyclic PP may be a relic of the evolutionary past and exists today preferentially in lower plant organisms.

ACKNOWLEDGMENT

We wish to thank the many colleagues who sent reprints and preprints of their work to help in writing this review. We should also like to thank Dr. W. R. Ullrich for his valuable criticism and reading the manuscript. Special thanks are due to all who helped us in the final preparation of this article.

Literature Cited

1. Anderson, J. M., Boardmann, N. K., Spencer, D. 1971. *Biochim. Biophys. Acta* 245:253
2. Arisz, W. H., Sol, M. M. 1956. *Acta Bot. Neer.* 5:218
3. Arnon, D. I., Tsujimoto, H. Y., McSwain, B. D. 1964. *Proc. Nat. Acad. Sci. USA* 51:1274
4. Arnon, D. I., Tsujimoto, H. Y., McSwain, B. D. 1967. *Nature* 214:562
5. Arnon, D. I. et al 1961. *Proc. Nat. Acad. Sci. USA* 47:1314
6. Asahi, T., Jagendorf, A. T. 1963. *Arch. Biochem. Biophys.* 100:531
7. Avron, M. 1967. *Curr. Top. Bioenerg.* 2:1
8. Avron, M., Neumann, J. 1968. *Ann. Rev. Plant Physiol.* 19:137
9. Barber, J. 1968. *Nature* 217:876
10. Beevers, L., Hageman, R. H. 1969. *Ann. Rev. Plant Physiol.* 20:495
11. Ben-Amotz, A., Avron, M. 1972. *Plant Physiol.* 49:244
12. Bianchetti, R. 1963. *G. Bot. Ital.* 70:321
13. Bishop, N. I. 1962. *Nature* 195:55
14. Bishop, N. I. 1966. *Ann. Rev. Plant Physiol.* 17:185
15. Bishop, N. I. 1967. *Photochem. Photobiol.* 6:621
16. Bishop, N. I. See Ref. 51, 1:459
17. Bishop, N. I., Gaffron, H. 1962. *Biochem. Biophys. Res. Commun.* 8:471
18. Bishop, N. I., Gaffron, H. 1963. *Nat. Acad. Sci. Nat. Res. Counc. Publ.*, 158
19. Bishop, N. I., Wong, J. 1971. *Biochim. Biophys. Acta* 234:433
20. Boardman, N. K. 1970. *Ann. Rev. Plant Physiol.* 21:115
21. Böhme, H., Reimer, S., Trebst, A. 1971. *Z. Naturforsch.* 26b:341
22. Böhme, H., Trebst, A. 1969. *Biochim. Biophys. Acta* 180:137
23. Bomsel, J. L., Pradet, A. 1968. *Biochim. Biophys. Acta* 162:230
24. Bornefeld, T., Domanski-Kaden, J., Simonis, W. See Ref. 51, 2:1379
25. Bornefeld, T., Simonis, W. 1973. In preparation
26. Bothe, H. 1969. *Z. Naturforsch.* 24b:1574
27. Bothe, H. See Ref. 51, 3:2170
28. Buchanan, B. B., Arnon, D. I. 1966. *Protides Biol. Fluids Proc. Colloq.* 14:143
29. Butt, V. S., Peel, M. 1963. *Biochem. J.* 88:31 p
30. Carr, N. G., Hood, W., Pearce, J. See Ref. 142, 3:1565

31. Chulanovskaya, M. V., Zalensky, O. V. 1970. In *Methods of Investigation of Photophosphorylation*, ed. E. B. Kirichenko, 111. Pushchino
32. Collander, R. 1930. *Acta Bot. Fenn.* 6:1
33. Cook, J. R. 1967. *J. Protozool.* 14:382
34. Cook, J. R., Carrer, M. 1966. *Plant Cell Physiol.* 7:377
35. Coster, H. G. C., Hope, A. B. 1968. *Aust. J. Biol. Sci.* 21:243
36. Cox, R. M., Fay, P. 1969. *Proc. Roy. Soc. London Ser. B.* 172:357
37. Cramer, W. A., Böhme, H. 1972. *Biochim. Biophys. Acta* 256:358
38. Dainty, J. 1962. *Ann. Rev. Plant Physiol.* 13:379
39. Decker, M., Tanner, W. 1972. *Biochim. Biophys. Acta* 266:661
40. Domanski-Kaden, J., Simonis, W. 1972. *Arch. Mikrobiol.* 87:11
41. Donze, M., Haveman, J., Schiereck, P. 1972. *Biochim. Biophys. Acta* 256:157
42. Drechsler, Z., Nelson, N., Neumann, J. 1969. *Biochim. Biophys. Acta* 189:65
43. Emerson, R. L., Stauffer, J. F., Umbreit, W. W. 1944. *Am. J. Bot.* 31:107
44. Eppley, R. W., Gee, R., Saltmann, P. 1963. *Physiol. Plant.* 16:777
45. Epstein, E. 1972. *Mineral Nutrition of Plants: Principles and Perspectives*, 412. New York, London: Wiley
46. Eßl, A. 1968. PhD thesis. Univ. Würzburg. 162 pp.
47. Faust, R., Orcutt, A. R., Shearin, S. 1971. *Planta* 100:360
48. Fay, P. 1965. *J. Gen. Microbiol.* 39:11
49. Fay, P. 1970. *Biochim. Biophys. Acta* 216:353
50. Fay, P., Stewart, W. D. P., Walsby, A. E., Fogg, G. E. 1968. *Nature* 220:810
51. Forti, G., Avron, M., Melandri, A., Eds. 1972. *Proc. 2nd Int. Congr. Photosyn. Res.* The Hague: Dr. W. Junk N.V. 3 vols.
52. Forti, G., Jagendorf, A. T. 1961. *Biochim. Biophys. Acta* 54:322
53. Forti, G., Parisi, B. 1963. *Biochim. Biophys. Acta* 71:1
54. Forti, G., Rosa, L. 1971. *FEBS Lett.* 18:55
55. Forti, G., Zanetti, G. See Ref. 142, 3:1213
56. Frenkel, A. W. 1954. *J. Am. Chem. Soc.* 76:5568

57. Frenkel, A. W. 1970. *Biol. Rev.* 45: 569
58. Gaffron, H. 1940. *Am. J. Bot.* 27:273
59. Gaffron, H., Bishop, N. I. 1963. In *La Photosynthese*, 119:229. Paris: CNRS
60. Gest, H., Kamen, M. D. 1948. *J. Biol. Chem.* 176:299
61. Gibbs, M., Ed. 1971. *Structure and Function of Chloroplasts.* Berlin, Heidelberg, New York: Springer-Verlag
62. Gimmler, H. 1973. *Z. Pflanzenphysiol.* In press
63. Gimmler, H., Neimanis, S., Eilmann, I., Urbach, W. 1970. *Z. Pflanzenphysiol.* 64:358
64. Gimmler, H., Simonis, W., Urbach, W. 1969. *Naturwissenschaften* 56:371
65. Gimmler, H., Urbach, W., Jeschke, W. D., Simonis, W. 1968. *Z. Pflanzenphysiol.* 58:353
66. Gingras, G. 1966. In *Currents in Photosynthesis*, ed. J. C. Goedheer, J. B. Thomas, 187. Rotterdam: Donker. 487 pp.
67. Glagoleva, T. A., Zalensky, O. V. See Ref. 31, 89
68. Good, N. E., Izawa, S., Hind, G. 1966. *Curr. Top. Bioenerg.* I:76
69. Gould, E. S., Bassham, J. A. 1965. *Biochim. Biophys. Acta* 102:9
70. Goulding, K. H., Merrett, M. J. 1966. *J. Exp. Bot.* 17:678
71. Grünsfelder, M. 1973. *Planta.* In preparation
72. Gyldenholm, A. O., Whatley, F. R. 1968. *New Phytol.* 67:461
73. Hall, D. O. 1972. *Nature* 235:125
74. Hall, D. O., Evans, M. C. W. 1972. *Sub. Cell. Biochem.* 1:197
75. Halldal, P., Ed. 1970. *Photobiology of Microorganisms.* Wiley Interscience
76. Harold, F. M. 1966. *Bacteriol. Rev.* 30:772
77. Hattori, A., Myers, J. 1967. *Plant Cell Physiol.* 8:327
78. Hauska, G. A., McCarty, R. E., Racker, E. 1970. *Biochim. Biophys. Acta* 197:206
79. Healey, F. P., Myers, J. 1971. *Plant Physiol.* 47:373
80. Heber, U. 1969. *Biochim. Biophys. Acta* 180:302
81. Heber, U. 1973. In preparation
82. Heber, U., Santarius, K. A. 1970. *Z. Naturforsch.* 25b:718
83. Heldt, H. W., Rapley, L. 1970. *FEBS Lett.* 10:143
84. Hill, R., Bendall, F. 1960. *Nature* 186:136
85. Hirt, G., Tanner, W., Kandler, O. 1971. *Plant Physiol.* 47:841
86. Hoagland, D. R., Davis, A. R. 1929. *Protoplasma* 6:610
87. Hoare, D. S., Hoare, S. L., Moore, R. B. 1967. *J. Gen. Microbiol.* 49:351
88. Hoare, D. S., Hoare, S. L., Smith, A. J. See Ref. 142, 3:1570
89. Hoch, G. E. 1972. *Methods Enzymol.* 24:165
90. Hoch, G. E., Owens, O. V. H., Kok, B. 1963. *Arch. Biochem. Biophys.* 101:171
91. Hoch, G. E., Randles, J. 1971. *Photochem. Photobiol.* 14:435
92. Hodson, R. C., Schiff, J. A. 1971. *Plant Physiol.* 47:296
93. Holm-Hansen, O. 1970. *Plant Cell Physiol.* 11:689
94. Jackson, W. A., Volk, R. J. 1970. *Ann. Rev. Plant Physiol.* 21:384
95. Jagendorf, A. T. 1962. in *Surv. Biol. Progr.* 4:369
96. Jeschke, W. D. 1970. *Planta* 91:111
97. Jeschke, W. D. 1970. *Z. Pflanzenphysiol.* 62:158
98. Ibid 1972. 66:397
99. Jeschke, W. D. 1972. *Planta* 103:164
100. Jeschke, W. D., Simonis, W. 1965. *Planta* 67:6
101. Ibid 1969. 88:157
102. Joliot, P., Joliot, A., Kok, B. 1969. *Biochim. Biophys. Acta* 153:635
103. Kaiser, W., Urbach, W. 1972. *Ber. Deut. Bot. Ges.* In press
104. Kanai, R., Simonis, W. 1968. *Arch. Mikrobiol.* 62:56
105. Kandler, O. 1950. *Z. Naturforsch.* 5b:423
106. Ibid 1954. 9b:625
107. Ibid 1957. 12b:271
108. Kandler, O. 1960. *Ann. Rev. Plant Physiol.* 11:37
109. Kandler, O., Haberer-Liesenkötter, I. 1963. *Z. Naturforsch.* 18b:718
110. Kandler, O., Tanner, W. 1966. *Ber. Deut. Bot. Ges.* 79:48
111. Kessler, E. 1964. *Ann. Rev. Plant Physiol.* 15:57
112. Kessler, E. 1968. *Planta* 81:264
113. Ibid 1970. 92:222
114. Kiyohara, T., Fujita, Y., Hattori, A., Watanabe, A. 1962. *J. Gen. Appl. Microbiol.* 8:165
115. Klob, W., Tanner, W., Kandler, O. See Ref. 51, 3:1998
116. Knaff, D. B., Arnon, D. I. 1969. *Proc. Nat. Acad. Sci. USA* 64:715
117. Komor, E., Hass, D., Tanner, W. 1972. *Biochim. Biophys. Acta* 266: 449

118. Komor, E., Tanner, W. 1971. *Biochim. Biophys. Acta* 241:170
119. Krall, A. R., Bass, E. R. 1962. *Nature* 196:791
120. Krause, G. H., Heber, U. 1971. In *Proc. 1st Eur. Biophys. Congr.*, ed. E. Broda, A. Locker, H. Springer-Lederer, 79. Vienna: Verlag Wiener Med. Akad.
121. Kuhl, A. 1960. *Ergeb. Biol.* 23:144
122. Kulaev, I. S. 1971. In *Chemical Evolution on the Origin of Life*, ed. R. Buvet, C. Pernamperuna. Amsterdam: North-Holland
123. Kylin, A. 1960. *Physiol. Plant.* 13:366
124. Ibid 1966. 19:644
125. Ibid, 883
126. Kylin, A., Tillberg, J. E. 1967. *Z. Pflanzenphysiol.* 58:165
127. Latzko, E., Gibbs, M. 1969. *Plant Physiol.* 44:396
128. Levine, R. P. 1969. *Ann. Rev. Plant Physiol.* 20:523
129. Lewenstein, A., Bachofen, R. 1972. *Biochim. Biophys. Acta* 267:80
130. Lilley, R. McC., Hope, A. B. 1971. *Biochim. Biophys. Acta* 226:161
131. Lookeren Campagne, R. N. van 1957. *Acta Bot. Neer.* 6:543
132. Lüttge, U., Ball, E., von Willert, K. 1971. *Z. Pflanzenphysiol.* 65:326
133. Lüttge, U., Pallaghy, C. K. 1969. *Z. Pflanzenphysiol.* 61:58
134. Lüttge, U., Pallaghy, C. K., Osmond, C. B. 1970. *J. Membrane Biol.* 2:17
135. MacLachlan, G. A., Porter, H. K. 1959. *Proc. Roy. Soc. London B* 150:460
136. MacRobbie, E. A. C. 1966. *Aust. J. Biol. Sci.* 19:363
137. MacRobbie, E. A. C. 1970. *Quart. Rev. Biophys.* 3:251
138. MacRobbie, E. A. C. 1971. *Ann. Rev. Plant Physiol.* 22:75
139. Majumdar, K., Simonis, W. 1973. In preparation
140. Maree, E., Forti, G., Bianchetti, R., Parisi, B. 1963. *La Photosynthese* 119:557. Paris: CNRS
141. Mehler, A. H. 1951. *Arch. Biochem. Biophys.* 33:65
142. Metzner, H., Ed. 1969. *Progress in Photosynthesis Research*, Vol. 1-3. Proc. Int. Congr. Photosyn. Res., Freudenstadt 1968. Tübingen
143. Miginiac-Maslow, M. 1971. *Biochim. Biophys. Acta* 234:353
144. Miginiac-Maslow, M., Moyse, A. See Ref. 142, 3:1203
145. Miyachi, S., Tamiya, H. 1961. *Plant Cell Physiol.* 2:405

146. Morris, I., Ahmed, J. 1969. *Physiol. Plant.* 22:1166
147. Nissen, P. 1971 *Physiol. Plant.* 24:315
148. Nobel, P. S. 1967. *Plant Physiol.* 42:1389
149. Nobel, P. S., Chang, D., Wang, C. T., Smith, S. S., Barcus, D. E. 1969. *Plant Physiol.* 44:655
150. Nultsch, W. 1962. *Ber. Deut. Bot. Ges.* 75:443
151. Nultsch, W. See Ref. 66, 421
152. Nultsch, W. 1967. *Z. Pflanzenphysiol.* 56:1
153. Nultsch, W. 1969. *Photochem. Photobiol.* 10:119
154. Nultsch, W. See Ref. 75, 213
155. Nultsch, W. 1971. *Photochem. Photobiol.* 14:705
156. Nultsch, W., Hellmann, W. 1972. *Arch. Mikrobiol.* 82:76
157. Nultsch, W., Bai, J. 1966. *Z. Pflanzenphysiol.* 54:84
158. Oelze-Karow, H., Butler, W. L. 1971. *Plant Physiol.* 48:621
159. Ohmann, E. 1964. *Biochim. Biophys. Acta* 82:325
160. Paneque, A., Aparicio, P. J., Cardenas, J., Vega, M., Losada, M. 1969. *FEBS Lett.* 3:57
161. Park, R. B., Sane, P. V. 1971. *Ann. Rev. Plant Physiol.* 22:395
162. Paschinger, H. 1969. *Arch. Mikrobiol.* 67:243
163. Pedersen, T. A., Kirk, M., Bassham, J. A. 1966. *Biochim. Biophys. Acta* 112:189
164. Pedersen, T. A., Kirk, M., Bassham, J. A. 1966. *Physiol. Plant.* 19:219
165. Penth, B., Weigl, J. 1969. *Z. Naturforsch.* 24b:342
166. Penth, B., Weigl, J. 1971. *Planta* 96:212
167. Piskunkova, N. F., Pimenova, M. N. 1971. *Mikrobiologiya* 40:692
168. Powls, J., Wong, J., Bishop, N. I. 1969. *Biochim. Biophys. Acta* 180:490
169. Pratt, L. H., Bishop, N. I. 1968. *Biochim. Biophys. Acta* 153:664
170. Pringsheim, E. G., Pringsheim, O. 1959. *Biol. Zentralbl.* 78:937
171. Pringsheim, E. G., Wiessner, W. 1960. *Nature* 188:919
172. Raven, J. A. 1967. *J. Gen. Physiol.* 50:1627
173. Raven, J. A. 1968. *J. Exp. Bot.* 19:233
174. Raven, J. A. 1969. *New Phytol.* 68:45
175. Raven, J. A. 1970. *J. Exp. Bot.* 21:1
176. Ibid 1971. 22:420

177. Raven, J. A. 1971. *Chem. Ind. London* 31:859
178. Raven, J. A., MacRobbie, E. A. C., Neumann, J. 1969. *J, Exp. Bot.* 20:221
179. Rensen, J. J. S. Van 1971. *Meded. Landbouwhogesch. Wageningen* 71-9:1
180. Rensen, J. J. S. Van. See Ref. 142, 3:1769
181. Ried, A. 1968. *Biochim. Biophys. Acta* 153:653
182. Ried, A. See Ref. 142, 1:512
183. Ried, A. See Ref. 51, 1:763
184. Ried, A., Setlik, I. See Ref. 51, 3:2077
185. Ried, A., Setlik, I., Berkova, E., Bossert, U. 1972. *Photosynthetica* 7. In press
186. Rippka, R., Neilson, A., Kunisawa, R., Cohen-Bazire, G. 1971. *Arch. Mikrobiol.* 76:341
187. Robinson, J. B. 1969. *J. Exp. Bot.* 20:201
188. Ibid, 212
189. Rurainski, H. J., Randles, J., Hoch, G. E. 1970. *Biochim. Biophys. Acta* 205:254
190. Santarius, K. A., Heber, U. 1965. *Biochim. Biophys. Acta* 102:39
191. Schilde, C. 1966. *Planta* 71:184
192. Schmidt, A., Schwenn, J. D. See Ref. 51, 1:507
193. Schön, G., Bachofen, R. 1970. *Arch. Mikrobiol.* 73:34
194. Schürmann, P., Buchanan, B. B., Arnon, D. I. 1972. *Biochim. Biophys. Acta* 267:111
195. Senger, H. 1970. *Planta* 92:327
196. Senger, H., Bishop, N. I. See Ref. 142, 1:425
197. Sestak, Z. 1969. *Photosynthetica* 3:285
198. Setlik, I. 1957. *Biochim. Biophys. Acta* 24:434
199. Shieh, Y. J., Barber, J. 1971. *Biochim. Biophys. Acta* 233:594
200. Simonis, W. 1960. *Handb. Pflanzenphysiol.* 5:1966
201. Simonis, W. 1964. *Ber. Deut. Bot. Ges.* 77:5
202. Ibid 1967. 80:395
203. Simonis, W. 1972. In *Theoretical Foundations of the Photosynthetic Productivity*, ed. A. A. Nichiporovich, 75. Moscow: Nauka
204. Simonis, W., Eßl, A. 1973. In preparation
205. Simonis, W., Gimmler, H. See Ref. 142, 3:1155
206. Simonis, W., Grube, K. H. 1952. *Z. Naturforsch.* 7b:194
207. Simonis, W., Kating, H. 1956. *Z. Naturforsch.* 11b:707
208. Simonis, W., Kuntz, F. J., Urbach, W. 1962. *Ber. Deut. Bot. Ges. Vortr. Gesamtgeb. Bot.* 1:139
209. Simonis, W., Mechler, E. 1963. *Biochem. Biophys. Res. Commun.* 13:241
210. Simonis, W., Urbach, W. 1963. In *Studies on Microalgae and Photosynthetic Bacteria*, 597. Tokyo: Jap. Soc. Plant Physiol.
211. Simonis, W., Urbach, W. 1963. *Arch. Mikrobiol.* 46:265
212. Simonis, W., Weichart, G. 1958. *Z. Naturforsch.* 136:55
213. Sinclair, J. 1968. *J. Exp. Bot.* 19:254
214. Smith, A. J., London, J., Stanier, R. Y. 1967. *J. Bacteriol.* 94:972
215. Smith, F. A. 1966. *Biochim. Biophys. Acta* 126:94
216. Smith, F. A. 1967. *J. Exp. Bot.* 18:348
217. Ibid 1968. 19:442
218. Smith, F. A. 1970. *New Phytol.* 69:903
219. Stewart, W. D. P., Haystead, A., Pearson, H. W. 1969. *Nature* 224:226
220. Strehler, B. L. 1952. In *Phosphorus Symposium*, 2:491. Baltimore: Johns Hopkins Press
221. Strehler, B. L. 1970. In *Methoden der Enzymatischen Analyse*, ed. H. U. Bergmeyer, 2:2036. Weinheim: Chemie-Verlag
222. Strotmann, H., Heldt, H. W. See Ref. 142, 3:1131
223. Surikow, T. 1971. *J. Exp. Bot.* 22:526
224. Syrett, P. J. 1966. *J. Exp. Bot.* 17:641
225. Syrett, P. J., Merrett, M. J., Bocks, S. M. 1963. *J. Exp. Bot.* 14: 249
226. Tagawa, K., Tsujimoto, H. Y., Arnon, D. I. 1963. *Proc. Nat. Acad. Sci. USA* 49:567
227. Tanner, W. 1969. *Biochem. Biophys. Res. Commun.* 36:278
228. Tanner, W., Dächsel, L., Kandler, O. 1965. *Plant Physiol.* 40:1151
229. Tanner, W., Grünes, R., Kandler, O. 1970. *Z. Pflanzenphysiol.* 62:376
230. Tanner, W., Kandler, O. 1967. *Z. Pflanzenphysiol.* 58:24
231. Tanner, W., Kandler, O. See Ref. 142, 3:1117
232. Tanner, W., Löffler, M., Kandler, O. 1969. *Plant Physiol.* 44:422
233. Tanner, W., Loos, E., Kandler, O. See Ref. 66, 243
234. Tanner, W., Zinecker, U., Kandler, O. 1967. *Z. Naturforsch.* 22:3

235. Taylor, F. J. 1960. *Proc. Roy. Soc. London B* 151:483
236. Teichler-Zallen, D., Hoch, G. E. 1967. *Arch. Biochem. Biophys.* 120:227
237. Teichler-Zallen, D., Hoch, G. E., Bannister, T. T. See Ref. 51, 1:643
238. Trebst, A. 1970. *Ber. Deut. Bot. Ges.* 83:373
239. Trebst, A., Burba, M. 1967. *Z. Pflanzenphysiol.* 57:419
240. Trebst, A., Harth, E., Draber, W. 1970. *Z. Naturforsch.* 25b:1157
241. Trebst, A., Schmidt, A. See Ref. 142, 3:1510
242. Ullrich-Eberius, C. I. 1973. *Planta* 109:161
243. Ullrich-Eberius, C. I., Simonis, W. 1970. *Planta* 92:358
244. Ibid 1970. 93:214
245. Ullrich, W. R. 1970. *Planta* 90:272
246. Ibid 1971. 100:18
247. Ibid 1972. 102:37
248. Ullrich, W. R. 1972. *Arch. Mikrobiol.* 87:323
249. Ullrich, W. R., Simonis, W. 1969. *Planta* 84:358
250. Ullrich, W. R., Urbach, W., Santarius, K. A., Heber, U. 1965. *Z. Naturforsch.* 20:905
251. Urbach, W. 1973. *Ber. Deut. Bot. Ges.* In preparation
252. Urbach, W., Gimmler, H. See Ref. 142, 3:1274
253. Urbach, W., Gimmler, H. 1970. *Z. Pflanzenphysiol.* 62:276
254. Urbach, W., Gimmler, H. 1970. *Ber. Deut. Bot. Ges.* 83:439
255. Urbach, W., Kaiser, W. See Ref. 51, 2:1401
256. Urbach, W., Simonis, W. 1962. *Ber. Deut. Bot. Ges. Vortr. Gesamtgeb. Bot.* 1:149
257. Urbach, W., Simonis, W. 1964. *Biochem. Biophys. Res. Commun.* 17:39
258. Urbach, W., Simonis, W. 1967. *Z. Naturforsch.* 22b:537
259. Vallée, M., Jeanjean, R. 1968. *Biochim. Biophys. Acta* 150:599
260. Vernon, L. P. 1968. *Bacteriol. Rev.* 32:243
261. Vidaver, W., French, C. S. 1965. *Plant Physiol.* 40:7
262. Vose, J. R., Spencer, M. 1967. *Biochem. Biophys. Res. Commun.* 29:532
263. Voskresenskaya, N. P. 1972. *Ann. Rev. Plant Physiol.* 23:219
264. Wassink, E. C. 1957. In *Research in Photosynthesis*, ed. H. Gaffron, 333. Proc. Gatlinburg Conf. Photosyn. 1955
265. Wassink, E. C., Wintermans, J. F. G. M., Tjia, J. E. 1951. *Proc. Kon. Ned. Akad. Wetensch Ser. C* 54:41
266. Weaver, E., Bishop, N. I. 1963. *Science* 140:1095
267. Wedding, R. T., Black, M. K. 1960. *Plant Physiol.* 35:72
268. Weichart, G. 1961. *Planta* 56:262
269. Weigl, J. 1964. *Z. Naturforsch.* 19b:845
270. Weigl, J. 1967. *Planta* 75:327
271. Welsch, F., Smith, L. 1969. *Biochemistry* 8:3403
272. Wiessner, W. 1964. *Plant Physiol.* 39 (Suppl.): 45
273. Wiessner, W. 1965. *Nature* 205:56
274. Wiessner, W. See Ref. 142, 1:442
275. Wiessner, W. See Ref. 75, 95
276. Wiessner, W., Amelunxen, F. 1969. *Arch. Mikrobiol.* 67:357
277. Wiessner, W., Gaffron, H. 1964: *Nature* 201:725
278. Wiessner, W., Gaffron, H. 1964. *Fed. Proc.* 23:226
279. Wiessner, W., Kuhl, A. 1962. *Ber. Deut. Bot. Ges. Vortr. Gesamtgeb. Bot.* 1:102
280. Wintermans, J. F. G. M. 1954. *Proc. Kon. Ned. Akad. Wetensch. Ser. C* 57:574
281. Witt, H. T. 1971. *Quart. Rev. Biophys.* 4:365
282. Yordanov, I. T. 1971. *C. R. Acad. Bulg. Sci.* 24:1551
283. Zalensky, O. V., Glagoleva, T. A., Chulanovskaya, M. V. 1966. *Bot. Zh.* 51:1718

Ann. Rev. Plant Physiol. 1973. 24:115–28

GAMETOPHYTE DEVELOPMENT IN FERNS AND BRYOPHYTES

✤ 7544

Hans Brandes[1]

Plant Science Research, Eli Lilly GmbH, 638 Bad Homburg, West Germany

CONTENTS

A study of the growth and development of fern, liverwort, and moss gametophytes yields an interesting biological system for studying basic developmental processes. Since the last comprehensive review in this series on the control of differentiation in Archegoniatae by Bopp in 1968 (6), research in this field has been vigorous. This review will cover mainly the literature of the years 1967 to 1972. The emphasis will be on: progress in methodology, particular characteristics, and control mechanism of gametophyte development.

Progress to better understanding of cellular differentiation has been achieved in many areas of developmental biology. In multicellular higher organisms morphogenesis has often been studied by using plants with a "simple" morphology or by cultivating isolated tissues or single cells in vitro. In lower organisms unicellular, filamentous, two-dimensional, and three-dimensional systems have been used successfully. The gametophytes of most ferns, liverworts, and mosses combine the presence of unicellular, filamentous, two-dimensional, and three-dimensional growth stages during early development.

[1] I am greatly indebted to numerous colleagues, especially Dr. H. Schraudolf, for advice and for help with the literature search and supply of reprints.

PROGRESS IN METHODOLOGY

Culture Techniques

The adaption, improvement, and development of new, very effective techniques has led to a rapid progress in the field of gametophyte development. Sterile culture techniques have been developed for a wide range of fern, liverwort, and moss gametophytes. In many cases the way is now open for handling gametophytes like unicellular or filamentous microorganisms (1, 27, 30, 39, 72, 86). Gametophytes can be cultured in vitro on simple defined media in a fully controlled environment, and in many cases the in vitro growth equals that in vivo. An ideal situation is encountered in the germination of unicellular spores which can be induced "en masse" to yield specialized cells of one or two types, protonema and rhizoid cells (38, 45, 61, 63, 92, 96).

Synchrony of cell division in apical cells of fern gametophytes was reported by Ito (27) after he transferred red-grown cultures to blue light conditions. An improved liquid suspension culture technique was developed by Miller & Machlis (39) for the liverwort *Sphaerocarpos donnellii* which substantially improved the reproducibility of quantitative experiments.

An interesting system for studying antheridia formation was described by Schraudolf (72). The two polypodiacaeous ferns *Polypodium crassifolium* and *P. polycarpon* offer certain advantages for morphogenetical studies. The gametophytes of these two ferns are reduced to two or three cells respectively which make the application of new cytological techniques very promising.

Biochemical and Biophysical Techniques

The development and application of cytological techniques in combination with biochemical and biophysical techniques has provided promising results in many areas of gametophyte development. DNA and RNA metabolism of developing gametophytes have been studied by using autoradiography (8, 23, 33, 78), cytophotometry (14, 76, 86, 87), and biochemical methods (66, 75, 80, 83). Ultraviolet absorption spectroscopy and fluorescence microscopy in combination with autoradiography and biochemical extraction and photometric determination of nucleic acids have made a more critical analysis of qualitative and quantitative results possible (7, 8, 20, 28, 33, 59, 77, 78). Protein metabolism during the development of spores, filaments, thalli or antheridia, and archegonia was investigated by following uptake, incorporation, and metabolism of labeled precursors (38, 53, 54, 59, 61, 62, 64). The regulation of protein synthesis in fern gametophytes was also investigated in an in vitro system by following amino acid incorporating activity in isolated chloroplasts during light-induced morphogenesis (64, 65). A microphoresis technique in combination with photometry has been successfully employed to determine quantitative alterations in nucleic acid content and base composition of individual cells associated with the initiation of gametophores in moss protonemata (66).

Electron Microscopical Techniques

Although having been developed much earlier for other organisms, electron microscopical techniques have recently been applied and improved for studying ultrastructural changes of specific developmental steps during gametophyte development (34). Using young and aged *Mnium* leaves, Lüttge & Krapf followed the ultrastructural changes during development and their influence on transport phenomena (36, 37). In other cases electron microscopical techniques have been applied to the study of ultrastructural polarity changes during maturation of eggs in developing archegonia (3), light-mediated changes in the ultrastructure of chloroplasts from liverwort gametophytes (40), and ultrastructural changes of chloroplasts from liverworts during senescence of the gametophyte (18).

Bioassays

Qualitative and quantitative determinations of endogenous and exogenous growth regulators make specific bioassays necessary. Because of the important role that growth regulators play during all phases of plant development, the addition of purified endogenous and/or well-known exogenous growth regulators to the culture medium during specific phases of gametophyte development have proved to be an important tool for studying morphogenetic processes in this group of organisms (6, 8, 66, 68, 85, 98, 102).

An improved cytokinin bioassay based on bud formation in isolated protonema filaments of *Funaria hygrometrica* has been developed by Hahn & Bopp (23). The test is very sensitive (lowest detectable concentration about $1.5 \times 10^{-8} M$) and highly specific. Buds can be induced in this bioassay only by cytokinins, i.e. by substances showing a substitution of the purine ring similar to that in kinetin. Besides this the test can be completed in 2 days and exhibits linearity between cytokinin concentration and number of buds induced. This bioassay is suitable for detecting naturally occurring, purified cytokinins as well as cytokinins in crude extracts.

The use of exogenous growth regulators and endogenous phytohormones (2, 32, 48, 57, 58) in combination with specific inhibitors of nucleic acid and protein metabolism has permitted the analysis of specific morphogenetic steps (8, 29, 66, 69). However, it has been demonstrated in many cases that the direct effect of inhibitors on cell elongation and cell division has to be separated from the direct effect of these inhibitors on cell differentiation (6, 69). Separation of a general growth inhibition (or promotion) from a specific inhibition (or induction) of a morphogenetic step could be achieved in the protonema system of *Funaria hygrometrica*. Protonemata precultivated under "normal" conditions are transferred to the new medium containing a growth regulator and/or a specific inhibitor (8, 23). With this technique protonemata can be transferred back to normal medium at any time during and after induction (or inhibition) of morphogenetic steps (8). The same technique has also been successfully applied for treatments of total protonemata or isolated filaments with radioactive precursors of RNA, DNA, and protein synthesis (8, 33). A similar technique has also been used in experiments in which only parts of isolated filaments or even individual cells were treated with specific compounds during their development.

CONTROL MECHANISMS OF GAMETOPHYTE DEVELOPMENT

In most species of ferns and bryophytes a germinated spore produces first a filamentous (one-dimensional) protonema. This filamentous stage is then followed by either a prothallial (two-dimensional) stage, typical of most fern and liverwort gametophytes (*Anemia, Pteridium, Riella, Sphaerocarpos*), or a gametophoral (three-dimensional) stage, typical of mosses (*Funaria*). The interesting morphogenetic problems are connected with the transition from spore to filament, filament to prothallium or gametophore, prothallium to antheridium or archegonium (41), and some other morphogenetic steps. These distinctly characterized developmental steps provide convenient material for analysis of germination (47, 62, 63, 92, 93, 102), elongation (38, 45), division (82, 99–101), differentiation (8, 48, 66, 68, 85, 98), dedifferentiation and regeneration (19, 22, 33, 51, 75, 86), and senescence (17, 21).

Spore Germination, Rhizoidal, Protonemal, and Thallial Growth

The differentiation pattern of spore germination provides a convenient system for the analysis of plant development. The germination of the spore of many fern species is promoted by an exposure to light (25). In some fern species (*Pteridium aquilinum, Asplenium nidus*) the germination of the spores is promoted by a short exposure to red light and inhibited by far red light (38, 43, 45, 47, 91).

Spore germination normally starts with the rupture of the exine and formation of a rhizoid. This is then followed by the differentiation of the protonema. A temporal separation of the initiation and elongation phase of the rhizoid and the protonema was achieved by Raghavan (61–63), who demonstrated (63) that the protrusion of the rhizoid and protonema are two separate phytochrome-controlled processes. When spores were irradiated continuously with white, red, blue, or far red light and examined daily, red light was found most effective in initiating rhizoids (about 100% in 96 hr) and protonemata (about 100% in 120 hr). If red light-induced spores were returned to darkness, rhizoid and protonema were only induced when the previous exposure had been at least 72 hr. Initiation of the protonema was suppressed by far red immediately after red light irradiation, while rhizoid initiation was not inhibited. This is interpreted to mean that initiation of the protonema and rhizoid during germination of spores of *Asplenium nidus* are two separate phytochrome-controlled processes, perhaps with different P_{fr} requirements (63). Prolonged treatments with far red light (72 hr or more) resulted in rhizoid and pseudoprotonemata formation even without red light irradiation; however, subsequent growth in far red light was inhibited. The formation of pseudoprotonemata suggests an additional morphogenetic effect which is probably attributed to changes in the balance of P_r and P_{fr} during the irradiation sequence.

A model of red, far red-induced cell elongation has been developed by Miller & Stephani (45). The model is based on experiments showing the effects of colchicine and light on cell elongation and on direct measurements. The authors came to the conclusion that red and far red light promote cell elongation in *Onoclea sensibilis* gametophytes by causing a constriction of the cell diameter while leaving the rate of volume growth unaltered. The model proposes light-mediated changes in the structure of the cell wall which leads to a restriction of lateral cell expansion and enhancement of elongation.

Several hypotheses have been developed concerning the photoreceptors for red and far red light in the promotion and inhibition of this photomorphogenesis. One hypothesis is based on the assumption that phytochrome alone is the photoreceptor system for these light effects (67). In contrast, a second hypothesis postulates the existence of another pigment in addition to the phytochrome system (43). A third hypothesis has been developed by Sugai & Furuya (91, 92). They demonstrated a blue light-induced (action spectra have two peaks at 440 and 380 nm) inhibition of spore germination (*Pteris vittata*) and subsequent dark recovery. Dosages of blue light and recovery period were quantitatively similar. Thus, there are probably at least three photoreactions involved in germination and all three are trigger type reactions involving low-energy reactions. Two of these are mediated by a phytochrome system, while the third is possibly controlled by violaxanthin (carotenoid). Blue light may convert this pigment to a growth hormone. However, this pigment system has not yet been isolated.

Studies on fine structure of red and blue-light-treated sporelings of the moss *Ceratodon purpureus* showed differences at the ultrastructural level (96). Blue light chloroplasts remain richer in starch and have a denser stroma than red light chloroplasts. The mobilization of reserve lipids is increased in red light, but there are more mitochondria in blue light treated cells.

Regardless of the nature of the pigment(s) which perform the initial absorption of light, the subsequent physiological changes which then lead to morphogenetic changes in the gametophyte may be considered separately.

The relationship between changes in RNA and protein synthesis during rhizoid initiation and elongation has been examined (38, 52, 62). DeMaggio & Raghavan (38) examined radioactivity profiles of proteins extracted during rhizoid initiation and elongation (*Pteridium aquilinum*) after separation by acrylamide gel electrophoresis. No significant quantitative differences in the profiles could be detected during germination. However, when germinating spores were treated with actinomycin D during rhizoid elongation, parts of the protein profiles were changed. No changes in the pattern of ^{14}C-labeled proteins could be detected in the rhizoid. DeMaggio & Raghavan (38) have shown that actinomycin D begins to penetrate the spore as early as 4 hr after transfer to the drug. This is long before any measurable synthesis of proteins is initiated in the spores. Thus lack of penetration of the drug into spores can be ruled out. The lack of inhibition of rhizoid initiation and protein synthesis by actinomycin D during rhizoid initiation can be interpreted by the hypothesis that differential controls are exerted over protein synthesis during initiation and elongation of the rhizoid and protonema. Protein synthesis during rhizoid initiation may be regulated at the translation level, while a shift to the transcription level may occur during elongation of the rhizoid and initiation of the protonema.

Very interesting morphogenetic variations in the gametophyte of a 3X race of the fern *Hypodematium crenatum* as compared to X and 2X(2n = 82) plants have recently been described by Loyal & Paik (35). The first germ cell in 3X gametophytes failed to exhibit a polarity gradient but retained potentiality to undergo mitosis more than once, the plane of spindle being different at each division. The two cells of almost equal size grow into two separate filaments, and the differentiation of the rhizoid is bypassed (35).

Growth-regulating effects of ethylene on dark grown gametophytes of *Onoclea sensibilis* have been reported by Miller, Sweet & Miller (46). Evidence that gametophytes produce ethylene was substantiated by direct gas chromatographic demonstration of ethylene formation. Elongation of the rhizoid was decreased and cell division inhibited; however, elongation of the protonema was increased. Some mutual effects on their own growth of *Polypodium* gametophytes in culture may be due to a diffusible volatile substance secreted into the medium (81). A reversible inhibition of spore germination in *Onoclea sensibilis* under ethylene treatment has also been demonstrated. It has been concluded for both filament and rhizoid elongation that the growth regulating activities of ethylene and auxin are independent in filament and rhizoid growth. However, inhibition of elongation by supraoptimal auxin concentrations could not be attributed to an auxin-stimulated production of ethylene (46).

Transition from Filamentous Sporelings to Two-dimensional Thalli

The induction and growth of two-dimensional thalli of many ferns is controlled by blue light. The change of growth pattern in the protonema cells induced by light is a very rapid reaction. Immediately after transferring the gametophytes from red to blue light, the cell cycle of apical cells begins to progress from the early G1 phase. Red light, however, maintains continued further elongation and keeps the cells in the early G1 phase (26, 27). Synchronous cell division of apical cells could be induced by transferring red light protonemata to blue light conditions.

Our understanding of the biochemical control mechanism of this photomorphogenesis is still quite rudimentary. As is known from earlier work, blue light-induced morphogenesis is connected with an increase in RNA and protein synthesis (6). There are several reports that various antibiotics and RNA bases selectively prevent the transition from filamentous to two-dimensional growth (4, 9, 10, 42, 60, 74). On the basis of these results a hypothesis has been developed that the induction of biplanar growth requires the synthesis of new species of RNA and protein. An interaction of light quality and RNases during the development of the gametophytes of *Asplenium nidus* has been postulated by Raghavan (60). The initiation of two-dimensional growth could be inhibited in blue or white light by adding RNase A or B to the medium. The protonemata continued filamentous growth. The specificity of these enzymes seems to be rather satisfactory since length and cell number of RNase-treated protonemata exceed those of untreated controls without transition to two-dimensional growth which normally occurs when the protonema has reached a certain critical cell number. This is interpreted by the hypothesis that RNases may act by selectively hydrolyzing RNA essential for two-dimensional growth without affecting the capacity of the protonema cells to divide.

It is assumed that the new type of development requires the synthesis of new proteins which are formed on new RNA templates. In the presence of RNases filamentous growth is mediated by RNA insensitive to RNases, while two-dimensional growth requires the synthesis of new RNase-sensitive RNA (60). Similar results were obtained with gametophytes of *Pteridium* and *Dryopteris* (4, 9, 10, 12, 13). When gametophytes were cultured in red light they remained filamentous. A treatment with blue light induced two-dimensional growth. If two-dimensional prothalli induced by blue light were transferred back into red light, they returned to the filamentous growth. Reversal from two-dimensional to one-dimensional growth was enhanced by factors that inhibit cell division (83).

The metabolism of RNA and protein in the gametophytes of *Pteridium aquilinum* during light-induced morphogenesis has also been investigated by studying incorporation of ^{3}H-uridine and ^{3}H-leucine incorporation into RNA and protein respectively (59). Blue light enhanced the RNA and protein synthesis in subcellular fractions of gametophytes of *Pteridium*. Nuclei, chloroplasts, and the soluble cytoplasm are shown to serve as the principal sites of RNA and protein synthesis in the fern gametophytes. These data are in line with data on RNA and protein synthesis accompanying two-dimensional growth in the gametophytes of other species of ferns. A dramatic increase in the amino acid-incorporating activity of isolated chloroplasts was reported by Raghavan & DeMaggio (64). The effects of irradiation on the metabolism of the cells of the gametophytes were first visible in the nucleus and chloroplasts. Whether blue light affects the nucleus directly or through growth regulators released within the cell is still an open question.

Studies on several species have indicated, however, that the use of inhibitors often delays cell division. Since the transition from one-dimensional growth to two-dimensional growth occurs when the filament has reached a certain critical cell number (9, 42, 71, 74), a general inhibition of cell division has to be separated from a specific inhibition of two-dimensional growth.

Davis (10) could demonstrate that the critical cell number may be exceeded by certain treatments such as white-red cycles. For white-red cycles the rate of elongation was controlled by the intensity of red light. The increased elongation delayed the initiation of two-dimensional development. In both cases the rate of transition to two-dimensional growth was correlated with the amount of elongation per division. It can be concluded from these results that the reorientation of the mitotic spindle, and thus the transition from one-dimensional to two-dimensional growth, is also controlled by interactions between cell elongation and cell division. This hypothesis is supported by the results reported for other species by several authors (42, 60, 83).

The role of phytochrome action on the timing of cell division in gametophytes of *Adiantum* has been studied by Wada & Furuya (101). On the basis of their trials the involvement of at least two pigment systems is postulated for the control of this photomorphogenesis. Similar interaction between phytochrome and a blue light absorbing pigment has been reported for spore germination and cell elongation growth by other authors (45, 92). Although the photoreceptor for the blue light absorbing pigment system is still unknown, it has been postulated to be a flavoprotein, possibly riboflavin (104). On the basis of experiments with pyrimidine analogs and riboflavin, Yeoh & Raghavan (104) have postulated a correlation between pyrimidine metabolism, synthesis of riboflavin, and induction of two-dimensional growth. Reinvestigation of the same problem by other authors (11, 74) makes this interpretation somewhat questionable. Phytochrome, the photoreceptor for the red-far red light-absorbing pigment system, on the other hand, has been isolated by Taylor & Bonner (94) from gametophytes of the liverwort *Sphaerocarpos donnellii*. Since spectrum maxima for the *Sphaerocarpos* pigment were at 655 and 720 nm, it is concluded that substantial differences exist in the absorption spectrum of the pigment in plants widely separated phylogenetically (94).

Effects of the direction of irradiation on orientation of cell division have been reported by Wada & Furuya (99, 100). Two-dimensional gametophytes expanded in a plane which was at a right angle to any given direction of irradiation with continuous white light. However, the first longitudinal cell division of apical cells of filamentous gametophytes of *Adiantum* appears to be controlled not only by the direction of irradiation which actually induced the division but also by that inducing the proceeding transverse division (100).

Steiner (88–90) has studied polarotropism in *Dryopteris chloronema* (responding to short and long wavelength visible irradiation) and *Sphaerocarpos* germ tube (responding only to short wavelength visible irradiation). Polarotropism in *Dryopteris* is mediated by phytochrome and a photoreceptor system absorbing in the blue spectral region. The dose response curves show 2 maxima and 2 minima. However, in *Sphaerocarpos* the polarotropism action spectra were different. Germ tubes of *Sphaerocarpos* did not respond polarotropically to irradiation with wavelengths above 550 nm (88). The action spectrum obtained in *Dryopteris* looks similar to those action spectra described for a variety of other blue-ultraviolet-mediated photoresponses. A flavin-type photoreceptor has been postulated to be involved in this tropism together with phytochrome (89).

Action spectra for phototropism and (photo) growth in the protonema of *Physcomitrium* were studied by Nebel (49). The major peak of phototropic activity was found at 730 nm. The blue-absorbing system responsible for phototropism in other groups of plants could not be demonstrated in *Physcomitrium*. It has been suggested that phototropism is dependent upon an interaction between the photosynthetic and phytochrome systems (49). Evidence for disc-shaped phytochrome photoreceptors was reported by Nebel (50). Responses of protonemata to red or far-red polarized light indicate that light striking the apical cell surface perpendicularly is absorbed independently of the angle of polarization. However, irradiation striking the flank tangentially is absorbed if the plane of polarization is parallel to the cell surface. It can be concluded that the active photoreceptors (phytochrome) are dichroic and coin- or disc-shaped.

The discovery of an atypical strain of *Onoclea sensibilis* was reported by Miller & Miller (44). About 25% of the gametophytes derived from the spores of this strain were able to undergo two-dimensional development in darkness in contrast to normal plants which are filamentous in darkness. Under conditions of reduced ethylene concentration the proportion of two-dimensional plants rose to 75%. Atypical strains should be well suited for studying some of the basic problems of control mechanism involved (56).

Transition from Filamentous Protonema to Three-dimensional Gametophores

The moss protonema system provides a convenient system for studying hormonal induction of differentiation. Filamentous chloronema differentiates caulonema cells and these caulonema cells are the prerequisite for the formation of gametophores (moss plants). The differentiation leads from the one-dimensional protonema stage to the three-dimensional tissue stage (6). Under certain conditions, caulonema cells dedifferentiate to chloronema cells and consequently lose their potential to form gametophores.

Occurrence of an adenin-type cytokinin, bryokinin (2), indicates that hormonal substances (32) produced by the plant itself are regulating differentiation in mosses. Studies on the morphogenetic effect of applied cytokinins play an important role for a better understanding of plant development. Cytokinin activity is restricted to a specific morphogenetic change during protonema development, the formation of gametophores (buds) on certain caulonema cells. Brandes & Kende (8) could demonstrate that cytokinin-induced buds can be reversed to protonema if the cytokinin is withdrawn during bud induction. The cytokinin obviously does not act as a "trigger." It needs to be present for a critical period of time during which the differentiation is "stabilized." Similar observations are reported from other plant species. If the cytokinin is removed during a washing period and if the protonemata are then transferred back to nutrient agar containing a cytokinin, bud formation continues normally. This proves that it is the cytokinin itself which is removed from the cells and not some other factor(s) required for bud formation. Autoradiography of ^{14}C-benzyladenine-treated filaments showed that the label was accumulated by target cells. This label was loosely bound to the protonema cells. Rinsing with water removed the labeled hormone—and also caused dedifferentiation of the young buds. These data are interpreted by the hypothesis that the easily removable, probably noncovalently bound hormone is the physiologically active one (8). It should also be mentioned that the naturally occurring plant hormones of gametophytes (2, 24, 32, 46, 48, 73) and other organisms (31, 55, 84, 103) are also released into the surrounding medium. Continuous removal of these hormones also retards or reduces further development. Thus, the presence of binding sites may be the biochemical basis for the difference between responding and nonresponding cells. A similar model for the mechanism of cytokinin activity has been developed by Berridge, Ralph & Letham (5). They could demonstrate a direct relationship between binding of cytokinins to isolated ribosomes and biological activity. Thus, activity of cytokinins seems to be restricted to the level of translation rather than transcription.

Changes in nucleic acid and protein metabolism during bud induction and bud differentiation have been demonstrated earlier (7). The problem of measuring nucleic acids has been approached either with direct photometric measurements (75, 76) or microphoresis technique. Schneider, Lin & Skoog (66) have used a microphoresis technique to determine changes in nucleic acid metabolism associated with spontaneously and hormonally induced bud differentiation. The total RNA content of kinetin-induced bud cells was found to be nearly 15 times that of noninduced protonemal cells (22.0 $\mu\mu$g/cell compared to 1.6 $\mu\mu$g/cell). The same dramatic increase in total RNA was apparent in bud cells which developed spontaneously. The adenine (A) to guanine (G) ratio for DNA from bud and protonemal cells was identical. However, the A:G ratio for RNA from bud cells was much lower than from protonemal cells. The base composition of total RNA of kinetin-induced buds approached that of DNA while the base composition of DNA remained constant (66). As is known from earlier work, cytokinins appear in specific fractions of RNA (16, 79). However, the regulation of differentiation seems to be related to that part of cytokinin which is freely available in the cells (5, 8, 55, 84, 95, 103).

Formation of Antheridia and Archegonia

Hormonal control of antheridia and archegonia formation has been demonstrated in many gametophytes (6, 48). The investigations on antheridia formation led to the hypothesis that in Polypodiaceae antheridium formation is controlled by structurally similar substances (antheridiogens). Prothalli of other ferns (*Anemia* and *Hygodium* family:Schizaeaceae) also produce antheridiogens (48). Recently Endo et al (15) could isolate, purify, and chemically identify antheridiogen from *Anemia phyllitidis*. Spectroscopic and chemical studies on the isolated hormone have led to the conclusion that antheridiogen from *Anemia phyllitidis* (A_{An}) has a diterpenoid structure ($C_{19}H_{22}O_5$) which biogenetically can be derived from a gibberellin skeleton (15). Pure A_{An} is active to a dilution of 10 μg/l in the antheridium formation bioassay; furthermore, it induces dark-germination in *A. phyllitidis* to the very low concentration of 0.3 μg/l. It should be noted that gibberellins also induce antheridia in this species and some antheridiogens are active in some gibberellin specific bioassays (48, 68, 97, 98).

A second *Anemia*-antheridiogen has been isolated by Schraudolf (73). This new antheridiogen is characterized by its Rf-value and a low activity in the dark germination bioassay (spore germination of *Anemia* spores). The structural similarity (15) between A_{An} and gibberellin and the similarities in the activity in the antheridiogen bioassay (68) have led to the hypothesis that antheridiogens and gibberellins are closely related plant hormones (73). There are, however, also some indications that antheridia can be induced by a wide variety of compounds (70), indicating that the hormone induction of antheridia is less specific than the induction of buds on moss protonemata by cytokinins (8, 23). Although the hormonal induction of antheridia has been studied in many species, the control mechanism has not yet been clarified.

There are, however, some indications that antheridiogens (and gibberellins) act as realisators of inactive RNA and cannot be blocked by inhibitors like actinomycin D. This hypothesis has been supported by results obtained with gametophytes of *Polypodium crassifolium*. In this species antheridia can be initiated directly from the spore. Inhibition of antheridia formation cannot be achieved by a treatment with actinomycin D, whereas the spermatogene tissue was inhibited by the same treatment (69). Since actinomycin D has been shown to penetrate the spore before the onset of DNA and RNA synthesis, the missing activity cannot be explained by a lag period in uptake (38).

Very interesting results have been obtained with spores of an aberrant strain of *Onoclea sensibilis*. When the aberrant spores were cultured in darkness under conditions of reduced ethylene concentration, the proportion of two-dimensional plants rose to 75%, and up to 50% of the gametophytes produced antheridia. No antheridia were initiated under comparable conditions on "normal" (non-aberrant) gametophytes (44). Some data on the molecular basis of induction of differentiation has been reported by Jayasekera & Bell (28). By using developing archegonia of *Pteridium aquilinum* and following incorporation of tritiated uridine, these authors could demonstrate two phases of incorporation of the nucleoside into nuclei during oogenesis. The first is interpreted as indication of gene activation initiating oogenesis. The second phase is believed to be concerned with the differentiation of the egg itself. Part of the cytoplasmic RNA located in nucleus-like bodies may represent genetic information remaining untranslated until after fertilization (28).

CONCLUDING REMARKS

Significant progress has been made in our knowledge of the mechanisms controlling gametophyte morphogenesis. By far the greatest advances have been made in the identification and investigation of naturally occurring and synthetic regulators of plant development. The application of new techniques of nucleic acid and protein chemistry, high resolution autoradiography and spectroradiometry, along with the evaluation of specific induction and inhibition of "simple" morphogenetic steps, may rapidly lead to further insights into the processes of plant morphogenesis.

Literature Cited

1. Basile, D. V. 1970. *Science* 170: 1218–20
2. Bauer, L. 1966. *Z. Pflanzenphysiol.* 54:241–53
3. Bell, P. R. 1969. *Z. Zellforsch.* 96: 49–62
4. Bergfeld, R. 1968. *Planta* 81:274–79
5. Berridge, M. V., Ralph, R. K., Letham, D. S. 1970. *Biochem. J.* 119:75–84
6. Bopp, M. 1968. *Ann. Rev. Plant Physiol.* 19:361–80
7. Brandes, H. 1967. *Planta* 74:45–54
8. Brandes, H., Kende, H. 1968. *Plant Physiol.* 43:827–37
9. Burns, R. G., Ingle, J. 1968. *Plant Physiol.* 43:1987–90
10. Davis, B. D. 1968. *Am. J. Bot.* 55: 532–40
11. Davis, B. D. 1968. *Plant Physiol.* 43: 1165–67
12. Davis, B. D. 1968. *Bull. Torrey Bot. Club* 95:31–36
13. Davis, B. D. 1971. *Am. J. Bot.* 58: 212–17
14. Döhren, F.-R. v. 1966. PhD Thesis. Technical Univ., Hanover, Germany
15. Endo, M., Nakanishi, K., Näf, U., McKeon, W., Walker, R. 1972. *Physiol. Plant.* 26:183–85
16. Fox, J. E., Sood, C. K., Buckwalter, B., McChesney, J. D. 1971. *Plant Physiol.* 47:275–81
17. Fredericq, H., De Greef, J. A. 1968. *Physiol. Plant.* 21:346–59
18. Fredericq, H., De Greef, J. 1971. *Biol. Jaarb. Kon. Nat. Gen. Belgium* 39:338–42
19. Giles, K. L. 1971. *Plant Cell Physiol.* 12:447–50
20. Giles, K. L., Taylor, A. O. 1971. *Plant Cell Physiol.* 12:437–45
21. De Greef, J., Butler, W. L., Roth, T. F., Fredericq, H. 1971. *Plant Physiol.* 48:407–12
22. Grotha, R., Stange, L. 1969. *Planta* 86:324–33
23. Hahn, H., Bopp, M. 1968. *Planta* 83:120–23
24. Hartmann, E. 1971. *Planta* 101: 159–65
25. Hillman, W. S. 1967. *Ann. Rev. Plant Physiol.* 18:301–24
26. Ito, M. 1969. *Embryologia* 10:273–83
27. Ito, M. 1970. *Planta* 90:22–31
28. Jayasekera, R. D. E., Bell, P. R. 1971. *Planta* 101:76–87
29. Jordan, W. R., Skoog, F. 1971. *Plant Physiol.* 48:97–99
30. Kato, Y. 1970. *Bot. Gaz.* 131:205–10
31. Klämbt, D. 1967. *Wiss. Z. Univ. Rostock., Math. Naturwiss. Reihe* 16:623–25
32. Klein, B. 1967. *Planta* 73:12–27
33. Knoop, B., Hahn, H., Bopp, M. 1969. *Planta* 88:288–92
34. Lehmann, H., Schulz, D. 1969. *Planta* 85:313–25
35. Loyal, D. S., Paik, P. 1971. *Curr. Sci.* 40:384–85
36. Lüttge, U., Bauer, K. 1968. *Planta* 78:310–20
37. Lüttge, U., Krapf, G. 1968. *Planta* 81:132–39
38. DeMaggio, A. E., Raghavan, V. 1972. *Exp. Cell. Res.* 73:182–86
39. Miller, D. H., Machlis, L. 1968. *Plant Physiol.* 43:714–22
40. Ibid, 723–29
41. Miller, J. H. 1968. *Bot. Rev.* 34: 361–440
42. Miller, J. H. 1968. *Physiol. Plant.* 21:699–710
43. Miller, J. H., Miller, P. M. 1967. *Physiol. Plant.* 20:128–38
44. Miller, J. H., Miller, P. M. 1970. *Am. J. Bot.* 57:1245–48
45. Miller, J. H., Stephani, M. C. 1971. *Physiol. Plant.* 24:264–71
46. Miller, P. M., Sweet, H. C., Miller, J. H. 1970. *Am. J. Bot.* 57:212–17
47. Mohr, H. 1956. *Planta* 46:534–51
48. Näf, U. 1968. *Plant Cell Physiol.* 9:27–33
49. Nebel, B. J. 1968. *Planta* 81:287–302
50. Ibid 1969. 87:170–79
51. Ootaki, T. 1967. *Bot. Mag. Tokyo* 80:1–10
52. Payer, H. D. 1969. *Planta* 86:103–15
53. Payer, H. D., Mohr, H. 1969. *Planta* 86:286–94
54. Payer, H. D., Sotriffer, U., Mohr, H. 1969. *Planta* 85:270–83
55. Phillips, D. A., Torrey, J. G. 1972. *Plant Physiol.* 49:11–15
56. Pray, T. R. 1968. *Am. J. Bot.* 55: 951–60
57. Pryce, R. J. 1971. *Phytochemistry* 10: 2679–85
58. Ibid 1972. 11:1355–64
59. Raghavan, V. 1968. *Planta* 81:38–48
60. Raghavan, V. 1969. *Am. J. Bot.* 56: 871–79
61. Raghavan, V. 1970. *Exp. Cell Res.* 63:341–52
62. Ibid 1971. 65:401–7
63. Raghavan, V. 1971. *Plant Physiol.* 48:100–2

64. Raghavan, V., DeMaggio, A. E. 1971. *Plant Physiol.* 48:82–85
65. Raghavan, V., DeMaggio, A. E. 1971. *Phytochemistry* 10:2583–91
66. Schneider, M. J., Lin, J. C. J., Skoog, F. 1969. *Plant Physiol.* 44: 1207–10
67. Schnarrenberger, C., Mohr, H. 1967. *Planta* 75:114–24
68. Schraudolf, H. 1966. *Planta* 68: 335–52
69. Ibid 1967. 74:123–47
70. Ibid, 188–93
71. Ibid 1967. 76:37–46
72. Schraudolf, H. 1968. *Flora, B* 157: 379–85
73. Schraudolf, H. 1972. *Z. Pflanzenphysiol.* 66:189–91
74. Schraudolf, H., Legler, K. 1969. *Physiol. Plant.* 22:312–18
75. Schulz, R. 1971. *Z. Pflanzenphysiol.* 64:335–49
76. Schulze, L. 1968. *Biophysik* 4:252–65
77. Sigee, D., Bell, P. R. 1968. *Exp. Cell Res.* 49:105–15
78. Sigee, D., Bell, P. R. 1971. *J. Cell. Sci.* 8:467–87
79. Skoog, F., Armstrong, D. J. 1970. *Ann. Rev. Plant Physiol.* 21:359–84
80. Smith, D. L. 1972. *Protoplasma* 74:133–48
81. Smith, D. L., Rogan, P. G. 1970. *New Phytol.* 69:1039–51
82. Sobota, A. E. 1970. *Am. J. Bot.* 57:530–34
83. Sobota, A. E. 1972. *Plant Physiol.* 49:914–18
84. Spiess, L. D., Lippincott, B. B., Lippincott, J. A. 1971. *Am. J. Bot.* 58:726–31
85. Ibid 1972. 59:233–41
86. Stange, L. 1970. *Z. Pflanzenphysiol.* 63:84–90
87. Stange, L., Kleinkauf, H. 1968. *Planta* 80:280–87
88. Steiner, A. M. 1969. *Planta* 86: 343–52
89. Steiner, A. M. 1969. *Photochem. Photobiol.* 9:493–506
90. Ibid, 507–13
91. Sugai, M. 1971. *Plant Cell Physiol.* 12:103–9
92. Sugai, M., Furuya, M. 1968. *Plant Physiol.* 9:671–80
93. Szweykowska, A., Schneider, J., Prusinska, U. 1969. *Acta Soc. Bot. Pol.* 38:139–42
94. Taylor, A. O., Bonner, B. A. 1967. *Plant Physiol.* 42:762–66
95. Tegley, J. R., Witham, F. H., Krasnuk, M. 1971. *Plant Physiol.* 47: 581–85
96. Valanne, N. 1971. *Can. J. Bot.* 49: 547–54
97. Varner, J. E., Chandra, G. Ram 1964. *Proc. Nat. Acad. Sci. USA* 52:100–6
98. Voeller, B. 1964. *Science* 143:373–75
99. Wada, M., Furuya, M. 1970. *Develop. Growth Differentiation* 12: 109–18
100. Wada, M., Furuya, M. 1971. *Planta* 98:177–85
101. Wada, M., Furuya, M. 1972. *Plant Physiol.* 49:110–13
102. Weinberg, E. S., Voeller, B. R. 1969. *Proc. Nat. Acad. Sci. USA* 64:835–42
103. Wood, H. N., Braun, A. C., Brandes, H., Kende, H. 1969. *Proc. Nat. Acad. Sci. USA* 62:349–56
104. Yeoh, O. C., Raghavan, V. 1966. *Plant Physiol.* 41:1739–42

Ann. Rev. Plant Physiol. 1973. 24:129–72

PROTOCHLOROPHYLL AND ❖ 7545
CHLOROPHYLL BIOSYNTHESIS
IN CELL-FREE SYSTEMS
FROM HIGHER PLANTS

Constantin A. Rebeiz
Department of Horticulture, University of Illinois, Urbana-Champaign

and Paul A. Castelfranco
Department of Botany, University of California, Davis

CONTENTS

INTRODUCTION

Investigations about the chemical nature and properties of the green coloring substances found in higher plants were initiated in the nineteenth century. In

1870 Timiryazev postulated a structural relationship between the red pigment of blood and the green pigment of leaves (64). Ten years later the relationship between these two pigments was demonstrated by Hoppe-Seyler (69, 70). Hoppe-Seyler hydrolyzed hemoglobin in concentrated sulfuric acid and isolated a purple pigment with characteristic absorption spectrum. By the action of alkalies and acids, a product with a similar absorption spectrum was isolated from chlorophyll (69, 70). In 1890 Nencki and his collaborators established that porphyrins were made up of pyrrole nuclei (64). Kuster (89) in 1913 proposed a correct formula for the ring system of porphyrins in which four pyrrole rings were linked together into a macrocycle by four methene bridges. Willstätter & Stoll (176) contributed extensively to our knowledge of porphyrins derived from the chemical degradation of chlorophyll and their relationship to the porphyrins derived from hemin. They were first to recognize hemin as an iron salt and chlorophyll as a magnesium salt. In 1929 Fischer and collaborators (36, 37) achieved the complete chemical synthesis of hemin and protoporphyrin as well as a variety of related porphyrins. They also elucidated the nature of isomerism in porphyrins. In 1940 they proposed a structural formula for chlorophyll (36, 37). Final confirmation of the accepted structural formula of chlorophyll was provided when Woodward et al (178) and Strell et al (160) achieved the complete synthesis of chlorophyll a.

In comparison with the immense body of knowledge on pyrrole, porphyrin, and phorbin chemistry, our understanding of how plant tissues synthesize and accumulate chlorophyll is still primitive. The biosynthetic pathway of chlorophyll a in terms of terminal porphyrin and phorbin intermediates was postulated by Granick in 1950 (52). By 1950 Granick (50–52) had isolated protoporphyrin IX, Mg protoporphyrin, and protochlorophyllide[1] from X-ray irradiated *Chlorella* mutants. He had also demonstrated that the protoporphyrin IX isolated from *Chlorella* mutants was identical with the protoporphyin IX of blood heme (50). In 1950 Koski & Smith (87) had purified protochlorophyll[1] from etiolated barley seedlings and determined its spectral absorption properties. This made it possible to demonstrate its conversion in situ to chlorophyll a (Smith 154, Koski 85, Koski et al 86). Granick (52) arranged these porphyrins and phorbins by order of increasing structural complexity into a scheme, starting with protoporphyrin IX and leading to chlorophyll a. He also postulated a biosynthetic and functional relationship between chlorophyll and the iron protoporphyrin of blood (50, 52).

A number of workers contributed to our knowledge of the early steps of porphyrin biosynthesis. In 1952 Westall (175) crystallized porphobilinogen from the urine of a patient with acute porphyria and demonstrated its chemical conversion to uroporphyrin upon boiling over a wide pH range from 1 to 8. By 1952 Shemin and co-workers had traced back all the nitrogen and carbon atoms of the protoporphyrin of heme formed by duck erythrocytes to specific atoms of the substances: glycine and succinate (79). In 1953 Shemin & Russell (148) reported the

---

[1] The terminology proposed by Kirk & Tilney-Basset (84) is followed in this review.

conversion of ALA[2] to protoporphyrin by duck erythrocytes and showed that all the carbon and nitrogen atoms of this protoporphyrin were derived from ALA. They suggested that ALA probably was derived from glycine and succinyl-CoA (148).

In 1953 Bogorad & Granick (15) reported the conversion of porphobilinogen to protoporphyrin by colorless extracts of *Chlorella* cells. Falk et al (33) reported the same conversion in hemolysates of chicken erythrocytes. These experiments (15, 33) rightfully could be considered as the first attempts to study the biosynthesis of heme and chlorophyll at the cell-free level. They pointed out the possible role of porphobilinogen as a natural monopyrrole precursor of protoporphyrin. They also provided experimental evidence in support of Granick's contention that chlorophyll and heme share a common biosynthetic pathway as far as protoporphyrin. That same year (1953) Dresel & Falk (23) reported the conversion of ALA to porphobilinogen by hemolyzed chicken erythrocytes. A year later Granick (54) confirmed the findings of Dresel & Falk and reported the conversion of ALA to porphobilinogen by a colorless aqueous extract prepared from chicken erythrocytes.

Thus, by 1954 a considerable body of evidence suggested (*a*) that red blood cells and *Chlorella* extracts could convert ALA to protoporphyrin via porphobilinogen; (*b*) red blood cells and *Chlorella* probably shared a common biosynthetic pathway up to protoporphyrin IX. Moreover, porphyrin and phorbin intermediates that accumulated in X-ray irradiated *Chlorella* cells were organized into a logical pathway leading from protoporphyrin IX to chlorophyll *a* and involving a plausible series of oxidation-reduction and esterification steps. By analogy, higher plants were assumed a priori to possess the same chlorophyll biosynthetic chain. Thus for a detailed understanding of the step-by-step assembly of chlorophyll in lower and higher plants it became "necessary to isolate the different precursors from normal plants and to demonstrate their stepwise conversion into the next higher member of the sequence" (Löffler 98).

Since 1954 many papers on chlorophyll biosynthesis in lower and higher plants have been published. The field was competently and periodically reviewed (Granick & Mauzerall 62; Smith & French 156; Lascelles 90; Bogorad 12, 13; Granick 58; Jones 77; Marks 99; Kirk 81). In spite of this voluminous literature, the step-by-step enzymology of chlorophyll biosynthesis is still in its infancy. The middle portion of this biosynthetic pathway, i.e. the stepwise conversion of ALA to protoporphyrin IX, has been explored with moderate success in cell-free preparations from higher plants. However, the initial step (ALA biosynthesis) and the last steps (the conversion of protoporphyrin IX to chlorophyll) are still largely unknown. Progress in these two areas has been held back by the lack of cell-free systems able to catalyze these reactions. Cell-free systems from higher plants that are able to convert simple substrates (ALA, glycine, and succinate) to Mg protoporphyrin monoester, protochlorophyll, and chlorophyll have only very recently been described (Rebeiz 120; Rebeiz et al 122; Rebeiz & Castel-

---

[2] Abbreviation: ALA: δ-aminolevulinic acid.

franco 124, 125; Rebeiz et al 128, 130; Wellburn & Wellburn 174). It is hoped that these systems will prove useful in the stepwise elucidation of the initial and terminal portions of the chlorophyll biosynthetic chain.

In the last decade considerable efforts have been devoted to evaluating the reproductive, developmental, and nutritional autonomy of chloroplasts. The control of chlorophyll biosynthesis and the biosynthesis of thylakoid membranes have also received considerable attention. Excellent reviews dealing with these topics have been published (Bogorad 14; Kirk & Tilney-Basset 84; Leech 92; Lascelles 91; Kirk 81, 82; Schiff 136, 137; Goodwin 48; Woodcock & Bogorad 177; Givan & Leech 44). Here too the progress of the field was hindered by the lack of chloroplast preparations "capable of doing in vitro what they can do in vivo" (Woodcock & Bogorad 177). Techniques utilizing chlorophyll accumulation as a marker of chloroplast differentiation in vitro are being developed at present (Rebeiz et al 121, 127–130). In this article we shall consider recent reports of cell-free systems from higher plants, which are capable of carrying out one or more steps in the chlorophyll biosynthetic sequence. In the last part of this article a section will be devoted to chlorophyll accumulation and its relationship to chloroplast maintenance and differentiation in vitro. Figure 1 depicts the structures of the intermediates in the chlorophyll biosynthetic pathway. Roman numerals in the text designate these various intermediates according to Figure 1.

FORMATION OF δ-AMINOLEVULINIC ACID

The evidence that ALA (I) is the first intermediate of the heme and chlorophyll biosynthetic chains is compelling. In 1959 Granick (56) showed that etiolated barley seedlings treated with ALA accumulate large quantities of protochlorophyllide in the dark. That original report has been confirmed in many laboratories. The reader is referred to the comprehensive reviews by Granick & Mauzerall (62) and Bogorad (12, 13) for a summary of this evidence. Additional evidence has been provided recently by the conversion of ^{14}C-ALA to ^{14}C-porphyrins, Mg protoporphyrin monoester, protochlorophyll and chlorophyll a and b by cell-free systems from cucumber cotyledons (122, 124, 125).

In red blood cells from various animal sources, ALA is formed from succinyl CoA and glycine by the reaction:

$$(\text{COOH}) \cdot \text{CH}_2 \cdot \text{CH}_2 \cdot \overset{\times}{\text{C}}\text{O} \cdot \text{SCoA} + \text{H}_2\text{N} \cdot \overset{\square}{\text{C}}\text{H}_2 \cdot \overset{\text{O}}{\text{C}}\text{OOH} \rightarrow$$

$$\text{H}_2\text{N} \cdot \overset{\square}{\text{C}}\text{H}_2 \cdot \overset{\times}{\text{C}}\text{O} \cdot \text{CH}_2 \cdot \text{CH}_2 \cdot \text{COOH} + \overset{\text{O}}{\text{C}}\text{O}_2 + \text{CoASH}$$

This condensation reaction is catalyzed by ALA-synthetase (succinyl CoA—glycine succinyl transferase) in the presence of pyridoxal phosphate. The synthesis of ALA in mammalian red blood cells, the possible origin of succinyl CoA, and the properties of ALA synthetase were reviewed at length by Granick (58). ALA synthetase was purified about eightyfold from the chromatophores of

R. spheroides (Kikuchi et al 80), and a mechanism for the reaction was proposed. The enzyme has eluded isolation from green plants. Multiple efforts to detect it in higher plants have failed. Porra & Irving (116) could not demonstrate the enzyme in either etioplasts or chloroplasts, although the enzyme was detected in cell-free extracts of yeast.

Some indirect evidence has been published recently suggesting that ALA synthetase might be present in green plants. Wellburn & Wellburn (173) isolated *Avena* etioplasts by a novel technique and separated these organelles from other subcellular constituents and bacteria on a Sephadex G-50 column (173). This technique allowed them to isolate intact etioplasts in good yield, with about 0.2% and 0.1% mitochondrial and bacterial contamination respectively. The etioplast preparation was incubated in a medium previously shown to sustain ^{14}C-ALA incorporation into protochlorophyll and chlorophyll by isolated etioplasts (124, 125). Both glycine-2-^{14}C and succinate 2,3-^{14}C were incorporated into chlorophyll *a* (174). This incorporation could not be accounted for by the slight bacterial and mitochondrial contamination, and in the light, this incorporation was enhanced about twentyfold (174). These authors concluded that ALA synthetase is present in the etioplasts of higher plants. However, recent work by Beale and Castelfranco (unpublished) suggests that glycine-1-^{14}C is almost as good a precursor of ALA in greening cucumber cotyledons as glycine-2-^{14}C. This finding, which is inconsistent with the action of succinyl CoA-glycine succinyl transferase, points to the possibility that ALA may be synthesized in higher plants by some other route.

Evidence for the in vivo synthesis of ALA in green plants was also reported by Beale. Beale (4, 5) observed that in the presence of 10 mM levulinic acid, a competitive inhibitor of ALA dehydratase, autotrophically growing cultures of *Chlorella* produce about half the normal amount of chlorophyll and accumulate ALA in amounts equivalent to the deficit of chlorophyll. In order to exclude the possibility that the accumulation of ALA was due to direct amination of the levulinic acid, cultures were grown in the presence of 10 mM ^{14}C levulinic acid. The ALA that was formed by the cells was separated from the levulinic acid by passage through a cation exchange column. The crude ALA fraction contained an amount of radioactivity such that at most 10% of the ALA could have been formed by amination of the levulinic acid.

When cycloheximide was supplied to a culture of *Chlorella* which was accumulating ALA in the presence of levulinic acid, ALA accumulation stopped within one-half hour, whereas chlorophyll synthesis continued for an additional 6 hours at the expense of the ALA that had previously accumulated within the cells. This contrasted with the immediate cessation of chlorophyll synthesis when cycloheximide was supplied to cultures that were not previously incubated with levulinic acid and therefore had low endogenous levels of ALA. Beale concluded from these observations that in *Chlorella* the requirement for protein synthesis during chlorophyll formation is due to the need to synthesize continually a labile enzyme involved in the formation of ALA. Although no conclusions can be drawn regarding the rate of ALA synthesis or the nature of the reaction, the

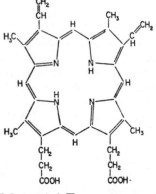

I. δ Aminolevulinic acid

II. Porphobilinogen.

III. Uroporphyrinogen III

IV. Coproporphyrinogen III

V. Protoporphyrinogen IX

VI. Protoporphyrin IX

Figure 1 Some intermediates of the chlorophyll biosynthetic pathway

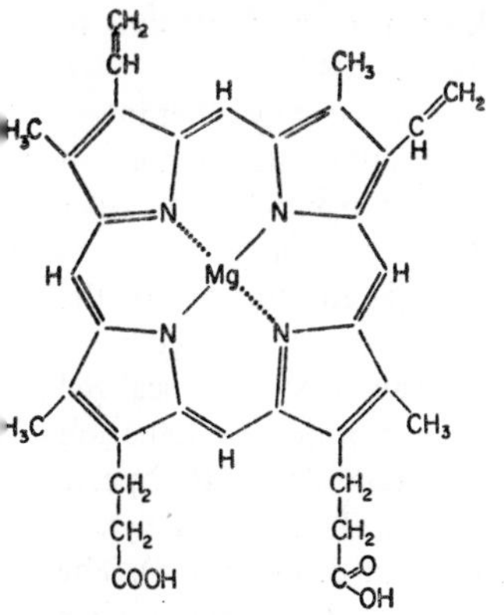

VII. Mg protoporphyrin IX

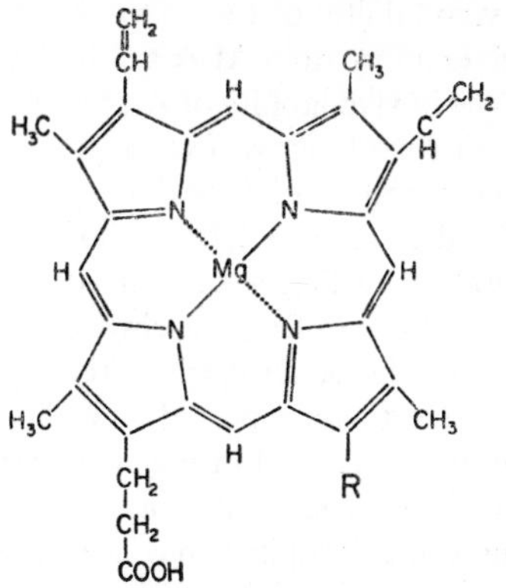

VIII. Mg protoporphyrin IX derivatives

a. R= $CH_2 \cdot CH_2 \cdot COO \cdot CH_3$; Mg protoporphyrin monomethylester

b. R= $CH:CH \cdot COO \cdot CH_3$; Acrylate derivative

c. R= $CHOH \cdot CH_2 \cdot COO \cdot CH_3$; Hydroxypropionate derivative

d. R= $CO \cdot CH_2 \cdot COO \cdot CH_3$; Ketopropionate derivative

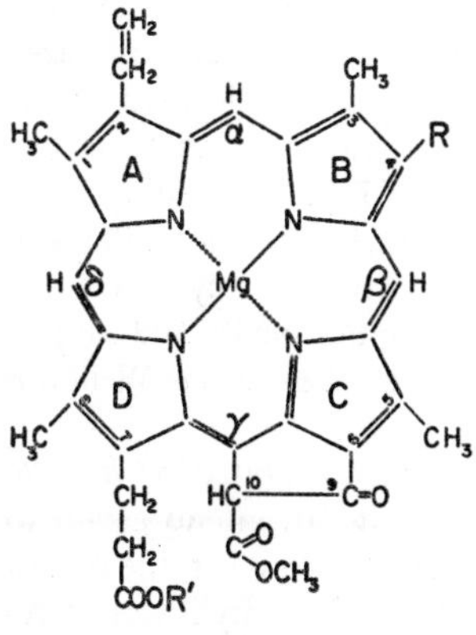

IX. Protochlorophyll derivatives

a. R= $CH:CH_2$; R'= H; Mg 2,4-Divinylpheoporpyrin a_5 (Vinylprotochlorophyllide)

b. R= CH_2-CH_3 ; R'= H; Protochlorophyllide

c. R= CH_2-CH_3 ; R'= $C_{20}H_{39}$; Protochlorophyllide phytyl ester

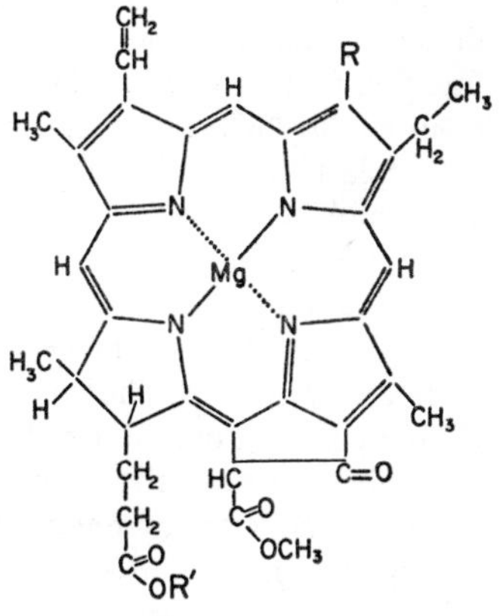

X. Chlorophyll derivatives

a. R= CH_3 ; R'= H; Chlorophyllide

b. R= CH_3 ; R'= $C_{20}H_{39}$; Chlorophyll a

c. R= CHO; R'= $C_{20}H_{39}$; Chlorophyll b

in vivo lability of the enzyme(s) can explain the lack of success of attempts to isolate an in vitro ALA synthesizing system from green plants.

The in vivo lability of ALA synthetase was likewise pointed out by other investigators working with vascular plants. Nadler & Granick (105) estimated a half life of about 1.5 hr for the enzyme in etiolated barley seedlings. Süzer & Sauer (164) determined a half life no longer than 10 min in the same tissue.

Assuming that a short-lived ALA synthetase, catalyzing the transfer of the succinyl group from succinyl CoA to the amino methylene group of glycine, is responsible for ALA synthesis in green plants, and that this process is localized in the plastids as reported by Wellburn & Wellburn (174), the origin of glycine and succinyl CoA used in the condensation reaction must be considered. It appears that chloroplasts contain a substantial amino acid pool (48). The origin of glycine found in chloroplasts has been discussed recently by Givan & Leech (44). The possible origin of chloroplast succinyl CoA has been discussed recently by Kirk (81). Given the availability of succinic acid in chloroplasts, the former could be converted to succinyl CoA by the action of succinyl thiokinase. Indeed, succinyl thiokinase, the enzyme that forms succinyl CoA from succinic acid and coenzyme A, was reported in crude chloroplast pellets isolated by differential centrifugation from wheat leaves (Nandi & Waygood 106). Very recently, however, Kirk & Pyliotis (83) tried to detect this enzyme in wheat and pea chloroplasts without much success. They isolated the chloroplasts by differential centrifugation in two different media and in sucrose gradients without pelleting. In all cases the low succinyl-CoA thiokinase activity could be accounted for in terms of mitochondrial contamination. Since succinyl CoA was mostly present in the mitochondrial fraction, Kirk & Pyliotis (83) suggested that in higher plants ALA is synthesized from succinyl CoA and glycine in the mitochondria and translocated to the chloroplasts for use in chlorophyll biosynthesis. It appears to us that succinyl CoA thiokinase should be looked for in the plastids of rapidly greening tissue before ruling out the presence of this enzyme in plastids. Furthermore, the high concentration of sucrose used in gradients is liable to inactivate many enzymes of the chlorophyll biosynthetic pathway (52, Rebeiz unpublished), and should therefore be avoided. The isolation technique developed by Wellburn & Wellburn (173) seems appropriate for this type of work.

An alternate route to the formation of ALA has been demonstrated in *Chlorella* extracts (57). Twelve amino acids were shown to donate an amino group to γ,δ-dioxovaleric acid (α-ketoglutaraldehyde) to form ALA. However, the role of this enzyme is not clearly understood in green plants as its activity is not correlated with the rate of chlorophyll formation (41, 57). In conclusion, in spite of many efforts, the origin of ALA in higher plants remains a mystery.

BIOSYNTHESIS OF PORPHOBILINOGEN

ALA dehydratase catalyzes the formation of one molecule of porphobilinogen (II) from two molecules of ALA; two molecules of H_2O are eliminated in the process:

$$2 \text{ ALA} \rightarrow \text{porphobilinogen} + 2 \text{ H}_2\text{O}$$

The enzyme has been partially purified from ox liver (Gibson et al 42), rabbit reticulocytes (Granick & Mauzerall 60), and *R. spheroides* (Shemin 147). The properties of these preparations were reviewed by Granick (58). From kinetic studies Granick & Mauzerall (60) suggested a reaction mechanism involving: (*a*) the firm binding of one ALA to the enzyme; (*b*) a looser binding of a second ALA molecule resulting in the spontaneous formation of a ketimine; (*c*) an ensuing aldol condensation; (*d*) and finally a hydrogen shift that results in pyrrole formation.

In green plants the enzyme was detected in gradient purified chloroplasts from *Euglena gracilis* and spinach leaves by Carell & Kahn (18). Nandi & Waygood (107) purified the enzyme from primary leaves of wheat and detected it also in crude chloroplast preparations. As with the animal enzymes, the wheat enzyme required activation by sulfhydryl compounds and was inhibited by EDTA. However, in contrast to the animal enzymes, it required a divalent cation for maximum activation. This observation is compatible with the mechanism proposed by Granick & Mauzerall since the presence of a metal at a β position of one of the ALA molecules should favor the formation of the enolate ion involved in the aldol condensation (60). Pyrophosphate, ATP, and levulinate were found to be inhibitory. The pH optima for the wheat enzyme were 7.5–7.6 in tris-HCl and 7.2–7.8 in phosphate buffer. Michaelis constants of $1 \times 10^{-3} M$, $3.1 \times 10^{-4} M$, and $3.7 \times 10^{-5} M$ were found for ALA, Mg^{++}, and Mn^{++} respectively.

Tigier, Batlle & Locascio (169) purified the enzyme sevenfold from soybean callus tissue by ammonium sulfate precipitation and Sephadex G-100 fractionation. The soybean enzyme exhibited a broad pH optimum between 8.4 and 8.8 and a K_m at pH 8.8 of $3.5 \times 10^{-4} M$ for ALA. It had a high optimal temperature of 55°C. In contrast to the wheat enzyme, the soybean enzyme did not require a thiol activation, although it was inhibited by typical thiol inhibitors. The authors suggested that this might be caused by a different tertiary structure for this enzyme, rendering the disulfide groups inaccessible to thiol activators. This suggestion, however, is unlikely since thiol inhibitors seem to gain easy access to these same functional sulfide groups.

ALA dehydratase has also been detected in crude chloroplast preparations derived from greening tissue cultures of *Kalanchoë crenata* (Stobart & Thomas 158). The enzyme activity increased with increasing chlorophyll content of the cultures but was not linearly related to the chlorophyll content of the culture. In *Euglena gracilis*, Ebbon & Tait (25) detected the enzyme in both pigmented and etiolated cultures. The cellular level of ALA dehydratase tripled when dark-grown cells of *Euglena* were induced to green. When the pigmented cells were placed back in the dark, the level of the dehydratase as well as that of chlorophyll fell off gradually. Growth of the cultures in the dark was not accompanied by increased dehydratase activity.

ALA dehydratase constitutes a good marker for the intracellular localization of porphyrin biosynthesis. Indeed, once porphobilinogen is formed it is unlikely to cross the plastid envelope in either direction (18). *Rhodopseudomonas spheroides* cells were also shown to be impermeable to porphobilinogen (65). There is little doubt that ALA dehydratase is present in chloroplasts. It has been found invariably in this organelle whenever looked for (18, 25, 107, 158). That chloroplasts are the site of porphyrin and phorbin biosynthesis is also amply documented (18, 124, 129). The intrachloroplastic localization of this enzyme, however, is more controversial. Carell & Kahn (18) showed that a substantial proportion of the chloroplastic enzyme remained bound to the membranes of lysed and washed mature *Euglena* chloroplasts. On the other hand, Rebeiz et al reported that most of the free porphyrin biosynthetic activity of greening etioplasts was found in the soluble fraction (supernatant of the centrifugation at 105,000 g) of the lysed plastids (127). The green membrane fraction exhibited very little porphyrin biosynthetic activity when incubated with ALA. These results suggest that in cucumber etioplasts the ALA dehydratase is either in the stroma or very loosely bound to the membranes. This difference in ALA dehydratase behavior between greening cucumber etioplasts and mature *Euglena* chloroplasts may be related to the taxonomic difference or to the difference in the physiological age of the organelles or both.

FORMATION OF PORPHYRINOGENS

Fruitful work on the stepwise biosynthesis of tetrapyrroles was delayed until the reduced nature of the porphyrin intermediates was recognized. Various observations (Bogorad 8, Granick 55) culminated in the work of Neve & Labbe (109), which demonstrated that the actual intermediates in the tetrapyrrole biosynthetic chain were not porphyrins but porphyrinogens. Porphyrinogens are reduced porphyrins, that is, hexahydroporphyrins, which lack a continuous conjugate double bond system.

Biosynthesis of Uroporphyrinogen III

Our knowledge of the enzymology of uroporphyrinogen III (III) formation in higher plants stems mainly from the early work of Bogorad (8–10, 15). It has already been adequately summarized (Bogorad 12, 13; Granick 58). In brief it appears that 4 moles of porphobilinogen are transformed into 1 mole of uroporphyrinogen III by the cooperative action of two enzymes: uroporphyrinogen I synthetase (formerly designated porphobilinogen deaminase) and uroporphyrinogen III cosynthetase (formerly designated uroporphyrinogen isomerase):

$$4 \text{ porphobilinogen} \rightarrow \text{uroporphyrinogen III} + 4 \text{ NH}_3$$

Lockwood & Rimington (97) suggested the convenient trivial name porphobilinogenase to designate this two-enzyme system.

Uroporphyrinogen I synthetase was isolated from aqueous extracts of acetone powder (Bogorad 9). It catalyzes the condensation of 4 moles of porphobilinogen to produce 1 mole or uroporphyrinogen I and 4 moles of ammonia. It is specific for porphobilinogen and requires sulfhydryl reagents for activity (9). Uroporphyrinogen III cosynthetase was isolated from wheat germ (Bogorad 10). This enzyme will not act on uroporphyrinogen I but will convert porphobilinogen to uroporphyrinogen III (and minor amounts of uroporphyrinogen I) in the presence of uroporphyrinogen I synthetase (10). The enzyme is inhibited by hydroxylamine. The joint mechanism of action of these two enzymes has been the subject of much speculation. Bogorad (10, 13) suggested that most probably uroporphyrinogen I synthetase acts on porphobilinogen to produce a noncyclized tripyrrylmethane. The latter and one more molecule of porphobilinogen are then assembled by the uroporphyrinogen III cosynthetase into a molecule of uroporphyrinogen III. A similar porphobilinogenase complex has also been demonstrated in cell-free extracts of *Rhodospyrillum spheroides* (65, 68). Recently, Stevens & Frydman (157) purified uroporphyrinogen III cosynthetase from wheat germ and described some of its properties. There is no doubt that a clear understanding of the coordinated mechanism of these two enzymes depends on the unequivocal identification of the product of uroporphyrinogen I synthetase. Llambias & Batlle (96) purified uroporphyrinogen I synthetase and the cosynthetase from soybean callus and studied their properties in some detail. The two enzymes were resolved into two separate bands on starch-gel electrophoresis. They observed that during incubation of the porphobilinogenase with porphobilinogen, uroporphyrinogen formation occurred after a lag phase of 3 to 4 hours. Subsequently the biosynthesis of uroporphyrinogen increased linearly with time. The uroporphyrinogen formed during the lag phase accounted for only part of the porphobilinogen consumed. The authors concluded that a precursor of uroporphyrinogen must have accumulated during the lag phase (95, 96). They looked for such an intermediate in the incubation mixture and detected a uv absorbing species similar in its absorption properties to monopyrroles. Elution from Sephedex G-15 indicated a probable molecular weight of 600 to 700, suggesting a tripyrrole (95). In the presence of porphobilinogen the intermediate was converted into uroporphyrinogen III by the cosynthetase (95).

The intermediate was labeled by incubating the porphobilinogenase complex with ^{14}C-porphobilinogen. The label was easily separated from the enzyme upon purification on a Sephadex G-25 column. This indicated that the ^{14}C intermediate was probably not tightly bound to the enzyme (96). However, purified uroporphyrinogen I synthetase produced also an intermediate when incubated with porphobilinogen. The authors claim that this intermediate has different properties than the one converted into uroporphyrinogen III. The purification of these intermediates and their rigorous characterization by physical organic techniques is eagerly awaited.

Llambias & Batlle (96) also performed detailed kinetic studies on the porphobilinogenase and partially purified uroporphyrinogen I synthetase. Their data showed a pronounced deviation from Michaelis-Menten kinetics. It is compatible, however, with the presence of two substrate binding sites on the porphobilinogenase. Ammonium ions appear to act competitively at the first binding site (96). From this type of data and inhibition studies Llambias & Batlle (96) suggested a scheme that would account for the kinetics of the formation of uroporphyrinogen I and uroporphyrinogen III. This scheme incorporates features of Bogorad's (13) and Cornford's (20) hypotheses. Essentially it postulates the formation of a uroporphyrinogen I synthetase-porphobilinogen complex which grows into an enzyme-dipyrryl methane complex. Uroporphyrinogen III cosynthetase becomes attached to this complex and catalyzes the addition of two more porphobilinogen molecules, one at a time, to produce uroporphyrinogen III.

Biosynthesis of Coproporphyrinogen III

Uroporphyrinogen III is decarboxylated to coproporphyrinogen III (IV) by the enzyme uroporphyrinogen decarboxylase:

$$\text{Uroporphyrinogen III} \rightarrow \text{Coproporphyrinogen III} + 4\ CO_2$$

The enzyme has not yet been purified from green plants. It is probably present in *Chlorella* since frozen and thawed *Chlorella* preparations do convert uroporphyrinogen III to coproporphyrin (11). Its presence in green plants is undoubtedly ubiquitous since various preparations have been shown to convert porphobilinogen (18) or ALA (122, 127) to copro- and protoporphyrin.

Our knowledge of the enzymology of this reaction is mainly derived from the work of Mauzerall & Granick (102), who isolated their preparation from rabbit reticulocytes. The animal enzyme exhibits a high affinity for its substrate. No intermediates to this reaction have yet been found. The decarboxylation appears to proceed stepwise and at random. Decarboxylation of the four uroporphyrinogen isomers occurs in the following order of decreasing rate: III > IV > II > I(102).

Biosynthesis of Protoporphyrinogen IX

The oxidative decarboxylation of coproporphyrinogen III to protoporphyrinogen IX (V) is catalyzed by coproporphyrinogen oxidase:

$$\text{coproporphyrinogen III} \rightarrow \text{protoporphyrinogen IX} + 2\ CO_2 + 4\ H$$

The enzyme was detected in cell-free extracts of *Euglena* (Granick & Mauzerall (61). It has not yet been isolated from higher plants. It was purified about twentyfold from acetone powders of beef liver mitochondria by Sano & Granick (135). The mechanism of the oxidative decarboxylation has not been completely elucidated. The animal oxidase requires molecular oxygen and attacks mostly coproporphyrinogen III. One intermediate with one vinyl and three propionic residues appears, then disappears. Since the enzyme was found to be inactive toward the porphyrinogens of hematoporphyrin ($CH_3.CHOH.$ at positions 2 and 4 of the macrocycle) of 2,4-diacetyl deuteroporphyrin [$CH_3.(CO).$ at positions 2 and 4] and trans-2,4-diacrylic-deuteroporphyrin [$(COOH).CH:CH.$ at positions 2 and 4], Sano & Granick eliminated the possible involvement of a conventional β-oxidation on the propionic acid residues at position 2 and 4 of the macrocycle (135). Instead they postulated the removal of a hydride ion accompanied by a simultaneous decarboxylation (135).

Recently Sano (134) revised this hypothesis. He synthesized 2,4-bis-(β-hydroxypropionic acid) deuteroporphyrinogen IX [that is, a protoporphyrinogen IX with $(COOH).CHOH.CH_2$. at positions 2 and 4 of the macrocycle] and studied the conversion of this compound under various incubation conditions. Although this porphyrinogen was unstable under acid conditions, it yielded insignificant amounts of protoporphyrin IX (VI) (2.4%) under the standard incubation conditions in the absence of enzyme. In the presence of enzyme the protoporphyrinogen was converted to protoporphyrin IX in 20–22% yield. Based on this observation and the absolute requirement for O_2, Sano proposed a mechanism for the reaction involving the β-oxygenation by molecular oxygen of the 2 and 4 propionic residues of coproporphyrinogen to yield 2,4-β-hydroxy propionic acid deuteroporphyrinogen. This intermediate is then dehydrated with simultaneous decarboxylation to yield protophyrin IX with two vinyl groups at positions 2 and 4 of the macrocycle.

Since deuteroporphyrinogen IX-4-propionic acid is converted by coproporphyrinogen oxidase into the 4-monovinyl derivative (Porra & Falk 115), it seems likely that the enzyme is unable to distinguish between the 2 and 4 positions and should therefore be able to decarboxylate either or both residues. This would explain the appearance and disappearance of the intermediate with three propionic residues (135). Indeed, under both enzymatic and nonenzymatic oxidation of the 2,4-dihydroxypropionicdeuterophorphyrinogen, Sano detected a variety of intermediates with only three propionic residues. More recently, French et al (38) isolated and tentatively identified cis-2 acrylic, 4 propionic deuteroporphyrin [$(COOH).CH:CH$. at position 2 and $(COOH).CH_2.CH_2$. at position 4 of the macrocycle] from Meconium (the accumulated products of development and metabolism of the fetus). Speculating about the role of this compound as an intermediate in the oxidative decarboxylation of coproporphyrinogen, the authors visualize the coproporphyrinogenase reaction as proceeding via an initial desaturation of the 2,4-propionate side chains. The putative diacrylic intermediate combines with the thiol groups of the enzyme to form a thioester from which CO_2 would be lost by decarboxylation. The concepts of Sano (134) and those of French et al (38) are not contradictory. It is not unreasonable to suppose that the O_2 requirement is actually involved in an initial hydroxylation as suggested by Sano (134). The hydroxylated propionic residues are then dehydrated stereospecifically to yield the cis-acrylic intermediate which is decarboxylated to yield protoporphyrin IX, as suggested by French et al (38).

Relationship of Porphyrinogen and Protoporphyrin Formation to Chlorophyll Biosynthesis

Recent results (Rebeiz et al 127) suggest the compartmentation of free porphyrins, Mg porphyrin, and phorbin biosynthetic activity in developing chloroplasts. Rebeiz et al (127) isolated developing chloroplasts from greening cucumber cotyledons by conventional differential centrifugation. The isolation medium was previously shown to sustain protochlorophyll and chlorophyll biosynthesis from ALA in vitro (Rebeiz & Castelfranco 125; Rebeiz et al 128, 130). This incubation medium was also shown to preserve the ultrastructure of the developing chloroplasts during prolonged periods of incubation in vitro (Rebeiz et al 129).

The isolated plastids were lysed and the components separated by centrifugation (105,000 g, 1 hr) into a water-clear soluble protein fraction and a compact green membrane fraction. Incubation of the soluble protein fraction with ALA and cofactors resulted in the production of uroporphyrin, coproporphyrin, and protoporphyrin, probably via their respective porphyrinogens (127). The green membranous fraction was capable of a very limited biosynthetic activity. Upon incubation of the soluble protein fraction with the green particulate fraction in the presence of ALA and cofactors, uro, copro, proto, and metal porphyrins were formed. One of the metal porphyrins was tentatively identified as Mg protoporphyrin monoester (VIIIa). It was concluded that the biosynthesis of free porphyrins, including protoporphyrin from ALA, is catalyzed by soluble enzymes, while the conversion of free porphyrins to metal porphyrins and Mg porphyrin and phorbins requires the membrane fraction (127). A final confirmation of these results with purer plastid preparations and a more thorough characterization of the Mg porphyrin intermediate is now in progress (C. A. Rebeiz unpublished). This suggested compartmentation during active chlorophyll biosynthesis raises several interesting questions about (a) the physical relationships of the enzymic components of the free porphyrin synthesizing machinery to one another, (b) the functional and spatial relationship of the soluble enzyme system to the membrane-bound system, (c) the nature of the last metabolic product of the stromal system (protoporphyrinogen or protoporphyrin) and its relationship to the membrane bound enzymes of the biosynthetic chain. This soluble stromal system is indeed enormously efficient in the sequential handling of ALA and the ensuing intermediates during protoporphyrin formation (127). Therefore it appears possible that the stromal system may exist as an organized macroenzymatic complex. However, the purification data available so far indicate that these enzymes are easily separated from each other by conventional protein fractionation techniques (96, 107, 135, 169). During purification of uroporphyrinogen I synthetase and uroporphyrinogen III cosynthetase, Llambias & Batlle (96) obtained definite evidence that these two enzymes enter into various association-dissociation reactions. These reactions appear to be affected by sodium concentration. During in vitro incubations, uro and coproporphyrins (oxidized porphyrinogen) are also produced. It is not clear whether these byproducts are excreted as porphyrinogens and then oxidized after leaving the enzyme complex or are first oxidized by the enzyme complex and then excreted. We have repeatedly noticed that under certain incubation conditions intact differentiating chloroplasts do not excrete free porphyrins; instead they accumulate chlorophyll for brief periods of incubation (Rebeiz et al 128). This observation derived from intact plastids suggests that the continued normal operation of the chlorophyll biosynthetic chain involves a very delicate physical and functional relationship between the stromal system and the membrane system. Whenever this relationship is disturbed by feeding excess ALA (122, 132), by injuring the membrane system, by withholding some essential cofactors (128), or by disrupting the plastid structure (127), oxidized porphyrins accumulate. The relationship of the stromal and membrane systems to each other is not understood. Is the stromal system very loosely bound to the membranes? Does the membrane system impose a specific conformation on the soluble system? A certain kind of physical proximity between the two systems must be assumed for an efficient transfer of the terminal metabolite of the stromal system to the membranes. However, this relationship need not remain unmodified when chloroplasts reach full maturity and chlorophyll biosynthesis ceases or slows down.

Although the detectable terminal product of the soluble stromal system is protoporphyrin IX, it is not certain that the oxidation of protoporphyrinogen to protoporphyrin is enzymatic. It is also not clear whether protoporphyrin IX or protoporphyrinogen IX is the natural substrate for the membrane-bound enzymes.

FORMATION OF MG PROTOPORPHYRIN MONOMETHYL ESTER

Mg protoporphyrin (VII) and Mg-protoporphyrin monomethyl ester (VIIIa) were isolated from X-ray *Chlorella* mutants and chemically identified by Granick (50, 57). Granick was first to postulate their probable role in the chlorophyll biosynthetic sequence. This possible biochemical function was strongly ascertained by the observation that etiolated barley seedlings accumulated Mg protoporphyrin monomethyl ester (and protochlorophyllide) upon immersion in a solution of ALA. The accumulation of Mg protoporphyrin monoester was also reported by Rebeiz et al (122, 132) in etiolated cucumber cotyledons incubated with ALA. When the etiolated cotyledons were incubated with small amounts of ^{14}C-ALA in the dark, ^{14}C-Mg protoporphyrin monoester was formed and then disappeared as it was metabolized further and most of the label accumulated in protochlorophyll (122, 132).

The next step after the production of protoporphyrinogen is probably the insertion of Mg into protoporphyrin(ogen) IX. Although the insertion of Mg has not been demonstrated as an isolated reaction, the following considerations suggest that the reaction sequence is protoporphyrin→Mg protoporphyrin→ Mg protoporphyrin monomethyl ester, rather than protoporphyrin→protoporphyrin monomethyl ester→Mg protoporphyrin monomethyl ester: (*a*) Mg protoporphyrin has been isolated from higher plants (56); (*b*) Mg protoporphyrin is esterified faster by various methyl transferase preparations than protoporphyrin (25, 31, 119, 165).

In 1970 Rebeiz et al (122) reported the biosynthesis of various free porphyrins and metal porphyrins by cell-free homogenates prepared from etiolated cucumber cotyledons. The major metalloporphyrin fraction was tentatively identified as Mg coproporphyrin. Although the ligand was unambiguously identified as coproporphyrin, no metal analysis was performed on this fraction. It was considered an artifact of incubation (122). Upon incubating the crude homogenates with ^{14}C-ALA, minor amounts of a fast moving ^{14}C-metalloporphyrin were also detected on silica gel H.

This porphyrin was unambiguously identified as a metal derivative of protoporphyrin monoester by chromatography in various solvents. The amounts recovered from silica gel H chromatography were inadequate, however, to permit identification of the metal. Since the ^{14}C-metalloprotoporphyrin monoester cochromatographed with standard Mg protoporphyrin monomethyl ester in several solvents, it was tentatively identified as such (122, 124).

Subsequent efforts were directed at increasing the rate of synthesis of this metal protoporphyrin monoester for the unambiguous identification of the metal. Recently Rebeiz et al (127) reported the accumulation of microgram quantities of this fraction when the stromal enzymes and membrane fractions of developing chloroplasts were incubated together with ALA and cofactors (Mg^{++}, K^+, P_i, CoA, GSH, ATP, NAD, methanol, O_2). After elution from silica gel H, the pooled metal protoporphyrin monoester eluates exhibited an absorption spectrum very similar to Mg protoporphyrin monoester. Other metal porphyrin intermediates were also formed but were not identified. More recently an improved cell-free system consisting of intact differentiating chloroplasts has been described (128). It is capable of faster rates of Mg-protoporphyrin monoester biosynthesis and accumulation. The unambiguous identification of the metal(s) in this fraction is now in progress (Rebeiz, unpublished).

The biosynthesis of Mg protoporphyrin monoester (pending final confirmation of the metal) from ALA in cell-free systems prepared from etiolated and greening cucumber cotyledons appears to be intimately linked to the formation of protochlorophyll (124). Incubation in the absence of cofactors (Mg^{++}, K^+, P_i, CoA, GSH, ATP, NAD, methanol, O_2) depresses the formation of both protochlorophyll and Mg-protoporphyrin monoester. Oxygen (aeration) is absolutely required for this biosynthesis. The biosynthesis from ALA takes place in isolated etioplasts (124) and developing chloroplasts (127). It has a pH optimum of 7.7 in tris-HCl and seems to proceed nearly as well at 20°C as at 28°C. It requires the cooperation of both the stromal and membrane system of the plastids, suggesting that the steps beyond protoporphyrin IX are membrane bound (127).

Although the biosynthesis of Mg protoporphyrin monoester from ALA has been accomplished in vitro, the insertion of Mg into exogenous protoporpoyrin IX to produce Mg protoporphyrin has not been demonstrated as an isolated reaction. The actual substrate for Mg insertion as well as the Mg donor for the reaction are not known. In 1947 Smith (154) reported an increase in Mg in an ether soluble fraction during the early stages of greening of barley seedlings. This observation suggested that the Mg donor to the chelation reaction might be a lipid-Mg complex. However, this finding has not been confirmed in other laboratories.

The tetrapyrrole substrate for the reaction is very likely a protoporphyrinogen or a protoporphyrin complexed to a protein or a lipoprotein carrier. The argument for such a carrier is based on the observation that Mg insertion appears to be a membrane-bound reaction. In such a case a lipoprotein carrier for the protoporphyrin(ogen) may facilitate the access of the substrate to the membrane-bound enzyme. It is not clear whether protoporphyrin IX or the corresponding porphyrinogen is the natural substrate for the chelation. It is generally assumed that the substrate for the reaction is an oxidized porphyrin, namely protoporphyrin IX. This assumption is based on the observation that porphyrinogens do not coordinate with ferrous ions in aqueous solutions at a neutral pH, while porphyrins do (102). The analogy may not apply in this case, since we are dealing with Mg rather than Fe and since a specialized membrane-bound enzymatic process is involved. However, the enzymatic chelation of an oxidized porphyrin appears to offer certain advantages over the chelation of a porphyrinogen from the viewpoint of metabolic regulation. This becomes especially true if the Mg inserting enzyme is specific with respect to the oxidation state of the tetrapyrrole ring, but not with respect to the number of its carboxylic functions. Indeed, the natural tetrapyrrole intermediates are hexahydroporphyrins. Under normal conditions the Mg inserting enzyme is presented with only one porphyrin to chelate, namely protoporphyrin IX; therefore, only one product will be formed in spite of the presence of various porphyrinogens. In addition to Mg protoporphyrin monoester, other metal porphyrins were also formed by the cucumber cell-free systems. These metal porphyrins were not identified. They might be Zn porphyrin complexes, Mg complexes, or mixtures of both. Preliminary absorption spectra rule out ferric and ferrous iron (114, 122, 127, 128). It is not clear whether the formation of these metal porphyrins is enzymatic or not, and if it is, it is not known whether Zn or Mg insertion is catalyzed by the same enzyme that produces Mg protoporphyrin or by different enzymes. If some of the metal porphyrins prove to be Mg complexes of tri and tetracarboxylic acid porphyrins, this would be strong evidence that the Mg inserting enzyme is not specific for protoporphyrin but can chelate other oxidized porphyrins that are artificially produced during in vitro incubations.

Zinc chelatases have been reported in crude extracts of etiolated barley leaves (94) and chromatophores of *Rhodopseudomonas spheroides* (108). Ferrochelatases have also been reported in spinach chloroplasts (Jones 76) and in chloroplasts and etioplasts of bean and oat (Porra & Lascelles 117). The functional relationship of these chelatases to the Mg inserting enzyme is obscure.

Gorchein (49) has recently reported the incorporation of Mg into protoporphyrin by whole cells of *R. spheroides*. He has unambiguously identified the metal porphyrin product as a Mg protoporphyrin monoester. The reaction was enhanced where protoporphyrin was first sonicated with a lipid extract from *R. spheroides* (or ox heart) before addition to a reaction mixture containing a chelator such as EDTA. Magnesium protoporphyrin was not detected as an intermediate. This led Gorchein to suggest that the insertion of Mg is obligatorily coupled with methylation and is catalyzed by a multienzyme complex. Although Gorchein's results do suggest that the substrate for the chelation is protoporphyrin, it is not certain that the protoporphyrin lipid complex was metabolized as such by the cells or first modified and then used in the reaction. A membrane-bound multienzyme system such as the one encountered here is probably so tightly coupled in situ as to obscure the sequence of catalytic events. Therefore, the appearance of Mg protoporphyrin monoester as the major product of the reaction catalyzed by whole cells and the absence of a Mg protoporphyrin intermediate need not conflict with the classical accepted sequence:

protoporphyrin IX → Mg protoporphyrin

→ Mg protoporphyrin monomethyl ester.

The *R. spheroides* system did not respond to added $MgCl_2$ even when the soluble endogenous Mg pool was depleted by dialysis of the cells (49). This raised again the question of the actual nature of the Mg donor to the chelation reaction. The chelation catalyzed by *R. spheroides* cells could not be shown either with *Rhodospirillum rubrum* and *Chromatium* species, *Euglena gracilis* strain Z, or etiolated barley shoots (49).

Esterification of Mg Protoporphyrin

The methyl esterification of exogenous Mg protoporphyrin has been reported by several investigators. The enzyme catalyzing this reaction has been named S-adenosylmethionine-magnesium protoporphyrin methyl transferase (25, 119).

The esterification of Mg protoporphyrin IX to Mg protoporphyrin mono-methyl ester was first reported by Tait & Gibson (165) and Gibson, Neuberger & Tait (43) in chromatophores from *Rhodopseudomonas spheroides*. The incorporation of label from S-adenosyl-L-methionine-methyl-^{14}C with a dilution factor of only 3 indicated that this compound was probably the immediate methyl donor for the reaction (43, 165). Exogenous ^{14}C formate contributed 90% of the chlorophyll methoxy group in *Chlorella vulgaris*. The radioactivity was diluted 42 times (63). In the cell-free system prepared from etiolated cucumber cotyledons, methanol was one of the cofactors needed for the production of Mg protoporphyrin monoester (and protochlorophyll) from ALA (124). Ethanol, isopropanol, and t-butanol could not satisfy this requirement (124). It is not known whether formate and methanol should be regarded as precursors of the S-adenosyl methionine methyl group.

The location of the methyl ester group at the sixth position of the macrocycle has not been unambiguously demonstrated. It is based upon analogy with the known structure of chlorophyll *a* (77). This in turn is based on the assumption that during chlorophyll formation the methyl esterification is accomplished only once at the level of Mg protoporphyrin IX.

Recently Ebbon & Tait (25) investigated the intracellular distribution of the methyl transferase in *Euglena gracilis* and studied its properties after solubilization. The enzyme was found in the chloroplasts. Its apparent presence also outside the chloroplasts is very likely due to its solubilization during chloroplast isolation. The chloroplast enzyme was solubilized by 0.5% Tween 80, suggesting that in situ it is membrane bound in a hydrophobic environment (25). It was purified 68-fold. In the soluble form, the enzyme exhibited a pH optimum of 8.0 in tris-HCl or phosphate buffer and K_m values of 10 μM for Mg-protoporphyrin and 20 μM for S-adenosylmethionine, in the absence of Tween 80. As with chromatophores of *R. spheroides*, the solubilized enzyme methylated protoporphyrin IX only very slowly.

In higher plants Radmer & Bogorad (119) detected the enzyme in chloroplast preparations of greening maize seedlings. Zinc protoporphyrin was less satisfactory as a substrate than Mg protoporphyrin, but better than protoporphyrin IX. More recently Ellsworth & Dullaghan (31) reported a densitometric assay for this enzyme in crude homogenates of etiolated wheat seedlings. The Mg protoporphyrin monoester produced at the end of incubation was converted to protoporphyrin monoester and separated on thin layers of silica gel. The amount of protoporphyrin monoester on the thin layers was estimated densitometrically. Although this assay is simple and rapid, thin layers of silica gel are notorious for their rapid degradation of porphyrins, phorbins, and carotenoids. Ellsworth & Dullaghan reported a pH optimum of 7.7 at 25°C and K_m values of 36 μM for Mg protoporphyrin and 48 μM for S-adenosylmethionine. The enzyme was assayed in the presence of 0.1% Triton X-100. Protoporphyrin IX was methylated at one-tenth the rate of Mg protoporphyrin.

In conclusion, all the evidence available so far derived from R. *spheroides* chromatophores (43, 165), the solubilized *Euglena* enzyme (25), chloroplast preparations (119), and crude homogenates (31) from higher plants, indicate that Mg protoporphyrin is the best substrate for the methylation reaction.

FORMATION OF PROTOCHLOROPHYLL

All the enzymes necessary for the formation of protochlorophyll from added ALA are obviously present in isolated etioplasts (Rebeiz & Castelfranco 124) and developing chloroplasts (Rebeiz et al 128, 130). It was shown that crude cotyledonary homogenates as well as isolated etioplasts were capable of converting ¹⁴C-ALA into ¹⁴C-protochlorophyllide (IXb) and ¹⁴C-protochlorophyllide ester (IXc) (124). Isolated developing chloroplasts were capable of synthesizing and accumulating microgram quantities of protochlorophyll from ALA (Rebeiz et al 128, 130). The formation of protochlorophyll by homogenates was recently confirmed by Shlyk et al (145) in a crude homogenate prepared from greening corn leaves.

The biosynthesis of protochlorophyll from ALA by isolated etioplasts is accomplished in a complex incubation medium. The system has a pH optimum of about 7.7 in Tris-HCl and performs nearly as well at 20°C as at 28°C. For the formation of protochlorophyllide ester, O_2, GSH, methanol, Mg, P_i, and NAD are required. For the formation of protochlorophyllide ATP, CoA and probably K are required in addition to the above cofactors (124). This differential cofactor requirement for the biosynthesis of protochlorophyllide and protochlorophyllide ester was interpreted by Rebeiz & Castelfranco (124) as an indication that protochlorophyllide ester is not formed from protochlorophyllide, but both are formed in parallel from a common precursor, probably Mg protoporphyrin monoester. This hypothesis had been formulated earlier on the basis of kinetic studies of ¹⁴C-ALA incorporation into protochlorophyllide and protochlorophyllide ester in whole etiolated cucumber cotyledons (Rebeiz et al 131, 132). These studies, carried out on seedlings from 1 to 5 days old, had shown that the incorporation of the label into both forms of protochlorophyll was only consistent with biosynthetic models involving the simultaneous production of protochlorophyllide and protochlorophyllide ester in parallel from a common precursor (132).

The complex cofactor requirement for the biosynthesis of protochlorophyll from ALA was tentatively interpreted as follows (124): O_2 as a requirement for coproporphyrinogen oxidase; Mg as a structural component of protochlorophyll and a cofactor for activation reactions; GSH as a general reductant providing a protective effect for functional sulfhydryl groups and porphyrinogens; P_i, as a structural component of membranes. Methyl alcohol was viewed as the methyl donor for the propionic acid residue at the sixth position of the macrocycle. A modified β-oxidation sequence without a thioester intermediate was visualized to take place on the methylated propionic residue. Since the carbon undergoing oxidation is conjugated with the Mg tetrapyrrole system, the removal of a hydrogen α to the macrocycle would be facilitated and could take place in the absence of a thio-ester linkage. NAD was therefore suggested as a cofactor for a β-oxidation type dehydrogenase acting on the methylated propionic residue at the sixth position of the macrocycle and preceding cyclopentanone ring formation.

ATP and CoA were considered to be required for the activation of the free propionic acid residue of protochlorophyllide at position 7 of the macrocycle. This activation was visualized as a prerequisite for positioning this molecule at a specific site of the prolamellar body membranes. Similarly, the fatty ester group of protochlorophyllide ester was considered to position this latter molecule at other sites of the membranes.

Additional understanding of the steps between Mg protoporphyrin monomethyl ester and protochlorophyll was also provided by in vivo investigations. Jones (73) isolated several tetrapyrrole intermediates from *Rhodopseudomonas* cultures to which 8-hydroxy quinoline had been added. One of these accumulated intermediates was identified as 2,4-divinylpheoporphyrin (IXa) [i.e. a protochlorophyllide with two vinyl groups at positions 2 and 4 of the macrocycle (74)]. Later Jones (75) demonstrated that the protochlorophyll of the inner seed coat of *Cucurbita pepo* was actually a mixture of two pigments. One of the pigments was identified as a protochlorophyllide phytyl ester and the other as 2,4-divinylpheoporphyrin a5. This prompted Jones to suggest that the 2,4-divinylprotochlorophyllide was probably an intermediate in chlorophyll biosynthesis in higher plants.

The possible involvement of a β-type oxidation mechanism in the formation of the cyclopentanone ring of phorbins was first suggested by Granick (53). The occurrence of such an activity outside the mitochondria is made more plausible by the discovery of an extramitochondrial β-oxidation system in etiolated peanut cotyledons which breaks down palmityl CoA to acetyl CoA (123, 126). Recently Ellsworth & Aronoff (29, 30) isolated from ultraviolet *Chlorella* mutants Mg protoporphyrin derivatives having the expected structure of the natural intermediates of a β-oxidation sequence, i.e. with methylated acrylate (VIIIb), hydroxy propionate (VIIIc), and ketopyruvate (VIIId) residues at the sixth position of the macrocycle. Individual intermediates or mixtures of intermediates were recovered after chromatography on sucrose columns. Mass and visible spectroscopy suggested that Mg protoporphyrin monoester as well as the hydroxy and keto intermediates contained from zero to two ethyl groups, while protochlorophyllide contained zero or one ethyl group. These results indicate that in these *Chlorella* mutants reduction of the vinyl group at the fourth position of the macrocycle was not specific and could take place at any intermediate step of the β-oxidation sequence. These observations prompted Ellsworth & Aronoff (30) to postulate a pathway beginning at Mg protoporphyrin monoester and leading to the formation in parallel of protochlorophyllide (Mg-vinyl pheoporphyrin a_5) and vinyl protochlorophyllide (Mg-divinyl pheoporphyrin a_5).

Considerable confusion still surrounds the chemical nature of protochlorophyll. It would be important to know whether both protochlorophyllide and protochlorophyllide ester occur in the monovinyl and divinyl forms. The nature of the fatty alcohol at the seventh position of the macrocycle needs also to be ascertained. The alcohol was identified as phytol in the protochlorophyllide ester from pumpkin seed coat (12, 156). The generalization of this observation to the protochlorophyllide ester of leaves has been questioned by Fischer & Bohn (34) and Fischer & Rüdiger (35). Recently we have failed to detect phytol in the protochlorophyllide ester of etiolated cucumber cotyledons by gas chromatographic techniques (Rebeiz, unpublished).

In studying the incorporation of ^{14}C-methanol into protochlorophyll formed from ALA in vitro, it was observed that the protochlorophyllide ester, segregated on silica gel H, was 25 times more radioactive than protochlorophyllide (Rebeiz & Castelfranco 124). This observation is consistent with the previously stated hypothesis that protochlorophyllide ester is not formed from protochlorophyllide but is formed from a common precursor by a parallel series of reactions. Because of the metabolic separateness of the two protochlorophylls, the C_{10} carboxyls could have been methylated by exogenous ^{14}C-methanol with different degrees of dilution in the two compounds. Another explanation is that C^{14}-methanol labels the alcohol in the propionic acid ester which is attached to position 7 of the macrocycle in protochlorophyllide ester. It should be noted also that the nature of the alcohol or alcohols attached to the C-7 propionic acid in protochlorophyllide ester has not been determined and might, in some cases, turn out to be other than phytol.

Recently Rebeiz et al (127, 128) have described cell-free systems which synthesize and accumulate nanomole quantities of Mg-protoporphyrin monoester, protochlorophyll, and chlorophyll. It is hoped that further research with these systems will demonstrate the accumulation of other biosynthetic intermediates in amounts large enough to permit their characterization by physicochemical means and the study of their biosynthesis in single-step reactions. The protochlorophyll formed in vitro showed absorption maxima in ether at 626 nm and true emission maxima at 633–634 nm at room temperature (Rebeiz, unpublished).

The multienzyme complex catalyzing the reactions between Mg protoporphyrin and protochlorophyll appears to be membrane bound. Therefore, it might be necessary to solubilize these enzymes for a study of their catalytic activity in single step reactions. So far no single step reaction between Mg protoporphyrin monoester and protochlorophyll has been studied under physiological conditions. Jones & Ellsworth (72) recently reported the hydrolysis of protopheophytin ester

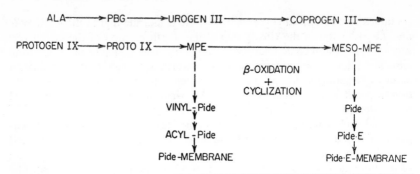

Figure 2 Proposed pathway to membrane-bound protochlorophylls. ALA: δ-amino-levulinic acid; PBG: porphobilinogen; UROGEN III: uroporphyrinogen III; COPROGEN III: coproporphyrinogen III; PROTOGEN IX: protoporphyrinogen IX; PROTO IX: protoporphyrin IX; MPE: Mg protoporphyrin monomethyl ester; MESO-MPE: Mg 2-vinyl-4-ethylprotoporphyrin IX monomethyl ester; VINYL-Pide: 2,4-divinyl pheoporphyrin a_5; Pide: protochlorophyllide; ACYL-Pide: acylprotochlorophyllide; Pide E: protochlorophyllide phytyl ester; Pide-MEMBRANE: membrane bound protochlorophyllide; Pide E. MEMBRANE: membrane bound protochlorophyllide phytyl ester (Rebeiz & Castelfranco 124).

(protochlorophyllide ester without its Mg) by an enzyme preparation from etiolated wheat seedlings. The reaction was performed in 80–90% aqueous acetone. The enzyme appeared to hydrolyze protopheophytin but not pheophytin *a*. Jones & Ellsworth concluded that this enzyme probably catalyzes the phytylation of protochlorophyllide but not of chlorophyllide, in situ. At present this conclusion does not appear to be entirely warranted by the experimental evidence available.

Figure 2 reproduces the working hypothesis of Rebeiz & Castelfranco (124) of protochlorophyll biosynthesis in higher plants. This hypothesis summarizes the knowledge so far acquired from in vivo and in vitro studies and takes into account standard concepts of protochlorophyll biosynthesis as well as the effect of the postulated β-oxidation upon Mg protoporphyrin monomethyl ester and its mesoderivative. No attempts are made to assign divinyl or monovinyl structures to the protochlorophyll species produced since no critical evaluation of the relative percentage of these forms is yet available.

Formation of Protochlorophyll Holochromes

Although we have some insight into the nature of the tetrapyrrole intermediates between protoporphyrin and protochlorophyll, we are completely in the dark as to the nature of the macromolecular carriers with which these chromophores are probably associated during these synthetic steps. It is now certain that in higher plants protochlorophyll must be attached to a protein, the holochrome protein, before the photoreduction of protochlorophyll(ide) to chlorophyll(ide) can take place (7).

The various possible modes of attachment of the protochlorophyll chromophore to the protein carrier have been discussed by Smith & French (156) and Boardman (7). Attachment of about 20–30% of the protochlorophyll, possibly by hydrogen bonding between the electronegative keto oxygen of the cyclopentanone ring and the hydroxyl group of a tyrosine residue of the holochrome, was suggested (7). The possible involvement of tyrosine rather than other amino acids was suggested by the ultraviolet absorption properties of tyrosine and its high pK values (7, 156). The pH values that caused inactivation of the phototransformation and change in the ultraviolet absorption spectrum of the holochrome were close to the pK values of tyrosine.

A 30% phototransformation of protochlorophyllide in etiolated *Chlorella* irradiated with ultraviolet light (280 nm) suggested that about 20–30% of the protochlorophyll was probably attached to a tyrosine residue. Tyrosine supposedly transferred the absorbed excitation energy to the protochlorophyll chromophore, thus making the photoreduction possible (7). However, Björn (6) reported that with purified protochlorophyll holochrome only the radiation absorbed by the protochlorophyll chromophore was effective in the photoreduction. Light absorbed by aromatic amino acid residues of the holochrome was ineffective. Oku & Tamita (113) reported essentially the same observation: irradiation of the protochlorophyll holochrome by ultraviolet light of 253.8 nm did not cause any measurable phototransformation. On the contrary, it caused inactivation of the holochrome. The ultraviolet irradiation also caused alterations to the chlorophyllide holochrome, suggesting that the light sensitivity of protochlorophyll and chlorophyllide holochrome might be different. These observations do not exclude the attachment of the protochlorophyll chromophore to an amino acid residue of the holochrome protein, but they cast some doubts about the transfer of excitation energy from the aromatic amino acid residues of the holochrome to the protochlorophyll chromophore.

Although our fundamental understanding of the protochlorophyll chromophore-holochrome protein interaction is still very primitive, considerable evidence points out that such an interaction is indeed real. Several forms of protochlorophyll holochromes have been reported in the literature. They are usually described by their absorption and fluorescence emission spectra as well as their phototransformability. These parameters allow us to make certain distinctions between the various types of holochromes which have been detected. The various protochlorophyll holochromes have not been physically separated. This has definitely hampered a deeper understanding of their macromolecular characteristics. Only in one instance was a partial separation reported between a phototransformable protochlorophyll holochrome and a nonphototransformable protochlorophyll-protein complex (104).

It has been recognized for some time that in vivo at least two types of protochlorophyll protein complexes are detectable: one is the phototransformable holochrome and absorbs maximally in visible light at about 650 nm; the other is nonphototransformable and shows a red absorption maximum of about 636 nm at room temperature. The evidence for these two forms has been reviewed by several workers (7, 12, 58, 82, 84, 156). Recently, Dujardin & Sironval (24) investigated the pigment protein interactions in the protochlorophyll holochrome both in situ and in vitro. They subjected etiolated bean primary leaves to various treatments: freezing and thawing cycles, extraction in dilute and viscous buffers, infiltration with HCl, and thermal denaturation, and monitored their absorption and emission spectra at 77 K. They were able to show that the protochlorophyll holochrome complex showing a red absorption maximum at 647–648 nm could undergo two kinds of shifts toward the blue: an 8 nm shift from 647–648 nm to 638–640 nm and a 20 nm shift from 647–648 nm to 627–628 nm. The protochlorophyll 638–640 was still phototransformable, while the protochlorophyll 627–628 was not. All these forms yield the same protochlorophyll chromophore upon extraction in an organic solvent. The authors suggested that the protochlorophyll 627–628 (emission at 631 nm) probably represents pigment molecules loosely linked to proteins, since the absorption was very similar to the absorption of protochlorophyll monomers in organic solvents. The protochlorophyll 639–640 (emission at 643 nm) was visualized as a complex with stronger pigment-protein interactions. Protochlorophyll 647–648 (emission at 657 nm) was considered to be an aggregated form involving both pigment-protein and pigment-pigment interactions. Although these conclusions are at best tentative, in part because the various forms of protochlorophyll holochromes were artificially induced, they do raise the possibility that more than two protochlorophyll holochrome forms might occur in vivo.

Evidence for the simultaneous presence of more than two protochlorophyll holochrome forms (protochlorophylls 650 and 636) in situ was reported by Kahn et al (78) and recently reviewed in this series by Kirk (81). Kahn et al (78) investigated the absorption, emission, and excitation (that is, energy transfer) spectra at 77 K of etiolated bean leaves, isolated prolamellar membranes, and protochlorophyll holochrome preparations. They came to the conclusion that protochlorophyll 636, one of the two protochlorophylls usually observed at room temperature in etiolated tissues, is probably made up of two components: a nonphototransformable component that absorbs at 628 nm and fluoresces at 630 nm and a phototransformable, nonfluorescent component, detected only as a 638–639 nm shoulder in excitation spectra of protochlorophyll fluorescence. However, no excitation spectra for the 630 nm fluorescence were shown to confirm that the protochlorophyll 628 nm form does actually fluoresce at 630 nm. The protochlorophyll 638–639 component was not fluorescent. This was ascribed to a transfer of its excitation energy to aggregated protochlorophyll 650. This latter form fluoresced at 655 nm. These results suggested that protochlorophyll 638–639 and protochlorophyll 650 must occur in groups in close proximity to each other to permit excitation energy transfer. Isolated prolamellar body membranes exhibited absorption and emission spectra that resembled the spectra of etiolated leaves, i.e. with substantial protochlorophyll 650 absorption and emission characteristics. However, the isolated holochrome preparations, although still phototransformable, exhibited spectra with a pronounced absorption maximum at 638 nm and a weak shoulder at 650 nm. Excitation spectra for protochlorophyll fluorescence were compatible with the absorption spectra and exhibited a major peak at 638 nm and a shoulder at 648 nm. The fluorescence had a peak at 652 nm and a shoulder at 642 nm. Taken together these results suggest that although most of the energy absorbed by the protochlorophyll 638 was still being transferred to protochlorophyll 650, a certain fraction was being re-emitted at 642 nm by the protochlorophyll 638 form. The difference in spectral characteristics between isolated protochlorophyll holochrome and the protochlorophyll in intact etiolated tissues has been reported by several investigators (24, 138, 139). It definitely indicates that the extraction of the holochrome partially disrupts the organization of the pigment and protein molecules. The reconstitution of the native state in vitro has not yet been achieved.

Additional insight into the possible structure of extracted protochlorophyll holochrome was recently contributed by Schultz (139) and Schultz & Sauer (140). Circular dichroism spectra of greening red kidney beans and barley leaves as well as protochlorophyll and chlorophyllide holochromes were determined. Circular dichroism spectra (CD) can be defined as the differential absorption of left and right circularly polarized light as a function of wavelength. They have the advantage of detecting close pigment associations via the interaction of intermolecular electronic transitions that are too close in energy level to be detected by conventional absorption spectrophotometry. Pigment-pigment interaction normally produces CD spectra whose number of components is equal to or less than the number of interacting molecules in a unit cell of the aggregate (Hochstrasser & Kasha, quoted by Schultz 139). The CD spectra of extracted protochlorophyll holochrome exhibited a minimum at 647 nm, a maximum at 637 nm, and another minimum at 613 nm. By reference to the magnetic circular dichroism spectra, the low energy transition was assigned to 643 nm and the high energy transition to 615 nm. In other words, according to the assignment of these electronic transitions, the low energy transition exhibited a splitting at 647 nm and 637 nm that was detected in the CD spectrum. This in turn suggested that the low energy transition consisted of two identical transitions belonging to two interacting groups of protochlorophyll molecules. In this respect the CD of the protochlorophyll holochrome resembled the CD of protochlorophyll dimers in anhydrous carbon tetrachloride. The authors concluded that the pigment chromophores in the phototransformable photochlorophyll holochrome were probably dimers. The CD spectra of inactive protochlorophyll were less simple. Circular dichroism bands similar to those of active protochlorophyll were detected in the CD spectrum of recently converted chlorophyllide holochrome but were shifted about 6 to 7 nm toward the blue. However, the assignment of the electronic transitions was not as obvious as in the case of transformable protochlorophyll. If the blue-shifted bands were actually those of inactive protochlorophyll, this might be an indication that this inactive protochlorophyll is also aggregated.

Interesting information about the relationship of phototransformable protochlorophylls and inactive protochlorophyll was recently reported by Horton & Leech (71), by Gassman (39), and by Akoyunoglou & Michalopoulos (2). Horton & Leech (71) observed that about 67% of the protochlorophyll 650 in isolated whole etioplasts and lysed etioplasts were converted to nonphototransformable protochlorophyll 630 after 5 hours of incubation in darkness. This conversion was only 24–25% in the presence of ATP and Mg. In another experiment, 90 percent of the protochlorophyll 650 was converted to protochlorophyll 630 by incubating in the dark for 5 hours. Upon addition of ATP and Mg a considerable proportion reverted to the 650 nm form as evidenced by phototransformation to chlorophyllide under a flash regime. These experiments raise the possibility that protochlorophyll 650 might be made up of activated protochlorophyll 630. An activated aggregate might facilitate the various conformational changes, triggered by illumination, via a stepwise dissipation of the trapped activation energy.

A reversible conversion of phototransformable protochlorophyll 650 to inactive protochlorophyll 633 at room temperature in situ was very recently reported by Gassman (39) when primary etiolated bean leaves were exposed for 3 minutes or less to H_2S gas. It was observed that after these exposures to H_2S, 95% of the H_2S-induced protochlorophyll 633 would spontaneously revert to the 650 form. Compounds such as HCN and NH_3 caused an irreversible conversion of protochlorophyll 650 to protochlorophyll 633. Gassman suggests that H_2S may alter the interaction between the hydrogen donor on the protein moiety of the holochrome and the protochlorophyll chromophore by reducing a disulfide bond in the protein, thereby causing a reversible conformational change in the complex.

An irreversible conversion of phototransformable protochlorophylls 650 and 637 to inactive protochlorophyll 628 was reported by Akoyunoglou & Michalopoulos (2). Freeze dried etiolated bean leaves showed the presence of both protochlorophyll 650 and 637. Upon addition of a trace of H_2O to the etiolated freeze dried tissue, both forms were converted into inactive protochlorophyll 628. In the opinion of the authors the addition of a trace of H_2O to the dried tissue was not likely to change the binding sites of the pigment molecules on their protein. The authors concluded that the difference in the position of the absorption peaks of the various protochlorophyll forms is most probably due to different orientations of the protochlorophyll chromophores in the prolamellar body rather than to different bindings on their holochrome apoprotein.

Recently Henningsen & Kahn (66) reported the solubilization of protochlorophyllide holochrome into photoactive subunits by the action of a buffer containing saponin and glycerol.

Sephadex G-100 chromatography suggested a molecular weight of 63,000 for the subunits from barley and 100,000 for those from bean leaves. The photoactive subunits exhibited an absorption maximum at 644 nm and a fluorescence emission maximum at 652 nm. An emission maximum at this wavelength is usually assigned to protochlorophyll 648 (78). However, the reason for the unusually large Stoke's shift between emission and absorption maxima (8 nm) is not known, and neither is the reason why this particular holochrome preparation has an absorption peak at 644 nm, i.e. at a wavelength 6 nm higher than most holochrome preparations (78). Photoconversion of the subunits yielded chlorophyll(ide) with an absorption maximum at 678 nm and a main fluorescence emission at 685 nm. The emission spectra of the converted subunits definitely indicated a pronounced 630 nm fluorescence maximum that could be attributed to inactive protochlorophyll 628. Since preliminary spectrofluoremetric analysis of partially photoconverted preparations gave no indication of energy transfer from protochlorophyll to chlorophyllide, the authors inferred the presence of a single protochlorophyll chromophore per subunit. If such were the case, it must be concluded that the 630 nm fluorescence is due to a population of inactive subunits. This system should be definitely useful for reactivation and reconstitution studies of inactive and photoactive subunits into more complex systems.

Very few data are available concerning the exact nature of the protochlorophyll chromophores in the various protochlorophyll holochromes. Such an insight must probably await the physical separation of the various protochlorophyll holochrome forms in quantities large enough for a detailed chemical characterization.

FORMATION OF CHLOROPHYLL

Reduction of Protochlorophyll Holochrome

Protochlorophyll is the precursor of chlorophyll in greening etiolated tissues (155). In higher plants this transformation requires a specific pigment-protein complex, the protochlorophyll holochrome. Protochlorophyll holochrome is photoreduced by the addition of two trans-hydrogens at the 7 and 8 positions of the macrocycle to produce chlorophyll(ide) (Xa). The photoreduction of protochlorophyll in vivo and in vitro has been reviewed at great length (7, 12, 58, 84, 156).

The phototransformation of protochlorophyllide ester into chlorophyll a (Xb) is still controversial. Godnev et al (45) and Rudolph & Bukatsch (133) had reported such a photoreduction in the past. Sironval et al (153) reported that immediately after a flash of actinic light at room temperature a net increase of about 10% was observed in the protochlorophyllide ester plus chlorophyll a pools. Their data did not indicate, however, whether the increase in esterified pigment was due to an increase in protochlorophyllide ester or chlorophyll a.

We have repeatedly observed in etiolated cucumber cotyledons the photoconversion of protochlorophyllide ester into chlorophyll a (Rebeiz, unpublished). Etiolated cucumber cotyledons containing both protochlorophyllide and protochlorophyllide ester were chilled to -10 to $-15°C$ and illuminated for 2 min with 240 fc of white light. The esterified and nonesterified pigments were segregated on cellulose MN 300 thin layers and their absorption spectra recorded. The esterified pigments exhibited both chlorophyll a and protochlorophyll red absorption maxima. This observation suggested that in etiolated cucumber cotyledons, protochlorophyllide ester is at least partially phototransformable.

Recently Godnev et al (46) revised some of their earlier statements concerning the photoactivity of protochlorophyllide ester. They illuminated etiolated barley sprouts with 628 nm light under various light-dark regimes. They could not establish the conversion of protochlorophyllide ester into chlorophyll a in the 628 nm light. The authors had probably assumed that the protochlorophyllide ester in situ is present only in the 628 nm form, thus the exclusive use of the 628 nm monochromatic light for the phototransformation. Since there is no evidence that this is the case, we must accept the conclusions drawn from this experiment, with some reservation.

Very recently Krasnovski et al (88) were able to achieve the chemical photoreduction of protochlorophyll to chlorophyll with a yield of about 12%. The photoreduction was carried in 0.01 to 0.02 M solutions of ascorbic acid in pyridine at 20°C. These results demonstrated the feasibility of effecting the quantitative photoreduction of the semi-isolated double bonds in solutions of protochlorophyll pigment. It is expected that further studies on this system might reveal more about the mechanism of the photoreduction.

It now appears that the photoreduction of protochlorophyll to chlorophyll(ide) in situ results in the formation of two additional intermediates between protochlorophyll 650 and chlorophyllide 684, exhibiting red absorbing maxima at about 668 and 678 nm. Evidence for the various intermediates occurring in vivo has been reviewed recently by Kirk in this series (81). Thorne (166) has recently confirmed the appearance of these various intermediates in etiolated bean leaves during the early stages of photoconversion. The following summary is that given by Thorne (166): Protochlorophyll (E650, F655)$\xrightarrow{\text{I}}$Intermediate (E668, F674)$\xrightarrow{\text{II}}$ Chlorophyllide (E678, F687)$\xrightarrow{\text{III}}$Chlorophyllide (E682, F694)$\xrightarrow{\text{IV}}$Chlorophyll (E672, F680), where E represents the excitation and F the emission peaks of these intermediates at 77 K. The Roman numerals designate the various shifts or reactions during which the intermediates are interconverted. Intermediate E668, F674 gave chlorophyll after dark ethanol extraction from the leaf. The nature of the intermediates and corresponding shifts will be discussed further in another section.

There is less agreement about the spectroscopic properties of the various intermediates formed during the conversion of extracted protochlorophyll holochrome. Some of the values reported for the red absorption maximum of the first photoproduct detected range from 678 nm to 680–682 nm (7, 138, 140). The reported values for the dark shifted intermediate range from 672–675 nm (7, 138, 140). Obviously more work is needed in this area before a comparison of the intermediates formed in vivo and those formed in vitro can be made. Since the spectroscopic characteristics of extracted holochrome are slightly different from those of protochlorophyll in vivo, some differences in the spectral characteristics of the photoconversion products might also be expected. The importance of understanding the molecular basis of these differences needs no further stressing.

Schultz (139) and Schultz & Sauer (140) reported that the phototransformation of protochlorophyll holochrome in the presence of 2 M sucrose yielded an absorption maximum at 678 nm. The photoreduction was carried in 2 M sucrose in order to prevent the occurrence of the dark shift and accompanying conformational changes. This allowed the study of the spectroscopic characteristics of the first intermediate in the absence of such a shift. The CD spectra showed a splitting of the low energy transition into a positive maximum at 678 nm and a weaker negative component at 687 nm. The high energy transition at 590 nm was not apparently split. These data, by analogy with the CD spectrum of protochlorophyll holochrome, indicated that the chlorophyllide thus formed in 2 M sucrose was also aggregated probably into dimers. Since the 678 to 674 nm dark shift was inhibited, the absorption maximum remained at 678 nm for 1 hr at 0 to 5°C.

Illumination of the protochlorophyll holochrome in the absence of high sucrose concentrations resulted in a chlorophyllide with a red absorption maximum at 677 nm. This absorption maximum started a very rapid dark shift to 674 nm via 677 and 676 nm absorbing forms. The shift was completed in 1 hr at 0 to 5°C and in 15 min at 22°C. The CD spectra obtained after the dark shift was completed resembled those of chlorophyll a monomers; i.e. they showed one unsplit minimum for the low energy transition at 681 nm and another minimum at 640 nm for the high energy transition. The same type of CD spectrum was obtained for homogenates of bean leaves that had received one hour of illumination. The authors concluded from their results that since sucrose blocked the dark shift of chlorophyllide, the CD spectra obtained in 2 M sucrose were representative of the initial photoproduct of the reaction. They concluded that this product was an aggregated form of chlorophyllide. In the absence of 2 M sucrose the dark shift took place very rapidly, and the CD spectra detected only monomeric, nonaggregated chlorophyllide.

More recently Mathis & Sauer (100) were able to stabilize the newly transformed chlorophyllide aggregate at temperatures below 1°C and to study its CD properties in the absence of high sucrose concentration. They came to the conclusion that the photoreduction process probably involved two light reactions. The initial one converted one member of the protochlorophyllide dimer and formed a chlorophyllide-protochlorophyllide holochrome intermediate with a weak CD spectrum. The second light reaction, which was considered temperature dependent, photoconverted the second member of the dimer. The net result was a chlorophyllide holochrome exhibiting the strong CD characteristics of a dimer. High concentrations of sucrose did not interfere with the photoreduction process but prevented the final dissociation into monomers. In the absence of sucrose and at temperatures above freezing the chlorophyllide dimer dissociated very rapidly into chlorophyllide monomers.

Kahn et al (78) and Thorne (166) investigated the state of aggregation of the photoconverted intermediates by spectrofluorimetry. They used etiolated bean leaves or holochrome preparations in which only part of the protochlorophyll had been converted to chlorophyll(ide) in order to monitor the excitation energy transfer between the protochlorophyll not yet transformed and the intermediate photoproducts. Energy transfer was monitored either via excitation spectra or by estimates of fluorescence yield efficiencies at definite excitation wavelengths. The demonstration of energy transfer between protochlorophyll and the various photoreduced intermediates at 10% or 40% photoconversion was interpreted as an indication that the protochlorophyll and the photoproducts were still in sufficiently close proximity to permit resonance energy transfer. The use of liquid nitrogen allowed a better resolution of excitation and emission maxima; it also stopped all photochemical and biochemical activity at the desired states of photoconversion and stabilized the very labile photoproducts so they could be detected. It must be stressed, however, that although experiments performed at 77 K may yield very interesting information, it is not always clear how such information is related to the biophysical state of the pigments at ordinary temperatures.

Kahn et al (78) demonstrated energy transfer between protochlorophyll and newly formed chlorophyll(ide) in partially converted leaves and protochlorophyll holochrome preparations. Thorne (166) demonstrated energy transfer in bean leaves between protochlorophyll and each of the photointermediates mentioned earlier with almost identical efficiency. Taken together these observations suggest that protochlorophyll and the newly formed chlorophyllide were probably existing together in an aggregated state; both CD and spectrofluorimetric studies reached essentially the same conclusion.

The number of protochlorophyll chromophores per aggregate is still a matter of controversy. Schultz & Sauer (140) estimated the pigment aggregate to be essentially a dimer. However, CD spectra might not detect pigment-pigment interactions with very small energy differences. In other words, the dimer suggested by the CD measurements of Schultz & Sauer (140) might well represent the average of two major interactions. On the other hand, Kahn et al (78) suggested a basic aggregate of 5 chromophores per unit, and Thorne (166) suggested an aggregate of 20 chromophores per unit. It therefore appears that for the moment the size of the protochlorophyll and newly formed chlorophyll(ide) aggregates is still an open question.

The presence of protochlorophyll in an aggregated state does not seem to be a prerequisite for the photoreduction to chlorophyllide. This is indicated by the recent findings of Henningsen & Kahn (66), who reported that protochlorophyll(ide) subunits of 60,000 to 100,000 molecular weight and presumably containing one protochlorophyll chromophore per subunit were photoconverted. The newly formed chlorophyllide exhibited red absorption maximum at 678 nm and a corresponding emission maximum at 685 nm. No spectroscopic data were reported for the behavior of the newly formed chlorophyll(ide) when left in the dark.

The observation of a dark shift accompanied by disaggregation of chlorophyllide holochrome in vitro (7, 140), presumably without concomitant phytylation, suggested that these two processes were not obligatorily coupled. This was recently confirmed by Akoyunoglou & Michalopoulos (2). They reported that in etiolated bean leaves the chlorophyllide 682 to chlorophyllide 672 dark shift preceeded that of phytylation.

Recent attempts have been made at defining more accurately the nature of some of the photoconverted intermediates encountered in vivo. Several articles still in press dealing with this topic were brought to our attention (Mathis & Sauer (101), Sironval (149), and Sironval & Kuyper (151)).

The kinetics of the photoconversion process in etiolated bean leaves was recently studied by Thorne & Boardman (168). Taking into account that proto-chlorophyll molecules occur in clusters which allow energy transfer between protochlorophyll and chlorophyll(ide), the true kinetics of the photoconversion was found to be of the first order. This conclusion was recently corroborated by Nielsen & Kahn (110), who found that in protochlorophyll(ide) holochrome sub-units, with no energy transfer between protochlorophyll and chlorophyll(ide) the photoconversion indeed obeyed first order kinetics.

The source of hydrogen for the photoreduction has not yet been elucidated. The photoreduction kinetics seem to favor the hypothesis that the hydrogen donor is in close association with the holochrome protein. Recently Oku & Tomita (112) detected a plastoquinone in close association with the protochlorophyll holochrome in a ratio of 1 plastoquinone to 2 protochlorophyll molecules. They suggest that this plastoquinone might serve as a hydrogen donor in the reaction.

The holochrome protein to which the protochlorophyll chromophore is at-tached has been visualized as a shuttling photoenzyme. Supposedly the photo-enzyme attaches to a photoinactive protochlorophyll 630 molecule to yield proto-chlorophyll 650, and then catalyzes the photoreduction of the chromophore. The chlorophyllide thus formed leaves the photoenzyme and the photoenzyme in turn attaches to another protochlorophyll 630 molecule, and so forth. This concept was suggested by the work of Gassman & Bogorad (40), Sironval et al (152), and Bogorad et al (16). It was emphasized by Sundqvist (161–163) and more recently developed further by Granick & Gassman (59) and Süzer & Sauer (164). Granick & Gassman (59) calculated that in etiolated bean leaves the loading of the photo-enzyme after a flash of light was accomplished with a half time of about 20–50 seconds. Süzer & Sauer (164) estimated a half time of about 20 seconds in etiolated barley. The concept of a shuttling holochrome protein has mainly evolved from observations describing the gradual photoconversion of a pool of "inactive" protochlorophyll by a flashing light regime in the presence of protein synthesis inhibitors (mainly cycloheximide and chlorampehnicol). Since chlorophyllide accumulated under these conditions, it was assumed that the holochrome protein which is necessary for the photoreduction could not have been synthesized de novo in the presence of the inhibitors but must have been used repeatedly in the photoreduction. However, it must be pointed out: (a) It is only an assumption that these protein synthesis inhibitors blocked the synthesis of all plastid proteins in vivo (Bogorad 16). (b) It is not known how fast protein biosynthesis may proceed in vivo under a given set of conditions. Therefore the great rapidity of protochlorophyll 650 regeneration after a flash of light cannot be used as an absolute proof for the absence of the de novo protein biosynthesis in vivo. (c) Recent energy transfer studies by Thorne (167) have shown that although molecular transposition of chlorophyllide does ultimately take place, it does so only after the elimination of the lag phase. During the initial photoconversion process, no transposition or diffusion of chlorophyll(ide) away from the proto-chlorophyllide complex was detected (Thorne 166).

Therefore the concept of the shuttling protochlorophyll holochrome must be still considered an open question.

Biosynthesis of Chlorophyll a

The synthesis of chlorophyll a (Xb) from ALA by crude cell-free homogenates and isolated differentiating chloroplasts has been studied by Rebeiz (120), Rebeiz & Castelfranco (125) and Rebeiz et al (128, 130). ^{14}C-chlorophyll a formed from ^{14}C-ALA was purified to constant specific radioactivity and degraded to ^{14}C-maleimides (125). These results were confirmed by Wellburn & Wellburn (174) with isolated etioplasts from barley. The biosynthesis of chlorophyll a took place in the light in the same incubation mixture which sustained protochlorophyll biosynthesis in the dark (124, 125). Although the direct transformation of added protochlorophyll to chlorophyll a was not achieved, the following observations indicate that chlorophyll a was probably synthesized via protochlorophyll. When the incubation was performed in the dark, protochlorophyll accumulated; when the incubation was performed in the light, chlorophyll a was also found. These results suggest that in either case functional protochlorophyll was formed from ALA, and that in the light, this functional protochlorophyll gave rise to chlorophyll a.

We have already noted that according to spectroscopic evidence, the photoreduction of aggregated protochlorophyll holochrome yields aggregated chlorophyllide which disaggregates rapidly at room temperature. The disaggregated monomeric chlorophyllide exhibits red absorption maxima at 672–674 nm. Esterification of the monomeric chlorophyllide in dark shifted holochrome preparations has not yet been reported. However, some of the protochlorophyllide ester (a few percent) might be converted to chlorophyll a during the photoreduction process.

The relationship between chlorophyllide disaggregation and esterification is not fully understood. However, the following experimental observations have been made: (a) Disaggregation into monomeric chlorophyllide can take place in vitro without accompanying phytylation; this indicates that the latter is not a prerequisite for the former. (b) Both chemical processes seem to be associated with the production of a "reacted" prolamellar body, i.e. with the transition from a highly regular, paracrystalline ultrastructure to a less orderly, more diffuse ultrastructure (Treffrey 170, 171; Murray & Klein 104). That paracrystalline structures are probably made up of pigment aggregates is indicated by recent observations reported by Murray & Klein (104): tissues containing protochlorophyll 650 (supposedly highly aggregated) contained crystalline prolammelar bodies; photoreduction yielded an initial chlorophyllide with the red absorption maximum of aggregated chlorophyll; on the other hand, tissues containing protochlorophyll 635 (not as highly aggregated as protochlorophyll 650) lacked prolamellar bodies; illumination yielded an initial chlorophyllide that exhibited the red absorption maximum of monomeric chlorophyllide.

Recently Treffrey (171) has observed that illumination of etiolated tissues at 0°C, i.e. under conditions that do not favor rapid disaggregation, reduced the protochlorophyll to chlorophyll(ide) ratio but prevented phytylation. The prolamellar bodies remained in the crystalline condition (170). Segregation of the plastid membranes by counter current distribution in a dextran-polyethylene glycol two-phase system revealed that the membranes had a high affinity for the more polar lower phase. When the photoreduction was performed at room temperature under conditions that normally favor disaggregation and phytylation, the membranes had a high affinity for the less polar upper phase. These results indicate that under conditions which favor the production of monomeric chlorophyllide, at least part of the chlorophyllide produced was converted to chlorophyll a and incorporated into stable membrane structures which could be separated by physical means.

Taken together, these observations suggest that (a) the transformation of the prolamellar body from the crystalline to the reacted state might be a prerequisite for the production of monomeric chlorophyll(ide); (b) the incorporation of chlorophyll(ide) into primary lamellae occurs in the monomeric state; (c) phytylation of monomeric chlorophyllide might be a prerequisite for the ordered chlorophyll accumulation in a thylakoid membrane.

It is not certain that the esterifying long chain fatty alcohol found in the chlorophyll of mature green tissues is always phytol. Although most of the esterifying fatty alcohol in chlorophyll a and b is phytol, minor amounts of other alcohols might also esterify certain chlorophyll molecules having specific structural roles. Such minor chlorophyll species might originate in the small amounts of photoreduced protochlorophyllide ester. As mentioned earlier, the esterifying alcohol in these cases may be different from phytol.

Various observations summarized in previous reviews (12, 13, 58) suggest that the production of phytol during greening is tightly coupled to chlorophyll biosynthesis and accumulation. More recent reports by Liljenberg (93), Stobart et al (159), and Brown & Lascelles (17) tend to confirm such observations.

In vivo labeling studies suggest that the immediate precursor of phytol might be geranylgeraniol (a precursor of carotenoids) (Costes 21) and more specifically all-trans geranyl geranylpyrophosphate (48). All the enzymes required for the synthesis of acyclic and cyclic carotenes from mevalonic acid have been demonstrated in plastids of tomato fruits (Porter 118). Chloroplasts isolated by nonaqueous techniques and disrupted by sonication will effectively incorporate 2-^{14}C-mevalonic acid into phytoene in the presence of ATP and Mg^{++} (Goodwin 47), but aqueously isolated chloroplasts will not. Geranylgeranyl pyrophosphate is formed by isolated carrot plastids during in vitro incubation but the hydrogenation of all trans geranylgeranyl pyrophosphate to yield phytol has not yet been reported in vitro (48). The mature plastids found in tomato fruit and carrot root tissues, though specialized for the accumulation of carotenoid pigments and no longer photosynthetic, are nevertheless homologous with chloroplasts.

The enzyme responsible for the phytylation of chlorophyllide is probably membrane bound. However, to the best of our knowledge, no studies have been made on the conversion of exogenous chlorophyllide or chlorophyllide holochrome to chlorophyll bound to plastid membrane fractions. On the other hand, numerous published reports describe the properties of solubilized enzyme preparations named "chlorophyllase."

Chlorophyllase preparations have been assayed under the following conditions: (a) Low organic solvent concentrations: in aqueous medium containing Triton X-100 (McFeeters et al 103). Only the hydrolysis of chlorophyll a, pheaphytin a, chlorophyll b, and pheophytin b were reported. (b) In the presence of high concentrations of acetone: 40–60% aqueous acetone, (Ogura & Takamiya 111, Chiba et al 19, Wellburn 172, Bacon & Holden 3). Two of these studies, Chiba et al (19) and Wellburn (172), reported the esterification of Mg-containing phorbins. (c) Under medium acetone concentration: 10% aqueous acetone (Ellsworth 26, 28).

Because of the extreme conditions used in the in vitro assays, it is difficult to relate these experiments to the actual synthetic reaction operating under physiological conditions. This objection has been raised in many reviews. In the ensuing section we will restrict our discussion to those reports where synthetic activity was observed. Chiba et al (19) extracted chlorophyllase from *Chlorella* cultures and performed the assay in a reaction medium containing about 40% aqueous acetone. Chlorophyll a was formed from methyl chlorophyllide but not from chlorophyllide. Some chlorophyllide was also formed probably from the methyl chlorophyllide. The authors suggested that in situ, chlorophyll a might be formed by transesterification of methyl chlorophyllide with phytol. Hines & Ellsworth (67) isolated small amounts of methyl chlorophyllide from expanding wheat leaves and passion flower leaves. They interpreted their observation in line with the transphytylation of methyl chlorophyllide suggested by Chiba et al (19). Wellburn (172) solubilized a chlorophyllase preparation from *Chlorella* and showed that (in 58.6% aqueous acetone) it esterified methyl chlorophyllide but not chlorophyllide with phytol and geranylgeraniol. Geranylgeranyl pyrophosphate was not a substrate.

Recently Ellsworth (26, 28) confirmed the observations of Chiba et al (19) and Wellburn (172). In addition, he reported that chlorophyllase preparations from wheat seedlings exhibited three types of activities: (a) hydrolysis of pheophytin a, (chlorophyll a without Mg); (b) phytyl esterification of pheophorbide a (chlorophyll without Mg and without phytol); (c) phytyl transesterification of methyl pheophorbide a. These activities were reported to have different optima under different acetone concentrations, buffer concentrations, pH, and temperature (30). In addition, slightly different elution profiles were noticed on different types of Sephadex (26). However, these three activities were not successfully separated from one another (26). Bacon & Holden (3) point out that during purification, chlorophyllase preparations have the tendency to enter reversible associations or aggregations [a pronounced property of structural proteins (Criddle 22)]. These association phenomena might explain the detection of more than one activity in Ellsworth's preparations.

In conclusion, on the basis of these in vitro studies, it appears possible that methyl chlorophyllide is an intermediate during the phytylation of chlorophyllide. The nature of the enzyme(s) catalyzing this process require further study under more physiological conditions.

Biosynthesis of chlorophyll b

The biosynthesis of chlorophyll *b* (Xc) was recently reviewed in this series by Shlyk (141).

The biosynthesis of ^{14}C-chlorophyll *b* from ^{14}C-ALA in vitro was reported by Rebeiz & Castelfranco (125). It was accomplished in cell-free crude homogenates prepared from greening cucumber cotyledons that had received 4.5 hr of white fluorescent light (240 fc at 28°C). The biosynthesis was carried out in the presence of the same cofactors that were shown to be essential for protochlorophyll and chlorophyll *a* biosynthesis. Cell-free homogenates prepared from etiolated cotyledons were capable of ^{14}C-chlorophyll *a* but not ^{14}C-chlorophyll *b* formation. Once the capacity for chlorophyll *b* biosynthesis was acquired by prior illumination of the tissue, it was maintained in the cell-free homogenate. The conversion of added chlorophyll *a* into chlorophyll *b* in this system was not attempted.

It now appears that chlorophyll *a* can be converted into chlorophyll *b* in vitro, in the dark. This conversion was first reported by Shlyk & Prudnikova in 1965 (142). Green barley shoots were allowed to incorporate ^{14}CO$_2$ under 5000–9000 lux. They were ground in phosphate buffer at pH 7–7.2 and the suspension was centrifuged. The precipitate was stirred into a saturated solution of sucrose and incubated in the dark. The specific activity of the chlorophyll *b* increased 1.5 times. The authors concluded that this was due to the formation of chlorophyll *b* from chlorophyll *a*. It should be noted that such a rise in specific radioactivity of chlorophyll *b* could also be caused by exchange of the ^{14}C phytol between high specific radioactivity chlorophyll *a* and lower specific radioactivity chlorophyll *b* (Ellsworth et al 32). In later experiments Shlyk & Prudnikova (143) showed that when exogenous unlabeled chlorophyll *a* was added to the reaction mixture, the increase in specific radioactivity of chlorophyll *b* was lowered. This was interpreted as an indication that the production of ^{14}C-chlorophyll *b* from ^{14}C-chlorophyll *a* was now diluted by the addition of unlabeled chlorophyll *a*.

The reports of Shlyk & Prudnikova were recently confirmed by Ellsworth et al (32). This group reported quantitative data for the conversion of ^{14}C-chlorophyll *a* into ^{14}C-chlorophyll *b*. The ^{14}C-chlorophyll *b* formed in vitro was degraded to pheophytin *b* and Rhodin *g* of constant specific radioactivity. The conversion of chlorophyll *a* into chlorophyll *b* proceeded best when NADP was added to the reaction mixture. The enzymatic activity tolerated high acetone concentrations and could not be destroyed completely by boiling for 5 minutes. During the incubation there was some indication that two processes occurred simultaneously in the homogenates: the transphytylation between labeled chlorophyll *a* and unlabeled chlorophyll *b*, and the biochemical conversion of the phorbin moiety of chlorophyll *a* to that of chlorophyll *b*. The enzyme has not been isolated and the nature of the enzymatic reaction oxidizing the methyl group at position 3 of the macrocycle to a formyl group is therefore still unknown.

More recently Perkins & Liu (quoted by Ellsworth 27) reported that the in vitro homogenates can also transform (at a slow rate) pheophytin *a* into pheophytin *b* and chlorophyll *b* and pheophytin *b* into chlorophyll *a* and pheophytin *a*. This observation, if confirmed, raises again the question of chlorophyll *b* conversion into chlorophyll *a*. The publication of the details of these experiments will be of great interest.

Very recently Shlyk et al (145) reported that chlorophyll *b* accumulated in microgram quantities in a crude homogenate prepared from greening corn leaves. The accumulation took place in the absence of added ALA. Chlorophyll *b* was detected spectrofluorimetrically in crude ether extracts prepared before and after the incubation; in this solvent, chlorophyll *b* exhibited a peak at 655 nm. However, in order to confirm this important observation, it must be shown that the species which fluoresces at 655 nm in this mixture is indeed chlorophyll *b* and not a degradation product of prolonged incubation.

It is not clear yet whether in situ chlorophyll *b* is formed from chlorophyll(ide) *a* before, during, or after incorporation into the thylakoid. Shlyk (146) has suggested that chlorophyll *b* is formed from newly formed (young) chlorophyll *a* molecules at specific sites named "reaction centers." However, it is not clear in Shlyk's hypothesis whether the chlorophyll *b* molecules would be subsequently translocated to other sites or not.

CHLOROPHYLL ACCUMULATION IN VITRO

Chlorophyll accumulation is defined here as the incorporation of chlorophyll *a* and *b* into stable, insoluble, functional structures, the thylakoid membranes. This definition, therefore, explicitly relates chlorophyll accumulation and thylakoid membrane assembly during chloroplast differentiation. Although in vivo both processes are obviously related, it is our intent to distinguish between the biosynthesis of a chlorophyll molecule and the process of incorporation of many such molecules into an ordered, stable, functional, insoluble structure.

A considerable amount of research has dealt with the various factors affecting chlorophyll accumulation in vivo: namely, nutritional deficiencies; effect of inhibitors, antibiotics, growth regulators; age of the tissue; light quality and intensity; alternating light-dark regimes; temperature; mutations; etc. For a review, the monograph by Kirk & Tilney-Basset is recommended (84). More recently the developmental (48, 81), reproductive (177), and nutritional (44) aspects of chloroplast autonomy, as well as the process of thylakoid membrane assembly (82), have received a good deal of attention. It is definitely beyond the scope of this review to discuss these processes in detail; however, we must point out that our fundamental understanding of all these phenomena suffers because of the difficulty of studying these processes in vitro, i.e. in the absence of extrachloroplastic influences. We concur with Woodcock & Bogorad (177) that "the problem is clear: plastids capable of doing in vitro what they can do in vivo need to be prepared." As a beginning toward this goal of a wholly self-sufficient plastid preparation, we would like to describe a system which is presently under investigation. This system is capable of sustaining chlorophyll accumulation and partial chloroplast differentiation in vitro (Rebeiz et al 128, 129).

Homogenates were prepared in tris-sucrose from etiolated and greening cucumber cotyledons, a tissue which had been thoroughly studied with respect to its ability to accumulate protochlorophyll and chlorophyll in vivo (120, 131, 132). Tissue homogenates and isolated plastids were incubated aerobically in the dark or in the light at pH 7.7 in the presence or absence of CoA, GSH, K, P_i, methanol, Mg, NAD, ATP. This medium was previously shown to sustain protochlorophyll and chlorophyll biosynthesis in vitro (124, 125). Ultrastructural changes were followed by electron microscopy (129). Chlorophyll, protochlorophyll, metal porphyrins, Mg protoporphyrin monoester, and free porphyrins were monitored spectrophotometrically and spectrofluorometrically (128; Rebeiz unpublished). It was observed that the plastids were destroyed very rapidly when incubated in the presence of all the other cellular constituents. However, rapid separation of the plastids from the rest of the cellular constituents greatly improved the maintenance of the plastids upon subsequent incubation (129). This indicated that crude homogenates contained factors responsible for plastid degradation. The cofactors previously shown to be essential for chlorophyll biosynthesis were required for the maintenance of the membrane structures during subsequent in vitro incubations. It was observed that the outer and inner plastid membranes were more fragile in etioplasts than in developing chloroplasts. When developing chloroplasts, rapidly isolated from greening cotyledons, were incubated with the cofactors in subdued light for 16 hr, their original structure was apparently perfectly well preserved and a limited extent of new grana growth could be observed (Figure 3; 129). Their prolamellar bodies at the end of the incubation looked "reacted," while whole cotyledons incubated in H_2O under the same temperature and light conditions for 16 hours contained paracrystalline prolamellar bodies.

The chlorophyll biosynthetic pathway was monitored at the level of several intermediates during incubations which achieved good structural stabilization and partial chloroplast differentiation (Rebeiz et al 128). In the presence of ALA and cofactors, microgram quantities of protochlorophyll and chlorophyll accumulated after 1 hr of incubation. Part of the accumulated chlorophyll was not extractable into organic solvents; it had to be estimated spectrofluorometrically in the precipitate obtained upon addition of acetone. We believe that this "bound chlorophyll" (128) may differ from "extractable chlorophyll" in terms of its association with the plastid membrane proteins in their native condition. The relative amount of bound chlorophyll formed before and after incubation depended on the extent of greening of the tissue from which the plastid preparation is obtained (128). This observation suggested that the accumulation of chlorophyll in situ and in vitro may be coupled to the biosynthesis of membrane-bound chlorophyll receptor sites.

Four types of plastid preparations were studied (128): (a) etioplasts isolated from etiolated cotyledons having a lag phase of chlorophyll biosynthesis; (b) etioplasts isolated from cotyledons in which the lag phase had been removed by preirradiation and dark incubation; (c) etioplasts isolated from cotyledons that were still in the lag phase but had received continuous irradiation for various periods of time; (d) developing chloroplasts isolated from cucumber cotyledons which had received 4.5 hours or more of continuous light. All etioplast preparations

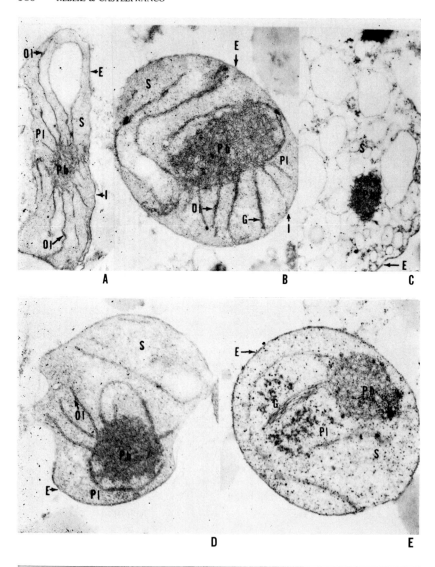

Figure 3 Etioplasts, isolated from cotyledons after irradiation for 2.5 hr and incubation in low light intensity for 16 hr. A, in situ etioplast; B, without cofactors, zero time; C, without cofactors, after 16 hr incubation; D, with cofactors, zero time; E, with cofactors, after 16 hr incubation. E, envelope; G, granum; I, invagination and/or interaction of inner component of envelope with peripheral lamellae; 01, overlapping of membranes in peripheral lamellae, possibly incipient grana; pb, prolamellar body, crystalline or reacted state; pl, peripheral lamellae; S, stroma (Rebeiz et al 129).

under subdued light accumulated nonphototransformable protochlorophyll and small amounts of extractable chlorophyll but little or no bound chlorophyll. Developing chloroplasts, on the other hand, accumulated both bound and extractable chlorophyll and lesser amounts of protochlorophyll. These results were interpreted as follows: (a) The biosynthesis and accumulation of chlorophyll in vitro appears to be coupled to the biosynthesis of membrane-bound chlorophyll receptor sites. (b) This inability of etioplasts to carry out continued phototransformation under these conditions may be due to: (i) insufficient quantity of reusable holochrome; (ii) insufficient quantity of reductant; (iii) the possibility that the holochrome is not a reusable catalyst but a protein which functions once in the phototransformation of protochlorophyll and then becomes an integral part of the chlorophyll receptor site.

When the differentiating plastids were incubated with ALA in subdued light in the absence of cofactors, chlorophyll accumulation did not take place, only porphyrins accumulated (128). After 16 hours of incubation, the results were not so clear cut. In the presence of cofactors, much of the chlorophyll produced was degraded and considerable quantities of pheophytin a and pheophorbide accumulated. Free porphyrins, metal porphyrins, Mg protoporphyrin monomethyl ester, protochlorophyll, and protopheophytin were also formed (Rebeiz, unpublished). Sixteen-hour incubation of crude homogenates with ALA in the absence of cofactors were previously shown to produce free porphyrins, metal porphyrins, a trace of Mg protoporphyrin monoester, but no protochlorophyll or chlorophyll (122).

The protochlorophyll accumulated in all these cell-free preparations exhibited an absorption maximum in ether at 626 nm. In the aqueous plastid suspension, it exhibited a fluorescence maximum at 636–637 nm at room temperature and at 632–633 nm, at 77 K (128). In these studies with developing chloroplasts it was noticed that the population as a whole was not synchronized (129): some plastids would show new grana formation in vitro, while others would begin to degenerate. The degenerating plastids would accumulate large amounts of nonphototransformable protochlorophyll, metal porphyrins, free porphyrins and their chlorophyll would begin to degrade (Rebeiz, unpublished).

Although the complete differentiation of etioplasts into chloroplasts in vitro has not yet been achieved, these preliminary studies indicate this can now be considered a realistic goal for future research.

CONCLUDING REMARKS

1. Cell-free studies indicate that the synthesis of chlorophyll from ALA is an enzymatic cofactor and energy requiring process localized in the etioplasts and developing chloroplasts. New techniques, more appropriate for the study of solid state biochemistry, need to be developed for the stepwise study of the terminal steps of chlorophyll biosynthesis.

2. Cell-free studies confirm (58) that the control of ALA production [be it by respression (91) or inhibition of ALA synthesis] plays a major regulatory role in

the production of nontransformable protochlorophyll. The availability of excessive amounts of ALA results in the accumulation of nontransformable protochlorophyll by the isolated plastids.

3. Separation of the plastid structure into a soluble stroma and an insoluble membrane system brought into focus the regulatory role of the membranous component in the process of protochlorophyll biosynthesis.

4. Cell-free studies of chlorophyll accumulation suggest that the latter is intimately related to structural protein biosynthesis and membrane assembly in vitro. It is not clear whether the chlorophyll receptor (binding) sites are all alike or are heterogeneous. Their relationship to chlorophyll biosynthetic "modes I and II" observed under flashing light-dark regimes is not clear (1, 150). Neither is their relationship to Shlyk's "reaction centers" (144, 146).

5. Cell-free studies, as well as other considerations, suggest that the concept of a shuttling protochlorophyll holochrome photoenzyme may have to be revised.

6. The number of variables affecting chlorophyll accumulation is so large that an adequate quantitative description of this complex process, although desirable, is not feasible at present.

ACKNOWLEDGMENT

We are greatly indebted to Mrs. Carole C. Rebeiz and Dr. Samuel I. Beale for reading the manuscript and making many valuable suggestions.

Literature Cited

1. Akoyunoglou, G., Argyroudi-Akoyunoglou, J. H., Michel-Wolwertz, M. R., Sironval, C. 1966. *Physiol. Plant.* 19:1101–4
2. Akoyunoglou, G., Michalopoulos, G. 1971. *Physiol. Plant.* 25:324–29
3. Bacon, M. F., Holden, M. 1970. *Phytochemistry* 9:115–25
4. Beale, S. I. 1970. *Plant Physiol.* 45:504–6
5. Ibid 1971. 48:316–19
6. Björn, L. O. 1969. In *Progress in Photosynthesis Research*, ed. H. Metzner, 2:618–29. Tübingen: Chem. Pflanzenphysiol.
7. Boardman, N. K. 1966. In *The Chlorophylls*, ed. L. P. Vernon, G. R. Seely, 437–79. New York: Academic
8. Bogorad, L. 1955. *Science* 121:878–79
9. Bogorad, L. 1958. *J. Biol. Chem.* 233:501–9
10. Ibid, 510–15
11. Ibid, 516–19
12. Bogorad, L. 1965. In *Chemistry and Biochemistry of Plant Pigments*, ed. T. W. Goodwin, 3–74. New York: Academic
13. Bogorad, L. 1966. See Ref. 7, 481–510
14. Bogorad, L. 1967. *26th Symp. Soc. Develop. Biol. Suppl.* 1:1–31. London: Academic
15. Bogorad, L., Granick, S. 1953. *Proc. Nat. Acad. Sci. USA* 39:1176–88
16. Bogorad, L., Laber, L., Gassman, M. 1968. In *Comparative Biochemistry and Biophysics of Photosynthesis*, ed. K. Shibata, A. Takamiya, A. T. Jagendorf, R. C. Fuller, 299–311. Univ. Tokyo Press
17. Brown, A. E., Lascelles, J. In press
18. Carell, E. F., Kahn, J. S. 1964. *Arch. Biochem. Biophys.* 108:1–6
19. Chiba, Y. et al 1967. *Plant Cell Physiol.* 8:623–35
20. Cornford, P. 1964. *Biochem. J.* 91:64–73
21. Costes, C. 1966. *Phytochemistry* 5:311–24
22. Criddle, R. S. 1969. *Ann. Rev. Plant Physiol.* 20:239–52
23. Dresel, E. J. B., Falk, J. E. 1953. *Nature* 172:1185
24. Dujardin, E., Sironval, C. 1970. *Photosynthetica* 4:129–38
25. Ebbon, J. G., Tait, G. H. 1969. *Biochem. J.* 111:573–82
26. Ellsworth, R. K. 1972. *Photosynthetica* 6:32–40
27. Ellsworth, R. K. 1972. In *The Chemistry of Plant Pigments*, ed. C. D. Chichester, 85–102. New York: Academic
28. Ellsworth, R. K. In press
29. Ellsworth, R. K., Aronoff, S. 1968. *Arch. Biochem. Biophys.* 125:269–79
30. Ibid 1969. 130:374–83
31. Ellsworth, R. K., Dullaghan, J. P. 1972. *Biochim. Biophys. Acta* 268:327–33
32. Ellsworth, R. K., Perkins, H. J., Detwiller, J. P., Liu, K. 1970. *Biochim. Biophys. Acta* 223:275–80
33. Falk, J. E., Dresel, E. J. B., Rimington, C. 1953. *Nature* 172:292–94
34. Fischer, F. G., Bohn, H. 1958. *Ann. Chem.* 611:224–35
35. Fischer, F. G., Rudiger, W. 1959. *Ann. Chem.* 27:35–46
36. Fischer, H., Orth, H. 1940. *Die Chemie des Pyrrols. II. Pyrrolfarbstoffe, Part 1.* Ann Arbor: Edwards
37. Ibid 1943. *Part 2*
38. French, J., Nicholson, D. C., Rimington, C. 1970. *Biochem. J.*, 120:393–97
39. Gassman, M. L. In press
40. Gassman, M., Bogorad, L. 1967. *Plant Physiol.* 42:781–84
41. Gassman, M., Pluscec, J., Bogorad, L. 1968. *Plant Physiol.* 43:1411–14
42. Gibson, K. D., Neuberger, A., Scott, J. J. 1955. *Biochem. J.* 61:618–29
43. Gibson, K. D., Neuberger, A., Tait, G. H. 1963. *Biochem. J.* 88:325–34
44. Givan, C. V., Leech, R. M. 1971. *Biol. Rev.* 46:409–28
45. Godnev, T. N., Akulovich, N. K., Khodosevich, E. V. 1963. *Dokl. Akad. Nauk SSSR* 150:920–23
46. Godnev, T. N., Galaktionov, S. G., Raskin, V. I. 1968. *Dokl. Akad. Nauk SSSR* 181:237–40
47. Goodwin, T. W. 1969. See Ref. 6, 2:669–74
48. Goodwin, T. W. 1971. In *Structure and Function of Chloroplasts*, ed. M. Gibbs, 215–76. New York: Springer-Verlag
49. Gorchein, A. 1972. *Biochem. J.* 127:97–106
50. Granick, S. 1948. *J. Biol. Chem.* 172:717–27
51. Ibid 1948. 175:333–42
52. Ibid 1950. 183:713–30
53. Granick, S. 1950. *Harvey Lect.*

Series XLIV 1948–1949:220. Springfield: Thomas
54. Granick, S. 1954. *Science* 120:1105–6
55. Granick, S. 1955. *Am. Chem. Soc. 128th Meet.*, p. 69C
56. Granick, S. 1959. *Plant Physiol.* 34: XVIII
57. Granick, S. 1961. *J. Biol. Chem.* 236: 1168–72
58. Granick, S. 1967. In *Biochemistry of Chloroplasts*, ed. T. W. Goodwin, 2:373–410. New York: Academic
59. Granick, S., Gassman, M. 1970. *Plant Physiol.* 45:201–5
60. Granick, S., Mauzerall, D. 1958. *J. Biol. Chem.* 233:1119–40
61. Granick, S., Mauzerall, D. 1958. *Fed. Proc.* 17:233
62. Granick, S., Mauzerall, D. 1961. In *Chemical Pathways of Metabolism*, ed. D. Greenberg, 2:525–616. New York: Academic
63. Green, M., Altman, K. I., Richmond, J. E., Salomon, K. 1957. *Nature* 179:375
64. Gurinovich, G. P., Sevchenko, A. N., Soloniev, K. N. 1968. *Spectroscopy of Chlorophyll and Related Compounds*, 1–52. Minsk: Izdatel'stvo Nauka i Tekknika. 520 pp. Transl. No. AEC tr-7199 available from Nat. Tech. Inform. Serv., U.S. Dep. Commerce, Springfield, Va.
65. Heath, H., Hoare, D. S. 1959. *Biochem. J.* 72:14–22
66. Henningsen, K. W., Kahn, A. 1971. *Plant Physiol.* 47:685–90
67. Hines, G. D., Ellsworth, R. K. 1969. *Plant Physiol.* 44:1742–44
68. Hoare, D. S., Heath, H. 1959. *Biochem. J.* 73:679–90
69. Hoppe-Seyler, F. 1879. *Z. Physiol. Chem.* 3:339–50
70. Ibid 1880. 4:193–205
71. Horton, P., Leech, R. M. In press
72. Jones, C. B., Ellsworth, R. K. 1969. *Plant Physiol.* 44:1478–80
73. Jones, O. T. G. 1963. *Biochem. J.* 88: 335–43
74. Ibid 1963. 89:182–89
75. Ibid 1966. 101:153–60
76. Jones, O. T. G. 1967. *Biochem. Biophys. Res. Commun.* 28:671–74
77. Jones, O. T. G. 1968. In *Porphyrins and Related Compounds*, ed. T. W. Goodwin, 131–45. New York: Academic
78. Kahn, A., Boardman, N. K., Thorne, S. W. 1970. *J. Mol. Biol.* 48:85–101
79. Kamen, M. D. 1957. *Isotopic Tracers in Biology.* New York: Academic
80. Kikuchi, G., Kumin, A., Talmage,

P., Shemin, D. 1958. *J. Biol. Chem.* 233:1214–19
81. Kirk, J. T. O. 1970. *Ann. Rev. Plant Physiol.* 21:11–42
82. Kirk, J. T. O. 1971. *Ann. Rev. Biochem.* 40:161–96
83. Kirk, J. T. O., Pyliotis, N. A. 1972. *Z. Pflanzenphysiol.* 66:325–36
84. Kirk, J. T. O., Tilney-Basset, R. A. E. 1967. *The Plastids.* London: Freeman
85. Koski, V. M. 1950. *Arch. Biochem.* 29:339–43
86. Koski, V. M., French, C. S., Smith, J. H. C. 1951. *Arch. Biochem. Biophys.* 31:1–17
87. Koski, V. M., Smith, J. H. C. 1948. *J. Am. Chem. Soc.* 70:3558–62
88. Krasnovski, A. A., Bystrova, M. I., Lang, F. 1970. *Dokl. Akad. Nauk SSSR* 194:1441–44
89. Kuster, W. 1913. *Z. Physiol. Chem.* 82:463–83
90. Lascelles, J. 1964. In *Tetrapyrrole Biosynthesis and Its Regulation*, 84–109. New York: Benjamin
91. Lascelles, J. 1968. See Ref. 77, 49–59
92. Leech, R. M. 1968. In *Plant Cell Organelles*, ed. J. B. Pridham, 137–61. London: Academic
93. Liljenberg, C. 1966. *Physiol. Plant.* 19:848–53
94. Little, H. N., Kelsey, M. I. 1964. *Fed. Proc.* 23:223
95. Llambias, E. B. C., Batlle, A. M. del C. 1970. *FEBS Lett.* 6:285–88
96. Llambias, E. B. C., Batlle, A. M. del C. 1971. *Biochem. J.* 121:327–40
97. Lockwood, W., Rimington, C. 1957. *Biochem. J.* 67:8 p
98. Löffler, J. E. 1955. *Carnegie Inst. Washington Yearb.* 54:159–60
99. Marks, G. S. 1969. *Heme and Chlorophyll*, 121–62. London: Van Nostrand
100. Mathis, P., Sauer, K. 1972. *Biochim. Biophys. Acta* 267:498–511
101. Mathis, P., Sauer, K. In press
102. Mauzerall, D., Granick, S. 1958. *J. Biol. Chem.* 232:1141–62
103. McFeeters, R. F., Chichester, C. O., Whitaker, J. R. 1971. *Plant Physiol.* 47:609–18
104. Murray, A. E., Klein, A. O. 1971. *Plant Physiol.* 48:383–88
105. Nadler, K., Granick, S. 1970. *Plant Physiol.* 46:240–46
106. Nandi, D. L., Waygood, E. R. 1965. *Can. J. Biochem.* 43:1605–14
107. Ibid 1967. 45:327–36
108. Neuberger, A., Tait, G. H. 1964. *Biochem. J.* 90:607–16

109. Neve, R. A., Labbe, R. F. 1956. *J. Am. Chem. Soc.* 78:691–92
110. Nielsen, O. F., Khan, A. In press
111. Ogura, N., Takamiya, A. 1966. *Bot. Mag.* 79:588–94
112. Oku, T., Tomita, G. 1970. *Photosynthetica* 4:295–301
113. Ibid 1971. 5:133–38
114. Phillips, J. N. 1963. *Compr. Biochem.* 9:34–72
115. Porra, R. J., Falk, J. E. 1964. *Biochem. J.* 90:69–75
116. Porra, R. J., Irving, E. A. 1970. *Biochem. J.* 116:42 p
117. Porra, R. J., Lascelles, J. 1968. *Biochem. J.* 108:343–48
118. Porter, J. W. 1969. *Pure Appl. Chem.* 20:449–81
119. Radmer, R. J., Bogorad, L. 1967. *Plant Physiol.* 42:463–65
120. Rebeiz, C. A. 1967. *Magon, Ser. Sci.* 13:1–21
121. Rebeiz, C. A. 1972. *Proc. 6th Int. Congr. Photobiol.*, 233
122. Rebeiz, C. A., Abou-Haïdar, M., Yaghi, M., Castelfranco, P. A. 1970. *Plant Physiol.* 46:543–49
123. Rebeiz, C. A., Castelfranco, P. A. 1964. *Plant Physiol.* 39:932–38
124. Ibid 1971. 47:24–32
125. Ibid, 33–37
126. Rebeiz, C. A., Castelfranco, P. A., Engelbrecht, A. H. 1965. *Plant Physiol.* 40:281–86
127. Rebeiz, C. A., Crane, J. C., Nishijima, C. 1972. *Plant Physiol.* 50:185–86
128. Rebeiz, C. A., Crane, J. C., Nishijima, C., Rebeiz, C. C. In press
129. Rebeiz, C. A., Larson, S., Weier, T. E., Castelfranco, P. A. In press
130. Rebeiz, C. A., Nishijima, C., Crane, J. C. 1971. *Plant Physiol.* 47:S-267
131. Rebeiz, C. A., Yaghi, M., Abou-Haïdar, M. 1969. *Proc. 11th Int. Bot. Congr.*, 178
132. Rebeiz, C. A., Yaghi, M., Abou-Haïdar, M., Castelfranco, P. A. 1970. *Plant Physiol.* 46:57–63
133. Rudolph, E., Bukatsch, F. 1966. *Planta* 69:124–34
134. Sano, S. 1966. *J. Biol. Chem.* 241:5276–83
135. Sano, S., Granick, S. 1961. *J. Biol. Chem.* 236:1173–80
136. Schiff, J. A. 1970. *Symp. Stadler* 3:89–113
137. Schiff, J. A. 1971. *Symp. Soc. Exp. Biol.* 24:277–301
138. Schopfer, P., Siegelman, H. W. 1968. *Plant Physiol.* 43:990–96
139. Schultz, A. J. Jr. 1970. *The develop-*
ment and organization of photosynthetic pigment systems. PhD thesis. Univ. California, Berkeley. 142 pp.
140. Schultz, A., Sauer, K. 1972. *Biochim. Biophys. Acta* 267:320–40
141. Shlyk, A. A. 1971. *Ann. Rev. Plant Physiol.* 22:169–84
142. Shlyk, A. A., Prudnikova, I. V. 1965. *Dokl. Akad. Nauk SSSR* 160:720–23
143. Shlyk, A. A., Prudnikova, I. V. 1967. *Photosynthetica* 1:157–70
144. Shlyk, A. A., Prudnikova, I. V., Fradkin, L. I., Nikolayeva, G. N., Savchenko, G. E. 1969. See Ref. 6, 2:572–91
145. Shlyk, A. A., Prudnikova, I. V., Savchenko, G. E., Grozovskaya, M. S. 1971. *Dokl. Akad. Nauk. SSSR* 200:222–5
146. Shlyk, A. A. et al 1967. *Stud. Biophys.* 5:17–24
147. Shemin, D. 1962. *Methods Enzymol.* 5:883–84
148. Shemin, D., Russell, C. S. 1953. *J. Am. Chem. Soc.* 75:4873–74
149. Sironval, C. In press
150. Sironval, C., Kirchman, R., Bronchart, R., Michel, J. M. 1968. *Photosynthetica* 22:57–67
151. Sironval, C., Kuyper, Y. In press
152. Sironval, C., Kuyper, Y., Michel, J. M., Brouers, M. 1967. *Stud. Biophys.* 5:43–50
153. Sironval, C., Michel-Wolwertz, M. R., Madsen, A. 1965. *Biochim. Biophys. Acta* 94:344–54
154. Smith, J. H. C. 1947. *J. Am. Chem. Soc.* 69:1492–96
155. Smith, J. H. C. 1948. *Arch. Biochem.* 19:449–54
156. Smith, J. H. C., French, C. S. 1963. *Ann. Rev. Plant Physiol.* 14:181–224
157. Stevens, E., Frydman, B. 1968. *Biochim. Biophys. Acta* 151:429–37
158. Stobart, A. K., Thomas, D. R. 1968. *Phytochemistry* 7:1313–16
159. Stobart, A. K., Weir, N. R., Thomas, D. R. 1969. *Phytochemistry* 8:1089–1100
160. Strell, M., Kaloyanoff, A., Koller, H. 1960. *Angew. Chem.* 72:169–70
161. Sundqvist, C. 1969. *Physiol. Plant.* 22:147–56
162. Ibid 1970. 23:412–24
163. Sundqvist, C. 1972. *Chlorophyll formation in dark grown plants. A study of the influence of δ-aminolevulinic acid on chlorophyll formation.* PhD thesis. Göteborg Univ., Göteborg, Sweden. 74 pp.
164. Süzer, S., Sauer, K. 1971. *Plant Physiol.* 48:60–63

165. Tait, G. H., Gibson, K. D. 1961. *Biochim. Biophys. Acta* 52:614-16
166. Thorne, S. W. 1971. *Biochim. Biophys. Acta* 226:113-27
167. Ibid, 128-34
168. Thorne, S. W., Boardman, N. K. 1972. *Biochim. Biophys. Acta* 267: 104-10
169. Tigier, H. A., Batlle, A. M. del C., Locascio, G. 1968. *Biochim. Biophys. Acta* 15:300-2
170. Treffrey, T. 1970. *Planta* 91:279-84
171. Treffrey, T. In press
172. Wellburn, A. R. 1970. *Phytochemistry* 9:2311-13
173. Wellburn, A. R., Wellburn, F. A. M. 1971. *J. Exp. Bot.* 22:972-79
174. Wellburn, F. A. M., Wellburn, A. R. 1971. *Biochem. Biophys. Res. Commun.* 45:747-50
175. Westall, R. G. 1952. *Nature* 170:614-16
176. Willstätter, R., Stoll, A. 1913. *Untersuchungen über Chlorophyll.* Berlin: Springer-Verlag
177. Woodcock, C. L. F., Bogorad, L. 1971. See Ref. 48, 89-128
178. Woodward, R. B. et al 1960. *J. Am. Chem. Soc.* 82:3800-2

Ann. Rev. Plant Physiol. 1973. 24:173–96

PROTEOLYTIC ENZYMES AND THEIR ❖ 7546
INHIBITORS IN PLANTS[1]

C. A. Ryan[2]

Department of Agricultural Chemistry, Washington State University, Pullman

CONTENTS

INTRODUCTION

Proteolytic activity in plants was reported over 170 years ago (173) but was probably recognized much earlier by inhabitants of tropical lands who used the

[1] Abbreviations used: ANA (α naphthylamide); BAEE (α-N-benzoyl-L-arginine ethyl ester); BANA (α-N-benzoyl-L-naphthylamide); BAPA (α-N-benzoyl-L-arginine p-nitro-analide); BTEE (N-benzoyl-L-tyrosine ethyl ester); pCMB (para-chloromercuriben-zoate); DFP (diisopropylfluorophosphate); EDTA (ethylenediaminetetraacetic acid); IAA (indoleacetic acid); LPA (L-leucine p-nitroanalide); PMSF (phenylmethane sulfonyl fluoride); TCA (trichloroacetic acid); TLCK (N-tosyl-L-lysine chloromethyl ketone); TPCK (N-tosyl-L-phenylalanine chloromethyl ketone); Z (carbobenzoxy-).

[2] The preparation of this report and work done at Washington State University was supported in part by U.S.P.H.S. Career Development Award Grant G.M. 17059. Scientific paper 3949, College of Agriculture, Washington State University, Project 1791.

latex and leaves of papaya plants as a meat tenderizer and as a vermifuge (104). Over the past 40 years a number of proteinases have been isolated from fruits and latex of several plants, but only recently have proteinases been studied in cellular processes in plant metabolism. A large number of reviews have been written through the years concerning proteinases in plants, usually emphasizing the properties of such well-known eyzymes as papain (12, 51), chymopapain (51, 88), ficin (51, 92), and bromelin (118). In addition to these enzymes, a number of other proteolytic enzymes from various plants have been studied extensively, particularly in germinating seeds where these activities have been associated with the events of seeding growth and development. Some enzymes have now been purified or partially purified, and studies of their substrate specificities and other properties have begun to provide an insight into the variety and complexity of plant proteolytic enzymes and of their physiological roles. It is the purpose of the first part of this review to discuss the more recent information available concerning these enzymes, their properties, and their roles, or possible roles, in seed germination and seedling development. In addition, a number of enzymes that are present in leaves, flowers, and fruits are reviewed. These enzymes are particularly interesting because of their possible roles in turnover and senescence in the tissues. Only a few of these enzymes have been highly purified and studied, but of those studied some have been found to possess unusual substrate specificities.

In many seeds and other storage organs, proteins that inhibit proteolytic enzymes (34, 91, 93, 135, 176) are often found in high concentrations. Inhibitor proteins are also found in virtually all animal tissues and fluids (91, 176) and have been the object of considerable research for many years because of their abilities to complex with and inhibit proteolytic enzymes from animals and microorganisms (176). Interestingly, they rarely inhibit proteolytic enzyme from plants. The inhibitors have become valuable tools for the study of proteolysis in medicine and biology. They are of particular interest because of their therapeutic potentials in controlling proteinases involved in a number of disorders such as pancreatitis (181), shock (181), and emphysema (114). Recently they have been under study as agents in regulating mammalian fertilization (188). Biochemists have also profitably studied plant proteinase inhibitors as model systems to explore the mechanism of action and inhibition of proteinases (40, 91). The presence of proteinase inhibitors in important plant foods has also made them the object of great interest nutritionally (93, 135). All of these aspects have been reviewed a number of times recently (34, 40, 81, 91, 93, 135, 176) and therefore will not be reviewed again here.

Although the interest in plant proteinase inhibitors has spanned over 50 years, only occasionally have their physiological function in plants been researched. In reviews of the inhibitors only a few paragraphs have been relegated to the relatively small amount of work done on their function. However, since 1960 a number of groups have approached the problem, either directly or indirectly, and the physiological importance of these plants constituents is becoming apparent. The inhibitors are part of a highly complex organization of proteins in plant cells that apparently are involved in the development and protection not only of the seeds and storage organs where they are concentrated, but also of aerial tissues as well. The second part of this review will be concerned primarily with recent information concerning the in vivo physiological functions of the proteinase inhibitors found in plant tissues.

PROTEINASES

Proteinases throughout all living systems have been classified into four major groups (57) based on their active site catalytic mechanisms. These include the serine proteinases, sulfyhydryl proteinases, metalloproteinases and acid proteinases. Within this broad generalized classification the enzymes can be more specifically classified with regard to their substrate specificities, i.e. endopeptidases, carboxypeptidases, or aminopeptidases, depending upon whether they attack internal peptide bonds of polypeptides or cleave single amino acid residues from the terminal ends. Although convenient and commonly used, this nomenclature cannot always be applied, particularly with regard to enzyme activities in plant extracts where full characterization of the enzymes has not been accomplished. Nevertheless, except for a number of peptide hydrolases that occur in plants that do not attack proteins but hydrolyze various small peptides, the enzymes included in this review have been viewed as belonging to the broad categories described above.

In germinating seeds the endopeptidases and carboxypeptidases are apparently involved primarily with breaking down reserve proteins. They usually have acidic pH optimums, and in this respect they support a concept presented by Matile (98), Yatsu (187), and St. Angelo & Ory (165) that they are part of a lysosomal-like digestive process that takes place in the aleurone grain at acid pH. On the premise that lysosomal digestion is occurring, it is possible that the aminopeptidases with acid pH optima might also be involved with this process, but not those having alklaline pH optima. If the latter enzymes have a role in the germination process, it is still unclear.

Following a discussion of the roles of the proteinases in germinating seeds and of the conditions that regulate the processes, the properties of some of the individual enzymes will be reviewed.

In many of the developmental studies reviewed herein, enzyme activities have been measured and not their absolute quantities. Therefore, in evaluating such data it should be kept in mind that several interpretations of increases and decreases in enzyme activities often can be made. These could include, for example, synthesis and degradation, activation or inhibition, or compartmentalization. Filner et al (42) have recently discussed in depth some of the difficulties encountered in interpreting activity data with special regard to a number of plant enzymes.

Proteolysis in Germinating Seeds

Studies of the proteinases of both dormant and germinating seeds has revealed that considerable complexity occurs among the proteolytic enzymes within tissues of seeds and developing seedlings. Some types of proteolytic enzymes appear to be tissue specific, while others are found in several tissues. Some are involved with the breakdown of storage proteins in the aleurone grains, whereas others apparently function in the turnover of cellular proteins in the growing seedling. The levels of some of these enzymes are under regulation by known hormones like gibberillic acid, abcissic acid, or cytokinins, while the levels of others apparently do not respond to these substances.

In many seeds considerable quantities of a few proteins are stored in the aleurone grains and are mobilized during germination by specific proteinases (102). The process of germination is highly organized to operate and be completed over a period of only a few days (102). Proteinase activities can change considerably during this period according to the needs of the plant in its unique internal and external environments. It is therefore not unexpected that the properties of proteinases from different plant genera, and even among species, may have evolved independently from each other and therefore exhibit different patterns of both regulation and activities. As we shall see, this indeed appears to be the case.

The variations in levels of proteolytic activities in seeds from different genera during germination suggest that there may be more than one way for the seeds to regulate proteolysis during the breakdown of storage proteins, although the overall proteolytic processes may be quite similar. Some seeds respond to the initiation of germination by producing large amounts of proteinase or proteinase activity, whereas other seeds apparently have considerable proteinase activity already present at the onset of germination. In germinating barley (17, 77), peas (15), squash (125), lettuce (155), cotton (69), and sorghum (47) there is an increase in proteolytic activity beginning with, or shortly after, the onset of germination. On the other hand, in germinating kidney beans (136), rice (68), or maize (106) there is a rapid liberation of amino acids from stored proteins early in the germination process, but during this period there is a slow decline in proteolytic activity.

In barley de novo synthesis of endopeptidase activity increased rapidly (77) during germination. In addition to the endopeptidase, in germinating barley several other proteolytic enzymes were present whose activities also changed during development. Eight peptidases were demonstrated (110), based on substrate specificity and pH optima. Present were three carboxypeptidases, three aminopeptidases, and two dipeptidases. The carboxypeptidase activities were in high concentrations in the aleurone layers, in the starchy endosperm, and in the scutellum. In these tissues their activities increased during germination. An interesting finding was that the starchy endosperm contained high activities of the three carboxypeptidase enzymes but was completely devoid of the five other enzymes. Along with the two dipeptidases, aminopeptidases were found in the scutellum, where their activities remained relatively constant during germination. In the seedling the aminopeptidases were highly active whereas the carboxypeptidases were only slightly active.

Prentice and associates have studied the activities of several peptide hydrolases in barley and wheat during germination (129, 131, 133, 134). Four peptide hydrolases were present in barley and wheat, designated A, B, C and LNAase

(24–26, 132). Enzymes A, B, and C were present in low levels in mature barley, whereas LNAase activity was much higher. During maturation the activities of enzymes A, B, and C decreased slowly while LNAase increased steadily. Wheat embryo contained peptide hydrolase A but no B enzyme (132). Germinating wheat contained both enzymes. These enzymes did not hydrolyze proteins and apparently were not responsible for the breakdown of storage proteins of the barley kernel endosperm during the development of the barley seedlings (24).

In germinating kidney beans (136) endopeptidase activity was initially quite high and decreased slowly until the eighth day, then increased sharply, followed again by a decline. Peptide hydrolase activity, measured with Z-tyrosine-p-nitroanalide, remained constant until day 8, then rapidly increased for several days. Peptide hydrolase activity, measured with BAPA, remained constant throughout the entire germination period. Leucine aminopeptidase activity, measured with L-leucine-naphthylamide, remained constant until day 8 and then increased slowly. By day 8 the cotyledons were green but the first leaves had not appeared. The significance of the sudden changes in enzyme concentrations at day 8 is not clear.

In germinating lettuce seeds the appearance of proteolytic activities (155) appeared to be regulated in a different way from the other seeds. Two proteinases were present, both having a pH optimum at 6.8 (154). One, a very small (MW 3000–4000) "trypsin-like" enzyme that was not synthesized de novo, developed during germination. It was apparently inhibited by an endogenous proteinase inhibitor that was destroyed during germination. It was postulated that a large amount of the "trypsin-like" enzyme may be present in the dormant seed either complexed with inhibitor or as a zymogen that was activated either autocatalytically or by other enzymes. The "trypsin-like" enzyme was purified 450-fold. It hydrolyzed BTEE and was inhibited by DFP. A carboxypeptidase with a pH optimum of 5.6 and an aminopeptidase, pH optimum 4.8, were also present in the seeds (155). The small size of the "trypsin-like" enzyme and its complex mode of activation in the germinating lettuce seeds certainly suggest that this system deserves more attention.

The profound changes in proteolytic activity in seeds in several cases can be directly associated with hormonal control. The most thoroughly studied system is that of barley. In 1967 Jacobsen & Varner (77) demonstrated that de novo proteinase and amylase syntheses in isolated barley aleurone layers were stimulated simultaneously by gibberellic acid. Synthesis of both enzymes was inhibited by abscisic acid, cycloheximide, and actinomycin D. Carlson (31) recently compared the ability of aneuploid aleurone tissues having normal triploid chromosomes to respond to gibberellic acid by synthesizing alpha amylase and proteinase. Trisomic lines, with additional copies of chromosomes 2, 4, and 6, produced an increased alpha amylase activity while additional copies of chromosomes 2 and 3 produced increased proteinase activity. Aleurone tissues with extra 2 chromosomes produced more alpha amylase activity than tetraploids at high gibberellic acid concentrations but not at low concentrations. This was interpreted to mean that chromosome 2 may be involved with the mediation of the gibberellic acid response at a post-transcriptional step. It is possible that this response may be involved with the early stages of gibberellic acid induction in barley reported by Evins & Varner (38), in which polysome formation preceded amylase production. Although these latter workers did not study proteinase synthesis, the close association of proteinase with the alpha amylase-inducing system certainly bears on the proteinase-synthesizing system. Gibberellic acid induced an increase in polysomes, the formation of ribosomes, and the synthesis of alpha amylase. All of these processes were inhibited by abscisic acid. In aleurone grains already synthesizing amylase, abscisic acid inhibited further synthesis but did not change the number of ribosomes. Therefore the post-transcriptional increase in the number of ribosomes and the percentage of polysomes are probably a prerequisite for alpha amylase synthesis and probably for proteinase synthesis as well.

In germinating barley, increases in carboxypeptidase activity in the starchy endosperm depended on the presence of the embryo (110). The inhibitory effect caused by excising the embryo could not be reversed by gibberellic acid. The authors considered the possibilities either of the existence of procarboxypeptidases that required the presence of the embryo for activation or that the enzymes were synthesized in the scutellum and transported to the endosperm. The possible role of proteinase inhibitors was not considered.

Hormonal influence also has been studied during the germination of cotton seeds (69, 70, 73). In cotton a carboxypeptidase was produced de novo 24 hours after germination and continued to increase until the fourth day, then disappeared. In this seed the mRNA was preformed and stored during embryogenesis, and its translation was inhibited by abscisic acid produced by tissues surrounding the embryo. When the embryo was severed from the mother plant, premature appearance of enzyme was noted, indicating that abscisic acid was probably diffusing from the ovule wall into the embryo.

In germinating squash the embryonic axis was required during the first 32 hours of germination for the de novo development of proteolytic activity in the cotyledons (124). The presence of cytokinin in the culture medium could replace the excised embryonic axis. During normal germination, proteolytic activity decreased after day 3. However, benzyladenine but not gibberellin in the culture medium prevented this decrease whether the embryonic axis was left intact or excised. The embryonic axis, therefore, may control both the induction and cessation of proteolytic activity in squash cotyledons.

A dipeptidase activity in the squash seedlings is also under regulation by the embryonic axis (167). This activity remains constant for the first 2 days of germination and then increases rapidly for the next 3 days. In two of three seedlots tested, the development of activity required the presence of the embryonic axis. In the seeds that required the embryonic axis, benzyladenine caused an increase in dipeptidase activity when the axis was removed. Cycloheximide inhibited the development of the dipeptidase activity.

Proteinases from Dormant and Germinating Seeds and Tubers

ENDOPEPTIDASES   Numerous endopeptidases have recently been purified or partially purified from developing, dormant, or germinating seeds or tubers. In several studies protein bodies were first isolated from the germinating seeds, and the endopeptidases associated with them were then studied.

Yatsu & Jacks (187) isolated membrane-bound protein bodies (aleurone grains) from cotyledonary tissues of ungerminated cotton seedlings. The bodies contained 100% of the endopeptidase of the cotyledons and 75% of the proteins in the tissues. The proteinase was not purified further nor extensively characterized but was shown to hydrolyze hemoglobin with two pH optimums of 1–2 and 3–4. Ory & Henningsen (122) isolated protein bodies from intact, dormant barley seeds. They found that there was no enzyme activity in the bodies specific for the substrate BAPA (i.e. no peptide hydrolase). An acid proteinase was present that hydrolyzed gelatin, but it was uncharacterized. An acid proteinase p(H optimum 5.2) was also identified in protein bodies isolated from sunflower seeds (152). It also was uncharacterized. These latter bodies did not contain amino peptidase activity (LPA as substrate). Protein bodies were also isolated from hempseed (165) with an endopeptidase activity associated with them. The enzyme was purified and called "edestinase" because it hydrolyzed edestin, the major storage protein of hempseed protein bodies. The enzyme had a pH optimum of 4.3 on pure edestin. It was not affected by cysteine, EDTA, pCMB, iodoacetate, or DFP.

The presence of acid proteinases in protein bodies supports the concept mentioned earlier that the bodies are lysosomal-like organelles in which intracellular digestion takes place. It is very likely that during the first stages of germination the endopeptidases found in the bodies are primarily responsible for initiating the process of fragmenting the storage proteins to polypeptides for further degradation and transport to the growing embryo, either as peptides or free amino acids.

Proteinases have also been isolated from total extracts of germinating seedlings. An acid proteinase with a high degree of specificity has been isolated from 5 to 6-day-old sorghum seedlings (47, 48). It was well characterized and had a molecular weight of 80,000 and a pH optimum of 3.6 on bovine serum albumin. It had no requirement for metal ions nor was it inhibited by DFP, sulfhydryl reagents, or by soybean trypsin inhibitor. It was inhibited by 1 mM N-bromosuccinimide (a tryptophan reagent). The enzyme had a narrow specificity requirement and hydrolyzed specifically peptide bonds between the alpha carboxyl of aspartic or glutamic acids and the amino group of the adjacent amino acid (48). Neither asparagine nor glutamine in these positions was hydrolyzed. This enzyme should prove to be a valuable regent in specifically fragmenting proteins for sequencing, and in identifying the positions of aspartic and glutamic acids in proteins. The major storage proteins of sorghum are high in glutamic and aspartic acids and should be excellent substrates for the enzyme during protein mobilization. During germination the activity of the enzyme remained constant for the first 2 days and then began to increase. Unlike barley (77), the increase did not parallel amylase increase, which began to rise almost immediately with the onset of germination.

An acid proteinase was purified 870-fold from cotyledons of germinating lotus seeds (158). The pH optimum was 3.8 on urea denatured casein and 2.4 on heat denatured casein. It was not a sulfhydryl enzyme nor was it inhibited by soybean trypsin inhibitor, potato trypsin inhibitor, or ovomucoid.

Five endopeptidases were partially purified from germinating barley (130) using carboxymethyl cellulose and gel filtration. Three of the enzymes were more active on hemoglobin at pH 3.8 than on casein at pH 6.0 and are apparently acid proteinases.

Other reports have been made of endopeptidases with acidic pH optimums in barley (36, 37), wheat (80, 163), and soybean (127). Little is known of the properties of any of these proteinases.

A sulfhydryl proteinase has been isolated from the tubers and fruit of *Carilla chocola*, a wild species of the Caricacaea (169). It has a pH optimum of 8–9.5 with casein and pH 6.5 with BAEE. Like papain, it clots milk.

Proteinase activity has been partially purified from developing vanilla pods (184). Activity was measured with hemoglobin and casein and had a pH optimum of 9. The enzyme was not a sulfhydryl enzyme and was only slightly affected by EDTA. Its activity was initially very high in newly developing pods but decreased as the pods matured.

CARBOXYPEPTIDASES   Only a few carboxypeptidases have been isolated from seeds, and they resemble those isolated from plant leaves (cf p. 184). It is apparent that these enzymes are considerably different from the mammalian carboxypeptidases in their specificities and mechanism of action. They are inhibited by DFP and are in this respect resemble the mammalian serine endopeptidases. It is anticipated that the active centers of these enzymes will be the object of considerable research in the near future, both from a mechanistic and an evolutionary point of view.

Germinating barley apparently contains three carboxypeptidases (111, 116, 175). One of them, purified to near homogeneity by Visuri et al (175), was free of endopeptidase, aminopeptidase, or dipeptidase activities. The purified enzyme liberated carboxyl-terminal amino acid residues from a wide range of Z-dipeptides. Peptides containing proline were not attacked. The enzyme had a pH optimum at 5.2 and was inhibited by DFP and pCMB. It had a molecular weight of about 90,000.

Another carboxypeptidase that was partially purified from germinating barley (116) hydrolyzed Z-dipeptides, preferably Z-ileu-phe, Z-gly-leu and Z-gly-tyr. It did not hydrolyze Z-gly-ala and in this respect differed from the enzyme described above by Visuri et al (175). The enzyme was not markedly affected by sulfhydryl reagents, EDTA, or $KBrO_3$.

The carboxypeptidase previously mentioned in cotton cotyledons was purified 2000-fold to homogenity by Ihle & Dure (71). The enzyme had a pH optimum of 6.7 and hydrolyzed several polypeptides sequentially from their C-terminal amino acids (72). Proline, asparagine, and glutamine were released, the latter two without deamination. The reaction rate with tripeptides was much less than with polypeptides and it did not hydrolyze dipeptides. The activity of the enzyme was inhibited by DFP, and it incorporated two moles per mole of enzyme. Inactivation studies supported the stoichiometric data and suggested that two active sites were present in the molecule. The enzyme had a molecular weight of 84,500 and was composed of three nonidentical subunits of molecular weight of 33,000, 31,000, and 24,000. Radioactive DFP was incorporated into the 33,000 and 31,000 molecular weight subunits but not into the smaller subunit. A carboxypeptidase isolated from French bean leaves (179) and another carboxypeptidase from citrus called carboxypeptidase C (190), are similar to this enzyme, particularly in their broad specificity; they will be discussed in a later section.

Potato tubers contained a carboxypeptidase (66) that was purified 200-fold (67). It inactivated bradykinin and kallidin, two kinins of mammalian origin. The enzyme had a molecular weight of 70,000 and an optimum activity at pH 7.0. Cobalt increased the activity severalfold after purification by isoelectric focusing but not before. Metal chelators, sulfhydryl reagents, or divalent metals other than cobalt did not affect the activity of the enzyme. This enzyme was also found in other edible tubers such as sweet potatoes, taros, and yams (67).

AMINOPEPTIDASES AND OTHER PEPTIDE HYDROLASES IN DORMANT AND GERMINAT-ING SEEDS   A number of enzymes that hydrolyze the amino terminal amino acid residues from dipeptides, dipeptidyl amides, and tripeptides have been reported recently. Some of the enzymes are clearly aminopeptidases, judging from their specificities with several substrates. However, in a few reports of peptide hydrolases it was not possible to make a distinction between aminopeptidases, esterases, amidases, and peptide hydrolases that have no specificity toward the amino or carboxyl end of dipeptidases or that hydrolyzed synthetic substrates having the terminal amino group blocked such as BANA. Only a few aminopeptidases have been isolated to a high degree of purity and well characterized.

An aminopeptidase from barley grains was purified 485-fold (84). It hydrolyzed L-phenylananyl-β-naphthylamide. The enzyme, molecular weight 65,000, preferred substrates with a hydrophobic N terminal side chain and it hydrolyzed dipeptides and a tripeptide (leu-gly-gly). Its pH optimum was 7.2 for acyl-β-naphthylamides and 5.8–6.5 for dipeptides. Neither DFP or EDTA affected its activity but pCMB was a potent inhibitor.

This enzyme was one of three aminopeptides in barley. It resembled the "arylamidases" of wheat germ (58) and was different from the barley peptide hydrolases which hydrolyze N-substituted p-nitroanalides or N-substituted ester substrates (24).

Buckner (23) isolated two aminopeptidases from 21-day-old pea leaves. One enzyme was purified 380-fold. It had a molecular weight of 70,000 and hydrolyzed only L-leucine-p-nitroanalide with a pH optimum of 7.7. The second enzyme was purified 1300-fold and had a molecular weight of 82,000. It hydrolyzed glycine-p-nitroanalide with an optimum pH of 8.2. This latter enzyme hydrolyzed to lesser estent L-leucine-p-nitroanalide and BAPA and exhibited a broad specificity with peptides. The former enzyme required N-terminal leucine or methioine in the dipeptides. Neither enzyme hydrolyzed casein or collagen. Both were inhibited by Hg^{++} or Cu^{++} and pCMB. It was suggested that these enzymes are not directly involved in germination but are more likely involved in turnover of cellular proteins.

An enzyme that hydrolyzed BAPA was isolated from roots of soybean seedlings (52). It was purified 45-fold and had a pH optimum of 8.0. It had low activities toward glycine-p-nitroanalide or L-leucine-β-naphthylamide. Several other nitro-analides were not hydrolyzed.

An aminopeptidase was purified 500-fold from germinating squash cotyledons (13). The enzyme hydrolyzed triglycine and tetraglycine but not pentaglycine, dipeptides, chloracetyl-gly-gly, hippuryl-L-leucine amide or hippuryl-DL-phenyl-lactic acid. It hydrolyzed slowly L-leucine amide, L-phenylalanine amide, and L-tyrosine amide. The enzyme was inhibited by Mn^{++}, Co^{++}, Sn^{++}, and EDTA. The pH optimum was 8.0.

Burger and associates have studied a group of peptide hydrolases in barley and wheat (24–26). These enzymes do not hydrolyze proteins but, depending on the enzyme, do hydrolyze several dipeptides, a few tripeptides, a number of N-benzoyl-L-amino acid p-nitroanalides, alpha-naphthylamide and leucine-naphthylamide. These enzyme activities are present in much smaller quantities than those of the proteinases.

These enzymes, active at neutral or alkaline pH, have been designated peptide hydrolases A, B, C, LNAase, and ANAases. Peptide hydrolase A (24) exhibited a broad substrate specificity with a variety of dipeptides and it readily hydro-lyzed BAPA. Peptide hydrolase B (24) did not hydrolyze BAPA but did hydro-lyze BAEE and exhibited a limited specificity on simple peptide substrates. Aro-matic residues on the carboxyl side of the dipeptide bond of dipeptides promoted hydrolases as did lysine residues at the amino side. Peptide hydrolase C (25) is remarkably similar to A, but possesses different elution characteristics from CM cellulose. Four enzymes were identified that hydrolyzed ANA and one that hydrolyzed LNA.

Arachin, a peptide hydrolase from ungerminated, shelled peanuts, noted earlier by Irving et al (75), was recently purified by Cameron & Mazelis (28). It is similar to peptide hydrolase A from barley and wheat. Its pH optimum is near 8. A simi-lar peptide hydrolase was purified from soybeans by Catsimpoolas et al that hydrolyzed BAPA with a pH optimum also near 8 (32).

An interesting peptidase activity that liberated β-naphthylamine from L-pyr-rolidonyl-β-naphthylamide was identified in seeds and sprouts of several plant genera (168). The enzyme is thought to catalyze the hydrolysis of N-terminal pyrrolidone-carboxylic residues from proteins. Several groups (35, 43, 74, 82) have reported that glutamyl tRNA cyclizes to produce pyrrolidonyl carboxyl-tRNA and then transfers it to the newly forming polypeptide chain. However, this reaction has not been confirmed in plants.

A L-gamma-glutamyl peptidase has been identified in a soluble extract from dormant onions (14). The enzyme catalyzed the liberation of p-nitroanaline from L-gamma-glutamyl-nitroanalide with a pH optimum of 9.0. The principal endogenous substrate of onion L-cysteine sulfoxide lyase (the enzyme largely responsible for the sensory attributes of onion) in freshly harvested onions is a

L-gamma-glutamyl peptide. The peptidase is capable of hydrolyzing this peptide.

Aminopeptidases and peptide hydrolases have been employed for genetic studies in plants and have yielded valuable information (151, 177, 182, 183). It is not apparent in some of these studies which enzyme or groups of enzymes are under study because the chromogenic substrates utilized to stain zymograms, such as alpha-naphthyl-acetate or L-leucine-β-naphthylamide, are not specific for just leucine aminopeptidase. Nevertheless, only a few bands usually react and the differences or similarities in these bands among species may be profitably utilized for inheritance studies. Scandalios (151) has recently reviewed the usefulness of these enzymes in providing information concerning the biochemical genetics of plants.

Proteinases from Leaves, Flowers, and Fruits

Recent studies into the possible role of proteinases in intracellular digestion of proteins during the growth of plant meristematic tissues as well as during senescence have suggested that compartmentalized proteinases may be important to these processes. In corn root tips (191) a carboxypeptidase called carboxypeptidase C (190) was found to be structurally bound in root tips (approximately 5 mm long). The enzyme was also found to be associated with both mitochondria and microsomes. The highest concentrations in the mitochondria were found in accompanying lysosomes, particularly light lysosomes, or meristematic vacuoles. The association of carboxypeptidase C with lysosomal organelles suggests that the enzyme may be involved with the intracellular digestive roles of the lysosomes.

The vacuole has also been considered to be a large lysosome that contributes to the intracellular digestion of proteins, and it has been shown to participate in senescence in the corolla of the morning glory. In the corolla of the morning glory, endopeptidase and aminopeptidase activities were present (99), but neither changed significantly during senescence although the protein content of the petals dropped sharply. This suggested that during senescence the bulk of the proteins came in contact with the existing proteinases. Electron micrographs of the process in the petals show that there is a rupturing of the central vacuole in some cells as well as an increase in the autophagic activity in other cells (99). This same process apparently occurred in senescing tobacco leaves or leaf disks (8). In senescence the vacuolar (lysosomal) enzymes apparently remain active through the death of the cells.

In detached oat leaves two proteinases increased in activity severalfold during senescence (97). Serine additions to the leaves enhanced senescence as well as the activities of the two proteinases, whereas both activities were significantly retarded by the addition of cytokinins or cycloheximide. Kinetin also retarded production of proteinase activity in *Brassica* (89) and oat leaves (157). In leaves of all three of the plants mentioned the suppression of proteolysis may be the major role of cytokinins in delaying senescence. However, if compartmentation of proteinases plays a role in leaf senescence, as it appears to do in flower petals, then an interesting new parameter must be considered in interpreting the full significance of the kinetin-induced suppression of senescence of leaf tissues.

A number of proteinases have been isolated from green and etiolated leaves, from flowers, and from fruits that apparently are involved in intracellular and, in the special case of insectivorous plants, extracellular digestion. The most thoroughly studied of these enzymes have been carboxypeptidase C (190) from citrus and another carboxypeptidase isolated from French bean leaves (174).

The first report of a carboxypeptidase from a plant tissue was made in 1964 by Zuber (189), who identified the enzyme in citrus fruit. It was called carboxypeptidase C (190). The enzyme has now been isolated from lemon leaves and orange leaves (164). The purified enzymes from these sources differed in some respects. The lemon enzyme had a molecular weight of 126,00, a pH optimum of 5.3, and was inhibited 89% by 10^{-3} DFP. It was unstable to lyophilization. The orange enzyme had a molecular weight of 175,000, had the same optimum pH as the lemon enzyme, but was only inhibited 18% with DFP. It was stable to lyophilization. Neither enzyme was affected by EDTA.

The specificity of carboxypeptidase C was similar to a mixture of carboxypeptidase A and B, but in addition it usually hydrolyzed C-terminal proline (190) and hydrolyzed peptides with proline as the penultimate amino acid. It also released carboxymethyl-cysteine from S-carboxymethylated peptides. Glycine at the C-terminus was attacked slowly. The enzyme has been demonstrated to be a potentially valuable tool for sequencing proteins (170) or polypeptides. The stability of the enzyme to pH extremes, its lack of autodigestion, and its broad specificity make it well suited for this role. The first six amino acids from the C-terminus of ribonuclease were determined unequivocally with the orange leaf enzyme (170), and ten C-terminal amino acids of insulin B chain were removed sequentially (164).

A search (191) for carboxypeptidase C in plants other than citrus demonstrated that it was present in several angiosperms but not in *Fagus* and *Magnolia*. It was proposed that the enzyme originated at the evolutionary level of the Pteridophyta and of the Angiospermae.

At least four proteinases have been isolated from French bean leaves (29, 30, 156, 179, 180). One, a carboxypeptidase called "phaseolain," has been purified 1180-fold. Its specificity was very broad, releasing acidic, neutral, and basic amino acids as well as proline from polypeptides (180). It resembled carboxypeptidase C from citrus (190) in its specificity and pH optimum (5.6). It was inhibited by DFP at the sequence of glu-ser-val. The other three enzymes hydrolyze casein and were named proteinase *a*, *b*, and *c* (29, 156, 180). Proteinase *b* had a molecular weight of 45,500 and was inhibited by DFP at the sequence thr-ser-met-ala (156). Proteinases *a* and *c* have not been well characterized.

An endopeptidase that hydrolyzed hemoglobin with a pH optimum of 9–10 was purified about 200-fold from bean leaves (138). The molecular weight was 100,000. The proteinase increased during leaf growth (ninth to sixteenth day). Like proteinase *b* from French bean leaves (180), the enzyme was inhibited by DFP. It is likely that all of the bean proteinases are involved in some aspects of protein turnover, but the high pH optimum of this enzyme suggests that it is not a lysosomal proteinase.

A peptide hydrolase was purified 4000-fold from tomato leaves (166). The enzyme hydrolyzed Z-L-tyrosine-p-nitrophenyl ester and BAPA. It was strongly inhibited by metal ions but not by SH reagents or soybean trypsin inhibitors. It had a molecular weight of 103,000, but this was an average of two isoenzymes of MW 90,000 and 113,000, separated on sucrose density gradients. The two isoenzymes had equal activity.

A proteinase was recently purified from dark grown oat shoots that was responsible for the degradation of phytochrome during purification (126). It was noted that two active forms of phytochrome could be detected in oat seedlings; a large protein was converted to the smaller species through proteolysis that could be prevented by the addition of PMSF (46). The phytochrome degrading enzyme was purified 600-fold from 5-day dark grown oat shoots (126). Its pH optimum was 6.4 and the enzyme was an endopeptidase. It hydrolyzed phytochrome, phytocyanin, casein, and collagen. It did not hydrolyze synthetic substrates or peptides. Inhibitors besides PMSF included Hg^{++} and TLCK but not soybean trypsin inhibitor nor TPCK. Rye shoots also contained the enzyme but in much smaller quantities. It has not been established whether the enzyme degrades phytochrome in vivo or if the phytochrome is a substrate for the enzyme only as a result of cell disorganization.

The fruit of the chinese gooseberry (*Actinidia chinesis*) contained an endopeptidase (103) that hydrolyzed BAEE, p-toleuene-sulfonyl-L-glutamic-p-nitrophenyl ester and gelatin and was enhanced by reducing agents and EDTA. The enzyme was crystallized. It had a molecular weight of 12,800 and exhibited a double pH optimum at 6.5 and 7.0. Its function is unknown.

A number of plant genera have evolved a mechanism for capturing and digesting insects. The proteinases of some of the genera have been studied.

An endopeptidase called "nepenthes" (119) was purified to electrophoretic homogeneity from the pitchers of tropical insectivorous plants of the genus *Nepenthes* (4). The enzyme exhibited an acid pH optimum of 2–3, hydrolyzing internal peptide bonds with aspartic acid, alanine, and tyrosine at the carboxyl residues (5). It was not inhibited by DFP or by pCMB. The enzyme was also purified from *Drosera pelata*, another insectivore. The enzymes from both sources were virtually identical in electrophoretic mobilities, specificities, and pH optimums, although *Nepenthes* and *Drosera* belong to different genera and have no morphological or ecological similarities.

The digestive fluid of the Venus fly trap (*Dionaea muscipula*) contained a proteinase (150) whose properties have not yet been investigated. The proteinase was secreted when the plant was fed small cubes of gelatin, and its activity doubled over a 3-day period following feeding, then subsided rapidly. This plant could prove to be an interesting system for studying the mechanism of induction and release of proteolytic enzymes from plant tissues.

PROTEINASE INHIBITORS

The physiological functions of proteinase inhibitors in plants has been somewhat of a puzzle. They were recognized in plant tissues about 35 years ago (141) and through the years have been considered as possible regulatory (96) or protective proteins (9, 176) and, because no direct evidence of these roles could be established, have also been considered as biochemical oddities. Until a few years ago, the physiological importance of proteinase inhibitors to the plant itself was not seriously considered. Now, within our present concepts of protein turnover, enzyme regulation, and plant protection through biological control, their possible in vivo functions cannot be overlooked.

The in vivo study of proteinase inhibitors has presented some special problems not ordinarily encountered in enzyme studies. The inhibitors must be measured indirectly if inhibitory activity is followed, or else they must be assayed immunologically if the inhibitors can be obtained in a high degree of purity. Several inhibitor species are often present in tissues (34, 91, 93, 135, 176); a few are closely related "isoinhibitors" (105, 113) and others are unrelated or so distantly related that their properties differ significantly. In some organs such as potato tubers (16, 63, 64, 76, 105) as many as 13 species of inhibitors can be present (16) that inhibit trypsin on chymotrypsin. To study any one inhibitor in vivo, therefore, it is necessary to distinguish it from among all of the proteins of the cell and observe its behavior. Only recently have such efforts been successful.

The occurrence of proteinase inhibitors in plant storage organs such as seeds and tubers is widespread. They have been isolated mainly from the Leguminosae, Graminae, and Solanaceae, but also from a number of other families (34, 91, 93, 135, 176). The plant proteinase inhibitors are generally small proteins having molecular weights under 50,000 and more commonly under 20,000 (91, 176). A number of the inhibitors from corn (60), potatoes (16, 63, 76, 105) and several legumes (33, 45, 59, 61, 79, 171) have minimum molecular weights of below 10,000 and are often present as dimers or tetramers. Nearly all plant inhibitors inhibit enzymes of animal or microbial origins having either trypsin-like or chymotrypsin-like specificities. Some are known to inhibit endogenous proteolytic enzymes of the plant from which they were derived, but in most instances inhibition of plant proteinases has not been reported.

The inhibitors contain "active sites" for the inhibition of proteolytic enzymes that apparently endow them with their specificity (40, 91). The "trypsin-specific" inhibitors always have either a lys-X or arg-X sequence at the binding site (91), whereas chymotrypsin-specific inhibitors usually have a leu-X at their active centers (19, 85, 153, 172). These sites are part of a very large binding area that is necessary for the proper structural alignments that confer the unusual stability to the enzyme-inhibitor complex (22). However, the mechanism of inhibition has not yet been saisfactorily explained. Laskowski & co-workers (91) have hypothesized a requirement for an acyl-intermediate between the enzyme and inhibitor. Alternatively, Ryan et al (1–3, 44) and Feeney et al (39, 41) have shown that inactive derivatives of the enzymes also bind inhibitors to form complexes that could not involve an acyl-intermediate as hypothesized by Laskowski.

The emerging picture from structural and specificity similarities among plant inhibitors from different sources indicates that the active inhibitor sites may have been conserved over millions of years of evolution and suggests that the inhibitory capacity is important for survival. This, together with recent advances into the physiology of inhibitors in plants, suggests that the inhibitors may have important roles (a) as regulatory agents in controlling endogenous proteinases, (b) as storage proteins, and (c) as protective agents directed against insect and/or microbial proteinases. Our present information concerning the status of these roles is reviewed in the following text.

Regulation of Endogenous Proteinases

Studies of the physiology of proteinase inhibitors in plants are relative recent. In 1955 Ofelt et al (121), after studying the presence of proteinase inhibitors in soybeans, concluded that in these seeds they do not function as inhibitors of endogenous proteinases. This conclusion was later affirmed by Birk (21). However, in 1965 Shain and Mayer (154) reported that a proteinase inhibitor in lettuce seeds apparently does inhibit an endogenous proteolytic enzyme activity. A more recent account (155) described the inhibited enzyme as an endopeptidase that was "trypsin-like" in its specificity (cf page 177). As the enzyme activity increased during germination, the inhibitor disappeared. This disappearance suggested that either the enzyme was present as an inactive form, complexed with inhibitor, or that the enzyme, activated from a precursor, titrated away the inhibitor as it increased in activity (155).

Ungerminated barley extracts are now known to contain three types of proteinase inhibitors (83): those that inhibit the endogenous endopeptidases of barley (108); those that inhibit microbial proteinases (101, 113); and those that inhibit trypsin (112). During germination, similar to the lettuce system (154), the inhibitors of endogenous proteinases disappear before endopeptidase activity begins to increase. These inhibitors apparently exist only in the resting barley grain, for they could not be found in any tissues after the onset of germination. Other inhibitors of endogenous proteinases do exist, however, in vegetative tissues of barley. Burger and Siegelman (27) reported in 1966 that a proteinase in barley kernels was inhibited by an endogenous, small molecular weight inhibitor. Extracts of endosperm had weak activity and the rootlet and seedling extracts produced strong inhibition. The inhibitor was dialyzable but was not isolated or identified.

In corn only one type of proteinase inhibitor has been found (60). It does not inhibit the endopeptidase from maize (106). In addition, the maize endopeptidase was not inhibited by lima bean or soybean trypsin inhibitors.

An inhibitor of mammalian carboxypeptidases A and B that is unique in inhibiting carboxypeptidases was isolated from potatoes (140, 146). It was found to inhibit strongly (65) the bradykinin inactivating carboxypeptidase that was also recently isolated from potatoes (66, 67). The inhibitor is the smallest proteinase inhibitor isolated to date and has a molecular weight of 3500. Nothing is known of the interaction between the proteinase and inhibitor in vivo, but in view of the importance of carboxypeptidases in breakdown of storage proteins in protein storage bodies, it is likely that it has a function in regulating the activity of the enzyme in dormant potatoes.

Storage

The storage role for the inhibitors is supported by their presence in large quantities in seeds and tubers. For example, they represent about 6% of the soybean's protein (137) and up to 10% of the soluble proteins of barley grains (109) or potato tubers (148). Inhibitor proteins are not confined to storage organs but are also found in aerial tissues, sprouts, roots, stems, and leaves of plants. Their presence in all cases is apparently transitory, indicating a possible temporary storage role. Ambe & Sohonie (6, 7) demonstrated the presence of trypsin inhibitors in all parts of double and field beans at all stages of growth. The inhibitors decreased in concentration in cotyledons and increased in sprouts during germination. The results of work by Kirsi & Mikola (83) and Shain & Mayer (155) in barley and lettuce, respectively, show a decrease in inhibitors in seeds upon germination. The former workers demonstrated that young rootlets and coleoptiles also contain inhibitors that disappear within 4–5 days. On the other hand, Birk & Waldman (21) reported that the inhibitor content in soybeans increased with maturation. The former workers demonstrated that in germinating barley all inhibitory activity disappeared from the endosperms within 4 to 5 days after the onset of germination. This same phenomenon occurred in maize (106) in which inhibitor began to disappear during imbibition and continued to disappear during germination.

In soybeans the inhibitors were concentrated in the cotyledons (21). Barley (109), wheat (63, 109), rye (62, 109), oats (109), and maize (106) endosperms contained trypsin inhibitors, and in barley, wheat, rye, and oats they were also present in embryos (83). In potatoes (148) inhibitors were found concentrated in the cortex of the tubers.

Using immunological techniques (123, 143), the existence of a single inhibitor, chymotrypsin inhibitor I (105), was studied during the life cycle of potato plants (144, 147). The techniques quantitatively distinguished the one inhibitor among all the inhibitors present. The inhibitor behaved as if it were a storage protein. Potato tubers' apical cortex always showed the highest levels of inhibitor I of any tissues tested (148). When seed pieces of potato tubers were planted, inhibitor I entirely disappeared within a few days as the new plants emerged and grew (148). The new stem and new aerial growth contained appreciable inhibitor I. When new tubers were established they accumulated inhibitor I as the concentration in aerial tissues diminished.

Inhibitor I could be induced to accumulate in leaflets within hours in the light by excising them from the potato plant (144). The accumulation was more effective in constant light than in diurnal light, while in total darkness it did not occur. The detached leaflet provided a convenient system for studying the process of protein accumulation in vegetative tissues.

The accumulated inhibitor I in detached leaves often reached about 1% of the total soluble protein present within 48 hours after detachment. However, in general, older detached leaflets accumulated less inhibitor I than younger leaflets.

The pattern of accumulation was polar in the leaflets, increasing at the base and proceeding toward the tip (144).

When lysine ^{14}C was fed to the detached leaflets in the light a significant amount of the exogenously supplied lysine ^{14}C was incorporated into the accumulated protein (144). The percentage of label incorporated into inhibitor I was over 12% with respect to all labeled TCA-precipitable proteins, suggesting that inhibitor I was being synthesized in detached leaflets in light preferentially compared to other proteins in the leaflets.

Indoleacetic acid was found to inhibit severely the accumulation, suggesting that this hormone may be involved with the regulation of the presence of inhibitor I in the plants (144). It is interesting that IAA did not significantly inhibit accumulation of inhibitor I in leaflets from plants already accumulating this protein before excision of the leaflets. This implies that IAA was not simply inducing or activating a system capable of destroying inhibitor I but was probably affecting protein synthesis at some step before messenger RNA for inhibitor I was translated.

The immunological and immunoelectrophoretic analyses have shown that the inhibitor I protein synthesized in the leaflets is not identical to that made in the tuber (144). The differences were seen in both antigenic determinants and electrophoretic mobility. The inhibitor is composed of subunits and it is possible, if not probable, that the differences between tissues represent differences in the concentration of subunits (105) present.

Among leaflets from several other plants of the Solanaceae family—including potato, pepper, eggplant, petunia, nightshade, and tobacco—containing no detectable inhibitor I protein, only potato leaflets accumulated inhibitor I when detached and supplied with water in constant light and dark for 48 hours (145), although other inhibitors did accumulate (56). In potato leaves the accumulation was found to be similar in nearly every respect to tomato leaves. The differences encountered among the different genera in accumulation are not understood.

Cycloheximide, but not chloramphenicol, was a potent inhibitor of the accumulation of inhibitor I in detached tomato leaflets (149). This indicated that cytoplasmic ribosomes may be involved in the synthesis and not chloroplast or mitochondrial ribosomes. An estimate of the kinetics of the protein synthesis machinery required for the process was made with the data from *E. coli* (117) and suggested that up to 400 redundant gene copies for inhibitor I subunits exist.

The synthesis and accumulation of this large amount of inhibitor I protein in detached potato and tomato leaflets was found to be associated with ultrastructural changes within the plant cells. A direct and consistent relationship was found between the presence of dense protein bodies in the vacuoles of the cells and the presence of inhibitor I (162). More direct evidence, utilizing specific antibodies labeled with ferritin, has confirmed that inhibitor I is a component of the vacuolar protein bodies (159).

Shumway et al (160) hypothesized that as a storage protein inhibitor I might be accumulating in vacuoles of meristematic tissues since its presence in these tissues could nearly always be detected immunologically. With electron microscopy they found that protein bodies did indeed accumulate in the many small vacuoles of very young developing cells of the apical meristems. As the cells matured the central vacuole became enlarged and the protein bodies disappeared.

Kirsi & Mikola (83) found that proteinase inhibitors also accumulated in young barley rootlets in high concentrations and then disappeared, a situation similar to that found in tomato apical meristems. This data implied that the inhibitors were synthesized in the meristems and then utilized for growth and development. New refinements in electron microscopy techniques involving the use of immunological markers with plant tissues should soon contribute significantly to the understanding of inhibitors, their compartmentation and storage, and their utilization in plant developmental processes.

Plant Protection

MICROORGANISMS   A number of reports have been published on the possible role of proteinase in facilitating microbial invasion of plant tissues (86, 87, 95, 128, 139, 142, 174), probably by hydrolyzing proteins in and between cell walls (50, 90, 120). It is therefore possible that large amounts of proteinase inhibitors in tissues directed against these microbial extracellular proteinase might be expected to severely retard proteolysis of cell wall and membrane proteins, reducing the disruption of cell organization.

Many of the proteolytic enzymes known to be secreted by microorganisms such as pronase and subtilisin are trypsin-like or chymotrypsin-like in their specificities (100). When these enzymes are challenged with inhibitors of trypsin or chymotrypsin they usually are strongly inhibited. Thus, the presence of inhibitors in plant tissues could prove to be an asset to the plant by arresting the extracellular proteinases of the pathogens (176). This area of research deserves much more study.

INSECTS   Interest in the effects of plant proteinase inhibitors toward insect attack by inhibiting their digestive proteinases began a quarter of a century ago. In 1947 Mickel & Standish (107) observed that larvae of certain pests were unable to develop normally on soybean products. These observations led Lipke et al in 1954 (94) to study the toxicity of soybean inhibitors on the complete development of *Tribolium confusum*, a common marauder of stored grain. Although these results were negative with respect to trypsin inhibitors, they revealed the presence of a specific inhibitor of *Tribolium* larval digestive proteolysis. Among all of the inhibitors in soybean, only two inhibitor fractions inhibited the midgut enzyme of *Tribolium castaneum* larvae (18). Birk et al (20) later isolated a *Tribolium* proteinase inhibitor from other proteinase inhibitors and showed that the inhibitor could completely inhibit larval gut proteolysis of both *Tribolium castaneum* and *Tribolium confusum*. A *Tribolium* inhibitor was also reported from wheat (11). Like the *Tribolium* inhibitor from soybeans, it was inactive toward trypsin and chymotrypsin.

This work was followed by a study on several inhibitors of the midgut proteinases of *Tenebrio molitor* (10), a common pest that also consumes stored grains. Lima bean inhibitor, ovomucoid, soybean inhibitor, and purified Bowman-Birk inhibitor inhibited *Tenebrio* larval trypsin. The *Tribolium* inhibitor from soybeans was ineffective.

The proteinases of *Tenebrio molitor* were studied in detail by Zwilling (192). Besides the presence of trypsin and chymotrypsin, previously reported by Applebaum et al (10), two new proteinases were reported, called alpha and beta protease. One enzyme was similar in molecular weight (24,000) to trypsin and chymotrypsin but did not hydrolyze typical ester substrates of either, nor was it inhibited by any natural proteinase inhibitors of either. The beta-protease had a molecular weight of about 60,000 and was inhibited by five naturally occurring plant and animal proteinase inhibitors of trypsin and chymotrypsin. This work suggested that insect protein digestion may be more complex than that of higher animals due to the variability in the kinds of digestive endopeptidases that are present.

Applebaum in 1964 (9) proposed that legumes did evolve proteinase inhibitors as a defense mechanism against insects and that digestion should be considered as a factor in host selection. It is clear from accumulating data that the digestive proteinases from a number of insect genera (49, 78, 106, 178, 185, 186, 192) are very similar to animal trypsin and chymotrypsin and are (49, 78, 106, 192) inhibited by plant proteinase inhibitors. The presence of inhibitors in the food of such plant-eating insects might be expected to have a significant effect upon the nutrition. If feeding continued upon such food, then the organism would face a serious shortage of amino acids and the possibility of protein starvation and death. From this point of view, proteinase inhibitors would be of great value to plant tissues when present to the extent of 6 to 10% of the soluble proteins, such as is found in some storage organs, or even the 1% that can accumulate in tomato and potato leaves (144).

The possible involvement of proteinase inhibitors as protective agents toward insects received strong support within the last year when Green & Ryan (53) found that wounding of the leaves of potato or tomato plants by adult Colorado potato beetles or their larvae induced a rapid accumulation of a proteinase inhibitor throughout the aerial tissues of the plants. This effect of insect damage could be simulated by mechanically wounding the leaves. Over 0.2 mg inhibitor protein accumulated per milliliter of leaf juice within 48 hours of wounding. The transport of a factor that has not yet been identified but is called PIIF (proteinase inhibitor inducing factor) was released from the damaged leaves and spread rapidly throughout the plants. In response to the factor, the level of proteinase inhibitor in both damaged and adjacent leaves rose strikingly within a few hours. Wounds inflicted on leaf tissues near midveins were more effective in inducing inhibitor in distal leaves than wounds near the outer edge of the leaves (55).

The half time of transmission of the inducing factor out of wounded leaves was found to be about $3\frac{1}{2}$ hours. The overall response was light and temperature dependent. Inhibitor accumulation increased linearly as a function of light intensity through 300 fc and approached a full response at intensities in excess of 500 fc (54). The temperature dependence was unusual (54). Between 22° and 32° the amount of inhibitor I that accumulated in leaves per unit time due to wounding increased ninefold. The optimum temperature for the response was 36°.

Much if not all of the data concerning the accumulation of inhibitor I in detached leaves reviewed under *Storage* is apparently due to the wound response. The synthesis of inhibitor is therefore de novo. In response to wounding, the inhibitor has been demonstrated to accumulate in the central vacuole of the cell where it remained for long periods of time (161).

The accumulation of a proteinase inhibitor in high concentration in leaves as a result of insect attack demonstrated that the insects can rapidly influence the protein synthesizing machinery in entire plants by attacking only a single leaf. The overall system is reacting in the manner of a primitive immune response. This type of response could be effective in repelling insects or their larvae, providing they have not evolved digestive systems to circumvent the inhibitory activity that accumulates. This response holds promise for designing new approaches to biological pest control.

CONCLUSIONS

Plant tissues, particularly germinating seeds, leaves, flowers, and fruits, are valuable systems for studying the roles of proteinases in the processes of development and senescence. One of the most interesting aspects of recent plant proteinase research is the principal role of plant carboxypeptidases in protein degradation in plant lysosomal-like organelles (aleurone bodies, protein bodies, and spherosomes), and their unusually broad substrate specificity. Also of interest are the many peptide hydrolases that occur in germinating seeds. The purification of these enzymes, including the endopeptidases, will continue to challenge the plant physiologists, molecular biologists, and geneticists to understand their regulation and functions and to apply this knowledge for man's benefit. Some of these rich new sources of enzymes will most certainly entice biochemists to study their structures and mechanisms as well as their biochemical evolution.

Related to the study of proteinases in a number of ways are the proteinase inhibitors whose presence provides more questions than answers. While the inhibitor's role as regulator of plant proteinases is becoming more strongly supported by evidence, this is probably only one of its functions in nature. Inhibitors are stored in tissues, utilized, and thus must be susceptible to proteolytic degradation. Virtually nothing is known of their mode of disappearance in tissues.

The most intriguing and complex—but perhaps the most potentially useful—function in plants is that of plant protection. The relationships between plants and insects and between plants and microorganisms are very complex, but the multiplicity of both proteinases and inhibitors complicates the picture even further. The recent discovery that insect wounding of a single plant leaf induces a proteinase inhibitor to accumulate throughout the plant presents a whole new concept of possible induced immunity toward marauding pests and possibly against some pathogenic microorganisms. From the accumulated data it is certain that the role of proteinase inhibitors in plants is not a passive one. It is also apparent that research concerning functions of proteinase inhibitors in plants has only just begun.

ACKNOWLEDGMENT

The author wishes to thank Dr. Milton Zucker for critically reading the manuscript and for his many helpful suggestions.

Literature Cited

1. Ako, H., Foster, R. J., Ryan, C. A. 1972. *Biochem. Biophys. Res. Commun.* 47:1402–7
2. Ako, H., Foster, R. J., Ryan, C. A. In review
3. Ako, H., Ryan, C. A., Foster, R. J. 1972. *Biochem. Biophys. Res. Commun.* 46:1639–45
4. Amagase, S., 1972. *J. Biochem.* 72: 73–81
5. Amagase, S., Nakayama, S., Tsugita, A. 1969. *Biochem. J.* 66:431–39
6. Ambe, K. S., Sohonie, K. J. 1956. *Experientia* 12:302–3
7. Ambe, K. S., Sohonie, K. J. 1956. *J. Sci. Ind. Res.* 15C:136–40
8. Anderson, J. W., Rowan, K. S. 1965. *Biochem. J.* 97:741–46
9. Applebaum, S. W. 1964. *J. Insect Physiol.* 10:783–88
10. Applebaum, S. W., Birk, Y., Harpaz, I., Bondi, A. 1964. *Comp. Biochem. Physiol.* 11:85–103
11. Applebaum, S. W., Konijn, A. M. 1966. *J. Insect. Physiol.* 12:665–69
12. Arnon, R. 1970. *Methods Enzymol.* 19:226–43
13. Ashton, F. M., Dahman, W. J. 1968. *Phytochemistry* 7:1899–1905
14. Austin, S., Schwimmer, S. 1970. *Enzymologia* 40:273–85
15. Beevers, L. 1968. *Phytochemistry* 7: 1837–44
16. Belitz, H. D., Kaiser, K. P., Santarius, K. 1971. *Biochem. Biophys. Res. Commun.* 42:420–27
17. Bhatty, R. S. 1969. *Cereal Chem.* 46:74–75
18. Birk, Y., Applebaum, S. W. 1960. *Enzymologia* 22:318–26
19. Birk, Y., Gertler, A. 1971. *Proc. Int. Res. Conf. Proteinase Inhibitors, 1st, 1970,* ed. H. Fritz, H. Tschesche, 142–48. Berlin: De Gruyter. 304 pp.
20. Birk, Y., Gertler, A., Khalef, S. 1963. *Biochim. Biophys. Acta* 67:326–28
21. Birk, Y., Waldman, M. 1965. *Qual. Plant. Mater. Veg.* 12:199

22. Blow, D. M. et al 1972. *J. Mol. Biol.* 69:137–44
23. Buckner, J. 1971. PhD thesis. Univ. North Dakota. 84 pp.
24. Burger, W. C., Prentice, N., Kastenschmidt, J., Moeller, M. 1968. *Phytochemistry* 7:1261–70
25. Burger, W. C., Prentice, N., Moeller, M. 1970. *Plant Physiol.* 40:860–61
26. Burger, W. C., Prentice, N., Moeller, M., Kastenschmidt, J. 1970. *Phytochemistry* 9:33–40
27. Burger, W. C., Siegelman, H. W. 1966. *Plant Physiol.* 19:1089–93
28. Cameron, E. C., Mazelis, M. 1971. *Plant Physiol.* 48:278–81
29. Carey, W. F., Wells, J. R. E. 1970. *Biochem. Biophys. Res. Commun.* 41:574–81
30. Carey, W. F., Wells, J. R. E. 1972. *J. Biol. Chem.* 247:5573–79
31. Carlson, P. S. 1972. *Nature, New Biol.* 237:39–41
32. Catsimpoolas, N., Funk, S. K., Wang, J., Kenney, J. 1971. *J. Sci. Food. Agr.* 22:79–82
33. Chu, H.-M., Chi, C.-W. 1965. *Sci. Sinica* 14:1441
34. Dechary, J. M. 1970. *Econ. Bot.* 24: 113–22
35. Doolittle, R. F., Armentrout, R. W. 1968. *Biochemistry* 7:516–21
36. Enari, T.-M., Mikola, J. 1967. *Eur. Brew. Conv. Proc. 11th Congr.*
37. Enari, T.-M., Puputii, E., Mikola, J. 1963. *Eur. Brew. Conv. Proc. 9th Congr.,* 37–44
38. Evins, W. H., Varner, J. E. 1972. *Plant Physiol.* 49:348–52
39. Feeney, R. E. 1971. See Ref. 19, 162–68
40. Feeney, R. E., Allison, R. G. 1969. *Evolutionary Biochemistry of Proteins,* 199–244. New York: Wiley. 290 pp.
41. Feinstein, G., Feeney, R. E. 1966. *J. Biol. Chem.* 241:5183–89
42. Filner, P., Wray, J. L., Varner, J. E. 1969. *Science* 165:358–67

43. Folkers, K., Enzmann, F., Boler, J., Bovers, C. Y., Shelley, A. V. 1969. *Biochem. Biophys. Res. Commun.* 37:123–36
44. Foster, R. J., Ryan, C. A. 1965. *Fed. Proc.* 24:473
45. Fratelli, V. 1969. *J. Biol. Chem.* 244: 274–80
46. Gardner, G., Pike, C. S., Rice, H. V., Briggs, W. R. 1971. *Plant Physiol.* 48:686–93
47. Garg, G. K., Virupaksha, T. K. 1970. *Eur. J. Biochem.* 17:4–12
48. Ibid, 13–18
49. Giebel, W., Zwilling, R., Pfleiderer, G. 1971. *Comp. Biochem. Physiol.* 38B:197–210
50. Ginzburg, B. Z. 1961. *J. Exp. Bot.* 12:85–107
51. Glazer, A. N., Smith, E. L. 1970. *The Enzymes*, ed. P. Boyer, 3:502–46. New York: Academic. 886 pp.
52. Graf, G., Hoagland, R. E. 1969. *Phytochemistry* 8:827–29
53. Green, T. R., Ryan, C. A. 1972. *Science* 175:776–77
54. Green, T. R., Ryan, C. A. 1972. *Plant Physiol.* 51:19–21
55. Green, T. R., Ryan, C. A. In review
56. Gurusiddaiah, S., Kuo, T., Ryan, C. A. 1972. *Plant Physiol.* 50:627–31
57. Hartley, B. S. 1960. *Ann. Rev. Biochem.* 29:45–72
58. Hasegawa, J. 1963. *J. Histochem. Cytochem.* 11:474–77
59. Haynes, R., Feeney, R. E. 1967. *J. Biol. Chem.* 242:5378–85
60. Hochstrasser, K., Illchmann, K., Werle, E. 1970. *Z. Physiol. Chem.* 351:721–28
61. Ibid, 1503–12
62. Hochstrasser, K., Werle, E., Schwarz, S., Siegelman, R. 1969. *Z. Physiol. Chem.* 350:249–54
63. Hochstrasser, K., Werle, E., Siegelman, S., Schwarz, S. 1969. *Z. Physiol. Chem.* 350:897–902
64. Hojima, Y., Moriya, H., Moriwaki, C. 1971. *J. Biochem.* 69:1019–25
65. Hojima, Y., Ryan, C. A. Personal communication
66. Hojima, Y., Tanaka, M., Moriya, H., Moriwaki, C. 1971. *Allergy* 20:755–62
67. Ibid, 763–69
68. Horiguchi, T., Kilagishi, K. 1969. *Nippon Dojo-Hisyogaka Zasshi* 40: 255
69. Ihle, J. N., Dure, L. S. III 1969. *Biochem. Biophys. Res. Commun.* 36: 705–10
70. Ibid 1970. 38:995–1001
71. Ihle, J. N., Dure, L. S. III, 1972. *J. Biol. Chem.* 247:5034–40
72. Ibid, 5041–47
73. Ibid, 5048–55
74. Ikenaka, Z., Smid, K. 1965. *Proc. Soc. Exp. Biol. Med.* 120:748–49
75. Irving, G. W. Jr., Fontaine, T. D. 1945. *Arch. Biochem. Biophys.* 6:351–64
76. Iwasaki, T., Kiyohara, T., Yoshikawa, M. 1971. *J. Biochem.* 70: 817–26
77. Jacobsen, J. V., Varner, J. E. 1967. *Plant Physiol.* 42:1596–1600
78. Kafatos, F. C., Tartakoff, A. M., Law, J. H. 1967. *J. Biol. Chem.* 242: 1477–87
79. Kakade, M. L., Simons, N. R., Leiner, I. E. 1970. *Biochim. Biophys. Acta* 200:168–69
80. Kiminski, E., Busuk, W. 1969. *Cereal Chem.* 46:317–24
81. Kassell, B. 1970. *Methods Enzymol.* 19:839–905
82. Kato, H., Iwanaga, S., Suziki, T. 1966. *Experientia* 22:49–50
83. Kirsi, M., Mikola, J. 1971. *Planta* 96:281–91
84. Kolehmainen, L., Mikola, J. 1971. *Arch. Biochem. Biophys.* 145:633–42.
85. Krahn, J. Stevens, F. C. 1970. *Biochemistry* 9:2646–52
86. Kuc, J. 1962. *Phytochemistry* 52:961
87. Kuc, J., Williams, E. B. 1962. *Phytopathology* 55:1285
88. Kunimitsu, D. K. Yasunobu, K. T. 1970. *Methods Enzymol.* 19:244–51
89. Kuraishi, S. 1968. *Physiol. Plant.* 21:78–83
90. Lamport, D. T. A. 1970. *Ann. Rev. Plant. Physiol.* 21:235–70
91. Laskowski, M. Jr., Sealock, R. W. 1971. See Ref. 51, 3:375–473
92. Leiner, I. E., Friedenson, B. 1970. *Methods Enzymol.* 19:261–73
93. Leiner, I. E., Kakade, M. L. 1969. *Toxic Constituents of Plant Foodstuffs*, ed. I. E. Leiner, 7–68. New York: Academic. 500 pp.
94. Lipke, H., Fraenkel, G. S., Leiner, I. E. 1954. *J. Agr. Food. Chem.* 2:410–14
95. Lumsden, R. D. 1968. *Diss. Abstr.* 78:17
96. Mansfield, V., Zieglehoffer, A., Horakova, J., Hladovec, J. 1959. *Naturwissenschaften* 46:172
97. Martin, C., Thimann, K. V. 1972. *Plant Physiol.* 49:64–71
98. Matile, Ph. 1968. *Z. Pflanzenphysiol.* 58:365

99. Matile, Ph., Winkenbach, F. 1971. *J. Exp. Bot.* 22:759-71
100. Matsubara, H., Feder, J. 1971. See Ref. 51, 3:721-95
101. Matsushima, K. 1955. *J. Agr. Chem. Soc. Jap.* 29:883-87
102. Mayer, A. M., Poljakoff-Mayber, A. 1963. *The Germination of Seeds.* New York: Macmillan. 236 pp.
103. McDowell, M. A. 1970. *Eur. J. Biochem.* 14:214-21
104. McKee, H. S. 1962. *Nitrogen Metabolism in Plants,* 329. Oxford: Clarendon. 728 pp.
105. Melville, J. C., Ryan, C. A. 1972. *J. Biol. Chem.* 247:3445-53
106. Melville, J. C., Scandalios, J. G. 1972. *Biochem. Genet.* 7:15-31
107. Mickel, C. E., Standish, J. 1947. *Univ. Minn. Agr. Exp. Sta. Tech. Bull. 178*
108. Mikola, J., Enari, T.-M. 1970. *J. Inst. Brew.* 76:182-88
109. Mikola, J., Kirsi, M. 1972. *Acta Chem. Scand.* 26:787-95
110. Mikola, J., Kolehmainen, L. 1972. *Planta* 104:167-77
111. Mikola, J., Pietila, K., Enari, T.-M. 1971. *Eur. Brew. Conv. Proc. 13th Congr.,* 21-28
112. Mikola, J., Soulinna, E.-M. 1969. *Eur. J. Biochem.* 9:555-60
113. Mikola, J., Soulinna, E. M. 1971. *Arch. Biochem. Biophys.* 144:566-75
114. Mittman, C., Ed. 1972. *Pulmonary Emphysema and Proteolysis.* New York: Academic. 562 pp.
115. Moeller, M., Burger, W. C., Prentice, N. 1969. *Phytochemistry* 8:2153-56
116. Moeller, M., Robbins, G. S., Burger, W., Prentice, N. 1970. *J. Agr. Food. Chem.* 18:886-90
117. Morse, D. E., Baker, R. F., Yanofsky, C. 1968. *Proc. Nat. Acad. Sci. USA* 60:1428-35
118. Murachi, T. 1971. *Methods Enzymol.* 19:273-84
119. Nakayama, S., Amagase, S. 1968. *Proc. Acad. Sci.* 44:358
120. Newcomb, E. H. 1963. *Ann. Rev. Plant. Physiol.* 14:43-64
121. Ofelt, C. N., Smith, A. K., Mills, J. M. 1955. *Cereal Chem.* 32:53-63
122. Ory, R. L., Henningsen, K. W. 1969. *Plant Physiol.* 44:1488-98
123. Ouchterloney, O. 1949. *Acta Pathol. Microbiol. Scand.* 26:507
124. Penner, D., Ashton, F. M. 1966. *Nature* 212:935-36
125. Penner, D., Ashton, F. M. 1967. *Plant Physiol.* 42:791-96
126. Pike, C. S., Briggs, W. R. 1972. *Plant Physiol.* 49:521-30
127. Pinski, A., Grossman, S. 1969. *J. Sci. Food. Agr.* 20:374-75
128. Porter, F. M. 1966. *Phytopathology* 56:1424-25
129. Prentice, N., Burger, W. C., Kastenschmidt, J., Huddle, J. D. 1967. *Physiol. Plant.* 20:361-67
130. Prentice, N., Burger, W. C., Kastenschmidt, J., Moeller, M. 1970. *Phytochemistry* 9:41-48
131. Prentice, N., Burger, W. C., Moeller, M. 1971. *Cereal Chem.* 48:587-94
132. Prentice, N., Burger, W. C., Moeller, M., Kastenschmidt, J. 1970. *Cereal Chem.* 47:282-87
133. Prentice, N., Burger, W. C., Wiederholt, E. 1969. *Physiol. Plant.* 22:157-60
134. Prentice, N., Moeller, M., Pomeranz, Y. 1971. *Cereal Chem.* 48:714-17
135. Puztai, A. 1967. *Nutr. Abstr. Rev.* 37:1-9
136. Puztai, A., Duncan, I. 1971. *Planta* 96:317-25
137. Rackis, J. J., Anderson, R. L. 1964. *Biochem. Biophys. Res. Commun.* 15:230-35
138. Racusen, D., Foote, M. 1970. *Can. J. Bot.* 48:1017-21
139. Ragozina, I. F., Schneider, Y., Lipsits, D. V. 1969. *Dokl. Akad. Nauk SSR* 184:242-45
140. Rancour, J. M., Ryan, C. A. 1968. *Arch. Biochem. Biophys.* 125:380-83
141. Read, J. W., Haas, L. W. 1938. *Cereal Chem.* 15:59-68
142. Reddy, M. N., Stuteville, D. L., Sorensen, E. L. 1971. *Phytopathology* 61:361
143. Ryan, C. A. 1967. *Anal. Biochem.* 19:434-40
144. Ryan, C. A. 1968. *Plant. Physiol.* 43:1859-65
145. Ibid, 1880-81
146. Ryan, C. A. 1971. *Biochem. Biophys. Res. Commun.* 44:1265-70
147. Ryan, C. A., Huisman, O. C. 1967. *Nature* 214:1047-49
148. Ryan, C. A., Huisman, O. C., Van Denburgh, R. W. 1968. *Plant Physiol.* 43:589-96
149. Ryan, C. A., Huisman, W. 1970. *Plant Physiol.* 45:484-89
150. Scala, J., Iott, K., Schwab, D. W., Samersky, F. E. 1969. *Plant Physiol.* 44:367-71
151. Scandalios, J. G. 1969. *Biochem. Genet.* 3:37-79
152. Schnarrenberger, C., Oser, A., Tolbert, N. F. 1972. *Planta* 104:185-94

153. Seidel, D. S., Leiner, I. E. 1972. *Biochim. Biophys. Acta* 258:303–9
154. Shain, Y., Mayer, A. M. 1965. *Physiol. Plant.* 18:853–59
155. Shain, Y., Mayer, A. M. 1968. *Phytochemistry* 7:1491–98
156. Shaw, D. C., Wells, J. R. E. 1967. *Biochem. J.* 104:5C
157. Shibaoka, H., Thimann, K. V. 1970. *Plant Physiol.* 46:212–20
158. Shinano, S., Kukushima, K. 1969. *Agr. Biol. Chem.* 33:1203–36
159. Shumway, L. K., Cheng, V., Ryan, C. A. In preparation
160. Shumway, L. K., Cheng, V., Ryan, C. A. 1972. *Planta* 106:279–90
161. Shumway, L. K., Rancour, J. M., Ryan, C. A. 1970. *Planta* 93:1–14
162. Shumway, L. K., Ryan, C. A. 1972. *Plant Physiol. Suppl.* In press
163. Skupin, J., Warchalewski, J. 1971. *J. Sci. Food Agr.* 22:11–15
164. Sprossler, B., Heilman, H.-D., Grampp, E., Uhlig, H. 1971. *Z. Physiol. Chem.* 352:1524–30
165. St. Angelo, A., Ory, R. L., Hansen, H. J. 1970. *Phytochemistry* 9:1933–38
166. Stern, M. S. 1970. *Plant Physiol.* 46S:16
167. Sze, H., Ashton, F. M. 1971. *Phytopathology* 10:2935–42
168. Szewczuk, A., Kwiatkowska, J. 1970. *Eur. J. Biochem.* 15:92–96
169. Tookey, H. L., Gentry, H. S. 1969. *Phytochemistry* 8:989–91
170. Tschesche, H., Kupfer, S. 1972. *Eur. J. Biochem.* 26:33–36
171. Tur-Sainai, A., Birk, Y., Gertler, A. 1970. *Isr. J. Chem.* 8:176 P
172. Uy, R. L., Feeney, R. E. 1971. *Fed. Proc.* 30:1130
173. Vauquelin, L. N. 1799. *Ann. Chim.* 43:267
174. Van Etten, H., Bateman, D. F. 1965. *Phytopathology* 55:1285
175. Visuri, K., Mikola, J., Enari, F. M. *Eur. J. Biochem.* 7:193–99
176. Vogel, R., Trautshold, I., Werle, E. 1966. *Natural Proteinase Inhibitors.* New York: Academic. 159 pp.
177. Wall, J. R. 1968. *Biochem. Genet.* 2:109–18
178. Ward, C. 1972. *Comp. Biochem. Physiol.* 42B:131–35
179. Wells, J. R. E. 1965. *Biochem. J.* 97:228–35
180. Wells, J. R. E. 1969. *Biochim. Biophys. Acta* 167:388–98
181. Werle, E. 1971. See Ref. 19, 23–27
182. West, N. B., Garber, E. D. 1967. *Can. J. Genet. Cytol.* 9:640–45
183. Ibid, 646–55
184. Wild-Altamirano, C. 1969. *J. Food Sci.* 34:235–38
185. Yang, Y. J., Davies, D. M. 1971. *J. Insect Physiol.* 17:117–31
186. Ibid 1972. 18:747–55
187. Yatsu, L. Y., Jacks, T. Y. 1968. *Arch. Biochem. Biophys.* 124:466–71
188. Zanefeld, L. J. D., Polakowski, K. L., Robertson, R. T., Williams, W. L. 1971. See Ref. 19, 236–46
189. Zuber, H. 1964. *Nature* 201:613
190. Zuber, H. 1968. *Z. Physiol. Chem.* 349:1337–52
191. Zuber, H., Matile, Ph. 1968. *Naturforscher* 23:663
192. Zwilling, R. 1968. *Z. Physiol. Chem.* 349:326–32

Ann. Rev. Plant Physiol. 1973. 24:197–224

SENESCENCE AND POSTHARVEST PHYSIOLOGY[1]

❖ 7547

Joseph A. Sacher[2]

Department of Biology, California State University, Los Angeles

CONTENTS

INTRODUCTION

Senescence of higher plants may be classified into three major types, namely: (*a*) population senescence (e.g. annual plants); (*b*) organism or individual plant senescence; and (*c*) determinate organ senescence, namely, leaves, fruits, and petals (134). This review will be concerned largely with the senescence of fruits, but with some comparisons made with other determinate organs. We shall use the definition of senescence as being the final phase in ontogeny of the organ in which a series of normally irreversible events is initiated that leads to cellular breakdown and death of the organ. Fruit ripening is regarded as a normally

[1] The following abbreviations will be used: ABA (abscisic acid); ADP and ATP (adenosine di- and triphosphate); BA (benzyl adenine); B-9 (succinic acid 2,2-dimethyl hydrazide); CEPA (2-chloroethylphosphonic acid); CH (cycloheximide); CK (cytokinin); 2,4-D (2,4-dichlorophenoxyacetic acid); DNA (deoxyribonucleic acid); GA (gibberellic acid); IAA (indole-3-acetic acid); RNA (ribonucleic acid); RNase (ribonuclease).

[2] Supported in part by grant GB-18071 from the National Science Foundation.

nonreversible process (21, 65, 76, 112, 131, 186) and was recognized as a senescent phase over 40 years ago by Blackman & Parija (29). Aesthetically and commercially, however, the term ripening is preferred to senescence, but the terms will be used interchangeably in this article. Under certain circumstances senescence may be experimentally reversed in leaves (70, 241), and to some extent in fruits (50), but this does not affect the use of the term as defined above.

Determinate organs appear to undergo a genetically programmed senescence of the type not dependent on specific "aging genes" (213). It has been stated previously that "there is no need to suppose any special genetic mechanism of senescence, as it can reasonably be explained by the general gene actions of normal development" (88). Although these views have not been proved, they are consistent with the absence of any evidence thus far for the presence of specific "aging genes" and the overwhelming evidence that the many enzymes called into play during organ senescence, as shown by their conspicuous increases in activity, are the same enzymes that function during development. Enzymes which have been investigated include respiratory enzymes (103, 108, 178), starch degrading enzymes (55, 153), and hydrolases which have the potential for catalyzing the degenerative aspects of senescence (59, 89, 92, 114, 124, 139, 177, 244). Others include the enzymes involved in both sucrose synthesis and hydrolysis during the sugar accumulation that attends grape ripening (107).

In this review attention will be given largely to endogenous factors and processes which affect senescence of fruits. Senescence in the whole plant has been discussed comprehensively (41, 134, 237); for general reviews of plant and animal senescence and theories of aging see (51, 54). Coverage will begin with developments since the most recent reviews on fruit ripening (95) and the biochemistry of senescence (224) in this series. Some of the past reviews on fruit ripening are the following (18–21, 23, 36, 64, 95, 170, 172, 207, 222). Of primary importance for fruit ripening are the two recent volumes on the physiology and biochemistry of fruits (118, 119). For the most recent reviews on the physiology and biochemistry of foliar senescence see (175, 205, 224, 225, 243) and especially the recent comprehensive review by Beevers (15); for ultrastructural changes in leaf and fruit senescence see (40). Following are some reviews relevant to senescence on lysosomes (150), growth regulators (80, 129, 210, 218, 238), growth retardants (3, 42, 127, 201), and ethylene (2, 36, 38, 145, 172, 208).

RNA METABOLISM

What is known about the metabolism of RNA during ripening indicates that there can be considerable variability among different fruits. In harvested yellow transparent apples (142) total RNA increased markedly (ca. 50%) during the climacteric, to such a large extent that the authors suggest the increase must be due largely to ribosomal synthesis. In another variety of apples (Cox's Orange Pippin) there occurs a sharp rise in RNase concommitant with the climacteric (177). During banana ripening there is no change in base composition nor the content of RNA in pulp tissue (232). From MAK column profiles of phenol-extracted RNA it was concluded that there are no significant variations in total RNA (181) during avocado ripening nor in the relative distribution of RNA among the different fractions. During the early part of the respiratory rise, ^{32}P was incorporated into a heavy ribosomal region as well as a mRNA fraction, but it declined thereafter and had ceased in mRNA by the climacteric peak. Coextensively there occurred a rapid increase of labelling into the sRNA fraction which could be attributed to products of RNA degradation, as observed by others (45, 143) during leaf senescence. Total RNA, extracted from avocado by a method of Smillie & Krotkov (206), declined during ripening (averaging 16% ± 3.9 in four experiments), paralleled by a decline in the rate of uridine incorporation (242). The disparity between these results and those of Richmond & Biale (181) on incorporation of precursors in avocado tissue may be due to the length of the tissue incubations in dilute solutions; i.e. impairment of RNA synthesis due to osmotic damage may have occurred in the 3-hour period used in the one instance (242) and not in the 1-hour period used in the other (181).

Using sucrose density gradient centrifugation, Ku & Romani (131) demonstrated that at the onset of the climacteric in pears there is a transient peak in the rate of synthesis of ribosomes, followed by a marked decline in such synthesis thereafter. Assay of synthesis was based on incorporation of radioactive uridine. The authors found no change in the relative amounts of monosomes and polysomes until after the climacteric peak, when polysomes declined. Ribosomes of the fig fruit undergo a gradual decline attending the climacteric rise, so that monomers predominated throughout the senescent phase (147). Some evidence of the dependency of ripening on synthesis of RNA is in the fact that ripening of pears (76) is inhibited by actinomycin D.

Peel tissue from preclimacteric apples which were treated with ethylene was used to follow ripening phenomenon (122, 178). Ethylene enhanced uridine incorporation into the disks prior to any detectable enhancement of incorporation of valine in Worcester Pearmain apples, which later paralleled the development of the malate decarboxylating system. From these data it was suggested that development of the malate system may be dependent upon prior RNA synthesis. In Cox's Orange Pippin apples, however, the increase in incorporation of valine and uridine occurred simultaneously.

Most of the evidence above, based on assays of the rate of incorporation of RNA precursors, indicated that RNA synthesis, including mRNA, was enhanced during the early part of the climacteric rise but declined rapidly thereafter and in avocado had ceased by the climacteric peak. In general this decline in synthetic capacity is not associated with any significant loss of RNA. The physiological significance of an increase in RNase is not clear, as an increase in this enzyme has been associated with a large increase in RNA in apples and a small decrease in avocados. Such differences must reside in differences in protoplasmic compartmentation.

In contrast to fruit, senescence of leaves is associated with a large degradation of RNA and protein (see review 15). The decline in RNA is correlated with a large increase in RNase (61, 67, 116, 209, 212, 220, 221), but exceptions have been reported (171). Dyer & Osborne (70) suggested that the capacity of a leaf for reversal of senescence may be related to metabolism of chloroplast RNA. They found that retention of chloroplast RNA in leaves that are yellow and senescent in appearance, and have already lost considerable cytoplasmic RNA and protein, allows regreening and thus reversal of senescence. This raises the possibility that a similar basis may exist for regreening of citrus fruits.

Strehler's (213) hypothesis that aging or senescence may be associated with a loss of certain species of tRNAs or their associated synthetases has been examined using soybean cotyledons as a model. Bick et al (26) found a sharp decrease in one species of tyrosyl-tRNA with cotyledon age and a concomitant increase in another species. They also observed that retardation of senescence by BA was associated with a 50% increase in two species (numbers 5 and 6) of leucyl-tRNA. They concluded that although the increase in these species could affect new synthesis of degradative enzymes associated with senescence, it was difficult to reconcile BA enhancing an increase in these tRNAs with its effect on retarding senescence. Although speculative, the increase in these tRNAs caused by BA would be consistent with BA retardation of senescence if they resulted in synthesis of proteins necessary for formation of repressors of senescence-associated hydrolases (e.g. RNase). In this connection DeLeo & Sacher (unpublished results) found evidence for development of repressors of RNase in plant tissues based on the observed stimulation of RNase synthesis by actinomycin D and the inhibition of this stimulation by cycloheximide. This regulation of RNase is very similar to that reported previously for RNase and other enzymes in bacteria (156) and animal tissues (81, 217).

PROTEIN METABOLISM

During ripening of apples (137, 177), pears (96, 97), and cantaloupe (187) there is a net increase in protein, but no change in protein content in bananas (33, 195, 196, 232) and grapefruit (183), and a decrease (16% $\pm$ 3.7) in total protein was found in assays of four different batches of Haas & Fuerte avocados (242), extracted according to the method of Smillie & Krotkov (206). Biale & Young (24) discuss the results of others and their own analyses and conclude that no increase in total protein is associated with ripening of avocado fruit. Davies & Cocking (55, 56) found that protein increases in tomato during the climacteric, associated with softening of the locule tissue, and decreases during the later stages of ripening. A review of the literature on changes in protein (115) in ripening tomato shows protein changes ranging from a decrease to an increase. Such differences may be due at least in part to changes in the physical and chemical characteristics of the tissue during ripening and the extraction procedures used. In this context it should be noted that many of the authors cited herein refer to methods used to counteract or ameliorate difficulties encountered during extraction of proteins from fruit tissues during different stages of ripening, particularly because of variable amounts of phenols or tannins (e.g. 48, 59, 63, 65, 77, 89, 107, 109, 185, 247). Mitochondrial protein of pears declines from the climacteric minimum through the peak (162), although the authors report a "slight though not statistically verifiable" increase in mitochondrial yield during the climacteric rise.

Polyacrylamide gel electrophoresis has been used to obtain profiles from several fruit tissues showing changes in protein bands or deletions or additions of isozymes of specific enzymes during ripening (48, 76, 89), a method that will doubtless prove more valuable in the future in more detailed studies of changes in activity of specific enzymes during ripening. It has been suggested (175) that each isozyme has a particular set of physical properties more favorably attuned to a particular cellular environment during growth or senescence. In this respect Haard (89) refers to preliminary studies during banana ripening which show a difference in substrate specificities among isozymes of peroxidase. By application of polyacrylamide gel electrophoresis for separation of radioactive proteins extracted from preclimacteric and early climacteric pears, Frenkel et al (76) found that some of the increase in activity of malic enzyme was a result of protein synthesis. From changes in the patterns of banana proteins detected by gel electrophoresis, Brady et al (33) conclude that a redirection of protein synthesis was involved in ripening.

Over the past several years there has been increasing use of radioisotopes for study of changes in protein synthesis during fruit ripening, these procedures having proved to be useful in studies of leaf senescence. In earlier studies Sacher (195) reported that in 1×5 mm disks of banana pulp tissue the rate of incorporation of three different concentrations of leucine and phenylalanine began to decrease before the onset of the respiratory climacteric and then declined rapidly up to the peak, but at different rates, with the rates being inversely proportional to the concentration. Comparable decreases were observed in avocado (242). Richmond & Biale (179, 180) conducted similar experiments (except the incubations of disks were shorter, 60 to 90 min) with disks of avocado pulp tissue from the preclimacteric, respiratory rise, and peak periods and found increased amino acid incorporation during the respiratory rise, which was inhibited by puromycin. Brady et al (33) studied protein synthesis during banana ripening with thick slices (3–10 mm) maintained in moist air, which underwent all aspects of normal ripening (169) that occurred in the intact fruit. Labeled leucine was provided by vacuum infiltration and the tissue slices incubated for 1 hour before extraction and assay. In slices that were treated with ethylene there was a strong stimulation of the rate of incorporation of lysine during the climacteric rise (at 13 to 24 hours); this dropped off sharply by 40 hours, which is about the middle of the climacteric rise.

The need for protein synthesis during ripening of bananas (33) and pears (76) was shown by the complete inhibition of ripening by cycloheximide. Of interest was the decrease in effectiveness of cycloheximide as it was administered later in the climacteric, indicating that enzymes essential for ripening had undergone synthesis earlier in the climacteric. This was verified by the enhanced incorporation of phenylalanine into proteins by pear tissue during the early part of the rise.

Hulme and co-workers (78, 79, 121, 178) have investigated the relationship of protein synthesis to other aspects of ripening using two methods. In the first method disks of peel are removed from preclimacteric apples, treated with ethylene, and then assayed over 24 hours for several biochemical phenomena associated with the climacteric. They also removed disks from apples as the climacteric develops in intact fruit. Using both methods they found that enhancement of protein synthesis is part of a ripening sequence in apple peel, which is preceded by stimulation of acetate incorporation into lipid and ethylene production and followed by the development of the malate effect.

Ripening of grapefruit, a nonclimacteric fruit, was associated with a marked and continuous decline in ^{14}C-leucine incorporation into flavedo (peel) disks, while uptake was unchanged (183). Despite this decline a large increase in activity of phenylalanine ammonia-lyase (PAL) occurred in the flavedo of intact fruit after gamma irradiation or ethylene treatment, and in freshly cut untreated flavedo disks (184). Since the increase in activity was inhibited by cycloheximide, it was ascribed to protein synthesis, which is consistent with the more recent demonstration of de novo synthesis of PAL in another tissue based on an increased buoyant density after incubation of the tissue in deuterium oxide (200).

Whether there is an increase or decrease or no change in total protein during ripening does not seem important, as it is apparent in all cases that increases in activity of many enzymes occur in each instance (Table 1). Also, the respiratory climacteric is unaffected by a decrease in mitochondrial protein during ripening of pears. Studies of the rate of incorporation of amino acids indicate that there is a period of accelerated protein synthesis early in the climacteric which is construed, but not proved, to be a reflection of the synthesis of enzymes associated with ripening. On the other hand, the marked decline in the rate of amino acid incorporation during ripening to the point of a complete loss by the climacteric (33, 180, 195), and the loss of capacity to repair damage from ionizing radiation (186), are clearly marks of a nonreversible senescence process. In a study of leucyl-tRNA synthetase changes in senescing cotyledons, Bick & Strehler (27) found that in senescing cotyledons the capacity of tRNA^{Leu} synthetase to aminoacylate tRNA^{Leu} species 1–4 had decreased to 10 to 16% of the values in the young cotyledons. Their data suggested that this was due to the loss of one of the two synthetase fractions for tRNA^{Leu} 1–4 during aging. Were a programmed change of this kind to occur during fruit ripening, it would be consistent with the rapid decline in protein synthesis observed in banana (33, 195) and avocado (180).

ENZYMES

In the past several years there have been a large number of investigations of changes in enzyme activity associated with ripening (Table 1). With few exceptions the hydrolytic enzymes investigated undergo increases in activity during ripening. There is increasing awareness of the effects of physical and chemical changes during ripening on extraction of enzymes and use of procedures that afford protection of enzymes against inactivation by phenols and tannins, as well as ensuring that changes in activity are not due to differential distribution of the enzyme during extraction (see citations in preceding section). The credibility of results obtained without such precautions may be questioned.

A number of enzymes which have the potential for catalyzing degenerative aspects of senescence undergo substantial increases during the climacteric, including cell wall enzymes such as polygalacturonase (114, 250), pectinmethylesterase (113, 124), and cellulase (92), increases which correlate with softening. Increases in chlorophyllase (139, 177) and lipase (177) have been associated with breakdown of chlorophyll and increases in free fatty acids respectively. It has been suggested that the fourfold increase in lipoxidase that occurs during apple ripening could be involved in ethylene biosynthesis (244). Although several metabolic roles for peroxidase are known (28, 80, 166), the function of its increased activity in fruit ripening is unknown (89, 130, 152). Haard (89) found peroxidase in both soluble and precipitate fractions in pear, but all of the increase in activity occurred in the particulate enzyme, which was associated with two isozymes not found in the soluble fraction. In this connection, Ku et al (130) report that during tomato ripening the threefold increase in soluble peroxidase is associated with loss of one and formation of three new isozymes. They suggest that the enzyme may be involved in ethylene synthesis during ripening, although this view is disputed (126). During ripening of mangoes the climacteric and ethylene production are attended by a large increase in activity of peroxidase and catalase associated with the disappearance of a heat-labile, nondialyzable inhibitor of these enzymes (152). An increase in these enzymes was also induced by treatment of mango tissue slices with ethylene.

Auxin retards ripening of banana (227) and grape (90, 91), and in the latter fruit endogenous auxin declines at the time that ripening commences. Since peroxidase has the capacity for IAA breakdown (204), it could provide a basis for control of ripening, i.e. by destroying endogenous IAA and thereby rendering the tissue sensitive to ethylene. Frenkel (74) investigated isozymes of peroxidase and IAA oxidase in pear, tomato, and the nonclimacteric blueberry and found an increase in from one to three isozymes of IAA oxidase in each of the fruits during ripening, but found an increase in peroxidase isozymes only in pear and tomato. He suggested that the increase in IAA oxidase in both climacteric and nonclimacteric fruits is consistent with the view that fruit ripening is accompanied by an increase in capacity for auxin degradation, which is necessary to make the fruit sensitive to ethylene. This view fits with the report below (52) of the disappearance of auxin as grape ripening is initiated and the inhibition of pear ripening by infiltration with auxin (75).

During apple ripening there occurs both an increase in soluble RNase (177) and an increase in total RNA (142). Perhaps the RNase is sequestered in the vacuole during the early period of ripening, as Matile and co-workers (150, 151) presented evidence for a "lysosomal" role of the vacuole in higher plants. A large increase occurs in RNase during senescence of bean endocarp (61, 198), associated with large decreases in nucleic acids and protein, but senescence of

Table 1 Behavior of some enzymes during ripening of some climacteric and nonclimacteric fruits

Kind of fruit	Enzyme	Fold increase
Banana peel (139)	Chlorophyllase	1.6
Apple peel (139, 177)	Chlorophyllase	2.8–3.0
Apple peel (177)	Lipase	1.6
Apple peel (244)	Lipoxidase	4.0
Apple peel (177)	Ribonuclease	1.6
Bean pod (198)	Ribonuclease	4.0
Banana peel[a]	Ribonuclease	1.7
Apple peel (177)	Acid phosphatase	1.5
Mango (154)	Acid phosphatase	2.0
Banana pulp (59)	Acid phosphatase	26.0
Banana peel[a]	Acid phosphatase	2.5
Avocado pulp[a]	Acid phosphatase	4.0
Banana pulp (89)	Peroxidase	2.7
Tomato (130)	Peroxidase	3.0
Mango (152)	Peroxidase	3.0
Pear (74)	Peroxidase	3 isozymes } increases in
Tomato (74)	Peroxidase	1 isozyme
Pear (74)	IAA oxidase	2 isozymes
Tomato (74)	IAA Oxidase	1 isozyme
Blueberry (74)	IAA Oxidase	1 isozyme
Mango (152)	Catalase	4.0
Tomato (92)	Cellulase	6.3
Avocado (250)	Polygalacturonase	13.0
Tomato (114)	Polygalacturonase	>100.0
Banana pulp (124)	Pectinesterase	increases
Tomato (113)	Pectinesterase	1.4
Avocado (250)	Pectinesterase	constant
Grapefruit peel (184)	Phenylalanine ammonia lyase	>100.0
Tomato (48)	Amylase	increases
Mango (153)	Amylase	2.0
Mango (153)	Invertase	3.0
Grape (107)	Invertase	>14.0
Grape (107)	Sucrose synthetase	4.5
Grape (107)	Sucrose phosphate synthetase	4.5
Grape (107)	Sucrose phosphatase	3.0
Grape (107)	Hexokinase	constant
Grape (108)	Phosphopyruvate carboxylase	decreases
Grape (107)	Glucose-6-phosphate dehydrogenase	constant

[a] Sacher, unpublished results.

Table 1—Continued

Kind of fruit	Enzyme	Fold increase
Cherry (103)	Glucose-6-phosphate dehydrogenase	constant
Pear (102)	Glucose-6-phosphate dehydrogenase	decreases
Mango (164)	Glucose-6-phosphate dehydrogenase	increases
Grape (108)	Pyruvate decarboxylase	decreases
Cherry (103)	Pyruvate decarboxylase	constant
Banana (247)	Pyruvate decarboxylase	constant
Pear (104)	Pyruvate decarboxylase	1.7
Apple peel (120)	Malic enzyme	4.0
Pear (104)	Malic enzyme	2.1
Cherry pulp (103)	Malic enzyme	constant
Grape (108)	Malic enzyme	decreases
Grape (108)	Malic dehydrogenase	decreases
Apple (100)	Mito. malic dehydrogenase	decreases
Banana (247)	Aldolase	constant
Cherry pulp (103)	Aldolase	decreases
Apple peel & pulp (101)	Aldolase	2.5–4.5
Pear peel & pulp (101)	Aldolase	2.8–2.9
Apple (100)	Mito. succinic dehydrogenase	decreases

this tissue resembles leaf rather than fleshy fruit senescence. Farkas and co-workers (221, 245) recently used Sephadex chromatography to demonstrate that during senescence of detached *Avena* leaves there are increases in two nucleases. One, a wound RNase (relatively purine specific RNase), increased to a peak and began to level off in 6 hours. Its development was enhanced by ABA and suppressed by BA. As the wound RNase began to decline, a sugar nonspecific nuclease increased, and its increase was accelerated greatly by ABA and inhibited by actinomycin D and cycloheximide. This sugar nonspecific nuclease increases during natural senescence in attached *Avena* leaves.

The role of the large increase in acid phosphatase, paralleling the climacteric, in apple (177), banana (59), and avocado (Sacher, unpublished results) is unknown, although its potential for catalyzing aspects of ripening has been discussed (59). In this connection, there is a sixfold increase in inorganic phosphate during banana ripening (T. Solomos, personal communication), which could be a product of acid phosphatase activity. Mattoo et al (154) reported a large increase in acid phosphatase during ripening of mango, which on the one hand controlled carotenogenesis by dephosphorylating the intermediates, and on the other hand the activity of the enzyme was stimulated by β-carotene. Chalmers & Rowan (43) reported evidence consistent with their hypothesis that an increase in orthophosphate in the cytoplasm during tomato ripening stimulates phosphofructo-kinase which contributes to the respiratory rise. The apple acid phosphatase is a soluble enzyme (177), while in banana and avocado this enzyme increases in both the soluble and particulate fractions (59). The increase in activity of both fractions of banana acid phosphatase was inhibited completely during a 5-hour treatment with cycloheximide or actinomycin D during the midclimacteric period.

All respiratory enzymes of the nonclimacteric fruits reported in Table 1 remain constant or decrease in activity during ripening. In cherry (103) these are glucose-6-phosphate dehydrogenase, pyruvate decarboxylase, malic enzyme, and aldo-lase; in grape (107, 108) they are glucose-6-phosphate dehydrogenase, pyruvate decarboxylase, malic enzyme, malic dehydrogenase, hexokinase, and phospho-pyruvate carboxylase. Among the climacteric fruits reported, some respiratory enzymes increase and others decrease in activity. Malic enzyme increases in apples (121) and pear (76, 104) during ripening, while decreases in mitochondrial malic dehydrogenase and succinic dehydrogenase occur in apple (100) and glucose-6-phosphate dehydrogenase in pear (102). Hartmann reported increases in aldolase in apple and pear (101) and pyruvate decarboxylase in pear (104). Subsequently, Young (247) investigated enzyme activity in banana during different stages of ripening, using a number of procedures which protect enzymes from inactivation by tannins during extraction. He found that the levels of aldo-lase and pyruvate decarboxylase remained constant, although a previous report (214) had shown an increase in these enzymes in banana in the absence of these precautions during extraction.

Hawker's (107) studies of enzymes involved in accumulation of reducing sugars during grape ripening show that large increases occur in activity of invertase, sucrose synthetase, sucrose phosphate synthetase, and sucrose phosphatase (Table 1). He pointed out that accumulation of reducing sugars is not simply a matter of translocation and hydrolysis of sucrose. His evidence bespeaks a process of sugar accumulation somewhat like that reported for reducing sugar accumulation in bean endocarp (194) and sucrose accumulation in sugar cane (83, 110, 199), which in both tissues involved the activities of the enzymes listed above in the free space, cytoplasmic and vacuolar compartment to varying degrees. Considering the significance of carbohydrates in fleshy fruits, Hawker's results should lead toward more work on the basic mechanism of their accumulation.

Although there has been an increase in our knowledge of the kind of enzymes that increase in activity during fruit ripening, more information is needed on the development of specific enzymes in a number of different kinds of fruits in order to further our understanding of the comparative biochemistry of ripening. There is general agreement that ripening involves synthesis of enzymes, based on stim-ulation of incorporation of amino acids only during the early phase of the climacteric (33, 65, 180), yet increases in the activity of most enzymes occur throughout the climacteric period (59, 114, 124, 130, 177, 244, 250). Little has been done, however, to determine for individual enzymes whether an increase in activity during ripening is due to activation or de novo synthesis (see 226). Since Filner & Varner (73) demonstrated de novo synthesis of α-amylase in barley aleurone by use of heavy isotope labeling coupled with isopycnic equilibrium density gradient ultracentrifugation, this method has been used to show de novo synthesis of other hydrolytic enzymes associated with senescence of barley endosperm (protease, 125), abscission zone tissue (cellulase, 136) and *Rhoeo* leaf sections (ribonuclease, J. A. Sacher & D. D. Davies, unpublished results). Progress in these areas, as well as investigations of the role of hormones in enzyme induction during ripening, will be important in the quest for understanding of programmed senescence.

An enhancement of activity of peroxidase and protease by certain amino acids has been demonstrated recently during leaf infection and senescence. Dezsi et al (62) found a positive correlation in bean leaves between the accumulation of pipecolic acid and increase in peroxidase activity in conjunction with leaf infection or wounding. Treatment of attached leaf halves with 20 different amino acids (0.15 M) showed that peroxidase activity was increased 64, 52, 44, 27, 25, and 24% in 7 days by glutamine, glycine, pipecolic acid, threonine, glutamic acid, and asparagine respectively. All the increases occurred in only one of two isozymes which increase during virus infection, and there was no change in the level of chlorophyll or protein, which indicated the effect was rather specific for peroxidase.

Shibaoka & Thimann (203) found that treatment of detached senescing oat leaves over a 3-day period with 0.03 M serine completely reversed cytokinin inhibition of RNase activity, while 0.2 M serine partially (ca. 50%) reversed the cytokinin inhibition of protease activity. Serine had little effect on protease activity in control leaves. Cysteine also acted as a modifier of hormone action in promoting protease activity in presence of kinetin, and the effect of either serine or cysteine was largely prevented by arginine.

A comparison of the results of Dezsi et al (62) and Thimann and co-workers (148, 149, 203) reveals that the amino acids glycine, glutamic acid, and threonine had a promotive effect on activity of both peroxidase and protease. However, the promotive effect of amino acids on peroxidase was independent of, and on protease dependent on, cytokinin treatment. The basis for this difference could be related to any one of a number of factors such as levels of endogenous hormones, leaf age, and particularly whether leaves are attached or detached. Regardless, the above reports indicate an aspect of control of enzyme synthesis which affected RNase, protease, and peroxidase, enzymes often associated with senescence. The question may be raised as to whether this aspect of control would apply also to other hydrolases which prevail during senescence such as esterase (10), cellulase (1), β-1,3 glucan hydrolase (165), and acid phosphatase (60). In previous reports cysteine (93), 1-phenylalanine (223), and glutamic acid (246) were shown to be effective accelerators of leaf abscission. Since each of these promoted synthesis of either protease (203) or peroxidase (62), the question arises whether promotion of synthesis of hydrolases could have been a factor involved in acceleration of abscission.

HORMONAL REGULATION OF RIPENING

Auxin

Auxin was recognized as a senescence retardant initially because of its effectiveness in preventing senescence of bean endocarp, abscission zone (191), leaf (94, 192), and cotyledon (11) tissues. More recently, the ubiquitous effect of auxin in retarding senescence and cellulase activity of the abscission zones of leaves and fruits has been recognized (1, 117, 202). Many investigations over the past few years indicate that auxin as well as cytokinins and gibberellin retard ripening wholly or in part.

There is evidence that under some field and laboratory conditions auxin can accelerate ripening of bananas (58), tomatoes (105), and apples (141). More recently the senescence-retarding effects of auxin have been clearly demonstrated in studies with bananas, pears, and the nonclimacteric grape. Hale and co-workers (90, 91) showed that the application of the auxin benzothiazole-2-oxyacetic acid during the lag phase in development of the grape berry prolonged this phase, thus retarding the onset of the succeeding phase (veraison)[3] when maturation and senescent changes occur. It is interesting that exogenous auxin has a senescent-retarding action at the time when endogenous auxin declines in this fruit (52).

The Australian groups at Sydney and Ryde have approached the study of banana ripening by using transverse slices (2–10 mm thick) of bananas and storing them in moist air. The cutting of slices induces an increase in respiration and production of ethylene that subsides within 2 days (160, 169), after which the slices ripen with all the normal changes in color, respiration, ethylene production, softening, etc, that occur in the intact fruit. Experimental solutions were infiltrated by aspiration as Frenkel et al (77) had found successful for whole pear fruits. Vendrell (227) found that infiltrating auxin (2,4-D or IAA) into banana slices delayed ripening for up to 41 days; the ripening that followed was normal. Auxin pretreatment delayed ethylene-induced ripening for several days; ripening subsequently occurred normally. They attributed the auxin retardation of ripening to maintaining the tissue in a juvenile state. When intact bananas were dipped into 2,4-D solution (229), ethylene production was stimulated and pulp ripening proceeded rapidly but peel ripening was delayed. These results were ascribed to auxin penetrating the peel, but hardly at all into the pulp tissue. Auxin in the peel retarded peel senescence, but the auxin-induced ethylene production in peel tissue penetrated the pulp tissue and initiated its ripening. Analyses showed that 24 hours after dipping a whole banana into radioactive 2,4-D, 80% of the radioactivity was recovered in the peel. After 4 days the proportion in peel declined only slightly. Treatment of 10 mm thick avocado slices by soaking in a solution of α-naphthaleneacetic acid (NAA) for a few hours delayed ripening for 48 hours

---

[3] Veraison (from the French) means the onset of the maturation stage of grape berries.

based on retardation of both softening and the increase in acid phosphatase that attends ripening of avocado (60; C. Grant & J. A. Sacher, unpublished results).

Frenkel & Dyck (75) report that infiltration of auxin (IAA or 2,4-D) into mature-green Bartlett pears stimulated ethylene production but delayed ripening, thus indicating that the inhibitory effect of auxin on ripening predominated over any effect of the auxin-induced ethylene. Consistent with their suggestion that endogenous auxin acts as a resistance factor that must be depleted before ripening can occur is the demonstration (see above, 74) of increases in IAA oxidase isozymes in pear, tomato, and blueberry during ripening.

Cytokinins

Since it was first shown that kinetin prevented leaf senescence by arresting degradation of protein and chlorophyll (182), numerous studies indicate that cytokinins are especially prominent as senescence retardants in leaves. Evidence from investigations of leaf disks of *Nasturtium* (14) and attached leaves of *Perilla* (98) indicated that the decline in protein during leaf senescence was attributable to a decline in the rate of synthesis. From several studies of the effect of cytokinin on leaf senescence, there was large agreement (132, 163, 196, 215, 219) that cytokinin maintained protein by retarding the rate of breakdown rather than enhancing the rate of synthesis. What has been reported on the effects of cytokinins on fruit ripening also indicates that they act as senescence retardants, particularly of the peel. Infiltration of kinetin into fresh banana slices enhanced the ethylene production and respiration induced by cutting (231). Addition of ethylene for 16 hr after kinetin infiltration resulted in normal ripening with respect to respiration, softening, and increase in sugar content, but degreening of the peel was delayed greatly. The effect of benzyl adenine (BA) in delaying color changes in mature green oranges (72) is associated with a maintenance of the level of endogenous gibberellin (86) and a prevention of an increase in ABA-like compounds. However, BA accelerated senescent color changes in freshly harvested young orange fruitlets (87). Little is known about the level of endogenous cytokinins in fruits during the green and ripening stages, but Gazit & Blumenfeld (82) demonstrated cytokinin activity in an extract from avocado during the period of cell division and growth, as well as an inhibitor towards which cytokinin is antagonistic. They (31) also showed that tissue cultures derived from avocado mesocarp had a cytokinin requirement.

Gibberellins

The first senescence retarding effect of gibberellin in fruit, retardation of chlorophyll loss in Navel oranges, was reported in 1958 by Coggins & Hield (49). Subsequently it was demonstrated that GA delays rind softening and accumulation of carotenoids (72, 135) as well as enhancing regreening of Valencia oranges (50). Thomson's (216) ultrastructural studies showed clearly that the regreening is due to a reversion of chromoplasts to chloroplasts. The inhibition of senescent changes by GA is overcome by ethylene. Lewis et al (135) also found that the effects of GA were correlated with increased oxygen consumption and a higher level of phosphorus, and thus suggested that the treatment maintained mitochondrial integrity. In contrast to the above, senescence of young citrus fruitlets after harvest is accelerated by GA just as it is by cytokinins (87). The above results correlate well with recent analyses of levels of endogenous hormones in citrus fruits. Goldschmidt et al (86) report that the rapid increase in ABA-like compounds after harvest and the natural decline in GA are inhibited by BA, while CEPA treatment has the opposite effect. They suggest that exogenous growth regulators act by affecting the balance of endogenous hormones in the senescent tissue.

The senescence-retarding effect of GA on other fruits also appears to predominantly affect color change. Dostal & Leopold (66) showed that treatment with GA retarded color changes of tomato fruit. The ethylene acceleration of color change but not the onset of the climacteric was prevented by GA, suggesting that development of the climacteric is not dependent on all ripening phenomena. Similarly, Vendrell (228) found that GA treatment of whole bananas delayed yellowing of the skin but all other aspects of ripening were normal. In contrast, GA accelerated normal ripening phenomena in slices (without stimulating ethylene production), except for some delay in yellowing of the skin. Unlike auxin (see above), GA did not counteract the ethylene stimulation of ripening. GA which acts as a senescence retardant is also effective in reducing low temperature breakdown in apples (239), thus showing a similarity between senescence and stress phenomena.

Kessler (128) reported an interesting effect of gibberellin on senescence during water aging of cucumber leaves, related to the loss of loop-forming ability of double-stranded DNA preparations (measured in presence of DNA ligase and other factors). In contrast, DNA from tissue aged in presence of cytokinin retained the capacity to form circles in response to attachment of gibberellin A_7 to the DNA at the site of internal single-stranded nicks. The significance of these results for senescence in vivo is not yet known but is dependent upon the correctness of the assumption that in plants the loop is the unit of DNA transcription.

Abscisic Acid

Recent analyses indicate that senescence is correlated with an increase in content of ABA in both climacteric and nonclimacteric fruits. Davis & Addicott (57) reported that a rapid rise in the level of ABA in the fruit wall of cotton is correlated with marked senescent changes in the fruit wall. Rudnicki and co-workers (188–190) have shown that ABA increases during ripening of apples, pears, and strawberries. In yellow transparent apples stored at 18°, ABA increased tenfold in 5 days, but not in attached fruit (190). Thus the fruit has the capacity to rapidly synthesize ABA or convert an inactive ABA-glucoside to ABA. Since enhancement of ethylene production by ABA has been reported (53), it was suggested that ABA may accelerate senescence via enhanced ethylene production. During cold storage of pears (188) a fourfold increase in ABA occurred prior to onset of ripening, which increased by 33% during ripening at 18° for 7 days. Controlled atmosphere storage inhibited ABA accumulation. Wang et al (236) found that in attached pears ABA increased only slightly at 16 to 24°, but at low temperatures

(7–18°) there was a marked increase which was associated with premature ripening. ABA also increased approximately threefold in attached strawberries from the white unripe to the red ripe stage (189).

Senescence of detached citrus fruits (86) involves an increase in ABA-like compounds, which occurs before other senescence phenomena, and this increase can be prevented by benzyladenine or enhanced by ethylene. Hale & Coombe (personal communication) have shown that a large increase in ABA generally precedes the onset of the ripening stage in grapes, and an earlier treatment with ABA (during the lag phase) hastens the onset of berry senescence.

The evidence from studies of both climacteric and nonclimacteric fruits clearly indicates that ABA has an important regulatory role in natural ripening of fruits. Support for this view may be deduced from the fact that (a) endogenous ABA increases preceding or attending fruit ripening, (b) treatments which hasten the onset of senescence cause an increase in endogenous ABA, and (c) treatment with ABA accelerates the onset of fruit ripening. It was reported (46) that treatments which delay senescence of nasturtium leaves retard the decline in endogenous gibberellin-like compounds and prevent the increase in endogenous ABA-like compounds. ABA accelerates senescence and increases the level of RNase in leaf (60, 211) and fruit (61) tissues, and acid phosphatase in leaf tissues (60). Auxin, cytokinins, and GA invariably have an effect opposite to that of ABA (47, 60, 61, 116, 211), whether ABA enhances synthesis of hydrolases in leaf (60, 211) and fruit tissues (61) or inhibits synthesis of hydrolases in barley aleurone tissue (47).

Ethylene

Since there have been several recent reviews on ethylene (2, 34, 36, 145, 172, 208), the main emphasis here will be on work during the last 3 years. It is no longer arguable whether ethylene is a ripening hormone (36) or a product (25) of ripening. Burg & Burg's (37) studies at below atmospheric pressure showed conclusively that without ethylene there is no ripening. Certainly the excess of ethylene produced during ripening can be called a product. Reid & Pratt (176) suggest that the respiratory climacteric may be caused directly by the large amounts of ethylene produced during ripening rather than by the energy demands of other ripening phenomena (172). This view was based on their demonstration that a respiratory rise comparable in magnitude and duration to the climacteric occurred in the nonclimacteric orange and potato tuber when each was treated with ethylene. They conclude that the critical difference between "climacteric" and "nonclimacteric" fruits is in the amount of endogenous ethylene produced during ripening.

The past few years have seen an amplification of evidence for the view that (35) as fruits mature there is an increase in sensitivity to ethylene. As maturation progresses in pears (235) and apples (99) there is a decline in the concentration of ethylene and duration of exposure to a given concentration of ethylene that is required to initiate ripening. The question of the nature of the substance that is translocated into the fruit (36) and confers a "resistance" (95) to ripening is not answered, but recent work suggests that one or more among auxin, cytokinin, and gibberellin could act in this capacity, varying perhaps with different kinds of fruit. This is suggested by the evidence discussed above that senescence is retarded by auxin in bananas (227, 229), grapes (90, 91), pears (75), and avocado slices (C. Grant & J. A. Sacher, unpublished results). Further support for the senescence-retarding action of auxin is that auxin treatment prevents induction of ripening of banana by exogenous ethylene (227), and the induction of ripening of pear (75) and grape (91) by the increase in endogenous ethylene caused by applied auxin. Frenkel (74) has shown that during ripening of pears, tomatoes, and cherries there is an increase in IAA oxidase; peroxidase increased only in pears and tomatoes. He suggested that such an increase in an IAA-destroying enzyme provides a basis for degradation of auxin, thus rendering the tissue sensitive to initiation of ripening by ethylene. It was speculated further that a principal role of ethylene is to induce IAA oxidase.

The studies of Vendrell (227–229) and Wade & Brady (231) indicate that auxin, cytokinin, and gibberellin could be involved in vivo as factors offering "resistance" to ripening of banana. The case is stronger for auxin as the principal senescence retardant in banana because it delays ripening in both pulp and peel and prevents ethylene-induced ripening (227). Gibberellic acid, in contrast, delays ripening in the whole fruit but accelerates it in slices. Unlike auxin, GA does not counteract the accelerating effects of exogenous ethylene on banana ripening (228). Direct application of kinetin had no effect on peel ripening, but pretreatment with kinetin delayed degreening of banana peel in response to a 16-hour treatment with ethylene (231).

In grapes endogenous ethylene does not increase with ripening (B. G. Coombe & C. R. Hale, personal communication), but the stage of maturation at which applied ethylene is effective in inducing ripening (91) coincides with the time when there is a decline in endogenous auxin (52). The picture in grape also includes an increase in endogenous ABA at veraison, and the fact that treatments which affected maturation affected the accumulation of ABA in a correlative manner. Also, treatment with ABA hastened veraison, and there is no correlation between the time of its onset and ethylene content following treatments which hasten or delay this event, as all treatments increase the level of extractable ethylene (B. G. Coombe & C. R. Hale, personal communication). Early treatments with the auxin, benzothiazole-2-oxyacetic acid, or CEPA delay the onset while CEPA applied late hastens veraison (91). This illustrates a sharp contrast in the nature of the response to ethylene at different stages of maturation, and it is consistent with their suggestion that an auxin-ethylene relationship is involved in the regulation of grape ripening.

Marei & Crane (146) more recently reported that the fig, a climacteric fruit, exhibits a 3-period pattern of development like the grape and shows the same two-phase response to exogenous ethylene, namely, inhibition of growth when applied in period I and stimulation of the climacteric and other aspects of ripening when applied in or preceding period III. It was proposed that the adverse effect of ethylene in period I may be due to inhibition of cell division. They also reported that a sudden burst of endogenous ethylene preceded or attended the onset of ripening and the respiratory climacteric. It is very interesting to see such striking similarities between nonclimacteric and climacteric fruits in respect of the pattern of growth and the two-phase response to ethylene. One difference is that in the fig applied auxin induces ripening by induction of ethylene (155), while at no stage in grape ripening have auxins been observed to accelerate ripening (91).

An increase in ABA-like growth inhibitors in the flavedo of citrus fruits precedes other aspects of senescence phenomena, and ethylene, which accelerates senescence, causes the largest increase in the ABA-like inhibitors (86). These results strongly suggest that ABA has a prominent role in enhancement of fruit senescence. It would be interesting to see if application of ABA under hypobaric conditions (37) will induce senescence of citrus flavedo or other fruits in which ABA accumulates.

Recent results reported by Mayak & Halevy (158) show a striking parallel to the foregoing. They found that during rose petal senescence there occurred a rise in endogenous ethylene followed by an increase in ABA. Consistent with this sequence is the fact that exogenous ethylene accelerated senescence and induced endogenous ABA. Also, applications of ABA accelerated senescence (see also 4) and suppressed ethylene formation. They suggested that these two hormones participate in regulation of senescence via the same pathway. The above could indicate that ABA is closer in the pathway to the mediation of senescence processes than ethylene, particularly as ABA-treatment promoted senescence while suppressing ethylene. They also reported (157) for the first time that endogenous cytokinin declined substantially in rose petals at the onset of senescence, and that there is a direct correlation between the level of endogenous cytokinin and the longevity of two rose varieties. Further studies (159) showed that the level of endogenous cytokinin peaked as flowers opened and then declined to a low level. The results of these studies of endogenous levels of hormones are consistent with previous reports of effects of treatment with hormones on senescence of intact flowers, such as ethylene acceleration of senescence of carnations (167, 168) and the delay in senescence of disks of narcissus perianth segments caused by BA or a combination of BA and 2,4-D (9).

The involvement of enzymes in the very rapid senescence of the corolla of *Ipomea* was reported recently by Winkenbach & Matile (151, 240). During a period of 60 hours beginning 24 hours before flower opening there are large increases in invertase, DNase, RNase, and β-glucosidase, which peak and then decline with wilting, accompanied by breakdown of ca. 65% of the protein, RNA, and DNA. It was suggested that the β-glucosidase is involved in hydrolysis of anthocyanins. Invertase is extracellular where it appears to function to hydrolyze sucrose that enters the flower, as hexose content is responsible for corolla turgescence. The suppression of invertase activity was associated with the development of a proteinaceous inhibitor.

It is clear that the necessary perspective for evaluation of the role of and interactions among auxin, cytokinins, gibberellins, and ethylene during ripening will be based on more information on the endogenous levels of these hormones at different stages of maturation and ripening. This information will make more valid any interpretations of the results of adding exogenous hormones. Investigations of the role of specific hormones during ripening in the absence of ethylene will doubtless be greatly assisted by use of Burg's (37) methods for maintaining plant tissues in a continuous flow system under hypoatmospheric conditions. Indeed it is surprising that this method has been used so little over the past 6 years. Looney (140, 141) ascribed the delay in apple ripening caused by application of succinic acid 2,2-dimethyl hydrazide (B-9) to a direct inhibition of autocatalytic ethylene production. Were B-9 to act similarly in other tissues, it could be a useful experimental tool for inhibition of production of ethylene, if there were no adverse side effects.

The designation of ethylene as the "trigger" of ripening infers that ethylene is not needed during the subsequent period of ripening. Freebairn and co-workers demonstrated that ripening was slowed down greatly in tomatoes (173) and bananas (112) initiated by ethylene treatment and stored at low ($2\frac{1}{2}$ and 5%) oxygen tension. This delay could be counteracted by providing ethylene, thus indicating that there is a continuous need for ethylene in order that ripening may proceed after being initiated. This interpretation is consistent with the results of Hulme et al (123), which indicated that in apples both production and full physiological action of ethylene require a relatively high concentration of oxygen. At least in banana it appears that part of the requirement for adding ethylene to counteract the delay in ripening caused by a low level of oxygen could be related to the fact that ethylene treatment sufficient to stimulate ripening in banana tissue partly and irreversibly suppresses endogenous production of ethylene (230). Others have shown that higher levels of ethylene are needed for initiation of respiration than for softening in pears (233, 234). In banana (230) a longer exposure to ethylene is needed for initiation of chlorophyll degradation than for starch degradation.

Summary of Hormonal Regulation

Considering the results reviewed above on both climacteric and nonclimacteric fruits, it seems that auxin is prominent as an endogenous senescence retardant, as it can counteract the stimulatory effect of ethylene or abscisic acid on senescence. Cytokinins and gibberellins are implicated to a greater or lesser extent, either separately or in concert with auxin, depending on the kind of fruit. The present evidence indicates that both cytokinin and gibberellin are particularly prominent as senescence retardants in citrus fruits.

The picture that is beginning to emerge is that auxin declines at about the time that the level of endogenous ABA begins to increase. At this time exogenous auxin and ABA are effective in retarding or hastening the onset of senescence respectively. Two enzymes which have the potential for breakdown of native auxin have been shown to increase at this time in several fruits (Table 1), namely IAA oxidase and peroxidase. At this critical period in natural ripening, the levels of endogenous ethylene and ABA increase, and application of either hastens the onset of senescence. The evidence is not yet complete enough to say what the sequential relationship is between ethylene and ABA during ripening. On the one hand there is evidence that treatment with ethylene increases the level of ABA, which could be construed to indicate that ABA is involved in mediating some senescence processes which have all heretofore been ascribed to ethylene action in both fruits and flowers. In this connection it is well established that ABA enhances the activity of some hydrolytic enzymes during leaf senescence (60, 61, 116, 211), and without exception ABA acts antagonistically in this respect with auxin, GA, or cytokinins. It is very likely that further work will show that other hydrolases which are associated with leaf senescence are similarly controlled by these four hormones. In the face of the evidence discussed above, it is reasonable to expect that in fruit tissue a decline in auxin (cytokinin or gibberellin) and increase in ABA could lead to an increase in hydrolases at the onset of ripening. On the other hand, the fact that enhancement of ethylene production by ABA has been reported (53) has led to the suggestion (190) that ABA may accelerate senescence by stimulating production of ethylene.

MEMBRANE PERMEABILITY

The possibility of permeability changes being involved in fruit ripening was proposed in 1928 by Blackman & Parija (29), who suggested that the climacteric was due to loss of organization resistance, thus leading to more ready access between enzymes and substrates. More recently the association of permeability changes with senescence of a nonclimacteric fruit (bean endocarp) tissue, leaf abscission zones, and leaf sections was reported, based on observed liquid-logging of air spaces, solute leakage, and an increase in free space (85, 191, 192). An increase in the rate of solute leakage from banana and avocado slices into water during ripening (193) and the extensive loss of ultrastructural organization in pear fruit (7) were also regarded as evidence consistent with the Blackman hypothesis. Other investigations showed increased leakage into water of ions from banana slices (12) and phosphate (248, 249) and total solutes (16) from avocado sections during ripening. The leakage of total solutes from avocado slices commenced prior to the onset of the respiratory climacteric. Ben-Yehoshua (16) and Richmond & Biale (180) reported large increases in free space in avocado pulp tissue during ripening. Since in the preceding studies tissue sections were incubated in dilute solutions, errors probably were introduced in measurements made during the climacteric due to susceptibility of the softening tissue to osmotic damage to membranes (39). Quantitative evidence was obtained indicating changes in permeability based on the inverse correlation observed between changes in respiration and electrical impedance during ripening of avocado (13) and pear (7). Ben-Yehoshua et al (17) reported that resistance to gaseous diffusion in avocado pulp was unchanged until tissue softening occurred, after which it increased markedly. This was attributed to the liquid logging of air spaces by cellular exudates, due to permeability changes.

Eilam (71) conducted a thorough study of permeability changes during senescence of bean cotyledons and *Arum* spadix. He concluded that the similarity of the increasing changes in leakage of K^+, apparent free space to glycerol and sucrose, and permeability changes across the membranes surrounding the osmotic volume in the two different tissues suggest that they are of general occurrence and are an early sign of senescence. In *Rhoeo* leaf sections (60) time course studies showed that a large increase in acid phosphatase was followed immediately by a rapid increase in free space, which was used as a reliable criterion of senescence.

Burg et al (39) measured the rate of leakage from banana slices into hypo- and hypertonic sugar or glycerol solutions and concluded that membrane permeability remains constant during banana ripening, and the increased leakage is proportional to the increase in concentration of endogenous sugars. They found that leakage could be prevented by a high concentration of glycerol or other suitable osmotic agent in the ambient solution. Sacher (195, 196) conducted further studies on permeability changes during banana ripening by measurements of the rate of leakage of amino acids (the concentration of which remains constant during ripening) into water, the rate of influx of $1\text{-}^{14}C$-mannitol ($0.1\ M$) into banana pulp disks (for measurement of the tissue free space), and changes in the capacity for active uptake of solutes. An increase in free space began 3 days before and had increased significantly by the onset of the climacteric, while an increase in the rate of leakage of amino acids began about 4 days before the onset of the climacteric. There was also a marked reduction in active uptake of amino acids 1 to 2 days before the beginning of the respiratory rest. These three measurements show that an increase in membrane permeability begins 3 to 4 days before there is any significant change in the endogenous sugar content, which Burg et al (39) showed does not begin until the onset of the climacteric. In fact, tissue disks from pre-climacteric bananas or avocados do not suffer any osmotic damage in water, as they undergo up to a fiftyfold increase in the capacity for active uptake from dilute radioactive solution during an incubation of 16 hours, which was measured after a 1-hr wash in running tap water (197). It is clear, however, from the work of Burg et al (39) that damage to membranes in some tissues can result from hypotonic solutions (see also 111). Lewis & Martin (138) demonstrated that during apple ripening there is an increase in the rate of K^+ leakage from apple disks bathed in isotonic solutions. They suggested that the increase in permeability may cause alterations in intracellular compartmentation which could promote enzyme activities deleterious to maintenance of normal cell organization.

The recent work of Brady et al (32) also showed that the permeability of banana pulp tissue increases before the onset of the climacteric and continues to increase during ripening. They confirmed the increase in amino acid leakage into dilute solutions reported earlier (195), but most important, their results demonstrated also that the increase in permeability during ripening occurs in the presence of 0.4 and 0.6 M mannitol and thus is not dependent on an increase in the soluble carbohydrate content of the tissue. Another important result of their work is the demonstration that permeability changes can be uncoupled from the respiratory climacteric. In ethylene-treated tissue slices the respiratory increase began within 8 hours and reached a half-maximum after 24 hours, while no increase in amino acid leakage nor in "apparent free space" to mannitol occurs before 32 hours, from which they concluded that the respiratory increase during ripening is not dependent on a change in total permeability. Yet they did not exclude the possibility of subtle changes in intracellular compartmentation. This observation casts some doubt on previous speculations on causal relationships between membrane permeability and the climacteric (7, 13, 16, 191, 195, 248), but as suggested also by Chalmers & Rowan (43), there is no reason to expect changes in the permeability of the plasmalemma and intracellular membranes to coincide.

The Sydney group (32) found that the free space of banana pulp tissue at the climacteric peak was 40%, while a free space of 60% was reported by Burg (34) for ripe bananas. Since the free space of green bananas is 20% (32, 195), it seems that the larger free space reported previously for banana (195) and avocado (180) was in part due to osmotic damage caused by hypotonic solutions. The susceptibility of ripening fruit tissue to membrane damage in hypotonic solutions is probably due in part to changes in the physical properties (softening) of the thin cell walls and the effect thereof on the capacity of the cell membranes to withstand the higher turgor pressure caused by the steep osmotic gradient. In this context it should be noted that disks of mature storage tissue of sugar beet, which has a sugar content of up to 20%, remain viable for extended periods of time in water (5) without membrane damage. Also apple peel discs (178) suffer no osmotic damage during 24 to 48 hours in dilute solutions.

It appears from the work with pulp tissue of banana, avocado, apples, and pears discussed above that permeability changes during fruit ripening may be ubiquitous. The cause of the changes in membrane properties that occur normally during senescence is not known. Abeles (2) and Burg (34) have discussed the evidence which indicates that ethylene does not affect permeability. The changes in permeability attending senescence of leaves are related to breakdown of membrane lipids (6, 68, 69). Senescence in apples and pears is associated with a marked decrease in particle-bound unsaturated fatty acids (185). Also, it is well known that a higher content of unsaturated than saturated fatty acids is associated with the resistance of membranes to the stress caused by chilling (144).

Because of technical difficulties, little or nothing is known about the extent to which changes in membranes during ripening affect intracellular activity of enzymes in vivo. Regardless, it is now clear that care should be given to the tonicity of incubation solutions when working with tissue sections of fruit tissues, as some are highly susceptible to osmotic damage (39) even in the preclimacteric stage (e.g. honeydew melon; D. H. Simon, J. A. Sacher & H. K. Pratt, unpublished results). Hawker (107) reported severe leakage of the contents of slices of grape berry. A safe way of providing inhibitors, isotopes, etc, without tissue damage seems to be the method of vacuum infiltration of whole fruits (76, 77) or thick (3–10 mm) slices of tissues (169). From studies (169) of the effect of infiltration on the rate of respiration of banana slices it was concluded that it did not affect gas exchange; thus there could not have been much water injection of the air spaces. In whole pear fruits (77) it was necessary to infiltrate materials in an isotonic carrier solution, as water infiltration inhibited ripening, which was attributed to membrane damage. In contrast, vacuum infiltration of dilute solutions into banana slices did not impair ripening (33, 169), but it was observed (169) that mannitol solutions of 12 atm pressure caused a 40% inhibition of respiration. Lee et al

(133) reported that large (3.0 cm) disks of pericarp from mature green tomato fruits ripened normally in moist air and could be vacuum infiltrated with aqueous solutions without adverse effects. It was reported earlier (8) that the use of isotonic (0.24 M) concentrations of mannitol and other carbohydrates to inhibit growth of pea epicotyl sections caused severe inhibition (ca. 65%) of cell wall metabolism, while use of $CaCl_2$ did not. It appears germane to note that assays of mature storage tissue of sugar cane (84, 106) demonstrated that the sucrose concentration of the free space approached that of the tissue, which was attributed to efflux from the sugar storage compartment. If this were to happen in fruit tissues that accumulate sugars during ripening (e.g. banana) it could ameliorate the osmotic gradient between protoplasts and free space in vivo.

THE CLIMACTERIC: RELATIONSHIP TO OTHER ASPECTS OF RIPENING

No aspect of ripening has been characterized as thoroughly as the respiratory behavior of climacteric fruits. Biale (22) concluded that "the machinery for energy generation is in full operation throughout the climacteric." In this section emphasis will be placed not on the respiratory machinery but on recent evidence bearing on the relation between the respiratory climacteric and other aspects of ripening. The view has been expressed that the respiratory climacteric is a response to the energy demands of the processes of ripening (172). One weakness of this hypothesis is the absence of evidence that the energy requirements for ripening of non-climacteric fruits are less than those of climacteric fruits. Actually, during the past few years considerable evidence has accumulated which indicates that other aspects of ripening are not dependent on the increase in respiration that occurs in climacteric fruits. Dostal & Leopold (66) reported that ethylene stimulation of color development in tomatoes was prevented by gibberellic acid, but the stimulation of the respiratory climacteric was not. Others (173) found that when tomato ripening was initiated with ethylene and the fruit given 2.5% O_2, normal ripening took place in the absence of a respiratory climacteric. These results suggested that ripening processes in tomato are not necessarily linked with or controlled by the climacteric burst of respiration. Similarly, Vendrell (228) reported that treatment of bananas with gibberellic acid retarded peel yellowing but not the respiratory climacteric or starch hydrolysis. Quazi & Freebairn (174) showed that at low oxygen tension bananas ripened normally, but the respiratory climacteric was greatly reduced. They suggested that part of the respiratory climacteric may represent wasted energy (see also 59).

McGlasson et al (161) studied the effect of a number of inhibitors on banana ripening. Cycloheximide inhibited ripening but not the respiratory climacteric, indicating that the climacteric rise is "not dependent upon and integrated with" other aspects of ripening. All inhibitors which suppressed the climacteric invariably inhibited ripening, suggesting that this dependence on the respiratory rise could involve the requirements for synthesis of proteins. Similarly, Frenkel et al (76) reported that cycloheximide inhibited ripening of pears but did not inhibit the respiratory climacteric. They concluded that protein synthesis is not a metabolic sink explaining the climacteric rise as suggested earlier.

The studies of Hansen and co-workers (97, 233, 235) showed that in fully mature Anjou pears the onset of softening slightly preceded the onset of the climacteric. Using pears at 71 and 86% of maturity, however, they were able to induce full ripening at low concentrations of exogenous ethylene prior to the onset of the climacteric, which began about 10 days after the initiation of softening. They concluded that respiration, ethylene production, and softening were differentially affected by ethylene treatment, It was subsequently shown (234) that in pears at 85% maturity an internal ethylene concentration of 0.08 $\mu l/1$ initiated softening, while 0.46 $\mu l/1$ was required for initiation of the climacteric. Also, at low oxygen ($2\frac{1}{2}$%) pears ripened without an increase in respiration (30). Hansen (97, 233) suggests that the climacteric may have no unique function in ripening that respiration does not have in all cells, and that its biochemical significance and relationship to senescence is in need of reconsideration. Biale (22) has recently suggested in this connection that "for metabolic impairments leading to senescence the search should be directed to protein and nucleic acid metabolism."

The consistency of results of many recent investigations with bananas, tomatoes, and pears (discussed above) certainly indicates that ripening is not dependent upon the climacteric increase in respiration. It has been suggested by Richmond & Biale (180) that the increase in incorporation of amino acids during the early stage of the climacteric reflects the induction of a set of enzymes which catalyze the respiratory climacteric and final cell breakdown. From other results discussed herein, however, it appears that the climacteric increase in respiration is not dependent upon protein synthesis (76, 161), whereas many other events in ripening are, which could mean that the increased respiration is due to activation or regulation of the activity of enzymes rather than de novo synthesis. In this connection, during banana ripening there occurs a large increase in fructose diphosphate and a sharp fall in phosphoenol pyruvate, positive and negative effectors respectively of phosphofructokinase (PFK). From this and other evidence, T. Solomos (personal communication) suggested that activation of pyruvate kinase and PFK at the climacteric rise in banana may be due to changes in the amount and distribution of enzymic effectors, rather than to a net increase in respiratory enzymes. As discussed above, Chalmers & Rowan (43, 44) came to a similar conclusion about the respiratory climacteric in tomato.

Literature Cited

1. Abeles, F. B. 1969. *Plant Physiol.* 44: 447–52
2. Abeles, F. B. 1972. *Ann. Rev. Plant Physiol.* 23:259–92
3. Addicott, F. T., Lyon, J. L. 1969. *Ann. Rev. Plant Physiol.* 20:139–64
4. Arditti, J., Flick, B., Jeffery, D. 1971. *New Phytol.* 70:333–41
5. Bacon, J. S. D., MacDonald, I. R., Knight, A. H. 1965. *Biochem. J.* 94: 175–82
6. Baddeley, M. S., Simon, E. W. 1969. *J. Exp. Bot.* 20:44–49
7. Bain, J. M., Mercer, F. V. 1964. *Aust. J. Biol. Sci.* 17:78–85
8. Baker, D. B., Ray, P. M. 1965. *Plant Physiol.* 40:360–68
9. Ballantyne, D. J. 1966. *Can. J. Bot.* 44:117–19
10. Balz, H. P. 1966. *Planta* 70:207–36
11. Basler, E., Nakazawa, K. 1961. *Bot. Gaz.* 122:228–32
12. Baur, J. B., Workman, M. 1964. *Plant Physiol.* 39:540–43
13. Bean, R. C., Rasor, J. P., Porter, G. C. 1960. *Yearb. Calif. Avocado Soc.* 44: 75–78
14. Beevers, L. 1968. See Ref. 238, 1417–35
15. Beevers, L. *Plant Biochemistry*, ed. J. Bonner, J. E. Varner. New York: Academic. In press
16. Ben-Yehoshua, S. 1964. *Physiol. Plant.* 17:71–80
17. Ben-Yehoshua, S., Robertson, R. N., Biale, J. B. 1963. *Plant Physiol.* 38: 194–201
18. Biale, J. B. 1960. *Advan. Food Res.* 10:293–354
19. Biale, J. B. 1960. *Handb. Pflanzenphysiol.* 12:536–92
20. Biale, J. B. 1962. *Food Preserv. Quart.* 22:57–62
21. Biale, J. B. 1964. *Science* 146:880–88
22. Biale, J. B. 1969. *Qual. Plant. Mater. Veg.* 19:141–53
23. Biale, J. B., Young, R. E. 1962. *Endeavour* 21:164–74
24. Biale, J. B., Young, R. E. 1971. See Ref. 119, 2:1–63
25. Biale, J. B., Young, R. E., Olmstead, A. J. 1954. *Plant Physiol.* 29:168–74
26. Bick, M. D., Liebke, H., Cherry, J. H., Strehler, B. L. 1970. *Biochim. Biophys. Acta* 204:175–82
27. Bick, M. D., Strehler, B. L. 1971. *Proc. Nat. Acad. Sci.* 68:224–28
28. Birecka, H., Galston, A. W. 1970. *J. Exp. Bot.* 21:735–45
29. Blackman, F. F., Parija, P. 1928. *Proc. Roy. Soc.* B 103:412–45
30. Blanpied, G. D., Hansen, E. 1968. *Am. Soc. Hort. Sci.* 93:813–16
31. Blumenfeld, A., Gazit, S. 1971. *Physiol. Plant.* 25:369–71
32. Brady, C. J., O'Connell, P. B. H., Smydzuk, J., Wade, N. L. 1970. *Aust. J. Biol. Sci.* 23:1143–52
33. Brady, C. J., Palmer, J. K., O'Connell, P. B. H., Smillie, R. M. 1970. *Phytochemistry* 9:1037–47
34. Burg, S. P. 1968. *Plant Physiol.* 43: 1503–11
35. Burg, S. P., Burg, E. A. 1965. *Bot. Gaz.* 126:200–4
36. Burg, S. P., Burg, E. A. 1965. *Science* 148:1190–96
37. Ibid 1966. 153:314–15
38. Burg, S. P., Burg, E. A. 1969. *Qual. Plant. Mater. Veg.* 19:185–200
39. Burg, S. P., Burg, E. A., Marks, R. 1964. *Plant Physiol.* 39:185–95
40. Butler, R. D., Simon, E. W. 1971. *Advan. Gerontol. Res.* 3:73–129
41. Carr, D. J., Pate, J. S. 1967. *Soc. Exp. Biol. Symp.* 21:559–99
42. Cathey, H. M. 1964. *Ann. Rev. Plant Physiol.* 15:271–302
43. Chalmers, D. J., Rowan, K. S. 1971. *Plant Physiol.* 48:235–40
44. Chalmers, D. J., Rowan, K. S. 1971. *Proc. Aust. Biochem. Soc.* 4:57
45. Cherry, J. H., Chroboczek, H., Carpenter, W. J. G., Richmond, A. 1965 *Plant Physiol.* 40:582–87
46. Chin, T. Y., Beevers, L. 1970. *Planta* 92:178–88
47. Chrispeels, M. J., Varner, J. E. 1967. *Plant Physiol.* 42:1008–16
48. Clements, R. L. 1970. See Ref. 118, 1:159–77
49. Coggins, C. W. Jr., Hield, H. Z. 1958. *Calif. Agr.* 12:(9)11
50. Coggins, C. W. Jr., Lewis, L. N. 1962. *Plant Physiol.* 37:625–27
51. Comfort, A. 1964. *Ageing. The Biology of Senescence.* London: Routledge and Kegan Paul. 365 pp.
52. Coombe, B. G. 1960. *Plant Physiol.* 35:241–50
53. Cracker, L. E., Abeles, F. B. 1969. *Plant Physiol.* 44:1144–49
54. Curtis, H. J. 1966. *Biological Mechanisms of Ageing.* Springfield, Ill.: Thomas. 133 pp.
55. Davies, J. W., Cocking, E. C. 1965. *Planta* 67:242–53
56. Davies, J. W., Cocking, E. C. 1967. *Biochem. J.* 104:23–33

57. Davis, L. A., Addicott, F. T. 1972. *Plant Physiol.* 49:644-48
58. Dedolph, R. R., Goto, S. 1960. *Bot. Gaz.* 121:151-54
59. DeLeo, P., Sacher, J. A. 1970. *Plant Physiol.* 46:208-11
60. Ibid, 806-11
61. DeLeo, P., Sacher, J. A. 1971. *Plant Cell Physiol.* 12:791-96
62. Dézsi, L., Pálfi, G., Farkas, G. L. 1970. *Phytopathol. Z.* 69:285-91
63. Dilley, D. R. 1966. *Plant Physiol.* 41: 214-20
64. Dilley, D. R. 1969. *Hort. Sci.* 4:111-14
65. Dilley, D. R., Klein, I. 1969. *Qual. Plant. Mater. Veg.* 19:55-65
66. Dostal, H. C., Leopold, A. C. 1967. *Science* 158:1579-80
67. Dove, L. D. 1971. *New Phytol.* 70: 397-401
68. Draper, S. R. 1969. *Phytochemistry* 8:1641-47
69. Draper, S. R., Simon, E. W. 1971. *J. Exp. Bot.* 22:481-86
70. Dyer, T. A., Osborne, D. J. 1971. *J. Exp. Bot.* 22:552-60
71. Eilam, Y. 1965. *J. Exp. Bot.* 16:614-27
72. Eilati, S. K., Goldschmidt, E. E., Monselise, S. P. 1969. *Experientia* 25: 209-10
73. Filner, P., Varner, J. E. 1967. *Proc. Nat. Acad. Sci. USA* 58:1520-26
74. Frenkel, C. 1972. *Plant Physiol.* 49: 757-63
75. Frenkel, C., Dyck, R. 1973. *Plant Physiol.* 51:6-9
76. Frenkel, C., Klein, I., Dilley, D. R. 1968. *Plant Physiol.* 43:1146-53
77. Frenkel, C., Klein, I., Dilley, D. R. 1969. *Phytochemistry* 8:945-55
78. Galliard, T., Rhodes, M. J. C., Wooltorton, L. S. C., Hulme, A. C. 1968. *Phytochemistry* 7:1453-63
79. Ibid, 1465-70
80. Galston, A. W., Davies, P. G. 1969. *Science* 163:1288-97
81. Garren, L. D., Howell, R. R., Tomkins, G. M., Crocco, R. M. 1964. *Proc. Nat. Acad. Sci. USA* 52:1121-29
82. Gazit, S., Blumenfeld, A. 1970. *Plant Physiol.* 46:334-36
83. Glasziou, K. T. 1961. *Plant Physiol.* 36:175-79
84. Glasziou, K. T., Gayler, K. R. 1972. *Plant Physiol.* 49:912-13
85. Glasziou, K. T., Sacher, J. A., McCalla, D. R. 1960. *Am. J. Bot.* 47: 743-52
86. Goldschmidt, E. E., Eilati, S. K.,
87. Goren, R. 1972. *Plant Growth Substances, 1970,* ed. D. J. Carr. New York: Springer. 837 pp.
87. Goldschmidt, E. E., Eilati, S. K., Riov, J., Bravdo, B. 1970. *Plant Physiol. Suppl.* 46:10
88. Györffy, B. 1965. *International Conference on Gerontology,* 101-9. Hung. Acad. Sci., Budapest
89. Haard, N. F. *Phytochemistry.* In press
90. Hale, C. R. 1968. *Aust. J. Agr. Res.* 19:939-45
91. Hale, C. R., Coombe, B. G., Hawker, J. S. 1970. *Plant Physiol.* 45:620-23
92. Hall, C. B. 1964. *Bot. Gaz.* 125:156-57
93. Hall, W. C., Herrero, F. A., Katterman, F. R. H. 1961. *Bot. Gaz.* 123: 29-34
94. Hallaway, M., Osborne, D. J. 1960. *Nature* 188:240-41
95. Hansen, E. 1966. *Ann. Rev. Plant Physiol.* 17:459-80
96. Hansen, E. 1967. *J. Am. Soc. Hort. Sci.* 91:863-67
97. Hansen, E., Blanpied, G. D. 1968. *J. Am. Soc. Hort. Sci.* 93:807-12
98. Hardwick, K., Wood, M., Woolhouse, H. W. 1968. *New Phytol.* 67: 79-86
99. Harkett, P. J., Hulme, A. C., Rhodes, M. J. C., Wooltorton, L. S. C. 1971. *J. Food Technol.* 6:3945
100. Hartmann, C. 1962. *Compt. Rend.* 255:996-98
101. Hartmann, C. 1963. *Phytochemistry* 2:407-11
102. Hartmann, C. 1968. *Physiol. Vég.* 6:289-97
103. Hartmann, C. 1971. *Qual. Plant. Mater. Veg.* 20:221-29
104. Hartmann, C., Lugon, M., Valade, D. 1968. *Physiol. Vég.* 6:279-87
105. Hartman, R. T. 1959. *Plant Physiol.* 34:65-72
106. Hawker, J. S. 1965. *Aust. J. Biol. Sci.* 18:959-69
107. Hawker, J. S. 1969. *Phytochemistry* 8:9-17
108. Ibid, 19-23
109. Ibid, 337-44
110. Hawker, J. S., Hatch, M. D. 1965. *Physiol. Plant.* 18:444-53
111. Helder, R. J. 1956. *Handb. Pflanzenphysiol.* 2:468-88
112. Hesselman, C. W., Freebairn, H. T. 1969. *J. Am. Soc. Hort. Sci.* 94:635-37
113. Hobson, G. E. 1963. *Biochem. J.* 86: 358-65
114. Ibid 1964. 92:324-32

115. Hobson, G. E., Davies, J. N. 1971. See Ref. 119, 2:437–82
116. Hodge, E. T. 1970. *Hormonal regulation of enzyme synthesis in leaf sections of Rhoeo discolor.* MA thesis. California State Univ., Los Angeles. 30 pp.
117. Horton, R. F., Osborne, D. J. 1967. *Nature* 214:1086–88
118. Hulme, A. C., Ed. 1970. *The Biochemistry of Fruits and their Products,* Vol. 1. New York: Academic. 620 pp.
119. Ibid 1971. Vol. 2, 788 pp.
120. Hulme, A. C., Jones, J. D., Wooltorton, L. S. C. 1965. *New Phytol.* 64: 152–57
121. Hulme, A. C., Rhodes, M. J. C., Galliard T., Wooltorton, L. S. C. 1968. *Plant Physiol.* 43:1154–61
122. Hulme, A. C., Rhodes, M. J. C., Wooltorton, L. S. C. 1971. *Phytochemistry* 10:749–56
123. Ibid, 1315–23
124. Hultin, H. O., Levine, A. S. 1965. *J. Food Sci.* 30:917–21
125. Jacobsen, J. V., Varner, J. E. 1967. *Plant Physiol.* 42:1596–1600
126. Kang, B. G., Newcomb, W., Burg, S. P. 1971. *Plant Physiol.* 47:504–9
127. Kefeli, V., Kadyrov, C. 1971. *Ann. Rev. Plant Physiol.* 22:185–96
128. Kessler, B. 1971. *Biochim. Biophys. Acta* 240:330–42
129. Key, J. C. 1969. *Ann. Rev. Plant Physiol.* 20:449–74
130. Ku, H. S., Yang, S. F., Pratt, H. K. 1970. *Plant Cell Physiol.* 11:241–46
131. Ku, L. K., Romani, R. J. 1970. *Plant Physiol.* 45:401–7
132. Kuraishi, S. 1968. *Physiol. Plant.* 21: 78–83
133. Lee, T. H., McGlasson, W. B., Edwards, R. A. 1970. *Radiat. Bot.* 10: 521–29
134. Leopold, A. C. 1961. *Science* 134: 1727–32
135. Lewis, L. N., Coggins, C. W. Jr., Labanauskas, C. K., Dugger, W. M. Jr. 1967. *Plant Cell Physiol.* 8:151–60
136. Lewis, L. N., Varner, J. E. 1970. *Plant Physiol.* 46:194–99
137. Lewis, T. L., Martin, D. 1965. *Aust. J. Biol. Sci.* 18:1093–101
138. Ibid 1969. 22:1577–80
139. Looney, N. E. 1967. *Nature* 214: 1245–46
140. Looney, N. E. 1969. *Plant Physiol.* 44:1127–31
141. Looney, N. E. 1971. *J. Am. Soc. Hort. Sci.* 96:350–53
142. Looney, N. E., Patterson, M. E. 1967. *Phytochemistry* 6:1517–20
143. Lynn, C. F., Pillay, D. T. N. 1971. *Z. Pflanzenphysiol.* 65:378–84
144. Lyons, J. M., Wheaton, T. A., Pratt, H. K. 1964. *Plant Physiol.* 39:262–68
145. Mapson, L. W., Hulme, A. C. 1970. *Progr. Phytochem.* 2:343–84
146. Marei, N., Crane, J. C. 1971. *Plant Physiol.* 48:249–54
147. Marei, N., Romani, R. 1971. *Biochim. Biophys. Acta* 247:280–90
148. Martin, C., Thimann, K. V. 1972. *Plant Physiol.* 49:64–71
149. Ibid, 432–37
150. Matile, Ph. 1969. *Lysosomes in Biology and Pathology,* ed. J. T. Dingle, H. B. Fell, 406–30. New York: Elsevier
151. Matile, Ph., Winkenbach, F. 1971. *J. Exp. Bot.* 22:759–71
152. Mattoo, A. K., Modi, V. V. 1969. *Plant Physiol.* 44:308–10
153. Mattoo, A. K., Modi, V. V. 1969. *Proc. Int. Conf. Subtrop. Fruits,* London. 111 pp.
154. Mattoo, A. K., Modi, V. V., Reddy, V. V. R. 1968. *Indian J. Biochem.* 5: 111–14
155. Maxie, E. C., Crane, J. C. 1967. *Science* 155:1548–50
156. May, B. K., Walsh, R. L., Elliott, W. H., Smeaton, J. R. 1968. *Biochim. Biophys. Acta* 169:260–62
157. Mayak, S., Halevy, A. H. 1970. *Plant Physiol.* 46:497–99
158. Ibid 1972. 50:341–46
159. Mayak, S., Halevy, A. H., Katz, M. 1972. *Physiol. Plant.* 27:1–4
160. McGlasson, W. B. 1969. *Aust. J. Biol. Sci.* 22:489–91
161. McGlasson, W. B., Palmer, J. K., Vendrell, M., Brady, C. J. 1971. *Aust. J. Biol. Sci.* 24:1103–14
162. Miller, L. A., Romani, R. 1966. *Plant Physiol.* 41:411–14
163. Mizrahi, Y., Amir, J., Richmond, A. 1970. *New Phytol.* 69:355–61
164. Modi, V. V., Reddy, V. V. R. 1967. *Indian J. Exp. Biol.* 5:233
165. Moore, A. E., Stone, B. A. 1972. *Planta* 104:93–109
166. Morgan, P. W., Fowler, J. L. 1972. *Plant Cell Physiol.* 13:727–36
167. Nichols, R. 1968. *J. Hort. Sci.* 43: 335–39
168. Ibid 1971. 46:323–32
169. Palmer, J. K., McGlasson, W. B. 1969. *Aust. J. Biol. Sci.* 22:87–99
170. Pentzer, W. T., Heinze, P. H. 1954. *Ann. Rev. Plant Physiol.* 5:205–24
171. Phillips, D. R., Horton, R. F., Fletcher, R. A. 1969. *Physiol. Plant.* 22:1050–54

172. Pratt, H. K., Goeschl, J. D. 1969. *Ann. Rev. Plant Physiol.* 20:541–84
173. Quazi, M. H., Freebairn, H. T. 1968. *Pak. J. Sci. Ind. Res.* 11:337–38
174. Quazi, M. H., Freebairn, H. T. 1970. *Bot. Gaz.* 13(1):5–14
175. Racusen, D., Foote, M. 1966. *Can. J. Bot.* 44:1633–38
176. Reid, M. S., Pratt, H. K. 1970. *Nature* 226:976–77
177. Rhodes, M. J. C., Wooltorton, L. S. C. 1967. *Phytochemistry* 6:1–12
178. Rhodes, M. J. C., Wooltorton, L. S. C., Galliard, T., Hulme, A. C. 1968. *Phytochemistry* 7:1439–51
179. Richmond, A., Biale, J. B. 1966. *Arch. Biochem. Biophys.* 115:211–14
180. Richmond, A., Biale, J. B. 1966. *Plant Physiol.* 41:1247–53
181. Richmond, A., Biale, J. B. 1967. *Biochim. Biophys. Acta* 138:625–27
182. Richmond, A., Lang, A. 1957. *Science* 125:650–51
183. Riov, J., Goren, R. 1970. *Radiat. Bot.* 10:155–60
184. Riov, J., Monselise, S. P., Kahan, R. S. 1969. *Plant Physiol.* 44:631–35
185. Romani, R. J., Breidenbach, R. W., van Kooy, J. G. 1965. *Plant Physiol.* 40:561–66
186. Romani, R. J., Ku, I. K., Ku, L. L., Fisher, L. K., Dehgan, N. 1968. *Plant Physiol.* 43:1089–96
187. Rowan, K. S., McGlasson, W. B., Pratt, H. K. 1969. *J. Exp. Bot.* 20:145–55
188. Rudnicki, R., Machnik, J., Pieniażek, J. 1968. *Bull. Acad. Pol. Sci.* 16:509–12
189. Rudnicki, R., Pieniażek, J., Pieniażek N. 1968. *Bull. Acad. Pol. Sci.* 16:127–30
190. Rudnicki, R., Pieniażek, J. 1970. *Bull. Acad. Pol. Sci.* 18:577–80
191. Sacher, J. A. 1957. *Science* 125:1199–1200
192. Sacher, J. A. 1959. *Plant Physiol.* 34:365–72
193. Sacher, J. A. 1962. *Nature* 195:577–78
194. Sacher, J. A. 1966. *Plant Physiol.* 41:181–89
195. Ibid, 701–8
196. Sacher, J. A. 1967. *Soc. Exp. Biol. Symp.* 21:269–303
197. Sacher, J. A. 1967. *Z. Pflanzenphysiol.* 54:410–26
198. Sacher, J. A. 1969. *Plant Physiol.* 44:313–14
199. Sacher, J. A., Hatch, M. D., Glasziou, K. T. 1963. *Plant Physiol.* 38:348–54
200. Sacher, J. A., Towers, G. H. N., Davies, D. D. 1972. *Phytochemistry* 11:2383–92
201. Schneider, G. 1970. *Ann. Rev. Plant Physiol.* 21:499–536
202. Scott, P. C., Leopold, A. C. 1966. *Plant Physiol.* 41:826–30
203. Shibaoka, H., Thimann, K. V. 1970. *Plant Physiol.* 46:212–20
204. Siegel, B. Z., Galston, A. W. 1967. *Science* 157:1557–59
205. Simon, E. W. 1967. *Soc. Exp. Biol. Symp.* 21:215–30
206. Smillie, R. M., Krotkov, G. 1960. *Can. J. Bot.* 38:31–49
207. Spencer, M. 1965. *Plant Biochemistry*, ed. J. Bonner, J. E. Varner, 793–825. New York: Academic
208. Spencer, M. 1969. *Progr. Chem. Org. Natur. Prod.* 27:31–80
209. Sodek, L., Wright, S. T. C. 1969. *Phytochemistry* 8:1629–40
210. Srivastava, B. I. S. 1967. *Int. Rev. Cytol.* 22:349–87
211. Srivastava, B. I. S. 1968. *Biochim. Biophys. Acta* 169:534–36
212. Srivastava, B. I. S., Matsumoto, H., Chadra, K. C. 1971. *Plant Cell Physiol.* 12:609–18
213. Strehler, B. L. 1967. *Soc. Exp. Biol. Symp.* 21:149–77
214. Tager, J., Biale, J. B. 1957. *Physiol. Plant.* 10:79–85
215. Tavares, J., Kende, H. 1970. *Phytochemistry* 9:1763–70
216. Thomson, W. W., Lewis, L. N., Coggins, C. W. 1967. *Cytologia* 32:117–24
217. Tomkins, G. M., Gelehrter, T. D., Granner, D., Martin, D., Samuels, H. H., Thompson, E. B. 1969. *Science* 166:1474–80
218. Trewavas, A. 1968. *Progr. Phytochem.* 1:113–60
219. Tung, H. F., Brady, C. J. See Ref. 86, 588–97
220. Udvardy, J., Farkas, G. L., Marre, E., Forti, G. 1967. *Physiol. Plant.* 20:781–88
221. Udvardy, J., Farkas, G. L. 1972. *J. Exp. Bot.* 23:914–20
222. Ulrich, R. 1958. *Ann. Rev. Plant Physiol.* 9:385–416
223. Valdovinos, J. G., Muir, R. M. 1965. *Plant Physiol.* 40:335–40
224. Varner, J. E. 1961. *Ann. Rev. Plant Physiol.* 12:245–64
225. Varner, J. E. 1965. See Ref. 207, 867–71
226. Varner, J. E. 1971. *Soc. Exp. Biol. Symp.* 25:197–205

227. Vendrell, M. 1969. *Aust. J. Biol. Sci.* 22:601–10
228. Ibid 1970. 23:553–59
229. Ibid, 1133–42
230. Vendrell, M., McGlasson, W. B. 1971. *Aust. J. Biol. Sci.* 24:885–95
231. Wade, N. L., Brady, C. J. 1971. *Aust. J. Biol. Sci.* 24:165–67
232. Wade, N. L., O'Connell, P. B. H., Brady, C. J. 1972. *Phytochemistry* 11:975–79
233. Wang, C. Y., Hansen, E. 1970. *J. Am. Soc. Hort. Sci.* 95:314–16
234. Wang, C. Y., Mellenthin, W. M. 1972. *Plant Physiol.* 50:311–12
235. Wang, C. Y., Mellenthin, W. M. 1972. *J. Am. Soc. Hort. Sci.* 97:9–12
236. Wang, C. Y., Wang, S. V., Mellenthin, W. M. 1972. *J. Agr. Food Chem.* 20:451–53
237. Wareing, P. F., Seth, A. K. 1967. *Symp. Soc. Exp. Biol.* 21:543–58
238. Wightman, F., Setterfield, G., Eds. 1968. *Biochemistry and Physiology of Plant Growth Substances.* Ottawa, Canada: Runge. 1642 pp.
239. Wills, R. B. H., Patterson, B. D. 1971. *Phytochemistry* 10:2983–86
240. Winkenbach, F., Matile, Ph. 1970. *Z. Pflanzenphysiol.* 63:292–95
241. Wollgiehn, R. 1967. *Soc. Exp. Biol. Symp.* 21:231–46
242. Wong, B. 1967. *The relationship between permeability characteristics, active uptake, incorporation of amino acids and the amount of protein and RNA in senescing avocado fruit tissue.* MA thesis. Calif. State Univ., Los Angeles. 50 pp.
243. Woolhouse, H. W. 1967. *Aspects of the Biology of Ageing.* Cambridge Univ. Press. 634 pp.
244. Wooltorton, L. S. C., Jones, J. D., Hulme, A. C. 1965. *Nature* 207:999–1000
245. Wyen, N. V., Erdei, S., Udvardy, J., Bagi, G., Farkas, G. L. 1972. *J. Exp. Bot.* 23:37–44
246. Yager, R. E., Muir, R. M. 1958. *Science* 127:82
247. Young, R. E. 1965. *Arch. Biochem. Biophys.* 111:174–80
248. Young, R. E., Biale, J. B. 1967. *Plant Physiol.* 42:1357–62
249. Young, R. E., Bieleski, R. L., Biale, J. B. 1961. *Plant Physiol. Suppl.* 36: xxx
250. Zauberman, G., Schiffmann-Nadel, M. 1972. *Plant Physiol.* 49:864–65

Ann. Rev. Plant Physiol. 1973. 24:225–52

PHOSPHATE POOLS, PHOSPHATE TRANSPORT, AND PHOSPHATE AVAILABILITY[1,2]

❖ 7548

R. L. Bieleski

Plant Diseases Division, D.S.I.R., Auckland, New Zealand

CONTENTS

[1] Abbreviations used: The widely accepted abbreviations for inorganic phosphate (Pi), adenosine triphosphate (ATP) and other nucleotides, nicotide adenine dinucleotide (NAD) and other cofactors, ribonucleic acid (RNA) and deoxyribonucleic acid (DNA) are used having their usual meanings. Phosphorus is P; used in combination it can also mean an esterified phosphate group. Thus, P-deficiency (phosphorus deficiency), P-esters (acid-soluble phosphate esters) but ester-P (esterified phosphate), glucose-6-P (glucose-6-phosphate) and similar abbreviations for specific esters, P-lipid (phospholipid), P-choline (phosphatyl choline).

[2] Throughout this article, solution concentrations will be given as molarities. The tissue content will usually be expressed as millimoles (mmoles), micromoles (μmoles) or nanomoles (nmoles) per gram fresh weight (g fr wt); or else in terms of molarities, where the tissue is taken to be an equivalent volume of water. This should help in comparing solution and tissue concentrations and the results of different studies. I have made the necessary conversions from other authors' data. Where values have been quoted in terms of dry weight alone, I have made the conversions by assuming dry wt/fresh wt ratios of 0.12, 0.14, and 0.17 for fleshy, normal, and woody tissues. They are sufficiently accurate for my comparisons. Unfortunately, authors do not always make clear the basis of their calculations. Calculations in terms of moles and molarities are free from ambiguity.

1. INTRODUCTION

To me, this review resembles an intercellular space, squeezed awkwardly between the nicely rounded cells of other reviews. I have not altogether avoided common ground covered by Läuchli in an excellent review on translocation of inorganic solutes (122). Other impinging reviews are those of Milthorpe & Moorby on vascular transport (152), Eschrich on phloem translocation (65), Laties on compartmentation in ion uptake and transport (119), Oaks & Bidwell on compartmentation of metabolites (169), Leggett on salt absorption (125), and Rowan on phosphorus metabolism (189). Earlier reviews devoted to phosphorus transport and nutrition are still well worth consulting (3, 143).

2. AVAILABILITY OF PHOSPHATE IN THE ENVIRONMENT

For the plant, the issue of Pi availability begins at the roots. Many soils throughout the world are P-deficient, and even in the more fertile soils, the Pi concentration of the soil solution is seldom higher than $10 \mu M$ (3, 70). In 135 representative U.S. soils, the Pi concentration of the soil solution was never higher than $8 \mu M$, and the modal concentration was $1.5 \mu M$ (11). In contrast, modal concentrations of K, Ca, and Mg were 90, 700, and 1000 μM respectively. In lake and pond waters, concentrations range from $0.05 \mu M$ to $2 \mu M$ or more (21, 40, 139). Surface layers of the sea have Pi concentrations below $0.1 \mu M$ (94, 150); and even the deeper layers, polar waters, or Pi-rich upwellings are $0.5-1.5 \mu M$ Pi. Organically bound P, mostly colloidal, tends to be highest in waters lowest in Pi, so that total P content is usually below 2 μmoles P/litre (21, 94). Lake and marine sediments are often high in P (40, 94, 180). These sediments can have a strong buffering effect on Pi concentrations of shallow estuarine waters (180), ponds, and small lakes (40). The roots of water plants, though often considered as simply anchoring the plants in place, may supply most of the Pi required by the plant, even when it is freely available in the water (42). Because ferric phosphate is hydrolyzed more readily at higher pH, the equilibrium concentration of Pi in seawater, about $2 \mu M$ (94) at pH 8, is higher than in most lakes where the water is often about pH 6 and the Pi concentration $0.04 \mu M$ (139).

Low solubility of Pi salts is also responsible for the low concentrations of Pi in soil solutions, but it does not explain the complete unavailability of much of the Pi in clay soils. Possibly Pi becomes locked specifically into the crystal lattice of the mica surfaces (233). About 98–99% of Pi in clay soils can be bound so firmly that it cannot be extracted by salt solutions, exchanged with added ^{32}Pi, or utilized by plants. A large part of soil P may be in the organic form: this is also largely insoluble, and the P may not be available to plants (223).

Thus soil solutions typically contain 2 μM Pi while plants contain 5–20 mM Pi (Section 9). A large part of this increase in concentration occurs within the root. since Pi in the xylem sap is 20–100 times more concentrated than in the soil solution (Section 3). Some Pi may be released into solution from soil particles, but in order to maintain this increase in concentration, it frequently has to move through the soil towards the root at a much greater speed than water. Most of the movement occurs by diffusion, although some bulk mixing of soil water does take place (70, 203). The role of diffusion is elegantly shown by autoradiography of roots growing in ^{32}Pi-treated soil (11, 130, 224). Around each root is a zone of depletion 1–2 mm wide: the distribution of ^{32}Pi across this zone fits a diffusion curve (224).

Diffusion constants for Pi have been measured. In water, the diffusion constant for Pi is 200 times greater than that in soils high in soluble Pi, but 2×10^5 greater than that in P-deficient soils $[D = 5-8 \times 10^{-6} \text{cm}^2/\text{sec}$ in water (173, 203), $2-4 \times 10^{-8}$ cm^2/sec in high-P soils (129), and $4 \times 10^{-11} \text{cm}^2/\text{sec}$ in P-deficient soils (11, 224)]. Two factors conspire to retard Pi movement in soils. In one of these, tortuosity, the fine soil particles increase the length of the diffusion path, and in the other, surface interaction, movement is retarded by a reversible binding of Pi to soil particle surfaces. Tortuosity has been both calculated (173) and measured in a model soil composed of fine glass beads (203). It reduces Pi movement to about 0.04 times the rate in water (i.e. $D = 2 \times 10^{-7} \text{cm}^2/\text{sec}$). Surface interaction is slight when soil Pi levels are high, Pi movement being reduced to about 0.1 times the rate in a model soil (i.e. $D = 2 \times 10^{-8} \text{cm}^2/\text{sec}$), but in P-deficient soils it can reduce Pi movement to 10^{-4} times (11, 129, 203, 224). Consequently, raising the Pi level of the soil will improve Pi supply to the root in two distinct ways—by increasing the amount available, and also by increasing its rate of diffusion. The second effect may be more important than the first.

Existence of Pi depletion zones in the soil shows that the root has the potential for taking up Pi faster than can be supplied by diffusion (203). This may be so even in culture solutions. These usually contain 1–5 mM Pi, since these concentrations are optimal under static culture conditions and with limited solution volumes (135). When the effort has been made to grow plants in large volumes of flowing culture solution, the optimal Pi concentrations, 1–10 μM, are similar to those of soil solutions (3, 6, 135). Even in stirred solutions, an unstirred layer up to 0.5 mm wide can surround the root and limit solute movement (56). There is an important consequence of this diffusion barrier. Beyond a certain point, there should be no advantage to a plant in having a more efficient Pi accumulation mechanism than another plant. This may be why species having widely different Pi requirements in the field show quite similar uptake rates per unit weight of root in the laboratory (6, 135). On the other hand, the geometry of the root system may be very important in determining the ability of the plant to take up Pi. The most obvious advantage is to have more roots. In competition experiments, workers have noted the ability of the better-rooted species (133) or varieties (84, 124) to take up more Pi. Also, finer roots with a larger surface area and greater length per unit weight should be able to draw Pi from a greater volume of soil and increase the uptake (16). High surface to volume ratio may be the key to the ability of phytoplankton to grow in Pi concentrations below 0.1 μM (94, 139).

3. UPTAKE OF PHOSPHATE

Almost all Pi uptake studies have used ^{32}Pi. This is taken into plant tissues from concentrations well below 0.01 μM (9, 190). Efflux of Pi by passive leakage and by backflow through the accumulation mechanism also occurs but has seldom been studied (227, 228), even though the balance between uptake and loss is critical in plant growth (47, 227). For example, *Nitella*, *Spirodela*, and *Lagarosiphon* take up ^{32}Pi from 0.1, 1, and 10 μM solutions, yet in each case the final equilibrium concentration of Pi in the solutions is 0.3–0.5 μM (my unpublished data); *Elodea* shows a similar behavior (227). Measured efflux rates [20–50 nmoles/g fr wt/hr (32, 227, 228)] are similar to uptake rates at these concentrations [10–50 nmoles/g fr wt/hr from 1–10 μM Pi (26, 227, 228)]. Reducing efflux could be more important to the plant than increasing the affinity of the uptake mechanism. There is a clear need for studies such as those of Pitman on influx and efflux of monovalent cations and anions (179).

The size of the concentration gradient to be surmounted when Pi enters the plant, 10^3–10^4, is greater than for most other ions. The potential difference between vacuole and medium [-100 mV (60, 99, 205)] or xylem sap and medium [-50 mV (39)] corresponds to a small gradient in Pi concentration in the opposite direction. Hence the observed gradient, 2000 instead of 0.02 (205), must arise through the operation of an effective Pi pump. The relationship between rate of uptake and Pi concentration is usually not simple in either plants (62, 109) or tissue slices. It can be changed markedly by "aging" tissue slices—that is, by washing them for $\frac{1}{2}$–7 days in water or 10^{-4} M CaSO$_4$. Aging increases the ability of many tissues to take up Pi from solutions of low concentration—storage tissue slices (137, 174), segments of epicotyl and hypocotyl (175), vascular bundles (26), leaf discs, and cells in culture.

The complex curve relating uptake (y) to concentration (x) has been analyzed as a composite of two or more simple curves. Two types have been invoked; a hyperbola $y = ax/(x+b)$, and a straight line $y = mx$. By analogy with enzyme reactions, the hyperbola is thought to represent active transport by the operation of a specific permease or enzyme system and is characterized by two parameters, the maximum rate $V_{max}(=a)$ and the "affinity" $K_m(=b)$. The straight line, on the other hand, is taken to show "diffusion kinetics," where rate of uptake is controlled by diffusion (7, 62, 119). It is sometimes overlooked that the hyperbola approximates to the line $y = mx$ at concentrations much below the K_m value (when $x \ll b$); i.e. a line that is straight up to 10^{-2} M could be the linear tail of a hyperbola with $K_m = 10^{-1}$ M (26, 109). Criteria other than a straight line relationship are required to show whether "diffusion kinetics" apply. A clear distinction should also be drawn between "uptake due to diffusion," where rate of movement is determined solely by physical parameters of concentration gradient and diffusion, and "uptake rate controlled by diffusion," where uptake is due to an active pump, but where the rate at which Pi can reach the pump is controlled by diffusion. For example, at 10 mM Pi, where diffusion kinetics apply, celery vascular bundles take up Pi at 1700 nmoles/g fr wt/hr (26). If this uptake were due to diffusion, then since the tissue itself contains 8 mM Pi, the efflux of Pi at low external concentrations should be about 1350 nmoles/g fr wt/hr, or sufficient to deplete the tissue of Pi in a few hours. Instead, the efflux rate is below 20 nmoles/g fr wt/hr. This anisotropic behavior shows that entry to the tissue at 10 mM Pi is still by means

of some sort of pump despite "diffusion kinetics" (76). Of course, the identification of diffusion-like processes does not depend solely on the straightness of the
line. Diffusion should be insensitive to metabolic inhibitors or low or high temperature (7, 48, 109). However, further clarification of this section of the Pi
uptake curve is needed.

Sometimes the tissue shows a "diffusion" curve alone, when uptake is proportional to concentration over the entire range (26, 137, 174, 175). Sometimes this
is so only at the higher concentrations, while at lower concentrations the curve
resembles a hyperbola (26, 62, 109). Uptake is then interpreted as being the sum
of a straight-line process and a hyperbolic process. Two pieces of evidence suggest that it is legitimate to split the curve in this way. (a) The same tissue can show
a single part or both components of the curve, depending on whether it is freshly
cut or aged. (b) Ca^{++} can activate one component of the curve and not the other
"diffusive" part (126). The reasonable assumption is that other complex curves,
even in the absence of aging phenomena, can be split up in this way.

Usually there are two components in the curve (26, 48, 109, 126), though sometimes there are three or more (62, 228). What do they represent? (a) Each could
be characteristic of a different cell type in the complex tissue being studied. (b)
Two components could result from the two anionic forms $H_2PO_4^-$ and $HPO_4^=$
which can coexist at physiological pHs (86). (c) Bacterial contamination of the
tissue could cause a second phase to appear. (d) There could be two separate
pumps present in the one cell—these could be both in the plasmalemma (in parallel) or separately in plasmalemma and tonoplast (in series). A full discussion of
this last aspect (119) lies outside the scope of my review.

Evidence that $H_2PO_4^-$ and $HPO_4^=$ are accumulated as separate individuals is
surprisingly sketchy. Clearly, $H_2PO_4^-$ can be accumulated: Pi uptake is most
rapid at a pH where over 95% is present in this form (86, 104). As the pH, and
consequently the proportion present as $HPO_4^=$, is increased, uptake of Pi declines.
Increased pH does not inhibit uptake of other ions in this way, and so it has been
argued that $HPO_4^=$ cannot be accumulated (4). In another study, uptake from
100 μM Pi at various pHs was proportional to the fraction of Pi present as
$H_2PO_4^-$ (86). However, at 1 μM Pi there was a more complex pattern, such that
the relationship of uptake rate to Pi concentration appeared biphasic at the higher
pHs. A separate accumulation mechanism of low V_{max} and high affinity (low K_m)
for $HPO_4^=$ was invoked. It was also necessary to propose that OH^- was a competitive inhibitor of both $H_2PO_4^-$ and $HPO_4^=$ uptake (86). Some of the results
could have been due to an effect of pH on active efflux of Pi into the medium via
the xylem transport system (228). Unfortunately, the behavior of other ions was
not followed. The question as to whether $HPO_4^=$ can be accumulated should be
reopened. However, presence of two ionic forms is probably not responsible for
the biphasic nature of most Pi uptake curves, which have been obtained at pH
6 or so, where $H_2PO_4^-/HPO_4^=$ ratio is 15 (86). Similar curves can be obtained
with ions of single ionic form (119).

Bacterial infection of root surfaces or tissue slices often increases Pi uptake.
The increased Pi taken up may be localized in the bacteria (10), but there can also

be a direct effect on the plant which results in increased transport into the xylem (38). At low Pi concentrations transport is reduced (9), presumably because bacteria preempt the limited supply. The contribution of bacteria towards uptake at different Pi concentrations has been estimated by comparing sterile and nonsterile tissue samples. Bacterial uptake fitted a hyperbola of low K_m (13 μM) (8). This does not mean that the low K_m phase that develops as tissue slices age is necessarily due to the bacterial flora that develops at the same time. Discs of beet aged under sterile conditions (7, 174) still show the aging response and complex uptake pattern. The aging phenomenon seems analogous to the behavioral changes that are encountered when bacteria are subjected to nutrient shifts, that is, a type of derepression. Increased Pi uptake and metabolism, including RNA and P-lipid synthesis, appear to be a consequence of this.

When bacterial and pH effects are excluded, two Pi accumulation paths can still coexist in one tissue, possibly in one cell. If they are in parallel, it is possible that they might represent accumulation mechanisms adapted to high and low Pi concentrations in the external medium. The explanation, however, is unsatisfactory: in nature the Pi concentration seldom rises above 10^{-5} M. The low K_m mechanism has ample capacity to supply the Pi requirements of the plant or tissue. If the two are in series, there is a problem as to how we can "see" the tonoplast mechanism, lying as it does behind that in the plasmalemma (119). This hypothesis provides one reasonable explanation for the existence of two mechanisms. Instead of there being one pump attempting to increase Pi concentration 10^4x from solution to cytoplasm, there would be two pumps, each increasing concentration 10^2x from solution to cytoplasm and then from cytoplasm to vacuole. If this were so, the Pi concentration of the cytoplasm would be much lower than that in the vacuole. Instead, a direct measurement on *Nitella* cytoplasm and vacuole found similar Pi concentrations (205); and two estimates made on higher plant tissues have indicated a Pi concentration rather higher in the cytoplasm than in the vacuole (121, 219). An alternative function of a tonoplast pump could be to segregate Pi in the cell (Section 8).

Factors other than concentration can affect Pi uptake. The cation present can change the Pi uptake rate by a factor of 3 (126, 206). The higher the valency of the cation, the greater the Pi uptake (69). Brief pretreatment of the tissue alters its subsequent behavior, suggesting that the cation binds to the cell surface and affects the ease with which Pi can approach the accumulation sites (69, 211).

After a period of drought stress, Pi uptake is decreased (58, 114), and transport into the xylem stream is so greatly affected that the Pi concentration there is no higher than in the external medium (114). Plasmolysis apparently breaks the plasmodesmata resulting in a change from accumulatory, symplasmic transport to passive, free space transport.

4. TRANSPORT ACROSS THE ROOT

Uptake into the root is followed by transport into the xylem. The two processes are not obligatorily linked; they can be inhibited to different degrees, and one can occur without the other (52, 80, 105). The root tip and the region initiating lateral roots take up Pi most actively, but the older parts of the root pass on a greater proportion to the transpiration stream (50, 188, 191). Consequently, transport of Pi (and water) into the xylem occurs fairly evenly along the terminal 10 cm or so of root (2, 225). Water and Pi thus move in through the root over a large area, about 1 cm^2, and then become channeled into a xylem of much smaller cross-sectional area, 5×10^{-5} cm^2. Even though Pi and water may flow through xylem at speeds above 200 cm/hr, across the cortex the flow will be 0.1–5 mm/hr, depending on whether the path lies through the whole cell, the wall, or the cytoplasm (235). When tritiated water is used to study water entry, it exchanges rapidly with most of the water there, with a half-time of $\frac{1}{2}$–1 hr (2), making it difficult to find the pathway of movement. As it is affected by temperature and metabolic inhibitors, some water movement must be through the interior of the cell, although the vacuoles do not appear to be in the direct path (75, 235). Ions, including Pi, do not move into the root in direct proportion to the water unless the whole cortex is removed from the stele (122). Most or all Pi movement is through the cytoplasm of the cortical cells rather than through the cell walls. Pi concentration in the xylem stream is up to 2 mM, 10–100 times that in the medium (39, 99, 216, 236) but less than that in the cortical cells, typically 10 mM (81, 99). It is increased by higher Pi concentration in the medium (216) and by water stress (80, 81), and decreased when water flow is increased by transpiration or by hydrostatic pressure (155, 172, 192).

Most of the Pi entering the root is taken up by the root hairs or outermost cell layers (121); the inner cortical cells contribute only when either the transpirational flow or the Pi concentration is very great (53). Thus when ^{32}Pi is supplied, the highest activity is initially found in the epidermis. With time it transfers to the cells of the stele, particularly the phloem and adjacent xylem cells (53, 236, 237). Distribution of total P in the root shows a similar pattern, with highest P concentrations in the root hairs and stelar tissues, particularly the phloem (121, 236). A high concentration of P in the xylem lumen can be shown by electron probe analysis (121) and by autoradiography (53) when special sectioning techniques are used.

The bulk of the evidence suggests that Pi travels across the root in the symplasm (122). Direct diffusional movement of Pi from soil to stele through the apparent free space is halted at the endodermis by the casparian band (118), a hydrophobic strip that extends from one plasmalemma to the next (36). One consequence of this is that excised roots show 1.3 times as much apparent free space as do intact roots, probably because this layer can be bypassed at the cut end (104, 118). Sometimes the endodermis does not act as a barrier, perhaps because it has been breached by the development of lateral roots. Where movement is solely through the symplasm, the endodermis would not be expected to act as a barrier unless the plasmodesmal connections at the endodermis restrict movement more than in other cells. Instead, cytoplasm and plasmodesmata appear to be normal (36, 49). The main function of the endodermis is probably to reduce backflow of accumulated ions from the xylem into the cortex and ultimately the soil (118).

It has been stated that for symplasmic movement to occur, a gradient in concentration would be expected from outside to inside. This was not found when the distribution of K^+ was studied (60), but these measurements were made on the cell vacuoles not the cytoplasm, and the two do not always have the same concentration. Also, this presumes that ions move along the symplasm by diffusion. It seems more likely that protoplasmic streaming plays a major part. As already noted, water movement is sensitive to inhibitors, and unless an active water pump is present, protoplasmic streaming would seem the most likely factor affected. Known rates of streaming are above 2 cm/hr. In roots, Pi appears to move at about 2 mm/hr, since ^{32}Pi applied to the outside takes 20–30 min to reach the xylem (54) and takes a similar time to cross cortical sleeves (76). These rates possibly could be achieved by diffusion alone, but it seems more likely that streaming contributes (235). The greatest barrier to movement through the symplasm is at the pits where the cytoplasm narrows down to plasmadesmata with a cross-section less than 2% of the cell wall (49). The flux across these strands is such that either no membrane can be present at that point (i.e. the plasmalemmae of adjacent cells are confluent through the pore) or the membrane is about 1000 times more permeable at that point than are other cell membranes (49). Electron microscopic evidence supports the first explanation (122). Complete continuity is necessary; if the strands have been broken by plasmolysis, the transport of Pi is stopped, even when the cells are no longer plasmolyzed (105, 114). The absence of membranes across the strands can also be inferred from data showing that depolarizing events in the epidermal cytoplasm are recognized within 8 sec in the lumen of the xylem (61). These data also suggest that the symplasm has a lower electrical resistance than has the apparent free space, implying that membranes across the strands are absent. The speed with which externally applied inhibitors affect xylem transport supports this view (123, 179).

During their passage across the root, ions exchange with those already present. This retards the apparent movement of a pulse of ^{32}Pi across the root [P seems to move as Pi rather than in ester form (54)]. The proportion of Pi which initially exchanges in this way, with a half-time under an hour, is only 1–5% of the total (54, 81). Also present is a much larger pool of Pi which exchanges with a half-time of about 3 days (81). These two pools probably lie in the symplasm and the vacuoles respectively. Taking into account the evidence for other ions as well (122), it seems that ion movement across the root occurs through the symplasm and that the vacuoles, although not inert, act as a buffering reservoir. If a nutrient has previously been in short supply, a bigger proportion of that entering can be diverted into storage. This means that the results of transport experiments can depend a great deal on the original state of the tissue. Steady-state experiments are thus preferable (123).

The final stage in passage of Pi from soil to xylem is the unloading into the xylem vessels. There are two related issues: where and how does the unloading occur? It must be somewhere inside the endodermal barrier. The original symplasm theory suggested that ions were lost into the xylem from the stelar symplasm by passive leakage. The balance of evidence (122) suggests that the movement is not passive but involves some sort of pump. The requirements for this pump are not the same as for the accumulating pump in the cortical cells, since it need not function against a concentration gradient. The Pi concentration of cortical cells, and their cytoplasm, is higher than that in the xylem stream (81, 99). There is no conclusive evidence as to which cells are involved in the unloading. The most probable candidates are the xylem parenchyma cells alongside the vessels. These show a marked ability to accumulate ions, including Pi, in vitro (27). Young xylem tissues have a high ATP content and respiration rate out of proportion to their supposedly passive function (116, 176). Unloading could occur by one of the transport processes already known to operate in plants. There could be a back flow through a normal type of uptake mechanism (227, 228) or an outwardly directed pump in the plasmalemma (122). Alternatively, the xylem vessels could retain a cytoplasmic lining along a considerable length of the young root (2), with the lumen acting as the equivalent of a vacuole, either open-ended like a test tube (208) or releasing its contents by breaking down. In this last case, P-esters might be expected to occur in the xylem stream.

In animal cells, secretion processes are often accomplished by exocytosis of microvacuoles which form in the cytoplasm. The process is not well documented in plant cells (138), though it does appear to operate in both sugar secretion in nectaries (67) and salt secretion in salt glands (214). Exocytosis would in many ways provide the most satisfying answer as to how transfer into the xylem occurs, but there is no evidence yet that the process operates in root cells. There seems room for studies using appropriate electron-microscopic techniques.

5. MYCORRHIZAS AND ROOT HAIRS

Physically, ectotrophic and endotrophic (*Endogone*) mycorrhizas are quite distinct, but it is becoming apparent that they have similar physiological roles. Ectotrophic mycorrhizas were first shown to increase growth two to three times, reduce P-deficiency symptoms, and increase the P content of the host (149, 207). Their effects were greatest on P-deficient soils (90). Recently, endotrophic mycorrhizas have also been found to increase growth three to six times, abolish P-deficiency symptoms, increase P content of the host (15, 78, 95, 96, 103), and to produce greatest effects on P-deficient soils (15, 55, 95, 103) [even in culture, higher Pi concentrations inhibit growth of *Endogone* (159)]. Occasionally the content of other nutrients in the host can also be affected slightly (15, 72), but P remains the limiting nutrient (149).

Possible explanations of this improved P nutrition have been discussed (90). (*a*) Mycorrhizas have an accumulation mechanism with a higher affinity than that in normal roots and can reduce the Pi concentration of the medium to a lower level. However, in P-deficient soils, diffusion rather than accumulation appears to limit Pi uptake, so that reducing Pi efflux would be more important to the plant (Section 2). The problem is one of the diminishing returns: a plant which reduces a 1 μM solution to 0.1 μM has already taken up 90% of the Pi, and reducing the concentration to 0.01 μM yields only another 9%. This of course applies to the ability of a single plant to obtain Pi from the environment, and in this situation mycorrhizas are able to make more Pi available. When two species are adjacent, a more efficient accumulation mechanism could be a considerable advantage if the two sets of roots were actually competing directly. The diffusion rate of Pi in soil is so low that either the roots would have to be very close together or the soil water would have to be repeatedly mixed to achieve this state. It seems that different affinities of accumulation systems for Pi are of only minor importance (135). (b) Mycorrhizas obtain Pi from insoluble salts in the soil. This would happen if the mycorrhiza could take up Pi from concentrations below the solubility product of the salt in question (this is like the first explanation, except that continued dissociation of the insoluble salt would continually replenish the Pi). The same thing would happen if mycorrhizas, like some soil bacteria, could release substances that dissolve the insoluble salts or could hydrolyze P-ester (158, 159). This does not appear to be important. (c) Mycorrhizal hyphae present a larger surface area for absorption or explore a greater volume of soil than do the roots alone (73, 96, 149). Most evidence supports this role. Root hairs are typically 0.7 mm long (129, 130) with 50 hairs per millimeter of root (59, 130). Mycorrhizal hyphae penetrate the surface of the root (sometimes through the root hairs themselves) at a lesser frequency, about 1–20 connections per millimeter of root (112, 158, 167), but they extend 10–40 mm out into the surrounding soil (112, 158, 166). The great length of these hyphae is more important than the lower frequency. The root hairs are spaced 100 μ apart at the base and 300 μ apart at the tips. If diffusion limits Pi uptake, each hair could be surrounded by a diffusion shell up to 0.5 mm wide. The shells would overlap (130), and the whole root would behave rather like a hairless root of the same diameter as the root hair envelope, increasing the effective diameter of the root from 0.2 to 1.6 mm and the rate of diffusion by about 2–3x (59). If diffusion were not limiting uptake, the root hairs would increase the surface area of the root, and thus the uptake, 3–5x (129, 130). In comparison, a root with 4 hyphal connections per millimeter would have the hyphae about 0.5 mm apart at the base, and the diffusion shells would not overlap to the same extent. With hyphae 20 mm long and 25 μ in diameter (158, 167) the increase in uptake would be about 60 times if diffusion were limiting, or 10 times if uptake were proportional to surface area. This assumes simple radial hyphae. In practice, the hyphae branch and anastomose so that the total hyphal length could be considerably greater (112, 158).

Four experiments suggest that this last explanation is the correct one. (a) Exchangeable Pi in soil was allowed to equilibrate with added [82]Pi. When mycorrhizal and nonmycorrhizal plants were grown on this soil, more [82]Pi was taken up by the mycorrhizal plants, but its specific activity was the same as that in the nonmycorrhizal plants (96, 195). If nonexchangeable Pi had been tapped, the specific activity of Pi in the mycorrhizal plants should have been lower. (b) Mycorrhizal and nonmycorrhizal plants were grown, and then the amount of exchangeable Pi remaining in the soil was measured. Soils supporting mycorrhizal seedlings showed a far greater loss of available Pi during the course of the experiment (72). (c) Sudan grass was grown in competition with mycorrhizal and nonmycorrhizal pines. A fourfold increase in Pi uptake in pine seedlings as a result of mycorrhizal infection was balanced by a corresponding decrease in Pi uptake by the grass. The hyphae had allowed the pines to compete more vigorously for existing Pi (207). (*d*) Five dissimilar species were compared. Plants with coarse roots and fewer root hairs showed greater responses to mycorrhizal infection, indicating that hyphae can substitute for root hairs (16). Autoradiographs should be used to compare depletion zones around mycorrhizal and nonmycorrhizal roots.

After Pi has entered the root hair or hypha, it moves into the root. The diffusion coefficient for Pi in the tissue should be similar to that in water or gelatine (130). Even though the total cross-sectional area of the root hairs (100 μ^2 each) is only 1% of the root surface area, Pi diffuses through P-deficient soil so slowly that total Pi movement through the hairs would be about 2500 times that through the whole area of soil (see Section 2). The root hairs would resemble conducting wires in an insulating matrix. Similarly, total movement through the hyphae (5 per millimeter of root, 500 μ^2 each) would be 1000 times as great as through the soil. Protoplasmic streaming (166) would probably overwhelm even this rapid diffusive movement, as ^{32}Pi travels along fungal hyphae at more than 2 cm/hr (134).

Movement between the mycorrhizal hyphae and the host is not well documented. In ectotrophic mycorrhizas, contact is made through the Hartig net, but the hyphae need not penetrate the host cell wall (90). In endotrophic mycorrhizas, hyphae penetrate the cell wall (73, 112, 158) so that the two protoplasts are separated by only the thin hyphal wall (85) or simply the two plasmalemmas (166). Though intracellular hyphae can be cut off and then digested by the host cell (72, 85, 166), transfer of nutrients in this way is likely to be unimportant. An informative study by freeze-etch electron microscopy of the host-hyphal interface could well be extended (85). This has shown that hemispherical wall protuberances similar to those of transfer cells increase the surface area in contact. The plasmalemmas remain intact so there is no continuity of symplasm. Whether transfer occurs by exocytosis or through the membrane is not yet known. Sucrose moves from host to hypha because it is metabolized in the hypha to mannitol and trehalose, forms that the host cannot use, and thus a sink is established (131). For Pi to be moved in the other direction, the root symplasm need only maintain itself as a sink by efficient unloading of Pi into the xylem, and so compete with the hypha for its phosphate. The P contents of the two are very similar (92). Though more than half the ^{32}P in a transporting hypha is esterified, only ^{32}Pi is transferred to the host (92). Only the Pi within the cytoplasmic pool of the hypha is transferred; in effect, the cytoplasm of the host and the vacuole of the hypha compete for the same Pi (92).

The fungus takes up Pi much more rapidly than does the host, but such uptake is also more sensitive to metabolic inhibitors, chilling, or low oxygen concentrations (90, 91). Transfer of Pi depends on both partners being metabolically active and is thus inhibited by chilling or anaerobic conditions (91). Because the hyphae form an intercepting layer around the root, and because transfer takes time, it can take mycorrhizal plants several days to reach the radioactivity of control plants when ^{32}Pi is supplied, despite their improved nutrition (154).

Here mycorrhizal effects have been discussed in terms of P nutrition. Nitrogen deficiency can sometimes stimulate mycorrhizal growth (159). Improved growth does not always depend on P nutrition, and other nutrients could be involved if their uptake were limited by diffusion.

6. TRANSPORT IN XYLEM AND PHLOEM

Two basic methods are used to study compounds that move in the xylem. When herbaceous plants are topped, the rooted stump often exudes a "bleeding sap" from the severed xylem (155). Apparently salt transport mechanisms in the root continue to function, water follows osmotically, and the resulting liquid which exudes is thought to be a reasonably accurate representation of the xylem sap in the intact plant. Alternatively, large amounts of sap ("suction sap") can be obtained by suction from xylem of woody stems (34). Compositions of the two sorts of sap are broadly similar, though when the two were directly compared in one plant, the root exudate was two to five times as concentrated as the xylem sap (155), presumably because transpirational water flow had been stopped.

Most of the P in xylem travels as Pi. Suction saps have little or no organic P (155, 200) but bleeding saps have up to 25% as organic P (217). This difference may be due to the method of sap collection rather than whether the plant is herbaceous or woody (155). The main compound present is P-choline (140), and its occurrence is something of a mystery. It occurs in quite high amounts in vascular tissues (26), enzymes for its synthesis are present (210), and it is subject to rapid turnover. Unlike other P-esters, it is a zwitterion, so it has a low net charge and relatively lipophilic character. It has been thought of as a form that can penetrate membranes readily; yet membrane transport processes for Pi are very effective and well documented. Why does the method of sap collection affect the presence of P-choline? It could possibly be produced in bleeding sap from P-lipid by phospholipase D, but not as any simple experimental artifact (217). Phosphatase is present in small amounts in bleeding saps (204), but appears to be virtually absent from suction saps (my unpublished data), so that it is unlikely that P-esters there have been hydrolyzed by phosphatase. The reason for the disparity between the two sorts of saps needs to be established. A difference in carbohydrate composition of bleeding and suction saps in grape has also been reported. In this case the suction sap has an extra component, sucrose, that is missing from the bleeding sap (89).

The concentration of Pi in suction saps from woody stems, as in bleeding saps from roots, can be greater than 1 mM, e.g. in apple (34) and in pear, willow, *Tecoma*, and bamboo (my unpublished data). In deciduous trees, there are spectacular changes at breaking of dormancy, when the Pi concentration of the sap can rise from 0.07 mM to 1 mM in 8 days (34, 46). In the earliest stages of bud break, Pi in the xylem sap seems to come from the wood parenchyma rather than from the roots (45). After the spring peak, concentration falls below 0.5 mM as the volume of water flowing through the plant increases (34). By this time most of the Pi must come from the soil. There may be diurnal changes in Pi concentration not paralleled by similar changes in K^+ and Ca^{++} concentration. A diurnal rhythm in bud development seems responsible; Pi is apparently taken into the phloem from the xylem along the length of the stem during the day and transported to the bud via the phloem (46). At onset of dormancy, Pi concentration in the xylem sap falls below 0.1 mM (34).

The xylem lumen is not simply an inert pipe, and both xylem parenchyma cells and the adjacent phloem take part in its function. When ^{32}Pi in solution is sucked through a stem section, it exchanges with Pi already present in the free space, but the living xylem cells also accumulate Pi actively (27, 156), changing the concentration and composition of the xylem stream (89). The movement out of the xylem vessels may sometimes be restricted. When xylem vessels of sugar cane are perfused with tritiated water or ^{14}C-sucrose solutions, the water can diffuse freely into the entire tissue volume, but sucrose movement is restricted to about 3% of the tissue volume (44). Apparently there is a barrier to sucrose diffusion close to the xylem vessels, and the apparent free space of the cortex is not accessible. These phenomena have not been properly studied.

Pi moves readily from xylem to phloem. When ^{32}Pi is perfused into willow xylem it can be recovered from the sieve tubes 80 min later (101). The time lag, due to exchange of the ^{32}Pi with Pi in the intervening cells, is reduced to 25 min when ^{32}Pi is applied directly to the cambial surface. The final entry into the phloem cells can involve movement against a gradient (27). The reverse movement from phloem cells to xylem sap has seldom been observed. It occurs in willow only in the spring (178) when the Pi concentration of the xylem sap rises suddenly. Nor does Pi move very readily from one vascular bundle to another (128) or laterally from one part of the phloem to another (25, 100). In the absence of a vascular anastomosis like that in grasses (44), any unevenness in the entry of Pi to the root or the vascular system can be reflected in the distribution pattern throughout the whole plant (100, 128). Physiologists using tracers in transport studies are used to seeing "hit-and-miss" distribution patterns that become logical when the vascular connections are sorted out.

Although lateral movement in the vascular system is restricted, this does not mean that P cannot move freely in the plant. P is one of the most mobile of nutrients, and an individual P atom can make several circuits through the plant. In contrast, Ca tends to go and stay where it is taken by the transpiration stream, so that changes in the P/Ca ratio of an organ can frequently indicate the amount of P redistribution that has occurred (145). Much of the redistribution occurs from the leaves. The process has been studied by feeding ^{32}Pi to the leaf through a flap or through the leaf surface. Over 50% of the Pi applied to the leaf surface can be taken up, to the extent that deficiency symptoms can be relieved or the P requirement of the plant supplied in this way (117, 215). It moves through the mesophyll cells and into the fine vascular bundles (13, 127, 163), the process being retarded by inhibitors, wilting, and low oxygen concentrations. Vein loading is a consequence of active accumulation and transport by the vascular tissues (26). If the vein is cut very close to the ^{32}Pi entry site, loading is prevented and ^{32}P moves slowly out into the surrounding leaf (163).

Once ^{32}Pi has entered the phloem in the veins, it is transported at speeds up to

80 cm/hr (24, 30) out of the leaf through the petiolar phloem. Not much transfer of ^{32}P into the petiolar parenchyma or xylem occurs at this stage. Further on, in the stem, up to half the ^{32}P can pass from the phloem into the xylem parenchyma cells (24, 25, 178), though it does not move so readily into the xylem sap (25, 178). There is a large area of contact between xylem and phloem, so that a small potential for movement between the two could move large amounts of Pi. Future work will probably show that the parenchyma cells of the woody xylem play an important role in the short-term storage of nutrients, particularly Pi, and that the extent to which Pi passes into or out of the phloem depends on the physiological state of the plant at that time.

With such a mobile element the pattern of redistribution seems to be determined by the properties of the source and sink rather than those of the transport system. The natural sinks for P in the plant are the meristematic and expanding tissues, while the sources are the mature and senescent ones. For example, when ^{32}Pi is applied to part of a mature leaf and is taken up into the vascular system, it will move into another part of the same leaf only if a sink is created there [e.g. by hormone treatment or fungal infection (161)]. As leaves become senescent, there is a large net loss of P so that the final amount in the leaf can fall to less than one half, even though Pi is continually entering the leaf through the xylem. In contrast, Ca levels increase until leaf fall (98, 171, 231). A small portion moves down the phloem into the root (23), from whence it can be remobilized to travel up the stem again (79, 145). Most of the P moves into the young developing leaves, flowers, or fruit, or into the bud itself (13, 23, 79). Very young leaves do not redistribute any ^{32}Pi which enters, but when they mature they in turn become sources.

Factors that influence behavior of potential sinks can be broadly grouped as either hormone or P-deficiency effects. It is not the hormone that makes the sink; it is the physiological state brought about by a number of interacting factors, one of which may be a hormone. Which hormone is effective will depend on the tissue and on existing hormone levels. For example, when a stem has the tip removed, it ceases to be an effective Pi sink, but if the cut end is treated with indoleacetic acid, a sink is regenerated within a few hours (57, 199, 202). Cytokinins and gibberellins do not replace indoleacetic acid but interact with it to some degree, depending on the state of the tissue (57, 202). On the other hand, in a senescing leaf, cytokinins cause treated tissue to become a sink (1, 161). Cytokinin treatment also retards or reverses senescence, and the Pi accumulating ability appears to be just one aspect of this rejuvenation (1). Aging of tissue slices generates sink-like behavior, and induction of potential meristematic activity is clearly involved in this phenomenon (Section 3).

It is easier to understand P-deficiency affecting Pi transport. The roots of a P-deficient plant usually retain more P and transport much less (190, 231). By preempting Pi, root growth is maintained at the expense of shoot growth, leading to the decreased shoot/root ratio that is a feature of P-deficiency (220). Correspondingly, P-deficient leaves sprayed with Pi solutions retain Pi, and the plant shows an increased shoot/root ratio (117, 215). The story is rather an oversimplification, and the shoot tip is still the ultimate sink; a major feature of P-deficiency is that bud growth is maintained at the expense of the rest of the plant (220).

As the phloem is the main pathway for P redistribution, any discussion of the detailed mechanism of P redistribution soon becomes one of the physiology of phloem tissues (65). However, some interesting features of P movement in phloem can be noted. The phenomenon of vein loading has similarities to the loading of xylem vessels in the root. Transfer cells, with their specialized structure and location at vein endings, are surely involved in some way, particularly since they are also prominent in vascular ansatomoses at nodes of grasses (170). Companion cells also have similarities with transfer cells and may mediate the movement of solutes between sieve tubes and the surrounding phloem in the main part of the vascular system. The ultimate result of loading is that Pi moves against an overall concentration gradient into the sieve tubes (26, 27), where a concentration as high as 30 mM has been recorded, though 3 mM may be more typical (115, 209). Nobody has investigated whether Pi concentrations in sieve tubes are susceptible to environmental factors.

Samples of sieve tube contents are obtained either by cutting the petiole or stem and collecting liquid which exudes from the phloem, or by obtaining liquid which exudes from aphid stylets. One problem is to obtain a sufficient volume of sap for anslysis. A second is that phosphatase activity is high in sieve tubes (238). A third and basic problem lies in understanding the nature of sieve tube exudates, since exudation often continues for a long time, and all the liquid obtained can clearly not have been in the column of sieve tubes at the time of cutting (65). Possibly the cells alongside the sieve tubes progressively burst as a result of the release of turgor and give up their contents to the sieve tube lumen. The characteristic solute and protein composition of the sap does not support this explanation. Probably a secretion of solutes into the sieve tubes by "loading" cells continues after cutting, in which case the sieve tube sap should be reasonably representative of the sieve tube contents. Enzymes, proteins, and RNA continue to appear in the exudate (17, 65) so that loading is not likely to be a simple transport of solutes across a membrane; again, exocytosis may be involved.

Whole phloem tissues have been analyzed and found to have a general P-ester pattern like that of other tissues (26, 30, 176, 177). Two characteristics are that UDP-glucose is a prominent nucleotide (176, 177), and ATP is present at a much higher concentration than in surrounding tissues (115, 116, 176, 177). In the past, opinion has been divided as to which P-ester, if any, is present within the sieve tubes themselves, but it now seems that the ester-P content can equal Pi content (30, 209). Glucose-1-P was one of the first esters found, but it may be an artefact arising from the hydrolysis of UDP-glucose (209). ATP is present in sieve tube saps in concentrations, 0.3–1.5 mM, that are about 5 times higher than those in most plant tissues (71, 115, 116) and more like those in cytoplasm or chloroplasts alone (Section 8). The proportion of ATP to total ester-P is also higher than in other tissues (30, 115), but the normal spectrum of P-esters is present; they are actively metabolized (30, 115) and the enzymes for their synthesis are present

(111). ³²Pi supplied to sieve tube exudate is soon incorporated into many P-esters (17).

Although half the P present is in the organic form, the transporting form in sieve tubes seems to be Pi. When ³²Pi is supplied to a leaf and the sieve tube contents are sampled either by aphid stylet or as exudates (30, 115), label appears first as ³²Pi. Subsequent incorporation of ³²Pi into ATP and other esters occurs very rapidly, but RNA becomes labeled more slowly (30, 115). The most reasonable explanation (65) is that in the intact sieve tube there is a stationary, cytoplasm-like phase which contains the P-esters, and a mobile vacuole-like phase which transports Pi; that ³²Pi passing down in the mobile phase exchanges into the stationary phase and is incorporated into P-esters there; and that both phases are swept out by turgor release when the phloem is cut or punctured by the aphid. If the sieve tube itself is not metabolically active, then its function must be closely integrated with that of another cell type (presumably the companion cell) which can behave in this way in order to give rise to the high ATP levels and the rapid P-ester turnover observed.

7. PHOSPHATASES AND PHOSPHATE AVAILABILITY

Any discussion of P availability would be incomplete without mention of phosphatases. Plant tissues invariably contain, in addition to low activities of phosphatases acting on specific substrates, a high activity of non-specific acid phosphatase. Typical levels are around 0.3–1.5 e.u./g fr wt (1 e.u. hydrolyzes 1 μmole test substrate/min at 30°C) although occasional activities as high as 35 e.u./g fr wt have been recorded (183 and my unpublished data). A typical plant tissue thus has the potential ability to hydrolyze about 100 nmoles P-ester/g fr wt/min, whereas the rate of total P-ester synthesis is less than 200 nmoles/g fr wt/min. However, it is unlikely that any significant part of the ester-P is normally hydrolyzed by nonspecific phosphatase. For example, P-deficiency in *Spirodela* causes a 20-fold increase in phosphatase activity (183) without causing any observable change in labeling kinetics or the pattern of P-esters (29). Thus P-esters must be in a separate compartment to phosphatase.

In animal cells, phosphatase is isolated in a well-characterized organelle, the lysosome. Lysosomes as such do not appear to exist in plant cells; instead the vacuole may take their place. Cytoplasmic material can be taken into the vacuole by invagination (147) and digested there (87, 238). Two types of lysosome-like particles have been obtained from meristematic cells: provacuoles (147) and analogous lipid-rich spherosomes (201). The provacuoles appear to be merely an early stage in a developmental sequence. In young cells, high membrane-bound phosphatase activity is associated with invaginations of either the dictyosome or the endoplasmic reticulum (20, 142, 147, 181, 238). These form phosphatase-containing vesicles which ultimately fuse to form the vacuole in the mature cell (147, 181, 238). Thus most of the phosphatase of the mature cell is in the vacuole, probably localized on the inner face of the membrane (181, 201) to be released into the cytoplasm only during senescence (20). These findings are based on electron microscopy of sections stained to reveal phosphatase. The specific staining procedures are not necessarily free from artifacts, but vacuolar localization has been shown more directly by isolating vacuoles and protoplasts from maize roots and comparing their enzymic activities; almost all the phosphatase was in the vacuole, along with other hydrolytic enzymes (146). Vacuolar localization of phosphatase is consistent with the view that much of the Pi is confined to the vacuole while P-esters are restricted to the cytoplasm (Section 8).

In P-deficient organisms, phosphatase appears to be localized in a different part of the cell. In *Escherichia*, yeast, *Neurospora*, and *Euglena*, P-deficiency causes a 5- to 50-fold increase in phosphatase, and in each case the increase is due to appearance of a new isoenzyme localized in the external cell membrane (references in 32). In higher plants, P-deficiency also increases phosphatase activity (132, 183, 218, 234). Again phosphatase becomes located so that it is accessible to substrates outside the cell (32, 234). The enzyme may be free in the external medium (218), but other evidence suggests it is generally bound in the cell wall or located on the outer face of the plasmalemma (32, 184). Phosphatase-containing vesicles derived from the dictyosomes can apparently move to the plasmalemma and be extruded, or the membrane may be interpolated into the plasmalemma (41). Although P-deficiency in higher plants causes a new isoenzyme to appear when activity increases (132, 183), it is not certain that this and the external phosphatase are the same (32).

The role of the external phosphatase has not been established. It might be a Pi transport agent, the phosphatase activity being incidental (234), or it might hydrolyze P-esters in the medium, converting them from a nonavailable form into Pi (32). Phosphatase activity in the external medium can produce Pi from P-esters (32, 213), but the significance of this under natural conditions, when concentrations of P-esters in the medium are very low, is uncertain.

8. COMPARTMENTS IN THE CELL

I use the terms "pool" and "compartment" to denote the situation where different portions of a compound are metabolically isolated from one another, whether or not this is due to their physical separation in cell organelles (169). The following procedures can show the presence of compartmentation. (*a*) Analysis of different cell organelles may reveal discontinuities in composition or concentration. (*b*) Comparing the rates of movement of various solutes into organelles can detect poorly permeable membranes. (*c*) Recognition of radioactive labeling anomalies can suggest the nature of the compartmentation. There can be unequal labeling in different parts of a cell, or assymetric labeling patterns in a molecule, or unexpected patterns of specific activity. Use of these procedures has revealed compartmentation of sugars, amino acids, and organic acids. The same criteria show compartmentation of Pi and P-esters. Direct analysis of isolated cell organelles has only occasionally been used to study P-ester pools. By using the large-celled alga *Nitella*, Smith showed that over 90% of the ester-P was in the cytoplasm, while most of the Pi was in the vacuole (205). Unfortunately, the experiments were rather cursory and did not yield as much information as their good design should have allowed. No attempt has been made to analyze P-ester levels in isolated vacuoles of higher plants (83), and few in chloroplasts that have been obtained by a good isolation procedure (113). Most information obtained has come as an adjunct to tracer-labeling studies (120, 196).

The uptake behavior of isolated mitochondria and chloroplasts has been studied. Uptake of Pi by mitochondria is rapid, against a concentration gradient, and metabolically active; half of the Pi supplied can be taken up in 10 min (102). The process requires substrate or ATP, can be inhibited, and follows enzyme kinetics (102, 151). Pi uptake is coordinated with that of Ca^{++} (63, 102) or Mg^{++} (151), the P/Ca or P/Mg ratio being in the region 1/1–1/2, suggesting that Ca or Mg phosphate is deposited within the mitochondrion (63, 151). A similar phenomenon has been reported for Pi entry into isolated chloroplasts (168). Because Pi precipitation may be involved (151), the biochemical role of this process is uncertain despite its metabolic characteristics. P-lipid can be taken into isolated mitochondria from a labeled microsome fraction (19).

A general problem in studying cell organelles is that their membranes can be affected easily by the isolating procedure. Much of the variation between different chloroplast studies undoubtedly depends on whether the outer membrane is intact (Class I) or absent (Class II) (110, 186). Class I chloroplasts are much less permeable to Pi and P-esters (14, 186) and reflect the in vivo state better (197, 230). The speed with which the chloroplast responds to an added substrate gives a measure of its permeability to that substrate. Generally, chloroplasts are found to be readily permeable to 3–P-glyceric acid, fructose-1,6-P, dihydroxyacetone-P, and possibly NAD, but they are much less permeable to fructose-6-P, ribulose-1,5-P, Pi, and NADP (14, 221). Opinion is divided as to whether or not ATP and ADP can penetrate (14, 113, 221, 230). The state of the original tissue may cause differences (110). UDP-glucose seems restricted to the cytoplasm (113).

As a result of these uncertainties, the most useful though indirect information has come from the study of inequalities in labeling patterns. In the simplest version, a pulse of ^{32}Pi is supplied to the tissue, and specific activities of Pi and P-esters are then compared (29, 33, 137, 226). In these studies, P-ester specific activity is 6–15 times that of Pi, indicating that most of the Pi in the tissue cannot be contributing to ester synthesis and is not equilibrating with the ^{32}Pi that has entered the tissue. This portion is termed the nonmetabolic pool. The proportion of Pi which is in the much smaller metabolic pool and contributing to ester synthesis will be 1/6–1/15 of the total (7–17%) if the Pi in the nonmetabolic pool has a specific activity near zero. This is a reasonable assumption, because the nonmetabolic pool takes more than a day to reach an equilibrium specific activity which is only about 1/10 the transient specific activity of the metabolic pool; therefore, after 1 hr its specific activity will be only about 1% of the metabolic pool.

Two results suggest that the metabolic pool is in the cytoplasm plus chloroplasts and the nonmetabolic pool is in the vacuole. In *Nitella*, cytoplasmic Pi becomes labeled much more rapidly than vacuolar Pi (205). In an elegant variation on this experiment, *Elodea* and spinach leaves were pulse-labeled with ^{32}Pi; then chloroplast and nonchloroplast fractions were isolated, and the specific activities of their Pi and P-esters were compared (219). The specific activity and amount of Pi in the vacuole was also estimated from the data. After 2 days, specific activity of vacuolar Pi was still less than half way to equilibrium. In spinach, about 72% of Pi was in the vacuole at 6 mM; 14% was in the chloroplast at 9 mM; and 14% was in the cytoplasm at 9 mM. In *Elodea*, 75% of Pi was in the vacuole at 5 mM, 6% in the chloroplast at 19 mM, and 19% in the cytoplasm at 19 mM. A similar estimate (196) showed *Elodea* and *Coleus* chloroplasts to contain 5% and 15% of the leaf Pi at 4–25 mM. The size of the nonmetabolic pool in *Spirodela*, 87% (29), corresponds to that of the vacuolar pool in *Elodea*. The nonmetabolic pool in potato is rather larger, 92% (33, 137), as is to be expected if chloroplasts are not present and contributing to the metabolic pool.

Similarly, pulse-labeling followed by nonaqueous isolation of chloroplasts has been used to show that 3-P-glyceric acid in vivo moves into and out of the chloroplast so rapidly (exchange half-time 15 sec) that it may function in transport of fixed carbon (97, 221). NADP and probably NAD penetrate slowly (193, 221), as do ribulose-1,5-P and sedoheptulose-1,7-P (97). Pi movement is much slower (exchange half-time 5–15 min) than ATP movement (197, 219). Thus in vivo and in vitro results consistently show that 3-P-glyceric acid penetrates the chloroplast membrane readily while Pi does not; and fructose-1,6-P and NAD penetrate more readily than fructose-6-P and NADP (14, 43, 110, 186, 193, 221). Differences between such closely related molecules suggest that movement of many of the P-esters is subject to specific transport processes. The chloroplast membrane, relative to tonoplast or plasmalemma, is very permeable to all P-esters, so the chloroplast is a very "leaky" pool as far as the P-esters are concerned. The vacuole, on the other hand, is a highly inaccessible pool of Pi, with an exchange half-time often in excess of 3 days (29, 33, 81, 219).

This isolation of the vacuolar Pi shows up in some whole-plant studies. For example, ^{32}P entering roots does not equilibrate with most of the Pi already there (54, 81). The equilibrating Pi is 1–5% of the total and is thought to be located in the symplasm. Similarly, [32]Pi entering mycorrhizal roots through ectotrophic mycorrhizal hyphae exchanges with less than 10% of the Pi in the hyphae (92), and again this is taken to be in the cytoplasm. Finally, an interesting situation is found in *Spirodela* (29). Normal plants behave as if 13% of the Pi were in the cytoplasm. This amount is 1.1 times the amount of ester-P in the cytoplasm. In P-deficient plants, all the Pi (0.82 times the amount of ester-P) behaves as if it were in the cytoplasm at a concentration of about 7 mM. Apparently, P-deficiency principally affects the Pi content of the vacuole. When Pi in the vacuole disappears, growth ceases because cytoplasmic Pi cannot be used for growth; it is an obligatory component, apparently necessary to maintain balance between different enzyme pathways (Section 9).

9. STORAGE POOLS

Pi, segregated in the vacuole, clearly serves a storage function. An alternative way to store Pi is for the cell to synthesize a storage compound. Polyphosphate occurs widely in lower plants (93) but has only occasionally been identified in angiosperms—in dodder (212), spinach leaf (153), maize root (222), *Urtica*, and *Deschampsia* (164). In *Banksia* it appears anomalously under P-deficient conditions (106). The procedures for identifying polyphosphate—solubility, acid lability, metachromatic staining—are equivocal, but they are reasonably convincing (153) when taken with chromatographic data. Polyphosphate may turn out to be a widely distributed storage product in higher plants, but it is likely to be less important than phytic acid. It is present in relatively small amounts [1–4% total P (153, 164, 212)], and in dodder where both occur, phytic acid predominates (212).

Phytic acid, on the other hand, is clearly established as a major P reserve, containing more than 70% total P in seeds (5, 64, 108, 160, 198) and present in large amounts in potato (194) and in small amounts in some actively growing green tissues (28, 185). The biochemistry of both synthesis (5, 185) and hydrolysis (22, 141) have been investigated. One pathway transfers phosphate from phytic acid to ADP, produces ATP, and so a potential role as a phosphagen has been discussed (157, 222). This role is unlikely to be important in seeds. The total content of phytic acid, 10 mg/g dry wt (160), utilized by the germinating seedling over a period of 6–10 days, corresponds to 12 μmoles $\sim$P per g fr wt seedling. This is the amount produced by respiratory metabolism in about 10 min. The role of phytic acid as a P-reserve is much more important (232).

In the earliest stages of germination, only free Pi is used for ester synthesis (232). Phytic acid is not transported into the developing seedling (64) but instead is hydrolyzed in the seed by the enzyme phytase. Only a part of the resulting Pi is esterified, Pi concentration stabilizes at 6–12 mM, and the pattern of the various P fractions comes to resemble that in a normal growing tissue (64, 88, 160). There is little or no phytase in the dry seed, but during germination its activity rises rapidly (141, 160). Appearance of phytase is prevented by inhibitors of protein synthesis (22, 141, 198) and repressed by Pi in the medium (22). If seeds are allowed to germinate in Pi solutions, hydrolysis of phytic acid is suppressed (198).

In potato, formation of phytic acid occurs during the later stages of tuber development. Most of the Pi which enters the tuber is converted to phytic acid, while Pi concentration remains constant at 5–6 mM (194). In developing seeds, phytic acid formation occurs at the same time as drying of the seed, so that again Pi concentration remains approximately constant (108). Apparently, changes in Pi concentration of the cytoplasm, as vacuolar function declines, triggers the conversion into phytic acid. Water stress hastens phytic acid formation (232), as does increasing Pi supply (5). There is no turnover of phytic acid in developing seeds, unlike other P-esters. Thus, during the life of the seed a major change occurs, from phytic acid formation and absence of phytase in the maturing, desiccating seed, to phytase synthesis and rapid phytic acid hydrolysis in the imbibing, germinating seed. The biochemical basis of this switch has not been studied.

The possibility that phytic acid, which can occur as the calcium salt phytin, may also function as a source of calcium has largely been ignored, even though germinating seeds frequently become limited in their growth by lack of calcium.

Phytic acid is also the major phosphate ester found in the soil, apparently because it is rapidly removed from solution as the highly insoluble Fe salt. Soil bacteria possessing phytase activity are known.

There is a coherent pattern in the behavior of Pi, phytic acid, and polyphosphate. Pi is either a substrate or a product of most key biochemical reactions. An increase in Pi concentration will enhance some and inhibit many others. However, Pi supply in the environment is often limited, and Pi supply to the cell is likely to fluctuate considerably. A means of buffering Pi concentration in the cytoplasm is needed. In the parenchyma cell, this is done by isolating Pi behind an impermeable membrane in the vacuole, so that the cytoplasmic concentration is kept at 5–20 mM, even in P-deficient plants (Section 10). In the developing seed the vacuole disappears, and this mechanism is no longer available. Instead a storage compound, phytic acid, is synthesized. This can readily remove Pi or return it to the cytoplasm without the need for a vacuole and membrane. The distribution of polyphosphate in maize roots also fits this pattern. Granules are present in the meristematic zone, where vacuoles are lacking or small, but absent from the elongating zone where proper vacuoles have developed (222). Similarly, the different properties and smaller proportions of vacuoles in bacterial and algal cells could make them much less capable of controlling Pi levels than the large vacuoles of parenchyma cells in higher plants. Polyphosphate synthesis would be an alternative control mechanism. Even so, polyphosphate can be located in a vacuole as well as in volutin granules in the cytoplasm (93).

10. ESTER POOLS

The four main organic-P fractions—DNA, RNA, lipid-P, and ester-P—can be regarded as four pools, each with its own characteristics [a "phosphoprotein" fraction probably represents a nonspecific binding of basic protein to highly electronegative P-compounds like phytic acid or polyphosphate (93)]. P in DNA is truly segregated, while at the other end of the scale terminal P groups of ATP turn over with a half-time of 2–20 sec (29, 33, 110, 137, 226). Most of the P-esters have half-lives of less than 30 min (29, 33). The different classes of RNA have only occasionally been distinguished, particularly with reference to their turnover. The same is true of P-lipids. It is not known whether there is a class of structural P-lipids that does not turn over. In actively growing *Spirodela*, in potato, and in cauliflower (19, 31), the half-lives of the P-lipids are about 1–5 hr; in apple fruit 3–10 hr (148); yet mature leaves may have rather a static P-lipid pool.

There is a basic difficulty in using ^{32}Pi as a precursor in turnover studies. The proper procedure is to pulse-label the tissue, then look at the way the labeling of the intermediate declines to zero. However, an inevitable endproduct of reactions involving P-esters is the precursor itself, Pi. Consequently, radioactivity of the intermediate cannot fall to zero. It is not even possible to study the time taken for an intermediate to reach the specific activity of the precursor since this cannot be calculated. The amount of ^{32}Pi which enters the tissue during the pulse is under

1% of the Pi already there, so the specific activity of ^{32}Pi in the tissue will be much lower than in the medium. But because the Pi in the tissue is present in two or more pools of unknown size, only an average value can be obtained, not the specific activity of the precursor Pi. As a result, the time taken for the intermediate to reach equilibrium labeling has been used as a criterion of turnover time. There must always be a small question mark attached to such measurements.

Depending on the P-nutrition of the plant, total P content falls in the range 2–40 mmoles P/Kg fr wt (P=0.05−1% dry wt) (37, 74, 165, 220). The level of Pi in the tissue is more sensitive to P nutrition than the levels of the various organic P fractions. Over the full range of P nutrition, ester-P, RNA, and P-lipid may change 5-fold (29, 187), while Pi content changes 50-fold (0.5–25 mM). Even the deficient plant contains 0.5–2 mM Pi (29, 165, 187). Levels of Pi above 30 mM are suggestive of unbalanced growth. Radiation damage (28, 144), Ca or N deficiency, or heterotrophic growth in the dark cause Pi concentration in *Spirodela* to rise from 30 mM to 70–80 mM (my unpublished data). In absence of other growth limitations, concentrations above 35 mM Pi or 45 μmoles total P/g fr wt may indicate P toxicity (68, 82, 136).

A typical ratio of DNA/RNA/lipid-P/ester-P is 0.2/2/1.5/1 (μmoles P/g fr wt) (28, 64). The size of the very important but seldom studied P-lipid fraction should be noted. Compared with the various P-lipids, each P-ester is present in much smaller amounts. Despite the different aims of different experiments and the range of techniques and tissues used, there has been a remarkable unanimity in results obtained, so that it is possible to write down a model P-ester pattern (Table 1). Published results can be fitted into it so easily that any marked departure is likely to be significant.

Many of the compounds have been identified by specific methods—identification of products, uv absorption and enzyme procedures. Even where the less rigorous chromatographic methods have been used, there is general agreement with the other procedures. The sugar phosphates and diphosphates have not been particularly well sorted out; because of their similarity, enzyme procedures will be required for better results (12). Problems caused by action of phosphatases have been generally recognized and circumvented. Those caused by specific adsorption of compounds during extraction are more insidious and have undoubtedly led to nucleotides being underestimated in some cases. Coenzymes have been satisfactorily determined only in experiments designed specifically for that purpose. UDP-glucose is frequently underestimated because it hydrolyzes rapidly under alkaline conditions; some UDP and most of the UMP may arise in this way. More attention should be paid to the nucleoside diphosphocompounds generally. Their concentrations appear to vary more than do those of other P-esters, and to be more informative of biochemical pathways operating in the tissue. For example, ADP-glucose levels are characteristically high in young seeds that are synthesizing starch (107, 162).

Two features of the generalized P-ester pattern require comment. The first is the ATP/ADP/AMP ratio of 10/3/1. Despite large changes in metabolism, changes in ATP level are often small and transient. Low ATP/AMP ratios can often be

Table 1   P-Ester Concentrations in a Model Tissue

Ester	Amount[a]	Ester	Amount[a]
phosphatidyl choline	630	phosphatyl choline	20
phosphatidyl ethanolamine	350	phosphatyl ethanolamine	20
phosphatidyl glycerol	300	1P-glycerol	10
phosphatidyl inositol	130	phosphatyl serine	5
phosphatidyl serine	50	ATP	100
phosphatidic acid	20	ADP	30
diphosphatidyl glycerol	20	AMP	8
3P-glyceric acid	85	UTP	40
P-enolpyruvate	15	UDP	50
6P-gluconate	5	UMP	5–40
phytic acid	15–300	CTP	10
glucose-6-P	200	CDP	3
fructose-6-P	60	CMP	1
mannose-6-P	40	GTP	10
glucose-1-P	15	GDP	5
ribose-5-P	3	GMP	1
ribulose-5-P	2	ADP-glucose	25
sedoheptulose-7-P	2	UDP-glucose	80
dihydroxyacetone-P	12	UDP-galactose	3
fructose-1,6-P	20	GDP-glucose	2
ribulose-1,5-P	8		
NAD	30	NADP	15
NADP	2	NADPH	5

[a] Amounts are in nmoles/g fr wt and apply to a young photosynthetic tissue. A meristematic tissue would have levels $2\times$ and a mature tissue 0.2–$0.5\times$ those shown. A nonphotosynthetic tissue would have much less phosphatidyl glycerol, 50 nmole/g fr wt, and rather more phosphatidyl ethanolamine. References: phospholipids (18, 28, 148); sugar phosphates (12, 28, 66, 120); nucleotides (28, 51, 66, 107, 113, 162, 229); cofactors (77, 113, 193). Because of space limitations, most of the sources used in compiling this table have not been cited.

explained by phosphatase action during extraction of the tissue. In some fine experiments, Bomsel & Pradet (35) have shown that a high ATP/AMP ratio is the normal state for plant tissues, and that only under extreme conditions (plants grown in N_2 atmosphere and darkness) are very low ATP/AMP ratios obtained (182). The ratios, including those reported by other authors, conform to the pattern expected if adenylate kinase were active and adenylate kinase is present. Bomsel & Pradet suggest that in plant cells as in bacteria, the ratio $\sim$P/total adenylate (the "energy charge") is a fundamental parameter controlling ATP-utilizing reactions in the cell.

Another consistent ratio is glucose-6-P/fructose-6-P/mannose-6-P, 4.5/1.5/1. This matches the ratio expected from the equilibrium constants of 6-phosphoglucose isomerase and 6-phosphomannose isomerase 4/1.5/1. Presumably these two enzymes are also active in plant tissues. Glucose-1-P does not fit a constant pattern. It is usually a minor ester, as to be expected from the equilibrium constant of phosphoglucomutase. When it is present in much higher amounts, starch phosphorylase has presumably been active; in some cases, this has apparently happened during extraction of the tissue.

Literature Cited

1. Adedipe, N. O., Fletcher, R. A. 1970. *J. Exp. Bot.* 21:968–74
2. Anderson, W. P., House, C. R. 1967. *J. Exp. Bot.* 18:544–55
3. Arnon, D. I. 1953. *Soil and Fertilizer Phosphorus in Crop Nutrition IV*, ed. W. H. Pierre, A. G. Norman, 1–42. N. Y.:Academic
4. Arnon, D. I., Fratzke, W. E., Johnson, C. M. 1942. *Plant Physiol.* 17:515–24
5. Asada, K., Kasai, Z. 1962. *Plant Cell Physiol.* 3:397–406
6. Asher, C. J., Loneragan, J. F. 1967. *Soil Sci.* 103:225–33
7. Barber, D. A. 1972. *New Phytol.* 71:255–62
8. Barber, D. A., Frankenburg, U. C. 1971. *New Phytol.* 70:1027–34
9. Barber, D. A., Loughman, B. C. 1967. *J. Exp. Bot.* 18:170–76
10. Barber, D. A., Sanderson, J., Russell, R. S. 1968. *Nature* 217:644
11. Barber, S. A., Walker, J. M., Vasey, E. H. 1963. *J. Agr. Food Chem.* 11:204–7
12. Barker, J., Jakes, R., Solomos, T., Younis, M. E., Isherwood, F. A. 1964. *J. Exp. Bot.* 15:284–96
13. Barrier, G. E., Loomis, W. E. 1957. *Plant Physiol.* 32:225–31
14. Bassham, J. A., Kirk, M., Jensen, R. G. 1968. *Biochim. Biophys. Acta* 153:211–18
15. Baylis, G. T. S. 1967. *New Phytol.* 66:231–43
16. Baylis, G. T. S. 1972. *Plant Soil* 36:233–34
17. Becker, D., Kluge, M., Ziegler, H. 1971. *Planta* 99:154–62
18. Ben Abdelkader, A. 1968. *Physiol. Veg.* 6:417–42
19. Ben Abdelkader, A., Mazliak, P. 1970. *Eur. J. Biochem.* 15:250–62
20. Berjak, P. 1972. *Ann. Bot. London* 36:73–81
21. Berman, T. 1970. *Limnol. Oceanogr.* 15:663–74
22. Bianchetti, R., Sartirana, M. L. 1967. *Biochim. Biophys. Acta* 145:485–90
23. Biddulph, O., Biddulph, S., Cory, R., Koontz, H. 1958. *Plant Physiol.* 33:293–300
24. Biddulph, O., Cory, R. 1957. *Plant Physiol.* 32:608–19
25. Biddulph, S. F. 1956. *Am. J. Bot.* 43:143–48
26. Bieleski, R. L. 1966. *Plant Physiol.* 41:447–54
27. Ibid, 455–66
28. Ibid 1968. 43:1297–1308
29. Ibid, 1309–16
30. Ibid 1969. 44:497–502
31. Ibid 1972. 49:740–45
32. Bieleski, R. L., Johnson, P. N. 1972. *Aust. J. Biol. Sci.* 25:707–20
33. Bieleski, R. L., Laties, G. G. 1963. *Plant Physiol.* 38:586–94
34. Bollard, E. G. 1953. *J. Exp. Bot.* 4:363–68
35. Bomsel, J.-L., Pradet, A. 1968. *Biochim. Biophys. Acta* 162:230–42
36. Bonnett, H. T. 1968. *J. Cell Biol.* 37:199–205
37. Bouma, D., Dowling, E. J. 1969. *Aust. J. Biol. Sci.* 22:515–21
38. Bowen, G. D., Rovira, A. D. 1966. *Nature* 211:665–66
39. Bowling, D. J. F., Macklon, A. E. S., Spanswick, R. M. 1966. *J. Exp. Bot.* 17:410–16
40. Boyd, C. E., Hess, L. W. 1970. *Ecology* 51:296–300
41. Branton, D., Moor, H. 1964. *J. Ultrastruct. Res.* 11:401–11
42. Bristow, J. M., Whitcombe, M. 1971. *Am. J. Bot.* 58:8–13
43. Bucke, C., Walker, D. A., Baldry, C. W. 1966. *Biochem. J.* 101:636–41
44. Bull, T. A., Gayler, K. R., Glasziou,

K. T. 1972. *Plant Physiol.* 49:1007–11
45. Burström, H. 1948. *Physiol. Plant.* 1: 124–35
46. Burström, H., Krogh, A. 1947. *Sv. Bot. Tidsk.* 41:17–44
47. Butler, G. W. 1953. *Physiol. Plant.* 6:637–61
48. Carter, O. G., Lathwell, D. J. 1967. *Plant Physiol.* 42:1407–12
49. Clarkson, D. T., Robards, A. W., Sanderson, J. 1971. *Planta* 96:292–305
50. Clarkson, D. T., Sanderson, J., Russell, R. S. 1968. *Nature* 220:805–6
51. Collins, G. G., Jenner, C. F., Paleg, L. G. 1972. *Plant Physiol.* 49:398–403
52. Crossett, R. N. 1968. *Aust. J. Biol. Sci.* 21:225–33
53. Ibid, 1063–67
54. Crossett, R. N., Loughman, B. C. 1966. *New Phytol.* 65:459–68
55. Daft, M. J., Nicolson, T. H. 1969. *New Phytol.* 68:945–52
56. Dainty, J. 1963. *Advan. Bot. Res.*1: 279–326
57. Davies, C. R., Wareing, P. F. 1965. *Planta* 65:139–56
58. Dove, L. D. 1969. *Planta* 86:1–9
59. Drew, M. C., Nye, P. H. 1970. *Plant Soil* 33:545–63
60. Dunlop, J., Bowling, D. J. F. 1971. *J. Exp. Bot.* 22:434–44
61. Ibid, 453–64
62. Edwards, D. G. 1970. *Aust. J. Biol. Sci.* 23:255–64
63. Elzam, O. E., Hodges, T. K. 1968. *Plant Physiol.* 43:1108–14
64. Ergle, D. R., Guinn, G. 1959. *Plant Physiol.* 34:476–81
65. Eschrich, W. 1970. *Ann. Rev. Plant Physiol.* 21:193–214
66. Farineau, J. 1969. *Planta* 85:135–56
67. Findlay, N., Mercer, F. V. 1971. *Aust. J. Biol. Sci.* 24:657–64
68. Foote, B. D., Howell, R. W. 1964. *Plant Physiol.* 39:610–13
69. Franklin, R. E. 1969. *Plant Physiol.* 44:697–700
70. Fried, M., Shapiro, R. E. 1961. *Ann. Rev. Plant Physiol.* 12:91–112
71. Gardner, D. C. J., Peel, A. J. 1969. *Nature* 222:774
72. Gerdemann, J. W. 1965. *Mycologia* 57:562–75
73. Gerdemann, J. W. 1968. *Ann. Rev. Phytopathol.* 6:397–418
74. Gerloff, G. C., Moore, D. G., Curtis, J. T. 1964. *Res. Rep. Univ. Wis. Coll. Agr. Exp. Sta. 14*
75. Ginsburg, H., Ginzburg, B. Z. 1970.

J. Exp. Bot. 21:580–92
76. Ibid, 593–604
77. Graham, D., Cooper, J. E. 1967. *Aust. J. Biol. Sci.* 20:319–27
78. Gray, L. E., Gerdemann, J. W. 1969. *Plant Soil* 30:415–22
79. Greenway, H., Gunn, A. 1966. *Planta* 71:43–67
80. Greenway, H., Hughes, P. G., Klepper, B. 1969. *Physiol. Plant.* 22:199–207
81. Greenway, H., Klepper, B. 1968. *Planta* 83:119–36
82. Greenwood, E. A. N., Hallsworth, E. G. 1960. *Plant Soil* 12:97–127
83. Gregory, D. W., Cocking, E. C. 1966. *J. Exp. Bot.* 17:57–67
84. Hackett, C. 1969. *New Phytol.* 68: 1023–30
85. Hadley, G., Johnson, R. P. C., John, D. A. 1971. *Planta* 100:191–99
86. Hagen, C. E., Hopkins, H. T. 1955. *Plant Physiol.* 30:193–99
87. Hall, J. L., Davie, C. A. M. 1971. *Ann. Bot. London* 35:849–55
88. Hall, J. R., Hodges, T. K. 1966. *Plant Physiol.* 41:1459–64
89. Hardy, P. J., Possingham, J. V. 1969. *J. Exp. Bot.* 20:325–35
90. Harley, J. L. 1969. *The Biology of Mycorrhiza.* London: Hill
91. Harley, J. L., Brierley, J. K. 1954. *New Phytol.* 53:240–52
92. Harley, J. L., Loughman, B. C. 1963. *New Phytol.* 62:350–59
93. Harold, F. M. 1966. *Bacteriol. Rev.* 30:772–94
94. Harvey, H. W. 1969. *The Chemistry and Fertility of Sea Waters.* Cambridge: Cambridge Univ. Press
95. Hayman, D. S., Mosse, B. 1971. *New Phytol.* 70:19–27
96. Ibid 1972. 71:41–47
97. Heber, U., Santarius, K. A., Hudson, M. A., Hallier, U. W. 1967. *Z. Naturforsch.* 22B:1189–99
98. Hes, J. W. 1958. *Acta Bot. Neer.* 7: 278–91
99. Higinbotham, N., Etherton, B., Foster, R. J. 1967. *Plant Physiol.* 42:37–46
100. Ho, L. C., Peel, A. J. 1969. *Ann. Bot. London* 33:743–51
101. Hoad, G. V., Peel, A. J. 1965. *J. Exp. Bot.* 16:742–58
102. Hodges, T. K., Hanson, J. B. 1965. *Plant Physiol.* 40:101–9
103. Holevas, C. D. 1966. *J. Hort. Sci.* 41:57–64
104. Jacobson, L., Hannapel, R. J.,

Moore, D. P. 1958. *Plant Physiol.* 33: 278–82

105. Jarvis, P., House, C. R. 1970. *J. Exp. Bot.* 21:83–90

106. Jeffrey, D. W. 1964. *Aust. J. Biol. Sci.* 17:845–54

107. Jenner, C. F. 1968. *Plant Physiol.* 43: 41–49

108. Jennings, A. C., Morton, R. K. 1963. *Aust. J. Biol. Sci.* 16:332–41

109. Jeschke, W. D., Simonis, W. 1965. *Planta* 67:6–32

110. Johnson, E. J., Bruff, B. S. 1967. *Plant Physiol.* 42:1321–28

111. Kennecke, M., Ziegler, H., Rongine de Fekete, M. A. 1971. *Planta* 98: 330–56

112. Kessler, K. J. 1966. *Can. J. Bot.* 44: 1413–25

113. Keys, A. J. 1968. *Biochem. J.* 108:1–8

114. Klepper, B., Greenway, H. 1968. *Planta* 80:142–46

115. Kluge, M., Becker, D., Ziegler, H. 1970. *Planta* 91:68–79

116. Kluge, M., Ziegler, H. 1964. *Planta* 61:167–77

117. Koontz, H., Biddulph, O. 1957. *Plant Physiol.* 32:463–70

118. Krichbaum, R., Lüttge, U., Weigl, J. 1967. *Ber. Deut. Bot. Ges.* 80:167–76

119. Laties, G. G. 1969. *Ann. Rev. Plant Physiol.* 20:89–116

120. Latzko, E., Gibbs, M. 1969. *Plant Physiol.* 44:396–402

121. Läuchli, A. 1967. *Planta* 75:185–206

122. Läuchli, A. 1972. *Ann. Rev. Plant Physiol.* 23:197–218

123. Läuchli, A., Epstein, E. 1971. *Plant Physiol.* 48:111–17

124. Lee, J. A. 1960. *Evolution* 14:18–28

125. Leggett, J. E. 1968. *Ann. Rev. Plant Physiol.* 19:333–46

126. Leggett, J. E., Galloway, R. A., Gauch, H. G. 1965. *Plant Physiol.* 40: 897–902

127. Leonard, O. A., Glenn, R. K. 1968. *Plant Physiol.* 43:1380–88

128. Levi, E. 1968. *Physiol. Plant.* 21:213–26

129. Lewis, D. G., Quirk, J. P. 1965. *Nature* 205:765–66

130. Lewis, D. G., Quirk, J. P. 1967. *Plant Soil* 26:445–53

131. Lewis, D. H., Harley, J. L. 1965. *New Phytol.* 64:256–69

132. Liedtke, M.-P., Ohmann, E. 1969. *Flora* 160A:378–90

133. Litav, M., Wolovitch, S. 1971. *Ann. Bot. London* 35:1163–78

134. Littlefield, L. J. 1966. *Physiol. Plant.* 19:264–70

135. Loneragan, J. F., Asher, C. J. 1967. *Soil Sci.* 103:311–18

136. Loneragan, J. F., Carroll, M. D., Snowball, K. 1966. *J. Aust. Inst. Agr. Sci.* 32:221–23

137. Loughman, B. C. 1960. *Plant Physiol.* 35:418–24

138. Lüttge, U. 1971. *Ann. Rev. Plant Physiol.* 22:23–44

139. Mackereth, F. J. 1953. *J. Exp. Bot.* 4:296–313

140. Maizel, J. V., Benson, A. A., Tolbert, N. E. 1956. *Plant Physiol.* 31:407–8

141. Mandal, N. C., Biswas, B. B. 1970. *Plant Physiol.* 45:4–7

142. Marinos, N. G. 1963. *J. Ultrastruct. Res.* 9:177–85

143. Marré, E. 1961. *Ann. Rev. Plant Physiol.* 12:195–218

144. Martin, R. P., Russell, R. S. 1954. *J. Exp. Bot.* 5:91–109

145. Mason, T. G., Maskell, E. J. 1931. *Ann. Bot. London* 45:125–73

146. Matile, P. 1966. *Z. Naturforsch.* 21B: 871–78

147. Matile, P., Moor, H. 1968. *Planta* 80: 159–75

148. Mazliak, P. 1967. *Phytochemistry* 6: 941–56

149. McComb, A. L. 1938. *J. Forest.* 36: 1148–54

150. Menzel, W. D., Ryther, J. H. 1960. *Deep-Sea Res.* 6:351–67

151. Millard, D. L., Wiskich, J. T., Robertson, R. N. 1965. *Plant Physiol.* 40: 1129–35

152. Milthorpe, F. L., Moorby, J. 1969. *Ann. Rev. Plant Physiol.* 20:117–38

153. Miyachi, S. 1961. *J. Biochem.* 50:367–71

154. Morrison, T. M. 1957. *Nature* 179: 907–8

155. Ibid 1965. 205:1027

156. Morrison, T. M., Heine, R. W. 1966. *Ann. Bot. London* 30:807–19

157. Morton, R. K., Raison, J. K. 1963. *Nature* 200:429–33

158. Mosse, B. 1959. *Trans. Brit. Mycol. Soc.* 42:439–48

159. Mosse, B., Phillips, J. M. 1971. *J. Gen. Microbiol.* 69:157–66

160. Mukherji, S., Dey, B., Paul, A. K., Sircar, S. M. 1971. *Physiol. Plant.* 25: 94–97

161. Müller, K., Leopold, A. C. 1966. *Planta* 68:186–205

162. Murata, T., Minamikawa, T., Akazawa, T., Sugiyama, T. 1964. *Arch. Biochem. Biophys.* 106:371–78

163. Nakata, S., Leopold, A. C. 1967. *Am. J. Bot.* 54:769–72

164. Nassery, H. 1969. *New Phytol.* 68: 21–23

165. Ibid 1971. 70:949–51
166. Neill, J. C. 1944. *N. Z. J. Sci. Technol.* 25:191–201
167. Nicolson, T. H. 1959. *Trans. Brit. Mycol. Soc.* 42:421–38
168. Nobel, P. S., Packer, L. 1965. *Plant Physiol.* 40:633–40
169. Oaks, A., Bidwell, R. G. S. 1970. *Ann. Rev. Plant Physiol.* 21:43–66
170. O'Brien, T. P., Zee, S.-Y. 1971. *Aust. J. Biol. Sci.* 24:207–17
171. Oland, K. 1963. *Physiol. Plant.* 16:682–94
172. O'Leary, J. W. 1965. *Bot. Gaz.* 126:108–15
173. Olsen, S. R., Kemper, W. D., Jackson, R. D. 1962. *Soil Sci. Soc. Am. Proc.* 26:222–27
174. Palmer, J. M. 1970. *Planta* 93:48–52
175. Palmer, J. M., Loughman, B. C. 1964. *New Phytol.* 63:217–31
176. Pavlinova, O. A. 1965. *Fiziol. Rast.* 12:606–17
177. Pavlinova, O. A., Afanas'eva, T. P. 1962. *Fiziol. Rast.* 9:133–41
178. Peel, A. J. 1967. *J. Exp. Bot.* 18:600–6
179. Pitman, M. G. 1972. *Aust. J. Biol. Sci.* 25:243–57
180. Pomeroy, L. R., Smith, E. E., Grant, C. M. 1965. *Limnol. Oceanogr.* 10:167–72
181. Poux, N. 1963. *J. Microsc.* 2:485–90
182. Pradet, A. 1969. *Physiol. Veg.* 7:261–75
183. Reid, M. S., Bieleski, R. L. 1970. *Planta* 94:273–81
184. Ridge, E. H., Rovira, A. D. 1971. *New Phytol.* 70:1017–26
185. Roberts, R. M., Loewus, F. 1968. *Plant Physiol.* 43:1710–16
186. Robinson, J. M., Stocking, C. R. 1968. *Plant Physiol.* 43:1597–1604
187. Roux, L. 1966. *Ann. Physiol. Veg.* 8:137–45
188. Rovira, A. D., Bowen, G. D. 1970. *Planta* 93:15–25
189. Rowan, K. S. 1966. *Int. Rev. Cytol.* 19:301–90
190. Russell, R. S., Martin, R. P. 1953. *J. Exp. Bot.* 4:108–27
191. Russell, R. S., Sanderson, J. 1967. *J. Exp. Bot.* 18:491–508
192. Russell, R. S., Shorrocks, V. M. 1959. *J. Exp. Bot.* 10:301–16
193. Ryrie, I. J., Scott, K. J. 1968. *Plant Physiol.* 43:687–92
194. Samotus, B., Schwimmer, S. 1962. *Plant Physiol.* 37:519–22
195. Sanders, F. E., Tinker, P. B. 1971. *Nature* 233:278–79
196. Santarius, K. A., Heber, U. 1965. *Biochim. Biophys. Acta* 102:39–54

197. Santarius, K. A., Heber, U., Ullrich, W., Urbach, W. 1964. *Biochem. Biophys. Res. Commun.* 15:139–46
198. Sartirana, M. L., Bianchetti, R. 1967. *Physiol. Plant.* 20:1066–75
199. Šebánek, J. 1967. *Planta* 75:283–85
200. Selvendran, R. R. 1970. *Ann. Bot. London* 34:825–33
201. Semadeni, E. G. 1967. *Planta* 72:91–118
202. Seth, A. K., Wareing, P. F. 1967. *J. Exp. Bot.* 18:65–77
203. Shapiro, R. E., Armiger, W. H., Fried, M. 1960. *Soil Sci. Soc. Am. Proc.* 24:161–64
204. Sheldrake, A. R., Northcote, D. H. 1968. *J. Exp. Bot.* 19:681–89
205. Smith, F. A. 1966. *Biochim. Biophys. Acta* 126:94–99
206. Splittstoesser, W. E., Beevers, H. 1964. *Plant Physiol.* 39:163–69
207. Stone, E. L. 1949. *Soil Sci. Soc. Am. Proc.* 14:340–45
208. Sutcliffe, J. F. 1959. *Biol. Rev.* 34:159–220
209. Tammes, P. M. L., van Die, J. 1964. *Acta Bot. Neer.* 13:76–83
210. Tanaka, K., Tolbert, N. E., Gohlke, A. F. 1966. *Plant Physiol.* 41:307–12
211. Tang, Van Hai, Laudelout, H. 1971. *J. Exp. Bot.* 22:830–36
212. Tewari, K. K., Singh, M. 1964. *Phytochemistry* 3:341–47
213. Thompson, E. J., Black, C. A. 1970. *Plant Soil* 32:161–68
214. Thomson, W. W., Berry, W. L., Liu, L. L. 1969. *Proc. Nat. Acad. Sci. USA* 63:310–17
215. Thorne, G. N. 1957. *J. Exp. Bot.* 8:401–12
216. Tiffin, L. O. 1970. *Plant Physiol.* 45:280–83
217. Tolbert, N. E., Wiebe, H. 1955. *Plant Physiol.* 30:499–504
218. Ueki, K., Sato, S. 1971. *Physiol. Plant.* 24:506–11
219. Ullrich, W., Urbach, W., Santarius, K. A., Heber, U. 1965. *Z. Naturforsch.* 20B:905–10
220. Ulrich, A., Berry, W. L. 1961. *Plant Physiol.* 36:626–32
221. Urbach, W., Hudson, M. A., Ullrich, W., Santarius, K. A., Heber, U. 1965. *Z. Naturforsch.* 20B:890–98
222. Vagabov, V. M., Kulaev, I. S. 1964. *Dokl. Akad. Nauk SSSR* 158:218–20
223. Van Diest, A. 1968. *Plant Soil* 29:248–56
224. Vasey, E. H., Barber, S. A. 1963. *Soil Sci. Soc. Am. Proc.* 27:193–97
225. Weibe, H. H., Kramer, P. J. 1954.

Plant Physiol. 29:342–48
226. Weigl, J. 1963. *Planta* 60:307–21
227. Ibid 1967. 75:327–42
228. Ibid 1968. 79:197–207
229. Weinstein, L. H., McCune, D. C., Mancini, J. F., van Leuken, P. 1969. *Plant Physiol.* 44:1499–1510
230. West, K. R., Wiskich, J. T. 1968. *Biochem. J.* 109:527–32
231. Williams, R. F. 1955. *Ann. Rev. Plant Physiol.* 6:25–42
232. Williams, S. G. 1970. *Plant Physiol.* 45:376–81
233. Wilson, A. T. 1968. *Aust. J. Sci.* 31:55–61
234. Woolhouse, H. W. 1969. *Brit. Ecol. Soc. Symp. 9th, 1968* 357–80
235. Woolley, J. T. 1965. *Plant Physiol.* 40:711–17
236. Yu, G. H., Kramer, P. J. 1967. *Plant Physiol.* 42:985–90
237. Ibid 1969. 44:1095–1100
238. Zee, S.-Y. 1969. *Aust. J. Biol. Sci.* 22:1051–54

Ann. Rev. Plant Physiol. 1973. 24:253–86

PHOTOSYNTHETIC CARBON FIXATION IN RELATION TO NET CO$_2$ UPTAKE[1]

❖ 7549

Clanton C. Black Jr.

Department of Biochemistry, University of Georgia, Athens, Georgia

CONTENTS

INTRODUCTION

Plant biology currently is enjoying a renaissance which has affected the research and thinking of plant scientists and teachers throughout the world. The resur-

[1] Abbreviations used are: C (carbon); C$_4$ (C$_4$ dicarboxylic acid); CAM (Crassulacean acid metabolism); Chl (chlorophyll); F-6P (fructose-6 phosphate); G-3P (glyceraldehyde-3 phosphate); G-6P (glucose-6 phosphate); K_m (substrate concentration to give half maximum velocity); OAA (oxaloacetate); pentose or C$_3$ (reductive pentose phosphate); PEP (phosphoenolpyruvate); 3-PGA (3-phosphoglyceric acid); 6-PGluA (6-phosphogluconic acid); R-5P (ribose-5 phosphate); RuDP (ribulose-1,5-diphosphate); Ru-5P (ribulose-5 phosphate).

gence of interest in the last decade gained tremendous thrust from the discovery of a new pathway for plants to assimilate CO_2 photosynthetically and from the acceptance of the fact that light respiration is quite different both in rate and in the pathway of carbon metabolism from dark respiration. When these metabolic activities were shown to have interrelationships with physiology, anatomy, morphology, ecology, and economic factors such as plant productivity and plant competition, there occurred a renovation of our thinking about plants. Sufficient work has been presented in the literature for many workers to realize that plants may be divided into large groups which have some specific characteristics; therefore this review is organized within the framework of characteristics which distinguish various groups of higher plants.

In order to summarize a massive amount of information that is available in the literature from diverse fields of study, Table 1 has been prepared. The information in Table 1 is an accumulation of research work covering nearly 10 decades of plant research by scientists from laboratories throughout the world. It is important to realize at the onset that these are broad and general classifications with ranges of data, thus specific exceptions can be cited to some of the pieces of data given in Table 1. However, even with these exceptions, the three groups of plants delineated in Table 1 are distinctly different. It is not within the scope of this review to dwell on each one of these characteristics. Rather these characteristics were collated so interested readers could follow any characteristic in more detail in the literature cited. In broad outline, this review will be concerned with net CO_2 uptake in these three major groups of higher plants: those which primarily utilize the reductive pentose phosphate cycle; those which utilize the C_4-dicarboxylic acid cycle; and those which exhibit Crassulacean acid metabolism (Table 1).

When one considers net CO_2 uptake in plants, one immediately is faced with the problem of defining net CO_2 uptake. Rather than give a strict definition for net CO_2 uptake, we will use the term in more of an operational sense of simply measuring the amount of CO_2 taken from the atmosphere under normal oxygen and CO_2 concentrations and at light intensities encountered naturally. Since plant research workers today are very conscious of the fact that dark respiration and light respiration are inadequately measured and various pools of CO_2 may exist in leaves, any definition will automatically become subject to the stated conditions. Thus an intuitive definition of net CO_2 uptake seems to be more proper in view of the current state of knowledge about photosynthesis and such problems as mitochondrial and peroxisomal respiration and the influence of light, dark, CO_2 pools, and other factors on these types of metabolism which are intimately concerned with net CO_2 uptake.

An examination of the characteristics listed in Table 1 leads one also to question whether this extensive list of characteristics is simply a fortuitous situation or if definite "cause and effect" relationships exist between all of these characteristics which will result, for example, in pentose plants being less productive than C_4 plants. To integrate these characteristics and determine whether "cause and effect" relationships exist will be approached in this review, but it also is certainly a problem that will continue to be the subject of intensive investigation in years to come. At the present time many of these characteristics have only initially been described and perhaps others remain to be discovered, thus integration into a complete understanding of such problems as net CO_2 uptake or plant productivity is just beginning.

It is pertinent to Table 1 and to subsequent sections for readers to realize that these major groups of higher plants are independent of the usual taxonomic criteria used to classify plants. For example: *Euphorbia corcollata* is a pentose plant; *Euphorbia maculata* is a C_4 plant; and *Euphorbia grandidens* is a CAM plant. Numerous other genera have species which fall into at least two of these groups of plants.

TOTIPOTENCY AND PLANT BIOLOGY

A fundamental tenet of plant biology is that each plant cell has the complete genetic information necessary to regenerate an intact plant. With the acceptance of totipotency in plants as basic, then it does not seem surprising that all green cells exposed to light will exhibit, for example, the pentose cycle and furthermore that the cycle will be expressed at varying levels of activity in the chloroplasts of specific green cell types. However, one continually encounters in the photosynthesis literature such statements as the following:
" .... agranal bundle sheath chloroplasts are inactive in the Hill reaction .... ";
or " .... bundle sheath chloroplasts .... lack grana and an active photosystem II." "In corn leaves, which have no photorespiration .... "; "Photosystem I and II are not linked in agranal chloroplasts .... "; or " .... the bundle sheath plastids, which in the tropical grasses appear to act as amyloplasts."

All of these statements imply that a green leaf cell has failed to express a portion of its genetic information as indicated by the lack of a certain feature or activity. Indeed, other work has shown all of the above statements to be false. Thus within any green plant cell one should expect to find genetic information relative to photosynthesis expressed at various levels of activity, and if the activity is not detected, statements should be made within the limitations of the assay sensitivities or preparation procedures.

When I propose that the major CO_2 fixation system in crab grass mesophyll cells is via a β-carboxylation reaction, that proposal should not be extrapolated to imply that RuDP carboxylation is absent. Rather the RuDP carboxylation system is a minor CO_2 fixation system in crab grass mesophyll cells just as, for example, β-carboxylation may be a minor contributor to net CO_2 fixation in spinach mesophyll cells in the light.

PHOTORESPIRATION

The contribution of photorespiration to net CO_2 uptake is an integral part of our consideration. However, photorespiration has been reviewed extensively in previous issues of this series (96, 164) and in other recent publications (76, 78,

Table 1 Characteristics Distinguishing Three Groups of Higher Plants

Characteristic[a]	Primary Type of Carbon Dioxide Fixation			References[b]
	Pentose	C_4	CAM	
Leaf anatomy in cross section	Diffuse distribution of organelles in mesophyll or palisade cells with similar or lower organelle concentrations in bundle sheath cells if present	A definite layer of bundle sheath cells surrounding the vascular tissue which contains a high concentration of organelles; layer(s) of mesophyll cells surrounding the bundle sheath cells	Spongy appearance. Mesophyll cells have large vacuoles with the organelles evenly distributed in the thin cytoplasm. Generally lack a definite layer of palisade cells	81, 84, 117
Theoretical energy requirement for net CO_2 fixation (CO_2:ATP:NADPH)	1:3:2	1:5:2	1:6.5:2	18, 32, 46
Major leaf carboxylation sequence in light	RuDP carboxylase	PEP carboxylase then RuDP carboxylase	Both PEP and RuDP carboxylase	18, 37, 84
CO_2 compensation concentration (ppm CO_2)	30 to 70	0 to 10	0 to 5 in dark; 0 to 200 with daily rhythm	84, 101, 129
Transpiration ratio (gm H_2O/gm of dry wt)	450 to 950	250–350	50 to 55	102, 148
Maximum rate of net photosynthesis (mg CO_2/dm^2 of leaf surface/hr)	15 to 40	40 to 80	Usually about 1 to 4; highest reported −11 to 13	28, 84, 161
Photosynthesis sensitive to changing O_2 concentration from about 1% to 21%	Yes	No	Yes	58, 74, 75
Leaf photorespiration detection: (a) exchange measurements (b) glycolate oxidation	Present Present	Difficult to detect Present	Difficult to detect Present	164, 168
Optimum day temperature for net CO_2 fixation	15 to 25°C	30 to 47°C	~35°C	53, 54

[a] Comparisons are at 21% oxygen and 0.03% CO_2, or otherwise at as near healthy physiological conditions as possible. Each characteristic generally applies to fully differentiated tissues from plants which have not been subjected to such treatments as moisture stress, defoliation, shading, or experienced aging. A dash indicates that data could not be located.

[b] For interested readers selected references are cited so they will have a beginning for detailed examination of each characteristic.

Table 1—(*continued*)

Characteristic[a]	Primary Type of Carbon Dioxide Fixation			References[b]
	Pentose	C₄	CAM	
Maximum growth rate: (gm of dry wt/dm² of leaf area/day; gm/m² of land area/ day	.5 to 2 19.5 ± 3.9	4 to 5 30.3 ± 13.8	.015 to .018 —	53, 54, 102
Leaf chlorophyll *a* to *b* ratio	2.8 ± .4	3.9 ± .6	2.5 to 3.0	33, 144
Leaf isotopic ratio δ¹³C/¹²C	−22 to −34	−11 to −19	−13 to −34	20, 144, 155
Requirement for sodium as a micro-nutrient	None detected	Yes	—	39
Leaf postillumination CO₂ burst sensitive to oxygen levels above about 1%	Yes	Not completely inhibited	—	38, 74, 75
Rapid leaf postillumination release of newly fixed ¹⁴CO₂ >5%	No	Yes	Yes	Unpublished data
Translocation of freshly fixed ¹⁴C from illuminated leaves (%/6 hr)	<50%	>50%	—	94, 130
Optimum day temperature for growth (dry matter production)	20 to 25°C	30 to 35°C	∼35°C	54, 84, 161
Response of net photosynthesis to increasing light intensity at temperature optimum	Saturation reached about ¼ to ⅓ full sunlight	Either proportional to or only tending to saturate at full sunlight	Uncertain, but apparently saturation is well below full sunlight	28, 31
Dry matter production (tons/hectare/ year)	22.0 ± 3.3	38.6 ± 16.9	Extreme variability in available data	53, 54

84, 177). But the following developments related to photorespiration are relevant to net CO₂ uptake in specific plants.

One problem that has been of considerable concern in considering glycolate as the sole substrate for photorespiration (176) is the lack of information on rates of glycolate synthesis. Various laboratories have been estimating that the rate of photorespiration is at least somewhere between 30 and 50% of the rate of leaf net CO_2 uptake (96, 164, 177), which means that the rates of photorespiration or glycolate synthesis have to be, for example in tobacco leaves, in the order of 50 to 75 μmoles/mg chl/hr. Only recently have rates of glycolate synthesis been achieved which are of the order of 50% of photosynthesis. Zelitch reported, with illuminated tobacco leaf disks, glycolate synthesis rates of 67 μmoles/g fr wt/hr in the presence of a glycolate oxidase inhibitor and rates of 13 μmoles in maize leaf disks (178); these rates compare favorably with his previous data on tobacco photosynthesis rates of up to 136 μmoles (177). Reported rates of glycolate or of serine and glycine synthesis in C_4 plants are uniformly low.

In addition to the problem of the rate of glycolate synthesis, the biochemistry of glycolate synthesis is uncertain. Some workers have been proposing for several years that glycolate synthesis involves carbon one and two of either a hexose or a pentose, possibly utilizing the transketolase reaction of photosynthesis, producing an additional complex which is cleaved in the presence of hydrogen peroxide to form phosphoglycolate which then is dephosphorylated by glycolate phosphatase (76, 147, 164). More recently, Ogren and co-workers have discovered (35, 36, 132), due to the fact that oxygen is an inhibitor of leaf CO_2 fixation, that ribulose diphosphate carboxylase may act as an oxygenase. With respect to CO_2, the enzyme is competitively inhibited by O_2 and may act as an oxygenase as shown.

$$RuDP + O_2 \xrightarrow[\text{carboxylase}]{\text{RuDP}} 3\text{-}PGA + \text{phosphoglycolate}$$

These experiments, however, are hindered by the very low rates (about 2 nmoles/min/mg protein) of glycolate production these workers observed. In related experiments using ^{18}O-oxygen to show that ^{18}O is incorporated into the carboxyl group of glycine and serine during photosynthesis in spinach leaves, it was concluded that glycolate arises from a two-electron oxidation of an intermediate of the pentose cycle (7). Isotope was not detected in glycerate or phosphoglycerate, which leaves the question still unanswered as to the mechanism of O_2 inhibition of photosynthesis or the precise pathway of glycolate synthesis. Zelitch has recently proposed, particularly in C_4 plants, an alternative pathway of glycolate synthesis utilizing organic acids. Label in C-2 or C-3 of acetate or pyruvate was incorporated primarily into C-2 of glycolate, hence he proposed multiple pathways for synthesis of glycolate in both maize and tobacco (178).

The entire question of photorespiration in C_4 plants has been a problem that is not yet completely resolved. Some workers can show clearly that CO_2 is released and O_2 is taken up by maize leaves when they utilize $^{18}O_2$ or $^{13}CO_2$ as markers (168), while other workers still report that measurements under a variety of conditions fail to detect CO_2 evolution in maize leaves (165). In consideration of photorespiration in C_4 plants, the very earliest experiments of Krotkov and co-workers, who were unable to detect CO_2 release from maize leaves, contained a proposal that the reason for this was an efficient re-utilization of CO_2 before it could escape to the atmosphere (75). Since that time almost all workers generally have embraced the internal leaf CO_2 cycling proposal; however, direct evidence to substantiate the proposal has been lacking.

Two types of experiments have been reported with isolated cells from C_4 leaves which demonstrate that photorespiration is active in C_4 plants. In one study isolated crabgrass bundle sheath cells released CO_2 when fed glycolate, and mesophyll cells in an adjacent chamber in a Warburg flask were able to take up the CO_2; hence free CO_2 was not released from glycolate oxidation when the two isolated cell types had access to the same atmosphere. Thus experimentally it is possible to demonstrate the cycling of CO_2 from glycolate oxidation between cells isolated from the leaves of a C_4 plant (122). In a second study isolated bundle sheath strands from a C_4 plant have been shown to exhibit the Warburg effect in that photosynthesis was inhibited by either raising the oxygen concentration or lowering the CO_2 concentration (49, 50). Both treatments will lower the level of glycolate production in leaves (76) and affect leaf photorespiration. However, if one does similar experiments on intact C_4 plant leaves, such effects are not observed easily. Thus one is led to conclude the bundle sheath cells exhibit the Warburg effect but the CO_2 which is released from photorespiration is effectively trapped by the mesophyll cells, and thus externally the Warburg effect is not exhibited by leaves of C_4 plants. Such proposals are further supported by the demonstration that the enzymes of the glycolate pathway are concentrated in the bundle sheath cells (122, 164) and that the bundle sheath strands synthesize glycolate (50). Although the activities in the isolated cell experiments are low, they are quite consistent with Krotkov's proposal (75).

In some other experiments on leaf CO_2 release, the base of excised leaves from maize and the rooted aquatic plant *Phragmites* were submerged in a solution of ^{14}C-labeled bicarbonate, and the leaves were illuminated at 600 fc at 25°C (79). Under these conditions corn released $^{14}CO_2$ at a rate between .017 to .12 mg of CO_2/dm^2 of leaf area/hr, whereas *Phragmites* released at a rate between .034 to .14 mg of CO_2/dm^2 of leaf area/hr. The rates of CO_2 release were independent of external O_2 levels of near 1% and 21%. Compare these CO_2 release rates with the rates of leaf photosynthesis in Table 1. Clearly these rates of CO_2 release from internal HCO_3^- moving up through the transpiration stream show that very little CO_2 can be released from the interior of either a C_3 or C_4 plant leaf. Using another rooted aquatic plant, cattail (128), other workers have demonstrated photosynthesis rates that varied from 44 to 68 mg of CO_2/dm^2 of leaf area/hr when illuminated at a light intensity of 2×10^6 ergs cm^{-2} sec^{-1}, which is about two times normal summer sunlight intensities. This work is quite similar to earlier work on cotton and sunflower which shows that some C_3 species under certain conditions may exhibit rates of photosynthesis comparable to C_4 plants even with photorespiration which is easily detected in these leaves. In another experiment with an aquatic angiosperm *Myriophyllum spicatum*, a biphasic temperature curve was demonstrated for photosynthesis with a tendency to approach an optimum around 25 to 30° with a sharp shoulder at 35° (156). In addition, the response of CO_2 fixation to CO_2 concentration indicated a nonlinear response which was extrapolated to zero to indicate that the compensation point of this plant may be near zero. The ability to remove virtually all of the CO_2 from the atmosphere of a sealed chamber by aquatic plants such as algae is well known, and perhaps rooted aquatic plants have an unusual metabolism, although both of these plants exhibited labeling characteristics for possessing the classical pentose cycle of photosynthesis (128, 156).

In other work related to photorespiration it is generally recognized that the rate of CO_2 release from serine formation has been too low to account for the expected rates of CO_2 evolution from glycolate oxidation via the glycolate pathway (164, 177), and alternative enzymic reactions have been sought such as a glycine decarboxylase (107, 109) which indeed is stimulated as the oxygen tension rises to 100% (108) as is photorespiration. A direct decarboxylation of glyoxylate forming CO_2 and formate and requiring chloroplasts, light, O_2, and manganous ions has been demonstrated in spinach at rates exceeding 50 μmoles/ mg chl/hr (179). These very interesting decarboxylation reactions may play a significant role in photorespiration and doubtless will be the subjects of more research. A glyoxylic acid reductase has been purified in a homogeneous form from spinach leaves, characterized in detail as to its physical and kinetic properties, and shown to exist in isozymic forms (112). However, any relationship between factors studied in this series of papers which affect enzyme activity and the in vivo enzyme role(s) is uncertain.

A physiological function for photorespiration remains speculative even though the glycolate pathway is a pathway for synthesizing amino acids, and as Tolbert formulates it (164), it is a means for conserving 75% of the carbon entering the pathway. But as Zelitch formulates photorespiration (177), it may result in either a 50 or 100% loss of the entering carbon depending upon the rather uncertain fate of formate. However one formulates photorespiration, the fact remains that its function is unknown, and its exact contribution has yet to be determined in regard to net CO_2 uptake, particularly in C_4 and in CAM plants.

PRIMARY ENZYMES OF CO_2 METABOLISM

The major enzymes currently considered to be involved in leaf CO_2 metabolism will be considered. The reader will note that most reactions involving mitochondria are omitted, simply because of the unfortunate lack of substantial work with these leaf systems, and that enzymes related to photorespiration and peroxisomes were just mentioned or have been reviewed recently.

Carboxylases

RIBULOSE DIPHOSPHATE CARBOXYLASE   The stoichiometric combination of RuDP with CO_2 to give 2 molecules of 3-PGA is the initial CO_2 fixation reaction of the pentose cycle. The carboxylation and hydration reaction proceeds toward triose formation in a practically irreversible manner. Since the enzyme was discovered in the early fifties and it became clear that Fraction I protein, discovered in the late forties, was identical to it, RuDP carboxylase has been the subject of continuous research (105). Currently much effort is being expended in an attempt to understand how its activity may be controlled since it is a particularly attractive idea to search for means of controlling a key enzyme such as an initial CO_2 fixing enzyme.

For a number of years it has been reasoned that the enzyme might be activated directly in the light and deactivated in the dark. There is a report on the isolation of a light activating factor (174), but its occurrence in numerous photosynthetic organisms has not been established. Generally most light activation effects on RuDP carboxylase are indirect in that light effects pH, Mg^{++} ion levels, NADPH/NADP$^+$ ratios, and RuDP or 6-PGluA concentrations (17, 105, 170), which in turn can regulate the enzyme as will be discussed. Thus no compelling evidence exists to support the idea of light directly influencing RuDP carboxylase activity.

An interesting regulation possibility can be presented by considering the recent work showing that RuDP carboxylase can be crystallized in the presence of Mg^{++} and HCO_3^- ions and reversibly solubilized by addition of the substrate RuDP (44, 115). Other compounds including R-5P, ATP, 3-PGA, P_i, and carbamyl-P would not solubilize the enzyme. These specific properties have been particularly useful in purifying and crystallizing tobacco RuDP carboxylase, but the implications as to regulation mechanisms seem exceptional. In green tissues it is known that the level of RuDP rises severalfold upon illumination, then falls in darkness (17), and Mg^{++} is known to move in chloroplasts upon illumination (121). It was estimated that the carboxylase-RuDP complex is 100 times more soluble than the carboxylase in the presence of excess HCO_3^- and Mg^{++} (115). Thus upon illuminating a leaf, the HCO_3^- level should drop as photosynthesis begins, and as the pentose cycle gains momentum the RuDP produced should "solubilize" and activate more carboxylase. Of course, this is an oversimplification. Walker, in a detailed discussion of RuDP carboxylase regulation, has used similar data from other laboratories to lucidly present a somewhat similar sequence of events associated with the well-known photosynthetic induction phenomena and other aspects of photosynthesis regulation (170).

Another mechanism for regulating RuDP carboxylase is its partial inhibition by 6-PGluA (51, 159). The mode of inhibition with respect to RuDP for the spinach enzyme in one report appears to be noncompetitive (51) while in the other reports it is competitive (159). Despite this difference, the data plainly indicate that an intermediate of the oxidative pentose phosphate cycle can regulate the entrance to the reductive pentose phosphate cycle. 6-PGluA levels rise in the dark in green tissues and fall in the light to near zero (17). 6-PGluA does not penetrate the chloroplast membrane, but data are available showing that enzymes of the oxidative pentose cycle are in chloroplasts (84, 114). In addition it has been observed that 6-PGluA does not inhibit (159) the lower molecular weight RuDP carboxylases found in *Rhodospirillum rubrum*, *Rhodopseudomonas spheroides*, and *Thiobacillus denitrificans* (4, 127). As noted in the photorespiration section, leaf RuDP carboxylases are sensitive to O_2 levels, which may be a regulatory mechanism. Whether O_2 levels effect RuDP carboxylases in bacterial systems is unknown.

One of the most vexing problems in photosynthesis has been the concentration of the "CO_2-species" required to give half-maximum velocity of the plant RuDP carboxylase with reports ranging up to 22 mM (124, 139). Walker (170) has illustrated the enigma with the following calculation: at pH 7.9 and 30°, and with a K_m for HCO_3^- of 22 mM, the partial pressure of atmospheric CO_2 in equilibrium with 22 mM HCO_3^- is approximately 1.8%. Since plants flourish at 0.03% CO_2 and the isolated enzyme requires 1.8% CO_2, for half-maximum velocity one clearly has a problem equating these values. The molecular species of CO_2 utilized by RuDP carboxylase was uncertain until 1969–70 when Cooper and co-workers (56, 57) presented evidence for a faster initial incorporation of radioactivity when $^{14}CO_2$ was used rather than when $H^{14}CO_3^-$ was used. The results are consistent with the idea of CO_2 being the molecular species used by RuDP carboxylase. The K_m HCO_3^- was corrected to a K_m CO_2 of about 0.45 mM (56). However, as Walker points out (170), this level of CO_2 still would be in equilibrium with an atmosphere of about 1.8% CO_2 which is about 60 times higher than normal air. The problem may not be as severe as these calculations appear if one considers that isolated, envelope-containing chloroplasts exhibit a K_m HCO_3^- near 0.4 mM (97) and isolated tobacco leaf cells have K_m HCO_3^- near 0.2 mM (98). Both of these values would be in approximate equilibrium with 0.03% CO_2. Thus in these isolated systems which have maintained a more normal environment for the enzyme, there is no problem as to the level of CO_2 needed for photosynthesis compared to the enzyme requirement.

PHOSPHOENOLPYRUVATE CARBOXYLASE   Until 1967 PEP carboxylase, which was first detected in plants by R. S. Bandurski in the early fifties, was considered primarily to have a role in dark CO_2 fixation reactions with only a minor role in leaf CO_2 fixation in light with the possible exception of CAM leaves (169). However, in 1967 (151) activity was detected at levels four to fivefold higher than the rates of leaf photosynthesis in specific plants such as sugarcane, maize, and sorghum which also are C_4 plants. This observation has been confirmed in numerous laboratories throughout the world. PEP carboxylase catalyzes the essentially irreversible carboxylation and hydrolysis reaction shown below.

$$PEP + CO_2 + H_2O \rightarrow OAA + P_i$$

The molecular "CO_2 species" utilized by PEP carboxylase seem to be HCO_3^- in most studies using such diverse tissue sources as avian, peanut cotyledons, and 12 to 14-day-old maize shoots (57, 73). However, there is a report, using similar methods, of CO_2 as the active species for the maize leaf enzyme (172). For most PEP carboxylases the apparent K_m for HCO_3^- or "total CO_2" is near .3 to .4 mM. However, there may not be any discrepancy in the "CO_2 species" used as the substrate by the enzyme from various laboratories since data are available on multiple forms of the enzyme in: cotton leaf tissue (131); *Chlamydomonas reinhartii* (45); leaves of a number of pentose, C_4, and CAM plant species; in nongreen root tissues (106, 162, 163); and in leaves of F_1, F_2, and F_3 hybrids of *Atriplex patula* spp. *hastate*×*Atriplex rosea* (85). It seems conceivable that a form of the enzyme from various portions of a plant may utilize a different CO_2 species.

PEP carboxylase also is an attractive enzyme to use in searching for regulatory functions. In bacterial systems it is subject to regulation by organic acids and nucleotides, but in leaves the usefulness of these regulators remains to be demonstrated. However, there is a report that the purified maize leaf enzyme is sensitive to OAA with 50% inhibition observed at .25 mM, while aspartate or pyruvate alone has little effect on the enzyme (123). Recently G-6P, F-6P, and R-5P have been shown to activate PEP carboxylase some two to threefold in C_4 leaf extracts (52, 162), while 3-PGA stimulates and inhibits depending upon its concentration (52). Sigmoidal shaped curves for the C_4 leaf enzyme have been presented, indicating that this enzyme may be an allosteric protein (52, 162). Thus in leaves of C_4 plants where PEP carboxylase is a major CO_2 fixing enzyme, regulatory mechanisms appear to be operating. PEP carboxylase apparently is insensitive to changing O_2 levels in contrast to RuDP carboxylase.

PEP carboxylase assays in leaf extracts from *Atriplex spongiosa*, *Atriplex hastata*, and *Sedum praeltum* show 80 to 90% inhibition of the enzyme in the presence of 5 to 10 mM levels of several bisulfite compounds (135). Extracts from C_4 plant leaves indicate a particular sensitivity to NaCl with an 80 to 90% inhibition by 100 to 200 mM salt, while other sources of the plant enzyme are very tolerant of anions (137).

In a recent study, plant PEP carboxylases have been divided into four types based on kinetic and physical characteristics (Table 2). It was further suggested that each form was associated with a different metabolic pathway in the respective tissues (162, 163). A thermostable, cyanide-sensitive PEP carboxylase from photosynthetic and a wide variety of other tissues has been reported (138), but this report has not been confirmed or extended.

PHOSPHOENOLPYRUVATE CARBOXYKINASE   In contrast to the previously discussed

Table 2   General Kinetic Properties of Plant PEP Carboxylases[a]

Tissue source of enzyme	K_m PEP[b]	K_m Mg^{++}[b]	V_{max}[c]
C_4 Photosynthetic	.59	.5	29
Pentose photosynthetic	.14	.097	1.5
CAM	.19	—	18
Nongreen or autotrophic	.19	~.09	1.2[d]

[a] References 162, 163.
[b] millimolar.
[c] μmoles/min/mg chl.
[d] μmoles/min/g fr wt.

carboxylases, PEP carboxykinase catalyzes a freely reversible carboxylation in which phosphate transfer also occurs:

$$PEP + ADP + CO_2 \leftrightarrow OAA + ATP$$

Little energy loss occurs in these changes, so the reaction can proceed readily in either direction, presumably dependent upon the respective substrate concentrations and other reactions which may be removing substrates. The plant enzyme was described briefly in 1957 (126), but it was not thought to play a major role in leaf photosynthesis until it was discovered in leaves of certain C_4 plants which have low levels of malic enzyme (68). The plant enzyme utilizes adenosine nucleotide derivatives more efficiently than other nucleotides and requires a divalent metal ion for activity. The C_4 leaf enzyme has not been intensively characterized as to its kinetic and physical properties or regulatory processes. Partially purified PEP carboxykinase from *Rhodospirillium rubrum* used CO_2 as the substrate (55), and based on the utilization of nucleotide di- and triphosphate by the C_4 leaf enzyme (68), it probably uses CO_2 as the substrate.

Transcarboxylase

Until early 1970 the Australian workers, Hatch and Slack, were proposing a transfer of carbon from a C_4-dicarboxylic acid to C-1 of 3-PGA via a transcarboxylation type reaction during C_4 photosynthesis (91). However, to date such an enzymatic activity has not been demonstrated. Currently one must conclude that the idea is viable but supporting data are lacking. As will be discussed later, the role of transcarboxylation has been replaced by a decarboxylation reaction in most C_4 plants, followed by another carboxylation.

Decarboxylases

As currently formulated, both C_4 and CAM plants depend upon a decarboxylation reaction to transfer carbon from a 4C compound into a 3C compound (see later sections). Only two decarboxylating enzymes have been detected for this role, and neither enzyme has been characterized in sufficient detail from C_4 or CAM leaves to propose regulation mechanisms other than compartmentalization. Malic enzyme catalyzes the following oxidative decarboxylation:

$$Malate + NADP^+ \leftrightarrow pyruvate + NADPH + CO_2$$

Equilibrium is favored to the right, but with excess levels of pyruvate and HCO_3^-, the reaction will reverse. Malic enzyme from maize leaves was partially purified and shown to be similar to the enzyme from other sources (100). Malic enzyme was $NADP^+$ specific, required a divalent metal ion, and as the pH varied the K_m for malate also varied. The preparation exhibited an OAA decarboxylase activity equal to half the malic enzyme activity at their respective optimum assay conditions (100).

PEP carboxykinase also will act as a decarboxylase, as noted in the previous section.

In CAM plants particularly, the oxidative decarboxylation catalyzed by 6-phosphogluconate dehydrogenase may play a role in converting carbon from starch to triose.

$$6\text{-PGluA} + NADP^+ \rightarrow Ru\text{-}5P + NADPH + CO_2$$

Extensive search in this and other laboratories for an aspartate or OAA decarboxylase activity has not detected appreciable activity as yet.

Carbonic Anhydrase

The hydration of CO_2 catalyzed by carbonic anhydrase at physiological pH values above 6, involves four components which are intimately associated with photosynthesis.

$$CO_2 + H_2O \leftrightarrow H^+ + HCO_3^-$$

Hence it is not surprising that research for a photosynthetic role has been pursued for several decades. However, the correct role for carbonic anhydrase never has been demonstrated. Renewed interest in the enzyme occurred when several workers assayed low levels of the enzyme in crude C_4 leaf extracts (47, 70, 71), which seemed to be contradictory to their high rates of photosynthesis (Table I). Seemingly these measurements comparing levels of carbonic anhydrase in C_4 versus C_3 plants are in error since later work has demonstrated no difference in enzyme activity between these two groups of plants (9, 84, 140, 141). Not only are no differences apparent in the activity of carbonic anhydrase in various plant groups in Table 1, but recent calculations with spinach reveal that there is sufficient enzyme present to hydrate over 100 times more CO_2 in chloroplasts than the rate of photosynthesis (142). It was assumed that the rate of photosynthesis equaled 200 μmoles/mg chl/hr and the enzyme activity data were obtained at 0°. Assuming carbonic anhydrase catalyzed reactions are readily reversible, and considering the usual temperature optima for photosynthesis, the activity of carbonic anhydrase—if the calculations are correct—is in great excess when compared to the maximum rates of leaf photosynthesis (Table 1). Unless the enzyme is compartmentalized, it can hardly be considered as rate-limiting for leaf photosynthesis.

Carbonic anhydrase is widely distributed in plants and exists in at least two electrophoretically separable types in higher plants. One type is found principally in dicots with a suggested molecular weight in the range of 140,000 to 190,000 whereas the monocot type is about 40,000. There is a suggestion of three isoenzymes in leaves of C_4 plants, and about one-third the activity in maize leaves can be isolated in a "particulate fraction" (9, 10). As much as 44% of the spinach leaf enzyme can be localized in the stroma region of the chloroplasts (142).

THE PENTOSE CYCLE

As outlined in the 1950s, the pentose cycle has withstood nearly two decades of testing and appears to be reasonably acceptable. A few questions remain concerning the low activity of some enzymes such as transaldolase and the diphosphatases. The high apparent K_m for HCO_3^- was a problem but appears to be settled if one accepts the data on intact chloroplasts or isolated leaf cells (see earlier section). Knowing now that trioses particularly can move in and out of chloroplasts, the asymmetric labeling in hexoses appears to be explainable. For a thorough discussion of the pentose cycle, including its supporting data and its deficiencies, the review a decade ago by Stiller still is pertinent (157).

The positive point to be made here is that the pentose cycle has been found in all photosynthetic organisms studied thus far. This point was first called into question by C_4 photosynthesis (86), but work in the last two years has shown the pentose cycle to be present in all C_4 plants (see next section). Indeed, it should be noted again that the NADP+ dependent G-3P dehydrogenase of the pentose cycle still is *the* enzyme that reduces net quantities of carbon to the level of sugars such that plants can grow and support the animal world. In all green photosynthetic cells studied to date, G-3P dehydrogenase is present.

Current research on the pentose cycle is concentrated upon mechanisms of its regulation, and such items as the apparent competitive inhibition by O_2 with respect to CO_2 of RuDP carboxylase will be the subject of future research (36). Hopefully, out of this work the exact mechanism of the RuDP carboxylase reaction will be elucidated, including the chemical properties of RuDP, and the still hypothetical intermediate(s) will be identified.

THE C_4 CYCLE

The beginning of the elucidation of a major new pathway for carbon flow during photosynthesis can be traced clearly to the efforts of the sugarcane industry to learn as much as possible about the sugarcane plant. Unquestionably the Hawaiian Sugar Planters' Association Experiment Station and similar organizations have laid a strong biochemical, physiological, and genetic foundation for understanding the sugarcane plant (42). Out of this basic work emerged the C_4 cycle, which has revolutionized current thinking about plant biology and catalyzed efforts in laboratories throughout the world to investigate numerous old and new aspects of plant biology such as the few listed in Table 1. The studies of $^{14}CO_2$ fixation with sugarcane leaves by Kortschak, Burr & associates cleanly demonstrated that in short-time tests of steady-state photosynthesis, a large percentage of the total counts were incorporated into malic and aspartic acids (41, 42, 113). In early reports ^{14}C could not be detected in 3-PGA in very short times (42), but in subsequent reports labeled 3-PGA was detected in photosynthesis periods as short as 1 second (41, 113). Scattered references to $^{14}CO_2$ fixation into apparent initial products other than 3-PGA also appeared in other literature, but the significance of these observations relative to pathways of carbon flow other than the pentose cycle was not mentioned (104, 160).

On the basis of this earlier work, sugarcane workers at another industrial laboratory in Australia initiated studies that allowed them to formulate in outline a major pathway of carbon flow, the C_4 cycle (86, 91). Hatch and Slack were the first workers to understand and to present a scheme for photosynthetic carbon flow which clearly did not utilize RuDP carboxylase (86). In the period from 1966 to 1969 these workers presented an amazingly thorough series of enzymic and ^{14}C kinetic labeling experiments to support the C_4 cycle. They reviewed this intensive period of research for Volume 21 of this series (91).

This review of carbon flow in the C_4 cycle will begin around 1969 and will evaluate its current status. However, to evaluate the current work, which is quite controversial in its interpretation, I will list the elementary criteria upon which I will base my evaluation. These criteria are needed to quantitatively establish a photosynthetic carbon pathway or cycle in a cell or an organelle in a fully differentiated tissue such as a C_4 leaf. The following criteria also must be compared to intact leaf studies. First, proof of separation of the cell or organelle must be furnished. This proof needs to be furnished in the form of photographs in sufficient detail to identify the cell or to determine the internal structure, for example, of a chloroplast. The level and type of contamination in the particular cell or organelle preparation should be assessed. The percent distribution of the cell or organelle in the intact leaf or cell should be determined. The second criterion would be the results of CO_2 fixation experiments with the isolated preparation. The third criterion would be to assay appropriate enzymes in the isolated preparation. For criteria two and three, the rate of the reactions should be compatible with the rate of net CO_2 fixation in the intact leaf and the total leaf or cell activity should be accounted for. The fourth criterion would be to identify the products of $^{14}CO_2$ fixation and to determine the internal distribution of ^{14}C in the products. The fifth criterion would be to feed to the cells or organelles suspected intermediates either labeled in known positions or unlabeled, and observe the products under various conditions such as plus CO_2 or light or dark. The expected or predicted products and labeling patterns should be observed. The sixth criterion would be to use known inhibitors of the enzymes in a suspected pathway to cause the accumulation or to prevent the formation of a compound. One who is familiar with the literature can observe that all such criteria have not been presented by any group of workers; however, such quantitative data will be required ultimately for the C_4 pathway to be described precisely and to become widely accepted.

In mid-1969 there were several apparent discrepancies in the supporting data for the C_4 cycle such as: a transcarboxylase type activity was proposed (86) for which no supporting enzymic data had been presented; the activity of RuDP carboxylase was about one-tenth of leaf photosynthesis, about 30 μmoles/mg chl/hr (151); malic enzyme was considered as a carboxylase with low activity, about 30 μmoles/mg chl/hr (151); during leaf photosynthesis 5 to 15% of the $^{14}CO_2$ was found immediately (under 5 second exposure) in 3-PGA (84, 86);

[14]C-labeling kinetics of malate were quite different from aspartate in pulse-chase and in light-dark experiments (86); and C-1 labeling data for glyceric acid indicated near linearity from time zero, unlike the labeling of a secondary product (86). Regardless of such discrepancies, it was quite clear by 1970 that Slack and Hatch, with their ^{14}C-kinetic labeling data and the discovery of the supporting enzymes such as PEP carboxylase, pyruvate P_i dikinase, and NADP[+] malic dehydrognease (5, 84–92, 99, 100, 149–152), had formed the framework for establishing a new pathway of photosynthetic carbon assimilation that occurred in a wide variety of genera.

In 1968 workers from at least four laboratories simultaneously published data correlating the presence of the C_4 cycle with a particular type of leaf anatomy (60, 64, 99, 116, and Table 1) in which morphologically quite distinct mesophyll cells and bundle sheath cells can be easily observed in a light microscope. Then as data on enzymes and biochemical activities continued to accumulate, the problem of C_4 photosynthesis expanded quickly and became involved with the fact that leaves of C_4 plants characteristically possess two types of chloroplasts separated into the two types of leaf cells.

In 1969 the first data showing substantial levels of RuDP in leaves of C_4 plants was published; furthermore it was evident that the previously low activities of other workers could be attributed in part to the lack of breakage of bundle sheath cells in the preparation of leaf extracts (26). Workers in numerous laboratories then realized that leaves of C_4 plants possessed two major carboxylases, PEP and RuDP, and two distinct photosynthetic cell types. Thus to understand leaf photosynthesis in C_4 plants it became imperative that the chloroplasts and/or the cell types be isolated and characterized.

A great deal of effort has been expended in attempting to separate, isolate, and characterize the mesophyll and bundle sheath chloroplasts. Historically the first report of an isolation attempt was grinding leaves in an aqueous medium followed by a one density step gradient centrifugation (16). Isolation of mesophyll chloroplasts apparently was achieved but not of the bundle sheath chloroplasts. Unfortunately, the chloroplasts were virtually inactive in fixing CO_2 (maximum of 2 μmoles/mg chl/hr) and the RuDP carboxylase activity in C_4 plants was unknown at that time. The second report was of the isolation of sugarcane chloroplasts in nonaqueous media, taking advantage of the fact that bundle sheath chloroplasts could, by manipulation of the environment, accumulate more starch than mesophyll chloroplasts (150, 152). The nonaqueous technique presumably separates the chloroplasts on the basis of density. This technique depends upon proteins being attracted to adjacent proteins within a cell when subjected to nonaqueous environments and then not exchanging or separating when mixed together by grinding. The third technique involved the differential grinding of leaves in which one theoretically can grind with increasing vigor and sequentially release the contents of the epidermal cells, the mesophyll cells, and then the bundle sheath cells, thus releasing preferentially one chloroplast or enzyme or cell content over another (26). Grinding leaves in progressively more vigorous fashion has been modified slightly in various laboratories and has been used subsequently on a number of C_4 plants (14, 22, 40, 95, 136, 141). As one might expect from the empirical nature of "vigorous grinding" and with the inherent differences in plants, much controversy has arisen from the use of this procedure.

A cell separation and isolation procedure was developed in which the theory essentially involved taking advantage of the fact that plant cell walls are exceptionally strong (65, 69). By utilizing plant cell walls to keep the contents of the mesophyll cell and bundle sheath cells intact, one could first isolate the cells. Then activity could be determined with each cell or on the contents, such as organelles, of the respective cells. The cell isolation procedure seems to be the most incisive technique developed to date to quantitatively study the roles of the two chloroplasts and the two cell types in C_4 photosynthesis.

In the following evaluation of current hypotheses on CO_2 fixation sites in C_4 leaves, it is important first to note some points of agreement. With intact leaves all workers agree that OAA, malate, and aspartate are the major products (over 50%) of short-term, light CO_2 fixation experiments with C-4 label observed, and that 3-PGA always is detected with C-1 label. In carefully prepared whole leaf extracts all C_4 plants have two carboxylases, namely PEP and RuDP carboxylase, with their respective activities in excess or equivalent to the rates of leaf photosynthesis on a chlorophyll basis. Microscopy reveals two distinct photosynthetic (or green) cell types, mesophyll and bundle sheath, in C_4 plant leaves. With these points of agreement as basic, these four hypothesis have appeared in recent literature as to the pathway and sites of carbon assimilation in C_4 leaves.

The current hypothesis for the pathway of carbon fixation from Hatch and Slack is that all of the CO_2 entering leaves is first carboxylated by PEP carboxylase in the mesophyll cells (6, 84). Malate or aspartate, depending upon the specific plant, then moves to the bundle sheath cells where decarboxylation occurs, allowing the pentose cycle to operate in the classical fashion finally to synthesize starch. Their ideas are schematically presented in Figure 1. To balance the carbon and nitrogen, either pyruvate or alanine would move from the bundle sheath cells to the mesophyll cells depending upon whether malate or aspartate had moved into the bundle sheath cell. Their scheme is strongly supported in a qualitative fashion by: ^{14}C studies on β-carboxyl labeling of C_4 acids and C-1 labeling of 3-PGA plus other pentose cycle intermediates during both steady-state photosynthesis and pulse-chase under a variety of conditions with intact leaves; detection of the appropriate enzymes at adequate levels, except for an OAA-decarboxylase; localization of the enzymes by nonaqueous separation and differential leaf grinding procedures; micrographs of their nonaqueously separated chloroplasts; and a strong inhibition of C_4 photosynthesis by a presumed PEP carboxylase inhibitor (135). Details of the evidence was reviewed in 1969 and 1970 (84, 91), and since that time the only other supporting data which these workers have presented is relative to CO_2 pool(s) (83) in C_4 plant leaves which will be considered later.

Positive enzymic data on a transcarboxylase type activity (82, 86) has not been presented, so the other major shifts in thinking about the C_4 cycle when one compares the 1969 review (91) with this review are now listed. RuDP carboxylase

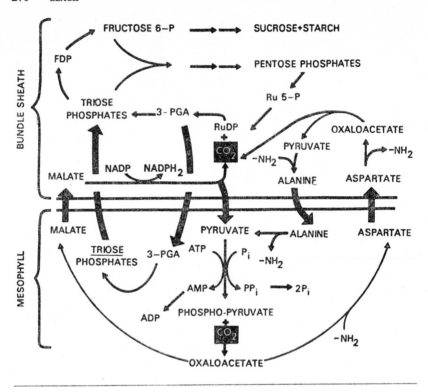

Figure 1 Probably reactions and intercellular movements of metabolites during the operation of the C₄-dicarboxylic acid pathway (6).

assumes a major role, with all of the incoming carbon dioxide ultimately being carboxylated by this enzyme. Malic enzyme moves from a carboxylating role to a decarboxylating role. Of course malic enzyme catalyzes an oxidative decarboxylation such that for each mole of CO_2 produced a mole of NADPH also is produced. Thus the scheme to the left of Figure 1 is supported consistently by the data cited above, particularly with sugarcane.

In the current literature one can find essentially this same scheme being supported by other workers (22, 25, 63, 72, 104, 117, 175) using various types of experimental approaches.

One of the points of greatest skepticism about the C₄ cycle as it is formulated in either Figure 1 or 2 is the idea and the supporting evidence for the movement of materials between cells to complete cycles of carbon flow. First it can be stated that electron microscopy data show numerous plasmodesmata connecting the cells in leaf cross-sections of all C₄ plants examined (34, 84, 117). The pathway of diffusion for the C₄ acids between chloroplasts in adjacent cells has been calculated, and the distances are such that diffusive processes could accommodate the rates of movement necessary to support C_4 photosynthesis (134). In the following discussion on isolated cells other data supporting cell-to-cell movements of materials will be cited.

One other point is noteworthy here, namely that the compounds which are postulated as moving between cells in C_4 photosynthesis—such as malate, aspartate, alanine, G-3P, and 3-PGA—also are compounds which penetrate chloroplasts. The majority of the chloroplast permeability studies are with pentose plant chloroplasts (17, 84, 121), but if one extrapolates these studies to C_4 plant chloroplasts, there is no experimental data available to question C_4 plant chloroplasts being permeable to these compounds. Hence the compounds which are postulated as moving between the two cell types in C_4 photosynthesis (Figures 1 and 2) also can penetrate the chloroplast envelope.

The second hypothesis for the major pathways of carbon flow in C_4 photosynthesis is based upon studies with isolated mesophyll cells and bundle sheath cells or strands principally from crabgrass leaves (65, 66, 69). Schematically the pathways are pictured in Figure 2. The technique of cell isolation has allowed a much more quantitative evaluation of the roles of cell types in leaf metabolism than any other available procedure. Thus the scheme in Figure 2 is partially in agreement with the scheme in Figure 1 in that the CO_2 entering a leaf which goes into the mesophyll cells follows a somewhat similar pathway in both schemes. However, a substantial amount of the CO_2 entering the leaf air spaces also enters the bundle sheath cells and is directly assimilated by the pentose cycle.

In the case of crab grass, about 15% of the CO_2 entering the leaf is fixed directly by the bundle sheath cells. This direct CO_2 fixation is supported by: the localization of most leaf RuDP carboxylase and other pentose cycle enzymes in the bundle sheath cells; the fixation of CO_2 by isolated bundle sheath cells when fed R-5P, ADP, or RuDP; microscopy studies on the internal leaf morphology which clearly show air spaces open to the bundle sheath cells such that physical barriers for gas distribution in the leaf are not evident; calculations for crab grass leaves showing that about 15% of the internal surface area exposed to gases may be bundle sheath cells; and in leaf CO_2 fixations, from numerous laboratories, label always is found in 3-PGA and the labeling kinetics for the C-1 position are not those of a typical secondary product (66). Data are not available to quantitatively evaluate the role of bundle sheath cells in directly fixing atmospheric CO_2 in the leaves of other C_4 plants. However, data on the levels of RuDP, PEP, 3-PGA, and malate in light and dark plus and minus CO_2 with maize leaves also can be interpreted as showing a direct carboxylation of RuDP using atmospheric CO_2 (118), although the specific cell in which these reactions occurred cannot be determined from these studies. The relative contribution of mesophyll and bundle sheath cells to net CO_2 uptake appears likely to be a characteristic of each plant.

Other supporting data for compartmentation of β-carboxylation in mesophyll cells and RuDP carboxylase in the bundle sheath cells in crab grass for about 85% of the leaf's atmospheric CO_2 supply to flow through the C_4 cycle, as shown in Figure 2, are as follows. In isolated whole mesophyll cells PEP carboxylase is

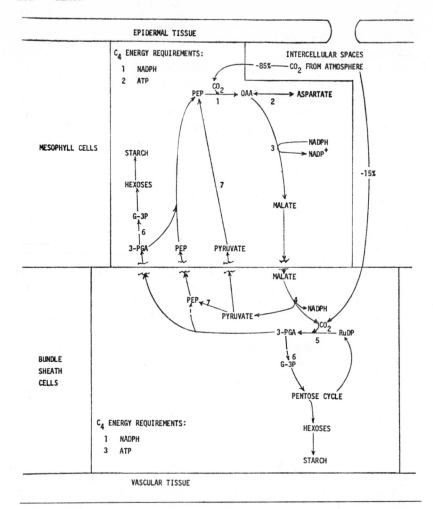

Figure 2 Scheme for the major pathways of carbon flow involved in net CO_2 uptake by the C_4 plant, crab grass. The reactions with numbers are catalyzed by: 1-PEP carboxylase; 2-aspartate transaminase; 3-malate dehydrogenase; 4-malic enzyme; 5-RuDP carboxylase; 6-G-3P dehydrogenase; and 7-pyruvate P_i diKinase.

quite active as it also is in extracts of these cells. We have assayed activity as high as 2000 μmoles, although the day-to-day activity usually varies from 700 to 1100 μmoles/mg chl/hr, which compares well with leaf photosynthesis rates of 300 to 450. It should be noted that PEP carboxylase at about 5 to 10% of this activity is present in crab grass bundle sheath cells just as it is present also in spinach mesophyll cells (66). NADP⁺-dependent malic dehydrogenase is localized in the mesophyll cells with activities ranging from 100 to 200 μmoles. The low level of RuDP carboxylase in mesophyll cells also is supported by the low 18 S component in the ultracentrifuge, which is typical of Fraction I protein in mesophyll cell extracts and the low activities of Ru-5P kinase and R-5P isomerase (32).

Several enzymes are about equally distributed in extracts from both bundle sheath cells and mesophyll cells and at adequate activities for leaf photosynthesis including G-3P dehydrogenase, adenylate kinase, pyrophosphatase, aspartate transaminase, and alanine transaminase. The level of transaminase activities seems to vary somewhat with the specific C_4 plant being studied (see section on variation in C_4 photosynthesis).

RuDP carboxylase, R-5P isomerase, Ru-5P kinase, malic enzyme, and aldolase are concentrated in crab grass bundle sheath cells. Pyruvate P_i dikinase has been a very difficult enzyme to assay with consistent day-to-date activity, but after much effort we now find activity between 100 and 200 μmoles/mg chl/hr in meso-phyll cells and 20 to 50 μmoles in bundle sheath cells. So the enzyme is active in both cells; thus pyruvate in the bundle sheath cells either can be metabolized (66) or phosphorylated or transported to the mesophyll cells (Figure 2).

The feeding of substrates to the isolated cells also support Figure 2. For example, pyruvate catalyzes CO_2 uptake with the mesophyll cells and is inactive with the bundle sheath cells, while R-5P plus ATP or RuDP catalyze CO_2 fixa-tion in bundle sheath cells without detectably influencing CO_2 fixation in the mesophyll cells. On the other hand, 2-PGA will stimulate CO_2 fixation by both cells. The products of light and dark CO_2 fixation in mesophyll cells are about 90% organic acids, while with bundle sheath cells fed PEP in the light, 3-PGA and sugars are formed (65–69, 103).

As outlined in Figure 2, one would predict that malate fed to the bundle sheath cells should transfer CO_2 to the pentose cycle. Feeding experiments with organic acids plus bundle sheath cells have not been convincing with crab grass under the conditions reported (32). However, by changing the ionic strength of the media, a transfer of carbon from malate has been achieved recently. In experiments with bundle sheath cells from *Panicum maximum* we have shown a very substantial CO_2 fixation rate at 320 μmoles/mg chl/hr which requires light, OAA and ATP or ADP (103). This is considered as proof for a light-dependent metabolism of an organic acid in the bundle sheath cells of a C_4 plant, such as could be predicted from both Figures 1 and 2.

The third hypothesis on leaf CO_2 fixation in C_4 plants is that all incoming CO_2 is fixed by PEP carboxylase either in nongreen cells, presumably bulliform cells, or in the mesophyll cell cytoplasm, and then the pentose cycle in the mesophyll cell chloroplasts is supplied with CO_2 by malic enzyme. The pentose cycle then operates in the classical fashion in the mesophyll chloroplast. Sucrose is excreted from the mesophyll chloroplast and translocated either via normal channels through the plant or to the bundle sheath chloroplasts which only act to synthe-size starch, presumably from sucrose, and as a storage site for starch (52).

This hypothesis was supported by enzyme data on leaf fractions released by the differential (or progressive) grinding technique using sugarcane, maize, or

Pennisetum purpureum (14, 40, 52). These authors had previously presented data showing phenols and phenoloxidase in sugarcane and spinach leaf extracts and proposed the use of β-mercaptoethanol or thioglycollate to prevent enzyme inactivation (13, 15). In attempting to isolate the chloroplasts, microscopic data was given (14, 46) on the leaf material remaining *after* grinding, but no data was given on the *contents* of each fraction released by progressive grinding. It should be obvious that the *contents* of the fraction which is used for assaying is the information that must be supplied. The maximum activity observed for PEP carboxylase in maize was 29 μmoles/mg chl/hr (40) and in sugarcane was 50 μmoles/ mg chl/hr (14), or less than 5% of the activity other workers with C_4 plants report. It was suggested that all other workers who failed to find substantial RuDP carboxylase activity in mesophyll cells may have inactivated the enzyme, but the large body of data from other laboratories on RuDP carboxylase in the bundle sheath cells was ignored by these workers.

As mentioned earlier, grinding a leaf in a progressively more vigorous fashion is an empirical technique, and each worker with each plant will of necessity be required to establish the separation and degree of purity of each fraction used. Without such care there is little doubt that with leaves of C_4 plants conflicting results can be obtained. A careful piece of work using the differential grinding technique with *Sorghum* and *Atriplex rosea* (95) failed to verify this work (14, 40) and the work on *Atriplex spongiosa* RuDP carboxylase activity (136) in which a differential grinding technique also was used. In another report using two progressive grinding techniques a 1:1 distribution on a chlorophyll basis of RuDP carboxylase in mesophyll extract:bundle sheath extract was reported for young maize leaves, again without proof that separation was achieved (141).

The fourth hypothesis for the site of CO_2 fixation in C_4 leaves is the suggestion that photosynthesis occurs primarily in the bundle sheath cells of maize (19). Three-week-old maize leaves were exposed to $^{14}CO_2$ for 1 min at 50 lx illumination intensity and then for either 10 sec or 1 min at 7000 lx. Then the leaves were killed with liquid nitrogen or isopentane and microautoradiograms were prepared. Analysis revealed that the labeling was restricted to the bundle sheath cells. Hence the hypothesis was proposed that the C_4 cycle in maize occurs in the structurally peculiar chloroplasts of the bundle sheath cells (19). I suggest that an exposure of leaves to $^{14}CO_2$ for 60 sec at low light is sufficient time for CO_2 fixation to occur, as in a preillumination type experiment (146), and the transport can occur in 10 sec. Doubts also exist as to the ability of the method used to hold water-soluble compounds precisely in place, although this subject is too involved to be considered here. Similar experiments have been presented in which substantial label was found in the mesophyll cells, and the results were interpreted simply to show rapid translocation (134). Finally, none of the criteria listed earlier have been fulfilled to support the bundle sheath cell fixation hypothesis; only the pattern of Ag grain deposition was presented as proof of the hypothesis.

CO_2 Assimilation Studies with Isolated C_4 Plant Chloroplasts

Only a limited number of CO_2 fixation studies are available with chloroplasts from C_4 plants. An important experiment was reported in 1970 with chloroplasts isolated from maize leaves which clearly demonstrated that the pentose cycle was operative (77). In these isolated maize chloroplasts the pentose cycle enzymes G-3P dehydrogenase, Ru-5P kinase, and RuDP carboxylase plus adenylate kinase and pyrophosphatase were present, while the key C_4 cycle enzymes PEP carboxylase, pyruvate P_i dikinase, and malic enzyme were absent or at low levels. Rates of up to 45 μmoles of CO_2 fixed/mg chl/hr were observed and the normal products of the pentose cycle were produced (77). These preparations did not metabolize malate or aspartate as one might predict from the enzyme content. Unfortunately, the mesophyll and bundle sheath chloroplasts were not separated (133, Fig. 2), so the crucial question remains from these studies as to whether the pentose cycle is highly active in both chloroplasts or only in one. With other isolated chloroplasts from sugarcane, maximum CO_2 fixation rates near 2 μmoles/ mg chl/hr were observed (12). Studies have been reported on the isolation of maize chloroplasts by a "laceration technique," which fixed CO_2 at about 1 μmole/mg chl/hr, but when supplemented with PEP, light, and a "supernatant," fixation rates up to 290 μmoles occurred (171). Again, in these preparations chloroplast separation was not achieved. From the carbon fixation experiments thus far with isolated C_4 plant chloroplasts one can conclude that the pentose cycle is operative, but information is not furnished on its location in mesophyll and/or bundle sheath chloroplasts, and no definitive information is furnished on whether PEP carboxylase and some other enzymes of the C_4 pathway are localized in chloroplasts. However, the data on adenylate kinase and pyrophosphatase (77) are very suggestive for their localization in a maize chloroplast.

Variation in C_4 Photosynthesis

Another quite unexpected aspect was added to C_4 photosynthesis also in 1970–71. Namely, with C_4 plants such as Bermuda grass it was observed that upon exposure of leaves to $^{14}CO_2$ for 5 sec, 33% of the label was in malate and 54% was in aspartate, while an identical experiment with sugarcane resulted in respective distributions of 62% and 22%. Following the exposure to $^{14}CO_2$, if the illuminated Bermuda grass leaves were transferred to $^{12}CO_2$, the label in aspartate was rapidly lost while malate maintained a fairly constant radioactivity (48, 66). The enriched labeling in aspartate over malate and the rapid loss of aspartate labeling relative to malate in $^{12}CO_2$ with Bermuda grass were directly opposite to the patterns originally reported in sugarcane (48, 86). At the same time it was noted that the level of malic enzyme in Bermuda grass and a number of other C_4 plants was much lower ($<5\%$) than in sugarcane, maize, or crab grass (6, 48, 61, 84, 100). In addition, it was noted that plants with a low malic enzyme content also exhibited an oxygen insensitive postillumination CO_2 burst, tended to have well-developed grana in the bundle sheath chloroplasts, and tended to have high levels of transaminases (61, 62). Hence Downton proposed that two patterns existed in C_4 photosynthesis which he termed "malate formers" and "aspartate formers" (61). Though this terminology may be convenient, it is misleading since all plants form malate and aspartate. Another example in which it could be misleading is illustrated in the demonstration that aspartate is the major product of

CO_2 fixation in pigweed leaves, while malate is the major product of CO_2 fixation in pigweed cotyledons, stems, and green callus induced from stem meristem (166).

Since malic enzyme had assumed a major role as a decarboxylase in C_4 photosynthesis, clearly another decarboxylase should be present in plants which appear to have low levels of malic enzyme. In search of such an activity, PEP carboxykinase was detected in *Panicum maximum, Panicum texanum,* and *Sporobolus poiretti* with activity two to seven times leaf photosynthesis. PEP carboxykinase is localized primarily in the bundle sheath cells of *P. maximum* along with enzymes of the pentose cycle (68, 103). It was reported that *P. maximum* fixed 99% of its $^{14}CO_2$ into malate plus aspartate in a 6 sec exposure (61), but we cannot repeat that observation. We find about 40% in aspartate, 40% in 3-PGA, and 14% in malate in a 5 sec exposure of *P. maximum* leaves to $^{14}CO_2$ (103). Since PEP carboxykinase is a freely reversible enzyme, we proposed that it acts as a decarboxylase in the bundle sheath cells of specific C_4 plants (68, 103). With *P. maximum* bundle sheath cells, we have shown CO_2 fixation which is dependent upon light, OAA, and ATP or ADP at rates equivalent to leaf photosynthesis; this is an experimental demonstration of the use of organic acids by bundle sheath cells of C_4 plants (103).

There is no information to show other mechanisms of decarboxylating C_4 acids in plants like Bermuda grass and pigweed which lack substantial levels of either malic enzyme or PEP carboxykinase in conventional assays. The present status of research leads us to tentatively form three categories of C_4 plants as shown in Table 3 (103). These three categories of C_4 plants also exhibit three characteristic types of postillumination CO_2 bursts (38, 103). In Table 3 a number of question marks are inserted where substantial uncertainty exists. Hopefully further research will clarify some of these questions.

ATP and NADPH Production in C_4 Plants

An examination of the light intensity response of photosynthesis with C_4 leaves (Table 1) indicates that the chloroplasts may exhibit a different or greater photo-

Table 3   Variations in Carbon Flow Pathways in C_4 Photosynthesis

	Sugarcane Maize Crab Grass	*Panicum maximum* *Panicum texanum*	Bermuda Grass Pigweed
Primary photosynthetic $^{14}CO_2$ fixation product accumulating (<10 sec exposures)	Malate	Aspartate, 3-PGA	Aspartate
Active intermediate from MC to BSC?[a]	Malate	OAA or aspartate	?
Decarboxylation enzyme	Malic enzyme	PEP carboxykinase	?
Active intermediate from BSC to MC?[a]	Pyruvate or PEP, 3-PGA	PEP	?
Carboxylase in MC[a]	PEP	PEP	PEP
Level of pyruvate P_i dikinase	Equals photosynthesis	<30% of photosynthesis	Equals photosynthesis

[a] Abbreviations: MC (mesophyll cells); BSC (bundle sheath cells).

chemical activity compared to pentose chloroplasts. The theoretical calculation showing a requirement of 5 molecules of ATP for CO_2 uptake to occur in C_4 plants (Table 1, 46) led to a study of the photochemical aspects of C_4 photosynthesis. In the early work with Bermuda grass, maize, sorghum, and sugarcane chloroplasts, noncyclic electron flow in the presence of $NADP^+$ and other Hill oxidants was coupled stoichiometrically with ATP formation and O_2 evolution just as with pentose plants (46, 93). PMS-catalyzed cyclic photophosphorylation also seemed to possess similar characteristics. However, in these studies the mesophyll chloroplasts were not separated from the bundle sheath. Further work with C_4 leaves also demonstrated that: ferredoxin is about twofold more concentrated in C_4 leaf extracts than in pentose leaves (119); the chlorophyll a to b ratio of C_4 plants is higher than pentose (Table 1, 33); and the content of P_{700} is about 50% higher in C_4 than in pentose leaves (33). These results, plus microscopic information on the variation in chloroplast morphology in mesophyll versus bundle sheath cells (34, 81, 84, 116, 117), made it important that the two chloroplasts from C_4 leaves be isolated to study their photochemistry. Again before discussing the photochemical studies with isolated chloroplasts it should be pointed out that proof of separation, contamination, morphological conditions, etc is just as important if these studies are to be convincing as is similar proof in the studies on CO_2 fixation. In most of the reports on photochemical activities from C_4 plants such proof has not been presented.

In an initial report on the photochemical activities in isolated bundle sheath chloroplasts from *Sorghum bicolor* and maize, it was reported that these chloroplasts did not exhibit the Hill reaction and lacked photosystem II (175). Subsequent reports have not substantiated the data on the absence of photosystem II and the Hill reaction in *Sorghum* and maize chloroplasts (2, 3, 23, 153). It also was reported that photosystems I and II were not linked in these bundle sheath chloroplasts (23), but more careful work demonstrated that the photosystems are linked (1, 24, 153, 154). Hence in the literature today there is agreement that the bundle sheath chloroplasts have an active photosystem II in which electrons flow from H_2O to $NADP^+$ which is sensitive to an inhibitor such as DCMU (3, 125, 154). All workers also agree that photosystem I is quite active in bundle sheath chloroplasts as indicated, for example, by a very active PMS catalyzed cyclic photophosphorylation (2, 3, 8, 143). The ratio of chlorophyll a to b in bundle sheath cells from crab grass have varied in numerous preparations from about 4.5 to 5.9 (66, 125), maize is about 5.4, and *Sorghum* is about 5.7 (2, 3) which also shows that photosystem I is likely to be quite active. Other C_4 plants, particularly dicots, seem to have chlorophyll a to b ratios which are only slightly higher or equal in bundle sheath chloroplasts compared to mesophyll. One unconfirmed report presents data on maize and *Sorghum* showing that the chlorophyll a to b ratios are 2.05 and 2.62 in bundle sheath chloroplasts and 1.64 and 2.10 in mesophyll chloroplasts (23). One could summarize and state that photochemical studies with bundle sheath chloroplasts show that photosystem I and II are present, they exist in nonstoichiometric quantities, electrons can flow from H_2O to $NADP^+$, and ATP synthesis can occur.

Mesophyll chloroplasts from the C_4 plants studied to date seem to be quite similar to spinach chloroplasts in chlorophyl content, in both photosystem I and II electron transport activity, and in ATP synthesis reactions (2, 3, 8, 66, 125, 143).

The theoretical implications and practical consequences of the altered ratios of photosystem I to II in bundle sheath chloroplasts of various C_4 plants appear quite important. The lack of a fixed stoichiometry between photosystem I and II in chloroplasts implies that the systems may exist separately and that various combinations may exist. The location within grana or appressed lamella or in a single lamellum of specific electron transport components or processes such as O_2 evolution or cyclic electron flow can be evaluated. The possibility of a variable photosynthetic unit seemingly is evident. There are practical consequences particularly as to carbon flow pathways in the various C_4 plants shown in Table 3 and illustrated in Figure 2 with crab grass. For example, if malic enzyme is utilized as a decarboxylase by one plant and PEP carboxykinase by another, or if aspartate versus malate, or pyruvate versus PEP, or 3-PGA versus G-3P, are the materials moving between cells, these plants would have quite different energy requirements in the respective cell types for net CO_2 fixation to occur. Seemingly the energy requirements in each cell for carbon flow can be reflected in the respective photochemical activity; however, sufficient data are not yet available with all of the C_4 plant types such as those categorized in Table 3 to reach such a broad conclusion.

CRASSULACEAN ACID METABOLISM

Interest in CAM plants was kindled markedly by correlating the fact that net CO_2 uptake involved the production of organic acids in both C_4 and CAM plants. Of course it is well known that organic acid production in CAM plants occurs at night primarily, and that the photosynthetic tissues are fleshy. Thus the analogy was made that in CAM plants there is a temporal separation of organic acid production from their degradation while in C_4 plants there is a spatial separation of the process. Then knowing the general leaf anatomy of both CAM and C_4 photosynthetic tissues, the statement that C_4 is "CAM mit Kranz" was a logical deduction (84, 117). Despite these external similarities, an examination of the characteristics listed in Table 1 reveals some marked differences.

In the history of carbon fixation research with CAM plants a landmark paper (37) demonstrated that malate was labeled 66% in C-4 and 33% in C-1 under a variety of dark or dark plus light conditions. These results were interpreted as indicating a double carboxylation sequence in forming malate with the 3-C precursor arising from starch. It was postulated that G-6P followed the oxidative pentose pathway to Ru-5P and then a carboxylation of RuDP occurred which would label one-half of the 3-PGA molecules in C-1. The 3-PGA then would form PEP, which upon carboxylation would result in a 2:1 labeling in the C-4:C-1 of malate. This sequence of events is pictured in the A portion of Figure 3. The 2:1 labeling distribution was verified in several laboratories (84, 161) with CAM plants.

The double carboxylation sequence proposal predicts that label should appear

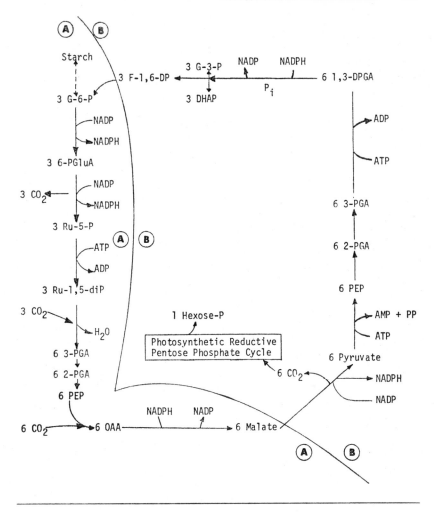

Figure 3 Scheme for net carbon dioxide fixation in Crassulacean acid metabolism plants in dark (A) and light (A plus B) (32, 144, 145).

in 3-PGA in the dark from $^{14}CO_2$ fixation (Figure 3, A portion; 37). The detection of labeled 3-PGA during dark $^{14}CO_2$ fixation by *Kalanchoë* was referred to in 1958 (37), but to my knowledge data on dark labeling of 3-PGA in CAM tissues has never been presented. In this laboratory we constructed the scheme in Figure 3 which allowed a calculation of the ATP and NADPH requirements for CO_2 fixation (Table 1). Using mesophyll cells isolated from the CAM plant *Sedum telephium* (144, 145), it has been demonstrated that label can be found in 3-PGA under a variety of dark $^{14}CO_2$ fixing conditions (unpublished data). We interpret the dark labeling of 3-PGA as supporting the double carboxylation sequence in Figure 3.

However, the malate labeling data upon which the double carboxylation sequence was proposed has been questioned with the demonstration of an apparent error in the degradation method which results in an underestimation of the C-4 label in malate (158). In dark $^{14}CO_2$ fixation studies with four CAM plant leaves, in exposure times under 30 sec, from 85 to 95% of the label was in C-4 of malate (158), and another report with *Kalanchoë* for a 1 min exposure indicates 82% of the label was in C-4 (120). In longer exposure periods to $^{14}CO_2$, up to 24 hr, as much as 54% of the label was found in C-4 (120, 158). Thus it was proposed that the β-carboxylation of PEP is the principal dark CO_2 fixation reaction in CAM plants (158). Presumably with only a single carboxylation the carbon from starch would flow through the glycolytic pathway and not through the oxidative pentose pathway as is pictured in Figure 3.

The light portions of CAM CO_2 fixation have traditionally been thought to involve a decarboxylation of malate and a refixation of the newly released CO_2 via the pentose cycle as shown in Figure 3, portion B. The question has been posed for some time as to the fate of pyruvate. By analogy with the C_4 cycle one could predict that pyruvate P_i dikinase might be utilized to generate PEP, since this enzyme is present in some CAM plants (111). The fate of PEP in the light is uncertain. In Figure 3, Portion B, PEP is shown returning to hexoses but PEP carboxylase also is operative in the light (11, 110). Future work on both the light and dark reactions of CAM metabolism should allow a quantitation of the contribution of the two carboxylases to net CO_2 uptake in light and in dark. Almost no quantitative information is available on peroxisomal and mitochondrial respiration in CAM plants, so little can be said about their contribution to net CO_2 uptake except that azide inhibits dark CO_2 uptake (120) and amytal inhibits light utilization of malate (59).

In studying CAM plants, the isotopic discrimination between ^{13}C and ^{12}C was examined since it had been demonstrated that C_4 and pentose plants have definite ranges of isotope discrimination (Table I). Apparently the major discrimination reaction in pentose plants is in the reaction catalyzed by RuDP carboxylase (173). Since PEP carboxylase has been assumed to fix most of the CO_2 in CAM plants, isotope discrimination values similar to C_4 plants were expected. However, values were found which spread over the range of all terrestrial higher plants (Table I, 144). The values also were independent of taxonomic criteria since the respective $^{13}C/^{12}C$ ratios for *Sedum rubrotintum*, *Sedum telephoides*, and *Sedum telephium* were -14, -22, and -29. Then it was learned that the values of a single CAM plant could be shifted by changing the environment from that typical of C_4 plants to that typical of pentose plants (unpublished data). Interpreted simply, the shift in isotopic ratios indicates that by changing the environment, CAM plants can be shifted from using PEP carboxylase as the primary carboxylase to using RuDP carboxylase as the primary carboxylase. We consider this to be a drastic change in CAM carbon metabolism and expect future research in these areas to offer many clear insights into CAM plants. Regulation of carbon flow pathways in CAM plants unquestionably will be complex since it is known that environmental as well as metabolic regulators such as substrate and product concentrations are operative.

CONCLUSIONS ON NET CO_2 UPTAKE

Information on photosynthesis with various plants is accumulating logarithmically, but the information has yet to be integrated sufficiently to explain, for example, why C_4 plants have a greater maximum capacity for net CO_2 uptake than other plants or why C_4 plants utilize H_2O more efficiently than pentose plants. Some approaches and ideas have emerged from the data that are focal points of some current thinking. For example, the successful crossing of a C_4 and pentose species of *Atriplex* has allowed work with their F_1, F_2, and F_3 progeny showing that they possess various physiological and biochemical characteristics of photosynthesis intermediate between the parents (27, 140). Though not yet complete, this approach seems to offer promise of finding the characteristic(s) that permit a plant to fix CO_2 or to use H_2O efficiently. Another idea has emerged from the literature that may yield some insight into C_4 photosynthesis. The literature shows that atmospheric levels of CO_2 are not saturating for leaf photosynthesis, measured with other conditions optimal. Thus one can think of photosynthesis as being rate limited by CO_2. When C_4 plants fix CO_2 in the mesophyll cells into C_4 acids, this apparently is a means for effectively trapping CO_2 for a few moments prior to releasing it in the bundle sheath cells. In the bundle sheath cell then the CO_2 concentration is effectively raised so that RuDP carboxylase can operate more efficiently than in a pentose leaf mesophyll cell, for example. Two types of experiments show that a pool(s) of CO_2 exists in leaves of C_4 plants, but direct data on the pool site inside of leaves are not yet available. Leaves of maize and *Amaranthus* were exposed to $^{14}CO_2$ for various times, and the "CO_2 pool" was trapped at low temperatures. The kinetics of the pool buildup and depletion indicate that it was derived from C_4 acids (83). Work in this laboratory has confirmed these experiments with crab grass, *Panicum maximum*, Bermuda grass, and nutsedge. In a second type of experiment (Table 1) we have found by investigating the substrate from which the CO_2 may be derived in the postillumination CO_2 burst (38) that C_4 plants release a burst of $^{14}CO_2$ in the dark in about 1 minute from newly fixed $^{14}CO_2$, which may be as much as 65% of the $^{14}CO_2$ fixed in the previous light period. Pentose plants only exhibit a dark release of $<5\%$ of their newly fixed $^{14}CO_2$, and the kinetics of buildup are not the same as in C_4 plants (Table 1). These two types of experiments are interpreted as showing that mesophyll cells in C_4 leaves act as "CO_2 harvesting antennas" for the bundle sheath cells to conduct photosynthesis efficiently.

OTHER GROUPS OF HIGHER PLANTS

In Table I only three major groups of higher plants are considered but the definite possibility exists as to defining other groups or subdividing groups. In fact, in Table 3 one can readily see that variations exist in the C_4 pathway enzymes with specific C_4 species. The carbon isotope ratio variation in CAM plants, for example, indicates that major differences exist in CAM plant primary carboxylases. Thus one can tabulate similar lists of characteristics for tree species, alpine plants, aquatic floating and rooted emergent higher plants, or yellow-green plants (21, 29, 30, 43). Relatively little information is available on some of these types of plants, but enough is available to warrant further investigation. For example, alpine plants exhibit rates of photosynthesis which are among the highest reported in the literature (29). Aquatic plants have very high rates of photosynthesis (see section on photorespiration), and some possess a highly developed chloroplast peripheral reticulum that was once thought to be unique to C_4 plants but that now has been found in many plants, particularly aquatic plants such as cattail, sea oats, and *Nymphoides indica* (80, 167). More detailed discussions on the possibility of other groups of plants are available (29, 30).

ACKNOWLEDGMENTS

A number of colleagues and students have contributed to this article through discussions, ideas, and the use of their unpublished data. These include Drs. Brown, Campbell, Chen, Dittrich, Fisher, Rouhani, Vines, and Wynn. I am grateful to them and to other students listed in the literature cited. However, responsibility for the inadequacies in this article and the lack of critical judgment in the choice of literature to be cited rests with the author.

Literature Cited

1. Andersen, K. S., Bain, J. M., Bishop, D. G., Smillie, R. M. 1972. *Plant Physiol.* 49:461–66
2. Anderson, J. M., Boardman, N. K., Spencer, D. 1971. *Biochim. Biophys. Acta* 245:253–58
3. Anderson, J. M., Woo, K. C., Boardman, N. K. 1971. *Biochim. Biophys. Acta* 245:398–408
4. Anderson, L. E., Fuller, R. C. 1969. *J. Biol. Chem.* 244:3105–9
5. Andrews, T. J., Hatch, M. D. 1969. *Biochem. J.* 114:117–25
6. Andrews, T. J., Johnson, H. S., Slack, C. R., Hatch, M. D. 1971. *Phytochemistry* 10:2005–13
7. Andrews, T. J., Lorimer, G. H., Tolbert, N. E. 1971. *Biochemistry* 10:4777–82
8. Arntzen, C. J., Dilley, R. A., Neumann, J. 1971. *Biochim. Biophys. Acta* 245:409–24
9. Atkins, C. A., Patterson, B. D., Graham, D. 1972. *Plant Physiol.* 50:214–17
10. Ibid, 218–23
11. Avadhani, P. N., Osmond, C. B., Tan, K. K. 1971. See Ref. 84, 288–93
12. Baldry, C. W., Bucke, C., Coombs, J. 1969. *Biochem. Biophys. Res. Commun.* 37:828–32
13. Baldry, C. W., Bucke, C., Coombs, J. 1970. *Planta* 94:124–33
14. Ibid 1971. 97:310–19
15. Baldry, C. W., Bucke, C., Coombs, J., Gross, D. 1970. *Planta* 94:107–23
16. Baldry, C. W., Coombs, J., Gross, D. 1968. *Z. Pflanzenphysiol.* 60S:78–81
17. Bassham, J. A. 1971. *Proc. Nat. Acad. Sci. USA* 68:2877–82
18. Bassham, J. A., Calvin, M. 1957. *The Path of Carbon In Photosynthesis.* Englewood Cliffs, N.J.: Prentice Hall. 104 pp.
19. Bednarz, R. M., Rasmussen, H. P. 1972. *J. Exp. Bot.* 23:415–21
20. Bender, M. M. 1968. *Radiocarbon* 10:468–72
21. Benedict, C. R. 1972. See Ref. 30, 7–25
22. Berry, J. A., Downton, W. J. S., Tregunna, E. B. 1970. *Can. J. Bot.* 48:777–86
23. Bishop, D. G., Andersen, K. S., Smillie, R. M. 1971. *Biochem. Biophys. Res. Commun.* 42:74–81
24. Bishop, D. G., Andersen, K. S., Smillie, R. M. 1972. *Plant Physiol.* 49:467–70
25. Björkman, O. 1971. See Ref. 84, 18–32
26. Björkman, O., Gauhl, E. 1969. *Planta* 88:197–203
27. Björkman, O., Nobs, M. A., Berry, J. A. 1971. *Carnegie Inst. Washington Yearb.* 70:507–11
28. Björkman, O., Pearcy, R. W., Harrison, A. T., Mooney, H. 1972. *Science* 175:786–89
29. Black, C. C. 1971. *Advan. Ecol. Res.* 7:87–113
30. Black, C. C., Ed. 1972. *Net Carbon Dioxide Assimilation In Higher Plants.* S. Sect. Am. Soc. Plant Physiol. 93 pp.
31. Black, C. C., Chen, T. M., Brown, R. H. 1969. *Weed Sci.* 17:338–44
32. Black, C. C., Edwards, G. E., Kanai, R., Mollenhauer, H. H. 1972. *Proc. 2nd Int. Photosyn. Congr.* 3:1745–57
33. Black, C. C., Mayne, B. C. 1970. *Plant Physiol.* 45:738–41
34. Black, C. C., Mollenhauer, H. H. 1971. *Plant Physiol.* 47:15–23
35. Bowes, G., Ogren, W. L. 1972. *J. Biol. Chem.* 247:2171–76
36. Bowes, G., Ogren, W. L., Hageman, R. H. 1971. *Biochem. Biophys. Res. Commun.* 45:716–22
37. Bradbeer, J. W., Ranson, S. L., Stiller, M. 1958. *Plant Physiol.* 33:66–70
38. Brown, R. H., Gracen, V. E. 1972. *Crop Sci.* 12:30–33
39. Brownell, P. F., Crossland, C. J. 1972. *Plant Physiol.* 49:794–97
40. Bucke, C., Long, S. P. 1971. *Planta* 99:199–210
41. Burr, G. O. 1962. *Int. J. Appl. Radiat. Isotop.* 13:365–74
42. Burr, G. O. et al 1957. *Ann. Rev. Plant Physiol.* 8:275–308
43. Carter, M. 1972. See Ref. 30, 54–74
44. Chan, P. H., Sakano, K., Singh, S., Wildman, S. G. 1972. *Science* 176:1145–46
45. Chen, J. H., Jones, R. F. 1971. *Biochim. Biophys. Acta* 229:208–15
46. Chen, T. M., Brown, R. H., Black, C. C. 1969. *Plant Physiol.* 44:649–54
47. Chen, T. M., Brown, R. H., Black, C. C. 1970. *Weed Sci.* 18:399–403
48. Chen, T. M., Brown, R. H., Black, C. C. 1971. *Plant Physiol.* 47:199–203
49. Chollet, R., Ogren, W. L. 1972. *Biochem. Biophys. Res. Commun.* 46:2062–66
50. Ibid 1972. 48:684–88

284 BLACK

51. Chu, D. K., Bassham, J. A. 1972. *Plant Physiol.* 50:224–27
52. Coombs, J., Baldry, C. W. 1972. *Nature* 238:268–70
53. Cooper, J. P. 1970. *Herb. Abstr.* 40: 1–15
54. Cooper, J. P., Tainton, N. M. 1968. *Herb. Abstr.* 38:167–76
55. Cooper, T. G., Benedict, C. R. 1968. *Plant Physiol.* 43:788–92
56. Cooper, T. G., Filmer, D., Wishnick, M., Lane, M. D. 1969. *J. Biol. Chem.* 244:1081–83
57. Cooper, T. G., Wood, H. G. 1971. *J. Biol. Chem.* 246:5488–90
58. Decker, J. P. 1955. *Plant Physiol.* 30: 82–84
59. Denius, H. R. Jr., Homann, P. H. 1972. *Plant Physiol.* 49:873–80
60. Downes, R. W., Hesketh, J. D. 1968. *Planta* 78:79–84
61. Downton, W. J. S. 1970. *Can. J. Bot.* 48:1795–1800
62. Ibid 1971. 49:1439–42
63. Downton, W. J. S. 1971. See Ref. 84, 3–17
64. Downton, W. J. S., Tregunna, E. B. 1968. *Can. J. Bot.* 46:207–15
65. Edwards, G. E., Black, C. C. 1971. *Plant Physiol.* 47:149–56
66. Edwards, G. E., Black, C. C. 1971. See Ref. 84, 153–68
67. Edwards, G. E., Gutierrez, M. 1972. *Plant Physiol.* 50:728–32
68. Edwards, G. E., Kanai, R., Black, C. C. 1971. *Biochem. Biophys. Res. Commun.* 45:278–85
69. Edwards, G. E., Lee, S. S., Chen, T. M., Black, C. C. 1970. *Biochem. Biophys. Res. Commun.* 39:389–95
70. Everson, R. G. 1970. *Phytochemistry* 9:25–32
71. Everson, R. G., Slack, C. R. 1968. *Phytochemistry* 7:581–84
72. Farineau, J. 1971. See Ref. 84, 202–10
73. Filmer, D. L., Cooper, T. G. 1970. *J. Theor. Biol.* 29:131–45
74. Forrester, M. L., Krotkov, G., Nelson, C. D. 1966. *Plant Physiol.* 41: 422–27
75. Ibid, 428–43
76. Gibbs, M. 1969. *Ann. NY Acad. Sci.* 168:356–68
77. Gibbs, M., Latzko, E., O'Neal, D., Hew, C. S. 1970. *Biochem. Biophys. Res. Commun.* 40:1356–61
78. Goldsworthy, A. 1970. *Bot. Rev.* 36: 321–40
79. Goldsworthy, A., Day, P. R. 1970. *Nature* 228:687–88
80. Gracen, V. E. Jr., Hillard, J. H.,

Brown, R. H., West, S. H. 1972. *Planta.* 107:189–204
81. Haberlandt, G. 1914. *Physiological Plant Anatomy.* Transl. M. Drummond. New York: Macmillan. 777 pp.
82. Hatch, M. D. 1971. See Ref. 84, 139–52
83. Hatch, M. D. 1971. *Biochem. J.* 125: 425–32
84. Hatch, M. D., Osmond, C. B., Slatyer, R. O., Eds. 1971. *Photosynthesis and Photorespiration.* New York: Wiley-Interscience, 565 pp.
85. Hatch, M. D., Osmond, C. B., Troughton, J. H., Björkman, O. 1972. *Carnegie Inst. Washington Yearb.* 71:135–41
86. Hatch, M. D., Slack, C. R. 1966. *Biochem. J.* 101:103–11
87. Hatch, M. D., Slack, C. R. 1967. *Arch. Biochem. Biophys.* 120:224–25
88. Hatch, M. D., Slack, C. R. 1968. *Biochem. J.* 106:141
89. Ibid 1969. 112:549
90. Hatch, M. D., Slack, C. R. 1969. *Biochem. Biophys. Res. Commun.* 34: 589–93
91. Hatch, M. D., Slack, C. R. 1970. *Ann. Rev. Plant Physiol.* 21:141–62
92. Hatch, M. D., Slack, C. R., Johnson, H. S. 1967. *Biochem. J.* 102:417–22
93. Hew, C. S., Gibbs, M. 1970. *Can. J. Bot.* 48:1265–69
94. Hofstra, G., Nelson, C. D. 1969. *Planta* 88:103–12
95. Huang, A. H. C., Beevers, H. 1972. *Plant Physiol.* 50:242–48
96. Jackson, W. A., Volk, R. J. 1970. *Ann. Rev. Plant Physiol.* 21:385–432
97. Jensen, R. G. 1971. *Biochim. Biophys. Acta* 234:371
98. Jensen, R. G. 1971. *Plant Physiol.* 48:9–13
99. Johnson, H. S., Hatch, M. D. 1968. *Phytochemistry* 7:375
100. Johnson, H. S., Hatch, M. D. 1970. *Biochem. J.* 119:273–80
101. Jones, M. B., Mansfield, T. A. 1972. *Planta* 103:134–46
102. Joshi, M. C., Boyer, J. S., Kramer, P. J. 1965. *Bot. Gaz.* 126:174–79
103. Kanai, R., Black, C. C. 1972. See Ref. 30, 75–93
104. Karpilov, Y. S. 1960. *Trans. Kazan. Sci. Chem. Inst.* 41:15
105. Kawashima, N., Wildman, S. G. 1970. *Ann. Rev. Plant Physiol.* 21: 325–58
106. Kerr, M. W., Robertson, A. 1971. *Biochem. J.* 125:34 p

107. Kisaki, T., Tolbert, N. E. 1970. *Plant Cell Physiol.* 11:247–58
108. Kisaki, T., Yano, N., Hirabayashi, S. 1972. *Plant Cell Physiol.* 13:581–84
109. Kisaki, T., Yoshida, N., Imai, Y. 1971. *Plant Cell Physiol.* 12:275–88
110. Kluge, M. 1971. See Ref. 84, 283–87
111. Kluge, M., Osmond, C. B. 1971. *Naturwissenschaften* 8S:414–15
112. Kohn, L. D., Warren, W. A., Carroll, W. R. 1970. *J. Biol. Chem.* 245:3821–30
113. Kortschak, H. P., Hartt, C. E., Burr, G. O., 1965. *Plant Physiol.* 40:209–13
114. Krause, G. H., Bassham, J. A. 1969. *Biochim. Biophys. Acta* 172:553–58
115. Kwok, S., Kawashima, N., Wildman, S. G. 1971. *Biochim. Biophys. Acta* 293–96
116. Laetsch, W. M. 1968. *Am. J. Bot.* 55: 875–83
117. Laetsch, W. M. 1969. *Sci. Progr. London* 57:323–51
118. Latzko, E., Laber, L., Gibbs, M. 1971. See Ref. 84, 196–201
119. Lee, S. S., Travis, J., Black, C. C. 1970. *Arch. Biochem. Biophys.* 141: 676–89
120. Levi, C., Gibbs, M. 1972. *Plant Physiol. Suppl.* 49:59
121. Lin, D. C., Nobel, P. S. 1971. *Arch. Biochem. Biophys.* 145:622–32
122. Liu, A. Y., Black, C. C. 1972. *Arch. Biochem. Biophys.* 149:269–80
123. Lowe, J., Slack, C. R. 1971. *Biochim. Biophys. Acta* 235:207–9
124. Maruyama, H., Easterday, R. L., Chang, H. C., Lane, M. D. 1966. *J. Biol. Chem.* 241:2405–12
125. Mayne, B. C., Edwards, G. E., Black, C. C. 1971. *Plant Physiol.* 47:600–5
126. Mazelis, M., Vennesland, B. 1957. *Plant Physiol.* 32:591–96
127. McFadden, B. A., Denend, A. R. 1972. *J. Bacteriol.* 110:633–37
128. McNaughton, S. J., Fullem, L. W. 1970. *Plant Physiol.* 45:703–7
129. Moss, D. N. 1962. *Nature* 193:587
130. Moss, D. N., Rasmussen, H. P. 1969. *Plant Physiol.* 44:1063–68
131. Mukerji, S. K., Ting, I. P. 1971. *Arch. Biochem. Biophys.* 143:297–317
132. Ogren, W. L., Bowes, G. 1971. *Nature* 230: 159–60
133. O'Neal, D., Hew, C. S., Latzko, E., Gibbs, M. 1972. *Plant Physiol.* 49: 607–14
134. Osmond, C. B. 1971. *Aust. J. Biol. Sci.* 24:159–63
135. Osmond, C. B., Avadhani, P. N. 1970. *Plant Physiol.* 45:228–30

136. Osmond, C. B., Harris, B. 1971. *Biochim. Biophys. Acta* 234:270–82
137. Osmond, C. B., Greenway, H. 1972. *Plant Physiol.* 49:260–63
138. Pan, D., Waygood, E. R. 1971. *Can. J. Bot.* 49:631–43
139. Paulsen, J. M., Lane, M. D. 1966. *Biochemistry* 5:2350–57
140. Pearcy, R. W., Björkman, O. 1971. *Carnegie Inst. Washington Yearb.* 69:632–40
141. Poincelot, R. P. 1972. *Plant Physiol.* 50:336–40
142. Poincelot, R. P. 1972. *Biochim. Biophys. Acta* 258:637–42
143. Polya, G. M., Osmond, C. B. 1972. *Plant Physiol.* 49:267–69
144. Rouhani, I. 1972. *Pathways of carbon metabolism in spongy mesophyll cells, isolated from Sedum telephium leaves, and their relationship to crassulacean acid metabolism plants.* PhD thesis. Univ. Georgia, Athens. 169 pp.
145. Rouhani, I., Vines, H. M., Black, C. C. 1973. *Plant Physiol.* 57:97–103
146. Samejima, M., Miyachi, S. See Ref. 84, 211–17
147. Shain, Y., Gibbs, M. 1971. *Plant Physiol.* 48:325–30
148. Shantz, H. L., Piemeisel, L. N. 1927. *J. Agr. Res.* 34:1093–1189
149. Slack, C. R. 1968. *Biochem. Biophys. Res. Commun.* 30:483–88
150. Slack, C. R. 1969. *Phytochemistry* 8:1387–91
151. Slack, C. R., Hatch, M. D. 1967. *Biochem. J.* 103:660–65
152. Slack, C. R., Hatch, M. D., Goodchild, D. J. 1969. *Biochem. J.* 114: 489–500
153. Smillie, R. M., Andersen, K. S., Bishop, D. G. 1971. *FEBS Lett.* 13: 318–20
154. Smillie, R. M., Andersen, K. S., Tobin, N. F., Entsch, B., Bishop, D. G. 1972. *Plant Physiol.* 49:471–75
155. Smith, B. N. 1972. *Bioscience* 22: 226–31
156. Stanley, R. A., Naylor, A. W. 1972. *Plant Physiol.* 50:149–51
157. Stiller, M. 1962. *Ann. Rev. Plant Physiol.* 13:151–70
158. Sutton, B. G., Osmond, C. B. 1972. *Plant Physiol.* 50:360–65
159. Tabita, F. R., McFadden, B. A. 1972. *Biochem. Biophys. Res. Commun.* 48: 1153–59
160. Tarchevskii, I. A., Karpilov, Y. S. 1963. *Fiziol. Rast.* 10:229–31
161. Ting, I. P., Johnson, H. B., Szarek, S. R. 1972. See Ref. 30, 25–53

162. Ting, I. P., Osmond, C. B. 1973. *Plant Physiol.* In press
163. Ibid. In press
164. Tolbert, N. E. 1971. *Ann. Rev. Plant Physiol.* 22:45–74
165. Troughton, J. H. 1971. *Planta* 100: 87–92
166. Usuda, H., Kanai, R., Takeuchi, M. 1971. *Plant Cell Physiol.* 12:917–30
167. van Steveninck, M. E., Goldney, D. C., van Steveninck, R. F. M. 1972. *Z. Pflanzenphysiol.* 67:155–60
168. Volk, R. J., Jackson, W. A. 1972. *Plant Physiol.* 49:218–23
169. Walker, D. A. 1962. *Biol. Rev.* 37: 215–56
170. Walker, D. A. 1972. *New Phytol.* In press
171. Waygood, R. E., Arya, S. K., Mache, R. 1971. See Ref. 84, 246–54

172. Waygood, E. R., Mache, R., Tan, C. K. 1969. *Can. J. Bot.* 47:1455–58
173. Whelan, T., Sackett, W. M., Benedict, C. R. 1972. *Assoc. Southeast Biol. Bull.* 19:109
174. Wildner, G. F., Criddle, R. S. 1969. *Biochem. Biophys. Res. Commun.* 37: 952–57
175. Woo, K. C., Anderson, J. M., Boardman, N. K., Downton, W. J. S., Osmond, C. B., Thorne, S. W. 1970. *Proc. Nat. Acad. Sci. USA* 67:18–25
176. Zelitch, I. 1964. *Ann. Rev. Plant Physiol.* 15:121–39
177. Zelitch, I. 1971. *Photosynthesis, Photorespiration, and Plant Productivity.* New York: Academic. 347 pp.
178. Zelitch, I. 1972. *Plant Physiol. Suppl.* 49:58
179. Zelitch, I. 1972. *Arch. Biochem. Biophys.* 150:689–707

Ann. Rev. Plant Physiol. 1973. 24:287–310

LIPID METABOLISM IN PLANTS[1]   ✦ 7550

Paul Mazliak

Laboratoire de Physiologie Cellulaire, Université de Paris VI, Paris, France

CONTENTS

Although numerous reviews devoted to lipid metabolism have appeared in recent years, relatively few articles (81, 103, 123, 128, 145, 160) are strictly restricted to plants. Some related topics concerning membrane lipids have been covered in special papers (18, 23, 146). This review will present the main pathways of lipid

---

[1] The following abbreviations are used: ACP (acyl-carrier protein); ADP, AMP, ATP (respectively adenosine di-, mono-, triphosphate); CDP, CTP (respectively cytidine di-, triphosphate; DGDG (digalactosyldigylceride); MGDG (monogalactosyldiglyceride); NAD$^+$, NADH (nicotinamide-adenine dinucleotide, oxidized and reduced forms); PA (phosphatidic acid); PC (phosphatidylcholine); PE (phosphatidylethanolamine); PG (phosphatidylglycerol); PGp (phosphatidylglycerol-phosphate); PGP (diphosphatidylglycerol); PI (phosphatidylinositol); PS (phosphatidylserine); SL (sulfolipid); UDP, UTP (uridine di-, triphosphate).

metabolism in plants,[2] and recent progress concerning higher plants will be stressed.

FATTY ACID BIOSYNTHESIS

Decisive advances in the study of the biosynthesis of long-chain saturated fatty acids have been made upon microorganisms or animal systems. Two biosynthetic pathways are classically distinguished:

1. A *de novo pathway*, catalyzed by a multienzyme complex (the fatty acid synthetase), localized in the cytoplasm of cells and amenable to fractionation in *Escherichia coli* (136), but apparently not dissociable without protein denaturation in yeast or mammalian tissues (4, 133). The following enzymic reactions contribute to one cycle of this de novo pathway in bacteria:

$$acetyl \sim S\text{-}CoA + ACP - SH \rightarrow acetyl \sim S\text{-}ACP + CoASH$$
$$malonyl \sim S\text{-}COA + ACP - SH \rightarrow malonyl \sim S\text{-}ACP + CoASH$$
$$acetyl \sim S\text{-}ACP + malonyl \sim S\text{-}ACP \rightarrow acetoacetyl \sim S\text{-}ACP$$
$$+ CO_2 + ACP\text{-}SH$$
$$acetoacetyl \sim S\text{-}ACP + NADPH + H^+ \rightarrow D(-)\beta\text{-hydroxybutyryl}$$
$$\sim S\text{-}ACP + NADP^+$$
$$D(-)\beta\text{-hydroxybutyryl} \sim S\text{-}ACP \rightarrow crotonyl \sim S\text{-}ACP + H_2O$$
$$crotonyl \sim S\text{-}ACP + NADPH + H^+ \rightarrow butyryl \sim S\text{-}ACP$$
$$+ NADP^+$$

Palmitic acid is the principal de novo product and the total reaction can be written:

$$acetyl \sim S\text{-}CoA + 7\ malonyl \sim S\text{-}CoA + 14(NADPH + H^+) \rightarrow$$
$$\rightarrow palmityl \sim S\text{-}CoA + 7\ CO_2 + 14\ NADP^+$$

Two recent papers (60, 186) suggest that the fatty acid synthetase is actually bound to the microsomes in animal cells.

2. An *elongation pathway*, catalyzed by enzymes linked to membranous structures and requiring a preformed fatty acid to which new carbon fragments are added, has also been demonstrated. In the mitochondria of animal tissues, Wakil (217) has proposed the following series of enzymic steps: acetyl$\sim$S-CoA+palmityl $\sim$S-CoA$\rightarrow\beta$-keto-stearyl$\sim$S-CoA+CoA-SH; β-keto-stearyl$\sim$S-CoA+NADH $+H^+\rightarrow\beta$-hydroxy-stearyl$\sim$S-CoA+NAD$^+$;β-hydroxystearyl$\sim$S-CoA$\rightarrow\alpha$-β-octadecenoyl$\sim$S-CoA+H_2O; α-β-octadecenoyl$\sim$S-CoA+NADPH+H$^+\rightarrow$stearyl $\sim$S-CoA+NADP$^+$. A microsomal elongation pathway has been described in rat liver (173), utilizing the malonyl-CoA as condensing unit:

$$malonyl \sim S\text{-}CoA + palmityl \sim S\text{-}CoA + 2(NADPH + H^+) \rightarrow$$
$$stearyl \sim S\text{-}CoA + CO_2 + 2\ NADP^+$$

The necessary enzymes for both the de novo pathway and the elongation mechanisms have been recently evidenced in plant cells.

[2] The metabolism of steroid and isoprenoid substances will be excluded.

Acetyl-CoA Synthetase

The enzyme has been partially purified from an homogenate of potato tuber (107): five activity peaks were obtained after DEAE-cellulose column chromatography and Sephadex column fractionation of the proteins. The major form of the potato acetyl-CoA synthetase followed an Iso Bi Uni Uni Bi Ping Pong mechanism, as judged by product inhibition studies. The global reaction can be written:

$$\text{acetate} + \text{ATP} + \text{CoA} - \text{SH} \xrightarrow{\text{Mg}^{++}} \text{acetyl} \sim \text{S-CoA} + \text{AMP} + (\text{P} \sim \text{P})_i$$

Acetyl-CoA Carboxylase

The limiting step in the de novo synthesis of fatty acids could be the formation of malonyl-CoA, catalyzed by the biotin-containing acetyl-CoA carboxylase:

$$\text{acetyl} \sim \text{S-CoA} + CO_2 + \text{ATP} \rightarrow \text{malonyl} \sim \text{S-CoA} + \text{ADP} + P_i$$

The presence of this enzyme was demonstrated in avocado mesocarp (7), wheat germ (89), or *Phaseolus* seedlings (28). Heinstein & Stumpf (94) described the properties of a purified acetyl-CoA carboxylase from wheat germ which could be an aggregate of a biotin-containing component and of a protein binding acetyl-CoA or malonyl-CoA. The localization in the cell of the acetyl-CoA carboxylase is not quite clear. The enzyme seems to be linked to mitochondria in avocado mesocarp (225), which would suggest some necessary cooperation between mitochondria and cytoplasm for the de novo fatty acid biosynthesis. In chloroplasts, Burton & Stumpf (27) have observed the presence of a heat-stable inhibitor of the enzyme which could well account for the low acetyl-CoA carboxylase activity in chloroplast preparations.

De novo System of Fatty Acid Biosynthesis

The pioneering researches of Stumpf and co-workers have demonstrated the existence of the de novo pathway for the biosynthesis of fatty acids in the cytoplasmic supernatant of avocado mesocarp cells (225). Palmitic acid was shown to be the exclusive product of the semipurified avocado fatty acid synthetase (87). The properties of a soluble fatty acid synthetase extracted from potato tuber have been also examined (108). Utilization of malonate or malonyl~S-CoA for fatty acid synthesis has been demonstrated in the chloroplasts isolated from lettuce (26) or spinach (161). Two plant ACPs have been prepared: from avocado mesocarp (196) and from spinach leaves (140). The two plant ACPs were very similar and differed principally from bacterial ACPs in containing about twice as many lysine residues. There is good evidence that there are at least two closely related ACPs in spinach (140).

Depending on growth conditions, *Euglena* contains either a single or two independent fatty acid synthetase activities (43). Grown in the light *Euglena* elaborates both an ACP-dependent chloroplast associated fatty acid synthetase (59) and an ACP-independent synthetase. These enzymes can be separated either by ammonium sulfate fractionation or by centrifugation in a sucrose density gradient. By contrast, only a single ACP-independent fatty acid synthetase was detected in etiolated cells.

Several recent papers describe carefully the regulation of fatty acid synthesis in chloroplasts (72, 116–119) or mitochondria (147, 149, 150). The cellular sites of fatty acid synthesis in avocado mesocarp were reexamined by Weaire & Kekwick (218), and it was found that neither the cytoplasm nor the mitochondria was able to synthesize fatty acids from acetate or malonate. The synthesis would thus occur in the chloroplasts only or other particles sedimenting therewith, and rupture of these particles during homogenization would give rise to apparent fatty acid synthesis capacities in the cytoplasm and mitochondrial fractions. Recent work of Harwood & Stumpf, however, distinguishes two separate types of fatty acid synthetases in avocado mesocarp (88). One is associated with chloroplast lamella fragments and the other with chloroplastic stromal or cytoplasmic proteins (85). On the other hand, fatty acid synthesis from acetate has been observed in the mitochondria (147, 149) or the cytoplasm (88, 108) of plant cells devoid of chloroplasts.

Elongation Systems of Fatty Acid Biosynthesis

Elongation mechanisms for the biosynthesis of the very long chain fatty acids of plant waxes have been suggested (128). Experiments with barley seedling slices (91) or germinating seeds of pea or safflower (134) have shown that large proportions (10 to 55% depending on the conditions) of fatty acids synthesized from acetate were saturated fatty acids in the range C_{20} to C_{28}. The labeling patterns of these fatty acids were consistent with a mechanism of chain elongation occurring in microsomes via malonyl-CoA (91). Similar results were recently obtained with microsomes extracted from the epidermis of pea leaves (129), thus confirming the importance of chain elongations in wax biosynthesis. In avocado cytoplasmic supernatant, Harwood & Stumpf (86, 87) characterized an arsenite-sensitive elongation system which formed stearate by malonyl-CoA addition to a C_{16} precursor, possibly palmityl-ACP.

Monounsaturated Fatty Acids

A general agreement seems now to have been reached for considering that the double bonds are introduced into formerly saturated fatty acids by desaturases requiring molecular oxygen and a reduced pyridine nucleotide (113, 114). Bloomfield & Bloch have shown that stearyl-CoA and palmityl-CoA were desaturated in *Saccharomyces cerevisiae* (21). In green *Eugena gracilis* and spinach chloroplasts, it was shown (164) that the ACP derivative of stearate was desaturated. The soluble stearyl-acyl-carrier-protein desaturase system from *Euglena* has been separated into three components: a NADPH-oxidase, the desaturase, and ferredoxin.

When saturated acids from C_{14} to C_{16} were incubated in *Chlorella* suspensions, they gave rise to Δ^7 and Δ^9 monoenes; C_{17} to C_{19} saturated fatty acids produced only the Δ^9 monoenes (25, 106). Sterculic acid is a potent inhibitor of the enzyme system desaturating stearic acid into oleic acid in *Chlorella vulgaris*. However, no inhibition of oleic formation from acetate is observed (115); it is suggested that the possible point of inhibition by sterculic acid is the transferase transforming stearyl-S-CoA into stearyl~S-ACP, the real substrate for the desaturase.

The particle bound acyl-CoA desaturase from bakers yeast was recently isolated by Schultz & Lynen (193). There was no indication that transfer of an acyl residue from CoA to the enzyme occurred in the reaction sequence. Moreover, the inhibition of the desaturase by palmitoleyl-CoA and oleyl-CoA was shown to be competitive, which strongly suggests that the thioesters are the acyl forms directly transformed. The enzyme converted decanoyl-CoA to a 9-decenoic acid derivative, therefore indicating no special chain length specificity. Baker & Lynen discussed the factors involved in stearyl-CoA desaturation by *Neurospora crassa* microsomes (6). Oxygen and NADH were required for the rapid desaturation of stearyl-CoA into oleyl phospholipid; NADPH is far less active than NADH. The data rule out the possibility of stearyl-phospholipid being an intermediate in the Δ^9 desaturation while they are consistent with oleyl-phospholipid being the substrate for the subsequent Δ^{12} desaturation.

In higher plants oleic acid is rapidly and heavily labeled when slices of various tissues are incubated in ^{14}C-acetate (30, 113, 142, 182); however, no precursor-product relationship can be clearly established in vivo between stearic acid and oleic acid. Moreover, we do not know whether the higher plant desaturases require the acidic forms, the oxygen esters, or the thioesters of saturated fatty acids, and we do not know if the enzymes present any chain length specificity.

When barley seedling slices synthesized labeled stearic and oleic acid from 1-^{14}C-acetate, Hawke & Stumpf (91) found that 53% of the total ^{14}C resided in the carboxyl carbon of stearic acid and 16.7% resided in the carboxyl carbon of oleic acid. These data suggest that stearic acid is not a direct precursor of oleic acid. On the contrary, when leaf preparations were incubated anaerobically with labeled acetate, stearic acid accumulated; on transferring the incubation to aerobic conditions, the radioactivity in stearic acid declined, giving place to an equal amount of oleic acid (84). Thus higher plants could also probably perform the direct desaturation of stearyl residues.

James (114) has observed that ^{14}C-fatty acids from C_2 to C_{14} could be incorporated entirely into the oleic acid of *Ricinus* leaves, while ^{14}C-palmitic or ^{14}C-stearic acid could not. Similarly, Fulco (63) has shown that slices of the ice plant *Carpobrotus chilense* can desaturate ^{14}C-myristate to ^{14}C-9-tetradecenoic acid and incorporate the labeled myristate into the oleic acid. All these data suggest that the introduction of the double bond in the acyl chain could occur when the acyl chain has acquired a specific length (14 carbons, for instance). This idea is reinforced by recent results of James' group on desaturases of four different origins (25).

Important quantities of petroselinic acid (6-octadecenoic acid) are synthesized during the maturation of ivy fruits and seeds (73). The precursor of the Δ^3-*trans*-hexadecenoic acid of photosynthetic tissues was shown to be palmitic acid (170), the reaction requiring not only oxygen but also light in both *Chlorella* and higher plant leaves. Reduction of ^{14}C-*trans*-3-hexadecenoic acid to ^{14}C-palmitic acid occurs rapidly in *Chlorella* (9).

Although desaturation of the common saturated fatty acids by subcellular fractions from animal tissues is well documented, similar efforts to study desaturation with comparable fractions from plant tissues have met with little success (76). Inkpen & Quackenbush have demonstrated a 9-desaturase activity in the 105,000 g supernatant of homogenates of cotyledons from soybean seeds (112). These cell-free preparations desaturate 1-^{14}C-palmitate, 1-^{14}C-stearate, 1-^{14}C-myristate, and 1-^{14}C-laurate to produce 9-10 unsaturated acids. In the best experiments, however, only 15% of the saturated acid was converted to monoenoic acid in 120 min, which is in contrast with the very rapid formation of oleic acid in entire cells.

Labeled 9-hexadecenoic, 9-octadecenoic, or 11-octadecenoic acids have been recently obtained by incubating isolated mitochondria from potato, cauliflower, *Lupinus*, or *Vicia faba* roots with ^{14}C-acetate (149). The kinetics of labeling of the various mitochondrial fatty acids indicated that the double bonds were introduced at the position 9 into the C_{16} or C_{18} saturated chains. Heavily labeled oleic acid was also obtained from ^{14}C-acetate with isolated chloroplasts (116) or microsomes (207).

No fatty acid desaturase has been isolated as yet from higher plants. Oleic acid was the major fatty acid formed from either acetyl-CoA or malonyl-CoA by a particulate subcellular fraction from developing castor oil seeds (53). While oleic acid was the major product formed when both NADH and NADPH were included in the incubation medium, stearic acid was the chief product in the presence of NADH alone. Particles sedimenting between 500 g and 15,000 g were first used as the source of enzymes, and it was very probable that in these conditions mitochondria were accompanied with other organelles (glyoxysomes, plastids, etc). The particulate fraction was stabilized and separated on a sucrose density gradient into several distinct protein bands (226). Unsaturated fatty acid biosynthesis from 1-^{14}C-acetyl-CoA was not associated with mitochondria (density 1.17–1.19 g/ml) but with one band of density 1.21 g/ml. The peroxisome was demonstrated as a distinct protein shoulder of density 1.24 g/ml on the denser side of the fatty acid synthetase band. It was recently proved that the enzyme systems for oleic acid synthesis in developing castor bean are located in proplastids (227).

Homogenates of developing soybean cotyledons, prepared with carbowax 4000 in the grinding medium, synthesize mainly stearic and oleic acid from malonyl-CoA. Kinetic studies indicate that stearic acid is a precursor of oleic acid. Fractionation of the fatty acid synthesizing system in a particulate system (sedimenting at 36,000 g for 10 min) and a supernatant fraction results, even when both fractions are recombined, into synthesis of only stearic acid (183). Thus the desaturase appears to be a very labile enzyme.

Polyunsaturated Fatty Acids

In slices of adult plant tissues, the polyunsaturated fatty acids, which are the more abundant fatty acids, are very poorly labeled from ^{14}C-acetate (30, 142); one must wait for as long as several days to observe a biosynthesis of linolenic acid quantitatively significant (15, 82, 83, 210). During the greening process of etiolated tissues (211–213) or during the maturation of some seeds (55), the biosynthesis of polyunsaturated acids is far more active, and in these conditions the kinetics of labeling of the different fatty acids from labeled sucrose or labeled acetate (54, 195) clearly indicate that a progressive desaturation chain is operating:

Oleic acid → linoleic acid → linolenic acid

Good percentages of desaturation (up to 82%) of 1-^{14}C-oleic acid into 1-^{14}C-linoleic and 1-^{14}C-linolenic acids have been obtained with actively growing mycelia of *Penicillium chrysogenum* (17).

The development of many plants or algae in total darkness has been shown to provoke a stop in chloroplast differentiation and an accumulation of galactolipids devoid of linolenic acid in plastids (171). The effect of light on fatty acid synthesis in chloroplasts, however, is poorly understood. With adult clover leaves, Tremolières (213) found that light stimulated radioactive acetate incorporation indifferently into all fatty acids and all the lipid classes without special effect on trienoic fatty acid or galactolipids. With growing pea leaves (211), on the contrary, a specific increase of acetate incorporation into linolenic acid was noted under illumination. Newman has shown that the desaturation process in the synthesis of linolenate is not phytochrome mediated during the greening of etiolated barley leaves (168).

It has been suggested that the fatty acids could only be desaturated when esterified to certain lipid molecules, e.g. monogalactosyldiglycerides (172), phospholipids (78, 185), or when bound to enzymes (114). Gurr & Brawn (77) have found that dioleyl-phosphatidylcholine is specially involved in the desaturation of oleate into linoleate in *Chlorella vulgaris*. The blue-green alga *Anabaena variabilis*, which does not contain lecithin, utilizes MGDG as an intermediate for linoleate synthesis (3). The apparent relative importance of the lecithin pathway in *Chlorella* is best observed when 1-^{14}C-oleate is directly given to the alga as a precursor; when linoleic acid is formed by *Chlorella* from ^{14}C-acetate, both MGDG and lecithin are good intermediates in the polyunsaturated fatty acid synthesis.

Researches with subcellular preparations do not seem to confirm the necessity for oleyl residues to be integrated in complex lipid molecules to be desaturated. A plastid fraction prepared from maturing safflower seeds (*Carthamus tinctorius*) catalyzed the conversion of oleyl-CoA to linoleyl-CoA (151, 152) as does a similar preparation from *Chlorella vulgaris* (82). Microsomes from maturing safflower seeds were also shown to contain an oleyl-CoA desaturase. Although there is a very high acyl transferase activity in these microsomes, which results in the rapid transfer of the oleyl moiety to an endogenous acceptor to form β-oleyl-phosphatidylcholine, this oleyl-oxygen ester does not serve as a substrate for the microsomal desaturase (215). Any modification of the substrate—that is, a *trans* double bond, a shift in the position of the *cis* double bond, chain length, and the substitution of ACP for CoA as the thioester moiety—resulted in complete loss of activity (216). Similarly it was found with *Chlorella vulgaris* and *Ricinus* seeds that among a wide variety of monoenoic fatty acids only those presenting a double bond in the ninth position (counted either from the carboxyl group or from the methyl end of the chain) could be transformed into methylene—interrupted dienoic fatty acids (105). Thus two kinds of desaturases could exist, differing by their "substrate recognition system": one kind would accept 9-monoenoic fatty acids; the other kind would accept ω-9-monoenoic acids. It is remarkable that oleic acid could be a substrate for both kinds of desaturases.

Unlike mature chloroplasts, chloroplasts isolated from immature spinach leaves incorporated 1-^{14}C-acetate into polyunsaturated fatty acids [these acids contained at the best 20% of total fatty acid radioactivity (119)]. Potassium cyanide completely inhibited linoleate synthesis but did not alter oleate nor linolenate synthesis: these facts suggest independent pathways for linoleate and linolenate synthesis in chloroplasts.

More work on polyunsaturated fatty acid synthesis by subcellular fractions is expected since, in vivo, linolenic acid respresents, for instance, 70% of the weight of total fatty acids in chloroplasts.

The fatty acid desaturases require oxygen. Usually the polyunsaturated fatty acids accumulate in nonphotosynthetic plant tissues at low temperature, and Harris & James (80) have shown that this accumulation could be explained by increased solubility of oxygen in cell fluids at low temperature. With green tissues, because of the production of oxygen by the Hill reaction, the temperature effect is not so marked (2).

A study was made of the stereospecificity of hydrogen removal in the sequential desaturations performed by intact cells of *Chlorella vulgaris* in the biosynthesis of oleic, linoleic, and α-linolenic acids. By use of *erythro-* and *threo-* 9,10^{2}H$_2^-$, -12,13-^{2}H$_2$-and-15-16-^{2}H$_2$ labeled precursors, it was demonstrated that the pair of hydrogen atoms removed from each of these positions had the *cis* relative configuration (159).

Some algae (*Euglena gracilis, Ochromonas danica, Porphyridium cruentum*) and mosses contain arachidonic acid. The distribution and biosynthesis of long chain tri- and tetraenoic acids were discussed by Nichols & Appleby (169).

In recent books on plant lipid metabolism (103, 145), the reader will find data upon the metabolic pathways concerning the substituted (hydroxylated, epoxyacids, etc) or the acetylenic fatty acids occurring in plant tissues.

FATTY ACID CATABOLISM

One can distinguish three catabolic pathways:

1. *The β-oxidation pathway* is the major one. The degradative sequence (dehydration, hydratation, dehydration, and release of acetyl-CoA) was demonstrated in plants by Stumpf & Barber in isolated mitochondria from germinating peanut cotyledons (202). Presumably, the fatty acids which are oxidized by the mito-

chondrial system must be first esterified by the coenzyme A. The thiokinases for long chain fatty acids are associated with microsomes (8, 180, 222, 223). As is the case in animal mitochondria, it seems logical to infer that the permeability barrier of the mitochondria can only be transgressed by the carnitine-esters of long chain fatty acids. The acyl-CoA-carnitine transferases of plant mitochondria have not yet been demonstrated; however, carnitine has been found in various plant tissues (177). This substance is abundant in embryos or tissues rich in mitochondria, while it is practically absent of tissues like endosperm or cotyledons where the β-oxidation sequence is normally followed by the glyoxylate cycle.

In germinating seeds, the β-oxidation pathway is working very actively extra-mitochondrially. The enzymes of the degradative cycle are located in microsomes (179, 180, 222, 223) or in glyoxysomes (38, 110, 111). The acetyl-CoA produced are utilized in the glyoxysomes (75, 156) to be converted into malate (and subsequently into sucrose) by the glyoxylate cycle (24, 31, 199). In castor bean seeds the fat is localized in the endosperm and the carbohydrate produced, principally sucrose, is absorbed by the cotyledons of the growing embryo. The rate of fat conversion reaches its maximum at the fifth day of germination (130). All the enzymes of the glyoxylate cycle are housed in the glyoxysome in vivo throughout the germination period, and the rise and fall in enzyme activities in phase with fat breakdown correspond to the net production and destruction of this organelle in *Ricinus* endosperm (71). Glyoxysomes, isolated on sucrose gradients, have been shown to contain a thiokinase which in the presence of CoA, $MgCl_2$, and ATP (GTP is ineffective) activates free fatty acids to their cognate fatty acyl-CoA derivatives (37). This thiokinase is specific for fatty acids of chain length greater than C_{10}. Glyoxysomes contain also catalase and glycolic acid oxidase (156).

Cytoplasmic particles, bounded by a single membrane similar in appearance to glyoxysomes, are often found in green leaf cells. These are the *peroxisomes* (or microbodies) which possess the enzymes of photorespiration and apparently participate in photorespiration through oxidation of glycolate, a product of photosynthesis (208). In those fatty cotyledons that convert stored lipids to carbohydrate in the early postgerminative stages and then expand and become photosynthetic organs, a unique relationship could exist involving both glyoxysomes and leaf peroxisomes (75). Sunflower, cucumber, and tomato cotyledons contain numerous glyoxysomes (characterized morphologically and enzymatically) during the first 4 days of germination; at this stage the cells are rapidly converting lipids to carbohydrates. Seven days after germination, numerous peroxisomes are appressed to chloroplasts, but it is yet to be determined whether these new organelles are derived from preexisting glyoxysomes or arise as a separate population of cytoplasmic particles.

The enzymes necessary to adapt the β-oxidation sequence to unsaturated fatty acids have been evidenced in the slime mold *Dictyostelum discoideum* (39). The β-oxidation of ricinoleic acid has been studied thoroughly (110, 223).

2. *The α-oxidation pathway* is a secondary catabolic sequence operating in plant tissues (139), especially in young leaves (99, 101). Two reaction sequences have been reported. The first, working in peanut cotyledons (139), involves a peroxida-

tion of the attacked fatty acid, a decarboxylation, and a subsequent reduction of the newly formed fatty aldehyde:

$$R-CH_2-CH_2-COOH \xrightarrow{+2H_2O_2} R-CH_2-CHO + 3H_2O + CO_2$$
$$\xrightarrow{+H_2O \ +NAD^+} R-CH_2-COOH + NADH + H^+$$

Recent reexamination of the α-oxidative system in peanut cotyledons (137) has shown, however, that D-2-hydroxypalmitic acid accumulates during the oxidation of palmitic acid in the presence of a H_2O_2 generating system. The second reaction system, operating in leaves, involves a direct α-oxidation by molecular oxygen of the attacked fatty acid with formation of a transitory 2-hydroxyacid (102). Sastry & Kates (188, 189) showed that 2-hydroxyacids were present in a cerebroside fraction from runner bean leaves. Hitchcock & collaborators (102, 104) showed clearly that the 2-hydroxyacids which accumulate in plant tissues have the absolute D-configuration while the 2-hydroxyacids which appear as intermediates of α-oxidation (catalyzed by acetone-dried pea-leaf powders) have the absolute L-configuration. The latter authors finally assume that both enantiomers are formed during α-oxidation, though presumably the rate of the biosynthesis of the L-α-hydroxyacids is greater than that of the D-α-type, since the main product is CO_2 rather than hydroxyacids (158). Finally, the authors proposed (100) the following reactions:

$$
\begin{array}{l}
\text{D-}\alpha\text{-R-CHOH-COOH} \xrightarrow[-CO_2-2H]{} R\text{-CHO} \longrightarrow R\text{-COOH (slow)} \\
+O \\
\text{R-CH}_2\text{-COOH}+1/2O_2 \\
\text{L-}\alpha\text{-R-CHOH-COOH} \xrightarrow[-CO_2-2H+O]{} R\text{-COOH (fast)}
\end{array}
$$

It is possible that α-oxidation could be a component of the basal respiration of potato slices (131), especially during the respiratory rise which is insensitive to cyanide, accompanying the "aging process."

3. *The peroxidation pathway* for fatty acid catabolism is catalyzed by the enzyme lipoxidase or lipoxygenase (E.C. 1.13.1.13). It consists of the oxidation by molecular oxygen of *cis, cis*-1-4-pentadiene systems to 5-hydroperoxy compounds. In plants the attacked systems reside mainly in linoleic and linolenic acids; the produced hydroperoxides often give rise to scission products, mainly aldehydes, which are strongly odorous. The lipoxygenase can be found in a number of seeds and has been obtained in a purified state from soybean (154) and peas (58) as well as from potato tubers (67).

The enzyme is highly specific for the oxidation of fatty acids containing the *cis, cis*-1-4-pentadiene unit with the formation of conjugated *cis, trans*-diene hydroperoxides (79). Dolev et al (47) have observed that the only product formed after soybean lipoxidase is incubated with linoleic acid is 13-hydroperoxy-9,11,-octadienoic acid, while spontaneous autoxidation of the unsaturated fatty acid gives a mixture of the 13-hydroperoxy and 9-hydroperoxy isomers. On the other hand, Galliard indicates that the 9-hydroperoxide is practically the unique product of potato lipoxygenase (67). The same isomer is obtained with the *Zea mays* germ lipoxidase (69). Finally, Christopher & Axelrod (35) have studied the products of the peroxidation of linoleic acid by two isozymes of soybean lipoxidase, and they have found that one isozyme formed a 1:1 mixture of 9- and 13-hydroperoxide while the other formed only the 13-hydroperoxide.

Heath & Packer (92) have observed that lipid peroxides increase in isolated chloroplasts in the light. The biological significance of the peroxidation of fatty acids is not clear. Fatty peroxides are very toxic substances, inhibiting many enzymes (192). The peroxidation of fatty acids must be a strictly controlled process in cells. It has been suggested that the aging of cells could result from a release of this control, thus permitting many metabolic disturbances by the newly produced fatty acid peroxides.

The cutin hydroxyacids have been considered as secondary products of a special peroxidation process of unsaturated fatty acids in plant epidermis (93).

PHOSPHOLIPID BIOSYNTHESIS

The main biosynthetic pathways evidenced in animal cells or microorganisms (126) are summarized in Figure 1. The mass of information available for these cells is in high contrast with the paucity of data concerning higher plants.

With entire organs or slices of various tissue (11, 56, 143, 153), *Avena* coleoptiles (228), or algae (34), it has been shown that the following elementary precursors of glycerolipids are integrated in newly synthesized molecules: ^{32}P-orthophosphate (34, 56, 143), $Na_2H^{33}PO_4$ (198), 1-^{14}C-acetate (56, 143, 198), ^{14}C-glycerol (143), ^{14}C-ethanolamine (153, 220), ^{14}C-serine (220), ^{14}C-choline (153, 219, 220). Individual enzymic steps have been evidenced in some cases, but rarely with subcellular fractions. Pure enzymes have practically never been isolated.

Phosphatidic Acid

The progressive acylation of glycerophosphate to give lysophosphatidate and subsequently phosphatidate has been demonstrated in isolated cauliflower mitochondria or corn etioplasts (52), with spinach leaves microsomes (125, 190), or in *Chlorella* (34, 190). The nature of the acyl donor (acyl-ACP or acyl-CoA) has not been explained. Presently, no information is available on the eventual utilization of dihydroxyacetone-phosphate as a precursor of phosphatidic acid in plants. The transformation of diglycerides into phosphatidic acids, catalyzed by a diglyceride kinase, has been suggested, in a mitochondrial preparation from peanuts (22).

In 1955, Mazelis & Stumpf (141) studied the incorporation of ^{32}P from orthophosphate into the phospholipid fraction of peanut cotyledon mitochondria. The first step is the esterification of ^{32}PO$_4$H$_3$ into AT ^{32}P. In 1960, Bradbeer & Stumpf (22) presented evidence that the only labeled phospholipid of peanut mitochondria from AT ^{32}P was a phosphatidic acid. Similar results were obtained recently with cauliflower mitochondria (49).

The enzymic biosynthesis of phosphatidic acid has been obtained by Cheniae

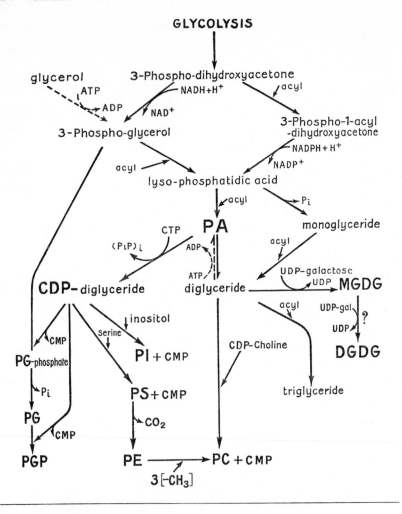

Figure 1 Biosynthetic pathways for glycerolipids (secondary redistributions of acyl groups are not indicated).

(34) with spinach leaf microsomes; acylation of ^{14}C-glycerophosphate was demonstrated with this system as well as the subsequent formation of mono-, di- or triglycerides. Incorporation of ^{32}P-glycerophosphate into phosphatidic acid by isolated spinach microsomes has also been obtained by Kates & Sastry (125). Barron & Stumpf (8) had also previously shown that the microsomes were the major site of triglyceride synthesis in avocado; ^{14}C-glycerol was incorporated by microsomal proteins first into glycerophosphate, then into phosphatidic acid, and

finally into glycerides. Evidence had been presented that the monoglyceride arose from the action of phosphatidic acid phosphatase on lysophosphatidic acid.

Phosphatidylglycerol, Diphosphatidylglycerol

Utilization of phosphatidic acid as a precursor for the biosynthesis of other phospholipids and glycerides has been demonstrated in apple tissues (144) or spinach (34) and avocado microsomes (8). The direct formation of CDP-diglyceride from phosphatidic acid and CTP has been demonstrated in isolated cauliflower mitochondria (98, 203, 204) and yeast (109, 201). In 1968, Douce (5, 48) furnished CTP-γ-^{32}P to physiologically active purified cauliflower mitochondria and observed first the formation of CD^{32}P-diglyceride and subsequently the formation of phosphatidylglycerol. In 1968 also, Sumida & Mudd (203, 204) independently demonstrated the incorporation of ^{14}C-CTP in CDP-diglyceride of cauliflower mitochondria. Part of the phosphatidic acid synthesized by isolated mitochondria can thus be utilized in the organelles to form phosphatidylglycerol (50), diphosphatidylglycerol (50), or phosphatidylinositol (204). The biosynthesis of phosphatidylglycerol is particularly active during the greening of etiolated cells (191). Experiments using $sn[1,3\text{-}^{14}C_2]$ glycerol-3$[^{32}P]$ phosphate have shown that in spinach leaves phosphatidylglycerol is synthesized by the microsomes (and probably not by the chloroplasts) via phosphatidylglycerolphosphate (124, 138).

Phosphatidylserine, Phosphatidylethanolamine, Phosphatidylcholine

It is surprising to report that very few papers deal with the biosynthesis of the major plant membrane phospholipids. Willemot & Boll (220), working on excised tomato roots, presented evidence from studies using ^{14}C-serine and ^{14}C-ethanolamine for the incorporation of serine and ethanolamine into the corresponding phosphatides and for the occurrence of a pathway whereby phosphatidylserine is decarboxylated to phosphatidylethanolamine, which is further methylated to phosphatidylcholine. Indirect evidence for the methylation of phosphatidylethanolamine to phosphatidylcholine was also brought by the kinetics of labeling from ^{32}P-orthophosphate of the phospholipids in etiolated seedlings of mung bean; in the swollen shelled seeds of the same plant, the biosynthetic pathway for phosphatidylcholine would be by CDP-choline (120). The existence of a curious N-acyl-phosphatidylethanolamine has been reported in pea seeds (178). The transformation of 3-phosphohydroxybutyrate into phosphorylserine (for which vitamin B_6 is a cofactor) was studied in extracts of etiolated epicotyls of *Pisum sativum* (200). The phosphorylcholine (135) was identified as an important constituent of the saps of higher plants. A choline kinase studied by Tanaka et al (205) was found in several plant tissues, but was not active enough to explain significant biosyntheses of phosphatidylcholine. This enzyme was also studied in *Cuscuta reflexa* where it could be localized in mitochondria (194).

The incorporation of ethanolamine-1,2-^{14}C into phosphatidylethanolamine is catalyzed primarily by the microsomal fraction from etiolated pea seedlings (214). The enzyme system has a pH optimum of 8.5 and no requirement for ATP, CTP, or CMP could be demonstrated. The Ca^{++} requirement was not met by Mg^{++} or

Mn^{++}. L-serine and choline competitively inhibited the incorporation of ethanolamine and were themselves incorporated into phosphatidylserine and phosphatidylcholine, respectively. This suggests that a single enzyme system would be responsible for the incorporation of any one of the three nitrogen bases into corresponding phospholipids. This system would not be a phospholipase D (having similar transferase activity as in cabbage) because the alcohols, with the exception of glycerol, were not effective inhibitors of substrate incorporation into phospholipids.

The fermentative production of CDP-choline by brewer's yeasts has been studied (127). Recently, Devor & Mudd (45) studied the enzymic incorporation of choline-1-2-^{14}C from CDP-choline into phosphatidylcholine by spinach leaf preparations. The enzyme appears similar to that of animal systems with regards to pH optimum (8.0), metal requirements (Mg or Mn), sensibility to SH poisons, and K_m for CDP-choline. The enzyme did not show any diglyceride specificity when exogenous diglycerides were added, which is markedly different from the synthesis of galactolipids in spinach (44, 46). The control of the fatty acid distribution in phosphatidylcholine molecules was realized by a microsomal acyltransferase: the acylation reaction was followed by the incorporation of ^{14}C-labeled fatty acids from the respective CoA-thioesters into phosphatidylcholine. When saturated and unsaturated fatty acyl-CoA derivatives where used, the saturated derivatives were incorporated primarily into the 1-position of the glycerol moiety and the unsaturated fatty acids went primarily to the 2-position. This pattern of incorporation agrees with the fatty acid distribution in vivo.

Purified Golgi apparatus fractions from onions contained the highest phosphorylcholine-cytidyl transferase activity on a protein basis when compared with other cell-fractions (157).

PHOSPHOLIPASES

Phospholipase D is the most thoroughly studied type of phospholipases (40, 51, 90, 122, 224). This enzyme degrades the major plant phospholipids to give an alcohol moiety (choline, ethanolamine, etc) and phosphatidic acid (often appearing in plant lipid extracts as an artifact) or various products of transphosphatidylation: phosphatidylmethanol (51), phosphatidylethanol (224), or phosphatidylglycerol (41). Phospholipase D has been found to be associated in some tissues with plastid preparations (121) but appears to be widely distributed in soluble from (95). The enzyme is stimulated by ethyl-ether, butyl-acetate, or anionic detergents (42). Calcium ions appears to be essential for the action of phospholipase D (92). In spinach leaves (121), Kates has demonstrated the activity of a phosphatidic acid phosphatase and a phospholipase C, both liberating diglycerides in the medium. With homogenates of the same leaves, Bartels & Van Deenen (10) have demonstrated the existence of enzymes carrying out the conversion of lysophosphoglycerides to corresponding diacyl-phosphoglycerides. These enzymes would provide cellular organelles with a means of adjusting the final fatty acid composition of glycerophosphatides after de novo synthesis.

An enzyme preparation that catalyzes the deacylation of mono and di-acyl phospholipids as well as galactolipids has been partially purified by Galliard from potato tuber homogenates (64, 65). Gel filtration, DEAE-cellulose chromatography, and free flow electrophoresis failed to achieve any separation of the acyl-hydrolase activities towards different classes of acyl lipids. The hydrolysis of sonicated aqueous dispersions of phosphatidylcholine required a detergent and was stimulated thirty to fortyfold by Triton X-100. In the crude homogenate, free fatty acids (particularly long-chain *cis*-unsaturated fatty acids) were the natural cofactors of the lipolytic acyl-hydrolase: PC, PE were hydrolyzed in the presence of these activators, but no stimulation was observed for the hydrolysis of lyso-compounds or acidic phospholipids or galactolipids (66). A tentative explanation of the difference in behavior of the enzyme towards the different classes of lipids is suggested by Galliard: the enzyme would attack easily only the micellar structures of lipids, and therefore PC, which forms lamellar spherulites in water, would only be attacked when a detergent or a free fatty acid would help to disorganize the regular structure of the spherulite.

GLYCOLIPID METABOLISM

The biosynthetic pathways of the major galactosyldiglycerides are summarized in Figure 1 (123). $^{14}CO_2$ was incorporated into the galactolipids of runner bean leaves (56) and *Chlorella* (61). In adult clover leaves, Tremolières (213) showed that $^{14}CO_2$ is more actively incorporated into the sugar moiety of galactolipids than into the fatty acid parts of these molecules. In green tissues, one can label easily the fatty acids of phospholipids or neutral lipids (185, 200) from $^{14}CO_2$ or ^{14}C-acetate, but it has proved difficult to find experimental conditions to get the labeled precursors incorporated into glycolipid fatty acids (185, 209, 210). Consequently, the assembly of the diglyceride and sugar moieties of plant galactolipids is well documented, while the formation of the typical dilinolenyl-diglyceride moiety of the molecule remains to be explained.

^{14}C-UDP-galactose was successfully incorporated into the galactolipids of isolated chloroplasts (132, 165, 176). Detailed optimum conditions and the effect of metal ions have been studied with a soluble preparation from spinach chloroplasts (33). In these chloroplasts (175) synthetic diolein could serve as an efficient acceptor for the MGDG synthesis but not for that of DGDG; on the other hand, MGDG isolated from the chloroplast is the acceptor for the second galactosylation. With acetone extracted chloroplasts as a source of enzymes, the highly unsaturated diglycerides were found by Mudd and collaborators to be the most active in stimulating the biosynthesis of chloroplast lipids (presumably mainly MGDG) from ^{14}C-UDP-glactose (163). On the contrary, Eccleshall and Hawke (57) found that synthetic diglycerides differing in fatty acid composition gave the same incorporation of galactose from UDP-galactose.

^{14}C-glycerophosphate was recently found (52) to be incorporated into MGDG by isolated corn etioplasts or spinach chloroplasts, provided that UDP-galactose is furnished in the medium. Without UDP-galactose the labeled precursor remained in lyso- and phosphatidic acid, di- and triglycerides. No formation of labeled DGDG was observed. It appears likely, therefore, that the first galactosylation step, giving MGDG in the chloroplast, is catalyzed by a separate enzyme from that which catalyzes the second galactosylation giving DGDG. In relation with this fact, it has been demonstrated that in spinach chloroplasts the fatty acid composition of MGDG and DGDG differed, particularly by the content in hexadecatrienoic acid. Ozone is an inhibitor of the metabolism of UDP-galactose in spinach chloroplast preparations (162).

We have only a few indications on the metabolism of the diglyceride moiety of galactolipids. Cell-free extracts of green *Euglena gracilis* catalyze the transfer of stearoyl and oleoyl groups from thioesters of *E. Coli* ACP to MGDG; this reaction is stimulated by α-glycerophosphate (181). In the presence of CoA and ATP, spinach leaf homogenates can effect the enzymic conversion of 2-acyl-3-monogalactosylglycerols (monogalatosylmonoglycerides) to 1,2-diacyl-3-monogalactosylglycerols (MGDG) (187). This reaction parallels the conversion of lysophospholipids to diacyl analogs (10).

Proposed pathways for the biosynthesis of the plant sulfolipid have been discussed recently (123).

The data on glycolipid metabolism so far presented concern photosynthetic tissues grown in normal conditions. Various environmental conditions (e.g. light and mineral nutrition) modify strongly this metabolism (166, 167). The light effect is the most important. The greening of etiolated tissues is accompanied by an increase in galactolipids (184) and accompanying linolenic acid and by the formation of transhexadecenoic acid in phosphatidylglycerol (208, 212). Manganese or nitrogen (20, 36) deficiency caused a reduction of linolenic acid and galactolipids in the chloroplasts of higher plants or algae.

Galactolipases from runner bean leaves have been studied by Sastry & Kates (188, 189). Galactolipids are deacylated with the liberation of free fatty acids and the corresponding mono- and di-galactosyl glycerols. Separate enzymes, each specific for the mono and digalactosyl diglycerides, have been postulated (97, 189) mainly from the fact that storage of enzyme preparation at 4°C for several days resulted in considerable decrease in activity towards MGDG (96). The demonstration of the existence of two different enzymes was furnished by Gatt (70). Attempts have been made to affect a separation of the two enzymes, nevertheless, Helmsing (97) ended with only a single protein showing hydrolytic activity towards MGDG and DGDG. Galliard (64–66) has isolated from potato tuber a hydrolytic enzyme which apparently does not distinguish between phospholipids or galactolipids.

PROGRAMMATION OF LIPID SYNTHESIS

When examining the rhythm of fat accumulation in oily seeds or fruits, one is struck by the fact that lipid biosynthesis is not a continuous process developing all through the growth and maturation of the organs. On the contrary, the major part of fat deposits is synthesized during a short period generally restricted to some days or weeks in the course of the seed or fruit development. The capacity of slices of flax or safflower seeds (197) to synthesize linoleic and linolenic acids in vitro from acetate is considerably stimulated during a short period (20th to 40th day after flowering for flax, 14th to 18th day for safflower) corresponding to the time of the natural great fat deposition in the seeds. The in vivo incorporation of $^{14}CO_2$ into the lipids of oat grains is much higher between the 11th and 18th day after flowering than in any other period (19). Miwa et al (155) have observed that the peculiar acids (12–13-dihydroxyoleic and 12–13-epoxy-oleic) appear in the seeds of *Vernonia anthelmintica* only during the period of great fat deposition. Between the 12th and the 36th day after pollination (29), 90% of the final content of ricinoleic acid is deposited in castor been seeds. Erucic acid accumulates rapidly in rapeseeds only after about the 20th day after flowering (1, 62). During the period of great fat deposition, the enzymes of lipid synthesis function very actively and produce the most characteristic fatty acid of the plants. Before or after this period, the physiological conditions in seed tissues are such that lipid biosynthesis is very low. Similarly, α-linolenic acid biosynthesis is only active in green tissues during the period of chloroplast differentiation (211).

LIPID SYNTHESIS DURING AGING

The physiological conditions required for the biosynthesis of polyunsaturated fatty acids can be induced artifically in a plant storage tissue by aging slices of this tissue in aerated conditions. Willemot & Stumpf (221) and later Ben Abdel-kader (11) demonstrated the rapid labeling of linoleic acid from ^{14}C-acetate in slices of potato parenchyma floating on water and aged in air during 24 hours. The precursor incorporation into polyunsaturated fatty acids was blocked if the aging process was realized with cycloheximide in the bathing solution (11, 221). In fact, it was recently found that the aging process induces the biosynthesis of a microsomal oleyl-CoA desaturase in the formerly dormant tissues (13). During the aging of potato slices, newly synthesized fatty acids are mainly integrated into phospholipids (11, 206). Galliard et al (68) studied similarly the development of a lipid synthesizing system during the aging of peel disks from preclimacteric apples. They found incorporation of ^{14}C-acetate into phospholipids and free fatty acids.

The aging of potato tuber slices induces an increase in protein, lipid phos-phorus, and fatty acid content of the mitochondrial (16) as well as the microsomal (12) fractions of parenchyma cells. These results suggest a synthesis of new mitochondrial and microsomal membranes during aging. Castelfranco and col-laborators (32) have shown that in fresh potato slices exogenous ^{14}C-choline is incorporated mainly into the lipids of microsomes. After 9 hours of aging, most of the radioactivity occurs in an intermediate band on gradients, while after 24 hours of aging, a considerable portion of the label is found in a heavy band which overlaps succinoxidase activity. The incorporation of ^{14}C-acetate into the fatty acids of mitochondria and microsomes increase strongly after aging, particularly when the organelles are left in the cells during the incubation period (in vivo conditions). During pulse-chase experiments, labeled lipids (namely phospholipids and linoleic acid) accumulate first in the microsomes and then in the mitochondria (14). This accumulation could not be observed in vitro with isolated mitochondria originating from aged tissue. One may think that the newly synthesized lipids are formed in a cellular compartment other than mitochondria and are subsequently transferred to new mitochondrial membranes.

LIPID TRANSFERS IN VITRO

Ben Abdelkader & Mazliak (143, 150) proposed the hypothesis of an exchange of lipids between the different cellular compartments or organelles to account for the lipid composition of the cellular membranes. The following experimental arguments have been presented (14): when potato tuber or cauliflower microsomes containing phospholipids labeled from 1-^{14}C-acetate, ^{32}P-phosphate, or 1-^{14}C-glycerol are incubated in vitro with unlabeled mitochondria and cytoplasmic supernatant, the radioactivity of the microsomes decreases strongly within 2 hours; the lipids of the mitochondria and those of the supernatant become radioactive, taking up to 40% of the initial radioactivity. The same transfer of labeled lipids is observed when one incubates labeled mitochondria with unlabeled microsomes. During these experiments one can observe a marked decrease of the specific radioactivities of the phospholipids present in the initially labeled fractions. This may be explained by a departure of labeled phospholipids from the radioactive fraction towards the other cellular fraction and by the arrival of unlabeled lipids from the initially unlabeled fraction to the radioactive fraction. The cytoplasmic supernatant acts as an intermediate in these exchanges. Phosphatidylcholine and phosphatidylethanolamine represent the most important fraction of the exchanged phospholipids.

However, it has been shown that these exchange phenomena concern also the various fatty acids engaged in the lipids of the different cellular fractions (148). ^{14}C-fatty acids are transferred in vitro from microsomes to mitochondria, or reciprocally, without any apparent metabolic alteration.

MEMBRANE LIPID TURNOVER

It has been known for some time that membrane lipids of mitochondria or microsomes do present a metabolic turnover in animal cells (74, 174). Data on turnover rates of membrane lipids in plant cells are very rare. Recently, "half-lives" of membrane lipids from potato or cauliflower cells have been calculated by measuring, during several days, the decrease in specific radioactivities of these lipids following an incubation of tissue-slices in 1-^{14}C-acetate solutions and subsequent transfer of the slices into a medium without any radioactive precursor (15). The phospholipids present various turnover rates in the different cellular fractions of the same tissue. The measured "half-lives" vary from 3 to 32 days, which is very similar to the data obtained with animal cells. The greatest turnover rates are found in cytoplasmic supernatants and the lower rates in mitochondria. The various fatty acids from the same cellular fraction present different turnover rates; some acids are metabolically stable (palmitic acid for instance), but oleic acid, on the contrary, is very rapidly metabolized. The polyunsaturated fatty acids which are the major constituents of membrane lipids present a slow turnover in cauliflower florets and potato as well as in clover leaf (210). These and other similar studies suggest that cell lipids may be divided in several metabolic pools. Some lipids would be submitted to active turnover [for instance, PC and MGDG in *Chlorella* (171) and oleic acid in higher plants (15)], while others would be rather metabolically inert and could possess more structural functions [DGDG, SL, and PE in *Chlorella* (167) and polyunsaturated fatty acids in membranes (15)].

Literature Cited

1. Appleqvist, L. A. 1968. *Physiol. Plant.* 21:455–65
2. Ibid 1971. 25:493–502
3. Appleby, R. S., Safford, R., Nichols, B. W. 1971. *Biochim. Biophys. Acta* 248:205–11
4. Ayling, J., Pirson, R., Lynen, F. 1972. *Biochemistry* 11:526–33
5. Bahl, J., Guillot-Salomon, T., Douce, R. 1970. *Physiol. Veg.* 8:55–74
6. Baker, N., Lynen, F. 1971. *Eur. J. Biochem.* 19:200–10
7. Barron, E. J., Squires, C., Stumpf, P. K. 1961. *J. Biol. Chem.* 236:2610–14
8. Barron, E. J., Stumpf, P. K. 1962. *Biochim. Biophys. Acta* 60:329–37
9. Bartels, C. T., James, A. T., Nichols, B. W. 1967. *Eur. J. Biochem.* 3:7–10
10. Bartels, C. T., Van Deenen, L. L. M. 1966. *Biochim. Biophys. Acta* 125:395–97
11. Ben Abdelkader, A. 1968. *Physiol. Veg.* 6:417–42
12. Ibid 1969. *C. R. Acad. Sci.* 268:2406–9
13. Ibid 1972. *C. R. Acad. Sci.* 275:51–54
14. Ben Abdelkader, A., Mazliak, P. 1970. *Eur. J. Biochem.* 15:250–62
15. Ben Abdelkader, A., Mazliak, P. 1971. *Physiol. Veg.* 9:227–40
16. Ben Abdelkader, A., Mazliak, P., Catesson, A. M. 1969. *Phytochemistry* 8:1121–33
17. Bennett, A. S., Quackenbush, F. W. 1969. *Arch. Biochem. Biophys.* 130:567–72
18. Benson, A. A. 1964. *Ann. Rev. Plant Physiol.* 15:1–16
19. Beringer, H. 1971. *Plant Physiol.* 48:433–36
20. Bloch, K. E., Chang, S. B. 1964. *Science* 144:560
21. Bloomfield, D. K., Bloch, K. 1960. *J. Biol. Chem.* 235:337–45
22. Bradbeer, C., Stumpf, P. K. 1960. *J. Lipid Res.* 1:214–20
23. Branton, D. 1969. *Ann. Rev. Plant Physiol.* 20:209–38
24. Breidenbach, R. W., Beevers, H. *Biochem. Biophys. Res. Commun.* 27:462–69
25. Brett, D., Howling, D., Morris, L. J., James, A. T. 1971. *Arch. Biochem. Biophys.* 143:535–47
26. Brooks, J. L., Stumpf, P. K. 1966. *Arch. Biochem. Biophys.* 116:108–16
27. Burton, D., Stumpf, P. K. 1967. *Arch. Biochem. Biophys.* 117:604–14
28. Cailliau-Commanay, L., Cavalie, G. 1972. *C. R. Acad. Sci.* 274:2045–48
29. Canvin, D. T. 1963. *Can. J. Biochem. Physiol.* 41:1879–85
30. Canvin, D. T. 1965. *Can. J. Bot.* 43:71–74
31. Canvin, D. T., Beevers, H. 1961. *J. Biol. Chem.* 236:988–95
32. Castelfranco, P. A., Tang, W. J., Bolar, M. L. 1971. *Plant Physiol.* 48:795–800
33. Chang, S. B., Kulkarni, N. D. 1970. *Phytochemistry* 9:927–34
34. Cheniae, G. M. 1965. *Plant Physiol.* 40:236–43
35. Christopher, J., Axelrod, B. 1971. *Biochem. Biophys. Res. Commun.* 44:731–36
36. Constantopoulos, G. 1970. *Plant Physiol.* 45:76–80
37. Cooper, T. G. 1971. *J. Biol. Chem.* 246:3451–55
38. Cooper, T. G., Beevers, H. 1969. *J. Biol. Chem.* 244:3507–13
39. Davidoff, F., Korn, E. D. 1964. *J. Biol. Chem.* 239:2496–2506
40. Davidson, F. M., Long, C. 1958. *Biochem. J.* 69:458–66
41. Dawson, R. M. C. 1967. *Biochem. J.* 102:205–10
42. Dawson, R. M. C., Hemington, M. 1967. *Biochem. J.* 102:76–86
43. Delo, J., Ernst-Fonberg, M. L., Bloch, K. 1971. *Arch. Biochem. Biophys.* 143:384–91
44. Devor, K. A., Mudd, J. B. 1971. *J. Lipid Res.* 12:390–402
45. Ibid, 403–11
46. Ibid, 412–19
47. Dolev, A., Rohwedder, W. K., Dutton, H. J. 1967. *Lipids* 2:28–32
48. Douce, R. 1968. *C. R. Acad. Sci.* 267:534–37
49. Ibid 1971. 272:3146–49
50. Douce, R., Dupont, J. 1969. *C. R. Acad. Sci.* 268:1657–60
51. Douce, R., Faure, M., Marechal, J. 1966. *C. R. Acad. Sci.* 262:1549–52
52. Douce, R., Guillot-Salomon, T. 1970. *FEBS Lett.* 11:121–24
53. Drennan, C. H., Canvin, D. T. 1969. *Biochim. Biophys. Acta* 187:193–200
54. Dutton, H. J., Mounts, T. L. 1966. *J. Lipid Res.* 6:221–25

55. Dybing, C. D., Zimmerman, D. C. 1966. *Plant Physiol.* 41:1465–70
56. Eberhardt, F. M., Kates, M. 1957. *Can. J. Bot.* 35:907–21
57. Eccleshall, T. R., Hawke, J. C. 1971. *Phytochemistry* 10:3035–45
58. Erikson, C. R., Svensson, S. G. 1970. *Biochim. Biophys. Acta* 198:449–59
59. Ernst-Fonberg, M. L., Bloch, K. 1971. *Arch. Biochem. Biophys.* 143:392–400
60. Favarger, P., Gerlach, J. 1971. *FEBS Lett.* 13:285–89
61. Ferrari, R. A., Benson, A. A. 1961. *Arch. Biochem. Biophys.* 93:185–92
62. Fowler, D. B., Bowney, R. K. 1970. *Can. J. Plant Sci.* 50:233–49
63. Fulco, A. J. 1965. *Biochim. Biophys. Acta* 106:211–12
64. Galliard, T. 1970. *Phytochemistry* 9:1725–34
65. Galliard, T. 1971. *Biochem. J.* 121:379–90
66. Galliard, T. 1971. *Eur. J. Biochem.* 21:90–98
67. Galliard, T., Phillips, D. R. 1971. *Biochem. J.* 124:431–38
68. Galliard, T., Rhodes, M. J. C., Wooltorton, L. S. C., Hulme, A. C. 1968. *Phytochemistry* 7:1453–63
69. Gardner, H. W., Weisleder, D. 1970. *Lipids* 5:678–83
70. Gatt, S., Baker, E. A. 1970. *Biochim. Biophys. Acta* 206:125–35
71. Gerhardt, B. P., Beevers, H. 1970. *J. Cell Biol.* 44:94–102
72. Givan, C. V., Stumpf, P. K. 1971. *Plant Physiol.* 47:510–16
73. Grosbois, M. 1971. *Phytochemistry* 10:1261–73
74. Gross, N. S., Getz, G. S., Rabinowitz, M. 1969. *J. Biol. Chem.* 244:1552–62
75. Gruber, P. J., Trelease, R. N., Becker, W. M., Newcomb, E. H. 1970. *Planta* 93:269–88
76. Gurr, M. I. 1971. *Lipids* 6:266–73
77. Gurr, M. I., Brawn, P. 1970. *Eur. J. Biochem.* 77:19–22
78. Gurr, M. I., Robinson, M. P., James, A. T. 1969. *Eur. J. Biochem.* 9:70–78
79. Hamberg, M., Samuelsson, B. 1967. *J. Biol. Chem.* 242:5329–35
80. Harris, P., James, A. T. 1969. *Biochim. Biophys. Acta* 187:13–18
81. Harris, R. V. 1966. *Rep. Progr. Appl. Chem.* 51:430–42
82. Harris, R. V., Harris, P., James, A. T. 1965. *Biochim. Biophys. Acta* 106:465–73
83. Harris, R. V., James, A. T. 1965. *Biochim. Biophys. Acta* 106:456–65
84. Harris, R. V., James, A. T., Harris, P. 1967. In *Biochemistry of Chloroplasts*, ed. T. W. Goodwin, 2:239. New York: Academic. 776 pp.
85. Harwood, J. L., Sodja, A., Stumpf, P. K. 1971. *Lipids* 6:951–54
86. Harwood, J. L., Stumpf, P. K. 1971. *Arch. Biochem. Biophys.* 142:281–92
87. Ibid 1972. 148:282–90
88. Harwood, J. L., Stumpf, P. K. 1972. *Lipids* 7:8–19
89. Hatch, M. D., Stumpf, P. K. 1961. *J. Biol. Chem.* 236:2879–85
90. Haverkate, F., Van Deenen, L. L. M. 1965. *Biochim. Biophys. Acta* 106:78–92
91. Hawke, J. C., Stumpf, P. K. 1965. *J. Biol. Chem.* 240:4746–52
92. Heath, R. L., Packer, L. 1968. *Arch. Biochem. Biophys.* 125:850–57
93. Heinen, W., Brand, I. V. D. 1963. *Z. Naturforsch.* 18:67–79
94. Heinstein, P. F., Stumpf, P. K. 1969. *J. Biol. Chem.* 244:5374–81
95. Heller, M., Aladjem, E., Shapiro, B. 1968. *Bull. Soc. Chim. Biol.* 50:1395–1408
96. Helmsing, P. F. 1967. *Biochim. Biophys. Acta* 144:470–72
97. Ibid 1969. 178:519–33
98. Hitchcock, C., James, A. T. 1964. *J. Lipid Res.* 5:593–99
99. Hitchcock, C., James, A. T. 1965. *Biochem. J.* 97:1c–3c
100. Hitchcock, C., Morris, L. J. 1970. *Eur. J. Biochem.* 17:39–42
101. Hitchcock, C., Morris, L. J., James, A. T. 1968. *Eur. J. Biochem.* 3:419–21
102. Ibid, 473–75
103. Hitchcock, C., Nichols, B. W. 1971. *Plant Lipid Biochemistry.* London: Academic. 388 pp.
104. Hitchcock, C., Rose, A. 1971. *Biochem. J.* 125:1155–56
105. Howling, D., Morris, L. J., Gurr, M. I., James, A. T. 1972. *Biochim. Biophys. Acta* 260:10–19
106. Howling, D., Morris, L. J., James, A. T. 1968. *Biochim. Biophys. Acta* 152:224–26
107. Huang, K. P., Stumpf, P. K. 1970. *Arch. Biochem. Biophys.* 140:158–73
108. Ibid 1971. 193:412–27
109. Hutchinson, H. T., Cronen, J. E. 1968. *Biochim. Biophys. Acta* 164:606–8
110. Hutton, D., Stumpf, P. K. 1971. *Arch. Biochem. Biophys.* 142:48–61

111. Hutton, D., Stumpf, P. K. 1971. *Plant Physiol.* 44:508–16
112. Inkpen, J. A., Quackenbush, F. W. 1969. *Lipids* 4:539–43
113. James, A. T. 1962. *Biochim. Biophys. Acta* 57:167–69
114. Ibid 1963. 70:9–19
115. James, A. T., Harris, P., Bezard, J. 1968. *Eur. J. Biochem.* 3:318–25
116. Kannangara, C. G., Henningsen, K. W., Stumpf, P. K., Appelqvist, L. A., Von Wettstein, D. 1951. *Plant Physiol.* 48:526–31
117. Kannangara, C. G., Stumpf, P. K. 1971. *Biochem. Biophys. Res. Commun.* 44:1544–52
118. Kannangara, C. G., Stumpf, P. K. 1972. *Plant Physiol.* 49:497–501
119. Kannangara, C. G., Stumpf, P. K. 1972. *Arch. Biochem. Biophys.* 148:414–24
120. Katayama, M., Funahashi, S. 1969. *J. Biochem.* 66:479–85
121. Kates, M. 1954. *Can. J. Biochem. Physiol.* 32:571–83
122. Ibid 1956. 34:967–80
123. Kates, M. 1970. *Advan. Lipid Res.* 8:225–65
124. Kates, M., Marshall, M. O. 1971. *J. Am. Oil Chem. Soc.* 48:88A
125. Kates, M., Sastry, P. S. 1964. *10th Int. Bot. Cong.* Abst. 60
126. Kates, M., Wassef, M. K. 1970. *Ann. Rev. Biochem.* 39:323–58
127. Kimura, A., Morita, M., Tochikura, T. 1971. *Agr. Biol. Chem.* 35:1955–60
128. Kolattukudy, P. E. 1970. *Ann. Rev. Plant Physiol.* 21:163–91
129. Kolattukudy, P. E., Buckner, J. S. 1972. *Biochem. Biophys. Res. Commun.* 46:801–7
130. Kriedemann, P., Beevers, H. 1967. *Plant Physiol.* 42:161–73
131. Laties, G. G., Hoelle, C. 1967. *Phytochemistry* 6:49–57
132. Lin, M. F., Chang, S. B. 1971. *Phytochemistry* 10:1543–49
133. Lynen, F. 1967. *Biochem. J.* 102:381–400
134. Macey, M. J. K., Stumpf, P. K. 1968. *Plant Physiol.* 43:1637–47
135. Maizel, J. V., Benson, A. A., Tolbert, N. E. 1956. *Plant Physiol.* 31:407–8
136. Majerus, P. W., Vagelos, P. R. 1967. *Advan. Lipid Res.* 5:1–33
137. Markovetz, A. J., Stumpf, P. K., Hammerström, S. 1972. *Lipids* 7:159–64
138. Marshall, M. O., Kates, M. 1972. *Biochim. Biophys. Acta* 260:558–70
139. Martin, R. O., Stumpf, P. K. 1959. *J. Biol. Chem.* 235:2548–59
140. Matsumura, S., Stumpf, P. K. 1968. *Arch. Biochem. Biophys.* 125:932–41
141. Mazelis, M., Stumpf, P. K. 1955. *Plant Physiol.* 39:237–43
142. Mazliak, P. 1965. *C. R. Acad. Sci.* 261:2716–19
143. Mazliak, P. 1967. *Phytochemistry* 6:941–56
144. Ibid, 957–61
145. Mazliak, P. 1968. *Le métabolisme des lipides dans les plantes supérieures.* Paris: Masson. 223 pp.
146. Mazliak, P. 1971. *Les membranes protoplasmiques.* Paris: Doin. 195 pp.
147. Mazliak, P., Ben Abdelkader, A. 1970. *Rev. Gen. Bot.* 77:53–71
148. Mazliak, P., Ben Abdelkader, A. 1971. *Phytochemistry* 10:2879–90
149. Mazliak, P., Oursel, A., Ben Abdelkader, A., Grosbois, M. 1972. *Eur. J. Biochem.* 28:399–411
150. Mazliak, P., Stoll, U., Ben Abdelkader, A. 1968. *Biochim. Biophys. Acta* 152:414–17
151. McMahon, V. A., Stumpf, P. K. 1964. *Biochim. Biophys. Acta* 84:359–61
152. McMahon, V. A., Stumpf, P. K. 1966. *Plant Physiol.* 41:148–56
153. Miedema, E., Richardson, K. E. 1966. *Plant Physiol.* 41:1026–30
154. Mitsuda, H., Yasumoto, K., Yamamoto, A., Kusano, T. 1967. *Agr. Biol. Chem.* 31:115–18
155. Miwa, T. K., Eable, F. R., Miwa, G. C., Wolff, I. A. 1961. *139th Congr. Am. Chem. Soc. St. Louis,* Abstr. 29c
156. Mollenhauer, H. H., Totten, C. 1970. *Plant Physiol.* 46:794–99
157. Morre, D. J., Nyquist, S., Rivera, E. 1970. *Plant Physiol.* 45:800–4
158. Morris, L. J. 1970. *Biochem. J.* 118:681–93
159. Morris, L. J., Harris, R. V., Kelly, W., James, A. T. 1968. *Biochem. J.* 109:673–78
160. Mudd, J. B. 1967. *Ann. Rev. Plant Physiol.* 18:229–52
161. Mudd, J. B., McManus, T. T. 1964. *Plant Physiol.* 39:115–19
162. Mudd, J. B., McManus, T. T., Ongun, A., McCullogh, T. E. 1971. *Plant Physiol.* 48:335–39
163. Mudd, J. B., Van Vliet, H. H. D. M., Van Deenen, L. L. M. 1969. *J. Lipid Res.* 10:623–30
164. Nagai, J., Bloch, K. 1968. *J. Biol. Chem.* 243:4626–33
165. Neufeld, E. H., Hall, E. W. 1964.

Biochem. Biophys. Res. Commun. 14: 503–8
166. Newman, D. W. 1964. *J. Exp. Bot.* 15:525–29
167. Newman, D. W. 1966. *Plant Physiol.* 41:328–34
168. Ibid 1971. 48:300–2
169. Nichols, B. W., Appleby, R. S. 1969. *Phytochemistry* 8:1907–15
170. Nichols, B. W., Harris, P., James, A. T. 1965. *Biochem. Biophys. Res. Commun.* 21:473–79
171. Nichols, B. W., James, A. T., Breuer, J. 1967. *Biochem. J.* 104:486–96
172. Nichols, B. W., Moorhouse, R. 1969. *Lipids* 4:311–16
173. Nutgeren, D. H. 1965. *Biochim. Biophys. Acta* 106:280–89
174. Omura, T., Siekevitz, P., Palade, G. B. 1967. *J. Biol. Chem.* 242:2389–96
175. Ongun, A., Mudd, J. B. 1968. *J. Biol. Chem.* 243:1558–66
176. Overath, P., Stumpf, P. K. 196-. *J. Biol. Chem.* 239:4103–10
177. Panter, R. A., Mudd, J. B. 1969. *FEBS Lett.* 5:169–70
178. Quarles, R. H., Clarke, M., Dawson, R. M. C. 1968. *Biochem. Biophys. Res. Commun.* 33:964–68
179. Rebeiz, C. A., Castelfranco, P. 1964. *Plant Physiol.* 39:932–38
180. Rebeiz, C. A., Castelfranco, P., Breidenbach, R. W. 1965. *Plant Physiol.* 40:286–89
181. Renkonen, O., Bloch, K. 1969. *J. Biol. Chem.* 244:4899–4903
182. Rinne, R. W., Canvin, D. T. 1971. *Plant Cell Physiol.* 12:387–93
183. Ibid, 395–403
184. Rosenberg, A., Gouaux, J. 1967. *J. Lipid Res.* 8:80–83
185. Roughan, P. G. 1970. *Biochem. J.* 117:1–8
186. Rous, S., Aubry, L. 1971. *FEBS Lett.* 239:509–12
187. Safford, R., Appleby, R. S., Nichols, B. W. 1971. *Biochim. Biophys. Acta* 239:509–12
188. Sastry, P. S., Kates, M. 1964. *Biochemistry* 3:1271–80
189. Ibid, 1280–87
190. Sastry, P. S., Kates, M. 1966. *Can. J. Biochem.* 44:459–67
191. Schantz, R., Douce, R., Duranton, H. M. 1972. *FEBS Lett.* 20:157–62
192. Schauenstein, E. 1967. *J. Lipid Res.* 8:417–28
193. Schultz, J., Lynen, F. 1971. *Eur. J. Biochem.* 21:48–54

194. Setty, P. N., Krishnan, P. S. 1972. *Biochem. J.* 126:313–24
195. Simmons, R. O., Quackenbush, F. W. 1954. *J. Am. Oil Chem. Soc.* 31:441–43
196. Simoni, R. D., Criddle, R. S., Stumpf, P. K. 1967. *J. Biol. Chem.* 242:573–81
197. Sims, R. P. A., McGregor, W. G., Plessers, A. G., Mes, J. C. 1961. *J. Am. Oil Chem. Soc.* 38:273–76
198. Singh, H., Privett, O. S. 1970. *Biochim. Biophys. Acta* 202:200–2
199. Sinha, S. K., Cossins, E. A. 1965. *Can. J. Biochem.* 43:1531–41
200. Slaughter, J. C., Davies, D. D. 1969. *J. Exp. Bot.* 20:451–56
201. Steiner, M. R., Lester, R. L. 1972. *Biochim. Biophys. Acta* 260:222–43
202. Stumpf, P. K., Barber, G. A. 1956. *Plant Physiol.* 31:304–8
203. Sumida, S., Mudd, J. B. 1970. *Plant Physiol.* 45:712–18
204. Ibid, 719–22
205. Tanaka, K., Tolbert, N. E., Gohlke, A. F. 1966. *Plant Physiol.* 41:307–12
206. Tang, W. J., Castelfranco, P. A. 1968. *Plant Physiol.* 43:1232–38
207. Thibaudin, A., Mazliak, P., Catesson, A. M. 1968. *C. R. Acad. Sci.* 266:784–87
208. Tolbert, N. E., Yamazaki, R. K. 1969. *Ann. NY Acad. Sci.* 168:325–41
209. Tremolières, A. 1970. *Année Biol.* 9: 113–56
210. Tremolières, A. 1971. *C. R. Acad. Sci.* 272:2777–80
211. Tremolières, A. 1972. *Phytochemistry.* 11:3453–60
212. Tremolières, A., Lepage, M. 1971. *Plant Physiol.* 47:329–34
213. Tremolières, A., Mazliak, P. 1970. *Physiol. Vég.* 8:135–50
214. Vandor, S. L., Richardson, K. E. 1968. *Can. J. Biochem.* 46:1309–16
215. Vijay, I. K., Stumpf, P. K. 1971. *J. Biol. Chem.* 246:2910–17
216. Ibid 1972. 247:360–66
217. Wakil, S. J. 1961. *J. Lipid Res.* 2:1–24
218. Weaire, P. J., Kekwick, R. G. O. 1970. *Biochem. J.* 119:48–49 P
219. Wen-Jung, T., Castelfranco, P. A. 1968. *Plant Physiol.* 43:1232–38
220. Willemot, C., Boll, W. G. 1967. *Can. J. Bot.* 45:1863–76
221. Willemot, C., Stumpf, P. K. 1967. *Plant Physiol.* 42:391–97
222. Yamada, M., Stumpf, P. K. 1965. *Plant Physiol.* 40:653–58

223. Ibid, 659–64
224. Yang, S. F., Freer, S., Benson, A. A. 1967. *J. Biol. Chem.* 242:477–84
225. Yang, S. F., Stumpf, P. K. 1965. *Biochim. Biophys. Acta* 98:19–26
226. Zilkey, B., Canvin, D. T. 1969.

Biochem. Biophys. Res. Commun. 34: 646–653
227. Zilkey, B., Canvin, D. T. 1972. *Can. J. Bot.* 50:323–27
228. Zimmerer, R., Hamilton, R. H. 1965. *Plant Cell Physiol. Tokyo* 6:681–87

Ann. Rev. Plant Physiol. 1973. 24:311-52

DORMANCY IN MICROBIAL SPORES ❖ 7551

A. S. Sussman and H. A. Douthit

Department of Botany, University of Michigan, Ann Arbor

Dedicated to Professor Frederick K. Sparrow Jr. in the year of his retirement as professor of botany at the University of Michigan, and in honor of his distinguished service to the field of mycology.

CONTENTS

INTRODUCTION

To paraphrase the felicitous description of seeds given by Harper (106), "Some spores are born dormant, some achieve dormancy and some have dormancy thrust upon them." This is to acknowledge that several types of dormancy exist in microbial spores in common with those in seeds which are discussed by Roberts (228) and others. The definitions given below are taken from Sussman (278), unless otherwise noted.

Dormancy: Any rest period or reversible interruption of the phenotypic development of an organism.

Constitutional (constitutive) dormancy: A condition wherein development is delayed due to an innate property of the dormant stage, such as a barrier to the penetration of nutrients, a metabolic block, or the production of a self-inhibitor. This state is imposed as soon as the spore or embryo is formed. Other names given to this type of dormancy in seeds are *innate* (106), *primary* (57), or *endogenous* (245) *dormancy* or *rest* (313).

Induced dormancy: A situation whereby a nondormant or exogenously dormant spore is made constitutionally dormant. Although this type of dormancy has not been set apart from constitutional dormancy in the literature on microbial spores (281), the phenomenon exists, as in the case of uredospores of *Puccinia graminis tritici*, which require a heat shock to germinate only after storage in the cold (30). The term *secondary dormancy* (57) also has been applied to this form of dormancy in seeds; the usual case involves a seed that has lost its innate dormancy and may be induced to regain it by high temperatures and a limited supply of oxygen (228).

Exogenous dormancy: A condition wherein development is delayed because of unfavorable chemical or physical conditions of the environment. Such a condition in seeds has been called *enforced dormancy* (106), *environmental dormancy, imposed rest* (120), or *quiescence* (243).

Activation: A treatment which leads to an increase in germination.

Germination: The first irreversible stage which is recognizably different from the dormant organism, as judged by morphological, cytological, physiological, or biochemical criteria.

CHARACTERISTICS OF DORMANT STAGES

Because dormancy is virtually ubiquitous in living systems and, "Since during diapause (dormancy) the organism is subjected to widely differing and harsh conditions, adaptations of the diapause stages to climatic factors show greater specificity, and are in general more varied and complex than adaptations in the active stages (60)." Specially formed structures like spores are not the only means through which dormant stages of protistans survive; modified cells like the "dauerzellen" of yeasts, sclerotia, rhizomorphs, and modified hyphae of other fungi can resist environmental upsets. And even the vegetative stages of bacteria and other protistans can do as well (281).

This leads to a question that occasionally is raised in the literature (246) about the relationship between dormancy and spores. In the spores of *Lenzites sepiariae*, which were studied by these authors, there may be no "unique, irreversible event that transforms a metabolically inactive cell into a vegetative hypha." Indeed, their studies suggest that there is more similarity between the metabolism of young (apical) cells of hyphae and these spores than between spores and older cells. For example, there is no clear point of morphological transition between the spore and vegetative cell, and the former contain all of the organelles found in the latter. Moreover, no new metabolic patterns were detected in spores, and respiratory rates were high in ungerminated spores and increased only as a function of protoplasmic increase. Although there is a shift in the pattern of metabolism of pentoses, the authors suggest that the spores are simply "young" cells divorced from the parent mycelium. Clearly, these spores, and others which are not constitutively dormant, need not differ conspicuously from vegetative cells to be subject to exogenous dormancy. The critical point is whether a developmental arrest can be survived, not whether the organism is in the form of a spore or vegetative mycelium. In fact, the spores of *L. sepiariae* do differ from vegetative cells by being more resistant to desiccation (246), a factor which enhances resistance to a variety of environmental factors (281).

Given this diversity of dormant stages, what are the common elements that characterize them?

Ultrastructural Correlates

The most obvious and usual difference between spores and vegetative stages is in the wall. Although there is wide variation in the size of the wall of spores and in the number of layers, the walls of vegetative stages of fungi usually are thinner and less complex. This is the case in the walls of several species of terrestrial phycomycetes (79, 112, 113, 299, 333), slime molds (203), fungi imperfecti (27, 32, 40, 102, 112), basidiomycetes (116), and ascomycetes (7, 42, 187, 188, 226). On the other hand, the walls of basidiospores of *Lenzites sepiariae* (136), conidia of *Neurosyra* (186), and of the zoospores of some fungi are no thicker than those of vegetative stages and, in some cases, may be thinner (136). Where the difference has been examined in detail, as in the case of ascospores and conidia of *Neurospora*, those stages with the heaviest and most complex walls also are the most resistant and are constitutionally dormant (281). However, this seems not to be invariably the case, for despite their massive wall, ascospores of *Podospora* germinate without a dormant period.

The morphological differences between the walls of spores and vegetative stages are reflected by differences in their chemistry as well. For example, Tokunaga & Bartnicki-Garcia (299) have shown that there is a significant difference in the relative proportions of C-3- and C-4-linked residues of the glucans in the cysts and hyphae of *Phytophthora palmivora*, which might be due in part to changes in the cellulose content. In addition, the branching of these materials (C-3- and C-6-linked residues and corresponding nonreducing terminal units) also differ. Also, a uronic acid-containing layer and one of melanin are present in ascospores of *Neurospora* (187, 188) and not in hyphae. Moreover, although the ratio of S- to R-glucan is 0.4 in ungerminated basidiospores of *Schizophyllum commune*, values of 1.30 and 1.52 are obtained 16 and 30 hours after germination, respectively (2). Other cases are reviewed by Bartnicki-Garcia (18). More recently, it has been shown that 88% of the ^{14}C-glucose incorporated into germinating uredospores of *Puccinia graminis* var. *tritici* went into insoluble products, most of which could be found in the germ tube wall (34).

Proteins in spore coats of *Bacillus* spores have been correlated with morphological structures. Solubility studies (147) have shown the outer layers of *B. megaterium* to consist of alkali-soluble and alkali-insoluble material. In *B. cereus*, *B. coagulans*, and *Clostridium sporogenes* these proteins had high levels of acidic and basic amino acids and precipitated at pH 6.0 as a network of fibrils (99). In *B. megaterium* the alkali-soluble protein was extracted from the inner of two layers of the spore coat; the outer layer and keratin-like inner layer were resistant to alkali (146).

Cytological changes that occur as vegetative cells of bacteria develop endospores have been reviewed (206). Recent morphological findings have extended the general descriptions, which were based mainly on a few species of *Bacillus*, to others as well (22, 166, 205), and to the thermophilic actinomycete *Thermoactinomyces* (58).

The chemical complexity of the exosporium (281) is reflected in a complex ultrastructure as revealed by electron microscopy of thin sections, and after negative staining, and by freeze etching (19, 122, 244, 256). The hair-like nap which occurs on the outer surface of the exosporium (92) may have some medical significance since virulence in *B. anthracis* is correlated with a much reduced nap (160).

Pilus-like appendages have been observed on spores of *Bacillus* (19). Generally these are less substantial than the appendages on the surface of *Clostridium* spores (244, 332). The significance of spore appendages is unclear, but the finding that they remain unchanged upon the germination and outgrowth of *Clostridium* (244) suggests that their role in these processes is not significant.

There are notable differences between the mureins from spores and vegetative cells of *Bacillus subtilis* as well (318). Not only is there a large proportion of the muramic acid residues in the δ-lactam form (giving the effect of two fused rings in spores), but N-acetyl groups and alanine residues are found as substituents in others. The density of cross-linking is low, and because of this and of seryl residues, there is a high concentration of free carboxyl groups, conferring a more polyacidic character on the spore surface. Inasmuch as charge neutralization in flexible polymers results in contraction (217), an increase in ionic strength will elicit contraction which will also be favored by the elasticity resulting from the relative lack of cross-links. In turn, contraction will be accompanied by dehydration, which is related to the temperature resistance that characterizes the sporulating state (281). These data are consistent with the contractile core theory (see below) through which the relatively dehydrated state of the bacterial spore can be explained.

These differences in chemistry are underscored by shifts in the synthesis of wall components during morphogenetic transitions. Thus, Bartnicki-Garcia (18) points to the drastically different qualitative composition of the mass of spores of *Mucor rouxii* wherein a glucan appears de novo during sporogenesis and chitin/chitosan polymers are formed upon germination. Also, Budd et al (33) have shown the accumulation of a peptidoglycan in dormant ascospores of *Neurospora tetrasperma* which probably is a precursor of wall material formed only after germination begins.

Another correlate of the dormant state in fungi is a paucity of endoplasmic reticulum. This has been observed in conidia of *Cunninghamella elegans* (113) and *Botrytis cinerea* (32), ascospores of *Neurospora tetrasperma* (188), basidiospores of *Lenzites sepiariae* (136), and uredospores of *Puccinia graminis tritici* (283). And even when considerable endoplasmic reticulum is present in spores, as in those of the myxomycete *Arcyria cinerea* (203), it is in the form of short segments, the more usual longer ones appearing only after germination. That basidiospores of *Agaricus biporus* (216) and ascospores of *Neurospora tetrasperma* (179) have very little phospholipid, as compared to vegetative stages and conidia respectively, may be accounted for on the basis of the lack of membranes described above. That there is active synthesis of phospholipid during germination has been shown by Langenbach & Knoche (163) for uredospores of *Uromyces phaseoli*. Thus, phosphatidic acid was detected in germinating spores but not in dormant ones, and inasmuch as this substance is a key intermediate in phosphatide synthesis in animal and plant cells as well as in those of protistans, the evidence favors the formation of such lipids and membranes at this developmental stage. Furthermore, germinating uredospores of *U. phaseoli* are able to methylate successively phosphatidyl-ethanolamine to form phosphatidylcholine by using methionine as a methyl donor (164). These authors also were able to show the condensation of choline, ethanolamine, and serine with diglycerides to form phosphatidylcholine, phosphatidylethanolamine, and phosphatidylserine, respectively, through studies of incorporation. This pattern of levels of phospholipids is similar to that shown for conidia of *Aspergillus niger* by Nishi (213). In both these organisms, although the level of phospholipid is high in dormant cells, germination results in rapid turnover and resynthesis of these important membrane components, as well as in a shift in composition.

A well-defined aspect of spore ultrastructure is the presence of lipid bodies which usually disappear upon germination. This has been found in spores of *Arcyria cinerea* (203) and *Dictyostelium discoideum* (56), conidia of *Cunninghamella elegans* (113) and *Aspergillus fumigatus* (40) and in several other species as well. These observations support the generalization that lipid reserves are used during the germination of many, if not most, fungal spores (53, 280), along with carbohydrates.

Although there is at least one case in which a change in the number and size of mitochondria in dormant spores has been related to changes in oxidative enzymes (126, 188), the data in other cases are obscured by difficulties in fixation (136). Nevertheless, there is frequent mention of changes in the number of mitochondria during germination and of changes in shape, even when there is no accompanying change in size (32).

Water Content

A usual concomitant of the dormant state in protistans is a decrease in the proportion of water in the cell (281). Hypotheses like that of the "anhydrous core" assume that, in addition, the water may exist in the "bound" state or in other than the liquid state. Because of the difficulties in determining the exact amount of water in each of its states in living cells and in localizing each type, these hypotheses have been difficult to test. Nevertheless, data have been gathered on the water content of some protistan spores. Black & Gerhardt (23), for example, estimated the water content of dormant spores in aqueous suspension as being 65%, whereas that for vegetative cells was 77%. These figures are in reasonable agreement with calculations made from recent measurements of the dielectric constants of spores and vegetative cells (43). Furthermore, almost all of the water in spores (97%) is freely and rapidly exchanged with D_2O. The external exchange rate changes little, if at all, upon germination and outgrowth (199).

Thus, although the "anhydrous core" hypothesis now seems unlikely in bacterial spores, the hypothesis of a contractile cortex has received recent support. Conductivity measurements of dormant spores (43) reveal that in spite of the high content of ionizable compounds such as dipicolinic acid (223), glutamic acid (211), phosphoglyceric acid (212), sulfolactic acid (25), and inorganic phosphate (212), these compounds apparently are electrostatically bound in nonionized and immobile forms. The reasons for this are not yet clear but, as Carstensen and colleagues speculate (43), a likely mechanism would involve the formation of chelation complexes or networks of such complexes (211). This binding of ions could account for the seeming paradox of low conductivity in the presence of an almost normal content of cellular water. Furthermore, such binding would be expected to effect charge neutralization, which could lead to a more compact and dense structure as discussed above.

A small amount of water is bound extremely tightly in bacterial spores (210). This water, which was between 6–15% of that bound, could not be removed by drying over P_2O_5 at 50°C as was shown by Grecz, Smith & Hoffmann (100). In fact, removal required heating to 110°C in a vacuum. That hygroscopicity is inversely correlated with resistance is suggested by these authors, who also have shown that spores of *Clostridium botulinum* are much more hydrophobic than are the vegetative cells of this organism.

The well-known increase in resistance gained by dehydrated cells (see below) may not be acquired without penalty; several studies attest to increased mutation rates under these circumstances in bacteria and fungi (249, 334). Moreover, many of these mutations have been shown not to be reparable. On the other hand, it also can be argued that these are a source of variation and so might confer selective advantage.

Respiratory Capacity

Reduced respiratory capacity is characteristic of almost all dormant systems, including plants, animals, and protistans (280). This aspect of microbial dormancy is dealt with in detail elsewhere (278, 281), and lists of the respiratory characteristics of dormant stages of other organisms may be found in the book by Altman & Dittmer (8). It has been claimed that uredospores of *Puccinia graminis* are an exception to the rule given above (257). However, the respiration of these spores was studied at only one interval after germination (72 hours), and the more recent data of Maheshwari & Sussman (192) prove that there is an overall increase in respiratory capacity during germination, although the kinetics are complex and can be divided into four stages. In this connection it is pertinent to note that Langenbach & Knoche (164) have correlated changes in phospholipids with the respiratory phases mentioned above in *P. graminis*. Another aspect of the respiration of these spores that has been studied is the effect of the self-inhibitor (4) of uredospore germination. Thus, Farkas & Ledingham (83) found that respiratory rates are inversely proportional to spore concentration and that compounds which counteract self-inhibition prevent the observed decline in respiration. Similar results were obtained by Naito & Tani (208) with uredospores of *Puccinia coronata*. Therefore, these authors concluded that respiratory inhibition is the basis of action of the self-inhibitor. By contrast, Williams & Allen (324) and Bush (36) found that oxygen uptake is independent of spore concentration. Although uredospores respired at a faster rate in low concentrations, according to Maheshwari & Sussman (192), especially in Phase 3, the effects of the inhibitor on respiration and germination are not proportional, inasmuch as the latter is more sensitive to inhibition than the former.

An exception of another kind is germinating macroconidia of *Fusarium solani*, which, when provided with ethanol or mannose as exogenous substrates, show very little increase in respiratory rate. However, increases do occur when glucose, fructose, or trehalose are the substrates (53).

The suggestion has been made that these differences in the respiratory response to germination may relate to the role that particular spores play in nature (278). For example, the thick wall, relative impermeability, and great longevity of ascospores of *Neurospora*, which show a large increase in respiratory capacity upon germination (278), suggest that it functions as a resting spore. Consequently, there is obvious selective advantage in the low rate of metabolism of the dormant spore. This is to be contrasted with the macroconidia of *Fusarium*, which do not need to be activated, depend upon exogenous substrates, and respire at a relatively rapid rate soon after harvest (53).

Although minute amounts of oxygen suffice to permit the germination of some fungi (286), most require frankly aerobic conditions for this process to go to completion (37, 49, 94, 236, 330). It has been found in several cases that not all stages in the germination process are equally susceptible to inhibition by the lack of oxygen as, for example, ascospores of *Neurospora* (94) and spores of *Mucor* (327) and *Rhizopus* (37). In the case of *Neurospora*, ascospores can be reversibly deactivated over several cycles by the imposition of anerobiosis after heat activation (93). As a result of these and similar studies, stages in the germination of several types of spores have been delineated (278).

It has been suggested (280) that the reduced metabolic capacity of dormant spores and their relatively dehydrated condition may be related. Many data show that as dehydration progresses there is a reduction, sometimes proportional, in oxygen consumption. Furthermore, in many of these cases the process is reversible, for rehydration results in the recovery of higher respiratory rates (277). For example, Terui & Mochizuki (296) have shown that the Q_{O_2} of conidia of *Aspergillus niger* is greater than 5 at 100% relative humidity but falls to 1 at 60%. However, the mechanism of this effect is not yet understood, but it is well known that the functioning of mitochondria can be markedly affected by changes in tonicity. In addition, isolated enzymes have been studied in soil and on other surfaces and inhibitory effects of desiccation have been demonstrated (260). But much more remains to be learned about this correlation as well as about the reciprocal one wherein hydration is deleterious to the spores of powdery mildews (29) and rusts (193). At least, in the latter case, osmotic effects have been suggested as being paramount.

The literature on the relationship of oxygen to dormancy and germination in bacterial spores is mixed. Spores of *B. cereus* reportedly had no cytochromes (70), and if respiration occurred at all, oxygen uptake was below the level of detection of the manometric technique used (281). It seems quite certain from the experiments of Desser & Broda (65), who measured the release of $^{14}CO_2$ instead of oxygen uptake as a measure of respiration, that hydrated dormant spores do catabolize endogenous carbon sources to CO_2. The Q_{CO_2} ($\mu l\ CO_2/hr/mg$ spores) for hydrated spores was dependent on temperature, significantly lower than that for vegetative cells and strongly influenced by the thermal history of the spores.

Early failures to observe cytochrome pigments in bacterial spores were due probably to the refractility difference between spores and the suspending medium. Tochikubo (298) has isolated particles from *Bacillus subtilis* spores which apparently have a full complement of cytochrome pigments. Bahnweg & Douthit (11) have shown that spectral profiles typical of cytochromes are demonstrated in dormant spores of *B. cereus* if they are suspended in a medium whose refractive index approximates that of the spores. In dormant spores these spectra suggest that the cytochromes are reduced, and attempts to cause their oxidation have been unsuccessful.

However, it is still far from clear that the cytochrome chain is important in spore germination. The extent to which alanine is converted to pyruvate and reduced pyridine nucleotides during alanine-induced germination is unknown, but the recent finding (258) that veronal, in the undissociated form, noncompetitively inhibits alanine-induced germination suggests that this may be an important aspect of alanine germination. On the other hand, attempts to block germination by inhibitors of cytochrome oxidase, such as cyanide and azide (314), have been unsuccessful. Azide has been reported to increase the rate of the endogenous respiration of dormant spores (65), suggesting that permeation does not enter into the inability of these compounds to block germination.

Thus, while there is a suggestion of the involvement in germination of the first part of the electron transport chain, based on the effect of veronal, the presence of reduced cytochromes in spores and the inability to inhibit germination by compounds which act on cytochrome oxidase argue against the importance of the terminal portion of this chain in the breaking of dormancy in bacterial spores.

Calorimetric techniques have been applied to the study of germinating and dormant conidia of *Actinomyces streptomycini* (143). There is a strong maximum in heat production associated with the emergence of the germ tube, but the swelling of conidia did not result in a marked increase in heat production. Entirely different results were obtained with dormant conidia which showed a "negative thermal phenomenon" whose explanation at present is not clear.

Resistance

Heightened resistance is, of course, a concomitant of dormancy that is ubiquitous. However, this subject will not be reviewed herein, and the reader is referred to the several reviews of this subject that have appeared previously (136).

MACROMOLECULAR SYNTHESIS AND GERMINATION

Evidence and Criteria for the Involvement of Protein Synthesis

There is little doubt now that protein synthesis often is a concomitant of the germination of fungal spores, even though the rate may be very low, as in the case of spores of *Fusarium solani* (50). This subject has been reviewed by Allen (5), Gottlieb (96), Sussman (278), and Van Etten (309), and a number of spores in which protein synthesis is a concomitant of germination are listed in Table 1. In addition, there is evidence that the spores of *Ustilago maydis* (39), uredospores of *Uromyces phaseoli* (302), and conidia of *Glomerella cingulata, Neurospora sitophila* (272), and *Geotrichum candidum* (16) undergo protein synthesis during germination. Among bacterial endospores, those of *Bacillus subtilis* (144, 148, 252) and *B. cereus* (54, 133, 233, 274, 275) also undergo protein synthesis, as do the microcysts of *Myxococcus xanthus* (225). Nevertheless, there is contradictory evidence in the literature as well. Thus, Shu et al (257) and Reisener (227) claimed that there is no net synthesis of proteins in uredospores of the obligate parasite, *Puccinia graminis tritici*, and Staples et al (272) reported similarly for uredospores of *P. helianthi, P. sorgi*, and spores of *Uromyces appendiculatus*. These findings led to the generalization that spores of obligate parasites, like the rusts, are deficient in protein synthesis, in contrast to those of saprophytes. Yet, as was noted above, Trocha & Daly (302) have shown that this is not the case for uredospores of the bean rust *Uromyces phaseoli*, so it is pertinent to discuss, as

Table 1 Variation in the requirement of fungal spores for RNA synthesis as a
concomitant of germination. Protein synthesis is a concomitant
of the germination of all spores listed

Organism	Time required to germi-nate	Size[b]	Requirement for RNA synthesis	Reference
Alternaria solani conidia	60 min	200×15	−	124
Peronospora tabacina conidia	60 min	16×24	−	124
Rhizophylctis rosea zoospores[a]	4–5 hr	3	+(rRNA)	171
Neurospora crassa conidia	8 hr	4–8	+	124
Aspergillus niger conidia	6–7 hr	3	+	131, 215
A. nidulans conidia	4–8 hr	3	+	12, 124, 253
Fusarium solani conidia	8–10 hr	60	+	50
Microsporum gypseum conidia	12–24 hr		+	17, 169
Botryodiplodia theobromae conidia	5 hr	30×13	−	28
Uromyces phaseoli uredospores	3 hr	20×26	−	302
Puccinia graminis tritici uredospores	2–3 hr	18×32	−	75
Lenzites sepiariae spores	4 hr	8×3	+	246

[a] Only small amounts of protein are synthesized.

[b] The sizes given are average values for the ranges given in the literature. When only
one number appears, the spores are spherical or roughly so.

these authors did, the reasons for the contradictory results. These reasons, and
others not discussed by Trocha & Daly, which apply to all aspects of protein
synthesis, are discussed below:

1. The use of inhibitors alone as indicators of the need for protein synthesis
may not be reliable because of the well-known difficulties with permeability and
nonspecific effects of these substances (297).

2. Incorporation of label from exogenously added amino acids may be difficult to evaluate because of uncertainties concerning the contribution of endogenous pools of substrates and the amount of substrates which leach into the
germination medium. For example, Daly et al (59) have shown that carbohydrates
and other materials leach into the growth medium during the germination of
uredospores of *Puccinia graminis tritici*, so these must be taken in account. That

amino acids also are released during the germination of uredospores of the crown rust has been shown (142).

3. Protein may be secreted into the germination medium, thereby raising questions as to the adequacy of recovery of labeled products (59).

4. The level of protein synthesis measured in vitro should be a reasonable approximation of that in vivo, otherwise their relationship is in question.

5. The percentage germination should be high.

6. Breakage of spores and recovery should be good enough to validate measurement of enzyme, protein, or nucleic acid extraction (311).

7. Determination of spore proteins may be inadequate because of the possible presence of trichloroacetic acid-soluble proteins in some spores (328). These workers have found that as much as 50 mg of protein nitrogen per gram of ungerminated spores may occur in this form, bound to glucans and mannans, perhaps as a reservoir of storage carbohydrate. Consequently, much protein might escape detection during the usual washing procedures which employ hot water or ethanol.

8. As Trocha & Daly (302) suggest, an appreciable amount of glycoprotein may be bound to the walls of spores and other developmental stages of fungi. That enzymatic activity is found associated with spore surfaces has been shown by several workers (121, 195) so, especially when wall enzyme synthesis is derepressed, a significant fraction of protein may be undetermined.

9. The occurrence of amino acids in macromolecules other than proteins may give misleading results when incorporation into spore residues is measured. Thus, it is possible that at certain developmental stages, like spore germination, significant amounts of "charged" tRNA exist.

10. The specificity of the usual means of estimating protein is inadequate to exclude certain nonproteinaceous components from being included in the analysis. For example, Johnson et al (140) showed that about 50% of the biuret-positive material found in tissues infected by a fungal parasite was not protein. Similar difficulties exist for the Kjeldahl, Lowry, and ninhydrin procedures. An alternative approach involving the hydrolysis of cellular residues by acids, followed by analysis of amino acids (151, 227, 257, 272), is promising but, as has been pointed out before (302), the hexosamines of the walls of spores and vegetative stages often are not considered, thereby leaving some uncertainty in the analysis. The use of enzymatic hydrolysis overcomes some of these difficulties but is harder to employ because of the introduction of contaminant amino acids.

11. The preparation of the spores themselves can lead to contradictory results. For example, their age, degree of hydration (278), treatment before use (193), and whether they have been washed free of certain nutrients (52) can affect the data. Moreover, mere suspension in medium can cause germination, so some workers have found it necessary to use dry spores to ensure that the dormant state is preserved (200).

12. The release of hydrolytic enzymes such as proteases and RNAases (50, 123) should be taken into account where macromolecular syntheses are being studied.

13. If the use of exogenously provided amino acids reveals no incorporation

into proteins, these precursors may be provided by endogenous pools or through the turnover of proteins (262).

14. When in vitro systems must be incubated for long intervals, the possibility of contamination must always be considered and means found to overcome the difficulty, such as the use of antibiotics which have no effect upon the system (125).

The careful study by Trocha & Daly (302) suggests that, at least in the case of rust spores, the generalization that obligate parasites are unable to synthesize protein during germination does not hold. This conclusion is further supported by the data of Staples and his group (269–271), which demonstrate the existence of a functional apparatus for protein synthesis in uredospores of *Uromyces phaseoli* in vitro, and of Dunkle et al (75), which show that uredospores of *Puccinia graminis tritici* fail to germinate in the presence of inhibitors of protein synthesis but did so when inhibitors of RNA synthesis are present. The use of labeled inhibitors to overcome the objection that the ineffective inhibitors did not penetrate would help to strengthen this argument. Another obligate parasite which has been shown to be capable of protein synthesis during germination is *Peronospora tabacina* (123, 125), but the rate is low.

An interesting case where no protein synthesis seems to be required is the formation of chlamydospores from germinated macrospores of *Fusarium solani*. Thus, it has been shown that chlamydospores can be formed in the presence of enough cycloheximide to inhibit protein synthesis, as measured by the incorporation of leucine-U-^{14}C. By contrast, ungerminated macrospores are prevented from forming chlamydospores under the same conditions, thereby reinforcing the notion that dormant cells must synthesize protein before development can be consummated.

The criteria used for deciding whether protein synthesis occurs during spore germination have been very diverse. Thus, the effect of inhibitors (75), incorporation of radioactive markers into protein (302), induced formation of enzymes (39), net increase in protein (227), and the presence of a functional ribosomal apparatus (269) have been applied separately and in combination as being diagnostic of such synthesis. Clearly, the use of inhibitors in the absence of concomitant studies on their effect upon incorporation is insufficient, and even were such studies done, side effects of the poison may confound the data. As for the use of induced enzyme formation alone, de novo synthesis is only one possible means whereby increased activity can appear, for "activation" of a precursor, as in the case of the trypsinogen-trypsin transformation, or removal of an inhibitor or degradative enzymes are alternative explanations. Even the presence of a functional synthetic system in vitro does not mean necessarily that it functions in vivo, because inhibitors may be present in dormant spores (10, 176, 247) which interfere with protein synthesis in the cell. And limitations in the resolving power of the existing techniques also must be considered inasmuch as the synthesis of a single key protein may be essential for development, yet it may be so small a part of the total protein that its synthesis would not be perceptible on the basis of the usual measurements. Even incorporation of label may be subject to errors of this kind, as is illustrated by the contradiction between the data of Lovett (185), who found no incorporation during the germination of zoospores of *Blastocladiella*, and those of Soll & Sonneborn (262, 263), who found a little. As the latter have pointed out, when rates of incorporation are low, as in certain pathogens (50, 123), contamination by cells of other stages in the life history of the same organism must be considered. Therefore, synchronous systems are advantageous, but the degree of synchrony has only rarely been determined in germinating fungal spores, although much more commonly in those of bacteria. Another possibility is that a small proportion of the cells under study incorporate at a normal rate, whereas none of the others do at all.

In at least one case which exhibits reasonable synchrony, that of germinating spores of *Bacillus cereus*, an interesting case has been made (54) for the synthesis of a protein, or proteins, which partially controls transcription during the germination or postgermination period. The synthesis of this protein(s) appears to be necessary for a greatly increased rate of RNA synthesis which is characteristic of this time in the life cycle. Although the nature of the protein(s) is unknown it has been assumed to be an RNA polymerase molecule, or part thereof, and thus would be unlikely to be noticed in assays of radioactive amino acid incorporation into total protein.

Sequence of Syntheses of Macromolecules Involved in Protein Synthesis

The sequence of syntheses of macromolecules undertaken by bacterial spores upon germination begins with RNA synthesis (transcription), and is followed by the formation of proteins (translation), and finally DNA replication (105, 155).

A similar sequence appears to be followed during the germination of zoospores of *Blastocladiella emersonii* (185), conidia of *Aspergillus niger* (329), *A. oryzae* (287), and *A. nidulans* (12), and the spores of *Fusarium solani* (50) and *Rhizopus stolonifer* (Brambl & Van Etten, personal communication). By contrast, some of the spores listed in Table 1 do not undergo RNA synthesis during germination but do synthesize protein, so the sequence in these cases probably is translation——→transcription——→DNA synthesis. These include conidia of *Peronospora tabacina, Alternaria solani* (124), and *Botryodiplodia theobromae* (28), and uredospores of the rusts *Uromyces phaseoli* (302), *Puccinia graminis tritici* (75, 123, 124), and *P. helianthi* (251). Other instances where the sequence of synthesis begins with translation include spores of *Geotrichium candidum* (16) and *Dictyostelium discoideum* (10). On the other hand, proflavine inhibited the germination of conidia of *Neurospora crassa* and *Aspergillus nidulans* at concentrations inhibiting RNA synthesis as well, suggesting that such synthesis is required for the germination of these fungi.

The data in Table 1 lead to a question as to why some spores require RNA synthesis for germination whereas others do not. It seems clear from the table that size alone cannot be the determinant. On the other hand, there is a fair correlation with the time required for germination; those spores germinating most rapidly do not require RNA synthesis during this time. In addition, all of these are spores of parasites but so are those of *Fusarium solani* and *Microsporum gyp-*

seum, which do undergo RNA synthesis during germination, so further data are required.

Is Protein Synthesis a Necessary Concomitant of Germination?

One of the most influential generalizations in developmental biology has suggested that differentiation is linked with differential gene action; thus, morphological change has been assumed to be determined by the synthesis of proteins which are specific and necessary to certain stages in development. And inasmuch as dormant spores upon germination differentiate by forming a new stage in the developmental process, differential gene activation has been assumed to be part of, or to control, the process. In fact, as the data reviewed above show, numerous examples support this hypothesis for fungal and bacterial spores (232, 274, 275), seeds (198), and animal systems (295). But is protein synthesis a *necessary* concomitant of spore germination, and if it is for some, can it be assumed to be so for all?

The data reviewed above provide abundant examples of fungal spore systems wherein protein synthesis occurs during germination and which contain a functional apparatus for this process. But, as was noted above, the mere presence of the apparatus and the fact that synthesis occurs under the usual conditions of germination does not establish that such synthesis must occur. The means whereby this question usually has been answered is through the use of inhibitors like cycloheximide for eucaryotes and chloramphenicol for procaryotes. In fact, there are only a few critical studies of this kind using fungal spores, including that of Hollomon (123), who found that cycloheximide inhibited both germination and protein synthesis. This is in contrast to the results with inhibitors of DNA and RNA synthesis which, although very effective in stopping the incorporation of precursors, failed to inhibit germination. The failure of other inhibitors of protein synthesis like puromycin, *p*-fluorophenylalanine, canavanine, and ethionine to inhibit germination, even when used at high levels (160–200 μg/ml), also was noted, but no data on their effect upon incorporation were reported. These results illustrate one of the difficulties in such experimentation, namely the possible failure of potential inhibitors to penetrate. That permeability barriers are a real deterrent to the study of the effect of inhibitors and other substances, especially in experiments with some fungal spores, is suggested in several studies (281). Another report in which inhibitors of protein synthesis have been shown to arrest germination is that of Dunkle et al (75) on uredospores of *Puccinia graminis tritici*. However, the effect of the poisons used upon incorporation was not determined, although effects upon subsequent developmental stages were observed.

The study with bacterial spores mentioned above (54) suggests that protein synthesis is necessary during some early stage in the germination-outgrowth sequence. The fact that these spores also possess polyribosomes (73, 84) supports this hypothesis. However, spores of *B. cereus* have a "defective" protein synthesizing system which is still only partially corrected by the end of germination (137, 155). This would argue that germination probably does not require ribosome-mediated peptide synthesis, a supposition that is strengthened by the experiments of Kanamitsu (144), who investigated spores of a strain of *B. subtilis* which had a temperature-sensitive defect in ribosome function. At the nonpermissive temperature these spores germinated normally but did not undergo outgrowth, suggesting that it is not germination but outgrowth for which protein synthesis is required.

However, Khan et al (148) reported that spores of a streptomycin-dependent strain of *B. subtilis* germinated very poorly unless supplied with streptomycin. Although these authors did not speculate on the relationship of protein synthesis to these processes, these results are consistent with a role for ribosomes in germination.

By contrast, it is believed by some that protein synthesis is not an essential prerequisite for germination, as in the case of zoospores of *Blastocladiella emersonii* (41, 262, 263). These workers have suggested that most of the structural events taking place during the germination of these cells are controlled by mechanisms other than protein synthesis. These events, which have been studied in detail by Cantino and co-workers (207, 261), reveal that the following changes occur during germination: retraction of the flagellum; formation of a cell wall and a spherical shape; development of adhesiveness and resistance to certain agents which can lyse nongerminating zoospores; disappearance of the nuclear cap of ribosomes and their dispersal into the cytoplasm; extensive alteration of the shape of the single mitochondrion; disappearance of gamma particles [one of the three types of DNA found in these cells (41)] and of the flagellar axoneme; and the formation of the germ tube. Of these extensive changes, only the disappearance of the flagellar axoneme and the formation of the germ tube fail to occur after treatment with 0.1 μg/ml cycloheximide, so almost all of these developmental events can proceed with little or no protein synthesis (262). These manifold events proceed in the presence of an amount of inhibitor which arrests all but about 20% of the low rate of incorporation by germinating cells, which is only about 1/1000 that of germinated ones. Thus, cycloheximide inhibits both amino acid incorporation from exogenously added ^{14}C-leucine or ^{3}H-phenylalanine as well as the completion of germination, but it does not prevent the transformation of zoospores to round cells which is accompanied by the changes listed above. Moreover, even when zoospores have been pretreated with cycloheximide, the kinetics of formation of round cells are unaffected. That endogenous sources of precursors are not being used in the synthesis of protein is suggested by the small size of the amino acid pools that are found prior to protein synthesis. Furthermore, no extensive protein breakdown occurs until *after* protein synthesis begins and turnover itself requires such synthesis.

Consequently, Soll & Sonneborn (262, 263) suggest that only "rearrangements in preexisting structures" are needed to permit the development of zoospores of *B. emersonii* to the round-cell stage. The rearrangements they mention include "assembly phenomena," amplification, and rapid physiological changes, none of which require concurrent protein synthesis. Assembly phenomena include situations where the self-assembly of polypeptide or protein monomers forms isozymes through random collisions and the union of concurrent products of translation or more complex systems like the cortical structures of *Paramecium* (266), wherein existing molecular assemblages participate in, and are essential for, other steps in assembly. Sonneborn (266) describes these systems in detail and discusses the role of allosteric transitions, homonucleation, and heteronucleation. Other protozoa demonstrate similar mechanisms, as do viruses and bacteria, the latter forming flagella by self-assembly. Other post-translational changes have been described for cytochrome *c* (87) and aldolase (162), both of which are altered in primary structure by deamidation of asparagine and glutamine residues, thereby yielding new subunits. Also, one of the subunits of bovine lens polymers is not the product of direct genetic translation but, like the above enzymes, is formed by slow post-translational changes from another subunit (63).

Amplification mechanisms are processes whereby a "trigger" or signal elicits a large response in the receptor organism. Such mechanisms exist at the level of the gene (31), but others are known to involve cellular structures like membranes rather than direct gene function (139). It is possible that the γ-particles of zoospores of *Blastocladiella emersonii* represent such a mechanism, for they are thought to release chitin synthetase for wall synthesis (303), an essential aspect of differentiation in this organism. Another such device may be the nuclear cap of this organism, which may help to suppress protein synthesis until its dispersal when development proceeds (63).

Finally, rapid physiological changes provide another means of effecting developmental changes in the absence of alterations in the regulation of gene function. Such phenomena as muscle contraction (135), changes in the form and size of mitochondria (109, 222), and the movements of microfilaments (322) fit in this category. Also, allosteric changes in proteins have been discussed by Koshland & Neet (159) in connection with development. An instance of a rapid change that is connected to the breaking of dormancy in spores of *Phycomyces blakesleeanus* is that involving trehalase, whose activity increases ten to fifteenfold immediately after the heat treatment that is required for activation of the spores (307). Soon after the heat treatment the enhanced activity decays and the increase and decline can be repeated several times (Figure 1). This remarkable cyclical phenomenon is not inhibited by cycloheximide, so protein synthesis and its arrest cannot explain the data. Other examples of heat-activated enzymes have been reported (284), as has regulation of activity by solubility (161), metal ions (13), and hormones (64).

These data lead to the conclusion that differentiation, and spore germination as an example of this process, need not require concomitant protein synthesis. In the case of some organisms like the protozoans studied by Sonneborn (266), "these differences depend not on differences in the kinds or amounts of macromolecules present, but on self-perpetuating differences in the location, orientation and arrangement of molecular groupings," Or the various other means through which preformed peptides and proteins are rearranged, aggregated, or otherwise modified to produce active enzymes could lead to developmental changes of wide scope, as suggested above. But it is legitimate to ask, as Ursprung (306) has done, whether multicellular organisms use mechanisms of differentiation found in uni-

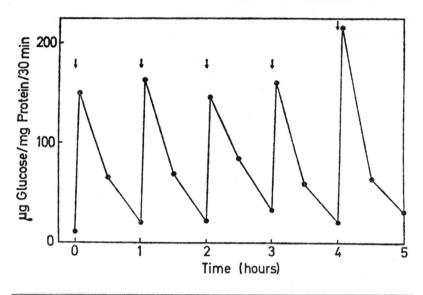

Figure 1   Increase and decay in trehalase activity in spores of *Phycomyces blakesleeanus* following repeated heat treatments at 50°C for 3 min (arrows), interrupted by periods of 1 hour at 18°C. Data of Van Assche, Carlier & Dekersmaeker (307).

cellular ones, because the development of differences *between* cells has permitted the former to exploit functional nuclear differences during differentiation. That multicellular organisms may employ these mechanisms is suggested by the early stages of sea urchin development, which proceed in the absence of protein synthesis (81, 82), and in other cases including *Stenostomum*, a turbellarian worm (266). But these conclusions should not be construed to mean that protein synthesis is not required at all by these systems, for it is only the time when these syntheses take place that is in question, not whether they occur. Also a nagging difficulty in all of these cases must be kept in mind for, as in the case of *Blastocladiella emersonii* (262), not all protein synthesis is inhibited by cycloheximide, and it must still be considered possible that the synthesis of a small amount of a key protein could be responsible for triggering developmental events in these spores and perhaps in others as well.

Finally, Barbara Wright and co-workers (103, 149) have called attention to the danger of using the results of inhibitor studies alone to conclude "that genes and messengers are sequentially transcribed and translated according to a developmental program" (103). They have shown that some enzymes like trehalose-6-phosphate synthetase in *Dictyostelium discoideum* may show "in vitro unmasking" such that their activity does not become manifest until special methods are applied. Among the possibilities considered to explain these effects is the presence of inhibitors, one of which has been shown to exist extracellularly in the case of trehalose-6-phosphate synthetase in *D. discoideum* (149).

Controls over Protein Synthesis

Although considerable attention has been devoted above to the possibility that protein synthesis need not be essential to the germination of some spores, control over the germination of others may be exerted by this process. Thus, Horikoshi et al (132) claim that the apparatus for protein synthesis is not complete in conidia of *Aspergillus oryzae* until 60 min after the breakage of dormancy, and a restriction on net protein synthesis has been noted by LéJohn & Lovett (171) in zoospores of *Rhizophlyctis rosea* as well. In addition, a 4-hour lag in protein synthesis has been observed in lyophilized spores (167), and washing appears to speed the rate of incorporation of label into protein in vitro in preparations from conidia of *Peronospora tabacina* (125) and *Fusarium solani* (50). Therefore, possible loci for such control will be examined using a modification of the outline suggested by Van Etten (309).

DOES DNA CHANGE DURING GERMINATION?   That replication of DNA is one of the last events during germination has been discussed above. Moreover, no qualitative changes in DNA have been detected during germination in the two species of this macromolecule found in spores of *Botryodiplodia theobromae* and in the single one of *Rhizopus stolonifer* (76). Experiments with *B. theobromae* showed that labeling of nuclear and mitochondrial DNA occurred at the same time during germination and that both components are synthesized during the rest of the germination period. In addition, it has been shown that mitochondrial DNA synthesis is not required for germination but is for vegetative growth to occur (75). Nor were any changes detected in the sedimentation and electrophoretic patterns or base composition of the DNA of conidia of *Aspergillus oryzae* (157, 288) during their germination. On the other hand, Myers & Cantino (207) have described the disappearance of the γ-particles of zoospores of *Blastocladiella emersonii* during encystment. Thus, one of the four species of DNA found in these cells undergoes changes as a result of this developmental process.

Doi (68) recently has reviewed the state of DNA in spores and vegetative cells of the sporulating bacteria. Papers that have appeared since then reiterate the likelihood that the primary sequence of bases and base-pairing of the DNA in spores is identical to those in vegetative cells (80, 242). Reports that DNA as extracted sometimes exists in an altered state, which shows up in physical parameters like buoyant density or melting temperature, are occasionally still seen (285), and in at least one case (80), that of DNA from spores of *Streptomyces venezuelae*, the cause of the shift in both of these parameters is known. Thus, a pigment extractable from the surface of these spores with ethanol and acetone binds tenaciously to the DNA, apparently causing a decrease in the buoyant density and an increase in the melting temperature. It can be removed from the DNA duplexes by treatment with high salt concentrations.

Considerable evidence exists supporting the idea that DNA in spores exists in an environment which renders it less sensitive to radiation (68). Experiments that have been published in the past 5 years have strengthened this hypothesis and to a certain extent have narrowed the list of spore structures that contribute to the "spore-specific state" of DNA. Utilizing the technique of transforming spores into osmotically sensitive spheroplasts, these experiments indicate that even in the absence of the cortex DNA is still less sensitive to ionizing (291) and ultraviolet (289) irradiation, less sensitive to strand breakage (292), and nonphotoreactivable (290). Upon lysis of these spheroplasts, DNA is converted into normally sensitive material, reflecting the change ordinarily encountered in the transformation from the dormant to the vegetative state. Sakakibara & Ikeda (241) have presented evidence that osmotically sensitive spore spheroplasts are similar to unaltered dormant spores in other ways as well. Thus, although they will not form colonies, they do respond to germinants in a way similar to unaltered spores. Macromolecular synthesis commences after treatment with alanine, and the order of appearance of RNA, protein, and DNA is the same as that seen after germination of dormant spores (68). As Sakakibara & Ikeda (241) point out, at least these properties of the dormant state do not derive from the presence of the cortex, which is mostly removed in the spheroplasting steps, but are due to some arrangement of chemicals and structures internal to it. Perhaps the analysis of radiosensitive mutants like those reported by Munakata & Ikeda (204), which are much like the vegetative cells from which they were produced, will help to clarify this aspect of dormancy.

As bacterial spores germinate and undergo outgrowth, their DNA undergoes remarkable transitions in sensitivity to irradiation. Stafford & Donnellan (268) described a peak in resistance to ultraviolet light in spores of *Bacillus subtilis* approximately 3 min after the initiation of germination. They correlated this decrease in radiation sensitivity with a decrease in thymine-containing photoproducts of the DNA. More recently, Nagatsu & Matsuyama (209) have described similar findings with spores of *Streptomyces cacaoi*. In these experiments, the increase in resistance to uv light is correlated in time with an increased sensitivity to gamma irradiation.

The assumption has been made that the differences in response of spores to irradiation during germination and outgrowth is evidence that the DNA encounters different environments at these times (268). That this is undoubtedly the case seems clear, but there are alternative explanations for the phenomenal recovery of bacterial spores damaged by irradiation. For example, Sonenshein (264), in studying the fate of phage DNA trapped within spores of *B. subtilis*, was led to the conclusion that a powerful repair mechanism comes into play during the germination period at a time when replication of the genome has not yet begun. Thus, the situation may be like that encountered in bacteria when irradiation is followed by exposure to nongrowing conditions. However, there may be differences among bacteria in the repair mechanism, as Terano et al (294) have shown, and other of their experiments with spores of *B. subtilis* (293) reveal that the period of most importance to the repair of damaged DNA is the initial stage of germination, which they were able to block by treatment of the spores with reducing agents. Since repair occurred in their system in the absence of DNA synthesis, they believed that repair was not of the excision type. However, as mentioned above, impermeability to radioactive DNA precursors may have led to this conclusion.

IS DNA IMMEDIATELY TRANSCRIBABLE IN DORMANT SPORES? Several studies of fungal spores reveal that synthesis of some types of RNA begins immediately in dormant conidia, such as those of *Aspergillus oryzae* (288, 304), and basidiospores of *Lenzites sepiariae* (246). When the types of RNA are analyzed, it is sometimes found that rRNA and tRNA are formed early in germination, as in the case of conidia of *Aspergillus oryzae* (215) and *Neurospora crassa* (119, 124). By contrast, pulse-labeling experiments revealed that although tRNA is formed in germinating spores of *Peronospora tabacina*, an unstable heterodisperse RNA appeared which is not rRNA (124). In addition, as will be discussed below, there is some variation in the nature of the mRNAs of dormant and germinating spores, as well as in the time when they are synthesized. So in the fungi at least there is considerable diversity in the time when the different RNAs are synthesized during germination.

A number of studies have shown that the synthesis of RNA commences very shortly after the initiation of germination of bacterial spores (217). For example, recent work reported by Setlow & Kornberg (250) shows that RNA synthesis is under way 2 min after the initiation of germination. Also, the finding (259) that low concentrations of 6-azauracil, an inhibitor of the synthesis of uridylic acid, blocks spore development after the depolymerization stage of germination, argues for the immediacy of RNA synthesis after germination. Further, exogenous nutrients appear to be unnecessary for the synthesis of these early RNA molecules (250), suggesting that the testing for early RNA synthesis by incorporation of exogenously added radioactive precursors should be viewed with caution. It seems likely, therefore, that in bacterial spores DNA is transcribed immediately upon germination.

DOES THE RNA POLYMERASE(S) IN DORMANT SPORES PERMIT THE SYNTHESIS OF THE VARIOUS TYPES OF RNA IN THE PRESENCE OF TRANSCRIBABLE DNA? Regrettably, only two detailed studies of isolated RNA polymerases of fungal spores were available to us, and one revealed that zoospores of *Blastocladiella emersonii* have at least three such enzymes (130). Their properties are much like those from other eucaryotes, and no general or specific inhibitors of these polymerases were found, nor were the ungerminated spores deficient in any of these. The other work shows that there are both qualitative and quantitative changes in the RNA polymerases of *Rhizopus stolonifer* as a function of germination (95). As in the case of zoospores of *B. emersonii*, at least three RNA polymerases are found in germinated spores of *R. stolonifer*, but ungerminated spores show only two such enzymes. In addition, the properties of RNA polymerase I from ungerminated spores differ from those of the equivalent enzyme in germinated spores in its response to Mn^{++}

and Mg^{++} and in its elution position of DEAE-cellulose columns. It is not yet known whether the RNA polymerase I of ungerminated spores is converted to the analogous enzyme in germinated spores or whether the latter is synthesized de novo.

The changes that occur in RNA polymerase during sporulation of bacteria have been reviewed recently (61). There were early suggestions of participation of RNA polymerase in this form of cellular differentiation when Kerjean et al (147a) reported that vegetative cells have two such enzymes whereas spores have only one, based upon heat-inactivation and sedimentation velocity studies. Recently the specificity of RNA polymerase for DNA templates has been shown to change during sporulation (182), and this apparently occurs concomitantly with alterations in the molecular weight of a subunit of the holoenzyme (183, 194). Evidence has accumulated implicating proteases (168, 202) in this alteration in molecular weight, and there are well-documented genetic links between protease production and sporulation (197, 202). Furthermore, there is evidence for the production of proteins typical of spores by means of proteolytic action on vegetative cell proteins (240). Thus, there is support for the hypothesis that the structural alteration of RNA polymerase is important in differential gene action in this system.

If this model is correct, then it would seem reasonable that the RNA polymerase present in sporulating cells would also be packaged in the spore. When the spore then germinates and begins to undergo outgrowth, one of the first requirements would be the alteration of its RNA polymerase to achieve the specificity characteristic of vegetative cells. Cohen et al (55) did a comparative study of the RNA polymerase found in vegetative cells and spores of *B. cereus* and found the two to be different with respect to their ability to transcribe DNA from bacteriophage CP51. This is a property of the vegetative cell polymerase which is lost during sporulation. The ability to transcribe this DNA is regarded as one of the earliest events after germination and so is consistent with the hypothesis stated above.

Many of these experiments were conducted on the RNA polymerase taken from mutants resistant to rifampin, which fail to sporulate and to degrade their polymerase (183, 265). However, Haworth (114) has isolated a number of RNA polymerase mutants of *B. subtilis* which are resistant to rifampin, streptovaricin, or streptolytigin. Although altered in their RNA polymerase, mutants resistant to the first two of these antibiotics were not affected in their ability to sporulate. The mutants resistant to streptolytigin not only had a reduced efficiency of sporulation but were altered in generation time as well. While these findings are not inconsistent with the hypothesis discussed above, they do indicate that the phenomenon of commitment to sporulation is not yet fully explained.

DOES THE LACK OF mRNA IN UNGERMINATED SPORES CONTROL PROTEIN SYNTHESIS DURING GERMINATION? The presence of stable mRNA seems to be established in several fungal spores, including zoospores of *Blastocladiella emersonii* (185), conidia of *Peronospora tabacina* (123), *Fusarium solani* (50), *Botryodiplodia theo-*

bromae (28), and *Microsporum gypseum* (16), and in uredospores of *Uromyces phaseoli* (331) and *Puccinia graminis tritici* (75). In all of these except the last, direct tests of the ability of RNA fractions to support protein synthesis were undertaken.

Evidence from electron microscopy alone, wherein the appearance of poly-ribosomes is taken to indicate the presence of mRNA, is less convincing because of the absence of data on whether the mRNA actually is functional. Such evidence has been used to supplement other work on the presence of stable mRNA in uredospores of *Uromyces phaseoli* (271), but is the only support for the assertion that it occurs in ungerminated basidiospores of *Lenzites sepiaria* (136).

On the other hand, dormant ascospores and conidia of *Neurospora crassa* contain only monoribosomes, but germination results in the formation of poly-ribosomes (119). After studying the sedimentation coefficients and nucleotide composition of the ribosomes at different stages, these authors (117) concluded that the ungerminated spores contained tRNA and rRNA but lack mRNA. Furthermore, dormant conidia of *Aspergillus oryzae* (131, 215) form only 80 S ribosomes, but polyribosomes appear 60 min after the breaking of dormancy, and similar conclusions have been reached for lyophilized basidiospores of *Schizophyllum commune* (167), conidia of *Fusarium solani* (224), and spores of *Dictyostelium purpureum* (85). The lack of mRNA also has been postulated from electron micrographs of the nuclear cap of *Blastocladiella emersonii* and dormant conidia of *Penicillium chrysogenum* (158).

Therefore, there are differences among fungal spores with respect to the presence of stable mRNA. Moreover, these differences do not seem to be due to the methods used, although it must be emphasized that the amount Mg^{++} used, for example, often determines whether functional polyribosomes will be recovered. Also, as Van Etten (309) has suggested, mRNA might be stored in spores in places other than on polyribosomes, but no evidence has yet been presented to confirm or deny this possibility.

The absence of mRNA in bacterial spores long has been assumed from experiments such as those of Kobayashi et al (156), which showed that protein synthesis during germination and outgrowth of *B. cereus* is sensitive to actinomycin D. This and the absence from spores of sedimentation profiles typical of polysomes (137) have contributed to the widely held assumption that bacterial spores are devoid of mRNA (156). However, it has been shown recently and independently by Feinsod et al (73, 84) and by Kobayashi (153), using different methods of breaking the spores, that polysomes can be isolated from dormant bacterial spores. Difficulties in spore breakage and problems with such things as germination during ribosome extraction (45, 137) probably have contributed to the earlier negative results. Presently it appears, at least in *B. cereus* and *B. megaterium* and perhaps in *B. subtilis* (73), that a sizable proportion of the ribosomes packaged in bacterial spores are bound in the form of polysomes. Thus, at least by inference, mRNA is present in these spores. Whether these messages are remnants of those present during sporulation or become translated during germination and outgrowth remains to be seen.

ARE THE VARIOUS tRNAS PRESENT AND FUNCTIONAL IN DORMANT SPORES AND ARE THEY IDENTICAL TO THOSE IN GERMINATING SPORES? That the sedimentation characteristics and nucleotide composition of the unfractionated tRNA found in conidia, ascospores, and hyphae of *Neurospora crassa* are similar has been shown (117, 118). This is also the case for conidia of *Aspergillus oryzae*, for which it has been shown that the sedimentation coefficients, nucleotide content, and thermal denaturation profiles of the unfractionated tRNA from dormant and germinated spores are similar (288). Although the biological activity of both preparations was low when assayed with a mixture of amino acids, they were of the same order of magnitude. It is not known if these tRNA preparations are active with more than one amino acid.

On the other hand, it has been suggested that the RNA in ungerminated teliospores of *Ustilago maydis* is complexed with a protein in an inactive form which is activated during germination (97). Thus, these workers were unable to isolate tRNA from ungerminated spores, and the total RNA fraction from these is bound to particles which do not adhere to a methylated kieselghur column and showed an unusual ultraviolet absorption spectrum. But tRNA fractions of the usual kind are obtainable from germinated spores, and a normal rRNA preparation is obtained from dormant spores if sodium chloride or the soluble fraction of germinated spores is used during extraction. However, recent work suggested that the profile of RNAs that can be obtained from these spores is identical to that in germinating ones so that the significance of these observations is in question (174).

The possibility that changes in the isoaccepting species of tRNA might play a significant role in morphogenesis has led to the isolation and characterization of these in a few fungus spores. For example, among the 13 amino acids examined, a qualitative change in the isoaccepting species of tRNA for lysine and a quantitative change in those for methionine have been reported for the germination of conidia of *Aspergillus oryzae* (132). Moreover, it has recently been demonstrated that tRNA methylase activities from ungerminated conidia of *Neurospora crassa* are lower than those from germinated ones (326). Although some methylation of conidial tRNA was catalyzed by enzymes from germinated spores, these same enzyme preparations had no activity on the tRNA preparations from hyphae of *Neurospora crassa*.

But the most detailed study of tRNA in fungal spores has been performed by Van Etten and his group (200). Thus, 10 aminoacyl-tRNAs were prepared with tRNA and aminoacyl-tRNA synthetases from ungerminated and germinated spores of *Rhizopus stolonifer*. Then the differences between these aminoacyl-tRNAs were analyzed by cochromatography on benzoylated-diethylaminoethyl cellulose columns. Although no significant differences during germination were detected in the case of 7 of the amino acids examined, quantitative changes in one of the two isoaccepting species of lysyl-tRNA and valyl-tRNA were observed, as well as a qualitative change in isoleucyl-tRNA. These workers also showed that tRNAs for at least 18 amino acids are present in ungerminated spores. That the acceptor activity of the tRNA from ungerminated spores of *R. stolonifer* is greater than that from germinated ones also has been shown (200), but the significance of this finding cannot be evaluated until each of the separate molecules is characterized and its properties studied.

Therefore, there are at least two cases known where qualitative as well as quantitative differences exist in the isoaccepting tRNAs of dormant and germinating fungal spores. It will be of interest to learn whether these differences are due to the increased synthesis of particular isoaccepting species of tRNA upon germination or to modification of preexisting species. As Merlo et al (200) suggest, it is still not clear whether these changes are a cause or a result of spore germination, and much remains to be done to clarify this issue. For example, does the change in a particular isoaccepting species reflect a change in a particular codon if different isoaccepting species have different coding specificities? Furthermore, it is to be expected that more isoaccepting tRNAs exist than have been found in species studied so far and that the function of some of these should include the initiation of translation. Given these numerous different compounds, many opportunities for control of protein synthesis exist through their lack or differences in specificity, but few studies of fungi have been detailed enough to explore these possibilities.

As is the case with the fungi, alterations in the species of transfer RNA have been documented during sporulation-germination cycles in the bacteria. Doi and associates (71, 145) and Lazzarini (165) were among the first to report on the nature of tRNA in sporulating bacilli. The major initial finding was of a quantitative shift in isoaccepting tRNA for valine as cells undergo the shift from vegetative growth to sporulation (72). An extensive recent study has been published by Vold & Minatogawa (315), who compared tRNA from vegetative cells and spores which had been broken in one of two ways: in their "physical" disruption technique, spores were disrupted in a Braun cell homogenizer with glass beads added; a "nonphysical" method involved reduction of the spores with dithiothreitol and treatment with lysozyme and sodium laural sulfate to cause lysis. Transfer RNA preparations made from spores broken by these methods were not identical. For example, saturation curves for aminoacyl-tRNA synthetase were similar whether the tRNA was from vegetative cells or "physically" disrupted spores. Preparations from "nonphysically" disrupted spores, however, required greater than tenfold more of the synthetase preparation for saturation. This difference in the properties of tRNA from spores could not be attributed to an easily removed substance which might alter the configuration of the tRNA molecules, nor were the authors sure which of the two forms exists in the spore.

These authors (315) also presented elution profiles for lysyl and leucyl-tRNA species from spores which were different from those of the aminoacyl-tRNA isolated from vegetative cells. The profiles were obtained from reverse-phase chromatography (RPC-5), and the resolution was excellent. The authors add that when 19 aminoacyl-tRNAs were investigated, they fell into three categories: (a) no differences between spores and vegetative cells; (b) a difference in the ratio of existing peaks of spores and vegetative cells; and (c) the appearance and disappearance of unique peaks. Only four of the aminoacyl-tRNAs fell into the third category: lysyl, glutamyl, tryptophanyl, and glycyl-tRNAs. These elution profiles were independent of the method of spore breakage. Although it continues to be interesting, the meaning of the differences in elution patterns of isoaccepting tRNA species of spores and vegetative cells remains obscure. However, the implied participation of these molecules in the transition from spore to vegetative cell is clear and clarification will be important.

ARE THE REQUISITE AMINOACYL-tRNA SYNTHETASES AND TRANSFERASES PRESENT AND FUNCTIONAL IN DORMANT SPORES? The synthesis of polyphenylalanine from phenylalanine in a system containing ribosomes and polyuridylic acid is stimulated by enzymes extracted from ungerminated uredospores of *Uromyces phaseoli* (271). It can probably be assumed that phenylalanyl-tRNA synthetase and transferases are responsible for this effect, especially in view of the work of Yaniv & Staples (331), who showed the latter enzymes to be present in extracts of dormant spores of this organism. In addition, this system appears to be exceptionally effective in transferase activity because 81% of the ^{14}C-phenylalanine added was transferred to protein in extracts of dormant spores and 99% when extracts of germinated spores were used. Moreover, on the basis of experiments on the kinetics of formation of aminoacyl-tRNA, it is likely that ungerminated spores of *Rhizopus stolonifer* contain at least 18 aminoacyl-tRNA synthetases (200). Ungerminated conidia of *Botryodipodia theobromae* contain at least 13 of these enzymes, but their specific activity increases during spore germination (311). However, these enzymes cannot be detected in extracts from dormant spores unless a special technique with a mechanical homogenizer is employed. In this system, enzymes from ungerminated spores, in the presence of ribosomes, polyuridylic acid, and tRNA, were less active in stimulating incorporation of phenylalanine into polyphenylalanine than were enzymes from germinated spores (311), although the two systems were similar in their response to inhibitors, energy supply, Mg^{++} and NH$_4^+$ ions, spermine, tRNA, polyuridylic acid, and ribosomes (309). In addition, both ungerminated and germinated spores of *B. theobromae* contain enzymes which transfer ^{14}C-phenylalanyl-tRNA to polyphenylalanine, a reaction which requires ribosomes, polyuridylic acid, GTP, and Mg^{++} and NH$_4^+$ ions (311), and this reaction is inhibited by RNase or GDP and stimulated by spermine. Although the transferases from ungerminated spores are not as active as those from germinated ones, their requirements and other properties are identical and resemble those of other systems as well.

Although it has been reported (69) that aminoacyl-tRNA synthetases are present in functional form in spores of *B. subtilis*, these authors are not aware of extensive study of these molecules from bacterial spores.

The presence of inhibitory factors is indicated by the initial failure of both Staples et al (270) and Van Etten (308) to extract active enzymes from ungerminated spores to effect the incorporation of ^{14}C-phenylalanine into polyphenylalanine unless special methods are applied (28, 271, 331). It is significant that in both cases the soluble fraction from ungerminated spores inhibited active in vitro incorporating systems. These data are reminiscent of those of Schmoyer & Lovett (247), who showed that a fraction prepared from the nuclear cap of zoospores of *Blastocladiella emersonii* was similarly inhibitory. The mechanism of these effects is unknown but could be accounted for as an effect of RNase. However, recent work by Bacon (10) establishes in the case of spores of *Dictyostelium discoideum* that a self-inhibitor of germination (dimethylguanosine) is responsible for preventing incorporation of precursors into protein at the level of the ribosome.

Kobayashi & Halvorson have reported that ribosomes of *B. cereus* do not bind aminoacyl tRNA (155), and it is likely that this inability is due to deficiencies in the ribosomes themselves (165). The presence in dormant spores of a complete set of soluble factors has yet to be shown, although it is likely that they are present. Mixing experiments (155) show that the supernatant from dormant spores has no positive action on protein synthesis when added to an amino acid incorporation system from vegetative cells. However, inasmuch as the activity of the supernatant fraction is demonstrable after heat activation of the spores, during which time biosynthetic reactions presumably do not occur, it is likely that initiation factors are present but in an inactive form in dormant spores.

ARE THE PROPERTIES OF RIBOSOMES FROM DORMANT SPORES SIMILAR TO THOSE FROM GERMINATING SPORES? The potential for morphogenetic regulation through the apparatus for protein synthesis is, of course, determined during the time when spores are formed, so this process must be considered before a full understanding of differentiation in such systems is reached. For example, "gene amplification" of the kind described by Brown & David (31) might be a means whereby ribosomes and their complement of rRNA are stored in large numbers in certain spores. Thus extra genes which help to synthesize extra 28 S and 18 S rRNA, as well as large amounts of ribosomes for storage, function only during oogenesis in certain animals and become inactive after the oocyte matures. No such information exists for fungal or bacterial spores at this time.

No differences appear to exist in the ribosomes of dormant and germinating fungal spores. Thus, the sedimentation coefficients of monosomes and the rRNA of ascospores, conidia, and hyphae of *Neurospora crassa* have been shown to be similar (117, 118). Although the ratios of protein to RNA in the different fractions were 33:67, 45:55, and 42:58, respectively, the nucleotide composition of rRNA was the same in extracts from these several stages. These data are supported by some which show that ribosomes from ungerminated conidia and vegetative stages of *N. crassa* are similar immunologically and that the proteins from these ribosomes run similarly on acrylamide gels (235). The same conclusion has been drawn for dormant and germinating conidia of *Aspergillus orzyae* on the basis of analyses of the nucleotide composition of rRNA, the sedimentation coefficients of monosomes (132, 150), and stability (150). However, deoxycholate was essential for the maximal recovery of ribosomes from dormant conidia of this organism but was not needed in the case of germinating ones (150). Other evidence for differences in ribosomes during germination include the instability of these in extracts from dormant conidia of *A. oryzae* (131) and lower phenylalanyl-tRNA binding capacity as compared with ribosomes from vegetative cells. On the other hand, the ribosomes of germinating and ungerminated cells of *Fusarium solani* were equally unstable (224).

Ribosomes which function in vitro have been successfully isolated from spores of *Dictyostelium discoideum* (10), conidia of *Botryodiplodia theobromae* (308, 312), *Aspergillus oryzae* (131, 150), *Fusarium solani* (224), zoospores of *Blastocladiella emersonii* (247), and uredospores of *Uromyces phaseoli* (271, 331). Those isolated from ungerminated spores of *Botryodiplodia theobromae* (308, 310) were less active than from germinated ones, but there was no such difference found in the case of uredospores of *Uromyces phaseoli* (331), *Fusarium solani* (224), and from spores of *Dictyostelium discoideum* from which the self-inhibitor has been washed away (10).

Experiments on bacterial spore ribosomes have been conducted for a number of years by Kobayashi, Halvorson, and associates (153–156, 173). The ribosomes isolated from dormant spores of *B. cereus* show low but definite activity in an in vitro system using polyuridylic acid and an artificial messenger (155). It should be emphasized that these experiments were conducted using ribosomes extracted in such a way that they did not appear as polysomes. Consequently, native mRNA, although probably degraded, probably was present on these ribosomes. Preincubation would have eliminated this native mRNA only if conditions were adequate and the ribosomes active. Since preincubation in a system designed for vegetative ribosomes did not increase the activity of these ribosomes, the authors reasonably concluded that they were inactive.

Kobayashi has recently extended these findings (153) by analyzing sedimentation profiles of spore ribosomes and subunits from them, and by analyzing their proteins by electrophoresis through polyacrylamide gels. The major finding from these experiments is that "defective subunits" can be isolated from dormant bacterial spores. Not only do they sediment through magnesium-deficient (0.1 mm) sucrose gradients with unusual sedimentation coefficients (42 S, 38 S, 26 S, and 21 S), but spore ribosomes appear to lack one or more proteins. These ribosomes become active if incubated with ribosomal proteins prepared from vegetative cells.

Thus it would appear that, at least in *B. cereus*, spore ribosomes as presently isolated are significantly different from vegetative ones. In addition, this "defect" is not repaired very rapidly upon germination. Ten minutes after the initiation of germination (germination may take 5–10 min for completion in this species) the activity of isolated ribosomes had increased to approximately 65% of that of vegetative ribosomes (155). However, care has to be exercised in the interpretation of these results. For example, the ionic environment of ribosomes may significantly alter their activity and structure (for example, 201). At least for a short time during isolation, ribosomes from dormant spores probably are subjected to a high concentration of calcium and other ions contained in the spore. Calcium is known to exchange with magnesium in ribosomes and to cause the appearance of extra sedimentation peaks (46). Also, it has been reported that the level of ribonuclease activity in the spore as a whole (45, 66), or on the ribosomes in particular (137), may be high with respect to the vegetative cell. Ribonuclease and protease activity might be particularly damaging if conditions in the spore or during isolation of the ribosomes were to cause an unfolding of the ribosomal structure exposing the substrate to hydrolytic activity. Thus it is probably premature to conclude that bacterial spores *contain* "defective" ribosomes but that they are inactive as presently isolated cannot be contested.

DO DORMANT SPORES LACK THE INITIATION FACTORS THAT ARE NECESSARY FOR THE TRANSLATION OF NATURALLY OCCURRING tRNAS? Although we know of no studies which deal with these factors directly, the subject is included as a reminder of its potential importance as a control in protein synthesis.

DOES THE APPARATUS FOR PROTEIN SYNTHESIS IN DORMANT SPORES HAVE ACCESS TO AN ADEQUATE POOL OF FREE AMINO ACIDS? Several fungal spores require the addition of amino acids before germination will take place (281): for example, those of *Rhizopus arrhizus*, to which proline must be added (321), and the microspores of *Trichophyton mentagrophytes*, which can use any of several amino acids to germinate (111). Three amino acids are lacking in the pool of dormant spores of *Dictyostelium discoideum* but are synthesized as soon as germination commences (10). Therefore, this form of control may obtain in these cases at least.

ARE ALL OF THE COMPONENTS OF THE APPARATUS FOR PROTEIN SYNTHESIS IN DORMANT SPORES IN CONTIGUITY, OR ARE THEY SEPARATED SPATIALLY? It is conceivable that although the components of the apparatus for protein synthesis are active in dormant spores, they might be separated in such a way as to prevent such synthesis from occurring. Activation then might result from bringing into contiguity these separated components of the system. The best documented of the cases among fungal spores where such regulation of protein synthesis might occur is the nuclear cap of zoospores of *Blastocladiella emersonii* (184), although tRNA and aminoacyl-tRNA synthetase are present in the cap as well (247).

A number of excellent electron micrographs of thin sections of spores from several species of bacteria have appeared recently (23, 62, 110, 232, 316, 317). Although these and others provide ample evidence that ribosomes appear in the central protoplast region (core) of these spores, there is no indication from them that spatial separation is involved in the failure of protein synthesis to take place in these dormant structures. In view of the chemical alterations in the machinery of protein synthesis discussed above, it appears likely that prokaryotic organisms utilize mechanisms other than spatial separation to ensure the cessation of protein synthesis in spores.

MECHANISMS OF GERMINATION

This general subject has been reviewed for fungal spores by Cochrane (51), Allen (5), and Sussman (278, 279), and for bacterial spores by Sussman & Halvorson

(281), Lewis (172), Gould & Dring (98), and Gould et al (99), and relevant papers can be found in the report of the Fifth International Spore Conference.

Constitutional Dormancy

SELF-INHIBITORS The early history of this field has been covered in detail in several reviews (5, 6, 51, 278) and will not be reviewed herein. The presence of self-inhibitors may be suspected if high concentrations of spores germinate less well than lower ones (26, 175). And the varying sensitivity to γ-irradiation, due to the length of time the conidia of *Aspergillus nidulans* have been held in buffer, has been ascribed to changes in the amount of a postulated self-inhibitor (248). But such observations alone are not fully diagnostic because it has been shown in the case of macroconidia of *Fusarium solani* that whereas only exogenous carbon is required at 3×10^5 conidia/ml, both nitrogen and carbon are required at 1×10^6 conidia/ml (101). This supports previous observations on the increased requirements of the same organism for carbon compounds at higher concentrations of spores (49). Furthermore, competition for oxygen has been invoked as an explanation in other cases (191). Another possible source of difficulty is determining the source of the inhibitor, especially in the case of pathogens and other microorganisms not cultured axenically. Thus, the "self-inhibitor" of conidia of *Peronospora tabacina* (253) later was shown to be a product of the leaves of its host (254). Moreover, substances formed by the host may reverse the action of self-inhibitors so that the latter must be studied in culture as shown for conidia of *Rhynchosporium secalis* (9).

Nevertheless, there are well-documented cases where self-inhibitors appear to be present, including those reported in the reviews mentioned above and the following more recent examples: conidia of *Glomerella cingulata* (175), *Microsporum gypseum* (169), *Phoma medicaginis* (47), *Leveillula taurica* (48), *Fusarium oxysporum* (91, 229) and *Eremethecium ashbyii* (214); spores of *Dothistroma pini* (138) and *Dictyostelium discoideum* (44, 239); arthrospores of *Geotrichum candidum* (219); pycnidiospores of *Mycosphaerella lignicola* (24); and uredospores of *Puccinia graminis* and *P. striiformis* (300).

Although self-inhibitors may function to prevent the germination of spores near the parent mycelium (48, 214), this cannot be assumed to be the only role in all cases. Thus, the broad activity range of the volatile sporostatic factors from *Fusarium oxysporum* (229, 230) and of the stable inhibitor from this organism argues for other roles for these substances. Moreover, as the above workers have shown, more than one self-inhibitor may be formed, as has also been found for pycnidiospores of *Mycosphaerella lignicola* (24), wherein different ones are formed depending upon the temperature at which the spores have been produced.

Studies of the mechanism of action of self-inhibitors have been handicapped by difficulties in their identification. Unfortunately, the reports that the inhibitor from uredospores of *Puccinia graminis* var. *tritici* was trimethylethylene (88) and that from *Uromyces phaseoli* was aspartate and/or glutamate (325) have been shown by Bell & Daly (20) to be erroneous. However, more recently the self-inhibitor from the wheat stem rust has been proven to be the *cis*- and *trans*-isomers of methyl 4-hydroxy-3-methoxycinnamate (methyl ferulic acid) (189) and that from the bean rust the *cis*- and *trans*-isomers of methyl 3,4-dimethoxycinnamate (190). That the rusts produce a family of cinnamic acid inhibitors is suggested by further work by Macko and his collaborators (unpublished) which shows that uredospores of the snapdragon and sunflower rusts produce 3,4-dimethoxycinnamate as a self-inhibitor. The large number of cinnamic acid derivatives formed by the hosts of rusts suggests that there is ample opportunity for the interaction of self-inhibitors and analogs formed by higher plants which may serve as activators by virtue of their ability to compete for the sites at which self-inhibitors act.

A series of volatile inhibitors of fungal spore germination have been identified from cultures of *Fusarium oxysporum*, including acetaldehyde, propionaldehyde, *n*-butyraldehyde, *n*-propanol, iso-butanol, ethyl acetate, iso-butyl acetate, and acetone (229–231). Although spores of *F. oxysporum* are themselves subject to inhibition by culture filtrates or the pure compounds, they are less so than are spores of other species like *Rhizopus stolonifer*. Therefore, these substances will be considered among the causes of environmental dormancy as well because of their broad specificity and the large number of species which may produce them in natural soil associations.

The self-inhibitor of the germination of spores of *Dictyostelium discoideum* has been identified as N^2-dimethylguanosine (2-dimethylamino-6-oxypurineriboside) and its mechanism of action studied in detail (10). Whereas incorporation of ^{14}C-leucine into proteins is strongly inhibited by the small concentrations of this substance that inhibit germination, respiration and incorporation into other substances are not much affected. Studies of a homologous in vitro system from *D. discoideum* reveal that a step or steps following the binding of aminoacyl-tRNA to the ribosomes probably is affected by the self-inhibitor.

That regulation of protein synthesis is involved in the action of the unidentified self-inhibitor of the germination of conidia of *Glomerella cingulata* also has been reported (176) so that this locus of action may be a general one for this group of substances.

Another self-inhibitor that has been identified is that of macroconidia of *Microsporum gypseum*. According to Page & Stock (218), phosphate ions are released into the medium by these spores, thereby inhibiting an alkaline protease which is essential to the germination of these spores (170). This is a case which illustrates the interdependence of two of the mechanisms discussed herein because the protease is thought to be located in lysosomes which presumably release the enzyme into the spore surface after heat activation, thereby opening up the wall to permit emergence of the germ tube. Thus this mechanism would fit into the next section as well as into this one.

Self-inhibitors of bacterial spores have been described for *Bacillus globigii* (273) and one has been identified as D-alanine (86). These have been reviewed elsewhere (281). That alanine racemase may be important to this type of autoinhibition has been suggested recently by data of Jones & Gould (141), who have shown that inhibition of this enzyme by analogs of alanine promotes more rapid germination. Presumably this inhibition allows less of the added L-alanine to be converted to D-alanine.

Dring & Gould (74) have speculated that the hydrolysis of peptidoglycan in the cortex of spores is one of the earliest steps in germination. According to this view, calcium and DPA are released very shortly after the murein fragments with perhaps a small lag in the release of calcium. If this timing is correct, then the large variety of germinating agents (320) could be interpreted as initiating the depolymerization of the cortex, one of the major results of which is to cause the release of calcium and DPA. That these compounds are involved in autoinhibition is suggested by experiments by Harrell and co-workers (107, 108), which show a positive correlation between the amount of DPA removed from spores during germination and the rate of oxidation of glucose. Also, the addition of DPA to cell-free extracts of spores stimulates glucose oxidation. Since this can be mimicked by EDTA, this probably is due to the well-known ability of DPA to chelate calcium. Thus it is likely that calcium rather than DPA is the inhibitor of the aerobic catabolism of glucose. However, Hachisuka et al (104) have published data which show that DPA *inhibits* the anaerobic catabolism of glucose. Thus the possibility exists that calcium and DPA, both of which exist in high concentration in the spore, are inhibitory to different phases of glucose oxidation. The germination reactions which bring about the hydrolysis of the cortex and subsequent release of calcium and DPA could thus be considered to release energy-yielding catabolic reactions from autoinhibition. The model of spore germination as caused by calcium chelates of DPA analogs, proposed recently by Lewis (173), is not inconsistent with this, except in detail.

COMPARTMENTALIZATION   The coexistence of enzymes in active form and their substrates in the same cell, even though the latter are not utilized, has led to the suggestion that these entities are separated from each other in vivo. Such lack of contiguity might take the form of separation by membranes or sequestration in different organelles or "channeling" wherein a common intermediate may be segregated into multiple pools (21).

A mechanism of this kind was proposed for zoospores of *Blastocladiella emersonii* (184, 185) whose nuclear cap has been mentioned previously. The original hypothesis was that the "packaging" of ribosomes in this way denied enzymes, and perhaps substrates, access to these organelles, thereby regulating protein synthesis. However, subsequent work has revealed that there is an inhibitor of ribosomes which functions even when they are dispersed in vitro so that aggregation of these organelles might serve another purpose, perhaps to prevent their turnover (247; Lovett, personal communication). Moreover, it is worth recalling that regulation of protein synthesis by self-inhibitors has now been established, and the substance described in spores of *B. emersonii* may fall in this category.

Another case where compartmentalization may take place is that of ascospores of *Neurospora*, which contain more than 10% trehalose (dry weight) and trehalase

(279). Yet the dormant spores utilize only their endogenous supply of lipids and have been found not to hydrolyze trehalose even after years of storage. But immediately upon activation by heat or chemicals, trehalose is utilized rapidly and is incorporated into a wide diversity of metabolites (33). The presence of active trehalase in dormant cells rules out the need for de novo synthesis of this enzyme as a result of activation, so other explanations must be sought. That furfural, an activator of germination, binds mainly to the walls of ascospores and to membranes has been shown (78), as well as the tendency of phenethyl alcohol, another activator of these spores (177), to bind to membranes (35). Furthermore, Hecker (115) has shown that the properties of the trehalase of *Neurospora* are such that the enzyme would probably be inactivated were it in the cytoplasm of dormant ascospores, and indeed by the use of immunofluorescent labeling he has localized this enzyme in the innermost major wall layer of these cells. Moreover, in vitro studies with wall materials show that no other materials from among the several tested protected trehalase better from heat inactivation, an essential property because of the need for high (60°C) activation temperatures. Consequently, it is likely in this system that activation involves the removal of trehalase from its association with the ascospore wall, along with an increase in the permeability of the plasma membrane, thereby allowing the cytoplasmically located trehalose to become accessible to its hydrolase. Support for this possibility comes from the work of Mandels, Vitols & Parrish (196) on spores of *Myrothecium verrucaria*, in which organic solvents like toluene or exposure to 60°C for 20 min result in the utilization of endogenous stores of trehalose. In addition, Souza & Panek (267) found that toluene disrupts the membranous barrier between trehalase and its substrate in yeast, thereby allowing the precocious hydrolysis of the sugar. But the fact that the addition of glucose alone to ascospores of *Neurospora* does not activate them has argued against this mechanism as the explanation of dormancy in this system (237, 238). However, glucose is absorbed only very slowly by intact ascospores (282), so this criticism probably does not hold in this case, although it would in that of spores of *Phycomyces blakesleeanus* which Rudolph and his co-workers have studied. Another difference between these systems is in the intermediates accumulated upon activation, for ascospores of *Neurospora* form considerable amounts of glucose 6-phosphate upon activation (77), whereas the spores of *Phycomyces* do not (238), so different metabolic fluxes and pathways probably exist in these organisms.

CHANGES IN PREFORMED PROTEINS   As was noted above, allosteric and other physical changes in proteins are of potential importance in regulating developmental events. An interesting example of such a change is to be found in the spores of *Phycomyces blakesleeanus*, wherein the heat treatment needed to break dormancy causes a very rapid ten to fifteenfold increase in trehalase activity (307). This increase soon decays but can be renewed by another heat shock, and several cycles of increase and decay can be elicited by this means (Figure 1). Inasmuch as cycloheximide does not prevent the rise in enzyme activity, the authors conclude that a physical change in the enzyme is responsible for the effect. However, the increased utilization of trehalose is only one of the requirements for the breakage of dormancy in these spores (237) so that changes of other kinds also are necessary in this system.

In spores of bacteria it now seems likely that more than one mechanism is used to protect enzymes from environmental stresses. Thus experiments from Kornberg's laboratory indicate that in the case of such enzymes as inorganic pyrophosphatase (301) and purine nucleoside phosphorylase (89), the polypeptide chains present in the vegetative cell are to be found in the spore as well. On the other hand, it is clear from the report of Sadoff (240) that some of the enzymes present in the spore have been modified during sporulation. For example, fructose-1, 6-diphosphate aldolase of the spore is lower in molecular weight than is that of vegetative cells. Further, it becomes more resistant to heat inactivation upon the addition of calcium ions, rather than more sensitive as is the case with the vegetative enzyme. Similarly, conditions close to those presumably found in the spore cause the disaggregation of glucose dehydrogenase. Thus, smallness may be advantageous to some enzymes for surviving the rigors to which spores may be subjected.

DE NOVO SYNTHESIS OF ENZYMES   Although the need for protein synthesis during germination is not universal, as was indicated before, there are many cases where this process is essential. Therefore, it is to be expected that there will be quantitative differences at least in the enzymatic constitution of many dormant and germinating spores. Although a large number of quantitative enzymatic changes have been described in such germinating fungal spores, it is not clear how they relate to the breaking of dormancy. Many of these changes have been summarized in reviews (5, 52, 278, 281) and will not be covered exhaustively herein. Attention is called to a previous section on *Ultrastructural correlates* in which it is indicated that a considerable increase in the synthesis of wall material occurs during spore germination, but in most cases the enzymes concerned have not been studied.

That qualitative changes in the enzyme complement of spores occur as a result of germination is suggested by the patterns of proteins found in *Verticillium albo-atrum* and *V. dahliae* (220, 221). Thus, for a particular isolate, the gel electrophoretic patterns of proteins from conidia were clearly different from those of the mycelium, some proteins occurring only in the spores. In fact, one protein was found only in nongerminated conidia and another after only 12 or 24 hours' incubation. Similar results have been found for *Drechslera*, in which it has been found that the protein patterns from the mycelium of two species differed less than the pattern from spores and mycelium of a single species (255). A greater than hundredfold increase in the NADP-specific glutamic dehydrogenase of germinating conidia of *Neurospora crassa*, which is arrested by cycloheximide, suggests the need for de novo synthesis of this enzyme. However, as the authors themselves point out (305), incorporation of ^{14}C-leucine into the enzyme during the period of increase was not demonstrated, so the data are equivocal.

In closing this section it is worth pointing out the pitfalls in extrapolating from in vitro evidence alone to the situation in vivo, a point made by Barbara Wright and co-workers in recent work on the differentation of *Dictyostelium discoideum*. They point out that the maximum specific activity of trehalose synthetase measured in vitro is about 1/100th that which must be present in vivo and that this enzyme becomes active in the organism only after in vitro measurements show a decline in activity. The other examples they provide to illustrate this point are sobering reminders of the need at all times to be concerned about the situation inside the cell.

PERMEABILITY   Permeability changes often are invoked to explain the transition from the dormant to germinating state, but few data exist to establish this connection in fungal spores. The effect of dimethylsulfoxide in activating spores of the slime mold *Lycogala epidendrum* (38) and those of *Bacillus pantothenicus* (323) has been used as an argument in favor of this possibility (38) because the former spores germinate in distilled water, but more slowly than in the chemical. But even were changes in permeability induced, it is doubtful that causality could be assigned unambiguously because of the other changes that accompany these (281).

Environmental Dormancy

The several factors involved in environmental dormancy have been discussed in detail previously (281), so only recent developments will be discussed. These relate mainly to the widespread soil fungistasis which was discovered by Dobbs & Hinson (67). Two principal hypotheses have occupied the attention of workers in the field (for reviews see 14, 180, 181, 319): the first assumes the presence of inhibitory substances in soil, and the second that the growth of competing microorganisms depletes the soil in the vicinity of spores of essential nutrients. The latter hypothesis was adopted because of the failure of several workers to extract inhibitory materials from soil (181, 319). Lockwood and co-workers (152, 178, 276) have provided evidence for the development of a nutritional insufficiency in soil because of the metabolic activities of neighboring organisms which act as "sinks" for these substances. Thus they claim that extracts of sterilized soil contain enough nutrients for the germination of nutritionally dependent spores but extracts of natural soil do not. Furthermore, germination in aqueous extracts of sterilized soil made fungistatic was high for non-nutrient requiring spores and low for others, and the increased fungistasis in sterilized soils which were amended with glucose and reinoculated was correlated with the loss of glucose.

On the other hand, the opposing view has been fortified by work on nonnutrient requiring propagules like sclerotia of *Sclerotium cepivorum* (3) and sporangia of *Pythium* (1), which are affected by the fungistatic principle. Still, it has been argued that such spores do not germinate because of the strong diffusion gradient set up by competing microorganisms (152), and it has been shown that continuous leaching of some non-nutrient requiring spores does prevent their germination. But recent work by Baker and co-workers and by Balis & Konyeas (15) has shifted attention to the first hypothesis, for they have been able to show the presence of a volatile fungistatic factor in soil (127) where contact with soil was not necessary for inhibition to be manifested. That some of the properties of this factor are similar to those described for soil fungistasis was established because both are water-soluble or inactivated in water, both are adsorbed on activated charcoal, and both are formed in greatest amounts in the upper layers of alkaline soils (127, 234). A difference between these is that soil fungistasis decreases immediately after wetting of dried soil but increases thereafter (67), whereas the volatile factor decreases with time (127). It is possible that the earthy smelling compounds formed by some actinomycetes are a source of the inhibitor (128). Moreover, a large number of fungi are inhibited by the factor (129). As these authors observe (129), the isolation and identification of these materials must be accomplished, along with proof that they inhibit at concentrations found in soil. The finding of volatile sporostatic substances in the cultures of several fungi (229), and the fact that acetaldehyde, n-propanol, and nonanoic acid accumulate in soil after inoculation with any of 16 species of fungi and inhibit the germination of test organisms (90), supports the inhibitor hypothesis.

Therefore, although soil fungistasis may "appear to be explicable in terms of nutrient relations and (does) not require an assumption that fungistatic substances are involved" (134), such substances have now been found and may explain the phenomenon, at least in part.

ACKNOWLEDGMENTS

We thank Drs. James Van Etten, James Lovett, David Smith, George Spiegelman, and Lyle R. Brown for helpful discussion and for the provision of unpublished data. Also, A. S. Sussman expresses his gratitude to Professor John Burnett of the Department of Agricultural Science, Oxford University, for providing the facilities which were used for much of the writing of this review.

Literature Cited

1. Agnihotri, V. P., Vaartaja, O. 1967. *Phytopathology* 57:1116–20
2. Aitken, W. B., Niederpruem, D. J. 1970. *J. Bacteriol.* 104:981–88
3. Allen, J. D., Young, J. M. 1968. *Plant Soil* 29:479–80
4. Allen, P. J. 1955. *Phytopathology* 45:259–66
5. Allen, P. J. 1965. *Ann. Rev. Phytopathol.* 3:313–42
6. Allen, P. J., Dunkle, L. D. 1971. *Proc. Conf. Morphological and Biochemical Events in Plant-Parasite Interaction, 1970,* ed. S. Akai, S. Ouchi, 22–58. Phytopathol. Soc. Japan, Tokyo
7. Alloway, J. M., Wilson, I. M. 1972. *Trans. Brit. Mycol. Soc.* 58:231–36
8. Altman, P. L., Dittmer, D. S. 1964. *Biology Data Book.* Fed. Am. Soc. Exp. Biol., Bethesda, Md.
9. Ayres, P. G., Owen, H. 1970. *Trans. Brit. Mycol. Soc.* 54:389–94
10. Bacon, C. 1972. *The self-inhibitor of the germination of spores of Dictyostelium discoideum.* PhD thesis. Univ. Michigan, Ann Arbor
11. Bahnweg, G., Douthit, H. A. 1972. *Cytochrome pigments in spores of B. cereus.* In preparation
12. Bainbridge, B. W. 1971. *J. Gen. Microbiol.* 66:319–25
13. Bais, R., Keech, B. 1972. *J. Biol. Chem.* 247:3255–61
14. Baker, K. F., Snyder, W. C. 1965. *Ecology of Soil-born Plant Pathogens.* London: Murray
15. Balis, C., Konyeas, V. 1968. *Ann. Inst. Phytopathol. Benaki, N. S.* 8:145–49
16. Barash, I. 1968. *Phytopathology* 58:1364–71
17. Barash, I., Conway, M. L., Howard, D. H. 1967. *J. Bacteriol.* 93:656–62
18. Bartnicki-Garcia, S. 1968. *Ann. Rev. Microbiol.* 22:87–108
19. Beaman, T. C., Pankratz, H. S., Gerhardt, P. 1972. *J. Bacteriol.* 109:1198–1209
20. Bell, A. A., Daly, J. M. 1962. *Phytopathology* 52:261–68
21. Bernhardt, S. A., Davis, R. H. 1972. *Proc. Nat. Acad. Sci. USA* 69:1868–72
22. Black, S. H. 1968. *J. Invertebr. Pathol.* 12:148–57
23. Black, S. H., Gerhardt, P. 1962. *J. Bacteriol.* 83:960–67
24. Blakeman, J. P. 1969. *J. Gen. Micro-*
biol. 57:159–68
25. Bonsen, P. P. M., Spudich, J. A., Nelson, D. L., Kornberg, A. 1969. *J. Bacteriol.* 98:62–68
26. Boyd, A. E. W. 1952. *Ann. Appl. Biol.* 29:322–29
27. Bracker, C. E. 1967. *Ann. Rev. Phytopathol.* 5:343–74
28. Brambl, R. M., Van Etten, J. L. 1970. *Arch. Biochem. Biophys.* 137:442–52
29. Brodie, H. J., Neufeld, C. C. 1942. *Can. J. Res. Sect. C* 20:41–61
30. Bromfield, K. R. 1964. *Plant Physiol.* 42:1633–42
31. Brown, D. D., David, I. B. *Science* 160:272–80
32. Buckley, P. M., Sjaholm, V. E., Sommer, N. F. 1966. *J. Bacteriol.* 91:2037–44
33. Budd, K., Sussman, A. S., Eilers, F. I. 1966. *J. Bacteriol.* 91:551–61
34. Burger, A., Prinzing, A., Reisener, H. J. 1972. *Arch. Mikrobiol.* 83:1–16
35. Burns, V. W. 1971. *Exp. Cell Res.* 64:35–40
36. Bush, L. 1968. *Phytopathology* 58:752–54
37. Bussel, J., Sommer, N. F., Kosuge, T. 1969. *Phytopathology* 59:946–52
38. Butterfield, W. 1968. *Nature* 218:494–95
39. Caltrider, P. G., Gottlieb, D. 1966. *Phytopathology* 56:479–84
40. Campbell, C. K. 1971. *Trans. Brit. Mycol. Soc.* 57:393–402
41. Cantino, E. C., Truesdell, L. C., Shaw, D. S. 1968. *J. Elisha Mitchell Sci. Soc.* 84:125–46
42. Carroll, G. C. 1967. *Arch. Mikrobiol.* 66:321–26
43. Carstensen, E. L., Marquis, R. E., Gerhardt, P. 1971. *J. Bacteriol.* 107:106–13
44. Ceccarini, C., Cohen, A. 1967. *Nature* 214:1345
45. Chambon, P., Deutscher, M. P., Kornberg, A. 1968. *J. Biol. Chem.* 243:5110–16
46. Choi, Y. S., Carr, C. W. 1967. *J. Mol. Biol.* 25:331–45
47. Chung, H. S., Wilcoxson, R. D. 1969. *Phytopathology* 59:440–42
48. Clerk, G. C., Ayesu-Offei, E. N. 1967. *Ann. Bot.* 31:749–54
49. Cochrane, J. C., Cochrane, V. W., Simon, F. G., Spaeth, J. 1963. *Phytopathology* 53:1155–60
50. Cochrane, J. C., Rado, T. R.,

Cochrane, V. W. 1971. *J. Gen. Microbiol.* 65:45–55

51. Cochrane, V. W. 1960. Review in *Plant Pathology, an Advanced Treatise,* ed. J. G. Horsfall, A. E. Dimond, 2:167–202. New York: Academic

52. Cochrane, V. W. 1966. In *The Fungus Spore,* ed. M. F. Madelin. Colston Papers No. 18. London: Butterworths

53. Cochrane, V. W., Cochrane, J. C., Collins, C. B., Serafin, F. G. 1963. *Am. J. Bot.* 50:806–14

54. Cohen, A., Keynan, A. 1970. *Biochem. Biophys. Res. Commun.* 38:744–49

55. Cohen, A., Silberstein, Z., Mazor, Z. 1972. In *Spores V,* ed. H. O. Halvorson, R. Hanson, L. L. Campbell, 247–53. Am. Soc. Microbiol.

56. Cotter, D. A., Miura-Santo, L. Y., Hobl, H. R. 1969. *J. Bacteriol.* 100:1020–26

57. Crocker, W. 1916. *Am. J. Bot.* 3:99–120

58. Cross, T., Davies, F. L. 1971. In *Spore Research 1971,* ed. A. N. Barker, G. W. Gould, J. Wolf, 175–80. New York: Academic

59. Daly, J. M., Knoche, H. W., Wiese, M. V. 1967. *Plant Physiol.* 42:1633–42

60. Danilevskii, A. S. 1965. *Photoperiodism and Seasonal Development of Insects.* Edinburgh: Oliver and Boyd

61. Dawes, I. W., Hansen, J. N. 1972. In *CRS Critical Reviews in Microbiology,* ed. A. I. Laskin, H. Lechevalier, 479–520

62. Dean, D. 1973. *Biochemical and Physiological Bases of Density Heterogeneity in Populations of Bacillus subtilis W23 Spores.* PhD thesis. Univ. Michigan, Ann Arbor

63. Delcour, J., Papaconstantinou, J. 1972. *J. Biol. Chem.* 247:3289–95

64. DeLuca, C., Gioeli, R. P. 1972. *Can. J. Biochem.* 50:447–56

65. Desser, H., Broda, E. 1969. *Arch. Mikrobiol.* 65:76–86

66. Deutscher, M. P., Chambon, P., Kornberg, A. 1968. *J. Biol. Chem.* 243:5117–25

67. Dobbs, C. G., Hinson, W. H. 1953. *Nature* 172:197–99

68. Doi, R. H. 1969. In *The Bacterial Spore,* ed. G. W. Gould, A. Hurst, 125–66. New York/London: Academic

69. Doi, R. H., Bishop, H. L., Migita, L. K. 1969. In *Spores IV,* ed. L. L. Campbell, 159–74. Am. Soc. Microbiol.

70. Doi, R. H., Halvorson, H. 1961. *J. Bacteriol.* 81:51–58

71. Doi, R. H., Kaneko, I. 1967. *Cold Spring Harbor Symp. Quant. Biol.* 31:581–82

72. Doi, R. H., Kaneko, I., Igarashi, R. T. 1968. *J. Biol. Chem.* 243:945–51

73. Douthit, H. A., Kieras, R. A. 1972. See Ref. 55, 264–68

74. Dring, G. J., Gould, G. W. 1971. *J. Gen. Microbiol.* 65:101–4

75. Dunkle, L. D., Maheshwari, R., Allen, P. J. 1969. *Science* 163:481–82

76. Dunkle, L. D., Van Etten, J. L. 1972. In press

77. Eilers, F. I., Ikuma, H., Sussman, A. S. 1970. *Can. J. Microbiol.* 16:1351–56

78. Eilers, F. I., Sussman, A. S. 1970. *Planta* 94:265–72

79. Ekundayo, J. A. 1966. *J. Gen. Microbiol.* 92:283

80. Enquist, L. W., Bradley, S. G. 1971. *Develop. Ind. Microbiol.* 12:225–36

81. Epel, D. 1967. *Proc. Nat. Acad. Sci. USA* 57:899–906

82. Epel, D., Pressman, B. C., Elsaesser, S., Weaver, S. 1969. In *The Cell Cycle: Gene-Enzyme Interactions,* ed. G. M. Padella, G. L. Whitson, I. L. Cameron, 279–98. New York: Academic

83. Farkas, G. L., Ledingham, G. A. 1959. *Can. J. Microbiol.* 5:141–51

84. Feinsod, F. M., Douthit, H. A. 1970. *Science* 168:991

85. Feit, I. N., Chu, L. K., Iverson, R. M. 1971. *Exp. Cell Res.* 65:439–44

86. Fey, G., Gould, G. W., Hitchins, A. D. 1964. *J. Gen. Microbiol.* 35:229–36

87. Flatmark, T., Sletten, K. 1968. *J. Biol. Chem.* 243:1623

88. Forsyth, F. R. 1955. *Can. J. Bot.* 33:363–73

89. Gardner, R., Kornberg, A. 1967. *J. Biol. Chem.* 242:2383–88

90. Garrett, M. K. 1972. *Plant Soil* 36:723–27

91. Garrett, M. K., Robinson, P. M. 1969. *Arch. Mikrobiol.* 67:370–77

92. Gerhardt, P., Ribi, E. 1964. *J. Bacteriol.* 88:1774–89

93. Goddard, D. R. 1935. *J. Gen. Physiol.* 19:45–60

94. Goddard, D. R. 1939. *Cold Spring Harbor Symp. Quant. Biol.* 7:362–76

95. Gong, C-S., Van Etten, J. L. 1972. *Biochem. Biophys. Acta* 272:44–52

96. Gottlieb, D. 1966. See Ref. 52, 217–23
97. Gottlieb, D., Rao, M. V., Shaw P. D. 1968. Phytopathology 58: 1593–97
98. Gould, G. W., Dring, G. J. 1972. See Ref. 55, 401–8
99. Gould, G. W., Stubbs, J. M., King, W. L. 1970. J. Gen. Microbiol. 60: 347–56
100. Grecz, N., Smith, R. F., Hoffmann, C. C. 1970. Can. J. Microbiol. 16: 573–79
101. Griffin, C. J. 1970. Can. J. Microbiol. 16:733–40
102. Gull, K., Trinci, A. P. J. 1971. J. Gen. Microbiol. 68:207–20
103. Gustafson, G. L., Wright, B. E. 1972. See Ref. 61, 453–78
104. Hachisuka, Y., Tochikubo, K., Murachi, T. 1965. Nature 207:220–21
105. Halvorson, H. O., Vary, J. C., Steinberg, W. 1966. Ann. Rev. Microbiol. 20:169–88
106. Harper, J. L. 1957. Proc. 4th Int. Congr. Crop Prot. Hamburg 1:415–20
107. Harrell, W. K. 1956. Bacteriol. Proc., 44
108. Harrell, W. K., Doi, R. H., Halvorson, H. O. 1957. J. Appl. Bacteriol. 20:xiii
109. Harris, R. A., Penniston, J. T., Asai, J., Green, D. E. 1968. Proc. Nat. Acad. Sci. USA 59:830–37
110. Hashimoto, T., Conti, S. F. 1971. J. Bacteriol. 105:361–68
111. Hashimoto, T., Wu, C. D., Blumenthal, H. J. 1972. Proc. Ann. Meet. Am. Soc. Microbiol., 56
112. Hawker, L. E. 1966. See Ref. 52, 151–61
113. Hawker, L. E., Thomas, B., Beckett, A. 1970. J. Gen. Microbiol. 60:181–89
114. Haworth, S. R. 1973. Genetic mapping and physiological characterization of RNA polymerase mutants of Bacillus subtilis. PhD thesis. Oregon State Univ., Corvallis
115. Hecker, L. I. 1972. The heat stability and localization of trehalase in ascospores of Neurospora. PhD thesis. Univ. Michigan, Ann Arbor
116. Heintz, C. E., Niederpruem, D. J. 1971. Mycologia 63:745–66
117. Henney, H. R. Jr., Storck, R. 1963. Science 142:1675–77
118. Henney, H. R. Jr., Storck, R. 1963. J. Bacteriol. 85:822–28
119. Henney, H. R. Jr., Storck, R. 1964. Proc. Nat. Acad. Sci. USA 51:1050–55
120. Hill, A. G. G., Campbell, G. K. G. 1949. J. Exp. Agr. 17:259
121. Hill, C. P., Sussman, A. S. 1964. J. Bacteriol. 88:1556–66
122. Hoeniger, J. F. M., Stuart, P. F., Holt, S. C. 1968. J. Bacteriol. 96: 1818–1834
123. Holloman, D. W. 1969. J. Gen. Microbiol. 55:267–74
124. Ibid 1970. 62:75–87
125. Holloman, D. W. 1971. Arch. Biochem. Biophys. 145:643–49
126. Holton, R. 1960. Plant Physiol. 35: 757–66
127. Hora, T. S., Baker, R. 1970. Nature 225:1071–72
128. Hora, T. S., Baker, R. 1972. Phytopathology 62:1475–76
129. Hora, T. S., Baker, R. 1972. Trans. Brit. Mycol. Soc. 59:491–500
130. Horgen, P. A. 1971. J. Bacteriol. 106:281–82
131. Horikoshi, K., Ikeda, Y. 1968. Biochim. Biophys. Acta 166:505–11
132. Horikoshi, K., Ohtaka, Y., Ikeda, Y. 1965. Agr. Biol. Chem. Japan 29: 724–27
133. Hoyhem, T., Rodenberg, S., Douthit, H. A., Halvorson, H. O. 1969. Arch. Biochem. Biophys. 125:964–74
134. Hsu, S. C., Lockwood, J. L. 1971. Phytopathology 61:1355–62
135. Huxley, H. E. 1969. Science 164: 1356–66
136. Hyde, J. M., Walkinshaw, C. H. 1966. J. Bacteriol. 92:1218–27
137. Idriss, J. M., Halvorson, H. O. 1969. Arch. Biochem. Biophys. 133:442–53
138. Ivory, M. H. 1967. Trans. Brit. Mycol. Soc. 50:563–72
139. Jaffe, L. F. 1969. Symp. Soc. Study Develop. Growth 28:83–111
140. Johnson, L. B., Brannaman, B. L., Zscheile, Z. P. 1966. Phytopathology 56:1405–10
141. Jones, A., Gould, G. W. 1968. J. Gen. Microbiol. 53:383–94
142. Jones, J. P., Snow, J. P. 1965. Phytopathology 55:499
143. Kalakutskii, L. V., Pozharitskaya, L. M. 1968. J. Gen. Appl. Microbiol. 14:209–12
144. Kanamitsu, O. 1971. J. Gen. Appl. Microbiol. 17:207–14
145. Kaneko, I., Doi, R. H. 1966. Proc. Nat. Acad. Sci. USA 55:564–71
146. Kawasaki, C., Nishihara, T., Kondo, M. 1969. J. Bacteriol. 97:944–46
147. Kawasaki, C., Nishihara, T., Kondo, M. 1969. Jap. J. Bacteriol. 25:209–14
147a. Kerjan, P., Marchetti, J., Szulmajster, J. 1967. Bull. Soc. Chim. Biol. 49:1139–58

148. Khan, K. P., Sengupta, S., Banerjee, A. B. 1968. *Acta Microbiol. Pol.* 17:305–8
149. Killick, K. A., Wright, B. E. 1972. *J. Biol. Chem.* 247:2967–69
150. Kimura, K., Ono, T., Yanagita, T. 1965. *J. Biochem.* 58:569–76
151. Klöker, W., Ledingham, G. A., Perlin, A. S. 1965. *Can. J. Biochem.* 43:1387–95
152. Ko, W-H., Lockwood, J. L. 1967. *Phytopathology* 57:894–901
153. Kobayashi, Y. 1972. See Ref. 55, 269–76
154. Kobayashi, Y., Halvorson, H. O. 1966. *Biochim. Biophys. Acta* 119:160–70
155. Kobayashi, Y., Halvorson, H. O. 1968. *Arch. Biochem. Biophys.* 123:622–32
156. Kobayashi, Y., Steinberg, W., Higa, A., Halvorson, H. O., Levinthal, C. 1965. *Spores III*, ed. L. L. Campbell, H. O. Halvorson, 200–12. Am. Soc. Microbiol.
157. Kogane, F., Yanagita, T. 1964. *J. Gen. Appl. Microbiol.* 10:61–68
158. Kornfield, J. M. 1961. *Structural and physiological aspects of germination of conidia of Penicillium chrysogenum.* PhD thesis. Univ. Wisconsin, Madison. *Diss. Abstr.* 22:396–97
159. Koshland, D. E. Jr., Neet, K. E. 1968. *Ann. Rev. Biochem.* 37:359–410
160. Kramer, M. J., Roth, I. L. 1968. *Can. J. Microbiol.* 14:1297–99
161. Kuczenski, R. T., Mandell, A. J. 1972. *J. Biol. Chem.* 247:3114–22
162. Lai, C. Y., Horecker, B. L. 1971. *J. Cell Comp. Physiol.* 76:381–88
163. Langenbach, R. J., Knoche, H. W. 1971. *Plant Physiol.* 48:728–34
164. Ibid, 735–39
165. Lazzarini, R. A. 1966. *Proc. Nat. Acad. Sci. USA* 56:185–90
166. Leadbetter, E. R., Holt, S. C. 1968. *J. Gen. Microbiol.* 52:299–307
167. Leary, J. V., Morris, A. J., Ellingboe, A. E. 1969. *Biochim. Biophys. Acta* 182:113–20
168. Leighton, T. J., Freese, P. K., Doi, R. H., Warren, R. A. J., Kelln, R. A. 1972. See Ref. 55, 238–46
169. Leighton, T. J., Stock, J. J. 1969. *Appl. Microbiol.* 17:473–75
170. Leighton, T. J., Stock, J. J. 1970. *J. Bacteriol.* 101:931–40
171. LéJohn, H. B., Lovett, J. S. 1966. *J. Bacteriol.* 91:709–17
172. Lewis, J. C. 1969. See Ref. 68, 301–58
173. Lewis, J. C. 1972. *J. Biol. Chem.* 247:1861–68
174. Lin, F. K., Davies, F. L., Tripathi, R. K., Raghu, K., Gottlieb, D. 1971. *Phytopathology* 61:645–48
175. Lingappa, B. T., Lingappa, Y. 1965. *J. Gen. Microbiol.* 41:67–75
176. Lingappa, B. T., Lingappa, Y., Bell, E. 1972. *Abstr. Ann. Meet. Am. Soc. Microbiol.*, 40
177. Lingappa, B. T., Lingappa, Y., Turian, G. 1970. *Arch. Mikrobiol.* 72:97–105
178. Lingappa, B. T., Lockwood, J. L. 1961. *J. Gen. Microbiol.* 26:473–85
179. Lingappa, B. T., Sussman, A. S. 1959. *Plant Physiol.* 34:466–72
180. Lockwood, J. L. 1964. *Ann. Rev. Phytopathol.* 2:341–62
181. Lockwood, J. L. 1968. In *The Ecology of Soil Bacteria*, ed. T. R. G. Gray, D. Parkinson, 44–65
182. Losick, R., Sonenshein, A. L. 1969. *Nature* 224:35–37
183. Losick, R., Shorenstein, R. G., Sonenshein, A. L. 1970. *Nature* 227:910–13
184. Lovett, J. S. 1963. *J. Bacteriol.* 85:1235–46
185. Ibid 1968. 96:962–69
186. Lowry, R. J., Durkee, T. L., Sussman, A. S. 1967. *J. Bacteriol.* 94:1757–63
187. Lowry, R. J., Sussman, A. S. 1958. *Am. J. Bot.* 45:397–403
188. Lowry, R. J., Sussman, A. S. 1968. *J. Gen. Microbiol.* 51:403–9
189. Macko, V., Staples, R. C., Allen, P. J., Renwick, J. A. A. 1971. *Science* 173:835–36
190. Macko, V., Staples, R. C., Gershon, H., Renwick, J. A. A. 1970. *Science* 170:539–40
191. Magie, R. O. 1935. *Phytopathology* 25:131–59
192. Maheshwari, R., Sussman, A. S. 1970. *Phytopathology* 60:1357–64
193. Maheshwari, R., Sussman, A. S. 1971. *Plant Physiol.* 47:289–95
194. Maia, J. C. da C., Kerjan, P., Szulmajster, J. 1971. *FEBS Lett.* 13:269–74
195. Mandels, G. R., Norton, A. B. 1948. *Quartermaster Gen. Lab. Res. Rep., Microbiol. Ser.* 11:1–50
196. Mandels, G. R., Vitols, R., Parrish, F. W. 1969. *J. Bacteriol.* 90:1589–98
197. Mandelstam, J., Waites, W. M. 1968. *Biochem. J.* 109:793–801
198. Marcus, A., Feeley, J. 1966. *Proc. Nat. Acad. Sci. USA* 56:1770–77
199. Marshall, B. J., Murrell, W. G. 1970. *J. Appl. Bacteriol.* 33:103–29
200. Merlo, D. J., Roker, H., Van Etten,

J. L. 1972. *Can. J. Microbiol.* 18: 949-56
201. Meselson, M., Nomura, M., Brenner, S., Davern, C., Schlessinger, D. 1964. *J. Mol. Biol.* 9:696-711
202. Millet, J., Kerjan, P., Aubert, J. P., Szulmajster, J. 1972. *FEBS Lett.* 23:47-50
203. Mims, C. W. 1971. *Mycologia* 63: 586-601
204. Munakata, N., Ikeda, Y. 1968. *Biochem. Biophys. Res. Commun.* 33: 469-75
205. Murray, R. G. E., Hall, M. M., Marak, J. 1970. *Can. J. Microbiol.* 16:883-87
206. Murrell, W. G. 1967. In *Advances in Microbial Physiology*, ed. A. H. Rose, J. F. Wilkinson, 1:133-251. London/New York: Academic
207. Myers, R. B., Cantino, E. C. 1971. *Arch. Mikrobiol.* 78:252-67
208. Naito, N., Tani, T. 1967. *Ann. Phytopathol. Soc. Japan* 33:17-22
209. Nagatsu, C., Matsuyama, A. 1970. *Agr. Biol. Chem.* 34:860-69
210. Neihof, R., Thompson, J. K., Deitz, V. R. 1967. *Nature* (London) 216: 1304-6
211. Nelson, D. L., Kornberg, A. 1970. *J. Biol. Chem.* 245:1128-36
212. Ibid, 1137-45
213. Nishi, A. 1961. *J. Bacteriol.* 81:10-19
214. Nordstrom, K. 1969. *J. Gen. Microbiol.* 55:1-7
215. Ono, T., Kimura, K., Yanagita, T. 1966. *J. Gen. Appl. Microbiol. Tokyo* 12:13-26
216. O'Sullivan, J., Lösel, D. M. 1971. *Arch. Mikrobiol.* 80:277-84
217. Ou, L., Marquis, R. E. 1970. *J. Bacteriol.* 101:92-101
218. Page, W. J., Stock, J. J. 1971. *J. Bacteriol.* 108:276-81
219. Park, D., Robinson, P. M. 1970. *Trans. Brit. Mycol. Soc.* 54:83-92
220. Pelletier, G., Hall, R. 1971. *Phytoprotection* 52:131-42
221. Pelletier, G., Hall, R. 1971. *Can. J. Bot.* 49:1293-97
222. Penniston, J. T., Harris, R. A., Asai, J., Green, D. E. 1968. *Proc. Nat. Acad. Sci. USA* 59:624-31
223. Powell, J. F. 1953. *Biochem. J.* 54: 210-11
224. Rado, T. A., Cochrane, V. W. 1971. *J. Bacteriol.* 106:301-4
225. Ramsey, W. S., Dworkin, M. 1970. *J. Bacteriol.* 101:531-40
226. Reeves, F. Jr. 1967. *Mycologia* 59: 1019-33
227. Reisener, H. J. 1967. *Arch. Mikrobiol.*

55:382-97
228. Roberts, E. H., Ed. 1972. *Viability of Seeds*, 321-59. London: Chapman & Hall
229. Robinson, P. M., Garrett, M. K. 1969. *Trans. Brit. Mycol. Soc.* 52: 293-99
230. Robinson, P. M., Park, D. 1966. *Trans. Brit. Mycol. Soc.* 51:113-24
231. Robinson, P. M., Park, D., Garrett, M. K. 1968. *Trans. Brit. Mycol. Soc.* 51:113-24
232. Rode, L. J. 1968. *J. Bacteriol.* 95: 1979-86
233. Rodenberg, S. et al 1968. *J. Bacteriol.* 96:492-500
234. Romine, M., Baker, R. 1972. *Phytopathology* 62:602-5
235. Rothschild, H., Itikawa, H., Suskind, S. R. 1967. *J. Bacteriol.* 94:1800-1
236. Rudolph, H. 1961. *Planta* 57:284-312
237. Rudolph, H., Furch, B. 1970. *Arch. Mikrobiol.* 72:175-81
238. Rudolph, H., Ochsen, B. 1969. *Arch. Mikrobiol.* 65:163-71
239. Russell, G. K., Bonner, J. T. 1960. *Bull. Torrey Bot. Club* 87:187-91
240. Sadoff, H. L., Celikkol, E., Engelbrecht, H. L. 1970. *Proc. Nat. Acad. Sci. USA* 66:844-49
241. Sakakibara, Y., Ikeda, Y. 1969. *Biochim. Biophys. Acta* 179:429-38
242. Sakakibara, Y., Saito, H., Ikeda, Y. 1969. *Biochim. Biophys. Acta* 174: 752-54
243. Samish, R. M. 1954. *Ann. Rev. Plant Physiol.* 15:185-201
244. Samsonoff, W. A., Hashimoto, T., Conti, S. F. 1970. *J. Bacteriol.* 101: 1038-45
245. Schafer, D. E., Chilcote, D. O. 1969. *Crop Sci.* 9:417-19
246. Scheld, H. W., Perry, J. J. 1969. *J. Gen. Microbiol.* 60:9-21
247. Schmoyer, I. R., Lovett, J. S. 1969. *J. Bacteriol.* 100:854-64
248. Scott, B. R., Alderson, T., Papworth, D. G. 1972. *Radiat. Bot.* 12:45-50
249. Servin-Massieu, M. 1971. *Curr. Top. Microbiol. Immunol.* 54:119-50
250. Setlow, P., Kornberg, A. 1970. *J. Biol. Chem.* 245:3645-52
251. Shaw, M. 1967. *Can. J. Bot.* 45: 1205-20
252. Shaw, M. V., Armstrong, R. L. 1972. *J. Bacteriol.* 109:276-84
253. Shepherd, C. J. 1957. *J. Gen. Microbiol.* 16:i (Abstr.)
254. Shepherd, C. J., Mandryk, M. 1963. *Aust. J. Biol. Sci.* 16:77-87
255. Shipton, W. A., McDonald, W. C. 1970. *Can. J. Bot.* 48:1000-2

256. Short, J., Walker, P. D. 1971. See Ref. 58, 189–91
257. Shu, P. K., Tanner, G., Ledingham, G. A. 1954. Can. J. Bot. 32:16–23
258. Sierra, G., Bowman, A. 1969. Appl. Microbiol. 17:372–78
259. Skoda, J., Blazkora, L., Dyr, J., Honzora, H., Vinter, V. 1967. Folia Microbiol. 12:557–61
260. Skujins, J. J., McLaren, A. D. 1967. Science 158:1570
261. Soll, D. R., Bromberg, R., Sonneborn, D. R. 1969. Develop. Biol. 20: 183–217
262. Soll, D. R., Sonneborn, D. R. 1971. Proc. Nat. Acad. Sci. 68:459–63
263. Soll, D. R., Sonneborn, D. R. 1971. J. Cell Sci. 9:679–99
264. Sonenshein, A. L. 1970. Virology 42:488–95
265. Sonenshein, A. L., Losick, R. 1970. Nature 227:906–9
266. Sonneborn, T. M. 1970. Proc. Roy. Soc. London 176:347–66
267. Souza, N. O., Panek, A. D. 1968. Arch. Biochem. Biophys. 125:22–28
268. Stafford, R. S., Donnellan, J. E. Jr. 1968. Proc. Nat. Acad. Sci. USA 59: 822–28
269. Staples, R. C. 1968. Neth. J. Plant Pathol. 74:25–36
270. Staples, R. C., App, A. A., McCarthy, W. J., Gerosa, M. M. 1966. Contrib. Boyce Thompson Inst. Plant Res. 23:159–64
271. Staples, R. C., Bedigian, D. 1967. Contrib. Boyce Thompson Inst. Plant Res. 23:345–47
272. Staples, R. C., Syamananda, R., Kao, V., Black, R. J. 1962. Contrib. Boyce Thompson Inst. Plant Res. 21: 345–62
273. Stedman, R. L., Kravitz, E., Anmuth, M., Harding, J. 1956. Science 124: 403–5
274. Steinberg, W., Halvorson, H. O. 1968. J. Bacteriol. 95:469–78
275. Ibid, 479–89
276. Steiner, G. W., Lockwood, J. L. 1970. Phytopathology 60:89–91
277. Stocker, O. 1956. Encycl. Plant Physiol. 3:652
278. Sussman, A. S. 1966. The Fungi, ed. G. C. Ainsworth, A. S. Sussman, 2:733–64. New York: Academic
279. Sussman, A. S. 1969. Soc. Exp. Biol. 23:99–121
280. Sussman, A. S. 1969. See Ref. 68, 1–38
281. Sussman, A. S., Halvorson, H. O. 1966. Spores, Their Dormancy and Germination, 354. New York: Harper

& Row
282. Sussman, A. S., Holton, R., von Böventer-Heidenhain, B. 1958. Arch. Mikrobiol. 29:38–50
283. Sussman, A. S., Lowry, R. J., Durkee, T. L., Maheshwari, R. 1969. Can. J. Bot. 47:2073–78
284. Swartz, M. N., Kaplan, N. O. Frech, M. E. 1956. Science 123:50
285. Szulmajster, J., Arnaud, M., Young, F. E. 1969. J. Gen. Microbiol. 57:1–10
286. Tabak, H. H., Cooke, W. B. 1968. Bot. Rev. 34:126–252
287. Tanaka, K., Kogané, F., Yanagita, T. 1965. J. Gen. Appl. Microbiol. 11:85–90
288. Tanaka, K., Motohashi, A., Miura, K., Yanagita, T. 1966. J. Gen. Appl. Microbiol. 12:277–92
289. Tanooka, H. 1968. Biochim. Biophys. Acta 166:581–83
290. Tanooka, H. 1969. Photochem. Photobiol. 9:95–97
291. Tanooka, H., Sakakibara, Y. 1968. Biochem. Biophys. 155:130–42
292. Tanooka, H., Terano, H. 1970. Radiat. Res. 43:613–26
293. Terano, H., Tanooka, H., Kadota, H. 1969. Biochem. Biophys. Res. Commun. 37:66–71
294. Terano, H., Tanooka, H., Kadota, H. 1971. J. Bacteriol. 106:925–30
295. Terman, S. A., Gross, P. R. 1965. Biochem. Biophys. Res. Commun. 21:595
296. Terui, G., Mochizuki, T. 1955. Technol. Rep. Osaka Univ. 5:219–27
297. Tisdale, J. H., DeBusk, A. Gib 1972. Biochem. Biophys. Res. Commun. 48:816–22
298. Tochikubo, K. 1971. J. Bacteriol. 108:652–61
299. Tokunaga, J., Bartnicki-Garcia, S. 1971. Arch. Mikrobiol. 79:293–310
300. Tolenaar, H., Houston, B. R. 1966. Phytopathology 56:1036–39
301. Tono, H., Kornberg, A. 1967. J. Biol. Chem. 242:2375–82
302. Trocha, P., Daly, J. M. 1970. Plant Physiol. 46:520–26
303. Truesdell, L. C., Cantino, E. C. 1970. Arch. Mikrobiol. 70:378–92
304. Tsay, Y., Nishi, A., Yanagita, T. 1965. J. Biochem. Tokyo 58:487–93
305. Tuveson, R. W., West, D. J., Barratt, R. W. 1967. J. Gen. Microbiol. 48: 235–48
306. Ursprung, H. 1965. In Organogenesis, ed. R. L. DeHaan, H. Ursprung 3–27. New York: Holt, Rinehart & Winston
307. Van Assche, J. A., Carlier, A. R.,

Dekeersmaeker, H. I. 1972. *Planta* 103:327–33
308. Van Etten, J. L. 1968. *Arch. Biochem. Biophys.* 125:13–21
309. Van Etten, J. L. 1969. *Phytopathology* 59:1060–64
310. Van Etten, J. L. 1971. *J. Bacteriol.* 106:704–6
311. Van Etten, J. L., Brambl, R. M. 1968. *J. Bacteriol.* 96:1042–48
312. Van Etten, J. L., Brambl, R. M. 1969. *Phytopathology* 59:1894–1902
313. Vegis, A. 1963. In *Environmental Control of Plant Growth*, ed. L. T. Evans, 265–87. New York: Academic
314. Vinter, V. 1970. *J. Appl. Bacteriol.* 33:50–59
315. Vold, B. S., Minatogawa, S. 1972. See Ref. 55, 254–63
316. Walker, P. D., Baillie, A. 1968. *J. Appl. Bacteriol.* 31:108–13
317. Walker, P. D., Thomson, R. O., Baillie, A. 1968. *J. Appl. Bacteriol.* 30:444–49
318. Warth, A. D., Strominger, J. L. 1972. *Biochemistry* 11:1389–96
319. Watson, A. G., Ford, E. J. 1972. In press
320. Wax, R., Freese, E. 1968. *J. Bacteriol.* 95:433–38
321. Weber, D. J. 1966. *Phytopathology* 56:118–23
322. Wessels, N. K. et al 1971. *Science* 171:135–43
323. Widdowson, J. P. 1967. *Nature* 214:812
324. Williams, P. G., Allen, P. J. 1967. *Phytopathology* 57:656–61
325. Wilson, E. M. 1958. *Phytopathology* 48:595–600
326. Wong, R. S. L., Scarborough, G. A., Borek, E. 1971. *J. Bacteriol.* 108:446–50
327. Wood-Baker, A. 1955. *Trans. Brit. Mycol. Soc.* 38:291–97
328. Wynn, W. K., Gajdusek, C. 1969. *Contrib. Boyce Thompson Inst. Plant Res.* 24:123–38
329. Yanagita, T. 1957. *Arch. Mikrobiol.* 26:329–44
330. Yanagita, T. 1964. In *Synchrony in Cell Division and Growth*, ed. E. Zeuthen, 391–420. New York: Wiley
331. Yaniv, Z., Staples, R. C. 1968. *Contrib. Boyce Thompson Inst. Plant Res.* 24:103–8
332. Yolton, D. P., Pope, L., Williams, M. G., Rode, L. J. 1968. *J. Bacteriol.* 95:231–38
333. Young, T. W. K. 1968. *Proc. Linnaean Soc. London* 179:1–12
334. Zamenhof, S., Eickhorn, H. H., Rosenbaum, D. *Nature London* 220:818–49

Ann. Rev. Plant Physiol. 1973. 24:353–80

HORMONE METABOLISM IN DISEASED PLANTS

❖ 7552

Luis Sequeira

Department of Plant Pathology, University of Wisconsin, Madison

CONTENTS

INTRODUCTION

The exaggerated growth responses that are frequently associated with the invasion of plants by pathogenic or symbiotic microorganisms, or those that occasionally accompany the exposure of plants to unfavorable environmental conditions, have attracted the attention of plant pathologists and physiologists for many years. The general theories that have been proposed over the past 50 years to explain these growth abnormalities usually include hormonal imbalances, for it has been frequently demonstrated that affected tissues contain higher or lower than normal amounts of the substances now recognized as important growth regulators in plants. These studies, for the most part, have suffered from the usual limitations imposed by the rather nonspecific bioassays used to measure hormonal activity. Also, experimental procedures have not taken into account both morphological and physiological changes in the host. Most studies have attempted to interpret alterations in normal growth as the result of changes in a single type of growth regulator. There is now a growing realization that the complex interrelationships between auxins, cytokinins, gibberellins, ethylene, etc must be taken into account. Until very recently, however, there were no adequate chemical techniques that allowed quantitation of all endogenous regulators. An integrated view of the growth regulator changes that accompany even the simplest plant response to irritation by a pathogenic agent is still very much in the future.

Although the present state of the art in plant hormone-disease relationships is disappointing in many respects, it should be pointed out that the study of such systems has contributed immensely to our knowledge of natural growth regulators. The discovery of gibberellins, and the more recent finding that 6-(3-methyl-2-butenylamino) purine (2iP) is an important natural cytokinin, were possible partly because of observations concerning unusual growth responses of plants to parasitic fungi and bacteria (35, 54, 112, 116). The fact that plant pathogens produce compounds which are identical to the natural growth regulators of higher plants has been and continues to be the basis for exciting and extremely important findings in the field of plant physiology.

Almost a decade has elapsed since this reviewer had an opportunity to compile some of the information on the relationship of growth regulators to pathogenesis (98). No useful purpose would be served by reviewing this information. Additional reviews on this subject, and in ancillary areas, have appeared in the intervening years (120, 122). It is the purpose of the present review to bring more closely into focus the most recent information available on the synthesis and degradation of growth regulators as they relate not only to abnormal growth phenomena induced by plant pathogens, but to the entire problem of disease resistance in plants as well. Completeness is not a primary aim in this review. Only those examples which seem to illustrate the points to best advantage have been included.

AUXINS

The capacity of a large number of plant pathogens to direct the growth of the host into new and abnormal directions has led to the demonstration that auxin levels in the affected tissues are higher, or in some cases lower, than normal (120, 122). To date, none of these studies has provided a reasonable explanation as to how these changes come about. Very few attempts have been made to study auxin synthesis and degradation by both host and pathogen, and no one has seriously tackled the problem of separating the contributions of both members to the auxin changes occurring at various stages of pathogenesis. In particular, the importance of auxin alterations during initiation of infection and the presumed role of IAA in creating conditions which are more favorable for multiplication of the pathogen have remained largely unexplored.

Crown Gall

Problems related to auxin synthesis have been studied largely with bacterial diseases, and particularly with those that involve overt growth alterations that result in galls and other tumor-like structures. Continued attention has centered on the crown gall disease incited by *Agrobacterium tumefaciens* because of the similarities of this disease to animal cancer. Once the bacterium comes in contact with host cells at a particular stage after wounding, these cells become fully transformed and grow in disorganized fashion even in the absence of the incitant (9). Early studies by Link and co-workers (61, 62) showed that *A. tumefaciens* produced a growth substance, presumed to be IAA, when grown in a liquid medium containing tryptophan. Assays of tomato hypocotyls bearing galls showed that they contained 15 times more auxin than normal tissues, and the term "hyperauxiny" was coined to describe this condition.

The application of tissue culture techniques resulted in the discovery that bacteria-free gall tissues grew well in the absence of auxin, while normal tissues required considerable quantities of auxin and certain growth factors for optimal growth (57). The absolute necessity of IAA for growth of crown gall tissue, however, was not clearly demonstrated until 1956 (55).

Separation of auxins from crown gall tissues and the analysis of the growth activity of these substances was first accomplished by Bitancourt (6). He found that tobacco and willow tumor cells contained IAA and indole-3-acetonitrile (IAN) at $4 \times 10^{-8} M$ and $1 \times 10^{-8} M$, respectively, well within the range previously indicated for total auxin activity (57). These levels are approximately $10 \times$ those in normal callus tissue. Further work in Bitancourt's laboratory confirmed the presence of IAA, IAN, and an auxin at the R_f of indole-3-acetic acid ethyl ester (IAE) in tumor tissue of tobacco (96). Similarly, Dye et al (27) found IAA and IAN, but indole-3-carboxylic acid (ICA) rather than IAE, in both crown gall and normal stem tissue extracts from tomato. These auxins were present in greater amounts in gall tissues; the amount of free IAA in the latter tissues was estimated to be 8–12 μg/kg of tissue. That other unidentified auxins may be present in abnormal amounts in crown gall tissues may be inferred from histograms showing the biological activity of crude extracts from *Parthenocissus* (73). Although the presence of artifacts in these crude assays is a distinct possibility, and one can question the quantitative aspects of assays carried out with minute amounts of tissue, it seems possible that the shift of a normal cell to the tumorous condition involves a series of auxins, of which IAA may constitute only a minor component.

Although there seems to be agreement as to the fact that crown gall tissues contain higher than normal amounts of IAA, the quantitative aspects of this phenomenon have not been resolved. First, comparisons between healthy and diseased tissues have been made on a fresh weight basis in most instances. Since it is very likely that healthy tissues have a higher degree of hydration than crown gall tissues, the IAA content of the latter would tend to be overestimated. When assay results have been calculated on a dry weight basis and both free and bound IAA have been determined (36), the results in some instances have indicated a higher total content for stem or callus tissues as compared with crown gall tissues. It is difficult to reconcile these results with the high total auxin levels reported by Link & Eggers (61) for tomato crown gall tissue following long-term extractions and the use of proteolytic enzymes to release bound auxin. Second, discrepancies in the quantitative estimates of auxin content in tumorous tissue are common. For example. Nitsch (73) and Schwarz et al (96) differ by approximately 1000-fold in their auxin estimates. Although such striking differences may be due to the nature of the tissues, the extraction procedure, and bioassay techniques, it is apparent that both qualitative and quantitative aspects of auxin content in crown gall tissues deserve a great deal of additional attention.

Synthesis of IAA by the crown gall bacterium in culture is dependent on the presence of tryptophan in the medium (46). Indole-3-pyruvate (IPyA) was shown to be the main intermediate leading to IAA, indole-3-lactic acid (ILA), and tryptophol; IPyA presumably originated from simple transamination of tryptophan, since in cell-free preparations the reaction is dependent on the presence of the transamination cofactors, α-ketoglutarate and pyridoxal phosphate (111). Since crown gall tissues can continue to synthesize auxin in the absence of the bacterium, what, if any, is the significance of the ability of this pathogen to synthesize IAA? In an attempt to answer this question, Lippincott & Lippincott (64) used a strain of A. tumefaciens which induces tumors in certain resistant hosts only when auxin is added. Addition of auxin soon after infection increased the number of tumors initiated by A. tumefaciens on primary pinto bean leaves. In inoculations with low concentrations of virulent (B6) cells, the presence of avirulent (IIBNV6) cells of the bacterium increased tumor initiation. Since both strains can synthesize auxin in culture and the increase in tumor initiation depends on the presence of viable cells of IIBNV6, it was suggested that auxin was the substance limiting tumor formation by B6 cells. It is obvious that this approach does not provide direct evidence for the involvement of bacterial auxin in tumor initiation, but the results should stimulate further work on this problem.

It is surprising indeed that there is no available information on the selection of IAA-less mutants of A. tumefaciens which could be used to obtain more direct answers to the questions raised above. In fact, if transformation of normal host cells to tumor cells is due to the uptake, incorporation, and transcription of part of the bacterial genome (113), IAA synthesis by tumor cells may follow routes which are different from those of normal cells. Presumably these routes could be identified with the use of appropriate precursors.

Bacterial Wilt

The approach suggested above has been used by Phelps & Sequeira (77) to determine the contribution to the IAA pool of each member of the host-parasite complex represented by tobacco and *Pseudomonas solanacearum*, a vascular pathogen. A comparison of the synthetic capacities of the host, as well as of virulent (K60) and avirulent (B1) forms of *P. solanacearum* in cell-free systems, suggested that each group of organisms utilized widely different pathways in the synthesis of IAA from tryptophan. Avirulent forms synthesized IAA by transamination in the presence of pyridoxal phosphate and α-ketoglutarate; synthesis proceeded via IPyA and indole-3-acetaldehyde (IAA1d). With the pathogenic strain, on the other hand, IAA synthesis proceeded in the absence of added transamination cofactors. With the use of ring and chain ^{14}C-labeled tryptophan it was determined that IAA was synthesized from products of the kynurenine pathway of tryptophan breakdown, but the actual pathway was not determined. In contrast, in cell-free systems from tobacco terminal buds, IAA synthesis appeared to proceed via tryptamine (TNH$_2$) and IAAld. By feeding radioactive TNH$_2$, which only the host could convert to IAA, it was determined that the host contributes most of the IAA which accumulates during the early stages of pathogenesis (100).

Tryptophan in the free form is usually present in very low concentrations in plant tissues (ca 0.5 μg/ml in tobacco xylem sap). Considering the relatively low efficiency of tryptophan conversion to IAA exhibited by most plant pathogens and host tissues, the question may be asked as to whether sufficient amounts of the precursors would be available for IAA synthesis. One would expect that most of the tryptophan would be shunted to bacterial protein synthesis and that very little would be available for other functions such as IAA synthesis. In the *P. solanacearum*-tobacco system, Pegg & Sequeira (75) showed that during the first 48 hr of infection important metabolic changes occur which lead to selective increases in aromatic amino acids. This change in amino acid content occurred at a time when the bacterium was still confined to the vascular system. Within 48 hr after inoculation the tryptophan content had increased by approximately 600%, and the phenylalanine content showed increases of similar magnitude; this selective increase in aromatic amino acids was thought to result from increased activity of the shikimate pathway. Thus the dramatic changes in concentrations of tryptophan would suggest that there is no shortage of precursor for IAA synthesis during the critical early stages of infection. The change in tryptophan content occurs at a time when increased protein synthesis, rather than breakdown, is induced in host tissues as a result of infection.

Olive Knot

The tumorous overgrowths incited by *Pseudomonas savastanoi*, unlike crown gall tumors, are dependent on the presence of the bacterium. It has been shown that IAA accumulates in the affected tissues (66). The bacterium produces relatively large amounts of IAA in culture, and washed cells or cell-free preparations produce indole-3-acetamide (IAM) and IAA from L-tryptophan. The sequence of IAM and IAA formation in reaction mixtures suggested that *P. savastanoi* synthesizes IAA from tryptophan via IAM (67). Kuo & Kosuge (58) established that tryptophan oxidative decarboxylase and indoleacetamide hydrolase are involved in the conversion of L-tryptophan to IAA and are produced even in the absence of tryptophan in the medium. The IAM pathway appears to be the dominant route for IAA synthesis by the bacterium, but an alternative pathway is present via an aminotransferase which catalyzes the conversion of L-tryptophan to IPyA. Whole cells of *P. savastanoi* converted IPyA to IAA, but the rate of conversion was relatively slow (59). Because the aminotransferase had a broad range of activity on several amino donors and acceptors, it was concluded that its capacity to convert tryptophan to IPyA was not one of its major functions in the cell.

IAA and IAM inhibit the tryptophan oxidative decarboxylase of *P. savastanoi*, but IAA does not inhibit the IAM hydrolase; as a result, IAA and not IAM tends to accumulate in the medium. IAA is also converted to a lysine conjugate, and therefore the extent to which free IAA accumulates is dependent on the relative activities of the systems that synthesize it and those that form the conjugate. With this background knowledge on IAA synthesis by the pathogen, it should be possible now to look at the synthetic capacities of infected tissues to determine to what extent the pathogen contributes to the high levels of IAA observed in tumor tissues.

Root Nodules

The possible involvement of growth regulator imbalances has been studied extensively in the root nodules which arise in leguminous plants as a result of invasion by bacteria of the genus *Rhizobium*. The early literature on this subject was very adequately reviewed by Braun (8). Interest on the influence of growth regulators has centered on two aspects of host responses to infection: (*a*) the characteristic root hair deformation or "curling" associated with the initial presence of the bacteria at the tip of the root hair; and (*b*) the rapid division and enlargement of infected cells in the root cortex. In both instances, a diffusible factor, presumably auxin, has been implicated. This idea has been supported by the demonstration that the bacterium produces IAA abundantly in culture and that nodule tissues contain relatively large amounts of auxin.

Regarding point *a*, Yao & Vincent (128) have shown that the markedly curled condition of host root hairs is restricted to host-parasite combinations involving homologous rhizobia. Bacteria-free filtrates, however, failed to produce the markedly curled condition. Since the filtrates were prepared after suspending cells in distilled water for one hour, the possibility that a more highly concentrated preparation, acting in more localized fashion, could reproduce the normal curling symptoms was not excluded. Although the filtrates had a low order of specificity, those from homologous strains of bacteria had a greater effect than those from heterologous strains. Although IAA can elicit root hair deformation comparable to that produced by many types of rhizobia, numerous arguments have been presented against the interpretation that this auxin is the "curling factor." In particular, the lack of specificity has been frequently noted, for this auxin is produced by many soil bacteria which have no effect on the root hairs of leguminous plants (128). The most recent study by Hubbell (39) indicates that IAA is not responsible, or at least not solely responsible, for the curling of root hairs. A crude preparation of an extracellular polysaccharide (glycoprotein) produced by *Rhizobium* induced curling of clover root hairs which was comparable to that induced by the bacterium. More significantly, a noninfective strain of the same bacterial species produced a different polysaccharide which induced only slight deformation of root hairs. Just how this compound could affect root hair development was not established, although a possible role as a pectic enzyme inducer was suggested.

Regarding point *b*, elaborate explanations have been advanced to explain the development of the nodules in terms of auxin released by the symbiont in the root tissues (8). Clover roots, for instance, release very substantial amounts of tryptophan, and it was shown that this amino acid can be converted to IAA by *Rhizobium trifolii* growing in the root exudates (50). It was conjectured that localized production of IAA within the root cortex by bacteria in the infection thread could initiate cell division and elongation in cells having the required concentration of cytokinins, since both growth regulators would be required for cell division to be initiated. These and other hypotheses are interesting, but the fact remains that the nodule is not a shapeless mass of tissue but a highly organized structure. Explanations that do not take into account the morphological development of the nodule in relation to the balance of different growth regulators cannot be particularly useful.

Emphasis has been placed in recent years on the pathways for synthesis of IAA from tryptophan by various species of *Rhizobium*. Rigaud (88, 89) concluded that *R. meliloti* produced IAA at a relatively rapid rate for several days after stationary phase was reached. In addition to IAA, IAAld and tryptophol were identified, but only the aldehyde was considered a precurosr of IAA. A more detailed study of auxin synthesis by rhizobia has been provided by Dullaart (25), who reported the production of IAA and ICA from L-tryptophan in vitro by cell-free preparations from root nodules and roots of *Lupinus luteus* and from *R. lupini*. No particularly useful differences were found between the pathways utilized for synthesis of IAA by the bacterium and those of the nodules or normal roots. In all cases, transamination of tryptophan to IPyA, followed by decarboxylation to IAA, was presumed to be the major pathway since addition of pyridoxal phosphate and α-ketoglutarate enhanced the reaction markedly in all cases.

As in the case of the *P. solanacearum*-tobacco system discussed previously, it was concluded that a very substantial part of the large amounts of IAA present in the nodules is of host origin and that it is the result of metabolic alterations induced by the symbiont. That these alterations do not affect all plant indolics in the same fashion was shown by the fact that ICA, as opposed to IAA, was present in the nodules at much lower concentrations than in the normal roots (24).

The reasoning that led to the suggestion that most of the auxin in nodules is of host origin is not entirely clear. Rhizobial cells have a greater capacity to synthesize IAA than root tissues, but Dullaart (25) assumed that bacterial proteins would not constitute a significant part of the total protein in homogenized nodule enzyme preparations which could also synthesize IAA actively. This is merely an assumption, and because it is a critical point in the main argument that is presented, it should be investigated further.

Auxin synthesis by plant-pathogenic fungi appears to proceed in most cases along pathways which are similar to those described for bacterial pathogens (98). It would serve no useful purpose to include this information in this review, particularly in view of the fact that little or no effort has been made to examine the problem of auxin synthesis within diseased tissues. In the few instances in which new or unusual pathways of IAA synthesis from tryptophan have been suggested, further work has almost invariably failed to support the original suggestion. As an example, IAA synthesis from tryptophan via TNH_2 has been reported in *Dibotryon morbosum* and *Taphrina deformans* (16), but careful work by Perley & Stowe (76) showed that *T. deformans* produced only IPyA, tryptophol, and IAA when grown in a medium supplemented with tryptophan.

AUXIN DEGRADATION

Although nonenzymatic inactivation of IAA may occur in plants, it is clear that this substance is normally degraded by enzymes of the peroxidase type (85). Principal exceptions appear to be the IAA-oxidizing enzymes of certain basidiomycetes (e.g. *Polyporus versicolor*) which are phenoloxidases of the laccase type (28, 118). At least one plant pathogen, *Marasmius perniciosus*, produces both a peroxidase and a laccase, each capable of catalyzing the oxidative breakdown of IAA (56).

Although there are numerous reports of changes in the activity of enzymes involved in auxin degradation following infection by plant pathogens (98), it is difficult to make an assessment as to the importance of these changes in pathogenesis. The activity of IAA oxidase appears to depend on a precise balance of phenolic cofactors and Mn^{++}, and the frequent presence of other phenolic constituents which inhibit the activity of the enzyme complicates any attempts to correlate the activity in vitro with that in the tissues. Attempts to measure activity of the enzyme in vivo by addition of exogenous, radioactive IAA is frequently complicated by changes in permeability which alter the uptake of auxin by the infected tissues. Also, the activity of the enzyme is known to change under conditions of stress such as injury (72) or lowered water potential (19). It seems likely, therefore, that changes in IAA oxidase activity reflect physiological stress brought about by infection and may not be connected in any way with the ability or inability of the pathogen to multiply in host tissues.

IAA Oxidase and Disease Resistance

Perhaps the most interesting and thorough study on IAA oxidase activity in infected plants has been carried out with the wheat-*Puccinia graminis* system. Interest in this system stems from the initial report by Shaw & Hawkins (105) that $^{14}CO_2$ derived from IAA was released much more slowly from leaf disks of rust-infected tissues than from comparable healthy tissues incubated with radioactive indoleacetate. The initial interpretation of the data on $^{14}CO_2$ recovery during the course of infection suggested that oxidase activity was inversely correlated with IAA content (although IAA content was not determined), and they postulated that the rapidity of change from the initially high rate to a reduced rate of enzyme activity was correlated with resistance or susceptibility of the tissue to fungal invasion. This suggestion has been the subject of considerable controversy. In an initial study of factors that influence rates of decarboxylation of IAA in rusted tissues, Daly & Deverall (17) determined that there was, in fact, a significant increase in decarboxylation during the first 2 or 3 days after inoculation, and that rates fell appreciably only after sporulation was well under way. They concluded, therefore, that rates of decarboxylation, presumably due to the activity of IAA oxidase, could not account for the increases in IAA observed during the early stages of infection. Further, they suggested that the low rates of decarboxylation observed during sporulation could be due to failure of IAA to be taken up by the diseased tissues.

In attempts to compare levels of IAA degradation in resistant or susceptible wheat varieties, the genetic diversity of the varieties used made it extremely difficult to establish correlations between resistance to rust and IAA oxidation and/or IAA content at different stages of infection. In an attempt to remedy this problem, Antonelli & Daly (2) used nearly isogenic lines of wheat, differing only at the Sr_6 locus. The Sr_6 gene is temperature sensitive, providing high resistance (infection type 0) at 20°C and susceptibility (infection types 3–4) at 25°C. The results of IAA decarboxylation studies with the susceptible material essentially confirmed earlier findings: there was an initial burst of decarboxylation during early stages of infection, followed by a lower rate at the time of sporulation. The initial increase in decarboxylation was similar in both susceptible and resistant combinations, but after 2–3 days, rates of decarboxylation in resistant tissues increased to eight times the values for healthy tissues, although there were no visible signs of infection. It was thought that such high rates correlated with the onset of resistance. With this, as with other methods which are dependent on the uptake of exogenous IAA into the tissues, it is difficult to determine if changes in rates of decarboxylation are not merely a reflection of changes in permeability. Therefore, data on IAA uptake would have been useful. Also, it is assumed that IAA oxidase mediates decarboxylation, but it is clear that a number of other oxidative enzymes can accomplish this reaction. Increases in rate of decarboxylation may merely reflect participation of other oxidases.

In an attempt to determine the possible relation between resistance, changes in IAA decarboxylation, and content of peroxidases which could catalyze the reaction, Seevers & Daly (97) measured the activity of these enzymes in leaves of near-isogenic lines of wheat carrying the Sr_6 or sr_6 alleles. As in previous studies based on IAA decarboxylation, marked peroxidase increases seemed to be characteristic of leaves showing the resistant reaction at 20°C beyond 2 days after inoculation, while peroxidase in susceptible leaves increased only slightly or not at all. However, when inoculated resistant lines showing high peroxidase activity were transferred to 26°C, reversion to a susceptible reaction occurred, but peroxidase activities were maintained at a high level. Thus there appeared to be no correlation between resistance and peroxidase activity. Because IAA oxidase activity paralleled total peroxidase activity, it could not be argued that the catalytic properties of the peroxidases had changed in plants subjected to conditions that caused the reversal in reaction to the pathogen. It could be argued, as a final and fragile resort, that the change to 26°C placed the plants under conditions which made peroxidase-mediated resistance mechanisms inoperative. In any case, it seems likely that the observed increases in peroxidases capable of IAA decarboxylation are the consequence rather than the cause of the metabolic changes that result in a resistant reaction.

There have been additional attempts to correlate activity of IAA oxidase with resistance to various diseases, e.g. bean to *Pseudomonas phaseolicola* (22), *Nicotiana tabacum*, *N. glutinosa*, and *N. sylvestris* to two different strains of tobacco mosaic virus (80), but one obtains the general impression that these studies have not been carried out in sufficient detail to permit any definite conclusions.

IAA Oxidase and Tumor Formation

Perhaps the most controversial question regarding the possible relationship of IAA oxidase activity to pathogenesis concerns the crown gall disease incited by *Agrobacterium tumefaciens*. It was initially thought that the high auxin content of the gall tissues was correlated with decreased activity of the enzyme, but there have been numerous claims and counterclaims, alternatively supporting or dismissing this concept (for a review of the earlier literature, see 98).

Stonier (110) has supported the idea that natural "auxin protectors," substances with relatively high molecular weight, increase dramatically in sunflower internodes inoculated with a virulent strain of *A. tumefaciens*. Since auxin protectors are antioxidants which prevent the peroxidase-catalyzed oxidation of IAA in plant meristems, Stonier concluded that the presence of high levels of at least one of these compounds (protector A) in tumor tissue explains the similarity to young meristematic tissues.

That crown gall tissue cultures release a potent IAA oxidase into the growth medium is evident from the recent work of Rubery (93). Since the presence of exogenous IAA is frequently inhibitory to the growth of crown gall cells in culture, Rubery suggested that the IAA oxidizing activity of the medium could regulate growth by preventing the accumulation of excess auxin. As with earlier work with tobacco root enzymes (102), the IAA oxidizing activity in *Parthenocissus* was separable from peroxidase activity by gel filtration. That the two activities are separable is not usually taken into account.

In addition to crown gall, several other malformations incited by plant pathogens have received attention in relation to IAA oxidase activity. Raa (83) considered the relationship between high IAA levels and activity of IAA oxidase in cabbage roots infected by *Plasmodiophora brassicae*. Roots infected by this pathogen form tumorous overgrowths (clubroots) which may contain 50 to 100 times the levels of auxin in the normal root. From clubroot tissues, Raa separated two IAA-oxidizing fractions differing in kinetic properties. Because at low concentrations of IAA there was very little difference between clubroot and normal root extracts in their ability to decarboxylate IAA, Raa concluded that activity of the enzymes could not afford a possible explanation for the difference in IAA content between the two tissues. The only reasonable alternative is that clubroot tissues synthesize IAA at a rate faster than that of normal root tissues.

In contrast, Mennes (68) extracted IAA oxidase from rhizobial nodules and normal roots of *Lupinus luteus* and, after partial purification by ammonium sulfate fractionation and elution through a Sephadex column, concluded that IAA degradation in root nodules was significantly lower than that of normal roots. No differences in the kinetics of the enzymes from root nodules and roots were found. Mennes confirmed earlier reports that *Rhizobium* produces IAA oxidase in culture (88), but could not prove conclusively that the bacteroids present within the tissues also have this activity. It would seem likely that the large amount of bacterial protein that would be carried along in the IAA oxidase fraction could affect the calculations of specific activities in the nodule preparations and thus result in lower values than those for normal tissues. An additional complicating factor, as Mennes has pointed out, is the increased content of polyphenols in nodule tissue which gives greater opportunity for nonspecific binding of oxidized phenolic products with proteins. To what extent nonspecific binding affects IAA oxidase activity in diseased tissues is difficult to estimate, but the common observation that nodule tissues are very rich in phenolic compounds and the lack of precautions to prevent their oxidation during extraction would suggest that reports of low IAA oxidase activity may be erroneous in many cases.

IAA Oxidase Inhibition

Of greater interest has been the demonstration of specific phenolic inhibitors of IAA oxidase in extracts of diseased tissues. Ortho-dihydric phenols are well-known inhibitors of the enzyme, and the frequent accumulation of caffeic acid and of caffeic acid derivatives (chlorogenic acid) in diseased tissues has suggested that inhibition of IAA oxidase by these compounds, linked to increased synthesis of IAA, might explain abnormal growth phenomena associated with infection. In the case of the *Pseudomonas solanacearum*-tobacco system, it was shown that activity of IAA oxidase was significantly decreased within 48 hr after inoculation and that this inhibition was due to the presence of dializable inhibitors, particularly scopoletin and chlorogenic acid (99). Amounts of scopoletin in diseased tobacco tissues were shown to be sufficient to cause significant inhibition of the enzyme (101). In stem tissues there were threefold increases in scopolin (the 7-glucoside of scopoletin) by 48 hr after inoculation, and concentrations continued to increase exponentially. Increases in scopoletin followed a pattern similar to that of scopolin, but the exponential rise occurred approximately 24 hr later. The accumulation of scopoletin was much more rapid than that of scopolin, reaching 18 times normal by 120 hr after inoculation. It was thought that increases in scopoletin might reflect increased hydrolysis of scopolin, but the specific activity of β-glycosidase did not increase significantly over control values.

Attempts have been made to localize the changes in scopoletin in diseased tobacco tissues (15). Use was made of the high fluorescence characteristic of scopoletin under uv light which allowed the content of this compound in individual cells to be quantitated by cytofluorometry. Scopoletin increased in cells of the endodermis opposite, and in the pith and xylem parenchyma adjacent to, infected xylem elements. Eight- to eighteenfold increases in scopoletin were detected by 5 days after inoculation; concentrations in the vacuolar contents of endodermal cells reached 6 μg/ml and were still increasing by 7 days after inoculation. These concentrations would be expected to cause almost complete inhibition of IAA oxidase (106). Hypertrophy of xylem parenchyma cells frequently crushed protoxylem elements, and it was in these cells that the earliest and most pronounced changes in scopoletin occurred. Copeman hypothesized that proliferation of the xylem parenchyma could be due to increased levels of IAA synthesized by host tissues under the influence of the bacterium (100) and maintained at a high level via scopoletin-mediated inhibition of IAA oxidase.

Reduction in IAA Content

The preceding discussion has dealt almost exclusively with reduced degradation of IAA in infected tissues and the consequences of concomitant increases in IAA. It should be apparent that the reverse situation, i.e. increased activity of IAA oxidase and the corresponding decrease in IAA content, can result in alterations of the normal growth pattern which may be just as extraordinary as those caused by increased auxin levels. It can be hypothesized that lower than regulatory levels of IAA could result from (a) interference in the synthesis and transport of this auxin; (b) a pathogen-induced increase in IAA oxidase activity of host origin; and (c) secretion of IAA oxidase by the pathogen within the host tissues.

Certain "physiological" disorders of plants have been correlated with decreases in the content of auxin. Perhaps the best example is the "frenching" disease of tobacco, a common disease characterized by the development of numerous sword-shaped narrow leaves, the loss of apical dominance, and the failure of internodes to elongate. These symptoms are indicative of auxin deficiency, and Kefford (49) found that severely "frenched" tips contain an average 2.16 μg IAA/kg fresh weight, while normal tissues contain 13.72 μg/kg. Just how this imbalance is triggered and maintained is still a matter of considerable controversy (125).

Of greater interest is the production of IAA oxidase by plant pathogens, although it is extremely difficult to obtain quantitative information on the amounts, and localization of the enzyme in the infected tissues. For example, IAA oxidase produced by *Omphalia flavida* (*Mycena citricolor*) is apparently involved in the premature abscission of coffee and *Coleus* leaves (103) induced by this pathogen. Presumably, abscission is triggered by a reduction in the amount of auxin moving from the blade to the petiole. The presence of an active IAA oxidase in infected coffee leaf tissues was demonstrated by comparing rates of decarboxylation of IAA ($-^{14}$COOH) by disks from healthy and infected leaves (91). Decarboxylation of IAA increased markedly as the disease progressed. It was postulated that the enzyme diffuses out of the infected areas and contributes to the increase in oxygen consumption characteristic of the healthy areas surrounding the lesions. Although the production of IAA oxidase by the pathogen in vivo seems to be reasonably well documented, the effects on abscission are debatable. Certainly the possible involvement of ethylene as an inducer of abscission in this case deserves attention.

The possible role of an IAA oxidase system of pathogen origin has also been considered in the case of the witches' broom disease of cacao incited by *Marasmius perniciosus* (56). Infection results in inhibition of apical growth and proliferation of axillary buds (brooming). The fungus does not produce significant amounts of auxin or cytokinins in culture, but culture filtrates destroy IAA rapidly because of the formation of two enzymes. One is a peroxidase with a pH optimum of 3.0, and the other is a laccase with a pH optimum of 6.0. Direct demonstration of the presence of these enzymes in vivo was not obtained, but it seems likely that the very marked ability of the fungus to destroy IAA would be important in pathogenesis.

CYTOKININS

Infection by plant pathogens, particularly by those that are obligate in nature, frequently results in temporary, compatible associations in which the metabolism of host cells is stimulated, as evidenced by increased rates of respiration, increased protein synthesis, etc. Affected cells may divide and dedifferentiate, establishing around the parasite a zone of intense metabolic activity and a concomitant movement of metabolites towards the new meristematic areas. Those infection courts are often referred to as "metabolic sinks." The diversion of metabolites to these sinks results in eventual damage to the growth of the plant as a whole and abnormal development of the affected organs. Cytokinins are potent inducers of nutrient mobilization (71), and their roles in plant cell division and inhibition of senescence are well known. It is not surprising, therefore, that secretion of cytokinins by plant pathogens, or by host cells in response to infection, has been frequently suggested to account for altered transport and accumulation of metabolites associated with infection (20, 21, 52, 116, 117).

Cytokinins and Mobilization of Metabolites

Infection by powdery mildews often results in the development of "green islands," i.e. retention of chlorophyll around fungal colonies in detached or senescent leaves. Bushnell & Allen (10) first showed that water extracts of *Erysiphe graminis* spores induced green islands and starch accumulation in barley leaves. Similar islands were produced by yeast extract and by an extract from wheat stem rust uredospores. Although Bushnell & Allen did not demonstrate that movement of metabolites to the green islands did occur, it was suggested that a primary effect of the spore extract was a redirection of the transport system of the leaves. They suggested that the spore extracts contained growth substances, presumably cytokinins, which elicited the responses leading to green island formation.

Thrower (117) examined the way in which the rust of subterranean clover, *Uromyces trifolii*, obtains energy and nutrients from its host. After $^{14}CO_2$ was supplied to leaflets, there was a rapid and progressive movement of ^{14}C-glucose into the developing colonies; similar accumulation of labeled assimilate was obtained by applying droplets containing ATP or kinetin. It was envisaged that the process of accumulation, either induced by the fungus or by kinetin, involves ATP to provide energy for translocation. The analogous effects of rust infection and kinetin on long-distance movement of metabolites were confirmed by Pozsar & Kiraly (81), and assays of partially purified extracts of rust-infected bean leaves showed increased cytokinin activity as compared to healthy tissues (52). Bean rust uredospores were shown to contain relatively high levels of cytokinin-like compounds. Although these observations are interesting, it is unfortunate that the type of assays used (inhibition of root growth, retention of chlorophyll) are very nonspecific and therefore not entirely reliable. It is unfortunate also that no effort was made to identify the types of cytokinins involved.

A more thorough study of mobilization factors in rust-infected bean leaves and in bean rust uredospores was carried out by Dekhuijzen & Staples (21). Extracts of infected leaves produced almost four times more callus from tobacco pith tissues than did extracts from healthy plants. Purification of the extracts showed that total cytokinin activity increased in infected leaves and that this increase was due primarily to one fraction which stimulated mobilization of metabolites and supported the growth of callus tissue. Chemical identification of this substance, presumably a cytokinin, was not carried out. Uredospores and rust mycelium have a complex of compounds which are active mobilization inducers (including cytokinins from uredospore tRNA), but none of these compounds had R_f values similar to those of the compounds in extracts from infected leaves. It was implied, therefore, that cytokinin increases in infected leaves were strictly of host origin.

Cytokinins and "Fasciation"

One of the most interesting and detailed studies on the relationship of cytokinins to abnormal growth responses in diseased tissues stemmed from the elegant demonstration by Thimann & Sachs (116) that *Corynebacterium fascians*, incitant of the "fasciation" disease of peas, produced a chloroform-soluble cytokinin in culture. Infected pea tissue, but not uninfected tissue, yielded a cytokinin with similar biological and chemical properties. Results were based on two assays: retention of chlorophyll in senescing oat leaves, and the release of lateral pea buds from apical dominance. Because these assays are not very sensitive, or sufficiently specific, some doubt remained as to the nature of the presumed cytokinins. The fact that the characteristic symptoms of the disease in peas (loss of apical dominance, multiple bud development, thickening of internodes, etc) could be reproduced by the topical application of high concentrations of kinetin in carbowax added further credence to the role of a cytokinin in the disease syndrome.

Procedures for the isolation of the cytokinins produced by *C. fascians* were developed by Klämbt et al (54), and Helgeson & Leonard (35) identified the active fractions as 2iP, nicotinamide, and 6-methyl aminopurine. The first compound accounted for most of the biological activity and was found to be 10 times more active than kinetin in the tobacco callus test. The subsequent evidence for the widespread occurrence of the riboside of 2iP, 6-(3-methyl-2-butenylamino)-9-B-D-ribofuranosylpurine in the tRNA of many animal, bacterial, and plant cells (33) added further significance to the determination of the structure of the major cytokinin of *C. fascians*.

It is unfortunate that further work which might confirm the role of a cytokinin in the fasciation disease of peas has not been carried out. The fact that the bacterium remains mostly as a zoogleal mass on the surface of the affected tissues, and the relatively low amounts of cytokinin produced by the bacterium in culture (54), would suggest that there are factors other than cytokinins which contribute to the growth responses of host tissues. In this connection the amounts of IAA in the tissues in relation to the amounts of cytokinin released by the bacterium would be important. Two preliminary reports (51, 60) suggest that the bacterium can break down IAA. This may be one additional factor that would help to explain how the extremely low amounts of cytokinin which presumably are taken up by the tissues can exert such marked effects on the host cells.

Cytokinins and Tumor Formation

Because of the well-known effects of cytokinins on cell division, attention has been directed to the production of these compounds by organisms that cause tumors, galls, or other manifestations of disorganized growth. The problem is considerably simpler in cases where the pathogen can be grown separately from the host than in those involving obligate parasites. To the first category belong the rhizobia that cause nodule formation in leguminous roots, and *Agrobacterium tumefaciens*, the cause of the crown gall disease. A cytokinin is released in culture media by *Rhizobium japonicum* and *R. leguminosarum*, and the amounts released by the first during the logarithmic growth phase were thought to be sufficient to initiate the cortical cell divisions involved in nodule formation (78, 79). The cytokinin released by the bacterium had properties similar to those of zeatin, a naturally occurring cytokinin from higher plants. The authors point out that direct demonstration of the production of this cytokinin within the infection threads in a soybean root would be extremely difficult. The availability of gas liquid chromatographic methods, which allow identification of cytokinins in relatively crude plant extracts, should make it possible to demonstrate the presence of the rhizobial cytokinin in nodule tissues if soybean roots do not produce zeatin themselves. The use of similar methods has allowed a clear demonstration of the production of 2iP by *Agrobacterium tumefaciens* in culture (119). There had been prior reports that this bacterium produced a cytokinin in culture, but identification had not been carried out (53, 92). The gas-liquid chromatography procedure devised separated 2iP from all known naturally occurring cytokinins, and because it was applied after silylation of a simple ethyl acetate extract, the method would seem applicable for identification of these compounds in crown gall tissues.

That this information is badly needed is evidenced by the confusion as to the nature of the cell-division promoting factors (CDF) present in crown gall tissues. That tumor cells synthesize these factors is suggested by the fact that they do not require additional cytokinins in culture, while normal cells possess an absolute requirement for 6-substituted purines or for growth factors of the type described by Wood (126). The latter factor has a purinone chromophore, but the entire structure has not been elucidated. A third growth factor which promotes tumor growth (increase in mean tumor diameter) at day 3 after inoculation of pinto bean leaves with *A. tumefaciens* has been described (63). A partially purified preparation of this factor showed no cytokinin activity in the tobacco callus assay, however. It is clear that the central role of these various growth promoting factors cannot be elucidated until they have been characterized chemically in both tumor and normal tissues.

In the case of exaggerated growth responses caused by obligate parasites, it is obvious that the determination of a possible role of cytokinins is considerably more complicated because the parasite cannot be grown separately from the host. The abnormal growth of tissues infected with *Plasmodiophora brassicae*, the cabbage clubroot pathogen, has received some attention in terms of growth regulator content. Katsura et al (47) reported that an ethyl ether extract from clubbed roots contained considerably more cytokinin activity than a comparable extract from healthy roots. Results were based on the increase in (*a*) fresh weight of radish leaf disks; (*b*) chlorophyll retention of leaf tissues; and (*c*) weight of tobacco callus tissue in assays of extracts from clubbed roots. These conclusions have been strengthened by the reports that explants of tumor tissue (infected with *P. brassicae*), unlike those from healthy roots, do not require an exogenous source of cytokinin for growth in culture (43, 86). Extracts from infected tissue cultures grown in kinetin-free media contained relatively low, but significant, cytokinin activity (86). Extracts from turnip clubroots contained an active fraction which co-chromatographed in several systems with zeatin (20). Total cytokinin activity was estimated at 0.1–1.0 mg zeatin equivalent/100 kg fresh weight of clubbed tissue. Dekhuijzen & Overeem (20) have made the important observation that high auxin levels in clubbed tissues (47) would markedly affect the response to high levels of zeatin. The production of zeatin and zeatin riboside by mycorrhizal fungi (69), under conditions which do not result in tumorous overgrowths in spite of the close association of root cells with those of the symbiont, suggests that there must be factors other than cytokinins which result in the growth responses associated with infection by the clubroot pathogen.

Unlike root tissues infected with the clubroot pathogen, tobacco pith tissue infected with the root knot nematode *Meloidogyne incognita* fails to grow when either IAA or kinetin are omitted from the medium (95). The nematodes do not appear to secrete detectable amounts of auxin or cytokinin, but they promote growth of the tissues when both IAA and kinetin are supplied. The nematodes appear to increase the capacity of host tissues to respond to both growth regulators, but this effect is difficult to ascertain in tissue culture. Both nematode and plant tissue failed to grow in cytokinin-less media. It seems likely that very different results would have been obtained if the parasite had had an opportunity to become established.

ETHYLENE

The vast amount of literature on the multitudinous effects of ethylene on plant growth and its ubiquitous presence in plant tissue (82) leaves little doubt that this compound plays a primary role in the regulation of many growth processes. That ethylene may be involved in abnormal growth responses associated with invasion by plant pathogens may be derived from frequent demonstrations that a large number of bacteria and fungi produce considerable amounts of this gas in culture (37, 40), and from the equally frequent reports of increased amounts of ethylene emanating from diseased plants as compared to healthy ones (37). With ethylene, as with other growth regulators, there is very little quantitative information on rates of synthesis in relation to pathogen populations or progress of the disease. Therefore, it is not possible to conclude that ethylene increases precede rather than follow the appearance of disease symptoms. Since increased levels of ethylene are frequently associated with senescing tissues, or with those that have been injured, a primary role for ethylene as a disease determinant does not appear likely at first sight.

Early evidence for a possible role of ethylene in a number of bacterial, viral, and fungal diseases was the demonstration that greater amounts of epinasty developed in indicator plants confined with diseased plants than in those confined with healthy plants. This was followed by the demonstration that pathogens such as *Fusarium oxysporum* f. *lycopersici* (23) produced ethylene in culture. With this and with other pathogens the traditional method to demonstrate ethylene production has consisted of growing the organism in an enclosed vessel and then testing for the gas by chemical means, such as bromine addition or permanganate reduction, or by absorption with mercuric perchlorate followed by addition of hydrochloric acid to release the gas for measurement by manometric methods. None of these methods provides adequate quantitation of rates of ethylene production. Complications arise as a result of CO_2 accumulation and oxygen depletion in enclosed vessels. As a result, very little can be said about actual amounts of ethylene produced by plant pathogens. An exception to this situation is the recent work of Bonn et al (7) with the wilt-inducing pathogen *Pseudomonas solanacearum*. This bacterium is known to produce substantial amounts of ethylene in ordinary culture media and in diseased tissues (29). Bonn et al grew the bacterium in glass tubes coated on the inside with culture medium. The tubes were joined together, providing a system with adequate geometry (i.e. large surface area for bacterial growth in relation to air volume) for constant flow measurement of ethylene production. Continuous monitoring of the effluent air stream by means of gas chromatography showed that a virulent strain (K60) produced twice as much ethylene as an avirulent mutant of K60 (B1). Production of ethylene was detectable at early log phase, increased to a maximum during log phase (1.3×10^{-19} moles/min/cell for K60 vs 0.46×10^{-19} moles/min/cell for B1), and decreased after growth reached stationary phase. Sufficient information is not available at present to determine whether differences in ethylene production between virulent and avirulent strains have any significance in terms of their ability to grow in host tissues.

Bonn (unpublished information) has determined that similar constant air flow techniques can be used to obtain quantitative information on rates of ethylene production from infected plant tissues. Because the bacterium does not grow at a constant rate throughout the tissues, however, techniques will have to be developed to allow uniform exposure of host tissues to the pathogen.

Information on mechanisms of ethylene synthesis by plant pathogens is meager and at present not particularly helpful to our understanding of the relative role of host and pathogen in this process. Recent data by Swanson (114) supplies some light on the effects of substrate, cofactors, light, and temperature on ethylene production by *P. solanacearum*. When added to a basal medium, methionine (up to $10^{-3}M$) and $10^{-4}M$ flavine mononucleotide, alone or combined, appeared to increase production of ethylene by the bacterium (up to 4 μl/liter of medium). Results were confounded by nonenzymic production of ethylene in this system, but Swanson argued that production in the dark eliminated the nonenzymic portion. In a basal glucose-mineral salts medium, or in potato dextrose agar, production of ethylene by the bacterium reached a peak in about 10 days; in contrast, Bonn et al (7) reported that in their system the bacterium reached stationary phase in 48 hr and maximum ethylene production was reached at the end of log phase. It is unfortunate that Swanson did not provide conditions for rapid bacterial growth, for the protracted growth period (up to 21 days) would seem unusual. The controls he used would not eliminate nonenzymic synthesis as a complicating factor; for instance, the use of noninoculated medium as a control, rather than medium inoculated with heat-killed cells, may have provided erroneous estimates as to the amount of nonenzymic production of ethylene. Also, the use of stoppered flasks to allow gas accumulation presumably interfered with bacterial growth for the reasons previously listed. Hopefully, this work can now be reexamined by means of constant air flow systems.

Ethylene and Disease Resistance

In recent years, interest has centered on the production of ethylene by plant tissues that are either resistant or susceptible to specific pathogens. It is proposed that there are differences in rate of ethylene production in these two interactions. In the resistant reaction rapid ethylene evolution stimulates the synthesis of phytoalexins and other aromatic antifungal or antibacterial compounds via increased activity of a key enzyme in aromatic biosynthesis, phenylalanine ammonia lyase (90). In addition, ethylene treatment results in increases in peroxidase activity of plant tissues (30), and such increases have been frequently correlated with resistance to microbial invasion in plants.

Sweet potato root tissues, which are normally susceptible to attack by *Ceratocystis fimbriata*, can be made resistant by prior inoculation with nonpathogenic isolates of the fungus. Volatile materials from infected root tissue also increased resistance of uninfected susceptible root tissue (14). Stahmann et al (107) showed that the active volatile material produced by diseased sweet potato tissue was ethylene. Sweet potato slices exposed to low concentrations of ethylene (8 ppm) for 2 days became resistant to infection by *Ceratocystis*, and this was accompanied by increases in activity of peroxidase and polyphenoloxidase. Susceptible tissue inoculated with a nonpathogenic strain of the fungus liberated more ethylene than similar tissue inoculated with a strain that did not induce resistance. There is controversy as to the ability of the fungus to produce ethylene in culture (11, 41), but most of the ethylene produced by infected tissue appears to be of host origin.

Stahmann et al (107) concluded that the ability to induce or not to induce ethylene production was correlated with avirulence or virulence of the pathogen. They concluded that ethylene may be the trigger which diffuses from infected areas to adjoining tissues and results in metabolic changes that inhibit further spread of the pathogen. Since tissue was incubated in sealed chambers for up to 6 days, there is a good possibility that these results may not represent a comparable situation in nature. Also, the amounts of ethylene induced in the susceptible tissues by compatible strains (15 ppm) were sometimes higher than those produced by the incompatible strains (9 ppm) which presumably induce resistance to black rot.

Imaseki et al (41) have determined the time course of ethylene production of infected sweet potato slices incubated in sealed flasks: within 2 days, rates of ethylene production were proportional to the number of disks in the chamber; cessation of ethylene production resulted from oxygen depletion within the chamber. Because fungus growth on the disks was not quantitated, it is not clear to this reviewer whether the effect of oxygen depletion was on growth of the fungus or on ethylene synthesis by the host tissues. Ethylene was released mainly from the tissues adjoining the infected area, but the fungus-invaded portion did not produce ethylene by itself. The authors estimated that infected sweet potato tissue produced 195 mμl/g/hr, but the amounts of ethylene necessary for the induction of metabolic changes associated with infection are so small (42) that the mere effect of wounding would explain most of the changes observed. Concentrations as low as 1 ppm caused marked increases in the activity of peroxidase and phenylalanine ammonia lyase, in oxygen uptake, and in content of chlorogenic acid.

Preliminary results suggest that the mechanism of ethylene biosynthesis in diseased sweet potato tissue may be different from that in freshly cut tissue. In experiments with slices supplied with acetate-1-^{14}C or acetate-2-^{14}C, the two carbon atoms were incorporated into ethylene produced by fresh tissues, but that from diseased tissue was mainly derived from acetate-2-^{14}C (94). It was suggested that in diseased tissues acetate must enter the TCA cycle before ethylene is synthesized, and some support for this hypothesis was obtained by the demonstration that monofluoroacetate inhibited pyruvate-3-^{14}C incorporation into ethylene in diseased tissues but promoted its incorporation in fresh tissue. From a time course study on the incorporation of glucose-U-^{14}C into ethylene it was concluded that the conversion of some cellular component (possibly degradation products of cell wall constituents) to ethylene is triggered by inoculation and proceeds much more rapidly than the conversion of glucose to ethylene.

Although there is little question that increased ethylene production is a characteristic feature of numerous host-parasite interactions (37), the relationship of this phenomenon to metabolic alterations which may or may not result in containment of the parasite is not at all clear. The reported effects of ethylene treatment on increased resistance of sweet potato root tissues to *Ceratocystis* could not be confirmed by Chalutz & DeVay (11). To a large extent these discrepancies may be due to the lack of quantitative means of estimating the amount of fungus growth in the tissues. Also, inoculation with different isolates of the fungus results in widely different rates and amounts of ethylene evolution, and proper controls are difficult to devise. In the absence of solid quantitative information on the relation of pathogen growth and ethylene evolution in varieties which differ in resistance, or with isolates which differ in virulence, it is difficult to evaluate the proposed role of ethylene in resistance mechanisms.

Ethylene and Phytoalexins

A widely held view is that ethylene triggers increased activity and/or synthesis of key enzymes in the metabolism of aromatic compounds. Increased activity of phenylalanine ammonia lyase following ethylene treatment has been indicated (90), and the key role of this enzyme in aromatic biosynthesis is well known. Correlations between increased activity of this enzyme and phytoalexin production in pea tissues has been claimed (32), and the production of an isocoumarin and pisatin in carrot roots and pea pods, respectively, was demonstrated by prolonged exposure of plant tissues to relatively low concentrations of ethylene (12, 13). It is tempting, therefore, to suggest that ethylene is the triggering agent for phytoalexin production. The difficulty, of course, is that phytoalexin production can be elicited by a wide variety of chemical agents and, in fact, ethylene is not particularly effective as compared with other agents. Other products of the host-parasite interaction are far more effective, and there seems to be little reason to single out ethylene as the specific inducer of phytoalexin production. In fact, it is questionable that induction in the biochemical sense does take place.

Ethylene and Peroxidases

An additional role for ethylene in disease resistance mechanisms has been proposed, namely, the increased synthesis or activation of peroxidases. It is clear that even small concentrations of ethylene result in increased activity of peroxidases in a wide variety of plant tissues (30). That this increase results from increased synthesis of the enzymes is suggested by the reported requirement for RNA and protein synthesis in the pea system (87). Correlations between ethylene production and increased peroxidase activity have been proposed in the sweet potato-*Ceratocystis* system (107), and in cowpea mosaic virus-infected cowpea seedlings (65). That ethylene triggers de novo synthesis of peroxidase in some plant tissues is clear (104), but just how this increase results in resistance to infection is not. Induction of peroxidase by ethylene is not a generalized phenomenon, as shown by Hislop & Stahmann (38); barley seedlings incubated in an ethylene-containing atmosphere showed neither quantitative nor qualitative changes in peroxidases.

Hislop & Stahmann (38) have made a careful study of ethylene production and peroxidase changes in barley leaves infected with *Erysiphe graminis*, an obligate parasite that attacks the epidermal cells. Although penetration of susceptible and resistant cells is similar, development of haustoria is abundant only in the susceptible cells. Increases in peroxidase, associated with the appearance of a new isozyme band, occurred in both resistant and susceptible interactions, but these increases did not appear to be correlated with the degree of development of the parasite. Resistant leaves showed a peak in ethylene evolution within 24 hr after inoculation, but this was not always reproducible. Actually, in the later stages of infection, large amounts of ethylene were produced by the compatible combinations, while ethylene production declined in the incompatible ones. It seems clear that in this host-parasite combination peroxidase and ethylene synthesis are controlled by factors independent of the genes for resistance.

Perhaps the most critical evidence against the proposed relation between ethylene, peroxidase, and disease resistance was provided by Daly et al (18) in a careful evaluation of wheat stem rust resistance in nearly isogenic lines of wheat, some carying the Sr_6 allele. As indicated before, this gene is temperature sensitive, and plants possessing the dominant allele are resistant at 20°C but susceptible at 26°C. Daly and co-workers found that infection with race 56 of *Puccinia graminis* increased peroxidase activity in resistant lines of wheat and, similarly, that ethylene treatment could induce high peroxidase activity. In ethylene-treated susceptible leaves, however, peroxidase activity increased to levels higher than those obtained in untreated, inoculated resistant leaves. In addition, treated resistant leaves (at 20°C) became completely susceptible after ethylene treatment, in spite of the fact that high peroxidase levels were induced. The general conclusion was that peroxidase levels and/or ethylene levels were not causally related to resistance in this system.

Ethylene production by infected susceptible wheat leaves occurred at rates higher than those of resistant leaves, a situation comparable to that reported by Hislop & Stahmann (38) for the barley-*Erysiphe* system. In both systems, evolution of ethylene and subsequent increases in peroxidase would seem to result from injury to host cells, and this would be more extensive in the susceptible infected tissues.

It is apparent, even from this brief review of the more recent literature on the role of ethylene in plant disease resistance, that the initially attractive hypothesis proposed by Stahmann and co-workers in which ethylene was considered the key trigger of resistance mechanisms has not received support, even from further studies in the system for which it was proposed initially. It is unfortunate perhaps that so much emphasis has been placed on the role of ethylene in disease resistance, for this has diverted attention from its role in physiological and morphological phenomena associated with disease. Premature senescence of affected tissues, epinastic responses, adventitious root formation, premature abscission of plant parts, etc are all phenomena frequently associated with diseases incited by a wide variety of agents and presumably ascribable to the remarkable changes in ethylene that occur upon infection. Because of the extremely varied physiological effects of ethylene (82), abnormal growth phenomena previously ascribed to other growth regulators are more likely due to ethylene and should be reexamined.

GIBBERELLINS

The imposing volume of literature on gibberellins as natural growth regulators in plants had its origin in the discovery that *Gibberella fujikuroi* produced substances in culture that mimicked the exaggerated growth responses of rice plants infected by this pathogen (112). It is strange, therefore, that so little attention has been directed during the past 45 years to the study of gibberellins in relation to plant diseases. Most of the effort appears to have been in the direction of exogenous applications of gibberellins in order to relieve symptoms of individual diseases, and this approach has not provided any answers to basic problems concerning the role of gibberellins in plant disease. Even in the case of the bakanae disease of rice, the evidence for a specific role of gibberellins is still largely indirect. This indirect evidence is such that there is little reason to doubt that a causal relationship exists, but as far as the writer is aware, no actual assays for gibberellin content of diseased rice tissues have been made. Obviously, the effective assessment of the growth regulator status of the overgrowth in the bakanae disease requires not only the assay for gibberellins but that for auxins and other growth regulators as well. To the extent that this is yet to be done, it would seem unfortunate that the physiological implications of a discovery that had its inception in the work of plant pathologists have remained so very largely neglected.

Of considerable interest has been the demonstration that gibberellin production is not restricted to a few species in the genus *Gibberella*, but is common among microorganisms, including many plant pathogens. Admittedly much of this evidence is based on bioassays of relatively crude preparations, and none of the compounds has been identified chemically. It is clear that most fungi and bacteria produce only small amounts of gibberellin-like compounds as compared with *G. fujikuroi*. Katznelson & Cole (48) reported that a large number of rhizosphere bacteria and actinomycetes produced a gibberellic acid-like (GA$_3$) substance in amounts that varied from 1 to 14 μg/liter. The list included a number of plant pathogens, including *Agrobacterium tumefaciens* and three species of *Rhizobium*. Because gibberellins are rapidly broken down in soil, it was speculated that the most likely site for a potential effect on plant growth by microbially synthesized gibberellins is the rhizoplane, where both root and microbial cells are in intimate contact (48). That gibberellin production is common among nonrhizosphere fungi was shown by Aubé & Sackston (3), who determined that out of 114 isolates of *Verticillium* 21 produced significant amounts of gibberellin-like substances. The list included a number of pathogenic species, including *V. albo-atrum* and *V. dahliae*, but these produced only a fraction of the amounts of gibberellins recovered from *G. fujikuroi*. A number of endophytic bacteria, mostly pseudomonads, appear to be capable of synthesizing small amounts of gibberellins as well (70).

One of the first studies on gibberellins from infected plant tissues was that of Radley (84), who showed that ethanol extracts of root nodules from bean and pea contain substantially greater amounts of gibberellin-like substances than those from normal root extracts. Dullaart & Duba (26) have confirmed and expanded these observations. On the basis of lettuce hypocotyl and barley endosperm bioassays, they reported that the nodule tissue of *Lupinus luteus* contains as high as 3340 μg GA$_3$ equivalent/kg fresh weight, whereas gibberellins were very low or undetectable in the normal roots. The variability in the activity of extracts from nodule tissues was very high, however.

Plants of the creeping thistle *Cirsium arvense*, systemically infected with the rust fungus *Puccinia punctiformis*, undergo very striking morphological changes such as longer internodes, incurled leaf margins, reduced leaf area, and lack of inflorescences. These unusual symptoms have attracted the attention of plant pathologists for many years, and Bailiss & Wilson (4) have provided an interesting account of the growth regulator changes that accompany infection. They first determined that systemic infection caused a fourfold increase in stem length as compared with healthy plants, and that GA_3 treatment at 1.0 μg/plant/week produced comparable responses, while IAA treatment had no significant effect on stem elongation. In both infected and GA-treated stems, anatomical changes, as reflected by the increase in number and length of cortical cells and in the number of mitotic divisions in the apices, were very similar. Extracts from diseased plants contained slightly greater auxin activity than those from healthy plants, but extraordinary differences were found in the type and amounts of gibberellin-like substances. Based on results of lettuce hypocotyl assays, Bailiss & Wilson concluded that there was a maturity-related transition in the types of gibberellins found in the shoots, but greater activity was always found in the diseased shoots. In the early stages of infection most of the activity was due to gibberellins that separated at the R_f of GA_1 or GA_2, but as shoots became more mature, a second zone of activity appeared at the R_f of GA_3. On 8-week-old shoots, GA_3 was the predominant gibberellin. Diffusates from shoot apices, collected on agar blocks, showed essentially the same picture. It was concluded that the consistently higher level of gibberellins in diseased shoots was always correlated with the period when they grew faster than the healthy shoots. Whether the combined effects of increased IAA and gibberellin content would explain other morphological responses, not explicable on the basis of gibberellins alone, was not determined.

Related Compounds

Of considerable interest is the production by plant pathogens of substances which mimic gibberellins in their effects on plant growth. To this group belong the sesquiterpenoids, helminthosporol and helminthosporic acid. The first was isolated by Tamura et al (115) from culture filtrates of *Helminthosporium sativum*, the agent of crown rot of wheat, and the second was obtained as an oxidation product of helminthosporol. Both resemble gibberellins in their actions but are much less potent than GA_3 and are active on a narrower range of plants. For instance, they promote the elongation of the shoots of rice seedlings but not that of oat or wheat shoots (34). The inability of these two compounds to promote elongation of wheat seedlings was correlated with the fact that infection by *H. sativum* does not result in marked growth responses by the host plant.

The dialdehyde helminthosporal is also produced by *H. sativum* and is highly phytotoxic. In addition to these properties, it can stimulate the synthesis of amylase by embryo-less barley seeds, but it inhibits the GA_3-induced synthesis of amylase (123). It was hypothesized that this substance inhibits amylase synthesis by blocking ATP synthesis in cells of the aleurone layer. The evidence for the latter effect was not entirely convincing, however.

GROWTH INHIBITORS

Almost a decade ago, it was indicated that the study of plant growth inhibitors in relation to the growth alterations that accompany infection had been badly neglected and that this was regrettable in view of the ubiquitous presence of these compounds and the increasing evidence that they play a major role in plant growth regulation (98). The same statement could be made today. There has been little interest in these compounds from the pathological standpoint even though at least one, abscisic acid, has been generally recognized as an important growth regulator of wide distribution in plants. Its effects on dormancy, abscission, inhibition of extension growth, flowering, etc have been well established, and plant pathogens quite often affect these phenomena in a predictable manner.

Stunting is one of the earliest and most characteristic symptoms of invasion by vascular wilt fungi and bacteria and by systemic viruses. An increase in growth inhibitor content of affected tissue has been correlated with appearance of this symptom (5, 74, 108, 121), but there have been very few attempts to determine the nature of the inhibitors involved. Steadman & Sequeira (109) reported that in tobacco plants infected with *P. solanacearum* there was a correlation between inhibitor content, multiplication of the bacterium, and decrease in internode elongation. The inhibitor was identified as (+)-abscisic acid and it was indicated that all of the inhibitory activity in plant extracts could be attributed to this compound, although the possible presence of other inhibitory compounds was not excluded. Since the bacterium did not produce abscisic acid when cultured in a tobacco medium, it was concluded that the host was stimulated to produce the inhibitor at greater than normal amounts. Just how this stimulation takes place has not been determined. It is well known that abscisic acid increases significantly in plants placed under water stress. For instance, up to fortyfold increases in abscisic acid have been reported in excised wheat leaves allowed to wilt for a few hours (127). Increases in abscisic acid in tobacco infected with *P. solanacearum* occurred before there was noticeable wilting. Therefore, it seems possible that water stress per se does contribute to the observed increases in abscisic acid, but only in the later stages of infection.

Increases in growth inhibitors are also characteristic of plants infected with the vascular parasite *Verticillium albo-atrum*. Pegg & Selman (74) observed a marked increase in inhibitor content in tomato plants infected by this pathogen. Apical leaves of susceptible, inoculated tomato plants showed a fivefold increase in abscisic acid compared with resistant, inoculated plants on a fresh weight basis. However, the levels of abscisic acid required to produce a fraction of the natural disease stunting, by exogenous application, were 100-fold the normal tissue levels (Pegg, unpublished information).

More recently, Wiese & DeVay (124) reported that abscisic acid levels approximately doubled in cotton leaves infected with a defoliating isolate of *V. albo-atrum*. It was noted that inhibitor levels were not affected in leaves infected with a nondefoliating isolate. Because abscisic acid accelerates leaf abscission in cotton (1), Wiese & DeVay concluded that imbalances in this hormone were at least partly responsible for the defoliation of infected plants. They proceeded to search for other contributory factors and reported that alterations in rates of IAA decarboxylation and increases in ethylene production accompanied defoliation. It was not possible to determine how these various factors are interrelated and affect the normal control of abscission. Their work, however, represents one of the few attempts to elucidate the nature of the different components, from a hormonal standpoint, that influence symptom expression in plant disease.

The emphasis on abscisic acid changes in wilt diseases is well justified, for this growth regulator may be involved in water conservation in the plant. Although it increases the permeability of plant cells to water (31), it also brings about the closure of stomata (45). If, as reported, abscisic acid increases under water stress conditions, increased permeability to water in root cells and the closure of stomata would be of value in restoring turgor (45).

The work reviewed above has emphasized the relationship between stress imposed by infection with vascular parasites and the increased synthesis of abscisic acid by host tissues. Although synthesis of this compound by plant pathogens themselves has not been reported, it is likely that microorganisms produce substances which have effects similar to those of abscisic acid. For instance, extracts of conidia of *Peronospora tabacina* caused severe stunting of young tobacco seedlings (44). It is difficult, of course, to separate the purely toxic effects of plant metabolites from the reversible, nontoxic effect of substances such as abscisic acid. It is possible today, however, to make a more systematic attempt to identify such plant metabolites and to determine whether known growth substances are responsible for the effects of fungal extracts on plant growth.

CONCLUDING REMARKS

The evidence reviewed here points out clearly that hormonal changes almost invariably accompany the response of plant cells to pathogens. These changes cover the range from the relatively simple sequence of events that results in premature senescence to the complex interactions associated with tumor formation or the establishment of obligate parasites. Auxins, gibberellins, cytokinins, ethylene, and growth inhibitors appear to be implicated at various stages of the host-parasite interaction just as they are implicated at various stages of normal plant growth. The possible effects of constantly changing interactions among these regulators provide the investigator with a bewildering and almost forbidding array of variables. Yet all of these must be considered in any effort to unravel the role of these substances in the response of plants to disease. From an evolutionary point of view, the changes in growth regulator metabolism associated with invasion by plant parasites are probably not different from those that have evolved to cope with wounding and other stress conditions. The important question would seem to rest on the nature of the mechanism or mechanisms that trigger the complex changes that follow injury. The most serious gap in our knowledge, therefore, would appear to arise from our ignorance as to what transpires at a molecular level during the initial interaction between plant cells and potential pathogens. Progress has been limited also by the general lack of knowledge as to the mode of action of plant hormones.

It is essential that the next decade of investigations on disease-plant hormone investigations should seek new directions, new approaches. Continuation of present studies on the correlation between levels of endogenous hormones and the various manifestations of the diseased condition cannot hope to answer any basic questions regarding the origin of growth alterations. We need to differentiate between cause and effect rather than to seek mere documentation of the changes in various growth regulator levels that have occurred. The availability of new and extremely sensitive analytical techniques, and recent efforts at an integrated approach that considers both morphological and biochemical changes that accompany infection, remain our best hope that a serious attack on the problem will be made.

ACKNOWLEDGMENTS

The author is indebted to J. P. Helgeson for many helpful suggestions during preparation of this paper and for critically reading the manuscript. Certain of the author's studies included in this paper were supported in part by a research grant (GB-31584) from the National Science Foundation.

Literature Cited

1. Addicott, F. T., Lyon, J. L. 1969. *Ann. Rev. Plant Physiol.* 20:139–64
2. Antonelli, E., Daly, J. M. 1966. *Phytopathology* 56:610–18
3. Aubé, C., Sackston, W. E. 1965. *Can. J. Bot.* 43:1335–42
4. Bailiss, K. W., Wilson, I. M. 1967. *Ann. Bot. London* 31:195–211
5. Beauchamp, C. 1958. *Rev. Gen. Bot.* 65:477–517
6. Bitancourt, A. A. 1954. *Année Biol.* 30:361–70
7. Bonn, W. G., Sequeira, L., Upper, C. D. 1972. *Proc. Can. Phytopathol. Soc.* 39:26 (Abstr.)
8. Braun, A. C. 1959. In *Plant Pathology, An Advanced Treatise,* ed. J. G. Horsfall, A. E. Dimond, 1:189–248. New York: Academic
9. Braun, A. C. 1962. *Ann. Rev. Plant Physiol.* 13:533–58
10. Bushnell, W. R., Allen, P. J. 1962. *Plant Physiol.* 37:50–59
11. Chalutz, E., DeVay, J. E. 1969. *Phytopathology* 59:750–55
12. Chalutz, E., DeVay, J. E., Maxie, E. C. 1969. *Plant Physiol.* 44:235–41
13. Chalutz, E., Stahmann, M. A. 1969. *Phytopathology* 59:1972–73
14. Clare, B. G., Weber, D. J., Stahmann, M. A. 1966. *Science* 153:62–63
15. Copeman, R. J. 1969. *Histochemical and cytochemical changes in tobacco infected by Pseudomonas solanacearum.* PhD thesis. Univ. Wisconsin, Madison. 94 pp.
16. Crady, E. E., Wolf, F. T. 1959. *Physiol. Plant.* 12:526–33
17. Daly, J. M., Deverall, B. J. 1963. *Plant Physiol.* 38:741–50
18. Daly, J. M., Seevers, P. M., Ludden, P. 1970. *Phytopathology* 60:1648–52
19. Darbyshire, B. 1971. *Physiol. Plant.* 25:80–84
20. Dekhuijzen, H. M., Overeem, J. C. 1971. *Physiol. Plant Pathol.* 1:151–61
21. Dekhuijzen, H. M., Staples, R. C. 1968. *Contrib. Boyce Thompson Inst.* 24:39–52
22. Deverall, B. J., Daly, J. M. 1964. *J. Exp. Bot.* 15:308–13
23. Dimond, A. E., Waggoner, P. E. 1953. *Phytopathology* 43:663–69
24. Dullaart, J. 1967. *Acta Bot. Neer.* 16:222–30
25. Ibid 1970. 19:573–615
26. Dullaart, J., Duba, L. I. 1970. *Acta Bot. Neer.* 19:877–83
27. Dye, M. H., Clarke, G., Wain, R. L. 1962. *Proc. Roy. Soc.* 155:478–92
28. Fahraeus, G. 1961. *Physiol. Plant.* 14:171–76

29. Freebairn, H. T., Buddenhagen, I. W. 1964. *Nature* 202:313–14
30. Gahagan, H. E., Holm, R. E., Abeles, F. B. 1968. *Physiol. Plant.* 21:1270–79
31. Glinka, Z., Reinhold, L. 1971. *Plant Physiol.* 48:103–5
32. Hadwiger, L. A., Hess, S. L., von Broembsen, S. 1970. *Phytopathology* 60:332–36
33. Hall, R. H. 1968. In *Biochemistry and Physiology of Plant Growth Substances*, ed. F. Wightman, G. Setterfield, 47–56. Ottawa: Runge
34. Hashimoto, T., Sakurai, A., Tamura, S. 1967. *Plant Cell Physiol.* 8:23–34
35. Helgeson, J. P., Leonard, N. J. 1966. *Proc. Nat. Acad. Sci. USA* 56:60–63
36. Henderson, J. H. M., Bonner, J. 1952. *Am. J. Bot.* 39:444–51
37. Hislop, E. C., Hoad, G. V., Archer, S. 1971. In *Fungal Pathogenicity and the Plant's Response. Proc. 3rd Long Ashton Symp. 1971.* In press
38. Hislop, E. C., Stahmann, M. A. 1971. *Physiol. Plant Pathol.* 1:297–312
39. Hubbell, D. H. 1970. *Bot. Gaz.* 131:337–42
40. Ilag, L., Curtis, R. W. 1968. *Science* 159:1357–58
41. Imaseki, H., Teraniski, T., Uritani, I. 1968. *Plant Cell Physiol.* 9:769–81
42. Imaseki, H., Uchiyama, M., Uritani, I. 1968. *Agr. Biol. Chem.* 32:387–89
43. Ingram, D. S. 1969. *J. Gen. Microbiol.* 56:55–67
44. Izard, C. 1961. *C. R. Acad. Sci.* 253:2756–58
45. Jones, R. J., Mansfield, T. A. 1972. *Physiol. Plant.* 26:321–27
46. Kaper, J. M., Veldstra, H. 1958. *Biochim. Biophys. Acta* 30:401–20
47. Katsura, K., Egawa, H., Toki, T., Ishii, S. 1966. *Ann. Phytopathol. Soc. Jap.* 32:123–29
48. Katznelson, H., Cole, S. E. *Can. J. Microbiol.* 11:733–41
49. Kefford, N. P. 1959. *J. Exp. Bot.* 10:462–67
50. Kefford, N. P., Brockwell, J., Zwar, J. A. 1960. *Aust. J. Biol. Sci.* 13:456–67
51. Kemp, D. R., Steenson, T. I. 1971. *Biochem. J.* 125-65P
52. Kiraly, Z., El Hammady, M., Pozsar, B. I. 1967. *Phytopathology* 57:93–94
53. Klämbt, D. 1967. *Wiss. Z. Univ. Rostock* 16:623–25
54. Klämbt, D., Thies, G., Skoog, F. 1966. *Proc. Nat. Acad. Sci. USA* 56:52–59
55. Klein, R. M., Vogel, H. H. 1956. *Plant Physiol.* 31:17–22

56. Krupasagar, V., Sequeira, L. 1969. *Am. J. Bot.* 56:390–97
57. Kulescha, Z., Gautheret, R. 1948. *C. R. Acad. Sci.* 227:292–94
58. Kuo, T. T., Kosuge, T. 1969. *J. Gen. Appl. Microbiol.* 15:51–63
59. Ibid 1970. 16:191–204
60. Lacey, M. S. 1948. *Ann. Appl. Biol.* 35:572–81
61. Link, G. K. K., Eggers, V. 1941. *Bot. Gaz.* 103:87–106
62. Link, G. K. K., Wilcox, H. W., Link, A. D. 1937. *Bot. Gaz.* 98:816–67
63. Lippincott, B. B., Lippincott, J. A. 1970. *Plant Physiol.* 46:708–14
64. Lippincott, J. A., Lippincott, B. B. 1968. *Phytopathology* 58:1058 (Abstr.)
65. Lockhart, B. E., Semancik, J. S. 1970. *Phytopathology* 60:553–54
66. Magie, A. R., Wilson, E. E. 1962. *Phytopathology* 52:741 (Abstr.)
67. Magie, A. R., Wilson, E. E., Kosuge, T. 1963. *Science* 141:1281–82
68. Mennes, A. M. 1972. *Indole-3-acetic acid oxidase activity in root nodules and roots of Lupinus luteus L.* PhD thesis. Univ. Leiden, Netherlands. 96 pp.
69. Miller, C. O. 1968. See Ref. 33, 33–45
70. Montuelle, B. 1966. *Ann. Inst. Pasteur* 3 (suppl.):136–46
71. Mothes, K., Engelbrecht, L. 1961. *Phytochemistry* 1:58–62
72. Niemann, G. J. 1970. *Acta Bot. Neer.* 19:681–83
73. Nitsch, J. P. 1956. In *The Chemistry and Mode of Action of Plant Growth Substances*, ed. R. L. Wain, F. Wightman, 3–31. London: Butterworths
74. Pegg, G. F., Selman, I. W. 1959. *Ann. Appl. Biol.* 47:222–31
75. Pegg, G. F., Sequeira, L. 1968. *Phytopathology* 58:476–83
76. Perley, J. E., Stowe, B. B. 1966. *Plant Physiol.* 41:234–37
77. Phelps, R. H., Sequeira, L. 1968. See Ref. 33, 197–212
78. Phillips, D. A., Torrey, J. G. 1970. *Physiol. Plant.* 23:1057–63
79. Phillips, D. A., Torrey, J. G. 1972. *Plant Physiol.* 49:11–15
80. Phipps, J., Martin, C. 1968. *C. R. Acad. Sci.* 266:1796–99
81. Pozsar, B. I., Kiraly, Z. 1966. *Phytopathol. Z.* 56:297–309
82. Pratt, H. K., Goeschl, J. D. 1969. *Ann. Rev. Plant Physiol.* 20:541–84
83. Raa, J. 1971. *Physiol. Plant.* 25:130–34
84. Radley, M. 1961. *Nature* 191:684–85
85. Ray, P. M. 1958. *Ann. Rev. Plant Physiol.* 9:81–118

86. Reddy, M. N., Williams, P. H. 1970. *Phytopathology* 60:1463–65
87. Ridge, I., Osborne, D. J. 1970. *J. Exp. Bot.* 21:720–34
88. Rigaud, J. 1969. *Arch. Mikrobiol.* 66: 29–33
89. Rigaud, J. 1970. *Physiol. Plant.* 23: 171–78
90. Riov, J., Monselise, S. P., Kahan, R. S. 1969. *Plant Physiol.* 44:631–35
91. Rodrigues, C. J., Arny, D. C. 1966. *Phytopathol. Z.* 57:375–84
92. Romanow, I., Chalvignac, M. A., Pochon, J. 1969. *Ann. Inst. Pasteur* 117:58–63
93. Rubery, P. H. 1972. *Biochim. Biophys. Acta* 261:21–34
94. Sakai, S., Imaseki, H., Uritani, I. 1970. *Plant Cell Physiol.* 11:737–45
95. Sandstedt, R., Schuster, M. L. 1966. *Physiol. Plant.* 19:99–104
96. Schwarz, K., Dierberger, R., Bitancourt, A. A. 1955. *Arq. Inst. Biol.* 22: 93–118
97. Seevers, P. M., Daly, J. M. 1970. *Phytopathology* 60:1642–47
98. Sequeira, L. 1963. *Ann. Rev. Phytopathol.* 1:5–30
99. Sequeira, L. 1964. *Phytopathology* 54:1078–83
100. Ibid 1965. 55:1232–36
101. Ibid 1969. 59:473–78
102. Sequeira, L., Mineo, L. 1966. *Plant Physiol.* 41:1200–8
103. Sequeira, L., Steeves, T. A. 1954. *Plant Physiol.* 29:11–16
104. Shannon, L. M., Uritani, I., Imaseki, H. 1971. *Plant Physiol.* 47:493–98
105. Shaw, M., Hawkins, A. R. 1958. *Can. J. Bot.* 36:1–16
106. Sirois, J. C., Miller, R. W. 1972. *Plant Physiol.* 49:1012–18
107. Stahmann, M. A., Clare, B. G., Woodbury, W. 1966. *Plant Physiol.* 41:1505–12
108. Steadman, J. R., Sequeira, L. 1969. *Phytopathology* 59:499–503
109. Steadman, J. R., Sequeira, L. 1970. *Plant Physiol.* 45:691–97
110. Stonier, T. 1969. *Plant Physiol.* 44: 1169–74
111. Stowe, B. B. 1955. *Biochem. J.* 61:ix
112. Stowe, B. B., Yamaki, T. 1957. *Ann. Rev. Plant Physiol.* 8:181–216
113. Stroun, M., Anker, P., Gahan, P., Rossier, A., Greppin, H. 1971. *J. Bacteriol.* 106:634–39
114. Swanson, B. T. 1972. *Ethylene in relation to phytogerontology.* PhD thesis. Univ. Minnesota, St. Paul. 326 pp.
115. Tamura, S., Sakurai, A., Kainuma, K., Takai, M. 1963. *Agr. Biol. Chem.* 27:738–39
116. Thimann, K. V., Sachs, T. 1966. *Am. J. Bot.* 53:731–39
117. Thrower, L. B. 1965. *Phytopathol. Z.* 52:269–94
118. Tonhazy, N. E., Pelczar, M. J. 1954. *Science* 120:141–42
119. Upper, C. D., Helgeson, J. P., Kemp, J. D., Schmidt, C. J. 1970. *Plant Physiol.* 45:543–47
120. Van Andel, O. M., Fuchs, A. 1972. In *Phytotoxins in Plant Diseases,* ed. R. K. S. Wood, A. Ballio, A. Graniti, 227–49. London: Academic
121. Van Steveninck, R. F. M. 1959. *J. Exp. Bot.* 10:367–76
122. Veldstra, H. 1968. *Neth. J. Plant Pathol.* 74 (suppl. 1):55–66
123. White, G. A., Taniguchi, E. 1969. *Can. J. Bot.* 47:873–84
124. Wiese, M. V., DeVay, J. E. 1970. *Plant Physiol.* 45:304–9
125. Woltz, S. S., Littrell, R. H. 1968. *Phytopathology* 58:1476–80
126. Wood, H. N. 1970. *Proc. Nat. Acad. Sci. USA* 67:1283–87
127. Wright, S. T. C., Hiron, R. W. P. 1969. *Nature* 224:719–20
128. Yao, P. Y., Vincent, J. M. 1969. *Aust. J. Biol. Sci.* 22:413–23

Ann. Rev. Plant Physiol. 1973. 24:381–414

THE METABOLISM OF SULFATE   ❖ 7553

Jerome A. Schiff

Department of Biology, Brandeis University, Waltham, Massachusetts

Robert C. Hodson

Department of Biology, University of Delaware, Newark, Delaware

CONTENTS

THE SULFUR CYCLE IN NATURE

Of the elements required by living systems, carbon, nitrogen, and sulfur undergo extensive metabolic transformations. Sulfur, unlike carbon and nitrogen, can be utilized to some extent in its most highly oxidized naturally occurring form, sulfate; sulfate reduction is necessary for the formation of sulfur-containing amino acids and proteins (35).

Figure 1 shows the relation of sulfate to the many reactions which sulfur undergoes in the biosphere. Sulfate is utilized by living systems to form sulfate esters of polysaccharides, phenols, steroids, and other organic compounds through a series of activation and transfer reactions which will be discussed subsequently. It is also reduced by certain anaerobic organisms such as the bacterium *Desulfovibrio* in their energy metabolism where the process is called dissimilatory sulfate reduction and the product is hydrogen sulfide. Many organisms reduce sulfate to the thiol level found in the amino acids cysteine and methionine, and these serve as building blocks for proteins; this process is called assimi-

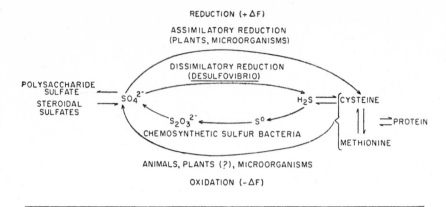

Figure 1 The sulfur cycle in nature showing the oxidation-reduction reactions that sulfur undergoes in various organisms (35).

latory sulfate reduction. Most organisms (including ourselves and higher plants) seem to be able to oxidize reduced sulfur compounds back to the level of sulfate, although only the chemosynthetic bacteria have been observed to couple the energy released (some 180 kcal per mole) to the reduction of carbon dioxide (35).

Reduced sulfur at the thiol level (R–SH or R–S–R') undergoes many reactions involving the transfer of the thiol group. These reactions have been considered in detail elsewhere and will not be discussed extensively in this review (138, 139). Our objective here will be to evaluate the present evidence for the pathway(s) of sulfate reduction and to present the emerging evidence for sulfate esterification, particularly of the sulfated polysaccharides. We have chosen to concentrate on these two areas because they represent important metabolic pathways in plants and are also areas of active research and discussion. Other areas which might be included have been ably reviewed very recently by others. For example, the entire field of inorganic sulfur metabolism has been thoroughly discussed in a monograph by Roy & Trudinger (110), while John Thompson (138) reviewed the area of sulfur metabolism with emphasis on sulfur amino acid metabolism and other transformations of reduced sulfur (138, 139). Challenger has described the interesting formation of organic sulfides by algae (16), and the sulfur metabolism of the algae has been reviewed (112, 114). The closely related subject of selenium metabolism has also been summarized (112, 125).

Compared with sulfate reduction, the metabolic transformations of reduced sulfur compounds are fairly well understood. In this review we will attempt to sort out the evidence and to present a unified pathway for sulfate reduction based on work with many different organisms. While sulfate esterification reactions are well understood in principle, the actual enzymology of sulfate transfer in the formation of sulfated polysaccharides is emerging slowly, and we include this field as one which shows signs of becoming extremely active in the near future.

PHYLOGENETIC DISTRIBUTION OF REACTIONS INVOLVING SULFATE TRANSFER AND REDUCTION

Reactions resulting in the esterification of organic compounds by sulfate are widely distributed (110). Sulfate esters of organic compounds are formed by higher animals in the form of steroid sulfates, phenol sulfates, and polysaccharide sulfates such as chondroitin sulfate and hyaluronic acid. The algae produce sulfated polysaccharides as wall constituents, frequently in copious amounts. The better known sulfated polysaccharides of commercial importance such as carrageenan and agar are produced by marine members of the red algae or Rhodophyta (86, 99). Sulfated polysaccharides have been found among the bacteria (137a), but we have been unable to find any reports of sulfated polysaccharides among the higher plants. It may be that these gelatinous materials have evolved in the algae as adaptations to periodic wetting and drying as water conservation mechanisms, a problem particularly in intertidal and terrestrial algae. In the higher animals, the sulfated polysaccharides are used to provide a matrix for cellular structures as in cartilage and skin (24). From our present knowledge, the occurrence of sulfated polysaccharides seems to be disjunct evolutionarily and probably represents the adaptation of particular groups of organisms to special needs.

Sulfate reduction (114), on the other hand, is found in prokaryotes such as bacteria and blue-green algae, except in those forms which seem to have lost it secondarily by mutation (see Figure 2). The eukaryotic algae also are able to reduce sulfate and grow on sulfate as the sole sulfur source as are the fungi and other plant-like protists. Sulfate reduction persists in the evolution of the higher plants, and all metaphytes whose sulfur nutrition has been studied seem to be able to utilize sulfate as the sole sulfur source for growth. The ability to reduce sulfate seems to have been lost in the evolution of the primitive animals or protozoa. Most protozoa and the entire animal kingdom are unable to reduce sulfate to the amino acid level and cannot grow on sulfate as far as is known, but allowance must be made for the ability of the gut flora of higher animals to reduce sulfate. A recently discovered exception is the parasitic protozoan *Leishmania tarentolae*, which reduces sulfate; Steele da Cruz & Krassner (133), who studied it, suggest that there are other reasons for relating this organism to the more plant-like flagellates. Although the absence of sulfate-reducing endosymbionts in this organism has not been conclusively demonstrated, none have been found so far, and this organism might represent a transitional form between algal flagellates and more specialized flagellated protozoa. Since nitrate reduction seems to have been lost just ahead of sulfate reduction [the Euglenophytes, judging from data for *E. gracilis* (17, 54), cannot reduce nitrate but can reduce sulfate], it is possible that these reactions have been lost because they are expensive energetically. The reduction of sulfate to the thiol level requires some 180 kcal

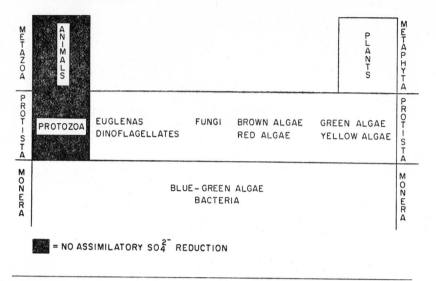

Figure 2   Phylogenetic distribution of the ability to carry out assimilatory sulfate reduction. Organisms in the unshaded areas are capable of carrying out assimilatory sulfate reduction (114).

per mole (35), and nitrate reduction is correspondingly demanding requiring about 100 kcal per mole. These reactions may have been deleted along with carbon dioxide fixation (requiring 110 kcal/mole) when photosynthetic reactions trapping light energy were lost. Evolving animals required energy for motility, and having lost photosynthesis, found themselves with a tight energy budget. Being predators, they could fulfill their architectural and energy requirements for reduced sulfur, nitrogen, and carbon by eating organisms such as microorganisms or plants which could perform these reductions. Contemporary animals obtain their reduced sulfur compounds by eating or harboring sulfate reducing plants or microorganisms, or by eating other animals which do.

Thus the problem of sulfate reduction must be studied in plants, plant-like protists, or in prokaryotes such as blue-green algae and bacteria.

SULFATE ACTIVATION

Sulfate is a relatively unreactive compound and must be activated before it can be used in biochemical reactions (19, 46, 93, 97, 108, 109). The enzymatic steps involved are shown in Figure 3.

Sulfate first reacts with ATP to form adenosine 5'-phosphosulfate (APS); the reaction is catalyzed by ATP sulfurylase (EC 2.7.7.4). This reaction proceeds with an unfavorable equilibrium for APS formation, undoubtedly because sulfuric is a stronger acid than phosphoric and the free energy of hydrolysis of the phosphate-

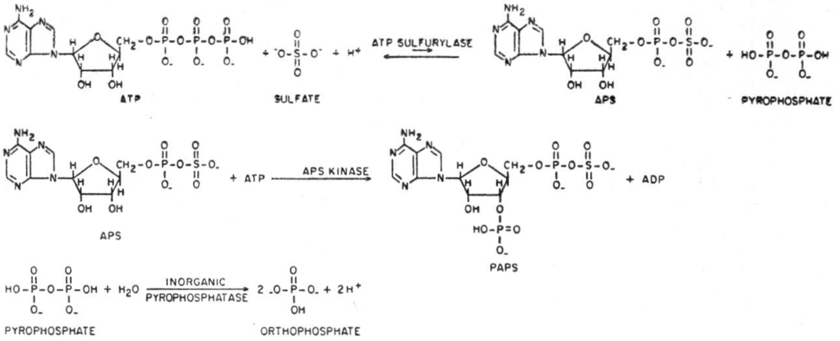

Figure 3   The enzymatic sequence in sulfate activation (114).

sulfate anhydride bond is higher than that of the phosphate-phosphate bond. Roy & Trudinger (110) estimate the group potential of phosphate-sulfate bond in APS to be about 19 kcal, compared with 8 kcal for the phosphate-phosphate bond of ATP. Consequently, the unfavorable equilibrium must be offset in order to accumulate an activated intermediate. This occurs through three mechanisms: (*a*) a second activating step [catalyzed by APS kinase (EC 2.7.1.25)] removes APS by reaction with another molecule of ATP to yield adenosine 3'-phosphate 5'-phosphosulfate (PAPS); (*b*) the APS kinase has a rather high affinity for APS (see Table 1); and (*c*) inorganic pyrophosphatase (EC 3.6.1.1) cleaves the pyrophosphate released in the sulfurylase reaction. Thus the activated intermediate PAPS can be accumulated efficiently since the first reaction is "pulled" by product removal through the second and third reactions. PAPS is the starting compound for sulfate ester formation in many systems (97, 110, 135). Assimilatory and dissimilatory sulfate reduction appear to use APS as the activated intermediate (95, 118, 145), and in these cases reactions 1 and 3 are involved together with whatever enzyme follows to use the APS and so help to pull the sulfurylase reaction.

Although the APS kinase reaction has been reported to be absent from extracts of pea shoots, tomato shoots, cauliflower florets, and *Lemna* plants (28), it has been found to be present in chloroplasts of french beans, maize, and spinach (78, 122). The properties and distribution of sulfate activating enzymes are shown in Tables 1 and 2.

SULFATE UPTAKE

Sulfate uptake and its control has been studied in bacteria (23, 60, 66), fungi (14, 79), the green alga *Chlorella* (151), roots and leaves of higher plants (32, 85), and in plant cell suspension cultures (41). Uptake seems to be accomplished through active transport mediated by a carrier with the properties of an enzyme. This car-

Table 1 Sulfate-Activating Enzymes from Plants and Microorganisms

Organism	Activity described	Enzymes assigned to activity	Enzyme purification	Repressors	References
Bacteria					
Bacillus subtilis	$SO_4^{2-} \rightarrow$ APS	Sulfurylase	None	Cystine	155
	$AP^{35}S \rightarrow PAP^{35}S$	Kinase	None	Cystine	155
Clostridium nigrificans	MoO_4^{2-}—dependent release of Pi from ATP	Sulfurylase	About 25-fold	...	3
Desulfovibrio desulfuricans	$^{35}SO_4^{2-} \rightarrow AP^{35}S$	Sulfurylase	None	None found	83, 87
	MoO_4^{2-}—dependent release of Pi from ATP	Sulfurylase	None	None found	87
Escherichia coli	$SO_4^{2-} \rightarrow$ APS	Sulfurylase	None	Cystine	155
	$AP^{35}S \rightarrow PAP^{35}S$	Kinase	None	Cystine	155
Nitrobacter agilis	$^{35}SO_4^{2-} \rightarrow AP^{35}S$	Sulfurylase	820-fold	...	149, 150
	$AP^{35}S \rightarrow PAP^{35}S$	Kinase	Partially	...	150
Nitrosomonas europaea	$^{35}SO_4^{2-} \rightarrow AP^{35}S + PAP^{35}S$	Sulfurylase+kinase	None	...	148
Salmonella pullorum	$^{35}SO_4^{2-} \rightarrow AP^{35}S + PAP^{35}S$	Sulfurylase+kinase	None	Cysteine	65
Salmonella typhimurium	$^{35}SO_4^{2-} \rightarrow AP^{35}S + PAP^{35}S$	Sulfurylase+kinase	None	Cysteine	65
	$SO_4^{2-} \rightarrow$ PAPS (in presence of excess kinase)	Sulfurylase	None	Cysteine	66
	APS→PAPS	Kinase	None	Cysteine	66
Fungi					
Aspergillus nidulans	$^{35}SO_4^{2-} \rightarrow PAP^{35}S$	Sulfurylase+kinase	None	...	55, 131, 132
	$AP^{35}S \rightarrow PAP^{35}S$	Kinase	None	...	55, 131, 132
	MoO_4^{2-}—dependent release of Pi from ATP	Sulfurylase	None	...	55
Aspergillus oryzae	$^{35}SO_4^{2-} \rightarrow PAP^{35}S$	Sulfurylase+kinase	None	...	131
Aspergillus sydowi	$^{35}SO_4^{2-} \rightarrow PAP^{35}S$	Sulfurylase+kinase	None	...	131
Eremothecium ashbyii	$^{35}SO_4^{2-} \rightarrow PAP^{35}S$	Sulfurylase+kinase	None	...	131
Nematospora gossypii	$^{35}SO_4^{2-} \rightarrow PAP^{35}S$	Sulfurylase+kinase	None	...	102
Neurospora crassa	MoO_4^{2-}—dependent release of Pi from ATP	Sulfurylase	None	...	131
Penicillium aurantio-brunneum	$^{35}SO_4^{2-} \rightarrow PAP^{35}S$	Sulfurylase+kinase	None	...	131
Penicillium chrysogenum	$^{35}SO_4^{2-} \rightarrow PAP^{35}S$	Sulfurylase+kinase	None	...	131
Penicillium citreo-roseum	$^{35}SO_4^{2-} \rightarrow PAP^{35}S$	Sulfurylase+kinase	None	...	131

Organism	Activity described	Enzymes assigned to activity	Enzyme Purification	Repressors	References
Penicillium notatum	$^{35}SO_4{}^{2-} \rightarrow PAP^{35}S$	Sulfurylase+kinase	None	…	131
Rhizopus stolonifer	$^{35}SO_4{}^{2-} \rightarrow PAP^{35}S$	Sulfurylase+kinase	None	…	131
Saccharomyces cecevisiae	$SO_4{}^{2-}$ or $MoO_4{}^{2-}$—dependent release of Pi from ATP	Sulfurylase	2000-fold	…	106, 107
Torulopsis utilis	$AP^{35}S \rightarrow PAP^{35}S$	Kinase	Partially	…	106, 107
	$^{35}SO_4{}^{2-} \rightarrow PAP^{35}S$	Sulfurylase+kinase	None	…	131
Algae					
Anacystis sp.	$^{35}SO_4{}^{2-} \rightarrow {}^{35}S_2O_3{}^{2-}$	Sulfurylase+kinase+ undefined enzymes	None	…	114
Chlamydomonas reinhardii	$^{35}SO_4{}^{2-} \rightarrow {}^{35}S_2O_3{}^{2-}$	Sulfurylase+kinase+ undefined enzymes	None	…	114
Chlorella protothecoides	$^{35}SO_4{}^{2-} \rightarrow PAP^{35}S$	Sulfurylase+kinase	None	…	114
Chlorella pyrenoidosa	$^{35}SO_4{}^{2-} \rightarrow AP^{35}S + PAP^{35}S$	Sulfurylase+kinase	None	…	114, 119
	$ATP\text{-}^{14}C \rightarrow PAPS\text{-}^{14}C$	Sulfurylase+kinase	None	…	114
	$MoO_4{}^{2-}$—dependent release of Pi from ATP	Sulfurylase	None	…	114
Chlorella vulgaris	$PAP^{35}S \rightarrow AP^{35}S$	3'-Nucleotidase	Partially	…	118, 145
	$^{35}SO_4{}^{2-} \rightarrow PAP^{35}S$	Sulfurylase+kinase	None	…	114
Euglena gracilis	$^{35}SO_4{}^{2-} \rightarrow PAP^{35}S$	Sulfurylase+kinase	Crude wholecell extract and chloroplast fragments	…	1, 18
Higher Plants					
Avena sativa (root tipes)	$^{35}SO_4{}^{2-} \rightarrow AP^{35}S$	Sulfurylase	Crude extract	…	8
Beta vulgaris (root)	$^{35}SO_4{}^{2-} \rightarrow AP^{35}S$	Sulfurylase	Crude extract	…	28
Brassica rapa (leaves)	$^{35}SO_4{}^{2-} \rightarrow AP^{35}S$	Sulfurylase	Crude extract	…	28
Lathyrus odoratus (shoots)	$^{35}SO_4{}^{2-} \rightarrow AP^{35}S$	Sulfurylase	Crude extract	…	8
Lemna gibba (leaves)	$^{35}SO_4{}^{2-} \rightarrow AP^{35}S$	Sulfurylase	Crude extract	…	28
Lycopersicum esculentum (leaves)	$^{35}SO_4{}^{2-} \rightarrow AP^{35}S$	Sulfurylase	Crude extracts	…	28
Phaseolus vulgaris (leaves)	$^{35}SO_4{}^{2-} \rightarrow AP^{35}S + PAP^{35}S$	Sulfurylase+kinase	Chloroplasts	…	78
Spinacea oleracea (leaves)	$^{35}SO_4{}^{2-} \rightarrow AP^{35}S + PAP^{35}S$	Sulfurylase+kinase	Chloroplasts	…	119
	$APS + ADP \rightarrow SO_4{}^{2-} + ATP$	Sulfurylase	Chloroplasts, leaves 53-fold, 112-fold	…	8, 118a
Zea mays (leaves)	$^{35}SO_4{}^{2-} \rightarrow AP^{35}S + PAP^{35}S$ of Pi from ATP	Sulfurylase+kinase	Chloroplasts	…	78

Table 2 Properties of Sulfate-Activating Enzymes from Plants and Microorganisms

Organism	Enzyme	pH Optimum	K_{eq}	K_m	MW	Remarks	References
Bacteria							
Nitrobacter agilis	ATP sulfurylase (EC 2.7.7.4)	7.4 (tris)	...	ATP 1.4 mM, APS 0.025 mM, PPi 0.12 mM, MgCl$_2$ 0.35 mM	700,000	Crude extract purified 820-fold, yielding one protein band on starch-gel electrophoresis.	149, 150
Nitrobacter agilis	APS kinase (EC 2.7.1.25)	7.6 (tris)		ATP 1.3 mM, APS 0.16 mM	280,000	Partially purified; PAPS reductase and a cytochrome c accompanied the kinase through purification. Final preparation had 5 protein bands (A "complex" on starch-gel electrophoresis).	149, 150
Fungi							
Saccharomyces cerevisiae	ATP sulfurylase (EC 2.7.7.4)	7.5–9.0 (tris)	1×10^{-8} (pH 8, 37°)	...	100,000	Purified approximately 2000-fold from crude extract. Requires ATP, Mg^{2+}; CTP, UTP, and GTP inactive.	44, 106
Saccharomyces cerevisiae	APS kinase (EC 2.7.1.25)	8.5–9.0 (tris)	...	APS <5 mM		Requires Mg^{2+} or Mn^{2+}. Partially purified.	44, 106
Higher Plants							
Spinacea oleracea	ATP sulfurylase (2.7.7.4)	...	...	APS 0.47 mM, PPi 3 mM	...	Purified 53-fold from chloroplasts; assayed in back reaction.	8

rier will mediate the transport of other group VI anions. Transport is unidirectional from the outside of the cell to the inside, with negligible exchange of internal sulfate with external sulfate or other anions. The absolute rate of transport is governed by pH, temperature, ionic strength, sulfate concentration, and the availability of metabolic energy. The apparent K_m for sulfate is in the range of 10^{-5} to $10^{-4}M$. Free intracellular sulfate depresses the rate of inward transport to some degree, but endogenous reduced sulfur, especially cysteine and methionine, has a pronounced inhibitory effect on the rate of sulfate uptake. Although this has been attributed to negative feedback inhibition or repression in bacteria (Figure 6) and higher plants, the exact mechanism of control in higher plants has not been established. Perhaps the most spectacular consequence of sulfate uptake is the high acidity of *Desmerestia*, a marine brown alga, which is due to the accumulation of high concentrations of sulfuric acid in the organism's vacuoles (see 112 for a review).

Pardee and co-workers (68, 87–91) have studied the control of sulfate transport in *Salmonella typhimurium*. They have isolated, purified, crystallized, and characterized a sulfate-binding protein from a membrane fraction, but the absence of the binding protein does not necessarily correlate with the absence of transport activity, although there is indirect evidence linking the binding protein to sulfate transport.

There is evidence that thiosulfate and sulfite are competitive inhibitors of sulfate uptake (20, 128) in *Salmonella* and *Penicillium*. It has also been shown in *Neurospora* and *Chlorella* mutants blocked for sulfate reduction that sulfate inhibits thiosulfate utilization (4, 103)

A sulfur source which can be sulfate is required for the successful completion of the *Chlorella* life cycle (42).

SULFATE AND SULFITE TRANSFER: THE FORMATION OF SULFATE ESTERS AND SULFONIC ACIDS

As far as is known, PAPS is the donor for sulfate transfer reactions resulting in the esterification of oxygen functions (110). The esters which are known include phenol sulfates, steroid sulfates, polysaccharide sulfates, choline sulfate, cerebroside sulfates, and flavonoid sulfates. Sulfation of nitrogen functions is also known and is represented by mustard oil glycosides and aryl sulfamates; in the case of sulfamate formation, PAPS has been shown to be the sulfate donor (110).

Perhaps the earliest sulfate transfer system to be investigated in vitro is the one which forms phenol sulfates in mammalian liver. The formation of such sulfates had been known as an important detoxification system for phenols in animals, and the first evidence for an activated sulfate compound (which was later identified as PAPS) was obtained from such a system (19). The enzyme involved is called phenol sulfotransferase (EC 2.8.2.1) and catalyzes the transfer of the sulfate group from PAPS to an acceptor phenol (15, 46). The red alga *Polysiphonia* forms 2,3-dibromobenzyl alcohol, 4,5-disulfate (47), and *Fusarium*, a fungus, secretes a hydroxynaphthoquinone sulfate (111).

Choline sulfotransferase is found in the fungi and results in the esterification of choline with the sulfate group of PAPS (62); mangroves (9) and red algae (72) can also form choline sulfate. Steroid sulfotransferases have been identified in animals which use PAPS as the sulfate donor (see 110).

Perhaps the sulfotransferases of greatest potential importance in the development and ecology of plants and animals are those which result in the formation of sulfate esters of polysaccharides. There are two potential mechanisms for these biosynthetic reactions. It is possible that the polymer backbone of the polysaccharide is constructed first and is then sulfated using PAPS as a donor. Another alternative is that the sugar monomers (probably as sugar nucleotides) become sulfated with PAPS as the sulfate donor, and these sugar sulfates are then polymerized to form the polysaccharide. There is suggestive evidence for both sets of reactions in work on polysaccharide sulfate formation in hen oviduct tissue and squid (40, 84, 135–137, 147). Sulfation of the already formed polysaccharide as well as formation of sulfated sugar nucleotides has been found.

Evidence from plant systems is still very fragmentary. The algae are major producers of sulfated polysaccharides which are present as cell wall constituents (86, 99). They give the plant thallus a considerable part of its structural integrity and protect it from desiccation; this is particularly important in intertidal and terrestrial algae. Percival's book on the marine algal polysaccharides serves as a convenient guide to these interesting compounds, and the reader is referred there since a more complete discussion of the carbohydrate chemistry of these interesting molecules would be out of place here (99).

Biosynthetic studies of some of these algal polysaccharides have been undertaken. PAPS occurs in algae (36) and $^{35}SO_4^{2-}$ incorporation of sulfate into the carrageenan of the red alga *Chondrus crispus* has been demonstrated in vivo (73); the biosynthesis of an extracellular polysaccharide of another red alga *Porphyridium* has been studied in vivo (104, 105). The biosynthesis and sulfation of fucoidin has been followed in plants and developing embryos of *Fucus* (11, 101) again in intact cells. Radioactive sulfate is incorporated into the polysaccharide fraction of *Chlorella pyrenoidosa* (Emerson strain 3); acidic polysaccharides can be localized in the wall and in golgi-like vessicles in this organism by electron microscopy of specifically stained preparations (38). *Prasiola stipitata* and the cells released by grinding the thallus secrete highly acidic polysaccharides (117). Although several laboratories have attempted to obtain cell-free extracts of these and other algae which will transfer the sulfate group from PAPS to free sugars, to a sugar nucleotide or to polysaccharide polymers, to date nothing positive has emerged in the form of an active cell-free system.

More highly reduced than the sulfated compounds are the sulfonic acids. In the sulfate esters there is an oxygen atom between carbon and the sulfur of sulfate, while in the sulfonic acids the sulfur is linked directly to carbon. Since the removal of oxygen from sulfur is difficult and costly in terms of energy, it is reasonable to think that the sulfur group is added in the form of sulfite or a compound closely related. In the case of the glucose-6-sulfonate (or 6 sulfoquinovose) found in the sulfolipid of plants (10), particularly in the chloroplast membranes, it has been shown that the sulfonic acid group comes more readily from sulfate than from reduced sulfur compounds such as cysteine (18). This would suggest that the mechanism does not involve the addition of a thiol to the sugar and a subsequent oxidation to the sulfonic acid group, but rather that the donor is formed more directly from sulfate, perhaps from PAPS or APS. It has been suggested (18) that sulfite or some similar molecule might add to enolpyruvate to form beta sulfonyl pyruvate, which might move through the glycolytic scheme in reverse in place of pyruvate yielding the six sulfonyl glucose in this way, although no direct evidence has been obtained for this series of reactions. Another alternative we might suggest is the formation of a 5–6 unsaturated glucose and the addition of a sulfite group across this double bond. Another possibility is the oxidation of the alcohol at carbon 6 of glucose to an aldehyde, followed by the addition of sulfite, dehydration, and reduction. Doubtless other mechanisms could be written, but in the absence of firm evidence none can be favored.

Other sulfonic acids of biological interest are cysteic acid, which will serve as a sulfur source for *Neurospora* (35, 53), and taurine, which is found in animal systems and can be formed from sulfate in chick embryos (74, 110). Taurine, N-methyl taurine, and N-dimethyl taurine have been found in various red algae (72). *Polysiphonia fastigiata* yields 2,L-amino-3-hydroxy-1-propane sulfonic acid (157), and N-(D-2,3-dihydroxy n-propyl) taurine or "glyceryl taurine" has been isolated from *Gigartina* (156). Homocysteic acid is formed by Sat_2^-, a mutant of *Chlorella pyrenoidosa* which is blocked for sulfate reduction (50).

COMPOUNDS CONTAINING OXIDIZED AND REDUCED SULFUR

Glucosinolates which contain divalent and hexavalent sulfur have been described, as well as alkaloids containing reduced sulfur and polyacetylenes containing oxidized and reduced sulfur (25, 27, 30, 31, 61, 134). Radioactive sulfate is rapidly incorporated into the sulfate group of 3-indolylmethylglucosinolate of *Sinapis alba* but only slowly into the divalent sulfur (124). Sulfur-labeled cysteine and methionine were not oxidized to sulfate appreciably in *Sinapis alba* or *Armaracia lapathifolia* respectively, and 90% of the radioactivity incorporated into p-hydroxybenzylglucosinolate of *Sinapis* or the allylglucosinolate of *Armaracia* was in the divalent sulfur (64, 152). Radioactive sulfur dioxide supplied to plants of the *Brassicaceae* rapidly labeled the sulfate group of glucobrassicin but only slowly labeled the divalent sulfur (67). The metabolic relationship of glucobrassicin, neoglucobrassicin, and glucobrassicin-1-sulfonate was studied using radioactive sulfate as a precursor (27); further studies have also appeared (77).

THE REDUCTION OF SULFATE IN ASSIMILATORY ORGANISMS

Since sulfate itself does not form a gas on acidification, but several inorganic species expected as reduction products of sulfate do (Figure 4), the formation of

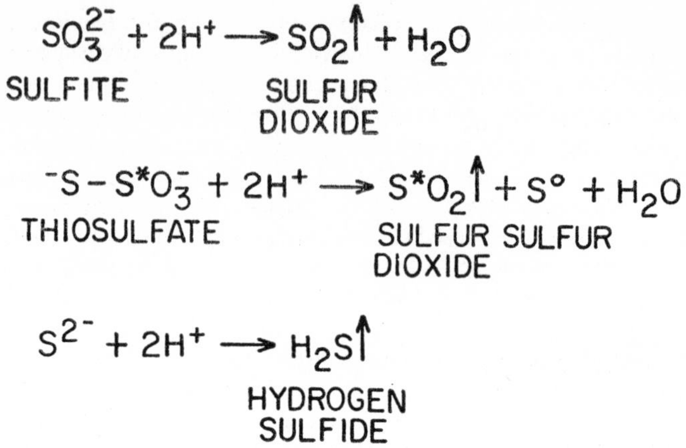

$$SO_3^{2-} + 2H^+ \longrightarrow SO_2\uparrow + H_2O$$

SULFITE   SULFUR
DIOXIDE

$$^-S-S^*O_3^- + 2H^+ \longrightarrow S^*O_2\uparrow + S° + H_2O$$

THIOSULFATE   SULFUR SULFUR
DIOXIDE

$$S^{2-} + 2H^+ \longrightarrow H_2S\uparrow$$

HYDROGEN
SULFIDE

Figure 4 Reactions forming acid-volatile radioactivity from possible compounds formed during sulfate reduction. Sulfate itself does not yield a gas on acidification.

acid-volatile radioactivity from $^{35}SO_4^{2-}$ was used as a convenient, sensitive and rapid assay from the very beginning of work on sulfate-reducing systems (45). Thus sulfite or thiosulfate would be expected to form sulfur dioxide on acidification while sulfide would yield hydrogen sulfide. In practice the gas is released from the enzyme reaction mixture with dilute acid and is trapped in dilute base or appropriate precipitants and counted. Using this assay, Hilz, Kittler & Knappe (44, 45) demonstrated that cell-free extracts of yeast suitably fortified with ATP, Mg^{2+}, and NADPH would form acid-volatile radioactivity from labeled sulfate; PAPS was implicated as the immediate precursor of the acid-volatile material. Although reduced thioctic acid appeared to act as a cofactor, this may be a reflection of the system's ability to react with thiols generally, as discussed further along. Subsequently, Wilson, Bandurski and co-workers fractionated this system (7, 141, 158) and showed that the gas produced on acidifying the reaction mixture was sulfur dioxide. They were able to fractionate the system into three enzyme fractions which they called "A," "B," and "C." They proposed that fraction A was reduced by NADPH. This reduced A in turn was thought to reduce fraction C (suggested to be a heat-stable protein disulfide). Enzyme B acted upon reduced fraction C and PAPS to yield sulfite, PAP, and oxidized fraction C. Fujimoto & Ishimoto (34) showed that extracts of *E. coli* could form sulfite from PAPS using NADPH, NADH, or reduced thioctic acid. Dreyfus & Monty (21, 22, 69) studied cysteine-requiring mutants of *Salmonella typhimurium* nutritionally and enzymatically in comparison with wild-type and concluded that sulfate was converted to PAPS by the usual activating reactions and that PAPS was reduced to free sulfite and then to free sulfide (Figure 6). They also demonstrated thiosulfate utilization and inferred that thiosulfate was not on the main path of reduction.

A cell-free system forming acid-volatile radioactivity was also obtained from *Chlorella pyrenoidosa* which required ATP, Mg^{2+}, and a reductant which could be either NADPH, NADH, or a thiol such as BAL, mercaptoethanol, gluta-thione, dithiothreitol, etc (48–50, 52, 71, 115). Using column chromatography, thiosulfate was identified as the product of the reductive reaction with NADPH or BAL as reductant. PAPS was shown to be an intermediate in the reaction sequence (52). A survey of other organisms such as yeast, *Salmonella* and *E. coli*, and *Chlamydomonas* using BAL as reductant also yielded thiosulfate as a product without the formation of sulfite (48). It later became apparent that the nature of the product depends on the thiol used (37, 146). BAL as reductant yields thiosul-fate, while all other thiols tested yield sulfite and/or the organic thiosulfate of the thiol. The unusual reaction of BAL is probably due to the hydroxyl group on the adjacent carbon to one of the thiol groups, which might facilitate the elimination of the α thiol by the β thiol to form the SH portion of thiosulfate, perhaps leaving behind the α,β episulfide of BAL. Whatever the thiol is in vivo, it too must be able to donate a sulfide moiety to thiosulfate formation since extracts from *Chlorella* without added thiol form thiosulfate using NADPH as reductant (71, 115).

As discussed above, the two steps of sulfate activation to yield APS and PAPS are widely distributed in plants and are employed in the esterification reactions which utilize PAPS. In the course of studies of assimilatory reduction in *Chlorella*, an enzyme was found which converts PAPS to APS (49, 145). APS, rather than PAPS, seems to be a substrate for reduction in *Chlorella* (37, 118, 145). This enzyme, called fraction "A" because it could be adsorbed to alumina c gamma gel (and is not to be confused with enzyme fraction A from yeast), is a 3' nucleotidase which is different from the widely distributed enzyme described by Shuster & Kaplan [3'-ribonucleotide phosphohydrolase (EC 3.1.3.6)] (129, 130) from a variety of tissues and purified by them from rye grass. The rye grass enzyme will cleave 3' nucleotides to yield inorganic phosphate without regard to substituents in the 5' position. "A" will only cleave the 3'-phosphate efficiently from nucleotides containing (as far as we know) a phosphate at the 5' position (37, 145). Unlike the Shuster & Kaplan enzyme, "A" will not act on adenosine 3'-phosphate very effectively, but like this enzyme "A" will cleave the 3'-phosphate from PAPS, CoA, and adenosine 3'5'-disphosphate (PAP). Its K_m for PAPS is in the vicinity of $10^{-4}M$.

As a source of APS, this reaction might provide several advantages over the ATP sulfurylase reaction alone. As mentioned before, the ATP sulfurylase forms APS against an unfavorable equilibrium (Figure 3). If there is much pyrophos-phate available to the enzyme, little APS will accumulate, especially in the presence of ATP given the low K_m of the APS kinase which converts APS to PAPS. It may be that it is more convenient for the cell to accumulate PAPS and to form APS as needed by the nucleotidase reaction. In addition, in cells which use PAPS for esterification and APS for reduction, the presence of "A" may provide a means of regulating the two processes separately by controlling the sizes of the two nucleotide pools. In any case, the substrate for reduction to sulfite or thiosulfate in *Chlorella* extracts is APS rather than PAPS (37, 118, 145). PAPS is inactive in reduction in the absence of "A"; the "S" fraction of *Chlorella* is specific for APS and reduces it to sulfite or thiosulfate. Wilson et al (158) found that APS would serve as a substrate for reduction in yeast extracts, but they regarded PAPS as the preferred substrate for reduction since it provided higher activities than APS.

It had been known from feeding experiments that leaves of higher plants could convert sulfate to sulfite (33) and that sulfite can be formed from PAPS in cell-free extracts from leaves. Schmidt & Trebst (119, 122, 142) have explored this further and have obtained results consistent with the *Chlorella* work reviewed above in that the substrate for the formation of sulfite in spinach chloroplast preparations can be PAPS. Schmidt later found that APS was the effective substrate for reduction and demonstrated the presence of a 3' nucleotidase converting PAPS to APS (118).

These reactions have become known as the "PAPS reductase system." Since we now know that APS is the preferred substrate (PAPS will not serve in the *Chlorella* system) perhaps "APS reductase complex" is a better term where APS preference has been demonstrated. (See Tables 3 and 4 for the distribution and properties of adenyl sulfate reductases.) The phylogenetic distribution of the "A" enzyme/or 3' nucleotidase, which acts on PAPS to form APS, is not yet known, but if "A" is distributed as widely as the adenyl sulfate reductases, all systems may well have an absolute specificity for APS, since PAPS reductase activity has been inferred by incubating PAPS with multienzyme preparations and measuring the formation of sulfite, thiosulfate, or acid-volatile radioactivity. Preliminary evidence indicates that "A" is present in *E. coli* and perhaps in yeast (146), although PAPS-specific transferase activity forming sulfite, thiosulfate, or organic thiosulfates with thiols has been identified in *E. Coli* (146). Work in the field has been moving rapidly, however, and articles now being published or in manuscript indicate that the APS reductase complex (or "S" fraction from *Chlorella*) can be fractionated still further into components. In this discussion, in addition to published papers, we will be guided by unpublished work of Schmidt, Schwenn, and Trebst and work of our own using the *Chlorella* system.

Schmidt, Schwenn, and Trebst (118, 120, 121, 126) have studied and purified to homogeneity (as judged by electrophoresis on acrylamide) two enzymes from *Chlorella* which they call APS transferase and thiosulfonate reductase. The transferase transfers the sulfate group of APS to a variety of acceptors; for example, it will transfer the sulfate group to glutathione to form S-sulphoglutathione. When incubated with the transferase and the reductase, the sulfate group of APS is ultimately reduced by the reductase in the presence of ferredoxin as a reductant and o-acetyl serine as an acceptor to yield cysteine. This reductase enzyme appears to have a low molecular weight component containing sulfhydryl groups which is necessary for activity. Abrams & Schiff (2) have shown that the sulfate group of APS can be transferred in the absence of reducing agents to a protein of the "S" fraction (or APS reductase complex) and that this binding activity is absent from mutants of *Chlorella* which lack active enzyme fraction "S." In the

...ons from Plants and Microorganisms

Organism	Activity Described	Electron Donor or Acceptor	Remarks	References
Bacteria				
Aerobacter aerogenes	PAP^{35}S→Acid-volatile radioactivity	Not specified	Crude extract. No activity with AP^{35}S	96
Aeromonas punctata	PAP^{35}S→Acid-volatile radioactivity	Not specified	Crude extract. No activity with AP^{35}S	96
Chlorobium limicola	^{35}SO$_4^{2-}$→AP^{35}S	Fe(CN)$_6^{3-}$	[Crude extracts	143
Chlorobium phaeoubrioides	AMP-dependent Fe(CN)$_6^{3-}$ reduction	Fe(CN)$_6^{3-}$	Crude extract	143
Chromatium vinosum	AMP-dependent Fe(CN)$_6^{3-}$ reduction	Fe(CN)$_6^{3-}$		143
	Methyl viologen reduction	Methyl viologen	Crude extracts	143
	AMP-dependent Fe(CN)$_6^{3-}$ reduction	Fe(CN)$_6^{3-}$		143
Clostridium kluyveri	PAP^{35}S→Acid-volatile radioactivity	Not specified	Crude extract. No activity with AP^{35}S	96
Clostridium nigrificans	AP^{35}S→Acid-volatile radioactivity	H$_2$+methyl viologen	Crude extract. No activity with PAP^{35}S	96
Clostridium pasteurianum	PAP^{35}S→Acid-volatile radioactivity	H$_2$+methyl viologen	Crude extract. No activity with AP^{35}S	96
Desulfovibrio desulfuricans	AP^{35}S→^{35}S^{2-}	H$_2$+methyl viologen	Crude extract	56, 94
Desulfovibrio vulgaris	^{35}SO$_3^{2-}$→AP^{35}S	Fe(CN)$_6^{3-}$	About 80–90% pure (ultracentrifugation)	81
Ectothiorhodospira mobilis	^{35}SO$_3^{2-}$→AP^{35}S	Fe(CN)$_6^{3-}$	[Crude extract	143
Escherichia coli	AMP-dependent Fe(CN)$_6^{3-}$ reduction	NADH; NADPH; dihydrollpoate	[Crude extracts. Slight activity with APS	34, 93
	PAPS→S^{2-}	NADPH; H$_2$+methyl viologen		34
Proteus mirabilis	PAP^{35}S→Acid-volatile radioactivity	Not specified	Crude extract. No activity with AP^{35}S	96
Proteus vulgaris	PAP^{35}S→Acid-volatile radioactivity	Not specified	Crude extract. No activity with AP^{35}S	96
Pseudomonas hydrophila	PAP^{35}S→Acid-volatile radioactivity	Not specified	Crude extract. No activity with AP^{35}S	96
Rhodopseudomonas spheroides	PAP^{35}S→Acid-volatile radioactivity	H$_2$+methyl viologen	Crude extract. No activity with AP^{35}S	96
Salmonella typhimurium	PAPS→S^{2-}	NADPH	Crude extract	21, 22, 66
Thiobacillus denitrificans	AP^{35}S→Acid-volatile radioactivity	H$_2$+methyl viologen	Crude extract. No activity with PAP^{35}S	96
Thiobacillus thiooxidans	AP^{35}S→Acid-volatile radioactivity	H$_2$+methyl viologen	Crude extract. No activity with PAP^{35}S	96
Thiobacillus thioparus	AP^{35}S→Acid-volatile radioactivity	H$_2$+methyl viologen	Crude extract. No activity with PAP^{35}S	96
Thiocapsa rosenpersicina	^{35}SO$_4^{2-}$→AP^{35}S	Fe(CN)$_6^{3-}$	Crude extracts from chromatophores and cytosol	143
Vibrio cholinicus (synonym of *Desulfovibrio desulfuricans*)	AMP-dependent Fe(CN)$_6^{3-}$ reduction	Fe(CN)$_6^{3-}$	Purified 60–80 fold; homogeneous in ultracentrifuge	143, 144
Fungi				
Saccharomyces cerevisiae	AP^{35}S→Acid-volatile radioactivity	H$_2$+methyl viologen	Crude extract. No activity with PAP^{35}S	96
	PAP^{35}S→^{35}S$_2$O$_3^{2-}$	NADPH; dihydrolipoate	Partially purified; requires 3 fractions (A, B, C). AP^{35}S supports some activity	7, 44, 158
	PAP^{35}S→^{34}SO$_3^{2-}$	NADPH+thioredoxin	Partially purified and free of activating enzymes	100
	PAP^{35}S→"bound sulfite ^{35}S"	NADPH	Partially purified; requires 3 fractions (A, B, C) in absence of carrier sulfite	140, 141
Algae				
Chlorella pyrenoidosa	^{35}SO$_4^{2-}$→^{35}S$_2$O$_3^{2-}$	NADPH; BAL	Crude extract	114
	AP^{35}S(PAP^{35}S)→S-sulfoglutathione-^{35}S	GSH	Partially purified. PAP^{35}S requires 3'-nucleotidase	118, 119
Chondrus crispus	AP^{35}S(PAP^{35}S)→^{35}SO$_4^{2-}$ ᵘᵣ ^{35}S$_2$O$_3^{2-}$	Thiol	Partially purified. PAP^{35}S requires 3'-nucleotidase	52, 145
Higher Plants				
Spinacea oleracea	^{35}SO$_4^{2-}$→Acid-volatile radioactivity	Thiol	Crude extract	146
	^{35}SO$_4^{2-}$→^{35}SO$_3^{2-}$	Endogenous+light	Chloroplast extract. Requires yeast-activating enzymes and fraction C	6
	AP^{35}S(PAP^{35}S)→S-sulfoglutathione-^{35}S	GSH	Chloroplast extract	119

Table 4 Properties of Adenyl Sulfate Reductases from Plants and Microorganisms

Organism	Enzyme	Electron Donor or Acceptor	pH Optimum	K_m	MW	Remarks	References
Bacteria							
Desulfovibrio vulgaris (called *D. desulfuricans* in ref. 43)	APS Reductase	$Fe(CN)_6^{3-}$	7.4	SO_3^{2-} 2 mM	220,000	Contains 1 mole FAD, 6-8 g atoms nonheme iron	81, 98, 144
Thiobacillus thioparus	APS Reductase	$Fe(CN)_6^{3-}$	7.4	SO_3^{2-} 2.5 mM, AMP 0.1 mM	170,000	Contains 1 mole FAD, 8-10 g atoms nonheme iron	75, 98, 144
	APS Reductase	Cytochrome c	9.5	SO_3^{2-} 0.017 mM, AMP 0.0025 mM	170,000		75, 98, 144
Thiobacillus denitrificans	APS Reductase	$Fe(CN)_6^{3-}$	7.2	SO_3^{2-} 1.5 mM, AMP 0.041 mM	...	Contains 1 mole FAD, 6-11 g atoms nonheme iron	13, 144
Thiocapsa roseopersicina	APS Reductase	$Fe(CN)_6^{3-}$	8.0	SO_3^{2-} 1.5 mM, AMP 0.073 mM	180,000	Contains 1 mole FAD, 4 g atoms nonheme iron, 2 g atoms heme iron. Purified 60 to 80-fold; homogeneous upon ultracentrifugation	144
	APS Reductase	Cytochrome c	9.0	SO_3^{2-} 0.093 mM, AMP 0.059 mM	180,000		144
Fungi							
Saccharomyces cerevisiae	PAPS Reductase	NADPH	7.5 (tris)	...	...	Partially purified into 3 fractions A, B, C. Some activity with APS. Fraction A purified 60-fold, fraction C to apparent homogeneity in ultracentrifugation	7, 44, 140, 141, 158
Algae							
Chlorella pyrenoidosa	APS Reductase	Thiol	...	...	330,000	Partially purified. PAPS is active in the presence of a 3'-nucleotidase	114, 118, 119, 145

presence of dithiothreitol and other thiols, this protein-bound radioactivity is released as the S-sulfonate of the thiol or as sulfite, while in the presence of sulfide or BAL it is released as thiosulfate. The radioactivity seems to be bound to a low molecular weight unit which can be relased by heat-denaturing the enzyme; this labeled low molecular unit releases acid-volatile radioactivity in the presence of suitable thiols. This is further evidence for a low molecular weight carrier in these reactions.

So far, the low molecular weight component seems to be similar to the enzyme fraction "C" from yeast already mentioned, and the APS transferase would be similar to components of fraction "B," and indeed Schwenn (126) found such an identity for fraction "C."

Taking together the earlier work already reviewed and these newer observations, we would suggest the following scheme which appears to be consistent with the known enzymology, the in vivo studies, and genetic investigations (Figure 5).

For reasons already presented, we indicate that sulfate is converted to APS by the usual ATP sulfurylase reaction and that APS can be converted to PAPS by the APS kinase reaction. Although the generality of the "A" reaction has not been proved, we would suggest that PAPS may be converted back to APS by "A" in many organisms as it is in *Chlorella* and spinach chloroplasts. We further assume that PAPS is the preferred substrate for esterification reactions, as already described.

We also assume that the APS reductase complex is composed of at least three components: a transferase, a low molecular weight sulfhydryl protein (or carrier, denoted "Car"), and a reductase. We further assume that the carrier Car-S$^-$ receives an SO_3^{2-} group from APS under catalysis by the transferase. Car-S-SO$_3^-$ would then be acted upon by the reductase in the presence of a suitable reductant (i.e. ferredoxin) and *o*-acetyl serine to yield cysteine. These assumptions are suggested not only by the known enzymology but because they also allow us to achieve a parsimonious explanation of the properties of mutants of sulfate reducing organisms, chiefly *Salmonella*, *E. coli*, and *Chlorella* (see Table 5); these will be described below.

The reactions we have described so far are shown in Figure 5 and constitute what we think is probably the main pathway in most assimilatory sulfate reducers.

It should be noted that the scheme we have presented here views the central intermediates as being bound to carriers, an idea which has been prevalent since the inception of work on sulfate reduction. Although sulfite, sulfide, and thiosulfate have been viewed as free intermediates at one time or other, the ready reaction of one sulfur compound with another and the close relation of these free inorganic compounds with organic thiosulfates and persulfides led many workers to suggest that the bound thiosulfates (R-S-SO$_3^-$) and persulfides (R-S-S$^-$) might give rise to free sulfite, sulfide, and thiosulfate by reactions incidental to the main pathway. Experiments with sulfur-labeled compounds have also led to ambiguous results in some instances because many sulfur compounds undergo exchange reactions. For example, sulfite exchanges with the SO$_3^-$ moiety of thiosulfate and organic thiosulfates quite readily (5, 71, 118), and many other exchange reactions are known

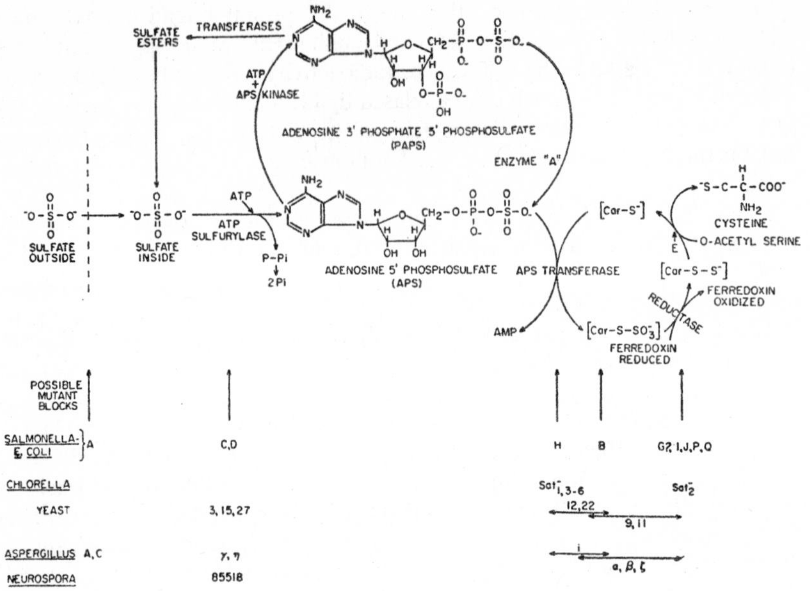

Figure 5   A unified scheme for assimilatory sulfate reduction. Sulfate is assumed to be converted to APS which is converted to PAPS to be used for sulfate esterification via various transferases; PAPS is assumed to be converted back to APS via a specific 3' nucleotidase (called fraction "A" in *Chlorella*). APS is assumed to be the substrate for further reduction as it is in *Chlorella*, yeast, and spinach chloroplasts, the sulfate being transferred from APS by an APS transferase. A carrier is assumed to exist (Car-S⁻) which may or may not be identical with thioredoxin; although only one SH group is shown, more than one is not precluded; thioredoxin has two which would permit internal S-S bonds to form and lead to the necessity for a reductive step to regenerate the reduced form. The carrier is thought to receive the sulfate group from APS via APS transferase; perhaps Car-S⁻ is the coenzyme for the transferase. Car-S-SO₃⁻, assumed to be the result of the transfer, is then assumed to be acted upon by the reductase which reduces the —SO₃ group with ferredoxin leading to the production of Car-S-S⁻; this presumably occurs when the carrier is complexed with the reductase Thus the carrier may act as a coenzyme for both the transferase and the reductase, a possible explanation for the pleiotrophy of certain mutations in *Salmonella* and *E. coli*. The Car-S-S⁻ would then donate its outer reduced sulfur to a reaction with *o*-acetyl serine to yield cysteine. This reductive pathway is composed largely of bound intermediates consistent with recent work. The accumulations of free compounds often seen can be explained as side reactions of the components of the main pathway, although independent pathways consisting of free intermediates cannot be ruled out. If we assume the pathway as shown is correct, free intermediates could arise through reductive cleavage of enzyme or carrier bound intermediates. In vitro thiols could release such compounds from either the transferase, from the carrier, or from the reductase. If we assume that the central inter-

(110). For this reason, the addition of carrier compounds to trap intermediates in the reduction of radioactive sulfate is very risky; one may obtain what he wishes to find by exchange of the free added nonradioactive carrier with the bound radioactive intermediate (71, 114).

A frequently written pathway is one which shows free sulfite and sulfide as intermediates (Figure 6; 21). It is true that sulfite accumulates in certain mutants blocked for sulfate reduction (21) and thiosulfate (50); in others sulfide is also found (21). But these could be formed from the breakdown of bound intermediates as well, or released from mutated enzymes in the pathway which normally retains the bound form. One reason why free sulfite has been an attractive possibility is the existence of sulfite reductases which reduce free sulfite to sulfide (110). Many of these enzymes are not specific; they will also reduce hydroxylamine, nitrite, and cytochrome c in some instances. It has been found that the reductase from *Chlorella* shown in Figure 5 for the bound intermediate (Car-S-SO_3^-) does not act on free sulfite (120, 121). These pieces of evidence suggest that sulfite reductase and free sulfite might not be obligatory participants in the main path of reduction in vivo. The question of whether sulfite reductase participates in sulfate reduction could be clarified if one could obtain mutants lacking this enzyme but possessing the reductase of Figure 5 for the bound intermediates and measure their ability to carry out sulfate reduction. It may be, however, that such mutants would be lethal even when grown on reduced sulfur if the other reactions catalyzed by "sulfite" reductases are necessary in other pathways. The question of the roles of bound and free intermediates needs to be further investigated. The possibility that there is more than one pathway of sulfate reduction has not been ruled out. It is conceivable that pathways employing free intermediates exist as well as those having bound forms—both paths might exist in the same cell. This point is discussed later in connection with cellular localization of sulfate reduction.

A word might be said here about the investment of reducing power in sulfate reduction. The early reactions of the pathway leading to Car-S-SO_3^- formation should not require reducing power per se. However, if the transferase or Car–S⁻

---

mediate is a bound thiosulfate (-S-SO_3^-), one example being the Car-S-SO_3^-, then thiols in vitro would release free sulfite (SO_3^{2-}) while BAL or sulfide would cause the release of thiosulfate ($S_2O_3^{2-}$). In addition, a variety of organic thiosulfates (R-S-SO_3^-) could be formed from added thiols, and these with excess thiol would also yield sulfite. Free sulfide (S^{2-}) could be formed from the reductive dismutation of Car-S-S⁻ or a similar enzyme-bound intermediate. The entry of exogenous sulfite into the scheme is assumed to be at the level of the carrier (Car-S⁻) either by addition or by cleavage of a disulfide form of the carrier by the sulfite; in either case Car-S-SO_3^- would result. Free exogenous sulfide or the reduced moiety of thiosulfate is assumed to enter by forming Car-S-S⁻ or by entering more directly into the formation of cysteine from o-acetyl serine. If these assumptions are correct, a unified scheme results upon which the mutant blocks can be placed of various strains of *Salmonella*, *E. coli*, *Chlorella*, yeast, *Aspergillus*, and *Neurospora*, as shown along the bottom of the figure (see Table 5). See text for further explanation and for references.

Table 5 Mutants with Impaired Sulfate Utilization

Salmonella / E. Coli	SO₄²⁻ Uptake	SO₄²⁻	Grows on: SO₃²⁻	Grows on: S₂O₃²⁻	S²⁻	Cyst(e)ine	Accumulates: SO₃²⁻ from SO₄²⁻	Accumulates: SO₃²⁻ from S₂O₃²⁻	S₂O₃²⁻ from SO₄²⁻	Homocysteic Acid from SO₄²⁻	SO₄²⁻ to APS
A		−	+	−	+	+	−	−			
A		−	+		+	+					
B		−	±	−	+	+	−	−			
B	−	−	−		+	+					
C		−	+	+	+	+					
C	+		+		+	+					+
D		−	+	+	+	+					
D		−	+		+	+					
E		−	−	−	−	+					
E	−	−	−		−	+					
G		−	−	+	+	+	+	+			
G	+	−	−		+	+					
H		−	+	+	+	+					
H	+	−	+		+	+					
I		−	−	+	+	+	+	+			
J		−	−	+	+	+	+	+			
P	−	−		−		+	+				
Q	+	−	−			+	+				

Whose Enzymology has been Studied

Converts: SO₄²⁻ to PAPS	SO₄²⁻ to SO₃²⁻	SO₄²⁻ to S²⁻	APS to PAPS	"A" Activity PAPS to APS	"S" Activity APS to SO₃²⁻	"S" Activity APSᵃ to S₂O₃²⁻	APS to S²⁻	PAPS to SO₃²⁻	PAPS to S²⁻	SO₃²⁻ to S²⁻	o-acetyl serine to cysteine	References
		+					+	+	+	+		21, 22, 59, 69
−						−		−	−	−		21, 22, 48, 59, 69
−		−					±	+	+	+		21, 22, 59, 69
		−					+	+	+	+		21, 22, 59, 69
−									±	−	−	21, 22, 58, 59, 69
+								+		−		21, 22, 59, 69
++		−				−	−	−	−	+		21, 22, 48, 59, 69
						+		+		−		21, 22, 48, 59, 69
										−		21, 22, 59, 69
+								+		−		21, 22, 59, 69
+									+	−		21, 22, 59, 69

ᵃ PAPS to S₂O₃²⁻ in case of *Salmonella* mutants, APS not tested.

Table 5 (*Continued*)

	SO_4^{2-} Uptake	SO_4^{2-}	Grows on:		S^{2-}	Cyst(e)ine	Accumulates:		$S_2O_3^{2-}$ from SO_4^{2-}	Homo-cysteic Acid from SO_4^{2-}	SO_4^{2-} to APS
			SO_3^{2-}	$S_2O_3^{2-}$			SO_3^{2-} from SO_4^{2-}	SO_3^{2-} from $S_2O_3^{2-}$			
Chlorella Sat⁻: 1,3–6		−		+							+
2		−		+					+	+	+
7R₁		+		+							+
Yeast 3		−	+	+							
12		−	+	+							
15		−	+	+							
22		−	+	+							
27		−	+	+							
9		−	−	+							
11		−	−	+							
Aspergillus γ	+	−	+	+		+					−
ι	+	−	+	+		+					+
η	+	−	+	+		+					+
α		−	−	+		+					+
β		−	−	+		+					+
ζ		−	−	+		+					+
A	−	−	+	±		+					+
C	−	−	+	±		+					+
Neurospora 85518											−

Converts: SO$_4^{2-}$ to PAPS	Converts: SO$_4^{2-}$ to SO$_3^{2-}$	SO$_4^{2-}$ to S^{2-}	APS to PAPS	"A" Activity PAPS to APS	"S" Activity APS to SO$_3^{2-}$	"S" Activity APS^a to S$_2$O$_3^{2-}$	APS to S^{2-}	PAPS to SO$_3^{2-}$	PAPS to S^{2-}	SO$_3^{2-}$ to S^{2-}	o-acetyl serine to cysteine	References
+				+	−	−						50
+				+	+	+						50
+				+	−	−						50
−	−											82
+	−											82
−	−											82
+	−											82
−	−											82
+	+											82
+	++											82
			+									55, 83, 127, 132
			+									55, 83, 127, 132
			−									55, 83, 127, 132
			+									55, 83, 127, 132
			+									55, 83, 127, 132
			+									55, 83, 127, 132
			+									55, 83, 127, 132
			+									55, 83, 127, 132
												102

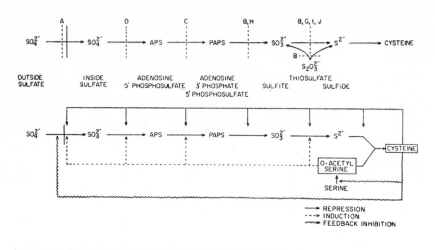

Figure 6 Pathways containing free intermediates which have been proposed for assimilatory sulfate reduction. The upper scheme is for sulfate reduction in *Salmonella* (22), the letters indicate the blocks inferred for various mutants (see Table 5). The lower scheme shows the similar suggestion for *E. coli*, together with the regulatory influences of cysteine and *o*-acetyl serine on the reduction pathway (60). It is not clear whether these pathways containing free intermediates exist in place of or in addition to the pathway of bound intermediates shown in Figure 5; on the other hand, the formation of free intermediates shown in this figure has also been attributed to side reactions in vivo or in vitro which are incidental to a main pathway of bound intermediates (such as the one in Figure 5).

undergo oxidation and must be kept in the reduced form for activity, a requirement for a reducing agent is understandable. Further, since what is frequently measured is the release of free sulfite or thiosulfate, a reducing agent would be necessary to form these compounds from $Car–S–SO_3^-$. Systems requiring NADPH or NADH for formation of free sulfite (21, 158) or thiosulfate (71, 115) are known, and in many of these thiols may be substituted as reductants [and as an S donor in the case of thiosulfate formation in presence of BAL (71, 115)]. It is probable that electron transport proteins are involved in passing the electrons from reduced pyridine nucleotides to $Car–S–SO_3^-$. Fraction A from yeast (7, 141, 158), for example, seems to serve this function, and it has been observed that thiosulfate formation in *Chlorella* extracts proceeds with pyridine nucleotides at pH 7.0, but thiols are the preferred reductant at pH 9.0 (71, 115). Perhaps an electron transport chain similar to that reducing thioredoxin (an SH-containing peptide), which is used for the reduction of ribonucleotides to desoxyribonucleotides, is involved (100); indeed, $Car–S^-$ in Figure 5, fraction C from yeast (7, 141, 158), and the low molecular weight acceptor for sulfate transfer from APS in *Chlorella* and spinach chloroplasts may be similar to or identical with thioredoxin.

When pyridine nucleotides are employed, the chain would operate to reduce the endogenous SH participants in the APS reductase complex, but thiols could preempt the chain by introducing electrons directly at the level of the thioredoxin-like compound.

If we regard the transferase as the enzyme plus Car-S⁻ (if our inferences are right, the transferase should have a high affinity for Car-S⁻ and a low affinity for Car-S-SO₃⁻, while the reverse should be true of the reductase), then various activities reported for the "PAPS reductase," "APS reductase," or transferase can be explained. It is known, for example, that the transferase will transfer the SO_3^- of APS to glutathione to form $G-S-SO_3^-$ (119). The same could occur with other thiols. It is likely that sulfite formation occurs by further sulfhydryl-catalyzed elimination from either the sulfhydryl-SO_3^- formed from the added thiol or from $Car-S-SO_3^-$. Similarly thiosulfate would be formed by the cleavage of $Car-S-SO_3^-$ by either free sulfide or by sulfide arising by extrusion of one of the sulfhydryl groups of BAL, as discussed above.

Thiosulfate dismutation and utilization might occur through a reductive cleavage to sulfide and sulfite as in the enzyme reaction studied by Kaji & McElroy (63), with the sulfite moiety appearing as $Car-S-SO_3^-$. Since evidence from wild-type and mutant *Chlorella* indicates that the sulfite moiety of exogenously fed thiosulfate must be oxidized to sulfate before further utilization (4, 51, 114), we assume that the sulfite formed from thiosulfate dismutation is immediately oxidized to sulfate and is then reduced. It is possible, of course, that thiosulfate is dismuted hydrolytically, which would yield sulfate and sulfide directly, but this reaction has not been observed as yet. Rhodanese activity has been found in extracts of *Chlorella* (2a).

Mutants of *Salmonella* (21, 22, 69), *E. coli* (59), *Chlorella* (50), yeast (82), and *Aspergillus* (55, 83, 127, 132) have been studied nutritionally and enzymatically (see Table 5); those of *Salmonella* and *E. coli* are rather similar in many respects (Tables 5, 6). Mutants in the A cistron are thought to be permease mutants with impaired sulfate uptake. Mutants C and D are thought to have blocks in sulfate activation; the placement of these mutants in our unified scheme is uncertain because we do not know if the "A" enzyme (or 3′ nucleotidase) converting PAPS to APS is present in these organisms, although there is preliminary evidence for its presence in *E. coli* (146). In the older linear schemes (Figure 6) where sulfate was assumed to flow through APS and PAPS respectively to the reductive steps, D was thought to block ATP sulfurylase while C blocked APS kinase.

A self-consistent unified scheme can be proposed which includes placement of blocks due to mutations involving the enzymes of the central reactions of sulfate reduction (Figure 5). We suggest that mutant H of *Salmonella* and *E. coli* is blocked in the transferase step. This would be consistent with its phenotypic characteristics shown in Table 5. Mutant H cannot grow on sulfate but can grow on sulfite, entering after the transferase step to form $Car-S-SO_3^-$ from $Car-S^-$; the rest of the pathway would not be blocked in this mutant. The sulfite moiety of thiosulfate would enter at the same place as sulfite in this scheme. *Chlorella* mutants Sat⁻₁,₃₋₆ are thought to be blocked in the transferase step since they lack

Table 6. Vocabulary of *Salmonella typhimurium* mutants
defective in sulfate metabolism[a]

| Cistron designation | | Presumed activity affected |
Old	New	
cys Aa	cys Aa	Sulfate permease
cys Ab	cys Ab	Sulfate permease
cys Ac	cys Ac	Sulfate permease
cys Ba	cys Ba	Regulation
cys Bb	cys Bb	Regulation
cys Bc	cys Bc	Regulation
cys Ca	cys C	APS-kinase
cys Cb	cys D	ATP-sulfurylase
cys Cc	cys J	Sulfite reductase
cys Cd	cys I	Sulfite reductase
cys Ce	cys H	PAPS reductase
cys Ea	cys Ea	Serine transacetylase
cys Eb	cys Eb	Serine transacetylase
cys G	cys G	Sulfite reductase

[a]References 21, 66.

"S" activity, the ability to convert APS to thiosulfate or sulfite, hence in the formation of $Car-S-SO_3^-$.

The B mutants of *Salmonella* and *E. coli* appear to be pleiotropic and affect several activities simultaneously (21, 22, 29, 34, 43, 57–60, 66, 69, 76, 92, 93, 153–155) In *Salmonella* these are conversion of PAPS to sulfite, the conversion of sulfite to sulfide, and the utilization of thiosulfate. In *E. coli* thiosulfate was not tested, but an effect of this locus on sulfate uptake was noted as well as on the activating enzymes. It has been suggested that the pleiotropy of the B locus arises from a regulatory gene which controls the structural genes concerned with the affected enzymes; the pattern of regulation is similar to the pattern of cysteine repression. It might be noted, however, that part of the pleiotropy can be explained by the scheme in Figure 5. If the B locus controls or codes for $Car-S^-$, the activity of both the transferase and the reductase in the main pathway will be affected along with any side reactions of $Car-S^-$.

It is thought that P and Q in *E. coli* and I and J in *Salmonella* influence sulfate reduction to sulfide. In terms of the scheme in Figure 5, we would place these mutants in the reductase which acts upon $Car-S-SO_3^-$; Sat_2^- of *Chlorella* is also blocked in this step (120a).

Locus E affects the production of serine transacetylase in *E. coli*. E also produces pleiotropic effects on other enzymes in the reductive pathway; the explanation generally offered is that *o*-acetyl serine is a regulator of the production or the activity of other enzymes in the pathway (see Figure 6). Therefore, we have placed E in OAS production in the scheme of Figure 5.

A linear scheme based on free intermediates has been proposed on the basis of this work (21; Figure 6). At present it cannot be decided whether two pathways exist (Figures 5 and 6) or whether the free intermediates arise from bound intermediates as side products in vitro or in mutants.

A distinct anomaly is mutant $Sat_7^-R_1$ from *Chlorella* (Table 5; 50). This mutant was obtained as a phenotypic revertant of Sat_7^-. Sat_7^- was selected for its lack of growth on sulfate, but the revertant $Sat_7^-R_1$ grows on sulfate albeit with an appreciable lag. $Sat_7^-R_1$ has not reverted all the way to wild type because it lacks significant levels of enzyme fraction "S" which converts APS to sulfite or thiosulfate. At first sight it would seem that enzyme fraction "S" is not on the main pathway of reduction in *Chlorella*, since this mutant lacks "S" but grows on sulfate. It is possible, however, that the enzyme extracts do not faithfully mirror the enzyme activity in vivo and that "S" is somehow lost or inactivated during extraction, but only in this mutant. On the other hand, it is also possible that there is more than one pathway of sulfate reduction in *Chlorella*, and that the one employing "S" has been blocked in $Sat_7^-R_1$ but not the parallel path. The same may be true of wild-type *Euglena*, which shows very low reduction activity in vitro yet grows well on sulfate (48). In this connection, we know very little about the cellular localization of the enzymes of sulfate reduction in eukaryotes. Spinach chloroplasts contain a sulfate-reducing system much like the one in *Chlorella* (121), but nothing is known of the nonchloroplast compartments in spinach. It is possible, of course, that the sulfate-reducing pathway is entirely chloroplastic both in higher plants and *Chlorella*, but nonphotosynthetic mutants of *Chlorella* grow as well on sulfate as wild type (39). The same is true for many other organisms; nonphotosynthetic mutants of *Chlamydomonas* (70) and *Scenedesmus* (12) grow on media containing sulfate as the only sulfur source, while mutants of *Euglena* which lack chloroplasts and plastid DNA grow as well as wild type on sulfate-containing media (116). There may be at least two pathways of sulfate reduction in chloroplast-containing cells, one in the chloroplast and one elsewhere in the cell, in the cytoplasm, or mitochondrion. It remains for future work to elucidate this. Until information of this sort is available, along with further studies, mutant Sat_7R_1 and wild-type *Euglena* remain anomalous and uninterpretable.

However, some recent work suggests that the two pathways may not be separate and distinct. Indeed, a useful working hypothesis can be made by assuming that the pathway in Figure 5 is the main pathway but that a separate and distinct sulfite reductase is also present which only acts upon free sulfite to reduce it to sulfide. (It is interesting that the free intermediate pathway shown in Figure 6, which employs a sulfite reductase of this type, has been postulated for organisms which can be conveniently grown on sulfite, i.e. in organisms which can be grown anaerobically to avoid the ready oxidation of sulfite to sulfate in the medium by molecular oxygen. This has facilitated the isolation of mutants which are blocked for sulfate utilization but which can utilize free sulfite from outside for growth.) If this hypothesis is correct, the usual pathway would be the one shown in Figure 3, but sulfite reductase would be available to reduce free sulfite should it be formed from bound intermediates along the way or if it is supplied from outside.

Certain *E. coli* and *Salmonella* mutants, reinterpreted in Figure 5 as thiosulfonate reductase mutants, may well be sulfite reductase mutants as originally suggested (see Figure 6, Tables 5, 6) if this hypothesis is true.

It has been known from the work of Schmidt & Trebst (121, 120) that gluta-thione-S-sulfonate (G–S–SO$_3^-$) will serve as a substrate for the thiosulfonate reductase in place of APS plus transferase. However, in studying mutant Sat$_2^-$ it has been found that APS and transferase yields no reductase activity while glutathione-S-sulfonate yields an appreciable amount (120a). Since Schmidt & Trebst showed that the thiosulfonate reductase acting on Car–S–SO$_3^-$ has no activity with free sulfite and since molecules of the type G–S–SO$_3^-$ are known to yield sulfite readily under certain conditions, the possibility that a separate sulfite reductase which acts on free sulfite is present was considered. Using a sulfite-generating system and the usual sulfite reductase assay, sulfite reductase activity was found in the extracts of this mutant which lacks thiosulfonate reductase activity. Added radioactive sulfite is also reduced. For this reason it is useful to think in terms of a sulfate-reducing system of the form of Figure 5 with bound intermediates, but with a parallel sulfite reductase step which would act upon free sulfite from outside the cell or from side reactions in wild type, or especially in mutants with impaired enzymes of the bound pathway.

DISSIMILATORY SULFATE REDUCTION AND THE EVOLUTION OF SULFATE REDUCING SYSTEMS

Dissimilatory sulfate reduction is found only in two genera of bacteria, as far as is known, *Desulfovibrio* and *Desulfotomaculum* (56, 94–97). These bacteria are strict anaerobes which utilize sulfate as an oxidizing agent in respiration in place of oxygen, thereby reducing sulfate to sulfide during oxidation of respiratory sub-strates. Oxidative phosphorylation of ADP to ATP is associated with these reactions. We will not be concerned with the detailed enzymology and physiology of the process, since this has been very recently reviewed (110) and is summarized in Figure 7.

In this pathway, sulfate is activated through the usual ATP sulfurylase reaction to form APS with the help of inorganic pyrophosphatase to pull the reaction. APS is the substrate for reduction to sulfite, the sulfite ultimately being reduced to sulfide. The immediate electron donor in these reactions is the unusually low potential cytochrome $c_3(Eo' = -205$ mv$)$ which ultimately receives electrons from molecular hydrogen via ferredoxin as a carrier. ATP formation from ADP is coupled to this process with P/H$_2$ ratios of 0.18. Although the sulfur intermediates in this pathway seem to accumulate as the free, unbound compounds, there is evidence for the formation of a flavin-sulfite adduct in the adenylyl sulfate reductase reaction of *Desulfovibrio* (80).

It was previously thought that the reduction of APS was restricted to the dissimilatory process while assimilatory reduction utilized PAPS. As discussed previously, APS has now been shown to be the substrate for reduction in assimilatory reducers such as *Chlorella* and spinach chloroplasts as well, the ap-

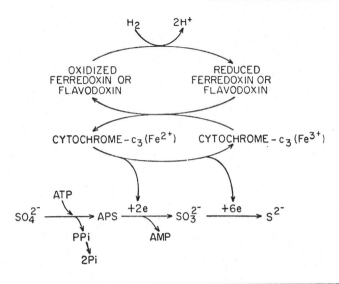

Figure 7 The dissimilatory pathway of sulfate reduction in *Desulfovibrio* (after Roy & Trudinger 110).

parent utilization of PAPS being due to a 3′ nucleotidase which converts PAPS to APS (37, 118, 145). If we add to this the fact that ferredoxin (120, 121) can act as a reductant in the assimilatory system (possibly via carriers as yet unknown), certain unities between asimilatory and dissimilatory sulfate reduction begin to emerge. It is quite possible that a primitive sulfate-reducing system similar to the system in *Desulfovibrio* evolved during the anaerobic phase of evolution before free molecular oxygen became available from the photolysis of water. With the appearance of oxygen-evolving photosynthesis, the assimilatory pathway presently found in aerobes may have evolved. As most people who work with enzymological systems reducing sulfate realize, the enzymes and intermediates in sulfate reduction are highly susceptible to inactivation by molecular oxygen, and at the very least, aerobic sulfate reducers would have to evolve mechanisms to protect the reduction process from molecular oxygen. When we consider the distribution of the free-intermediate pathway (Figure 6) (FIP) and the bound intermediate path (BIP) (Figure 5), it appears that the most anaerobic (and seemingly the most primitive) sulfate reducers are the dissimilatory ones, which so far seem to involve mainly unbound intermediates (Figure 7). The FIP may have evolved primarily as an energy-yielding process under anaerobic conditions. Under these conditions, enzymes and intermediates such as free sulfite and sulfide would need no protection from molecular oxygen. This pathway may have remained important in facultative organisms such as *E. coli, Salmonella,* and yeast, but at the same time the BIP may have evolved to provide greater protection of the intermediates of reduction from oxidation by molecular oxygen by binding them to carriers, and this may have become the predominant pathway in contemporary aerobic organisms. If we view the plastid and mitochondrion as having originated through endosymbiotic invasion of a nucleated cell by free-living procaryotes resembling blue-green algae and aerobic bacteria respectively (113, 123), then one can explain why contemporary chloroplasts have the BIP (118). The free-living blue-green algae-like plastid precursors may have been the first organisms to release molecular oxygen in photosynthesis, ending the anaerobic phase of the origin of life and compelling the selection of a reducing pathway more resistant to the presence of molecular oxygen. The same might be true of the earliest aerobic bacteria and contemporary mitochondria, although there seems to be no information which indicates whether sulfate reduction exists in mitochondria. Nonrespiratory mutants (petits) of yeast, however, are capable of growing on sulfate as the sole sulfur source, particularly "neutral" petits which lack mitochondrial DNA (James E. Haber, personal communication). This would suggest that the presence of functional mitochondria are not necessary for sulfate reduction, at least in yeast, and that energy derived from fermentation can be utilized for sulfate reduction.

The foregoing discussion has assumed that the FIP and BIP are real, distinct pathways. The fact that free intermediates can be readily obtained from bound intermediates by varying conditions in vitro has led many workers to suggest that the free intermediates found may not be on the main pathway of reduction but may be side products formed in vitro or by mutated enzymes in vivo. There seems to be no clear way to come to a firm conclusion at present. There may be a BIP which releases free intermediates under anomalous conditions, or there may be two sepatate or partially separate pathways. In addition, there may be other pathways of reduction about which we are completely ignorant. For example, mutant $Sat_7^- R_1$ of Chlorella (50) and wild type Euglena gracilis (48) fail to yield the usual sulfate-reducing activity in extracts but are able to grow perfectly well on sulfate. Is this due to difficulties in extraction or measurement of the enzyme, or does it point to an as yet undisclosed pathway of sulfate reduction?

In any case, considering the similarities between dissimilatory and assimilatory sulfate reduction and the possibility that the energy-coupled dissimilatory pathway arose first, perhaps we should look to this pathway for clues concerning the nature of the assimilatory pathway. The assimilatory pathway may have arisen from a loss of the energy-coupling function of the dissimilatory pathway combined with modifications necessary to aerobic existence.

CONCLUSIONS

This seems to be a good point at which to return to the bench. As the reader will appreciate, there are many problems in sulfate metabolism which require a great deal more work.

The area of sulfate transfer and esterification needs further investigation, especially of the biosynthesis of sulfated polysaccharides. Are the monomers sulfated before polymerization into the polysaccharide, or is the polysaccharide sulfated after polymerization? Do both processes occur and do they have developmental and regulatory significance?

How are sulfonic acids synthesized, especially glucose-6-sulfonate or 6-sulfoquinivose of the ubiquitous sulfolipid crucial for the formation of plastid membranes? Ideas exist but further studies are required.

Knowledge of the sulfate reducing pathway(s) is growing, but there is still a good deal of confusion. Is there more than one pathway of sulfate reduction? If so, how many? Are there bound or free intermediates in these pathways? Do sulfite reductases acting on free sulfite participate in sulfate reduction? Where are the pathways of sulfate reduction localized in the cell? What is the detailed enzymology of the reduction pathway(s), and what are the mechanisms of the reactions involved? Clearly there is much to do.

Sulfur biochemistry is plagued by promiscuous reactions of coy intermediates, and for this reason much of our time is spent in ruling out artifacts rather than ruling in reactions. Still, it is an exciting field for the stubborn and the patient, and the growing use of modern enzymological and genetic techniques has permitted highly significant advances to be made. These will continue, we feel, and there is every reason to be optimistic about the eventual solution of these problems.

ACKNOWLEDGMENTS

We would like to thank Dr. Ahlert Schmidt of the University of Bochum, who is presently on research leave at Brandeis, for his advice and valuable discussion and Dr. Achim Trebst of Bochum for providing Dr. Schwenn's thesis. Dr. William Abrams and Ms. Monica Lik-Shing Tsang also provided valuable assistance. The National Science Foundation through Grant GB 25920 to Jerome A. Schiff aided the original work reported herein and the preparation of the manuscript.

Literature Cited

1. Abraham, A., Bachhawat, B. K. 1963. *Biochim. Biophys. Acta* 70:104–6
2. Abrams, W., Schiff, J. A. 1973. *Plant Physiol. Abstr.* In Press
2a. Abrams, W., Schiff, J. A. Unpublished
3. Akagi, J. M., Campbell, L. L. 1962. *J. Bacteriol.* 84:1194–1201
4. Alt, F., Schiff, J. A. Unpublished
5. Ames, D., Willard, J. 1951. *J. Am. Chem. Soc.* 73:164–72
6. Asahi, T. 1964. *Biochim. Biophys. Acta* 82:58–66
7. Asahi, T., Bandurski, R. S., Wilson, L. G. 1961. *J. Biol. Chem.* 236:1830–35
8. Balharry, G. J. E., Nicholas, D. J. D. 1970. *Biochim. Biophys. Acta* 220:513–24
9. Benson, A. A., Atkinson, M. R. 1967. *Fed. Proc. Fed. Am. Soc. Exp. Biol.* 26:394
10. Benson, A. A., Shibuya, I. 1962. *Physiology and Biochemistry of Algae,* ed. R. Lewin, 371–83. New York: Academic
11. Bidwell, R. G. S., Ghosh, N. R. 1963. *Can. J. Bot.* 41:209–20
12. Bishop, N. I. 1971. *Methods Enzymol.* 23:130–43
13. Bowen, T. J., Happold, F. C., Taylor, B. F. 1966. *Biochim. Biophys. Acta* 118:566–76
14. Bradfield, G. et al 1970. *Plant Physiol.* 46:720–27
15. Brunngraber, E. G. 1950. *J. Biol. Chem.* 233:472
16. Challenger, F. 1959. *Aspects of the Organic Chemistry of Sulphur.* New York: Academic

17. Cramer, M., Myers, J. 1952. *Arch. Mikrobiol.* 17:384–402
18. Davies, W. H., Mercer, E. I., Goodwin, T. W. 1966. *Biochem. J.* 98:369–73
19. DeMeio, R. H., Wizerkaniuk, M., Schreibman, I. 1955. *J. Biol. Chem.* 213:439–43
20. Dreyfuss, J. 1964. *J. Biol. Chem.* 239:2292
21. Dreyfuss, J., Monty, K. J. 1963. *J. Biol. Chem.* 238:1019–24
22. Ibid, 3781–83
23. Dreyfuss, J., Pardee, A. B. 1966. *J. Bacteriol.* 91:2275–80
24. Dziewiatkowski, D. D. 1970. *Sulfur in Nutrition*, ed. O. H. Muth, 97–125. Westport, Conn.: Avi
25. Elliot, M. C., Stowe, B. B. 1970. *Phytochemistry* 9:1629–32
26. Elliott, M. C., Stowe, B. B. 1971. *Plant Physiol.* 47:366–72
27. Ibid. 48:498–503
28. Ellis, R. J. 1969. *Planta* 88:34–42
29. Ellis, R. J., Humphries, S. K., Pasternak, C. A. 1964. *Biochem. J.* 92:167–72
30. Epstein, E., Nabors, M. W., Stowe, B. B. 1967. *Nature* 216:547–49
31. Ettlinger, M. G., Kjaer, A. 1968. *Recent Advances in Phytochemistry*, ed. T. J. Mobry, R. E. Alston, U. C. Runeckles, 1:59–144. New York: Appleton
32. Ferrari, G., Renasto, F. 1972. *Plant Physiol.* 49:114–16
33. Fromageot, P., Perez-Milan, H. 1959. *Biochim. Biophys. Acta* 32:457–64
34. Fujimoto, D., Ishimoto, M. 1961. *J. Biochem. Tokyo* 50:533
35. Gibbs, M., Schiff, J. A. 1960. *Plant Physiol., A Treatise* 1B:279–319
36. Goldberg, I. H., Delbruck, A. 1959. *Fed. Proc. Fed. Am. Soc. Exp. Biol.* 18:235
37. Goldschmidt, E., Tsang, M., Schiff, J. A. In preparation
38. Gonzales, R., Schiff, J. A. 1972. *Abstr. Northeast. Sect. Meet., Am. Soc. Plant Physiol., Binghamton, N Y*
39. Granick, S. 1971. *Methods Enzymol.* 23:162–68
40. Itabuchi, O., Yamagata, T., Suzuki, S. 1971. *J. Biol. Chem.* 246:7357–65
41. Hart, J. W., Filner, P. 1969. *Plant Physiol.* 44:1253–59
42. Hase, E. 1962. See Ref. 10, 617–24
43. Henderson, R. J., Loughlin, R. E. 1968. *Biochim. Biophys. Acta* 156:195

44. Hilz, H., Kittler, M. 1960. *Biochem. Biophys. Res. Commun.* 3:140–42
45. Hilz, H., Kittler, M., Knappe, G. 1959. *Biochem. Z.* 332:151–66
46. Hilz, H., Lipmann, F. 1955. *Proc. Nat. Acad. Sci. USA* 41:880–90
47. Hodgkin, J. H., Craigie, J. S., McInnes, A. G. 1966. *Can. J. Chem.* 44:74
48. Hodson, R. C., Schiff, J. A. 1971. *Plant Physiol.* 47:296–99
49. Ibid, 300–5
50. Hodson, R. C., Schiff, J. A., Mather, J. P. 1971. *Plant Physiol.* 47:306–11
51. Hodson, R. C., Schiff, J. A., Scarsella, A. J. 1968. *Plant Physiol.* 43:570–77
52. Hodson, R. C., Schiff, J. A., Scarsella, A. J., Levinthal, M. 1968. *Plant Physiol.* 43:563–69
53. Horowitz, N. H. 1950. *Advan. Genet.* 3:33–71
54. Husisige, H., Sato, K. 1960. *Bull. Biol. J. Okayama Univ. Japan* 6:71–82
55. Hussey, C., Orsi, B. A., Scott, J., Spencer, B. 1965. *Nature* 207:632–34
56. Ishimoto, M., Fujimoto, D. 1959. *Proc. Jap. Acad.* 35:243–45
57. Jones-Mortimer, M. C. 1967. *Biochem. J.* 106:33
58. Ibid 1968. 107:51–53
59. Ibid 1968. 110:589–95
60. Ibid, 597–602
61. Josefsson, E. 1967. *Phytochemistry* 6:1617
62. Kaji, A., Gregory, J. D. 1959. *J. Biol. Chem.* 234:3007
63. Kaji, A., McElroy, W. D. 1959. *J. Bacteriol.* 77:630
64. Kindl, H. 1965. *Monatsh. Chem.* 96:527
65. Kline, B. C., Schoenhard, D. E. 1970. *J. Bacteriol.* 102:142–48
66. Kredich, N. M. 1971. *J. Biol. Chem.* 246:3474–84
67. Kutacek, M., Kefeli, U. I. 1968. *Biochemistry and Physiology of Plant Growth Substances*, ed. F. Wightman, G. Setterfield, 127. Ottawa: Runge
68. Langridge, R., Shinagawa, H., Pardee, A. B. 1970. *Science* 169:59–61
69. Leinweber, F. J., Monty, K. J. 1963. *J. Biol. Chem.* 238:3775
70. Levine, R. P. 1971. *Methods Enzymol.* 23:119–29
71. Levinthal, M., Schiff, J. A. 1968. *Plant Physiol.* 43:555–62
72. Lindberg, B. 1955. *Acta Chem. Scand.* 9:1323–26
73. Loewus, F., Wagner, G., Schiff, J. A., Weistrop, J. 1971. *Plant Physiol.* 48:373–75

74. Lowe, I. P., Roberts, E. 1955. *J. Biol. Chem.* 212:477–83
75. Lyric, R. M., Suzuki, I. 1970. *Can. J. Biochem.* 48:344–54
76. Mager, J. 1960. *Biochim. Biophys. Acta* 41:553
77. Mahadevan, S., Stowe, B. B. 1971. *Plant Growth Substances—1970*, ed. D. J. Carr. Canberra: Aust. Acad. Sci.
78. Mercer, E. I., Thomas, G. 1969. *Phytochemistry* 8:2281–85
79. Metzenberg, R. L., Parson, J. W. 1966. *Proc. Nat. Acad. Sci. USA* 55:629
80. Michaels, G. B., Davidson, J. T., Peck, H. D. Jr. 1970. *Biochem. Biophys. Res. Commun.* 39:321–28
81. Michaels, G. B., Davidson, J. T., Peck, H. D. Jr. 1970. *Biochem. Biophys. Res. Commun.* 39:321–28
82. Naiki, N. 1964. *Plant Cell Physiol.* 5:71–78
83. Nakamura, T. 1962. *J. Gen. Microbiol.* 27:221–30
84. Nakanishi, Y., Sonohara, H., Suzuki, S. 1970. *J. Biol. Chem.* 245:6046–51
85. Nissen, P. 1971. *Physiol. Plant.* 24:351–24
86. O'Colla, P. S. 1962. See Ref. 10, 337–52
87. Ohta, N., Galsworthy, P. R., Pardee, A. B. 1971. *J. Bacteriol.* 105:1053–62
88. Pardee, A. B. 1966. *J. Biol. Chem.* 241:5886–92
89. Pardee, A. B. 1967. *Science* 156:1627–28
90. Pardee, A. B., Prestige, S., Whipple, M. B., Dreyfuss, J. 1966. *J. Biol. Chem.* 241:3962–69
91. Pardee, A. B., Watanabe, K. 1968. *J. Bacteriol.* 96:1049–54
92. Pasternak, C. A. 1962. *Biochem. J.* 85:44–48
93. Pasternak, C. A., Ellis, R. J., Jones-Mortimer, M. C., Crichton, C. E. 1965. *Biochem. J.* 96:270–75
94. Peck, H. D. Jr. 1959. *Proc. Nat. Acad. Sci. USA* 45:701–8
95. Peck, H. D. Jr. 1962. *J. Biol. Chem.* 237:198
96. Peck, H. D. Jr. 1962. *Bacteriol. Rev.* 26:67–94
97. Peck, H. D. Jr. 1970. *Symposium: Sulfur in Nutrition*, ed. O. H. Muth, J. E. Oldfield, 61–79. Westport: Avi
98. Peck, H. D. Jr., Deacon, T. E., Davidson, J. T. 1965. *Biochim. Biophys. Acta* 96:429–46
99. Percival, E., McDowell, R. H. 1967. *Chemistry and Enzymology of Marine Algal Polysaccharides.* New York: Academic

100. Porqué, P. G., Baldesten, A., Reichard, P. 1970. *J. Biol. Chem.* 245:2371–74
101. Quatrano, R. S., Crayton, M. A. 1973. *Develop. Biol.* 30:29–41
102. Ragland, J. B. 1959. *Arch. Biochem. Biophys.* 84:541–42
103. Ragland, J. B., Liverman, J. 1958. *Arch. Biochem. Biophys.* 76:496
104. Ramus, J. 1972. *J. Phycol.* 8:97–111
105. Ramus, J., Groves, S. T. 1972. *J. Cell Biol.* 54:399–407
106. Robbins, P. W. 1958. *Methods Enzymol.* 5:964–77
107. Robbins, P. W. 1962. *Enzymes* 6:469–76
108. Robbins, P. W., Lipmann, F. 1957. *J. Biol. Chem.* 229:837–51
109. Ibid 1958. 233:681–85
110. Roy, A. B., Trudinger, P. A. 1970. *The Biochemistry of Inorganic Compounds of Sulfur.* Cambridge
111. Ruelius, H. W., Gaube, A. 1950. *Justus Liebigs Ann. Chem.* 570:121
112. Schiff, J. A. 1962. See Ref. 10, 239–46
113. Schiff, J. A. 1973. *Advan. Morphogen.* In press
114. Schiff, J. A., Hodson, R. C. 1970. *Ann. NY Acad. Sci.* 175:555–76
115. Schiff, J. A., Levinthal, M. 1968. *Plant Physiol.* 43:547–54
116. Schiff, J. A., Lyman, H., Russell, G. K. 1971. *Methods Enzymol.* 23:143–62
117. Schiff, J. A., Quatrano, R. S. 1972. *Biol. Bull.* 143(2):476
118. Schmidt, A. 1972. *Arch. Mikrobiol.* 84:77–86
118a. Schmidt, A. 1968. Dissertation. Gottingen
119. Schmidt, A. 1972. *Z. Naturforsch.* 276:183–92
120. Schmidt, A. Submitted for publication
120a. Schmidt, A., Abrams, W., Schiff, J. A. 1973. *Plant Physiol. Abstr.* In press
121. Schmidt, A., Schwenn, J. D. 1971. *11th Int. Congr. Photosyn.*, Stresa, 507
122. Schmidt, A., Trebst, A. 1969. *Biochim. Biophys. Acta* 180:529
123. Schnepf, E., Brown, R. M. 1971. *Origin and Continuity of Cell Organelles*, ed. J. Reinert, H. Ursprung. New York: Springer
124. Schraudolf, H., Bergmann, F. 1965. *Planta* 67:75–95
125. Shrift, A. 1969. *Ann. Rev. Plant Physiol.* 20:475–94
126. Schwenn, J. D. 1970. Dissertation, Univ. of Bochum

127. Scott, J. M., Spencer, B. 1960. *Biochem. J.* 106:471-77
128. Segal, I. H., Johnson, M. J. 1961. *J. Bacteriol.* 81:91
129. Shuster, L., Kaplan, N. O. 1953. *J. Biol. Chem.* 201:535
130. Shuster, L., Kaplan, N. O. 1955. *Methods Enzymol.* 2:551
131. Spencer, B., Hanada, T. 1960. *Biochem. J.* 77:305-15
132. Spencer, B., Hussey, C., Orsi, B. A., Scott, J. M. 1968. *Biochem. J.* 106: 461-69
133. Steele Da Cruz, F., Krassner, S. M. 1971. *J. Protozool.* 18:718-22
134. Stowe, B. B., Epstein, E., Vendrell, M. 1968. *Soc. Chem. Ind. London Monogr.* 31:102-10
135. Suzuki, S., Strominger, J. L. 1960. *J. Biol. Chem.* 235:257-66
136. Ibid 267-73
137. Ibid 274-76
137a. Taylor, B., Novelli, G. D. 1961. *Bacteriol. Proc.* 190
138. Thompson, J. F. 1967. *Ann. Rev. Plant Physiol.* 18:59-84
139. Thompson, J. F., Smith, I. K., Moore, D. P. 1970. See Ref. 24
140. Torii, K., Bandurski, R. S. 1964. *Biochem. Biophys. Res. Commun.* 14: 537-42
141. Torii, K., Bandurski, R. S. 1967. *Biochim. Biophys. Acta* 136:286
142. Trebst, A., Schmidt, A. 1969. *Progr. Photosyn. Res.* 3:1510-16

143. Truper, H. G., Peck, H. D. Jr. 1970. *Arch. Mikrobiol.* 73:125-42
144. Truper, H. G., Rogers, L. A. 1971. *J. Bacteriol.* 108:1112-21
145. Tsang, M., Goldschmidt, E., Schiff, J. A. 1971. *Plant Physiol.* 47:S-20
146. Tsang, M., Schiff, J. A. 1973. *Plant Physiol. Abstr.* In press; also unpublished work
147. Tsuji, M., Shimizu, S., Nakaishi, Y., Suzuki, S. 1970. *J. Biol. Chem.* 245: 6039-45
148. Varma, A. K., Nicholas, D. J. D. 1970. *Arch. Mikrobiol.* 73:293-307
149. Varma, A. K., Nicholas, D. J. D. 1971. *Biochim. Biophys. Acta* 227: 373-89
150. Varma, A. K., Nicholas, D. J. D. 1971. *Arch. Mikrobiol.* 78:99-117
151. Wedding, R. T., Black, M. K. 1960. *Plant Physiol.* 35:72
152. Wetter, L. R. 1964. *Phytochemistry* 3:57
153. Wheldrake, J. F. 1967. *Biochem. J.* 105:697-699
154. Wheldrake, J. F. 1967. *Biochem. J.* 105:697-699
155. Wheldrake, J. F., Pasternak, C. A. 1965. *Biochem. J.* 96:276-80
156. Wickberg, B. 1956. *Acta Chem. Scand.* 10:1097-99
157. Ibid 1957. 11:506-11
158. Wilson, L. G., Asahi, T., Bandurski, R. S. 1961. *J. Biol. Chem.* 236:1822

Ann. Rev. Plant Physiol. 1973. 24:415–44

CYTOKININS AS A PROBE OF
DEVELOPMENTAL PROCESSES[1]

❖ 7554

R. H. Hall

Department of Biochemistry, McMaster University, Hamilton, Ontario, Canada

CONTENTS

INTRODUCTION

A major effort of biological scientists centers on exploration of phenomena of growth and development at the molecular level. A model for mechanisms of control at the cellular level has been provided by advances in molecular biology, but a comparable model for explaining how multicellular events are organized has not emerged. Studies on multicellular organization are proceeding in many biological systems, and those scientists working with plants have already established

[1] Abbreviations used: ABA [abscisic acid, abscisin]; i⁶Ade [N⁶-(Δ²-isopentenyl) adenine]; i⁶Ado [N⁶-(Δ²-isopentenyl)adenosine]; Ade-CO-thr [N-(purin-6-ylcarbamoyl) threonine]; Ado-CO-thr [N-[9-(β-D-ribofuranosyl-9H)purin-6-ylcarbamoyl]threonine]; BAP [N⁶-benzylaminopurine]; GA [gibberellic acid]; IAA [indoleacetic acid]; RZ [ribosylzeatin]; Z [zeatin]; 2ms [2-methylthio].

415

several experimental approaches and a number of working hypotheses. These hypotheses have been drawn equally from knowledge of events in animal systems and from information peculiar to plant systems.

The "hormone" model of control, for example, has been borrowed largely from animal systems, and although it provides a useful model, it is not totally applicable to the plant situation. Differentiation can be thought to consist of two processes: the actual process of differentiation and the state of being differentiated. The well-defined hormonal systems of animals are part of the delicate control of function of mature tissues. The animal spends most of its life cycle in an adult state. In contrast, the plant's cycle is one of continuous growth and differentiation, and it is perhaps incautious to extend too far a model of hormonal control that explains events in mature tissues. I would even go so far as to suggest that the word "hormone" not be used in discussing control systems in plants because the word conjures up images of synthesis of specific agents, transportation of agents, target cells, receptor sites, etc. The model restricts what should at this stage in understanding be a free-moving exploration of growth and development in the plant.

> Whereas science links the present to the past via concepts, art leaps from the present to the future via percepts (McLuhan & Nevitt 116).

One of the difficulties in trying to formulate models of differentiation is the use of metaphors based on the mechanical sciences. These metaphors infer direct links between events and give rise to terms such as trigger, feedback, loops, etc. The very word "control" carries the connotation of a system of levers or wires hooked one to another which start and stop events. Events do indeed start and stop, slow down and speed up in developing tissue, but mechanical models of control provide little insight into such happenings.

In the plant there are compounds that appear to play a special role in development. I would like to suggest an alternative conceptual model of how such agents can influence the phenotypic expression of tissue. Phenotypic expression should be thought of as the total consequence of all the interactions of all the molecular processes of the cell or tissue. Expression comes not from the existence of the molecules of the cell as static entities, but rather from the resonant processes between molecules. When an agent such as a cytokinin enters and becomes part of a resonant system, every interaction is modified so that the total expression of the tissue is changed. It is much like adding a pinch of a powerful dye to an already colored paint—a perceptible shift in hue occurs. The model allows for what appears to be agent-agent interaction; as with colored paint, one can obtain infinite shading of hues by adding traces of two or three dyes in varying proportions.

This concept emphasizes the resonant interrelationships between molecules rather than the molecules themselves. Driesch (quoted by Goodman et al 64) suggested that such interrelationships, the entelechies, specify harmoniously the parts of the whole. The preliterate Greeks studied entelechies of nature, considering them to represent the true significance of nature, but when the Greeks became literate they ceased to study entelechies and turned to the study of causes, a tradi-

Figure 1   Fisherman and Hunter. Stone cut, Angasaglo, Amarook.

tion that has persisted for 2500 years. This tradition has produced an impressive body of knowledge in the physical sciences, but many scientists worry about the propriety of applying without questioning the concepts and experimental approaches of the physical sciences to a study of living phenomena. Polanyi, for example, writes that all manifestations of life cannot be explained by laws governing inanimate matter (131). The laws of physics and chemistry are based on concepts of direct cause and effect. Biological scientists should be seeking new approaches to an understanding of life processes, and they might well derive some inspiration from the preliterate Greeks' perception of the significance of the entelechy. The Eskimo artist also has a strong sense of the entelechy which is expressed in his art (Figure 1). His work expresses no point of view; its significance derives from the nature of all the interrelationships between himself and the happenings of the natural world.

Definitions

In reviewing the literature on cytokinins, I will try to use terms that have currency. Three key definitions are:

Cytokinin—A compound that promotes growth and differentiation in cultured callus tissue. Common cytokinins are shown in Figure 2.

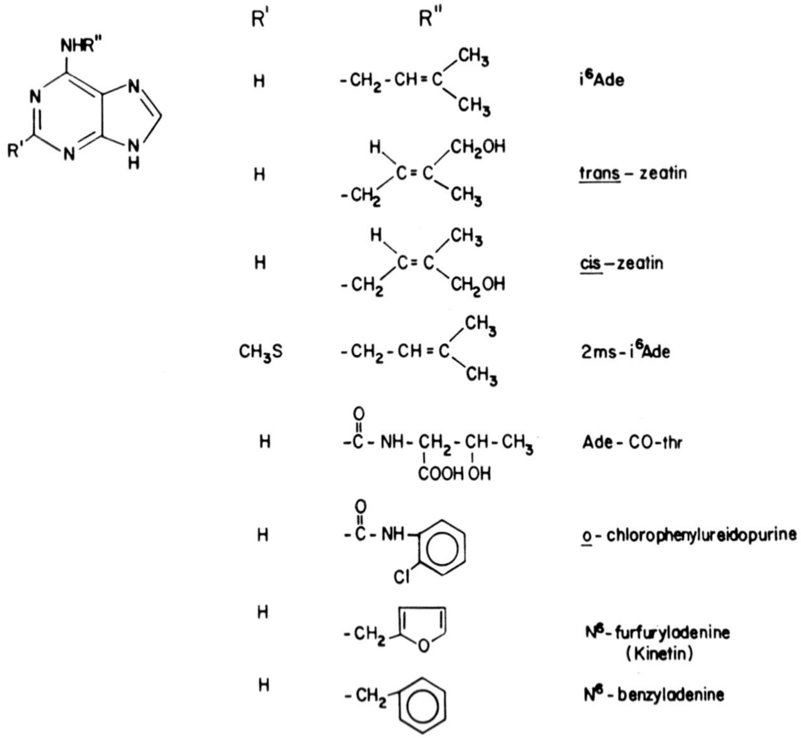

	R'	R''	
	H	$-CH_2-CH=C\begin{smallmatrix}CH_3\\CH_3\end{smallmatrix}$	i^6Ade
	H	(C=C with H, CH_2OH, $-CH_2$, CH_3)	*trans* – zeatin
	H	(C=C with H, CH_3, $-CH_2$, CH_2OH)	*cis* – zeatin
	CH_3S	$-CH_2-CH=C\begin{smallmatrix}CH_3\\CH_3\end{smallmatrix}$	$2ms-i^6Ade$
	H	$-\overset{O}{\overset{\|}{C}}-NH-CH_2-CH-CH_3$ ($\overset{\|}{COOH}$ $\overset{\|}{OH}$)	Ade - CO-thr
	H	$-\overset{O}{\overset{\|}{C}}-NH-$(phenyl with Cl)	o - chlorophenylureidopurine
	H	$-CH_2-$(furyl)	N^6-furfuryladenine (Kinetin)
	H	$-CH_2-$(phenyl)	N^6 - benzyladenine

Figure 2 Structures of cytokinins cited in review.

Cytokinin phenomenon—An aspect of growth and differentiation which involves cytokinins. It is not a precise definition and assumes that the laboratory demonstration of cytokinin activity reflects actual processes in plants.

Agent—I will use this in the generic sense to indicate a compound which is involved in bringing about a physiological event.

MORPHOGENESIS

The morphogenic events in the cycle of the plant—germination, branching, fruiting, senescence—share common features. The overall expression of the tissue may vary widely, but this is accomplished with a common set of molecules, biochemical processes, and, presumably, common control mechanisms. It is the infinite arrangement of the basic building materials and their dynamic interrelationships that give rise to the specific characteristics of the developed tissue. It is not surprising, therefore, that when the biochemical events which occur during morphogenesis are studied, the same general patterns are observed whether dealing with germination or senescence.

The cytokinin phenomenon appears to be involved at all stages of plant morphogenesis and can be manipulated experimentally in similar ways. Germination, branching, and fruiting represent accelerated growth, so if one measures gross biochemical parameters such as protein synthesis, nucleic acid synthesis, various enzyme levels, etc, quantitative changes will certainly be observed. Thus, when applying externally agents that stimulate natural development of tissue, to conclude that such agents cause a change in protein synthesis, nucleic acid synthesis, etc becomes meaningless. This is not to say such experiments cannot yield information, but one will not gain much insight by confusing cause and effect relationships with what really are correlations.

One correlation that emerges from all the experiments is that between rate of tissue growth and level of measurable cytokinin activity. Grape vines, for example, when stored at 1°C for about 1 year accumulate measurable cytokinin activity in the cambial tissue, apparently in anticipation of developing shoots in warmer weather (156). In developing cotton fruit, rice ears, and avocado seeds, the level of measurable cytokinin activity is much higher during the early stages of development, and as the tissue matures the cytokinin level also falls (57, 126, 142). The observed decrease in cytokinin activity could be due partly to formation of cytokinin inhibitors (16), but the important feature is that the level of free cytokinin seems to reflect the level of dynamic processes.

The developing seed and embryo of *Pisum arvense* L. have two maxima in growth rate, at 23 and 31 days after anthesis. Cytokinin activity that can be extracted from seeds shows two maxima in levels coinciding with these two maxima in growth rate. A third maximum in cytokinin level coincides with a maximum volume of endosperm of the seed (19).

Plant tissue, unlike animal tissue, is responsive to the external environment, and growth and development processes are sensitive to external circumstances. Plants subjected to stress such as limited water, salinity, or flooding, for example, show symptoms of premature senescence. Itai & Vaadia (91) noted that exudates of roots of sunflowers suffering from a lack of water contained significantly less cytokinin activity than those of control plants. As will be discussed in more detail below, processes of senescence seem to be associated with lower levels of observable cytokinin activity. In an analogous experiment, the premature aging of leaves of water-stressed sugar beets was reversed by application of BAP (149). The detached leaves of bean plants containing petioles generate adventitious roots when placed in an appropriate solution. The formation of roots from the petiole delays death of the leaf, but there is no need for the roots if the petiole has a callus (181). The interpretation is that either the developing roots or the callus produce cytokinins that help to maintain integrity of the biochemical processes of the leaf.

This interpretation suggests that the plant relies on active sites of morphogenesis for generation of cytokinin-like substances. Kende (95) has suggested that in the developing seedlings the preferred site of synthesis is first in the root tip; the site could then shift to the fruit at a later stage. I will discuss in more detail below the notion of translocation of cytokinins.

Differentiation at the cytological level has attracted attention. In higher plants the formation of tracheary elements offers a case of cytodifferentiation amenable to study. Vascular plants have conductive tissue and the distinctive cells of the xylem are tracheary elements. These cells are associated with other types of supporting cells. It is the development of the primary xylem into tracheary elements that can be followed, and this development requires the presence of cytokinins (172).

Cytokinins do not act alone, and there is evidence for many elements in the system for control of growth and differentiation. Steward & Rao (163) have studied the effect of various exogenous growth factors and nutrients on cultured explants of *Daucus carota* L. They conclude that when the quiescent cells with their endogenous endowment of growth factors and nutrients receive the appropriate external stimuli, the cells can then make use of available nutrients from both internal and external sources. The interaction of numerous factors in the growth control system means that in any experimental system most of the factors will already be present in the tissue and hence invisible to the experimenter. Only the externally applied factors can be readily studied and, in general, only one or two exogenous variables can be studied. This does not mean that the externally applied growth agents under study have any more biological significance than the ones already present internally. Thus in using exogenous growth factors as a tool for study of morphogenesis, one should not interpret complex phenomena in terms of the limited experimental system, but rather one should use the system as a probe of these complexities.

INTERACTION OF MULTIPLE AGENTS

The interaction of cytokinin with other plant growth factors has been demonstrated in many ways (see reviews 55, 97, 170). The opposing physiological effects of cytokinins and ABA have been studied by Hemberg (82). The rest of potatoes can be broken by administration of kinetin. ABA, considered to be one of the causes of the potato rest, disappeared under these conditions. However, Hemberg noted a problem in the design of the experiment: the observed disappearance of ABA could have been due to masking of its physiological activity in the subsequent analytical system. In a later paper (83), he presented evidence that disappearance of extractable ABA from the potato indeed resulted from the cytokinin treatment. The interaction of cytokinins and ABA is more complex than a simple plus/minus relationship. Blumenfeld & Gazit (17) observed that response of soya bean callus to low levels of kinetin (0.005 mg/l) is inhibited by ABA (10 mg/l), but when the concentration of the cytokinin is raised (approximately 0.5 mg/l) antagonism changes to synergism.

The relationships of cytokinin and auxin had been investigated by Went (179, 180), who showed that in tomato shoots growth is dependent upon hormone-like substances furnished by the root. Engelbrecht (44) and Mothes & Engelbrecht (122) showed that substances obtained from roots maintained chlorophyll, protein, and RNA levels in leaves, and that this activity could be replaced by cytokinins. Such observations led these workers and others (94, 103, 110) to the conclusion that roots supplied cytokinins needed for growth of the shoot. The apical region of the shoot synthesizes auxin, so these results suggest an interaction between auxin and cytokinin in the development of the shoot. Jordan & Skoog (92), in a study with coleoptile tips of *Avena sativa* L., showed that i⁶Ade and BAP stimulated synthesis of auxin. They also reported that in tobacco tissue cultures which normally require exogenous cytokinins and auxin for growth, sufficient auxin for growth will be synthesized in the presence of high levels of cytokinin.

It is probably more than just a question of synthesis. Hemberg (84), in studies on excised shoots of *Coleus blumei* Benth., found that kinetin caused an increase in bound auxin but not free auxin. The stimulation of auxin activity by cytokinins in excised shoots has some significance in interpretation of the *Avena* straight growth assay for auxin activity. Hemberg & Larsson (85) found that when the coleoptile segments were incubated in the presence of kinetin at low concentrations of IAA, the actual amount of auxin in the coleoptile was augmented over controls not using the cytokinin. The experiment does not permit us to distinguish between the possibility of synthesis of auxin in the coleoptile or of increased mobilization of auxin from the exogenous source. Regardless of the exact process, cytokinin does influence the amount of auxin, and the authors caution those who use this system as an assay to be sure the test extracts are free of cytokinin activity.

Some inkling as to how cytokinins can influence the auxin level comes from a study of enzymes that catalyze oxidization of IAA. Detachment of cotyledons from the lentil (*Lens culinaris* Med.) embryonic axes causes an increase in total peroxidase activity in the axes (56). Several isoenzymes are involved which can be resolved by starch gel electrophoresis. Although there is an increase in the total peroxidase activity, some of the isoenzymes diminish in quantity while others, particularly the cathodic ones, increase. The same result is obtained by application of kinetin to attached or detached axes. Moreover, application of cytokinin enhances activity of two anodic peroxidases and represses one cathodic isoenzyme. Cytokinin, therefore, has an effect on the qualitative distribution of the isoenzymes. Auxin also modulates the spectrum of isoenzymes, in some cases in a manner opposite to that of cytokinin and in other cases in an additive manner. The isoenzymes that are reduced in level by presence of auxin and cytokinin are those exhibiting the most powerful IAA oxidase activity.

These interrelationships have also been explored in tobacco callus tissue by Lee (106). He separated IAA oxidases into seven fractions by polyacrylamide gel disk electrophoresis. Either kinetin or Z increases the level of two of these isoenzymes. The increase was not obtained when the incubation was carried out in presence of actinomycin D^2 or cycloheximide, which interfere with RNA and protein synthesis, respectively. Lentil root growth of intact seedlings is inhibited by application of kinetin ($10^{-4}M$), and Darimont, Gaspar & Hofinger (29) noted that under these conditions indoleacrylic acid (the main auxin-like component of lentil roots) catabolism is enhanced, with a consequent drop in the in vivo level of this agent. As in the case of apical dominance, the observed physiological responses of auxin and cytokinin appear opposite (183), and application of auxin in the experimental system nullifies the inhibitory effect of cytokinin on lentil root growth. Although the physiological responses may seem to be opposite in these systems, the effects at the biochemical level as shown by the data on the oxidases are much more subtle. Moreover, there is evidence for involvement of ABA. This compound blocks the response of the peroxidase isoenzymes to cytokinin in the lentil root system, although it does not appear to influence the distribution of isoenzymes per se (56).

The involvement of growth agents can also be demonstrated in the detached cotyledons. In particular, an effect on starch metabolism can be demonstrated. For example, the α-amylase content of detached cotyledons of *Phaseolus vulgaris* is maintained in the presence of kinetin ($10^{-4}M$) (59). In the detached cotyledon of sunflower seedlings (*Helianthus annuus*), application of kinetin brought about a higher level of reducing sugars (61). These manifestations probably reflect changing metabolic events of the developing cotyledon, but it is not fair to conclude that the axis normally would produce a cytokinin necessary for orderly development of the cotyledon. Perhaps it would be more correct to conclude that the axis and cotyledon together produced the appropriate mixture of control agents necessary for harmonious development of both the axis and cotyledon.

When the cotyledon and axis are separated the axis makes an attempt to compensate. Gaspar, Khan & Fries (56), in noting the ability of the excised axis to recover from its injury, suggested that a wounding hormone is released whose properties resemble those of cytokinins.

In contemplating the significance of multiple interactions, one cannot even consider cytokinins as a single chemical agent. An indication of multiple cytokinins is obtained from the work of Radin & Loomis, who studied the changes in cytokinins during the development of radish roots (133). They separated chromatographically substances with cytokinin activity, including Z and its

---

[2] The use of actinomycin D as an inhibitor of the DNA-dependent RNA polymerase must be carried out with caution. The compound produces other effects unless the concentration is carefully adjusted.

derivatives. An unidentified cytokinin substance appeared only after extensive secondary thickening had taken place. In other words, as the root developed the ratio of detectable cytokinins changed.

APICAL DOMINANCE

Some measure of the complexities of the growth control system can be gained from a study of the phenomenon of apical dominance. The growing apex of a shoot inhibits outgrowth of lateral apices of the same plant. The inhibiting influence, at least in part, is exercised by auxin formed in the growing apex. In decapitated plants the inhibitory effect can be accomplished by application of auxin. The inhibiting effect of auxin, whether it comes from the growing apex or is applied exogenously, can be reversed by cytokinins (see for example 171, 183). In spite of the great deal of attention this phenomenon has attracted, little understanding of the basic processes has been attained. As would be expected in a developing bud, all biochemical events related to morphogenesis occur. General inhibitors of protein synthesis such as chloromycetin and 6-azauracil, for example, will inhibit bud growth stimulated by application of a cytokinin (143).

Gibberellins also seem to be involved in promotion of bud development. Ali & Fletcher (1) noted that BAP applied directly to inhibited buds will initiate growth in 16-day-old intact soya bean plants. And this growth was further enhanced when GA was applied 48 hours after the cytokinin treatment. These results led the authors to suggest that hormones have a sequential role in releasing buds from apical dominance. In keeping with the proposed model, these results are consistent with a changing relationship of cytokinin, auxin, GA, and other components of a growth control system.

Woolley & Wareing (189) have studied formation of lateral buds in *Solanum andigena*. The lateral bud has the potential to develop as a stolon or as a leafy orthotropic shoot. The presence of roots on the plant are necessary for conversion of the natural stolon to a leafy shoot, but BAP replaces the need for roots. Using $[\alpha-{}^{14}C]$-BAP, for example, they showed that radioactivity accumulated in the tip of the induced stolon prior to conversion to a leafy shoot caused by withdrawal of auxin. These results led them to suggest that the role of auxin in apical dominance is to control distribution and metabolism of cytokinins.

The absolute amount of an agent that can be detected in a tissue is no indication that the physiological activity normally associated with it will be expressed. Tucker & Mansfield measured the cytokinin activity of inhibited buds of *Xanthium strumarium* L. and found an accumulation with age; in fact, when the buds started to develop the concentration of cytokinin fell (173).

Consistent with the model proposed in the introduction, the concentrations and interactions of the controlling agents change sequentially during morphogenesis. Such changes have been shown experimentally by Bilderback (14), who studied the effects of growth agents on development of excised floral buds of *Aquilegia formosa* Fisch. Kinetin, at 10^{-6} or $10^{-7}M$, stimulated the excised buds to develop to a certain stage, a single whorl of many carpels. Application of gibberellin at this stage was necessary to stimulate development of the next stage—differentiation of the carpels into erect organs. A sequential effect of growth agents can also be demonstrated at the level of the single cell. Nagl (124) studied selective blockage of cell cycle stages in *Allium* root meristem and found that kinetin, ABA, IAA, and colchicine in different proportions stopped the cell cycle at either G_1, G_2, or at the mitotic cycle between karyokinesis and cytokinesis. The author suggests that each stage of the cell cycle is independently controlled by a specific balance of growth regulators and by synthesis of specific nucleic acids.

SENESCENCE

Senescence is as much a process of development as budding or germination and should be thought of as a positive process in the life of the plant. It has generally been investigated, however, as a reversal of morphogenesis ever since Chibnall (28) noted that when detached leaves underwent rapid aging a loss of protein and chlorophyll occurred. The process was prevented or even reversed if adventitious roots developed on the petiole. This observation led Chibnall to suggest that an agent originating in the roots was responsible for maintenance of balanced protein metabolism in the leaves. The suggestion was taken up by Richmond & Lang (138), who found that when the then newly described cytokinin kinetin was applied to detached leaves, it delayed the onset of senescence and at the same time slowed the decline of protein levels. This result was subsequently confirmed by several investigators (see review by Kende 95). Since that time the nature of senescence has been extensively studied, although most of the approaches have stemmed from Chibnall's original observation.

Recent studies have focused on the biochemistry of protein metabolism in the senescing leaf. Shibaoka & Thimann (153) noted that proteolysis is a dominant reaction in the yellowing of the leaf and that serine promotes proteolysis. They incubated the detached leaves of *Pisum sativum* in the presence of [^{14}C]-leucine. Incorporation of leucine into protein would indicate a certain measure of protein synthesis. Kinetin, however, did not increase incorporation of leucine into the protein, but it clearly prevented protein breakdown.

Tavares & Kende (169), in studying the effect of BAP on senescing disks of corn leaves, found that the specific activity of the protein in the leaf disk did not change over 48 hours. They came to the conclusion that cytokinins retard senescence of corn leaves primarily through inhibiting protein degradation. Mizrahi, Amir & Richmond (120) noted that an enhancement of protein content over controls in kinetin treated *Tropaeolum majus* leaves could not be accounted for by increased synthesis, and they concluded that the cytokinin acted by depressing protein degradation. They went further in suggesting that, although there may be a slight synthesis of protein under these conditions, the protein formed is a protein inhibitor.

There must be a delicate balance between the action of proteases and action of protease inhibitors, and in the normal course of senescence the action of the proteases begins to dominate. Martin & Thimann (114), on evidence that cyclo-heximide, a protein synthesis inhibitor, delays senescence, concluded that synthesis of proteases precedes senescence. Srivastava (quoted by Martin & Thimann 114) noted that when barley leaves are held in the dark, yellowing is significantly decreased, not only by cytokinins but also by chloramphenicol, puromycin, and tetracycline—antibiotics that inhibit directly or indirectly the synthesis of protein. Cycloheximide strikingly prevents senescence of oat leaves and at the same time prevents incorporation of leucine into protein. The activity of cytokinins added exogenously to detached leaves can be antagonized by amino acids, principally arginine and serine (114, 153). This observation suggests that some interaction between cytokinins and free amino acids occurs in the normal course of senescence.

Nucleic acids of plant leaves can be destroyed by several enzymes whose activity is sensitive to changes in environmental and metabolic conditions. Udvardy, Farkas & Marrè (174) found that the activity of ribonuclease I (this enzyme has a preference for purine nucleotide bonds) in *Avena* leaves increases over a $7\frac{1}{2}$ hour period following leaf detachment. The increase in activity is completely inhibited by cycloheximide. Other inhibitors of protein synthesis also effectively block the early increase of ribonuclease activity in detached or damaged leaves and roots. Treatment of excised leaves with either auxin or kinetin effectively blocks the increase in ribonuclease activity (see Dove 34).

ABA applied to leaf disks accelerates senescence in a variety of species, so it would be expected to accelerate changes at the molecular level (43). There is a ribonuclease associated with chromatin, and the activity of this particular ribonuclease increases in senescing leaves of barley (*Hordeum vulgare* L.) in presence of ABA. This effect of ABA is completely abolished by kinetin (160). Influence of cytokinins on ribonuclease activity is not limited to senescence. An RNase associated with microsomes has been obtained from the apical region of *Pisum sativum* L. (15). The activity of this enzyme seems to be enhanced by auxin and suppressed by cytokinins.

Again, as noted above, the interaction of ABA and cytokinin cannot be considered as a simple push-pull mechanism. Back, Bittner & Richmond (5) have carried out a study of the interaction of kinetin and ABA in senescing *Rumex pulcher* leaves. Their data show that ABA retards degradation of kinetin to adenine. This observation confirms an earlier observation of Mullins & Osborne (123), who found that ABA had a cytokinin sparing effect. Back, Bittner & Richmond (5) suggest that there is a rapid turnover of kinetin and that the senescence retarding action of cytokinins results from metabolism as a process. The relative accumulation of cytokinins effected by ABA may reflect interference by ABA of cellular interactions that results in an enhanced rate of senescence. The enzyme in the plant leaf that degrades kinetin to adenine presumably is the same enzyme described by Whitty & Hall (182) that catalyzes conversion of Z to adenine. Whitty & Hall noted an increased level of this enzyme in corn endosperm in the early stages of corn development which suggests that the activity of the enzyme parallels the amount of cytokinin activity in the tissue. Back & Richmond (7) have suggested that the seasonal variations in capability of a cytokinin or gibberellin to retard senescence of the detached leaves reflect changes in the endogenous quantities of ABA.

Gibberellin, as does cytokinin, retards the rate of senescence of detached leaves (10, 51). Gibberellin and cytokinins apparently do not retard senescence via the same mechanism, and Back & Richmond (6) suggest that one agent may affect the endogenous levels of the second agent by modifying its biosynthesis or rate of degradation.

Dyer & Osborne (36) have studied the nucleic acids in green and senescent leaves of three types of plants. High and low molecular weight RNA of the chloroplast is not present in senescent leaves of *Xanthium pennsylvanicum*, but both cytoplasmic and chloroplastic fractions can be found in the yellow leaves of *Vicia faba* and *Nicotiana tabacum*. Kinetin treatment temporarily arrests loss of chlorophyll and nucleic acid. Once the *X. pennsylvanicum* leaves turn yellow they cannot be regreened, whereas those of *N. tabacum* and to some extent those of *V. faba* can be rejuvenated by application of a cytokinin. The authors suggest that retention of chloroplast RNA in yellow leaves may be a major factor in determining the leaf's ability to regreen and that in those leaves that cannot be regreened, loss of chloroplast genomes is the key factor. They further make the point that the mechanism of the senescence process varies from plant to plant.

Richmond, Sachs & Osborne (139) studied the effect of kinetin on protein synthesis of isolated chloroplasts of *Nicotiana tabacum* L. They observed that cytokinins had a positive effect on incorporation of [^{14}C]-leucine into the protein, but they felt that a major effect of kinetin upon the chloroplast was not related to mechanisms of protein synthesis but rather to the permeability of the membrane of the chloroplast. There is other evidence that cytokinins have a selective effect on membrane permeability. Leaf disks and detached cotyledons of sunflowers in presence of kinetin show an increased uptake of K^+, Rb^+, Li^+, but not Na^+; thus, for example, the K^+/Na^+ ratios change (87, 88).

Berridge & Ralph (11) found that the principal effect of kinetin in excised Chinese cabbage leaves was to maintain ribosome levels. Interestingly, when the leaves were infected with turnip yellow mosaic virus, there was a decrease in the amount of virus in the presence of the cytokinin. Protein synthesis was maintained as evidenced by synthesis of viral protein, but formation of intact virus particles decreased. Under these conditions kinetin and BAP exhibited equilibrium binding to the ribosomes (12).

AUTONOMOUS TISSUE AND CROWN GALL

The crown gall disease is initiated by a tumor-inducing principle elaborated by the bacteria *Agrobacterium tumefaciens*. This as yet uncharacterized agent transforms normal plant cells to fully autonomous tumor cells over 3 to 4 days. Once the cellular transformation has been accomplished, the tumor cells grow autonomously in the absence of the initiating bacteria (see review by Wood 186). Our interest lies in the fact that autonomous tissue obtained from crown gall, or by other means, grown in culture does not require an exogenous source of auxin and cytokinin.

Autonomous tissue seems to be capable of synthesizing cytokinins. Dyson & Hall have identified i^6Ado in an autonomous strain of tobacco callus tissue (39), and Einset & Skoog (42) report the presence of i^6Ade and derivatives in an autonomous strain of tobacco callus. The interpretation of data showing the presence of these compounds in the free state has some pitfalls. Since i^6Ade is a component of the tRNA, it could be released under certain conditions of isolation. Rathbone & Hall (136) showed that in case of *Corynebacterium fascians*, which had previously been thought to secrete free i^6Ade, the base was being released from the tRNA during isolation. Normal cells cultured in vitro generally require the exogenous addition of both factors to the medium. Although under certain circumstances the capacity to form cytokinin can be induced, cultured sycamore cells, for example, will respond to exogenously added cytokinins, but in the absence of such factors they eventually develop the capacity to synthesize factors, one of which appears to be RZ (111, 112). These laboratory demonstrations show that the biosynthetic capabilities for auxin and cytokinin in the tissue used for bioassays are normally repressed.

The availability of cultured autonomous and normal tissue offers an experimental tool to compare biochemical events related to cytokinins. Wood (185) has described isolation of a cell division promoting factor (CDF) from the crown gall tumor tissue of *Vinca rosea* L. Chemical techniques suggest that the chromophore of CDF is a 3,7-dialkyl-2-alkylthio-6-purinone. The factor contains glucose but not ribose. CDF has also been found in dividing cells of several higher plant species that are taxonomically widely separated, such as tobacco and cactus. Wood & Braun (187) have obtained evidence that CDF is induced in the presence of 6-(substituted)-purines in normal plant cells. It apparently inhibits cyclic AMP phosphodisestase (188).

A group of compounds that are carcinogenic in animal tissue, the aminofluorenes, will induce formation of autonomous tissue when applied to hormone-dependent tobacco callus cultures (9). It is intriguing to speculate whether the cytokinin phenomenon is involved in animal tumorigenesis. The morphactin regulators of plant growth, methyl-2-chloro-9-hydroxyfluorene-9-carboxylate and *n*-butyl-9-hydroxyfluorene-9-carboxylate inhibit and stunt intact plants but do not act directly as herbicides at low doses. They inhibit seed germination and prevent expression of apical dominance. These two compounds also initiate formation of autonomous tissue nodules in hormone-dependent tobacco callus tissues (9).

A promising technological development in the study of autonomous strains results from the work of Syōno & Furuya, who isolated a temperature sensitive strain of autonomous tissue from a cytokinin-requiring callus strain, T2, of tobacco (166). This particular strain at 26°C grows rapidly on medium without added cytokinins, but at 16°C its growth is completely suppressed unless BAP or other cytokinin is added. This property is analogous to temperature sensitive mutants in somatic lines of cultured animal cells. It offers considerable potential for studying the autonomous process.

CYTOKININS AND PHOTOSYNTHESIS

There appears to be close involvement of cytokinins and light-sensitive processes; in some cases cytokinins appear to exert an action similar to that of phytochrome-dependent reactions such as chlorophyll synthesis and formation and maintenance of chloroplasts. Feierabend & Pirson (48) have investigated the photosynthetic enzymes of the Calvin cycle, carboxydismutase, NDP^+-dependent glyceraldehyde dehydrogenase, transketolase, and ribosephosphate isomerase. Although these enzymes can be formed independently of light and chlorophyll synthesis in germinating seedlings, their formation is promoted by far-red light. In the light-induced situation, a small but constant level of phytochrome is maintained by the far-red light (121), and this fact together with other evidence shows that formation of Calvin cycle enzymes is mediated by the phytochrome system (48). But in the case of dark grown seedlings, cytokinins are specifically required for development of the photosynthetic enzymes (45). Feierabend (45) feels that phytochrome and cytokinin act through different modes but are still interdependent; for example, the action of phytochrome demands a supply of cytokinins.

The interaction of kinetin and phytochrome seems to be fairly complex. Khudairi & Arboleda (98) noted an opposing effect of phytochrome and cytokinin on carotenoid biosynthesis. The ripening of tomatoes, which includes the synthesis of lycopene, is dependent on red light, thus showing a phytochrome dependency. Kinetin and gibberellin inhibit the synthesis of lycopene.

Tasseron-De Jong & Veldstra (168) have studied the interaction of phytochromes and cytokinins in *Lemna minor* L. When *L. minor* is grown under conditions of deficient light, cytokinins are able to supplement and partly overcome the deficiency. In the opinion of these workers, cytokinin seems to substitute for the effect of nonphotosynthetic light. The synthesis of betacyanins is also light controlled and involves phytochrome stimulation of amaranthin formation (in *Amaranthus tricolor* seedlings) (129). As in the case of the photosynthetic enzymes where phytochrome and cytokinins seem to promote similar effects, synthesis of this pigment is stimulated in darkness by kinetin (130). The stimulation does not appear to be related to activation of phytochrome. The effects of both light and cytokinin can be ascribed to gene activation since induction of amaranthin synthesis by either factor was prevented by actinomycin D, and also by inhibitors of protein synthesis.

Giudici de Nicola et al (63) have shown that when *A. tricolor* seedlings were stimulated to synthesize amaranthin in the dark by application of kinetin, light further enhanced amaranthin synthesis. These authors suggest that even though the genetic system has been activated by the cytokinin, the phytochrome system acts at a different level—for example, one system acting at the transcription and the other at the translation level. The enhancement of synthesis by action of light

is completely blocked by puromycin, a specific inhibitor of translation. The interactions, however, are a little more complex. The phytochrome system is activated
by a short period of irradiation (62). Under conditions of continuous illumination the photosynthetic system appears to be the photoreceptor for activation of
pigment synthesis. Thus, under continuous illumination in the presence of kinetin
the *A. tricolor* seedlings synthesize a large amount of amaranthin; this is said to
be due to stimulation of oxidative phosphorylation by the hormone and the
simultaneous induction of photophosphorylation. The fact that an increase in
oxidative phosphorylation under these conditions was noted does not mean any
direct relationship with cytokinin stimulation, for it is just a manifestation of a
general shift in metabolic activity.

Feierabend (46, 47) also investigated the oxidative pathway of the photosynthetic enzymes and found that the amount of available endogenous cytokinin can be
limiting for the rate of formation of glucose 6-phosphate dehydrogenase. The appearance of the photosynthetic enzymes in the presence of light is accompanied
by a decline in the rate of formation of glucose 6-phosphate dehydrogenase, but
the activity of this enzyme can be maintained at a high rate in presence of high
concentrations of cytokinin and auxin. The decline in the formation of glucose
6-phosphate dehydrogenase cannot be prevented by the application of cytokinins
alone, which is further evidence that cytokinins and auxins interact during the
orderly development of the plant tissue.

Cytokinins seem to be involved in promoting the synthesis of chlorophyll as
well as development of the chloroplasts (50). Etiolated barley seedlings lose the
ability to produce chlorophyll and soluble proteins with increasing age. Kinetin
retards this process (164). Etiolated cucumber cotyledons pretreated with BAP
and exposed to light synthesize 4.0% more chlorophyll than controls.

CYTOKININS IN NATURE

The cytokinin phenomenon has been associated with specific chemical structures,
and the bioassays devised to demonstrate activity reflect the idea that the phenomenon can be embodied in a structure. This concept, although limited in scope,
has provided some direction for research and has generated some insight. Several compounds occur naturally that give positive responses in the cytokinin bioassay systems. Whether each occurrence of such compounds relates to the cytokinin phenomenon cannot be assessed at the present level of understanding, but
as much information as possible about natural occurrence hopefully will allow us
to recognize patterns that will contribute to improved understanding. All the
known N^6-(substituted) adenines with cytokinin activity are derivatives of the
parent structure, i^6Ado. Table 1 lists the occurrence of i^6Ado and its known
derivatives. One of the more enigmatic features of the pattern of distribution is
the presence of these nucleosides in tRNA as well as in the free state.

Although i^6Ado occurs in the tRNA of all species, its derivatives segregate according to broad classes. The 2-methylthio derivative has been detected only in
the tRNA of plants and bacteria. The hydroxylated form, Z, has been found only

Table 1   Cytokinins—naturally occurring[a]

Compound	PLANT		ANIMAL		BACTERIA, FUNGI		YEAST	
	tRNA	free	tRNA	free	tRNA	free	RNA	free
i⁶Ado	20 22 70 176	39	70		4 21 70		70	
2ms-i⁶Ado	22 176				4 21			
trans-Z or *trans*-RZ	176	95 108				(fungi) 118 119		
cis-Z or *cis*-RZ	20 22 73 176							
2-methylthio-*cis*-RZ	20 22 176							
dihydrozeatin		101 102 159						
Ado-CO-thr or Ade-CO-thr	41		25 27	24	27 105 132	(fungi) 105	27 148	

[a] Numbers indicate references.

in plant tRNA, mainly in the *cis* form. On the other hand, tRNA from pea shoots, contains both the *cis* and *trans* forms; this is the only report that the *trans* form of Z occurs in tRNA. Perhaps during certain stages of growth the *trans* configuration is formed in the tRNA.

These tRNA components play a unique role in the function of tRNA. Both i⁶Ado and Ado-CO-thr occur only in the anticodon loop adjacent to the 3′ end of the anticodon, a position that enables them to help determine the conformation of the anticodon loop (see for example Ghosh & Ghosh 60). Fittler, Kline & Hall (49) observed that treatment of tRNA with aqueous iodine, a reaction specific for the Δ^2-isopentenyl side chain, resulted in loss of ability of tRNAs containing i⁶Ado to bind with ribosomes in the presence of the appropriate messenger; the ability to accept amino acids was not impaired. Gefter & Russell (58) performed a more elegant experiment in which they obtained mutants of *Escherichia coli* suppressor tRNA^{tyr} containing an adenosine residue in the position that normally would have a 2ms-i^{6}Ado residue. Kinetic studies of the amino acid accepting process showed no significant difference in the ability of the mutant forms to accept amino acids compared to the normal tRNA. However, the molecular species of tRNA^{tyr} containing unmodified adenosine was ineffective in in vitro tests of suppression, and its response in the ribosome binding tests was greatly reduced compared to the tRNA^{tyr} containing i^{6}Ado or 2ms-i^{6}Ado.

Miller & Schweizer (117) obtained tRNA from an *E. coli* threonine auxotroph grown under limiting growth concentrations of threonine, and those tRNAs known to contain Ado-CO-thr gave only 50% of the binding response to ribosomes in the presence of the appropriate messenger.

The distribution of i^{6}Ado and Ado-CO-thr in molecular species of tRNA follows a regular pattern (Table 2). Armstrong et al (2, 3) were the first to notice that i^{6}Ado and/or its corresponding analogs occur only in those molecular species of tRNA that respond to codons beginning with uridine. They fractionated yeast tRNA and found that all the tRNA fractions that contained cytokinin activity responded to codons beginning with letter U. This pattern was also observed in *Lactobacillus acidophilus*. *L. acidophilus* requires mevalonic acid as a growth factor and, consequently, when it is grown in the presence of radioactive labeled mevalonic acid, specific incorporation of the label into the Δ^2-isopentenyl side chain of the i^{6}Ado residues of tRNA takes place. Making use of this fact, Peterkofsky & Jesensky (128) fractionated the tRNA of *L. acidophilus* grown under these conditions and found that the radioactive label was confined to those tRNA molecules that accept leucine, tyrosine, cysteine, serine, and tryptophan.

Table 2 Genetic code showing distribution of i^{6}Ado and Ado-CO-thr, which occur only in tRNA species responding to UXX and AXX, respectively[a]

	U	C	A	G	
U	Phe 8	Ser 3, 90, 128, 161, 192	Tyr 3, 33, 64, 77, 78, 113, 128	Cys 3, 80, 128, 190	U
	Phe 8	Ser 3, 90, 128, 161, 192	Tyr 3, 33, 64, 77, 78, 113, 128	Cys 3, 80, 128, 190	C
	Leu 128, 190	Ser 3, 90, 128, 161, 192			A
	Leu 128, 190	Ser 3, 90, 128, 161, 192		Try 86, 128, 190	G
A	Ileu 99, 167	Thr 99	Asn 99	Ser 89	U
	Ileu 99, 167	Thr 99	Asn 99	Ser 89	C
	Ileu 99, 167	Thr 99	Lys 89, 158	Arg 104	A
	Met 89	Thr 99	Lys 89, 158	Arg 104	G

[a] Numbers indicate references.

These data have now been confirmed as the sequences of many of the specific tRNA molecules have been worked out. As more sequences became available, an analogous pattern began to emerge with respect to those tRNA species responding to codons with first letter A. In all these tRNA molecular species, Ado-CO-thr occurs adjacent to the 3′ end of the anticodon.

E. coli tRNA[phe] contains 2ms-i⁶Ado, but the tRNA[phe] of eukaryotic cells contains a different hypermodified nucleoside in this position, compound Y (35, 125, 134). Hecht et al (79) treated tRNA[phe] from wheat germ and yeast, both known to contain compound Y, with acid, under which conditions the base of compound Y is excised. The acid hydrolysate was active in promoting the growth of the cytokinin-requiring tobacco callus. Before one can make a final conclusion from these results, one would need to know the results of a bioassay performed on a chemically pure sample of compound Y.

It is clear from the available evidence that both i⁶Ado and Ado-CO-thr occupy a unique and important role in the known function of tRNA (amino acid transfer activity). The fact that these compounds and/or their derivatives exist in the free state and exhibit potent cytokinin activity and the fact that they occur in the tRNA may or may not be related. Nevertheless, investigators working in the general field of cytokinin phenomenology have been intrigued by the possibility that a significant relationship exists.

One suggestion has been that the endogenous cytokinins somehow or other exert their activity by incorporation into tRNA. For example, the synthetic cytokinin compound BAP was reported to be incorporated into the tRNA of callus tissue (20, 53). Richmond, Back & Sachs (137) specifically looked for incorporation of BAP into the RNA of senescing leaves but found no evidence for any incorporation. On the other hand, under similar conditions, kinetin was incorporated into the RNA.

It is difficult to make a case for the tRNA of cytokinin dependent tissue requiring an exogenous source of these particular nucleosides on the basis of available evidence. Chen & Hall (23) have shown that cytokinin dependent tobacco callus tissue contains the enzyme system that attaches the Δ^2-isopentenyl group to the appropriate adenine residue in tRNA. In the case of the tobacco callus tissue grown in the presence of radioactive BAP, even though this material was incorporated into the tRNA, the tRNA still contained its complement of i⁶Ado and derivatives (20). In general, various purine analogs can be incorporated into RNA in a rather nonspecific way. It is also quite possible that incorporation occurs in RNA fractions other than tRNA, and unless the nucleic acid fraction is rigorously characterized, one cannot exclude this possibility. Dyson (37), working in the laboratory of Fox, reexamined the RNA fraction used by Fox & Chen (53) and found that the radioactive BAP that had been incorporated resided in higher molecular weights of RNA and not in the tRNA. To establish whether incorporation of N⁶-(substituted)-adenine derivatives into tRNA has a biological significance, one would have to fractionate the tRNA and identify the incorporated material in a specific location in the tRNA sequence. Finally, studies on biosynthetic pathways of all modified nucleosides in RNA show that they are synthesized at the macromolecular level (see Hall 71).

N-[9-(β-D-Ribofuranosyl-9H)Purin-6-ylcarbamoyl]Threonine

The analogous status that Ado-CO-thr occupies in tRNA relative to i⁶Ado led Dyson et al (40) to investigate whether this compound has growth promoting properties in the cytokinin assay system. Ado-CO-thr or Ade-CO-thr per se has no growth promoting properties in a strain of soya bean callus tissue that requires cytokinin for growth. But since the test compounds are added to the test medium exogenously, they must penetrate the cell membrane, and in the case of Ado-CO-thr the polar carboxylic group might prevent entry into the cell. For example, analogs of BAP which contained an esterified carboxyl group attached to the α carbon of the benzyl group (the polar nature of the carboxyl group would be eliminated) were at least ten times as active as those derivatives with the corresponding free carboxyl group (109). With this in mind, Dyson et al (40) synthesized a series of Ade-CO-thr analogs and found that only those ureidopurine derivatives similar to Ade-CO-thr in size and configuration had cytokinin activity. Their most active analog was the ortho-chlorophenyl derivatives (Figure 2). A study of molecular models showed that the carbonyl group of the threonine residue of Ade-CO-thr and the chloro atom of the chlorophenyl derivative occupy the same spatial position relative to the carbonyl group of the urea portion of the molecule (Figure 3). This spatial arrangement is not possible for halogen atoms in the meta and para positions, and such analogs were relatively inactive as cytokinins. Replacement of the chloro group with a methyl group also nullified cytokinin activity. These data suggest that the strong cytokinin activity of the ortho-chlorophenyl derivative is due to its close similarity to Ado-CO-thr both in its three-dimensional shape and electronic configuration. The ortho-chlorophenyl derivative, unlike the naturally occurring Ado-CO-thr, is essentially a nonpolar compound and thus should have little difficulty in penetrating cell membranes. Dyson et al (40) concluded that the ureidopurine compounds represent a distinct class of growth promoting agents of which the archetype structure is Ado-CO-thr.

McDonald et al (115) also synthesized a series of ureidopurine analogs; the meta-chlorobenzyl and the ortho-methylphenyl analogs were weakly active in promoting growth of tobacco callus tissue, and 6-phenylureidopurine elicited quite a different budding response in the tissue compared to that of i⁶Ade. The growth promoting properties of symmetrical diphenylurea and related analogs have been known for many years (93, 150). The diphenylurea compounds, as is ortho-chlorophenylureidopurine, are flat molecules and their biological activity presumably can be attributed to a similarity in shape to that of Ade-CO-thr.

Ado-CO-thr appears to be synthesized at the macromolecular level, as are other components of tRNA. Chheda et al (25) and Powers & Peterkofsky (132) have shown that when a threonine auxotroph of *E. coli* is grown in presence of radioactive threonine the threonine is incorporated into the Ado-CO-thr component of tRNA. Threonine appears to be incorporated in an intact manner and, moreover, carbamoyl-phosphate is not the donor of the carbonyl group (132).

Laloue & Hall (105) have also shown that threonine is incorporated into the

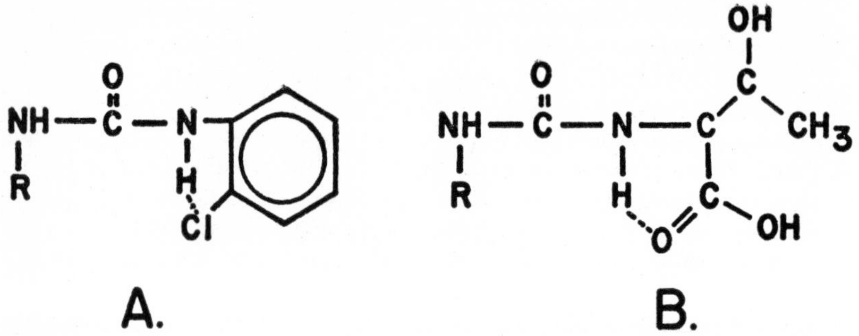

Figure 3 Diagram showing structural relationship and the possibility of hydrogen bonding indicated by dotted lines. R = purin-6-yl; A, ortho-chlorophenylureidopurine, B, Ade-CO-thr.

tRNA of the mycorrhizal fungus *Rhizopogon roseolus*. Their experiments suggest that a mechanism exists for synthesizing Ado-CO-thr independent of the tRNA. They carried out an experiment in which two radioactive labeled precursors, adenine and threonine, were added to the growing fungus. The ratio of radioactive adenine to radioactive threonine in the tRNA was substantially different from that of free Ado-CO-thr isolated from the medium. They concluded that the free Ado-CO-thr could not have been released from the tRNA; otherwise the ratio of the two radioactive components should have been the same from the two sources. The occurrence of Ado-CO-thr in the free state in extracts of *R. roseolus* together with free i⁶Ado and RZ (118, 119) lead one to speculate whether in some way Ado-CO-thr is part of the natural cytokinin activity of this fungus.

Metabolism of N⁶-(Δ²-Isopentenyl)Adenosine

As summarized in Figure 4, i⁶Ado and its hydroxylated derivative RZ undergo a number of metabolic reactions in plant tissue. Three enzyme systems have been found in plant tissue that catalyze conversion of i⁶Ado to different products. One enzyme system present in several tissues catalyzes its conversion to adenosine (127, 182). A powerful hydrolase exists in some tissue that converts it to the free base, i⁶Ade (182). Corn endosperm contains an enzyme that hydroxylates i⁶Ado to form Z (118). This particular reaction was first noted in *R. roseolus* by Miura & Miller (119). RZ itself undergoes degradative reactions similar to those of i⁶Ado. Whitty & Hall (182) found that the enzyme from corn endosperm catalyzed conversion of RZ to Z or to adenosine. Z can also be converted to adenine by means of the degradative enzyme in endosperm of corn. Sondheimer & Tzou (159) investigated the metabolism of Z in bean axes and among the metabolic products identified dihydrozeatin, RZ, ribosyldihydrozeatin, and the corresponding 5′ nucleotides.

Chheda & Mittelman (26) injected radioactive labeled i⁶Ado into human subjects and found that the nucleoside is rapidly metabolized. Although a large amount of radioactivity appeared in the urine, only 0.3% of it was in the form of i⁶Ado. Radioactive labeled uric acid was not recovered, suggesting that under these conditions i⁶Ado is not metabolized to uric acid. Their findings are consistent with the detection of a powerful enzyme in the bone marrow of animals, including humans, which catalyzes conversion of i⁶Ado to inosine (72, 74). Although the degradative enzyme system exists, free i⁶Ado can still circulate in the blood. Hall & Mintsioulis (74) injected a sample of i⁶Ado intravenously into a rabbit. There was an initial rapid clearing, but detectable levels (the order of 1 μg/ml) remained 4 hours after the initial injection.

The i⁶Ado residues in tRNA are synthesized by transfer of the Δ^2-isopentenyl group from Δ^2-isopentenylpyrophosphate. The enzyme that catalyzes this reaction has been partially purified from yeast and rat liver (100) and from plant tissue (23).

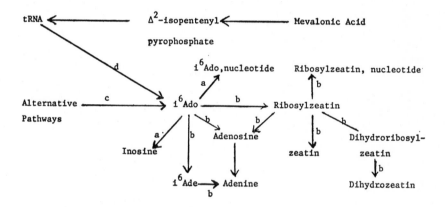

a   Animal tissue only

b   Plant tissue only

c   Speculative, no evidence obtained for alternative pathways.

d   Indirect evidence for this step

Figure 4 Scheme of known metabolic interactions.

The source of free i⁶Ado remains an open question. The tRNA represents a potential source, but presumably breakdown of the tRNA would be necessary to release the i⁶Ado. There is circumstantial evidence that alternative pathways might exist. Short & Torrey (155) measured the level of free cytokinins in the meristematic cells of pea root tips and found 27 times as much activity as could be obtained from the tRNA of these cells. In their experiment, however, they hydrolyzed the tRNA with acid, a treatment that partially destroys i⁶Ado (75, 141), so that the resultant biological activity might reflect a lower amount than actually present. In spite of this technical problem their data suggest the presence of a large excess of free cytokinin over that in the tRNA. Unless a rapid turnover of tRNA occurs in meristematic tissue, these data suggest an alternative source of i⁶Ado. In bacteria, another rapidly growing system, Mintsioulis & Hall (unpublished data) measured the turnover of i⁶Ado-containing tRNA molecules in the tRNA of *Acidophilus lactobacillus*. The Δ^2-isopentenyl side chains of the i⁶Ado residues can be specifically labeled by growing the bacteria in the presence of labeled mevalonic acid. There was no evidence of any tRNA turnover. Finally, another piece of evidence that bears on this question is the observation that 1-methyladenine, which is a gonad-stimulating compound in the starfish ovary, is synthesized independently of the tRNA (154).

It is quite possible that tissues make use of tRNA or of alternative pathways as a source of cytokinins, depending on circumstances. There is evidence for tRNA turnover in animal tissue (76), and in various parts of this review reference has been made to the degradative events that occur when plant tissue deteriorates. Sheldrake & Northcote (152) have advanced the hypothesis that agents involved in differentiation could arise from autolysis, and Torrey, Fosket & Hepler (172) feel that such a mechanism is plausible in the case of cambial activation.

The synthetic analogs used in cytokinin studies also can undergo metabolism. For example, when BAP was applied to *Lemna minor* the compound was taken up and within 24 hours had undergone extensive metabolism (13). One of the identified metabolic products was adenine. Dyson, Fox & McChesney (38) studied the metabolism of BAP in soya bean tissue and found that up to 25% of the added precursor was converted to a stable long-lived derivative.

The significance of the facile metabolism of the naturally occurring cytokinins does not derive from the existence per se of the various metabolites, but rather from their metabolism as a process. There is probably no one "active form" of cytokinin; all components of the metabolic system through their dynamic inter-conversions and interactions with other metabolic events express the ultimate response of the tissue. In studying the biochemistry of these metabolic happenings we can only keep adding to the mosaic with the hope that a pattern that provides insight will eventually emerge.

TRANSLOCATION

In the proposed model, cytokinin activity results from a dynamic system of inter-acting substrates and enzymes. Thus, in considering cytokinin participation in growth and development of an organ, we can ask how the particular tissue assembles all the needed components of the cytokinin system. Three general possibilities arise: 1. A given cell or group of cells can synthesize all elements of the cytokinin system. 2. With respect to the substrates, the tissue can synthesize only some of them; accordingly the missing compound(s) are obtained from another source. 3. Inability to synthesize missing substrate(s) may be due to lack of some enzyme activator that can be obtained from other tissues. [Varner (175) has commented on diverse ways of control of enzyme activity in plant systems.]

There is experimental evidence, already cited in this review, for translocation of some elements of the cytokinin system (see also review of Kende & Sitton 96). Briefly, the negative effects of removal of roots from a shoot can be offset by application of a cytokinin, and such data lead to the suggestion that roots produce a cytokinin which is translocated to other parts of the plant. Bui-Dang-Ha & Nitsch (18) were the first to positively identify such a compound, *trans*-RZ, which they isolated from chicory roots (*Cichorcium intybus* L.). More recently Dyson et al (unpublished data), working in our laboratory, obtained evidence that spring maple sap contains i^6Ado.

The RZ and i^6Ado isolated by the above workers were obtained under conditions which would yield only free nucleosides. There is always a possibility that elements of the cytokinin system are transported in a bound form. Kende (94), working with a root exudate, demonstrated that a factor inactive in the cell division assay was converted to an active form by acid hydrolysis, which suggests the presence of a bound form of cytokinin. Yoshida & Oritani (191) have obtained evidence for a bound form of Z in the roots of the rice plant. This particular compound was relatively inactive in the cytokinin bioassay but became much more active after acid hydrolysis or on treatment with β-glucosidase, suggesting it contains glucose.

The concept that circulating factors control the orderly development of plants no doubt gained currency because of the analogy with the well-characterized animal hormone systems [see for example the remarks of Steward (162)]. As discussed in the introduction, animal hormones are involved in functional control of mature tissues. The control mechanisms of growth and development of animal tissues remain as much a mystery as those of plant tissues. The danger of overemphasizing one model of control is that it discourages development of other models. Other models are indeed plausible; for example, experimental evidence supports the notion of a self-regulating coordinate system set up by the embryo (66). This model is based on a system of metabolic oscillators that can be synchronized throughout a tissue, and waves of activity propagated through the system provide the information which controls development (65). This model does not require the movement of any compounds between cells. Compounds of course move between cells but these are needed metabolic entities which do not necessarily have any more significance with respect to control than the movement of glucose. Electricity can be conducted from one place to another, but the significance in control of electricity lies not in how it arrives at the point of use but in how it is used.

STRUCTURE AND FUNCTION

The cytokinin phenomenon has been defined in terms of specific bioassay systems that in themselves are artificial. The systems have been devised so that a single compound will elicit a defined biological response. Although this experimental approach suggests a direct cause and effect relationship, we should avoid the trap of interpreting results in such a manner. Structure is tangible, function is intangible. Structure is a fact but function is a conception of a diffuse phenomenon. The two cannot possibly be equated; yet there is a tendency to consider function as defined by structure, a static entity. Rather, function arises from the resonant relationships between structures, and when one adds a single structure to an existing set of structures all relationships are modified to express a new function. Thus one can define function only in terms of total relationships. From the point of view of a single structure, any function one might wish to ascribe to it depends on the milieu the compound enters. The milieu of the cytokinin bioassay system has been standardized, and the value of using this system as a surrogate for the total plant lies in its use for study of naturally occurring compounds and their close derivatives.

I will confine myself to a few comments on activity as determined in the bioassays on variations of the naturally occurring cytokinins i⁶Ado and Z. Interpretation of data is also complicated by the fact that the agents are added exogenously and some structural feature that seems to be significant may only reflect the capacity of the tissue to utilize the agent in its added form.

Some have questioned whether the ribose group attached to the 9 position of the purine derivatives is of any value. A number of Z analogs with and without the 9-ribosyl group have been synthesized (81, 107). In every case the nucleoside was less active than the corresponding free base in the tobacco callus bioassay system. On the other hand, the 9-methyl derivative of Z is more active than Z in promotion of the growth of carrot callus (151). An analog of RZ in which the N-9 atom is replaced by a carbon (the formycin analog) gave only slight growth response (81). Note that the sugar-base bond is resistant to enzymic hydrolysis. One cannot conclude from these data that the ribose portion of RZ or i⁶Ado is unnecessary for activity because the cytokinin system is quite capable of attaching or removing the sugar residue, and the process of removal and/or attachment may in its own right be significant to the cytokinin phenomenon.

Both the *cis* and the *trans* forms of RZ occur naturally. Only the *trans* form has been found in the free state, whereas both the *cis* and *trans* forms have been found in the tRNA (Table 1). The *cis* form is less active than the *trans* form in promoting the growth of tobacco callus tissue (75). A more detailed account of structure-function relationships and analogs can be found in the review by Skoog & Armstrong (157).

CYTOKININS AND METHYLATION

Several workers have noted that the methylases responsible for catalyzing transfer of methyl groups from methionine to tRNA are more active in embryonic and neoplastic tissue than in normal adult tissues (see for example Riddick & Gallo 140). In a system involving cytokinin-stimulated growth and development one might expect to find an effect on cellular methylases. Schaeffer & Sharp (144, 146) found that during early phases of bud release from dormancy, induced by application of BAP, the rate of RNA methylation increased as evidenced by elevated levels of methylcytidine. In addition, there seem to be qualitative differences in the pattern of methylation of polar lipids, particularly the membrane component phosphatidylcholine (145, 147). Wainfan & Landsberg have found that cytokinins, including kinetin, i⁶Ado, Z, and BAP, inhibit the transfer of methyl groups from methionine to methyl-deficient tRNA (178). Analogs of Ado-CO-thr that are active in the soya bean callus assay are also effective inhibitors of the rate of tRNA methylation (177).

CYTOKININS IN NONPLANT SYSTEMS

In nonplant systems i⁶Ado exhibits a variety of physiological effects. For example, it inhibits growth of leukemic cells in vivo (165) and in vitro (135). It should be kept in mind that cytokinins also inhibit growth of plant cells above certain concentrations. Gallo, Whang-Peng & Perry (54) observed that i⁶Ado inhibits the mitosis of human lymphocytes stimulated by phytohemagglutinin at a concentration of about $10^{-6}M$. Their data suggest that in lower concentrations there might be stimulatory effects, although this result has not been confirmed.

Antibodies are manufactured mainly by plasma cells which are descendants of lymphocytes, but plasma cells are merely the end products of a long chain of events put into action by the body in response to invasion by foreign material (antigens). Primitive stem cells develop into virgin lymphocytes, and when the virgin lymphocyte encounters a specific antigen it responds by producing antibodies against that antigen. But first it divides and proliferates, producing a group of apparently specifically coded memory cells, and these memory cells may either proliferate to produce even more memory cells or differentiate ultimately to the antibody secreting plasma cells (184). Thus the immune mechanism involves cell division and differentiation, and the interference with this process by i⁶Ado makes one wonder whether the basic mechanisms of cell division and differentiation in plants may indeed share common biochemical features with those of the immune response.

Mature human lymphocytes cultured in vitro generally do not synthesize DNA or divide. Addition of phytohemagglutinin results in morphological transformation to lymphoblastoid cells capable of DNA synthesis and leads to eventual mitosis. These events are preceded by an increase in RNA and protein synthesis (54, 68). It is these events related to mitosis that seem to be inhibited by i⁶Ado.

Hacker & Feldbush (69) obtained evidence that inhibition of phytohemagglu-tinin-stimulated lymphocytes by i⁶Ado is due to inhibition of methylation of RNA. Not only is there inhibition but there may be a qualitative alteration in the methylation pattern. These data fit with those obtained by Wainfan & Landsberg, who observed that cytokinins inhibit methylation of tRNA in vitro (178).

Hacker & Feldbush (69) also studied the effects of i⁶Ado on phytohemaggluti-nin-transformed rat spleen cells. They found that the nucleoside inhibited incor-poration of uridine into RNA and thymidine into DNA, and they concluded that inhibition of these cells may reflect immunosuppression (68). These studies have been followed up by Diamanstein, Wagner & Bhargava (30), who showed that i⁶Ado inhibits the immune response which leads to formation of 19S hemoly-sin in mice immunized with optimal and suboptimal doses of sheep red blood cells. Their studies on incorporation of radioactive labeled leucine and uridine show that i⁶Ado interferes with the mechanism of cellular RNA and protein synthesis in thymus cells, but not spleen cells, from mice immunized with sheep red blood cells. These results confirm the immunosuppressive activity of i⁶Ado, but the data suggest that it acts only if given before the antigen is presented.

Cancerous cell lines other than leukemic lines are also sensitive to the action of i⁶Ado. Divekar & Hakala (32) studied the mechanism of inhibition of this com-pound in cultured sarcoma-180 cells. They found that for the nucleoside to be effective i⁶Ado must first be phosphorylated in the 5′ position. Interestingly enough, although i⁶Ado is phosphorylated by adenylate kinase to the 5′ mono-phosphate level, phosphorylation does not proceed further (31). A subline of sar-coma-180 cells resistant to i⁶Ado has a much lower capacity to phosphorylate this nucleoside (31) than have sensitive cells.

Rathbone & Hall (135) noted that a cell line derived from human myelogenous leukemia (Roswell Park 6410) contains an enzyme that hydrolyzes i⁶Ado to the free base i⁶Ade. The free base is much less active as an inhibitor of these cells than the nucleoside (52, 67).

It is difficult to know whether inhibition of various cell lines of cultured cells by i⁶Ado bears much relationship to the cytokinin phenomenon in plants; the mechanism of action may be quite independent. The immunosuppressant activity of i⁶Ado, however, is an intriguing observation and may be related to its capacity to inhibit cells under certain conditions. Since i⁶Ado is a natural component of mammalian tissues (at least in the tRNA), one can ponder whether it is involved in the normal mechanisms of the immune phenomenon.

CONCLUDING REMARKS

Plants have solved many metabolic problems peculiar to their existence, but when it comes to basic principles of control of development such principles probably hold for both animals and plants. Perhaps it is these common principles that we should be seeking to understand. The cytokinin phenomenon represents one aspect of our concept of mechanisms of growth and differentiation, and as such it provides a useful vehicle for probing these processes. It cannot, however, be considered an explanation of these processes.

Literature Cited

1. Ali, A., Fletcher, R. A. 1971. *Can. J. Bot.* 49:1727–31
2. Armstrong, D. J., Burrows, W. J., Skoog, F., Roy, K. L., Söll, D. 1969. *Proc. Nat. Acad. Sci. USA* 63:834–41
3. Armstrong, D. J. et al 1969. *Proc. Nat. Acad. Sci. USA* 63:504–11
4. Armstrong, D. J. et al 1970. *J. Biol. Chem.* 245:2922–26
5. Back, A., Bittner, S., Richmond, A. E. 1972. *J. Exp. Bot.* 23:744–50
6. Back, A., Richmond, A. 1969. *Physiol. Plant.* 22:1207–16
7. Ibid 1971. 24:76–79
8. Barrell, B. G., Sanger, F. 1969. *FEBS Lett.* 3:275–78
9. Bednar, T. W., Linsmaier-Bednar, E. M. 1971. *Proc. Nat. Acad. Sci. USA* 68:1178–79
10. Beevers, L. 1966. *Plant Physiol.* 41:1074–76
11. Berridge, M. V., Ralph, R. K. 1969. *Biochim. Biophys. Acta* 182:266–69
12. Berridge, M. V., Ralph, R. K., Letham, D. S. 1970. *Biochem. J.* 119:75–84
13. Bezemer-Sybrandy, S. M., Veldstra, H. 1971. *Physiol. Plant.* 25:1–7
14. Bilderback, D. E. 1972. *Am. J. Bot.* 59:525–29
15. Birmingham, B. C., Maclachlan, G. A. 1972. *Plant Physiol.* 49:371–75
16. Blumenfeld, A., Gazit, S. 1970. *Plant Physiol.* 46:331–33
17. Ibid 45:535–36
18. Bui-Dang-Ha, D., Nitsch, J. P. 1970. *Planta* 95:119–26
19. Burrows, W. J., Carr, D. J. 1970. *Physiol. Plant.* 23:1064–70
20. Burrows, W. J., Skoog, F., Leonard, N. J. 1971. *Biochemistry* 10:2189–94
21. Burrows, W. J. et al 1969. *Biochemistry* 8:3071–76
22. Burrows, W. J. et al 1970. *Biochemistry* 9:1867–72
23. Chen, C.-M., Hall, R. H. 1969. *Phytochemistry* 8:1687–95
24. Chheda, G. B. 1969. *Life Sci.* 8 (Part 2):979–87
25. Chheda, G. B., Hong, C. I., Piskorz, C. F., Harmon, G. A. 1972. *Biochem. J.* 127:515–19
26. Chheda, G. B., Mittelman, A. 1972. *Biochem. Pharmacol.* 21:27–37
27. Chheda, G. B. et al 1969. *Biochemistry* 8:3278–82
28. Chibnall, A. C. 1939. *Protein Metabolism in the Plant.* New Haven: Yale Univ. Press. 306 pp.
29. Darimont, E., Gaspar, T., Hofinger, M. 1971. *Z. Pflanzenphysiol.* 64:232–40
30. Diamantstein, T., Wagner, B., Bhargava, A. S. 1971. *FEBS Lett.* 15:225–28
31. Divekar, A. Y., Fleysher, M. H., Slocum, H. K., Kenny, L. N., Hakala, M. T. 1972. *Cancer Res.* 32:2530–37
32. Divekar, A. Y., Hakala, M. T. 1971. *Mol. Pharmacol.* 7:663–73
33. Doctor, B. P., Loebel, J. E., Sodd, M. A., Winter, D. B. 1969. *Science* 163:693–95
34. Dove, L. D. 1972. *Symp. Biol. Hung.* 13:180–85
35. Dudock, B. S., Katz, G., Taylor, E. K., Holley, R. W. 1969. *Proc. Nat. Acad. Sci. USA* 62:941–45
36. Dyer, T. A., Osborne, D. J. 1971. *J. Exp. Bot.* 22:552–60
37. Dyson, W. H. 1969. PhD thesis. Univ. Kansas, Lawrence. 113 pp.
38. Dyson, W. H., Fox, J. E., McChesney, J. D. 1972. *Plant Physiol.* 49:506–13
39. Dyson, W. H., Hall, R. H. 1972. *Plant Physiol.* 50:616–21
40. Dyson, W. H., Hall, R. H., Hong, C. I., Dutta, S. P., Chheda, G. B. 1972. *Can. J. Biochem.* 50:237–43
41. Dyson, W. H. et al 1970. *Science* 170:328–30
42. Einset, J. W., Skoog, F. 1972. *Plant Physiol.* 49(suppl.):61
43. El-Antably, H. M. M., Wareing, P. F., Hillman, J. 1967. *Planta* 73:74–90
44. Engelbrecht, L. 1964. *Flora* 154:57–69
45. Feierabend, J. 1969. *Progr. Photosyn. Res.* 1:280–83
46. Feierabend, J. 1970. *Z. Pflanzenphysiol.* 62:70–82

47. Feierabend, J. 1970. *Planta* 94:1–15
48. Feierabend, J., Pirson, A. 1966. *Z. Pflanzenphysiol.* 55:235–45
49. Fittler, F., Kline, L. K., Hall, R. H. 1968. *Biochem. Biophys. Res. Commun.* 31:571–76
50. Fletcher, R. A., McCullagh, D. 1971. *Can. J. Bot.* 49:2197–2201
51. Fletcher, R. A., Osborne, D. J. 1965. *Nature* 207:1176–77
52. Fleysher, M. H., Hakala, M. T., Bloch, A., Hall, R. H. 1968. *J. Med. Chem.* 11:717–20
53. Fox, J. E., Chen, C.-M. 1967. *J. Biol. Chem.* 242:4490–94
54. Gallo, R. C., Whang-Peng, J., Perry, S. 1969. *Science* 165:400–2
55. Galston, A. W., Davies, P. J. 1969. *Science* 163:1288–97
56. Gaspar, T., Khan, A. A., Fries, D. 1973. *Plant Physiol.* 51:146–49
57. Gazit, S., Blumenfeld, A. 1970. *Plant Physiol.* 46:334–36
58. Gefter, M. L., Russell, R. L. 1969. *J. Mol. Biol.* 39:145–57
59. Gepstain, S., Ilan, I. 1970. *Plant Cell Physiol.* 11:819–22
60. Ghosh, K., Ghosh, H. P. 1970. *Biochem. Biophys. Res. Commun.* 40:135–43
61. Gilad, T., Ilan, I., Reinhold, L. 1970. *Isr. J. Bot.* 19:447–50
62. Giudici de Nicola, M., Piattelli, M., Castrogiovanni, V., Amico, V. 1972. *Phytochemistry* 11:1011–17
63. Giudici de Nicola, M., Piattelli, M., Castrogiovanni, V., Molina, C. 1972. *Phytochemistry* 11:1005–10
64. Goodman, H. M., Abelson, J., Landy, A., Brenner, S., Smith, J. D. 1968. *Nature* 217:1019–24
65. Goodwin, B. C. 1971. *Control Mechanisms of Growth and Differentiation*, ed. D. D. Davies, M. Balls, 417–28. Cambridge Univ. Press. 498 pp.
66. Goodwin, B. C., Cohen, M. H. 1969. *J. Theor. Biol.* 25:49–107
67. Grace, J. T., Hakala, M. T., Hall, R. H., Blakeslee, J. 1967. *Proc. Am. Assoc. Cancer Res.* 8:23
68. Hacker, B., Feldbush, T. L. 1969. *Biochem. Pharmacol.* 18:847–53
69. Hacker, B., Feldbush, T. L. 1971. *Cancer* 27:1384–87
70. Hall, R. H. 1970. *Progr. Nucl. Acid Res. Mol. Biol.* 10:57–86
71. Hall, R. H. 1971. *The Modified Nucleosides in Nucleic Acids.* New York: Columbia Univ. Press. 451 pp.
72. Hall, R. H., Alam, S. N., McLennan,

B. D., Terrine, C., Guern, J. 1971. *Can. J. Biochem.* 49:623–30
73. Hall, R. H., Csonka, L., David, H., McLennan, B. 1967. *Science* 156:69–71
74. Hall, R. H., Mintsioulis, G. *J. Biochem. Tokyo.* In press
75. Hall, R. H., Srivastava, B. I. S. 1968. *Life Sci.* 7(Part 2):7–13
76. Hanoune, J., Agarwal, M. K. 1970. *FEBS Lett.* 11:78–80
77. Harada, F. et al 1968. *Biochem. Biophys. Res. Commun.* 33:299–306
78. Hashimoto, S., Miyazaki, M., Takemura, S. 1969. *J. Biochem. Tokyo* 65:659–61
79. Hecht, S. M., Bock, R. M., Leonard, N. J., Schmitz, R. Y., Skoog, F. 1970. *Biochem. Biophys. Res. Commun.* 41:435–40
80. Hecht, S. M. et al 1969. *Biochem. Biophys. Res. Commun.* 35:205–9
81. Hecht, S. M. et al 1971. *Biochemistry* 10:4224–28
82. Hemberg, T. 1970. *Physiol. Plant.* 23:850–58
83. Ibid 1972. 26:108–9
84. Ibid, 98–103
85. Hemberg, T., Larsson, U. 1972. *Physiol. Plant.* 26:104–7
86. Hirsh, D. 1970. *Nature* 228:57
87. Ilan, I. 1971. *Physiol. Plant.* 25:230–33
88. Ilan, I., Gilad, T., Reinhold, L. 1971. *Physiol. Plant.* 24:337–41
89. Ishikura, H., Yamada, Y., Murao, K., Saneyoshi, M., Nishimura, S. 1969. *Biochem. Biophys. Res. Commun.* 37:990–95
90. Ishikura, H., Yamada, Y., Nishimura, S. 1971. *FEBS Lett.* 16:68–70
91. Itai, C., Vaadia, Y. 1965. *Physiol. Plant.* 18:941–44
92. Jordan, W. R., Skoog, F. 1971. *Plant Physiol.* 48:97–99
93. Kefford, N. P., Zwar, J. A., Bruce, M. I. 1968. *Biochemistry and Physiology of Plant Growth Substances*, ed. F. Wightman, G. Setterfield, 61–69. Ottawa: Runge. 1642 pp.
94. Kende, H. 1965. *Proc. Nat. Acad. Sci. USA* 53:1302–7
95. Kende, H. 1971. *Int. Rev. Cytol.* 31:301–38
96. Kende, H., Sitton, D. 1967. *Ann. NY Acad. Sci.* 144:235–43
97. Khan, A. A. 1971. *Science* 171:853–59
98. Khudairi, A. K., Arboleda, O. P. 1971. *Physiol. Plant.* 24:18–22
99. Kimura-Harada, F., Harada, F., Nishimura, S. 1972. *FEBS Lett.* 21:71–74

100. Kline, L. K., Fittler, F., Hall, R. H. 1969. *Biochemistry* 8:4361–71
101. Koshimizu, K., Kusaki, T., Mitsui, T., Matsubara, S. 1967. *Tetrahedron Lett.* No. 14:1317–20
102. Koshimizu, K., Matsubara, S., Kusaki, T., Mitsui, T. 1967. *Agr. Biol. Chem.* 31:795–801
103. Kulaeva, O. N. 1962. *Sov. Plant Physiol.* 9:182–89
104. Kuntzel, B., Weissenbach, J., Dirheimer, G. 1972. *FEBS Lett.* 25:189–91
105. Laloue, M., Hall, R. H. *Plant Physiol.* In press
106. Lee, T. T. 1971. *Plant Physiol.* 47:181–85
107. Leonard, N. J., Hecht, S. M., Skoog, F., Schmitz, R. Y. 1969. *Proc. Nat. Acad. Sci. USA* 63:175–82
108. Letham, D. S., Williams, M. W. 1969. *Physiol. Plant.* 22:925–36
109. Letham, D. S., Young, H. 1971. *Phytochemistry* 10:23–28
110. Loeffler, J. E., Van Overbeek, J. 1964. *Régulateurs Naturels de la Croissance Végétale, 5ᵉ Colloque International sur les Substances de Croissance Végétale*, 77–82. Paris: Édition du Centre National de la Recherche Scientifique. 748 pp.
111. Mackenzie, I. A., Konar, A., Street, H. E. 1972. *New Phytol.* 71:633–38
112. Mackenzie, I. A., Street, H. E. 1972. *New Phytol.* 71:621–31
113. Madison, J. T., Kung, H. K. 1967. *J. Biol. Chem.* 242:1324–30
114. Martin, C., Thimann, K. V. 1972. *Plant Physiol.* 49:64–71
115. McDonald, J. J., Leonard, N. J., Schmitz, R. Y., Skoog, F. 1971. *Phytochemistry* 10:1429–39
116. McLuhan, H. M., Nevitt, B. 1972. *Take Today; The Executive As Dropout*. Don Mills: Longman Canada Ltd. 304 pp.
117. Miller, J. P., Schweizer, M. P. 1972. *Fed. Proc.* 31:450
118. Miura, G. A., Hall, R. H. *Plant Physiol.* In press
119. Miura, G. A., Miller, C. O. 1969. *Plant Physiol.* 44:372–76
120. Mizrahi, Y., Amir, J., Richmond, A. E. 1970. *New Phytol.* 69:355–61
121. Mohr, H., Wagner, E., Hartmann, K. M. 1965. *Naturwissenschaften* 52:209
122. Mothes, K., Engelbrecht, L. 1963. *Life Sci.* 2:852–57
123. Mullins, M. G., Osborne, D. J. 1970. *Aust. J. Biol. Sci.* 23:479–83
124. Nagl, W. 1972. *Am. J. Bot.* 59:346–51
125. Nakanishi, K., Furutachi, N., Funamizu, M., Grunberger, D., Weinstein, I. B. 1970. *J. Am. Chem. Soc.* 92:7617–19
126. Oritani, T., Yoshida, R. 1971. *Proc. Crop Sci. Soc. Japan* 40:325–31
127. Pačes, V., Werstiuk, E., Hall, R. H. 1971. *Plant Physiol.* 48:775–78
128. Peterkofsky, A., Jesensky, C. 1969. *Biochemistry* 8:3798–3809
129. Piattelli, M., Giudici de Nicola, M., Castrogiovanni, V. 1969. *Phytochemistry* 8:731–36
130. Ibid 1971. 10:289–93
131. Polanyi, M. 1966. *The Tacit Dimension*. New York: Doubleday. 108 pp.
132. Powers, D. M., Peterkofsky, A. 1972. *Biochem. Biophys. Res. Commun.* 46:831–38
133. Radin, J. W., Loomis, R. S. 1971. *Physiol. Plant.* 25:240–44
134. RajBhandary, U. L. et al 1967. *Proc. Nat. Acad. Sci. USA* 57:751–58
135. Rathbone, M. P., Hall, R. H. 1972. *Cancer Res.* 32:1647–50
136. Rathbone, M. P., Hall, R. H. 1972. *Planta* 108:93–102
137. Richmond, A., Back, A., Sachs, B. 1970. *Planta* 90:57–65
138. Richmond, A. E., Lang, A. 1957. *Science* 125:650–51
139. Richmond, A. E., Sachs, B., Osborne, D. J. 1971. *Physiol. Plant.* 24:176–80
140. Riddick, D. H., Gallo, R. C. 1970. *Cancer Res.* 30:2484–92
141. Robins, M. J., Hall, R. H., Thedford, R. 1967. *Biochemistry* 6:1837–48
142. Sandstedt, R. 1971. *Physiol. Plant.* 24:408–10
143. Schaeffer, G. W., Sharpe, F. T. Jr. 1969. *Bot. Gaz.* 130:107–10
144. Schaeffer, G. W., Sharpe, F. T. Jr. 1970. *Biochem. Biophys. Res. Commun.* 38:312–18
145. Schaeffer, G. W., Sharpe, F. T. Jr. 1971. *Physiol. Plant.* 25:456–60
146. Schaeffer, G. W., Sharpe, F. T. Jr. 1971. *Life Sci.* 10(Part 2):939–45
147. Schaeffer, G. W., St. John, J. B., Sharpe, F. T. Jr. 1972. *Biochim. Biophys. Acta* 261:38–43
148. Schweizer, M. P., Chheda, G. B., Baczynskyj, L., Hall, R. H. 1969. *Biochemistry* 8:3283–89
149. Shah, C. B., Loomis, R. S. 1965. *Physiol. Plant.* 18:240–54
150. Shantz, E. M., Steward, F. C. 1955. *J. Am. Chem. Soc.* 77:6351–53
151. Shaw, G., Smallwood, B. M.,

Steward, F. C. 1968. *Experientia* 24: 1089–90

152. Sheldrake, A. R., Northcote, D. H. 1968. *New Phytol.* 67:1–13

153. Shibaoka, H., Thimann, K. V. 1970. *Plant Physiol.* 46:212–20

154. Shirai, H., Kanatani, H., Taguchi, S. 1972. *Science* 175:1366–68

155. Short, K. C., Torrey, J. G. 1972. *Plant Physiol.* 49:155–60

156. Skene, K. G. M. 1972. *Planta* 104: 89–92

157. Skoog, F., Armstrong, D. J. 1970. *Ann. Rev. Plant Physiol.* 21:359–84

158. Smith, C. J., Ley, A. N., D'Obrenan, P., Mitra, S. K. 1971. *J. Biol. Chem.* 246:7817–29

159. Sondheimer, E., Tzou, D.-S. 1971. *Plant Physiol.* 47:516–20

160. Srivastava, B. I. S. 1968. *Biochim. Biophys. Acta* 169:534–36

161. Staehelin, M., Rogg, H., Baguley, B. C., Ginsberg, T., Wehrli, W. 1968. *Nature* 219:1363–65

162. Steward, F. C. 1971. *Ann. Rev. Plant Physiol.* 22:1–22

163. Steward, F. C., Rao, K. V. N. 1970. *Planta* 91:129–45

164. Stobart, A. K., Shewry, P. R., Thomas, D. R. 1972. *Phytochemistry* 11:571–77

165. Suk, D., Simpson, C. L., Mihich, E. 1970. *Cancer Res.* 30:1429–36

166. Syōno, K., Furuya, T. 1971. *Plant Cell Physiol.* 12:61–71

167. Takemura, S., Murakami, M., Miyazaki, M. 1969. *J. Biochem. Tokyo* 65: 489–91

168. Tasseron-De Jong, J. G., Veldstra, H. 1971. *Physiol. Plant.* 24:239–41

169. Tavares, J., Kende, H. 1970. *Phytochemistry* 9:1763–70

170. Thimann, K. V. 1965. *Recent Progr. Horm. Res.* 21:579–96

171. Thimann, K. V., Sachs, T., Mathur, K. N. 1971. *Physiol. Plant.* 24:68–72

172. Torrey, J. G., Fosket, D. E., Hepler, P. K. 1971. *Am. Sci.* 59:338–52

173. Tucker, D. J., Mansfield, T. A. 1972. *Planta* 102:140–51

174. Udvardy, J., Farkas, G. L., Marrè, E. 1969. *Plant Cell Physiol.* 10:375–86

175. Varner, J. E. 1971. See Ref. 65, 197–205

176. Vreman, H. J., Skoog, F., Frihart, C. R., Leonard, N. J. 1972. *Plant Physiol.* 49:848–51

177. Wainfan, E., Hong, C. I., Chheda, G. B. 1972. *Abstr. 164th Meet. Am. Chem. Soc. Div. Biol. Chem.* No. 206

178. Wainfan, E., Landsberg, B. 1971. *FEBS Lett.* 19:144–48

179. Went, F. W. 1938. *Plant Physiol.* 13: 55–80

180. Ibid 1943. 18:51–65

181. Wheeler, A. W. 1971. *Planta* 98: 128–35

182. Whitty, C. D., Hall, R. H. *Can. J. Biochem.* In press

183. Wickson, M., Thimann, K. V. 1958. *Physiol. Plant.* 11:62–74

184. Williamson, A. R., Askonas, B. A. 1972. *Nature* 238:337–39

185. Wood, H. N. 1970. *Proc. Nat. Acad. Sci. USA* 67:1283–87

186. Wood, H. N. 1972. *Progr. Exp. Tumor Res.* 15:76–92

187. Wood, H. N., Braun, A. C. 1967. *Ann. NY Acad. Sci.* 144:244–50

188. Wood, H. N., Lin, M. C., Braun, A. C. 1972. *Proc. Nat. Acad. Sci. USA* 69:403–6

189. Woolley, D. J., Wareing, P. F. 1972. *Planta* 105:33–42

190. Yamada, Y., Nishimura, S., Ishikura, H. 1971. *Biochim. Biophys. Acta* 247: 170–74

191. Yoshida, R., Oritani, T. 1972. *Plant Cell Physiol.* 13:337–43

192. Zachau, H. G., Dütting, D., Feldmann, H. 1966. *Hoppe-Seyler's Z. Physiol. Chem.* 347:212–35

Ann. Rev. Plant Physiol. 1973. 24:445–66

CHILLING INJURY IN PLANTS   ❖ 7555

James M. Lyons
Department of Vegetable Crops, University of California, Davis

CONTENTS

INTRODUCTION

Tropical and subtropical plants exhibit a marked physiological dysfunction when exposed to low or nonfreezing temperatures below about 10° to 12°C. This dysfunction is referred to as chilling injury and has been of great concern for many years with harvested plant parts because lowered storage temperatures are generally an effective means of extending the postharvest life of fruits and vegetables. The phenomenon has not been well understood or recognized for its importance even though plant sensitivity to low temperature has been recorded for centuries.

Molisch (117) cited several early studies demonstrating that a number of plant species were killed at low temperatures above the freezing point. He suggested (118) that this physiological harm should be referred to as "chilling injury" (Erkältung) to differentiate it from freezing damage (Erfrieren). Such injury to susceptible plants (or plant parts) has also been referred to as "low-temperature injury" (48), as "cold injury" (135), and in apples as "low-temperature break-down" (193). Chilling injury appears to be the preferable term (148) because it is not easily confused with freezing injury or with phenomena related to cold or winter hardiness (78, 190). Chilling-sensitive plant species appear to have a com-monality of temperature response, with the critical temperature below which injury occurs being most often around 10° to 12°C. This generalization does not apply in all cases, however, and species vary somewhat in tolerance with their region of origin. For example, the lower temperature limit is 0° to 4°C for tem-perate fruits such as apples, around 8°C for subtropical fruits such as citrus, avocado, and pineapple, and around 12°C for the more tropical banana (193).

As in many fields of plant physiology, instrumentation and techniques de-veloped in the last decade have provided the means of developing new insight on this elusive problem, leading to a more integrated concept of chilling injury. The most important recent discussions of chilling injury are by Levitt (78), in a book on environmental stresses, and by Fidler (48), in a review. General considerations of the effects of temperature on biological systems can be found in reviews by Bělehrádek (7, 8), Kayser (70), Langridge (77), and Luyet & Gehenio (90). Chill-ing injury has also been considered in conjunction with other topics in reviews on the storage and postharvest physiology of fruits and vegetables by Biale (10–12), Hansen (51), Miller (109, 111), Pentzer & Heinze (133), Ulrich (176), Wardlaw (184, 185, 187), and in a text by Ryall & Lipton (148). Additional discussion is found in general treatments of the biochemistry and physiology of specific com-modities such as citrus (13, 109, 171), avocado (14), banana (128, 183, 186), mango (61), pineapple (37), and tomato (56). Reviews on freezing injury in plants include some aspects of low-temperature effects in the range just above freezing (97, 98).

This review summarizes some of the striking horticultural manifestations of chilling injury, with particular emphasis on harvested crops, and relates these to the physiology of the tissue involved. The discussion considers how temperatures below around 10° to 12°C can so abruptly and dramatically affect physiological function in sensitive plant species.

HORTICULTURAL ASPECTS OF CHILLING INJURY

Symptoms

The symptoms of chilling injury vary with the plant tissue and the severity of in-jury, and they usually develop more rapidly if the tissue is transferred to a non-chilling temperature. Some generalities will be provided here without describing symptoms for all susceptible commodities. Specific information on various crops is given by Lutz & Hardenburg (89), Ryall & Lipton (148), and papers on the postharvest behavior of specific commodities. Although chilling injury has been recognized for many years, its severity is difficult to define quantitatively, being estimated essentially only qualitatively in terms of visual observations. Katz & Reinhold (69) experimented with changes in electrical conductivity for estimating injury in *Coleus* before external symptoms developed, but this method has not been extended widely to other plant material.

Perhaps the most apparent symptoms of general concern are surface pitting, necrotic areas, and external discoloration. As early as 1896, Molisch (117) described a brown spotting on sensitive tropical plants after brief exposure to chilling temperatures. The symptomology of chilling injury has since been observed and described for almost all commercially important horticultural commodities sensitive to low temperatures. If bananas, for example, are exposed to mild chilling when green, the peel develops a smoky or dull-yellow appearance (186), and with more severe chilling it turns dark brown or black. Cucumbers give an excellent example of typical surface pitting (42). The pitting results from injury and collapse of subsurface cells, followed rapidly by invasion of decay organisms. Such visible pitting can be related to rate of water loss in some commodities (120), and pitting has been reduced by maintaining a high relative humidity or waxing the fruit to decrease water loss. Some tissues become discolored both internally and externally as a result of chilling. Sweet potatoes, for example, develop a brown discoloration if the roots are cut and exposed to air for a short period after chilling (81).

Mature-green fruits commonly fail to ripen normally after chilling. Biale (9) reported that avocado fruits stored at chilling temperatures neither ripened normally nor developed the typical climacteric rise in respiration associated with ripening. Despite a lack of climacteric rise in respiration at 5°C, there was a cycle of ethylene production (as measured by the pea bioassay), though not until the fruit had been at the chilling temperature for 57 days (141). When avocado fruits were given a chilling treatment (5°C for 34 or 43 days) and then transferred to a warm temperature (20°C), CO_2 production rose but without the normal corresponding rise in ethylene (15). Abnormal patterns of ethylene evolution have also been shown in chilled bananas (121) and grapefruit, sweet oranges, avocados and leaves (29, 30). Because tomato fruits and bananas are harvested and shipped at a mature-green stage, their ripening behavior at various storage temperatures has been studied extensively (e.g. 79, 121, 147, 183–185).

Apple scald (superficial scald), a brown discoloration giving the skin a cooked appearance, is a symptom of chilling injury in some varieties of apples upon long-term storage at 0° to 4°C (108, 193). Low-temperature breakdown of apples, also caused by storage at such temperatures, occurs in the cortex, and in certain varieties involves most of the flesh (62, 193, 198). "Wooliness" of peaches and plums refers to a mealy texture and discolored appearance of the flesh caused by chilling injury in storage at 0° to 4°C (17, 34, 35).

Susceptibility to decay is increased by surface pitting, necrotic areas, and general weakening of the tissue. Injured tissue is readily infected by decay organisms such as alternaria rot (57, 99–104). Even without such ensuing decay, storage life would be much reduced by the pitting, necrosis, and lack of ripeness, so abbreviated storage life in itself can be considered a symptom of chilling injury in horticultural commodities (42). It is important that the symptoms described above are not necessarily specific to chilling injury. Surface pitting typical of chilling injury can be induced at high temperatures by low-oxygen atmospheres (120, 124, 125), which is relevant to the proposal of Nelson (124) that chilling injury is caused by the inability of tissue to obtain or use enough oxygen for normal respiration. Since, however, low oxygen can induce similar symptoms in both chilling-sensitive and chilling-resistant species (124), low oxygen does not appear to be a cause of chilling injury per se. Similarly, cucumbers at warm temperatures can become pitted as a result of mechanical injury and desiccation in the absence of chilling (3).

Symptoms of chilling in vegetative organs vary with the tissue involved, but under severe conditions the ultimate result is impairment of function and death. For example, if cotton seedlings are chilled at time of seed hydration, there will be abortion of the radicle tip, whereas chilling applied after germination but during early seedling growth will result in damage to the root cortex (20). These latter symptoms have also been described for muskmelon and pepper seedlings (53). Chilling of sensitive grass species has been shown to cause a reduction in photosynthesis (169) and changes in chloroplast ultrastructure (114, 167). Common symptoms of chilling injury in developing vegetative tissues are necrotic lesions, increased susceptibility to decay organisms, cessation of growth, and ultimately death.

Susceptibility

Chilling injury is a characteristic of plants of tropical or subtropical climates, although plants of the temperate zone can also have a sensitivity. Early works of Sachs (149), Ewart (44), and Molisch (117) described the influence of chilling temperatures on a number of species of tropical and semitropical plants. Sellschop & Salmon (153) demonstrated the sensitivity to chilling temperatures of a number of crop plants, including cotton, cowpea, peanut, corn, and rice. When these species were subjected at various stages of growth to chilling temperatures, 2° to 4°C for periods as short as 12 hours induced injury in some. Seible (151) described symptoms of chilling in several species exposed to low temperatures for only a few hours. Similarly, Spranger (163) described the symptoms of chilling injury on vegetative tissues from a number of ornamental plants. Since those early reports, susceptibility has been described in many horticultural crops of economic importance, and Lutz & Hardenburg (89) and Ryall & Lipton (148) have listed many of the susceptible commodities, their symptoms, and their safe low-temperature limits.

Varietal differences in susceptibility to chilling injury have also been reported. For example, Smith & Millet (160) showed that the average period for sprouting at 10°C ranged between 18 and 46 days among 10 tomato cultivars. Apeland (3) demonstrated varietal differences in cucumber responses to a 4-day chilling treatment at 5°C: decline to poor quality at 12.5°C took widely varied periods, from

10 days for 'Ohio M.R. 200', to 47 days for 'Marketeer'. Watada & Morris (188, 189) demonstrated that the symptoms and pattern of senescence were similar among 9 snap bean cultivars but that their storage lives differed significantly. Likewise, their rates and patterns of respiration were similar although they differed in severity of injury symptoms and reduction of storage life. Christiansen (26) demonstrated marked differences between 'Delta' and 'Acala' cotton selections in response to chilling injury during seed hydration, with the results suggesting that variation in resistance to chilling injury is a heritable factor.

Physiological age at chilling also affects susceptibility to injury. A good example is found in seedling growth and development. Although seedlings of chilling-sensitive species are susceptible, dry seeds of these same species experience no harm when stored for extended periods at chilling temperature. Wheaton (191) found that fully imbibed corn seeds (12 hours in aerated water) and 1-day-old seedlings were uninjured by 4 days at 1°C but that sensitivity appeared during the following 2 days and reached a maximum by the third day following imbibition. This transition in germinating seeds has also been studied in lima bean seeds (139, 140) and in cotton seeds (22, 23, 170). These studies indicated two critical periods of sensitivity: 1. Chilling applied coincident with imbibition as the initial step in germination causes an injury that can be avoided if the seed has been first allowed a brief hydration at elevated temperature before the chilling. 2. A second period of sensitivity appears sometime after 24 hours of germination. A precise determination of the metabolic events coincident with this transition in chilling sensitivity would be of immense value to an understanding of chilling injury in general. Physiological age certainly plays a role in the susceptiblity of plant tissues at other stages of growth as well. The less mature apical "hands" of bananas, for example, are more susceptible to chilling than the more mature basal hands (129).

Preharvest environment can influence susceptibility. Fidler (48) describes the influence on chilling sensitivity of the climate of origin. 'Cox's Orange Pippin' apples, for example, can be stored at 1.5°C when grown in mainland Australia, 2°C when in Tasmania, 2.5° to 3°C when in New Zealand, and 3° to 4°C when grown in the United Kingdom. Apeland (3) reported that greenhouse-grown cucumbers are more sensitive to 10°C than are similar cultivars grown under field conditions. Similarly, Palmer (128) cited studies indicating that banana fruits maturing at higher field temperatures are more susceptible than those maturing in a cooler climate.

Chilling Treatment

Exposure to chilling temperatures must be relatively long before cells of most sensitive plants are injured. The injuries observed most often require days or weeks of chilling temperatures, although Seible (151) described injury in several species (*Episcia, Achimines,* and *Gloxinia*) after only a few hours at 1° to 5°C, and bananas can be injured after a brief exposure below 10°C (183). In general, the severity of injury of sensitive plant tissues increases as temperature is lowered or as exposure is extended at any chilling temperature.

Since a reliable quantitative measure is lacking, chilling injury must be evaluated by qualitative visible symptoms as described above. It is a function of both physiological injury per se and rate of symptom development in the particular plant tissue. For example, van der Plank & Davies (182) reported that injury in plums and peaches was greater at 3°C than at 0°C. They explained these results on the basis of an "equilibrium" factor, the temperature increment below that critical for injury, and a "kinetic" factor which regulated the rate of chemical change and hence symptom expression. These factors operate in opposing directions as the temperature is lowered, and at some point (3° for plums and peaches) the kinetic factor allows a more rapid development of symptoms even though injury is ultimately greater at 0°C. Further, evaluating the injury that occurs during continued chilling (182) differs from that which develops at some elevated temperature (e.g. 25°C) after chilling for various periods (42, 79, 188). The latter is important as an experimental method since it not only standardizes the conditions for symptom expression, but reflects the usual practical situation, with commodities held at low temperatures during storage and transport followed by several days of higher temperatures during the marketing period. The injury that occurs during chilling is usually not immediately conspicuous but may very rapidly become visible during 2 or 3 days at the elevated temperature.

Much concern with chilling phenomena relates to the storage of fruits and vegetables, where chilling is most easily recognized as a problem. Chilling in the natural environment can be of equal concern, however. For example, chilling in the field as the season extends into the fall can cause tomato fruits to fail to ripen into marketable condition, even if held at proper temperature after harvest (119). And, of course, field temperatures will be of concern where chilling would prevent germination (52) or influence seedling development and growth after germination (20–22, 24, 53, 160), reduce photosynthetic rate (16, 169), and even affect reproduction in some temperate species (172).

Amelioration of Chilling

TEMPERATURE CONDITIONING   Acclimation and hardening are terms used to describe a change in woody species from a susceptible to a winter-hardy or resistant condition capable of withstanding freezing temperatures to around $-60°C$ (190). Similar "hardening" toward chilling injury is not possible, although a few experiments indicate that the sensitivity of some plant material can be reduced by exposure to temperatures slightly above the chilling range (3, 88, 101, 192). Wheaton & Morris (192) showed that exposing 5-day-old tomato seedlings grown at 25°C to a conditioning temperature of 12.5°C for as little as 3 hours provided some protection from a subsequent 2-day exposure to 1°C; maximum protection required 48 hours of such exposure. The protection was effective only against slight to moderate chilling, however, and none of the seedlings survived a 7-day exposure to 1°C. In similar attempts with sweet potatoes, conditioning treatments influenced respiration rates slightly but did not reduce chilling injury as measured by visible symptoms (192). The sensitivity of cucumber fruits chilled at 5°C for 4 or 6 days was decreased somewhat by a conditioning period at 12.5°C

(3). Bananas transferred directly from 21°C to 5°C showed more chilling injury than bananas whose temperature was lowered by steps of 3°C at 12-hour intervals (129). As with the tomato seedlings and sweet potatoes cited above, however, conditioning of cucumbers and bananas is effective only against slight chilling.

ALTERNATING TEMPERATURES It was demonstrated that injury in plums could be ameliorated by interrupting the chilling treatment with a warm (above 20°C) period of 2 or 3 days (158, 161). The basis of this system is the assumption that the cause of injury is accumulation of a toxic compound which can be metabolized, or depletion of a critical metabolite which can be restored at the warm temperature. This avoidance of chilling injury by applying a warm period has been observed in a number of plant tissues. Decreased ascorbic acid and increased chlorogenic acid in sweet potato roots during chilling could be reversed by a warm period at 15°C following 2 weeks' exposure to 7.5°C but not 4 weeks or longer (84). Cacao seed germination is completely destroyed by exposure to 4°C for as little as 10 minutes (64), but immediate immersion in 37°C water for 10 minutes restored germination to 50% (63, 65). In cacao seeds, the chilling exposure at 4°C that still allowed reversibility was 10 to 20 minutes. In cotton seedlings, a chilling-induced decrease in ATP levels was reversible after a 24-hour exposure to 5°C but not after a 48-hour exposure (164). In corn seedlings, visual leaf injury appeared in 36 hours at 0.3°C, but upon transfer to 21°C the leaves returned to normal and the leaf symptoms disappeared within 72 hours (33). Chilling injury in some of the seedlings could be partially reversed after 48 to 60 hours at 0.3°C, but not after 72 hours.

Thus, chilling injury can be avoided in many tissues if they are returned to a warming temperature before degenerative changes occur.

HYPOBARIC STORAGE Burg & Burg (18) described a system of fruit storage under reduced atmospheric pressures which greatly extended storage life at nonchilling temperatures. They (personal communication) have also shown that low-pressure storage at 50 to 200 mm Hg with continuous air changes to remove volatiles could relieve symptoms of chilling injury in avocados, bananas, grapefruit, peppers, and tomatoes. Tolle (173) applied the term "hypobaric" to the storage at subatmospheric pressure (low-pressure storage). Pantastico et al (129) showed that when bananas were held at 5°C and 220 mm Hg pressure, the green color of the fruit was maintained for about a month and symptoms of chilling injury were reduced. Pitting symptoms of chilling injury from storage of limes and grapefruit (130) at 4.5°C were lower under 220 mm Hg than under atmospheric pressure: 0.0% vs 65.4% in limes held 4 weeks, and 4.4% vs 23.5% in Marsh grapefruit held 7 weeks. Holding grapefruit at only 380 mm Hg did not give similar amelioration (49). Oudit et al (127) found no alleviation of chilling injury from hypobaric storage at 220 mm Hg in bananas, avocados, limes, cucumbers, and tomatoes. All the same, hypobaric storage appears able to alleviate symptoms of chilling injury under at least some conditions and therefore deserves further study.

MODIFIED ATMOSPHERES   The analogy between symptoms of injury induced by chilling temperatures and those induced by low oxygen (120, 124) has prompted investigations of the effect of modified atmospheres on chilling injury. Kidd & West (71) reported that increased CO_2 in modified-atmosphere storage increased the susceptibility of apples to low-temperature breakdown. Eaks (38) showed that atmospheres of compositions differing from 0% to 100% O_2 had very little influence on injury of cucumber fruits exposed to 5°C for 8 days, but that increased CO_2 concentrations intensified symptoms of chilling injury. Similarly, Tomkins (174) has shown that increased CO_2 in the storage atmosphere increases damage from chilling treatments in the tomato. Miller (109), in his review of citrus storage, described a number of studies in which modified-atmosphere storage was reported to decrease pitting and other symptoms of chilling injury in citrus. Despite some success reported for relatively high percentages of CO_2, such treatment generally resulted in some type of rind injury to citrus. Vakis et al (179) reported that 10% CO_2 in the storage atmosphere reduced chilling injury in grapefruit and avocados. Waxing of fruit and storage in pliofilm packages modified the internal atmosphere of citrus fruits and alleviated chilling injury in grapefruit (49) but aggravated it markedly in limes (55, 130).

OTHER TREATMENTS   Some evidence exists that chemical treatments can alleviate chilling injury. Recent studies in Australia have shown that superficial scald in apples, a chilling injury occurring after prolonged storage at 0°–4°C, was caused by injurious conjugated triene hydroperoxides, oxidation products of α-farnesene (58–60). It was further shown that oxidation of α-farnesene could be inhibited by the antioxidant diphenylamine, either as a coating on the apple skin or in wrappers around the fruit. Pantastico et al (130) could induce scald (an advanced stage of pitting) in limes by treatment with acetaldehyde at chilling temperatures, and this scald could be reduced approximately 12 to 50% by diphenylamine treatment. Another antioxidant which inhibits this scald (162) is ethoxyquin (6-ethoxy-2,2,4-trimethyl-1,2-dihydroquinoline). The fungicide thiabendazole (TBZ) has been shown to reduce pitting of grapefruit significantly during prolonged storage at around 8°C (150). It was concluded that TBZ possibly had some physiological effect on amelioration of chilling in addition to its effect as a fungicide. Amin (1) found that applications of IAA and vitamins increased the vegetative growth of chilled cotton plants over that of untreated controls.

Waxes applied to the surface of chilling-sensitive tissues have alleviated chilling symptoms in some instances. Waxing reduced surface pitting on chilling-sensitive fruits held at 0.5°C and 70% relative humidity, with the reduction in pitting related to slower water loss than in unwaxed fruits under similar conditions (120). Waxing was suggested by Platenius (137) as a method for reducing chilling injury. In contrast, Mack & Janer (96) observed that pitting was more severe on waxed than on unwaxed cucumber fruits held for one week at 2.2° to 3.3°C and 95% relative humidity. These results are difficult to assess since wax could slow desiccation, which has been associated with pitting (3, 120), or it could modify the internal atmosphere of a commodity (depending on the type of wax and thickness of coating). Which of these factors is operating in a given set of experimental conditions is not yet resolved.

Another disorder of apples called low-temperature breakdown has been shown to be caused by the accumulation of acetic acid during prolonged storage at low temperatures (195–199). Treatments that promoted water loss reduced acetic acid in the tissue by accelerating its removal as acetate esters (or the free acid), thereby reducing susceptibility to breakdown. Hypobaric storage or alternating temperatures (discussed above) could presumably accelerate removal or metabolism of acetic acid or its esters, thereby reducing low-temperature breakdown.

PHYSIOLOGICAL MANIFESTATIONS

Protoplasmic Streaming

Perhaps one of the more spectacular effects of chilling temperature on sensitive plant tissues is the effect on protoplasmic streaming. As early as 1864, Sachs (149) observed that protoplasmic streaming ceased at about 10°–12°C in root hairs of cucumber and tomato plants, and other workers of the time indicated that streaming continued down to or near 0°C in chilling-resistant species. Lewis (80) examined this differential response in a number of plant species. He found that streaming ceased or was just perceptible after 1 or 2 minutes at 10°C in petiole trichomes of chilling-sensitive plants and invariably ceased promptly at 5° or 0°C. If exposure to 0°C exceeded 24 hours, streaming failed to resume in most trichomes upon return to a warmer temperature. In contrast, streaming in chilling-resistant plants proceeded at temperatures down to 0°–2.5°C. Wheaton (191) confirmed these results with five sensitive and four resistant species.

The exact mechanism of protoplasmic streaming has not been delineated, but it is clear that the process requires energy and is dependent on the physical properties of the protoplasm and subcellular membrane system (68, 152). If, as discussed later, the ATP supply is greatly reduced in chilling-sensitive species, and the lipoprotein membranes undergo a phase transition from a flexible to a solid structure as an immediate effect of chilling, it follows that streaming would very quickly be greatly impeded or cease. Some evidence for the energy relationship can be found in the observation by Ewart (45) that streaming ceased at a higher temperature if the tissue was cooled under anaerobic rather than aerobic conditions. The impact of cessation of streaming on the events leading to injury is difficult to evaluate because the exact role of streaming in normal metabolism is little understood (152); however, it is likely that normal metabolism would be greatly upset by a complete cessation of the process.

Respiration

STUDIES WITH DETACHED ORGANS   Anomalous respiratory behavior during or after the chilling of sensitive plant tissues has been reported by a number of investigators (40, 41, 67, 71, 81, 121, 126, 129, 138, 188, 207). Except during the ripening of fruit of the climacteric type (10), the normal pattern of respiration is a gradual decline with time after harvest; and such is the case with chilling-resistant plant tissues at all temperatures (e.g. 148) and with chilling-sensitive plant tissues held above chilling temperatures. In contrast, Kidd & West (71) noted that respiration at 1°C was faster in apples injured by low-temperature breakdown, and similar reports of faster respiration (following an initial brief decline) have been made for sweet potato roots (81), tomatoes (79), cucumbers (41), peppers (126), and a number of other commodities. The faster respiration during storage at chilling temperatures usually precedes any external visible symptoms of injury (41, 81, 188). This respiration pattern was clear when data of Eaks & Morris (41) for cucumber fruits were presented as an Arrhenius plot (93). Their initial respiratory rates, at 24 hours of storage, indicated a discontinuity or "break" in the Arrhenius plot,[1] with a higher activation energy at the chilling than the nonchilling temperatures. The plot revealed an increased respiration after 5 days of storage in fruits stored at 0°C but not at 5° or 10°C, which correlates with the fact that upon transfer to a warm temperature for 3 days injury was observed in the 0°C fruit but not in the 5° or 10°C fruit. Upon 10 days of storage, injury was observed at 5° and 10°C as well as 0°C and respiration had increased at all three temperatures. This must reflect an injury phenomenon which has upset metabolism and induced accelerated respiration.

Discussions relative to the effect of temperature on Q_{10} values must consider the relationship between time and temperature. For example, if Q_{10} values were calculated at the 24-hour period for the cucumber data of Eaks & Morris (41) above, the Q_{10} values would be higher at the lower temperatures than at higher temperatures—the situation generally observed with most plant species, whether chilling-sensitive or not. In contrast, Q_{10} calculated from the respiration rates after 10 days would give values similar over the entire temperature range.

Another respiratory phenomenon of importance in relation to chilling injury is the greatly exaggerated respiration rate observed at warm temperatures after transfer from a chilling treatment. This respiratory stimulation upon transfer has been observed for many plant species (4), but with chilling-sensitive species it is greatly amplified and can be used as an index of the severity of chilling injury. If, for example, cucumber fruits are transferred to 25°C after 4 days at 5°C, there is a peak in respiration, but the rate quickly returns to that of fruits held continuously at 25°C, indicating that injury is not yet permanent (41). With chilling for 8 or 12 days, in contrast, a similar peak in respiration upon transfer to 25°C occurred but the rate did not return to a normal level. Further, symptoms of injury become apparent in the 8-day and 12-day treatments but not in the 4-day treatment. Responses have been similar in chilling-sensitive pepper (126), citrus (39), tomatoes (79), snap beans (188), and bananas (121). Thus, in evaluating the impact of chilling treatments on the respiratory process, it is important that the initial reversible

---

[1] A more complete discussion of the usefulness of presenting data as an Arrhenius plot, as well as the thermodynamic basis for such a plot, is found in reviews by Raison (142, 143) and Lyons (91).

response to brief exposures be distinguished from the response when chilling time has been long enough to cause irreversible degenerative changes in the tissue.

The respiratory quotient (RQ) at chilling temperatures has been examined in an attempt to demonstrate altered metabolism in sensitive plant tissues at chilling temperatures. Platenius (138) studied the effect of chilling temperatures on the RQ of a number of plant species but found no differences which could be related to chilling sensitivity. Eaks & Morris (41), on the other hand, demonstrated that the RQ for cucumbers held at 0°C was less than unity for the first 7 days of storage and then rose abruptly above unity. One can attempt to interpret RQ values in terms of the substrate available for the respiratory process, but, regardless of the inferences drawn, the sudden shift from less than unity to above unity correlates with rather severe injury and does reflect a major upset in normal metabolism of the tissue. In contrast, cucumbers held at the nonchilling temperature of 15°C exhibited an RQ of unity throughout the 14-day storage period.

Anomalous respiration of seedling tissues exposed to chilling temperatures has also been shown. For example, respiration in cucumber leaves (76, 200) declined for short periods at chilling temperatures but increased markedly upon transfer to warm temperatures. As with the other plant material studied, the magnitude and course of the respiratory pattern depended on the severity of previous chilling.

TISSUE-LEVEL STUDIES   Observation of the anomalous respiratory behavior associated with chilling injury has naturally led to a questioning of the impact of chilling on cellular respiration and oxidative phosphorylation. In one approach to this question, slices and segmented tissues are utilized to facilitate the use of respiratory inhibitors. Shichi & Uritani (154) prepared disks from sweet potato roots and showed that the tissue lost the ability to respond to 2,4-dinitrophenol (DNP) at 20°C after 10 to 15 days at 0°C. This led them to propose that the respiratory mechanism was uncoupled and ATP formation retarded as a result of the chilling treatment. Lewis & Workman (82) demonstrated a marked decline in the incorporation of ^{32}P at 20°C in tissue slices from mature green tomatoes following chilling at 0°C, an observation which also suggests a reduced capacity for oxidative phosphorylation. Wheaton (191), on the other hand, found that root segments from chilling-sensitive corn were highly responsive to DNP after 3 hours at 5°C. He did not extend the chilling treatment to the point where injury caused degenerative cellular changes, and his data suggest that the loss of phosphorylative capacity is not the primary step in chilling injury but follows upon some irreversible physiological breakdown. Creencia & Bramlage (33) demonstrated that DNP could stimulate respiration in segments of corn leaves if the tissue was tested before injury from chilling was irreversible. Upon irreversible injury the capacity to respond to DNP was lost and the respiration rate increased. Similarly, malonate inhibited the respiration of healthy banana slices but not that of chill-injured slices (121, 122). Tissue segments from cotton seedlings had a higher respiration rate at 25°C following chilling treatments up to 24 hours at 2.8°C (2). Respiratory activity was similar in tissue slices from both bananas and limes (180). Respiration of tissue slices from pepper fruits increased markedly after 7 weeks at 1°C, was relatively constant at 6°C, and declined slightly at 18°C (126).

MITOCHONDRIAL STUDIES   Changes in mitochondrial physiology have been of interest because of the central role of these organelles in the respiratory process. Lieberman et al (84) attempted further elucidation of the mechanism of chilling effects on the respiratory process by studying mitochondria isolated from sweet potato roots. Their basic approach was to compare activity at 25°C of mitochondria derived from sweet potato roots stored at chilling (7.5°C) and nonchilling (15°C) temperatures. They found essentially no difference resulting from the first 4 weeks of storage. After the fifth week, however, activity began to decline in the mitochondria from the chilled roots, and by the tenth week the chilling treatment yielded completely inactive mitochondria. Similarly, Minamikawa et al (115) found a decrease in oxidative activity at 25°C of mitochondria isolated from sweet potato roots chilled at 0°C. Uritani et al (177) also compared activity at a high temperature (28°C) of mitochondria from unchilled and chilled sweet potato roots and found that oxygen uptake and respiratory control (RC) were lower in the chilled tissues. If the sweet potato roots were chilled only slightly, cytochrome c added to the reaction mixture could restore the respiration rate to that of mitochondria from unchilled tissues. When the roots were severely injured, restoration of the rate by added cytochrome c was only partial, and they suggested that cytochrome c was released from the mitochondrial membranes during the chilling treatment. They also presented data indicating that malate dehydrogenase activity was impaired during storage of the roots at 0°C. Electron micrographs of mitochondria derived from the chilled sweet potato roots showed a large proportion in an extremely swollen form not observed in mitochondria from healthy tissue (203, 204). The swollen appearance resulted from degradation of mitochondria and the release of phospholipid from both the inner and outer membranes during storage at chilling temperatures (202). Furthermore, the capacity to bind added phospholipid (after removal by aqueous acetone treatment) was greatly reduced in mitochondria from chill-injured sweet potato roots in comparison to those from healthy roots (205). Pantastico et al (130) isolated mitochondria from limes and grapefruit after chilling treatment and found that the energy transfer system was impaired. Because of the many complicating factors operating during the procedures for isolation of plant mitochondria (144), distinguishing direct effects of chilling from indirect effects due to the preparation of mitochondria from injured tissue is most difficult.

In contrast to the approach used above, Lyons & Raison (93) isolated mitochondria from healthy plant tissues and studied the effect of a wide range of temperatures on oxidative activity of the healthy mitochondria. When the effect of temperature on oxidative activity was presented as an Arrhenius plot, the mitochondria from chilling-resistant plant species each exhibited a linear plot with a constant activation energy (Ea) over the entire temperature range from near 1°C to 25°C. In contrast, mitochondria from chilling-sensitive species each exhibited a discontinuity in plot at 10° to 12°C, with a marked increase in Ea below the break temperature, indicating that an immediate and direct effect of low temperature on these species is to suppress mitochondrial respiration. Their data also clearly showed that phosphorylative efficiency was not influenced at any temperature for either the chilling-sensitive or the chilling-resistant plant species, and therefore, although the rate of production would be decreased concomitantly with the suppressed mitochondrial respiration, the efficiency of energy production is not altered by low temperature.

Metabolic Changes

Numerous attempts have been made to correlate changes in cellular constituents or related enzyme activities with chilling injury. Tissue given a chilling treatment is invariably compared with healthy tissue, but this approach has not contributed significantly to an understanding of the mechanism of chilling injury. Cordner & Mathews (31) found essentially no change in sucrose, total sugar, acid-hydrolyzable substances, or starch in squash fruit after 8 days in storage at 2°–4°C. Lorenz (87) similarly found that all major components in summer squash remained practically constant during storage at chilling temperatures. Jones (67) found a slight decrease in hydrolysis of sucrose in chilled papaya fruits, with a concomitant increase in soluble solids. Roots of sweet potatoes injured by chilling decreased in ability to synthesize carotenoid pigments (46, 47) and had an accelerated loss of ascorbic acid (46, 47, 85). Accelerated loss of ascorbic acid as a result of chilling was observed also in pineapple (110, 112) and bananas (123), but not in guava (157) or tomato (32, 79). Barnell & Barnell (6) reported that the pulp of chilled banana fruit contained higher levels of tannins, and suggested that oxidation of these compounds gave rise to the dark color symptomatic of chilling injury in that fruit. Similarly, Murata & Ku (122) observed increased levels of tyrosine and dopa in chilled banana tissue which could produce the dark pigment upon polymerization and oxidation. Lieberman et al (85) showed an increased accumulation of chlorogenic acid in chilled sweet potato roots, a fact most probably instrumental in the decreased activity of mitochondria prepared from such tissues (83, 84). Both chlorogenic acid and total polyphenols in pepper seeds increased initially and then gradually decreased during storage at chilling temperatures (72). Compounds closely associated with intermediary metabolism—acetaldehyde, ethanol, and the keto acids—have been shown to increase in several tissues injured by chilling (113, 121, 122, 126, 130, 206). Chilling altered the amino acid content of sensitive grass species causing a sharp decrease in the content of those amino acids closely related to intermediates of the C_4-photosynthetic pathway within 1.5 days (168). As pointed out earlier, some primary responses to chilling temperatures must be clearly distinguished from events which are the result of degenerative tissue injury; many of the chemical changes listed above result from the latter event. Similarly, many enzyme systems isolated from chilling-injured tissue have shown an altered activity. For example, as a result of chilling treatments of sensitive tissues increases have been reported in the activity of invertase (19), polygalacturonase (181), phenylalanine ammonia-lyase and tyrosine ammonia-lyase (73), catalase (122), pyruvate decarboxylase, alcohol dehydrogenase, isocitrate dehydrogenase, glucose-6-phosphate dehydrogenase, and phosphoenol-pyruvate carboxykinase (206). Likewise, decreases have been observed in a number of systems including malate dehydrogenase (177), pectinmethylesterase (181), and amylase (19).

Studies of chilling effects on seed germination and seedling development have revealed the extreme sensitivity and short periods required for changes. For example, Guinn (50) showed that chilling decreased RNA, protein, and lipid-soluble phosphate in young cotton seedlings, and Stewart & Guinn (164, 165) showed a very early decline in ATP and other nucleotides as well. Some of these changes in the phosphorylating system were observed as early as 6 hours after exposure to chilling temperatures. Similarly, studies of enzymes isolated from cotton seed exposed to chilling temperatures have shown isocitratase levels sharply depressed but little or no effect on malate synthetase (116, 159). Christiansen (25), in a recent review of his own and co-workers' studies on germinating cotton (20–23, 27, 28), presented a table listing a number of aberrations in the metabolism of cotton seed subjected to chilling temperatures. This comparative analysis indicated that chilling changed the entire metabolic system of the cell, with some processes recovering quickly and others only slowly. He concluded that the initial impact of chilling temperatures was a rapid physical change in membranes to a freely permeable state which allowed secondary events to alter metabolism (this is discussed in more detail in the next section).

Membrane Phenomena

PERMEABILITY   Change in membrane permeability in response to chilling temperatures has often been investigated as a possible cause of chilling injury. Some studies to develop information on permeability covered the uptake and transloca-tion of water and solutes by intact plants in response to temperature. Kramer (74) investigated the effects of soil temperature on water absorption by cool- and warm-season crops, and showed that low temperature reduced water absorption in all species but much more in the chilling-sensitive warm-season crops. He con-cluded that the reduced water uptake was caused by an increased viscosity and a decreased permeability in response to low temperature. Studies on the transport of water and ions through chilling-sensitive plants (36, 66, 200) indicated that transport also was reduced in response to low temperatures. Similarly, Hartt (54) reported that translocation in sugar cane ceased completely at 5°C. Although it is evident that these water uptake and translocation phenomena are greatly de-creased by chilling temperatures, it is difficult to derive specific knowledge on membrane permeability from these studies.

Studies in which permeability changes are measured by solute leakage or ion accumulation have provided some direct evidence for an increased membrane permeability in response to chilling. Lewis (79) reported no difference between chilled and healthy tomato tissue disks in electrolyte leakage into tap water. In contrast, Lieberman et al (84) showed that ion leakage at 20°C was five times as great from chilled sweet potato root tissue as from healthy tissue. Wheaton (191) observed increased leakage of potassium ions in beans and corn root tips at chilling temperatures but no increased leakage in chilling-resistant pea or wheat. These differences were apparent in as little as 3 to 6 hours after exposure to 1°C. This increased leakage of electrolytes has also been shown for *Coleus* (69), orange, grapefruit, and lime (130), cotton roots (27), cotyledons (50), and cucumber leaves (200). Christiansen et al (27) found an increased exudation from roots of cotton seedlings in response to chilling temperatures, but also showed that the exudation could be completely prevented (or even reversed, once started) by adding calcium or magnesium. These studies were also extended to show that comparison of root exudation among several genetic lines at chilling temperatures revealed differences which may provide basis for genetic selection to reduce chilling injury in cotton (26).

PHASE TRANSITIONS   Early workers noted that plants (and animals) originating in warm climates tended to have more saturated fatty acids in their lipids (132), and it was suggested that solidification of the protoplasmic lipids could account for observed death or injury at low temperatures (above 0°C) (131, 166). This concept was extended in recent studies focused on the mitochondria (75, 92, 93, 95, 145, 146), clearly showing that the membranes did undergo a physical-phase transition from a flexible liquid-crystalline to a solid-gel structure at 10° to 12°C—which correlates precisely with the temperature below which injury occurred in the sensitive species of tropical origin. Electron spin resonance (ESR) studies using spin-labeled compounds with intact mitochondria and with extracted phospholipids indicated that the lipids controlled the physical state of the membranes (107, 145). This change in the physical state of the membrane lipids imparts a conformational change in membrane-bound enzymes and can account for the discontinuities observed in the Arrhenius plots of mitochondrial oxidation discussed earlier (93); for discontinuities in phosphoenolpyruvate carboxylase in chilling-sensitive C_4 species (134), and tomato and corn chloroplasts (143, 155); and for ^{14}C-incorporation by microsomes (175) and ATPase systems (106) in warm-blooded animals. None of these phase transitions of membranes or membrane-bound enzymes are observed in chilling-resistant species. It was also recently reported that mitochondria from different apple cultivars undergo phase transitions at temperatures ranging from 3° to 10°C (105). Several of these cultivars, though not all, are susceptible to low-temperature breakdown during storage at these temperatures, and the existence of the phase transition indicates the same mechanism for low-temperature breakdown and chilling injury in sensitive species of tropical origin.

    Studies on the physical phase transition stemmed from the observation that membrane lipids from chilling-sensitive plant species tended to have a higher proportion of saturated to unsaturated fatty acids than do their resistant counterparts (95). The correlation between fatty acid composition and chilling sensitivity is not precise (92, 178, 201), but it has been shown for microorganisms that the fatty acid composition can determine the existence of a temperature-induced phase transition (43). In higher plants and animals this has been more difficult to elucidate; nevertheless it has been shown that dietary modifications can alter the fatty acid composition of mitochondrial membranes but have little effect on shifting the temperature of the phase transition in rat liver mitochondria (194), though it did shift the transition temperature considerably in sheep liver mitochondria (J. K. Raison, unpublished data). It is not yet completely clear whether the fatty acid composition of the membrane lipids determines the physical state in response to decreasing temperatures, as originally proposed, or whether other membrane components, such as sterols, exert an additional influence, as has been shown in model membranes for cholesterol (86).

THE MECHANISM OF CHILLING INJURY

A number of mechanisms have been proposed to accommodate the physiological and biochemical changes associated with chilling injury. Since a review of these changes indicates that chilling disrupts the entire metabolic and physiological process, however, it would almost appear futile to explain the various phenomena by some primary single change or "master reaction" controlling chilling injury (7, 26). Nevertheless, a single controlling response is found in the evidence that cellular membranes in sensitive plants undergo a physical-phase transition from a normal flexible liquid-crystalline to a solid gel structure at the temperature critical for chilling injury (93, 143, 145).

Figure 1 [modifying a previous presentation by Lyons & Raison (94) and incorporating some proposals by Levitt (78)] presents in schematic fashion the consequences resulting from the phase transition in cellular membranes and how this accommodates the known events in chilling. As the temperature is lowered in chilling-sensitive species, the membrane lipids solidify at the critical temperature (e.g. 10°–12°C in sensitive species of tropical origin), and the change in state would be expected to bring about a contraction that causes cracks or channels, leading to increased permeability. This immediate effect on permeability would lead to an upset in ion balance as well as account for the ion leakage that results from chilling in some tissues. The phase transition not only increases permeability, but also increases the Ea of membrane-bound enzyme systems, leading to a suppressed reaction rate and establishing an imbalance with nonmembrane-bound enzyme systems. For example, as temperature is decreased the rate of reaction of the soluble enzyme systems such as glycolysis will decrease with a constant Ea and a Q_{10} of about 2. Similarly, the rate of reaction of the membrane-associated enzymes of mitochondrial respiration will decrease with a Q_{10} of about 2 until the critical temperature at which the phase transition occurs. Below this critical temperature the membrane-bound enzyme system exhibits a marked increase in Ea, establishing a major imbalance in the two systems. Then metabolites such as pyruvate, acetaldehyde, and ethanol would be expected to accumulate at the interface between glycolysis and the mitochondrial system, and these compounds do indeed accumulate very early in chilling (121, 126, 130). Similar events can be projected for the chloroplast, where the phase transition leads to suppressed activity (155) and changes in metabolites after brief chilling (168). The external

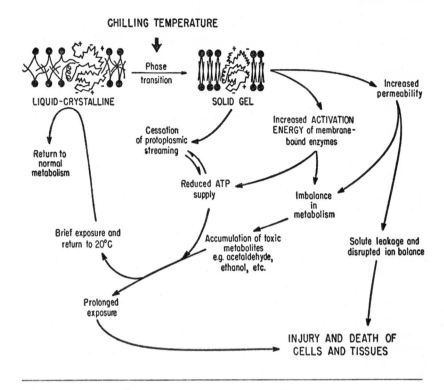

Figure 1 Schematic pathway of the events leading to chilling injury in sensitive plant tissues. The membrane model is an adaptation of that proposed by Singer (156).

symptoms of injury and ultimate death of the tissue would reflect the cell's inability to withstand increasing concentrations of these metabolites as a function of time. Apparent differences among species or cultivars in chilling sensitivity could be explained by different tolerances in withstanding or metabolizing these toxic compounds, even though the primary response is the same. For example, Watada & Morris (188) found that a number of snap bean cultivars were similar in respiratory pattern and were all injured by chilling temperatures but differed widely in visible symptoms. Similarly, McGlasson & Raison (105) showed a phase transition in all apple cultivars studied although some failed to develop symptoms, apparently being resistant to injury.

Also, a greatly reduced energy supply accompanying the suppressed mitochondrial respiration, along with the possibility of altered activity of the membrane-bound ATPase system (106), would greatly upset the normal energy balance of the cell. This altered energy supply, coupled with the rigidity of the membrane system following the phase transition, very easily accounts for the cessation of protoplasmic streaming observed in trichomes of sensitive species. It is also probably a critical factor in germination and seedling injury under chilling stress.

The temperature-induced phase change in the lipid portion of the membranes is completely reversible, though the effect on the whole organism is reversible only until the system incurs some degenerative injury. Thus, with a short chilling treatment followed by a warmer temperature, respiration increases sharply but only transiently, with the normal metabolism soon reestablished. If chilling temperatures continue long enough for degenerative changes to occur, however, the respiration rate remains elevated, reflecting a disrupted metabolism. These effects are the basis of the success that intermittent warming has given in ameliorating chilling effects, as discussed previously.

It is of interest that the symptoms of chilling injury are not specific to chilling but are identical to symptoms induced by suboxidation (124). It was suggested that suboxidation injury resulted from the accumulation of toxic products from disrupted metabolism, and Barker & Mapson (5) found an accumulation of fermentation products, including acetaldehyde and ethanol, in plant tissues held under anaerobic conditions. Thus, the visible symptoms of chilling and suboxidation injury are similar because they are both caused by the same mechanism—an accumulation of toxic metabolites—even though the accumulation stems from two entirely different effects.

The mechanism for chilling injury described here encompasses several elements operating independently or simultaneously: imbalances in metabolism, accumulation of toxic compounds, and increased permeability. Many reports (90, 135, 136, 153, 182) have focused on the formation of toxic substances, differential effects of temperature on the rate of vital enzymatic processes, and disequilibrium of reactions and changes in the velocities of interrelated chemical reactions. Levitt (78) proposed that all types of chilling injury can be the result of a change in cell permeability, and Christiansen (25) concluded that the initial impact of chilling on cotton is a rapid physical change in membranes to a freely permeable state. Similarly proposed as the mechanism of injury has been reduced ATP supply (97, 164, 165) or possibly a breakdown in the ATP-ADP transfer system leading in some cases to an accumulation of unavailable energy (130, 179).

These concepts can all be accommodated by the common event of a temperature-induced phase transition in the cellular membranes as the primary response in chilling injury.

Literature Cited

1. Amin, J. V. 1969. *Plant Soil* 31:365–73
2. Amin, J. V. 1969. *Physiol. Plant.* 22: 1184–91
3. Apeland, J. 1966. *Int. Inst. Refrig. Bull. 46, Annexe* 1:325–33
4. Appleman, C. O., Smith, C. L. 1936. *J. Agr. Res.* 53:557–80
5. Barker, J., Mapson, L. W. 1963. *Proc. Roy. Soc. London, Ser. B* 157(968):383–402
6. Barnell, H. R., Barnell, E. 1945. *Ann. Bot.* 9:77–99
7. Bělehrádek, J. 1935. *Protoplasma Monogr. 8.* 277 pp.
8. Bělehrádek, J. 1957. *Ann. Rev. Physiol.* 19:59–82
9. Biale, J. B. 1941. *Proc. Am. Soc. Hort. Sci.* 39:137–42
10. Biale, J. B. 1950. *Ann. Rev. Plant Physiol.* 1:183–206
11. Biale, J. B. 1960. *Advan. Food Res.* 10:293–354
12. Biale, J. B. 1960. *Handb. Pflanzenphysiol.* 12:536–92
13. Biale, J. B. 1961. *The Orange*, ed. W. B. Sinclair, 96–130. Berkeley: Univ. Calif. Div. Agr. Sci. 475 pp.
14. Biale, J. B., Young, R. E. 1971. See Ref. 61, 2:2–63
15. Biale, J. B., Young, R. E., Olmstead, A. J. 1954. *Plant Physiol.* 29:168–74
16. Björkman, O., Pearcy, R. W. 1971. *Carnegie Inst. Yearb.* 70:511–20
17. Boyes, W. W. 1952. *Deciduous Fruit Grow.* 2:13–17
18. Burg, S. P., Burg, E. A. 1966. *Science* 153:314–15
19. Chhatpar, H. S., Mattoo, A. K., Modi, V. V. 1971. *Phytochemistry* 10:1007–9
20. Christiansen, M. N. 1963. *Plant Physiol.* 38:520–22
21. Christiansen, M. N. 1964. *Crop Sci.* 4:584–86
22. Christiansen, M. N. 1967. *Plant Physiol.* 42:431–33
23. Ibid 1968. 43:743–46
24. Christiansen, M. N. 1969. *Proc. Nat. Cotton Counc. Beltwide Cotton Prod. Res. Conf.*, 127–30
25. Ibid 1971. 71–72
26. Ibid 1972. In press
27. Christiansen, M. N., Carns, H. R., Slyter, D. J. 1970. *Plant Physiol.* 46: 53–56
28. Christiansen, M. N., Thomas, R. O. 1969. *Crop Sci.* 9:672–73
29. Cooper, W. C., Rasmussen, G. K., Waldon, E. S. 1969. *Plant Physiol.* 44:1194–96
30. Cooper, W. C., Reece, P. C. 1969. *Proc. Fla. State Hort. Soc.* 82:270–73
31. Cordner, H. B., Matthews, W. A. 1930. *Proc. Am. Soc. Hort. Sci.* 27: 520–23
32. Craft, C. C., Heinze, P. H. 1954. *Proc. Am. Soc. Hort. Sci.* 64:343–50
33. Creencia, R. P., Bramlage, W. J. 1971. *Plant Physiol.* 47:389–92
34. Davies, R., Boyes, W. W., Beyers, E. 1936. *S. Afr. Dep. Agr. Forestry Rep. Low Temp. Res. Lab. Capetown 1934–35*, 78–114
35. Davies, R., Boyes, W. W., de Villiers, D. J. R. 1937. *S. Afr. Dep. Agr. Forestry Rep. Low Temp. Res. Lab. Capetown 1935–36*, 130–60
36. Drew, M. C., Biddulph, O. 1971. *Plant Physiol.* 48:426–32
37. Dull, G. G. 1971. See Ref. 61, 2: 303–24
38. Eaks, I. L. 1956. *Proc. Am. Soc. Hort. Sci.* 67:473–78
39. Eaks, I. L. 1960. *Plant Physiol.* 35: 632–36
40. Eaks, I. L. 1965. *Proc. Am. Soc. Hort. Sci.* 87:181–86
41. Eaks, I. L., Morris, L. L. 1956. *Plant Physiol.* 31:308–14
42. Eaks, I. L., Morris, L. L. 1957. *Proc. Am. Soc. Hort. Sci.* 69:388–99
43. Esfahani, M., Limbrick, A. R., Knutton, S., Oka, T., Wakil, S. J. 1971. *Proc. Nat. Acad. Sci. USA* 68: 3180–84
44. Ewart, A. J. 1895–97. *J. Linn. Soc. London Bot.* 31:364–461
45. Ewart, A. J. 1903. *On the Physics and Physiology of Protoplasmic Streaming in Plants.* Oxford: Clarendon. 131 pp.
46. Ezell, B. D., Wilcox, M. S. 1952. *Plant Physiol.* 27:81–94
47. Ezell, B. D., Wilcox, M. S., Crowder, J. S. 1952. *Plant Physiol.* 27:355–69
48. Fidler, J. C. 1968. *Low Temperature Biology of Foodstuffs*, ed. J. Hawthorne, E. J. Rolfe. *Rec. Advan. Food Sci.* 4:271–83
49. Grierson, W. *Proc. Am. Soc. Hort. Sci. Trop. Reg.* In press
50. Guinn, G. 1971. *Crop Sci.* 11:262–65
51. Hansen, E. 1966. *Ann. Rev. Plant Physiol.* 17:459–80
52. Harrington, J. F. 1963. *Proc. 16th Int. Hort. Congr.* 2:435–41
53. Harrington, J. F., Kihara, G. M. 1960. *Proc. Am. Soc. Hort. Sci.* 75: 485–89
54. Hartt, C. E. 1965. *Plant Physiol.* 40: 74–81

55. Hatton, T. T. Jr., Reeder, W. F. 1967. *Proc. Am. Soc. Hort. Sci. Trop. Reg.* 11:23-32
56. Hobson, G. E., Davies, J. N. 1971. See Ref. 61, 2:437-82
57. Hruschka, H. W., Smith, W. L., Baker, J. E. 1967. *Plant Dis. Rep.* 51:1014-16
58. Huelin, F. E., Coggiola, I. M. 1970. *J. Sci. Food Agr.* 21:44-48
59. Ibid, 82-86
60. Ibid, 584-89
61. Hulme, A. C., Ed. 1971. *The Biochemistry of Fruits and Their Products,* 2:233-54. New York: Academic. 788 pp.
62. Hulme, A. C., Smith, W. H., Wooltorton, L. S. C. 1964. *J. Sci. Food Agr.* 15:303-7
63. Ibañez, M. L. 1963. *Turrialba* 13: 127-28
64. Ibañez, M. L. 1964. *Nature, London* 201:414-15
65. Ibañez, M. L. 1963. *Turrialba* 13: 31-32
66. Jansen, R. D., Taylor, S. A. 1961. *Plant Physiol.* 36:639-42
67. Jones, W. W. 1942. *Plant Physiol.* 17:481-86
68. Kamiya, N. 1962. *Handb. Pflanzenphysiol.* 17:979-1035
69. Katz, S., Reinhold, L. 1965. *Isr. J. Bot.* 13:105-14
70. Kayser, C. 1957. *Ann. Rev. Physiol.* 19:83-120
71. Kidd, F., West, C. 1927. *Gt. Brit. D.S.I.R. Food Invest. Board. Spec. Rep. (1925-26),* 37-41
72. Kozukue, N., Ogata, K. 1971. *J. Jap. Soc. Hort. Sci.* 40:300-4 (in Japanese)
73. Ibid, 416-20
74. Kramer, P. J. 1942. *Am. J. Bot.* 29: 828-32
75. Kumamoto, J., Raison, J. K., Lyons, J. M. 1971. *J. Theor. Biol.* 31:47-51
76. Kushnirenko, S. V. 1961. *Sov. Plant Physiol.* 8:268-74
77. Langridge, J. 1963. *Ann. Rev. Plant Physiol.* 14:441-62
78. Levitt, J. 1972. *Responses of Plants to Environmental Stresses.* New York: Academic. 697 pp.
79. Lewis, D. A. 1956. *Physiological studies of tomato fruits injured by holding at chilling temperatures.* PhD thesis. Univ. California, Davis. 153 pp.
80. Lewis, D. A. 1956. *Science* 124:75-76
81. Lewis, D. A., Morris, L. L. 1956. *Proc. Am. Soc. Hort. Sci.* 68:421-28
82. Lewis, T. L., Workman, M. 1964. *Aust. J. Biol. Sci.* 17:147-52
83. Lieberman, M., Biale, J. B. 1956. *Plant Physiol.* 31:420-24
84. Lieberman, M., Craft, C. C., Audia, W. V., Wilcox, M. S. 1958. *Plant Physiol.* 33:307-11
85. Lieberman, M., Craft, C. C., Wilcox, M. S. 1959. *Proc. Am. Soc. Hort. Sci.* 74:642-48
86. Long, R. A., Hruska, F. E., Gesser, H. D. 1971. *Biochem. Biophys. Res. Commun.* 45:167-73
87. Lorenz, O. A. 1951. *Proc. Am. Soc. Hort. Sci.* 57:288-94
88. Lutz, J. M. 1945. *US Dep. Agr. Circ. 729.* 8 pp.
89. Lutz, J. M., Hardenburg, R. E. 1968. *US Dep. Agr., Agr. Handb. 66.* 94 pp.
90. Luyet, B. J., Gehenio, P. M. 1940. *Biodynamica* 3:33-99
91. Lyons, J. M. *Cryobiology.* 9:341-50
92. Lyons, J. M., Asmundson, C. M. 1965. *J. Am. Oil Chem. Soc.* 42:1056-58
93. Lyons, J. M., Raison, J. K. 1970. *Plant Physiol.* 45:386-89
94. Lyons, J. M., Raison, J. K. *Proc. 13th Int. Cong. Refrig.* 3:167-72
95. Lyons, J. M., Wheaton, T. A., Pratt, H. K. 1964. *Plant Physiol.* 39:262-68
96. Mack, W. B., Janer, J. R. 1942. *Food Res.* 7:38-47
97. Mayland, H. F., Cary, J. W. 1970. *Advan. Agron.* 22:203-34
98. Mazur, P. 1969. *Ann. Rev. Plant Physiol.* 20:419-48
99. McClure, T. T. 1959. *Phytopathology* 49:359-61
100. McColloch, L. P. 1953. *US Dep. Agr. Yearb.,* 826-30
101. McColloch, L. P. 1962. *US Dep. Agr. Mkt. Res. Rep. 518.* 19 pp.
102. McColloch, L. P. 1962. *US Dep. Agr. Mkt. Res. Rep. 536.* 16 pp.
103. McColloch, L. P. 1966. *US Dep. Agr. Mkt. Res. Rep. 749.* 5 pp.
104. McColloch, L. P., Worthington, J. T. 1952. *Phytopathology* 42:425-27
105. McGlasson, W. B., Raison, J. K. *Plant Physiol.* Submitted 1972
106. McMurchie, E. J., Raison, J. K., Cairncross, K. D. *Comp. Biochem. Physiol.* In press
107. Mehlhorn, R. J., Keith, A. D. 1972. *Membrane Molecular Biology,* ed. C. F. Fox, A. D. Keith, 192-227. Stamford, Conn.: Sinauer. 525 pp.
108. Meigh, D. F. 1970. See Ref. 61, 1: 555-69
109. Miller, E. V. 1946. *Bot. Rev.* 12:393-423

110. Miller, E. V. 1951. *Plant Physiol.* 26: 66–75
111. Miller, E. V. 1958. *Bot. Rev.* 24:43–59
112. Miller, E. V., Heilman, A. S. 1952. *Science* 116:505–6
113. Miller, E. V., Schomer, H. A. 1939. *J. Agr. Res.* 59:601–8
114. Millerd, A., Goodchild, D., J. Spencer, D. 1969. *Plant Physiol.* 44:567–83
115. Minamikawa, T., Akazawa, T., Uritani, I. 1961. *Plant Cell Physiol.* 2: 301–9
116. Mohapatra, N., Smith, E. W., Fites, R. C., Noggle, G. R. 1970. *Biochem. Biophys. Res. Commun.* 40:1253–58
117. Molisch, H. 1896. *Sitzungsber. Kaiserl. Akad. Wiss. Wien., Math.-Naturwiss. Kl.* 105:82–95
118. Molisch, H. 1897. *Untersuchungen über das Erfrieren der Pflanzen.* Jena: Fischer. 73 pp.
119. Morris, L. L. 1954. *Proc. Conf. Transp. Perishables, Univ. Calif., Davis,* 101–5
120. Morris, L. L., Platenius, H. 1938. *Proc. Am. Soc. Hort. Sci.* 36:609–13
121. Murata, T. 1969. *Physiol. Plant.* 22: 401–11
122. Murata, T., Ku, H. S. 1966. *J. Food Sci. Technol.* 13:466–71 (in Japanese)
123. Murata, T., Ogata, K. 1966. *J. Food Sci. Technol.* 13:367–70 (in Japanese)
124. Nelson, R. 1926. *Mich. Agr. Exp. Sta. Tech. Bull. 81.* 38 pp.
125. Nelson, R. 1933. *J. Agr. Res.* 46: 695–713
126. Ogata, K., Kozukue, N., Murata, T. 1968. *J. Jap. Soc. Hort. Sci.* 37:65–71 (in Japanese)
127. Oudit, D. D., McGlasson, W. B., Lee, T. H., Hall, E. G., Edwards, R. A. *J. Hort. Sci.* Submitted 1972
128. Palmer, J. K. 1971. See Ref. 61, 2: 65–105
129. Pantastico, E. B., Grierson, W., Soule, J. 1967. *Proc. Am. Soc. Hort. Sci. Trop. Reg.* 11:82–91
130. Pantastico, E. B., Soule, J., Grierson, W. 1968. *Proc. Am. Soc. Hort. Sci. Trop. Reg.* 12:171–83
131. Payne, N. M. 1930. *Ecology* 11:500–4
132. Pearson, L. K., Raper, H. S. 1927. *Biochem. J.* 21:875–79
133. Pentzer, W. T., Heinze, P. H. 1954. *Ann. Rev. Plant Physiol.* 5:205–44
134. Phillips, P. J., McWilliam, J. R. 1971. *Photosynthesis and Photorespiration,* ed. M. D. Hatch, C. B. Osmond,

R. O. Slatyer, 97–104. N. Y.: Wiley 565 pp.
135. Plank, R. 1938. *Food Res.* 3:175–87
136. Plank, R. 1941. *Planta* 32:364–90
137. Platenius, H. 1939. *J. Agr. Res.* 59: 41–58
138. Platenius, H. 1942. *Plant Physiol.* 17: 179–97
139. Pollock, B. M. 1969. *Plant Physiol.* 44:907–11
140. Pollock, B. M., Toole, V. K. 1966. *Plant Physiol.* 41:221–29
141. Pratt, H. K., Biale, J. B. 1944. *Plant Physiol.* 19:519–28
142. Raison, J. K. 1972. *J. Bioenergetics* 4:357–81
143. Raison, J. K. *Soc. Exp. Biol. Symp., 27th.* In press
144. Raison, J. K., Lyons, J. M. 1970. *Plant Physiol.* 45:382–85
145. Raison, J. K., Lyons, J. M., Mehlhorn, R. J., Keith, A. D. 1971. *J. Biol. Chem.* 246:4036–40
146. Raison, J. K., Lyons, J. M., Thomson, W. W. 1971. *Arch. Biochem. Biophys.* 142:83–90
147. Rosa, J. T. 1926. Ripening and storage of tomatoes. *Proc. Am. Soc. Hort. Sci.* 23:233–42
148. Ryall, A. L., Lipton, W. J. 1972. *Handling, Transportation, and Storage of Fruits and Vegetables,* Vol. 1: *Vegetables and Melons.* Westport, Conn.: Avi. 473 pp.
149. Sachs, J. von 1865. *Handbuch der Experimental-Physiologie der Pflanzen.* Leipzig: Engelman. 514 pp.
150. Schiffmann-Nadel, M., Chalutz, E., Waks, J., Latter, F. S. 1972. *J. Hort. Sci.* 7:394–95
151. Seible, D. 1939. *Beitr. Biol. Pflanz.* 26:289–330
152. Seifriz, W. 1943. *Bot. Rev.* 9:49–123
153. Sellschop, J. P. F., Salmon, S. C. 1928. *J. Agr. Res.* 37:315–38
154. Shichi, H., Uritani, I. 1956. *Bull. Agr. Chem. Soc. Jap.* 20:284–88
155. Shneyour, A., Raison, J. K., Smillie, R. M. *Biochim. Biophys. Acta.* In press
156. Singer, S. J. 1971. *Structure and Function of Biological Membranes,* ed. L. I. Rothfield, 145–222. N. Y.: Academic. 486 pp.
157. Singh, K. K., Mathur, P. B. 1954. *Indian J. Hort.* 11:1–5
158. Smith, A. J. M. 1950. *J. Hort. Sci.* 25:132–44
159. Smith, E. W., Fites, R. C., Noggle, G. R. 1971. *Proc. 1971 Nat. Cotton Counc. Beltwide Cotton Prod. Res. Conf.,* 45–47

160. Smith, P. G., Millet, A. H. 1964. *Proc. Am. Soc. Hort. Sci.* 84:480–84
161. Smith, W. H. 1947. *Nature* 159:541–42
162. Smock, R. M. 1957. *Proc. Am. Soc. Hort. Sci.* 69:91–100
163. Spranger, E. 1941. *Gartenbauwissenschaft* 16:90–128
164. Stewart, J. McD., Guinn, G. 1969. *Plant Physiol.* 44:605–8
165. Ibid 1971. 48:166–70
166. Tait, J. 1922. *Am. J. Physiol.* 59:467
167. Taylor, A. O., Craig, A. S. 1971. *Plant Physiol.* 47:719–25
168. Taylor, A. O., Jepsen, N. M., Christeller, J. T. 1972. *Plant Physiol.* 49:798–802
169. Taylor, A. O., Rowley, J. A. 1971. *Plant Physiol.* 47:713–18
170. Thomas, R. O., Christiansen, M. N. 1971. *Crop. Sci.* 11:454–56
171. Ting, S. V., Attaway, J. A. 1971. See Ref. 61, 2:107–69
172. Toda, M. 1962. *Proc. Crop Sci. Soc. Jap.* 30:241–44
173. Tolle, W. E. 1969. *US Dep. Agr. Mkt. Res. Rep. 842.* 9 pp.
174. Tomkins, R. G. 1963. *J. Hort. Sci.* 38:335–47
175. Towers, N., Kellerman, G., Raison, J. K., Linnane, A. W. *Biochim. Biophys. Acta.* In press
176. Ulrich, R. 1958. *Ann. Rev. Plant Physiol.* 9:385–416
177. Uritani, I., Hyodo, H., Kuwano, M. 1971. *Agr. Biol. Chem.* 1248–53
178. Uritani, I., Yamaki, S. 1969. *Agr. Biol. Chem.* 33:480–87
179. Vakis, N., Grierson, W., Soule, J. 1970. *Proc. Am. Soc. Hort. Sci. Trop. Reg.* 14:89–100
180. Vakis, N., Grierson, W., Soule, J., Albrigo, L. G. 1970. *Hort. Sci.* 5:472–73
181. Vakis, N., Soule, J., Biggs, R. H., Grierson, W. 1970. *Proc. Fla. State Hort. Soc.* 83:304–10
182. Van der Plank, J. E., Davies, R. 1937. *J. Pomol. Hort. Sci.* 15:226–47
183. Von Loesecke, H. W. 1949. *Bananas; Chemistry, Physiology, Technology,* 59–66. N. Y.: Interscience. 189 pp.

184. Wardlaw, C. W. 1937. *Trop. Agr. Trinidad* 14:131–39
185. Ibid 1938. 15:171–73
186. Wardlaw, C. W. 1961. *Banana Diseases, Including Plantains and Abaca.* London: Longmans. 648 pp.
187. Wardlaw, C. W., Leonard, E. R., Baker, P. E. D. 1934. *Trop. Agr. Trinidad* 11:196–200
188. Watada, A. E., Morris, L. L. 1966. *Proc. Am. Soc. Hort. Sci.* 89:368–74
189. Ibid, 375–80
190. Weiser, C. J. 1970. *Science* 169:1269–78
191. Wheaton, T. A. 1963. *Physiological comparison of plants sensitive and insensitive to chilling temperatures.* PhD thesis. Univ. California, Davis. 93 pp.
192. Wheaton, T. A., Morris, L. L. 1967. *Proc. Am. Soc. Hort. Sci.* 91:529–33
193. Wilkinson, B. G. 1970. See Ref. 61, 1:537–54
194. Williams, M. A., Stancliff, R. C., Packer, L., Keith, A. D. 1972. *Biochim. Biophys. Acta* 267:444–56
195. Wills, R. B. H. 1968. *J. Sci. Food Agr.* 19:354–56
196. Wills, R. B. H., McGlasson, W. B. 1968. *Phytochemistry* 7:733–39
197. Wills, R. B. H., McGlasson, W. B. 1969. *J. Sci. Food Agr.* 20:446–47
198. Wills, R. B. H., Scott, K. J. 1971. *Phytochemistry* 10:1783–85
199. Wills, R. B. H., Scott, K. J., McGlasson, W. B. 1970. *J. Sci. Food Agr.* 21:42–44
200. Wright, M., Simon, E. W. *J. Exp. Bot.* In press
201. Yamaki, S., Uritani, I. 1972. *Agr. Biol. Chem.* 36:47–55
202. Yamaki, S., Uritani, I. 1972. *Plant Cell Physiol.* 13:67–79
203. Ibid, 795–805
204. Yamaki, S., Uritani, I. *Agr. Biol. Chem.* In press
205. Yamaki, S., Uritani, I. *Plant Physiol.* In press
206. Yoshioka, K., Honda, K. 1970. *J. Food Sci. Technol.* 17:489–95 (in Japanese)
207. Ibid 1972. 19:131–38

Ann. Rev. Plant Physiol. 1973. 24:467–92

THE FATE OF PESTICIDES IN
THE ENVIRONMENT

❖ 7556

Donald G. Crosby[1]

Department of Environmental Toxicology, University of California, Davis

CONTENTS

Contrary to popular opinion, chemicals have been used for thousands of years to control unwanted animals, plants, and microorganisms—the "pests." Well into the age of synthetic organic compounds, mankind was so preoccupied with the effectiveness of his so-called "pesticides" that there was little thought of what eventually would happen to the steadily increasing volume and number of these chemicals released into the environment.

Although preceded by a limited interest (e.g. 214), the controversial book *Silent Spring* (31) represented a landmark of published concern over the environmental fate of pesticides. While overshadowed by a prodigious volume of semitechnical literature, a few technical reviews of the general subject finally have appeared in recent years (75, 157, 161, 190, 225, 234); more specific reviews will be mentioned

[1] Preparation of this review was supported in part by a research grant (GB-33723X) from the National Science Foundation.

later. Rather than attempt the monumental task of an exhaustive coverage or emphasize some particular aspect of that fate, the present review offers only some available possibilities and key references. However, it should be mentioned at the outset that there is no instance in which we know the total fate of any pesticide or even its fate in a single environmental compartment.

The term "pesticide" encompasses chemical agents used to control a wide variety of creatures inimical or just bothersome to man—insects, mites, rodents, nematodes, weeds, fungi, and others. The individual active agents generally are designated by accepted common names (110, 129); their chemical names also may be found in several other current compilations (84, 151, 221). In deference to the special interests of readers of this series, plant-related aspects and plant-regulating chemicals will receive particular attention, but the discussion could as well describe a host of other xenobiotics—substances foreign to our natural environment—and even the usual distinction of pesticides according to use (herbicides, insecticides, etc) generally is not crucial to considerations of their environmental fate.

It is claimed that more than 900 different chemicals are used as pesticides in the United States (174). Fortunately for those concerned with environmental studies, the majority can be classified into about 25 basic chemical types. Armed with knowledge of type reactions and the influence of the limited number of substituent variations superimposed upon them, investigation and prediction of their environmental fate may yet prove feasible. However, these "active ingradients" are formulated into thousands of mixtures— dusts, wettable powders, emulsifiable concentrates, etc—which are the products of commerce. Although the type of formulation can exert a powerful influence on the dispersal and breakdown of its pesticides (5, 72, 85), this complex subject must be all but ignored here because of space limitations.

For proper perspective, one must appreciate the volume of pesticides used now and in the past. Table 1 lists the applications of some common insecticides, herbicides, and fungicides reported to the California Department of Agriculture in 1970 (28), the year preceding drastic restriction of several of them; actual use undoubtedly was substantially larger, but these data currently are as comprehensive and indicative as any in our country. Estimates of past use in the state are even more inaccurate, but certainly at least 20 million pounds of each of the principal pesticides must have been injected into the California environment. California's total use of pesticides—including large quantities of inorganic chemicals—totaled more than 140 million pounds in 1970 (28).

Despite the widespread application to crops in past years, scattered monitoring data indicate that, with several notable exceptions, pesticide levels in soil (215), water (13, 19), and biota (148) have not built up. However, application efficiency often has been very low. Published estimates (43, 45) have suggested that as little as 20% of an application reaches its target, but the actual efficiency can be shown to be far less than that. For example, residue analysis for methoxychlor applied to alfalfa at a rate of 1.5 lbs/acre revealed only 24 ppm (about 3% efficiency) reaching the leaf surface (83). Between application and harvest, this amount declined to below 1 ppm, leaving less than 0.2% available for insecticidal action. Since as little as 10 μg of the insecticide provides a lethal dose for many insect pests (177), probably less than 500 mg of methoxychlor actually would be utilized—a pesticidal efficiency of less than 0.05%.

Where did the rest go? It was distributed between air, soil, water, and other biota, but eventually *all* of it—even the minute proportions which killed the insects—was subjected to biological and nonbiological transformations in the environment.

NONBIOLOGICAL TRANSFORMATIONS

Pesticides are ordinary compounds that possess some reactive or structural feature which confers antibiotic activity. Otherwise, they exhibit the expected chemical reactions (155) when provided with the necessary reagents and energy. Our environment suffers no shortage of such reagents: oxygen; the reducing power of both inorganic and hydrogen-rich organic compounds; water and powerful nucleophiles such as hydroxide ions, thiols, and amines; and a variety of substances recognized for their catalytic properties. Consequently, oxidation, reduction, hydrolysis and other nucleophilic reactions, and certain elimination and isomerization reactions can be expected to play an important part in nonbiological transformations.

Energy, too, is available in several forms. For example, the summer temperature of a metal or mineral surface can approach the boiling point of water; soil surfaces in the Sacramento Valley often exceed 50°C; and even a leaf surface can reach 40°C. Sunlight, however, appears more important than heat for pesticide transformations, and several reviews of pesticide photochemistry have appeared (51, 52, 57, 192). The ultraviolet (uv) light which reaches the earth's surface

Table 1 Typical 1970 use of pesticides in California[a]

Active ingredient (AI)	Use[b]	Pounds of AI	Acres treated
Toxaphene	I	2,659,274	753,725
Parathion/Methylparathion	I	2,102,734	3,221,083
DDT	I	1,164,700	643,899
Carbaryl (Sevin)	I	936,059	440,302
Disulfoton	I	212,551	219,948
2,4-D and esters	H	1,217,378	1,322,630
Simazine	H	312,280	211,755
Trifluralin	H	240,906	255,949
D-D/Telone	N	5,726,005	37,019
Sulfur	F	14,191,375	851,813
Maneb	F	503,248	232,027

[a] Data selected from (28).

[b] Insecticide (I), herbicide (H), nematicide (N), fungicide (F).

presents a wavelength range delimited by atmospheric absorption (290 nm) at one end and by low energy (about 450 nm) at the other (136) but capable of driving many types of reactions. Ultraviolet intensity is affected by season (about a six-fold increase at 290–313 nm between December and June in Washington, D.C.), latitude (Barrow, Alaska, receives half again as much total daily energy in June as does Miami, Florida), altitude (approximately doubling for each km of elevation), and climate. As beachgoers will testify, clouds are rather transparent to uv, and shaded areas may be exposed to as much uv light dispersed by the open sky as comes from direct sunlight (136).

Most pesticides are benzenoid compounds which absorb appreciable energy within the 290–450 nm region. However, photochemical sensitization (29) also may allow transformations of substances which otherwise would be stable to light. The notorious unreliability of sunlight and other difficulties in conducting environmental experiments with pesticide photolysis have resulted in a variety of techniques and equipment aimed at environmental simulation (49, 57). In fact, most of the examples in this section represent artificial situations; the photochemistry is valid enough, but there still is little evidence of its operation in the real world.

Oxidation

With the exception of a few locations such as certain soils and perhaps the ocean deeps, oxygen is an ubiquitous force for environmental transformations of pesticides. Most common reactions of oxygen reflect the free-radical character of its usual (triplet) electronic state, but in recent years the reactions and environmental significance of the more reactive double-bonded (singlet) oxygen produced by both light and chemical means has received increasing attention (191).

Pesticide photooxidation at aliphatic carbons is common (51). The first product usually is a peroxy radical which can abstract a hydrogen atom from its surroundings to form a hydroperoxide, rearrange to a carbonyl group, or be reduced (again by a hydrogen-rich medium) to form an alcohol. Methoxychlor underwent photooxidation to 4,4'-dimethoxybenzophenone, p-anisic acid, and p-methoxyphenol (48, 141), exactly the products to be expected from decomposition of a peroxy intermediate. Its close relative DDT formed the analogous dichlorobenzophenone and p-chlorobenzoic acid (101), but there is evidence both for and against a similar peroxy intermediate (64, 162, 173).

DDT normally is quite resistent to oxidation—photochemical or otherwise—under environmental temperatures and wavelengths of light, but a number of photosensitizers promoted appreciable breakdown of DDT in water (138) and on leaf surfaces (113). The extent to which photosensitization operates in nature is entirely unknown at present, but the action of a variety of common natural products including riboflavin, chlorophyll, anthracene, anthraquinone, xanthone, and even the natural insecticide rotenone (112) to induce photodecomposition of pesticides which normally would resist or be incapable of this process (147), suggests that sensitized photooxidations may be of major importance.

Rotenone itself was rapidly C-oxidized in both light (38) and the dark (144), a property which has limited the utility of this and other natural insecticides including the pyrethrins (37). A number of herbicidal acids and their salts—naphthaleneacetic acid (NAA) (59), fenac (55), and the phenoxy compounds 2,4-D and 4-CPA (62, 63)—also underwent photooxidation, although the initial reaction in this case may be loss of the carboxyl group as CO_2 with generation of a solvated electron (118).

Carbon atoms adjacent to hetero atoms seem especially prone to photooxidation. Typified by monuron (60), the result usually is dealkylation, and the photolysis of trifluralin (139), zectran (1), and diphenamid (200) attest to the diversity of N-alkyl groups which can be removed photochemically under mild conditions.

Epoxidation of the isolated double bond of aldrin, heptachlor, and trichloroethylene is one of the most obvious of the environmental transformations of pesticides, although its mechanism is poorly understood. Oxidation at an allylic position adjacent to a double bond by either radical (triplet) (67) or singlet (91) oxygen is commonplace, but apparently cannot operate with these pesticides. However, their epoxidation by electropositive oxidants such as peracids also is well known (67) and may provide a mechanism by which oxygen is able to convert even highly hindered olefinic pesticides to epoxides with ease (231).

Aromatic rings may react with oxygen in the presence of light; for example, monuron was hydroxylated in the position *ortho* to the urea sidechain (60). Again, this type of oxygenation resembles the action of peracids (231), although radical-generating systems such as Fenton's reagent also formed similar products.

The hetero atoms present in many pesticides often are oxidized nonbiologically. For example, the thioether (sulfide) link in aliphatic pesticides such as demeton (42) or in aromatic compounds such as abate (203) was oxidized to the corresponding sulfoxides and sulfones by air, and the cyclic sulfite insecticide endosulfan formed the corresponding sulfate (101). The environmental conversion of phosphorothionates and phosphonothionates to the corresponding oxygen analogs (oxons) has been widely observed and is due in part to light and air (66, 135). Simple dithiocarbamates were oxidized to thiuram disulfides, while the ethylene bis-dithiocarbamates formed elementary sulfur, et¹ ylenethiuram monosulfides, and polymers (146).

Amines undergo air oxidation, and the photolysis of monuron in water produced 4-chloronitrobenzene (48) via its hydrolysis product 4-chloroaniline. The photooxidation of other aromatic amines resulted in the formation of azobenzenes (168), especially when sensitized (202), and the phosphit. herbicide Falone was converted to the phosphate as was the phosphorotrithioite defoliant Folex (155).

Reduction

The abstraction of hydrogen atoms from organic solvents by free radica¹· is well known. In recent years, photochemical study of pesticides revealed that energetic irradiation in methanol, for example, led to replacement of aromatic ring halides with hydrogens. Chlorinated (192) and iodinated (114) benzoic acids and their esters were easily reduced [the halogen adjacent (*ortho*) to the sidechain was the most labile], dichlobenil (2,6-dichlorobenzonitrile) was converted to o-chloro-benzonitrile and benzonitrile (192), and pentachlorophenol (PCP) formed principally 2,3,5,6-tetrachlorophenol (53). Methylthiotriazines such as prometryne were reductively desulfurized (184), and aromatic nitro compounds such as trifluralin were reduced to amines (193).

However, similar reductions occur even in water. Irradiation of aqueous solutions of some of the same compounds similarly resulted in photoreduction; chlorinated benzoic (54) and phenylacetic (55) acids were dechlorinated, PCP formed tetrachlorophenols (244), and dexon lost its side chain to give dimethylaniline (58, 166). These reactions are difficult to interpret as free-radical hydrogen abstractions: there would seem to be insufficient energy available for homolytic cleavage of the H–OH bond, and the reactions take place in the presence of excess oxygen which is a good radical scavenger.

Strictly chemical reductions—by transition metal ions, for example—probably are widespread, but few have been reported. A particularly significant instance was the reduction of DDT to DDD by iron porphyrins (33, 163), a reaction which surely must be in general operation considering the wide distribution of both the chlorinated hydrocarbons and the porphyrins.

Hydrolysis and Other Nucleophilic Reactions

Like air, water is present to some extent almost everywhere. Its heat capacity, polarity, solvent properties, and transparency to uv light make it an important medium for chemical reactions. Of course, it also is chemically reactive, especially as it dissociates into hydroxide and hydronium ions; the pH normally varies between about 5 and 8, but a significant proportion of U.S. cities receive drinking water with a pH greater than 9 (232).

Pesticide degradation by alkaline hydrolysis often occurs readily, and it is surprising that the extent of these environmental transformations has received so little scientific attention. Esters generally undergo hydrolysis: dinitrophenol esters such as binapacryl were rapidly converted to their more toxic phenols; the cyclic sulfite endosulfan formed the cis-diol; carbamate insecticides gave the phenol, amine, and CO_2; and organophosphorus esters formed substituted phosphoric acids as well as a variety of alcohols, phenols, and thiols (155). Phosphates were more reactive than phosphorothionates; for example, demeton was 9.4% hydrolyzed in 60 min at pH 7.9 and 37°C, while its oxygen analog was 39% hydrolyzed under the same conditions (87). The hydrolysis of phosphate esters was catalyzed by many natural amino acids, metal ions, and chelates; Cu^{+2} at 3×10^{-7} M caused a twentyfold increase in the hydrolysis rate of 10^{-4} M parathion (130).

The thiadiazine ring of DMTT (Mylone) was opened by mild hydrolysis, and a subsequent elimination reaction formed methyl isothiocyanate (68). Carbothion (Vapam) also decomposed to methyl isothiocyanate in water at a rate increasing with dilution (155). Other typical decomposition reactions in dilute aqueous alkali include elimination [malathion to diethyl fumarate (180)], reverse aldol [dicofol to dichlorobenzophenone and chloroform (207)], decarboxylation [trichloroacetic acid (TCA) to chloroform (155)], and Michael addition of hydroxide ion to paraquat and diquat (followed by oxidation) (218).

In dilute base, a number of halogenated pesticides were hydrolyzed to the corresponding hydroxy compound; chloroacetic acid formed glycolic acid, dalapon provided pyruvic acid, and chlorotriazines gave hydroxytriazines (155). However, under the influence of light, even the normally stable ring chlorines of many aromatic pesticides underwent nucleophilic displacement of halide ion with formation of the corresponding phenol. Chlorinated phenoxyacetic (62, 63), benzoic (54), and phenylacetic acid herbicides (86), propanil (168), and metobromuron (205) gave hydroxy analogs, and multiple chlorines were replaced stepwise; pentachlorophenol (PCP) provided tetrachlororesorcinol, chloranil (presumably from tetrachlorohydroquinone), and chloroanilic acid (175, 244). Other substituents, including the diazosulfonate group of dexon (58, 166) and the nitrophenate group of nitrofen (58), were replaced by hydroxide upon sunlight irradiation in water.

Of course, environmental substrates are rich in other reactive nucleophiles. Ross (206) estimated that 99.9% of the guanine amino group, 76–99% of the imidazole group of histidine, 1–6% of the sulfhydryl of glutathione, and 0.1–1% of the amino endgroup of lysine are in nucleophilic form at pH 7.5. In water, captan reacted rapidly with sulfhydryl compounds with displacement of the side chain as thiophosgene (134, 183), and other pesticides which reacted with the sulfhydryl group under mild conditions included dichlone (183), dyrene (26), and organophosphorus esters (105). Natural amines reacted with methyl bromide (241) and ethylene oxide (240).

Other Nonbiological Reactions

Quite a variety of other pesticide reactions have been reported to take place under so-called environmental conditions. Dieldrin underwent an internal cyclization in light to form photodieldrin, and aldrin, endrin, and heptachlor reacted similarly (112, 201, 204, 250) under appropriate conditions. Under the influence of light or mild heat, carbaryl (56), DDT (95), and many other pesticides underwent elimination reactions.

Many organophosphorus esters, especially the phosphorothionates, isomerize under the influence of heat or light. For example, the conversion of parathion to Hs O,S-diethyl O-(p-nitrophenyl) isomer occurred slowly upon storage at room temperature (but was 50% complete in 5 hours at 140°C) (160), while it was rapid in light (135). On the other hand, demeton isomerizes so rapidly that it is sold as an equilibrium mixture which contains 65% thiolo isomer (221). Demeton and its methyl homolog (104) underwent self-alkylation in water to form sulfonium salts as much as 10^4 times more toxic than the original insecticide. The unsaturated phosphate insecticide mevinphos (Phosdrin) showed *cis-trans* isomerization to an equilibrium mixture containing 60% *cis*-isomer (221).

BIOLOGICAL TRANSFORMATIONS

If one were to examine elementary texts on the laboratory practice of organic chemistry and attempt to summarize the principal types of organic reactions from them, the classification might well narrow down to oxidations (oxygenations), reductions (hydrogenations), ionic reactions involving both initiation and reception of nucleophilic attack, and a few other less common transformations including eliminations and isomerization.

The nonbiological transformations of pesticides in the environment obviously fit such a classification despite a limited selection of reagents—oxygen, water, a few nucleophiles, and "reducing power"—generally restricted to dilute solutions at ambient temperature. The biochemical complex represented by the living process must operate under the same restrictions, and it is logical that chemical transformations within both the environment and its coevolving creatures often should be much the same.

In the course of normal life, survivors have developed the ability to transform or "metabolize" foreign substances, pesticides included. In fact, most animals, higher plants, and microorganisms utilize only relatively few biochemical pathways of metabolism, and recent reviews on pesticide transformations carried out by living organisms (32, 124, 156) can be consulted for details. Rather than summarize metabolism according to phylogeny, it is more pertinent to our consideration of the environmental fate of pesticides to examine what should by now be familiar basic processes.

Oxidation

Most, if not all, biochemical oxygenations of pesticides and other xenobiotics are carried out by low-specificity, mixed-function oxidases often localized in microsomes or similar structures. The oxidizing power of this system is demonstrated by its ability to oxidize alkyl groups: C-methyl groups of landrin (217), binapacryl (79), and pyrethrins (248) were hydroxylated and eventually oxidized to carboxyl groups; the methylene group of furadan (158) was oxidized stepwise to the ketone; and the methylene and methine carbons of such diverse substances as DDT (209), dieldrin (82), and rotenone (247) were hydroxylated.

Oxidative dealkylation at hetero atoms is a general reaction. Stepwise N-demethylation of amines [Zectran (2)], amides [diphenamid (140)], ureas [monuron (219)], and carbamates [carbaryl (137)] has been widely reported, as well as the loss of the N-ethyl group of atrazine (216) and the N-propyl group of trifluralin (90). O-Dealkylation is represented by the metabolism of methoxychlor (119) and dicamba (23); S-dealkylation is uncommon, perhaps due to a greater tendency for S-oxidation, although there is evidence that prometryne can be demethylated to the mercaptotriazine (17).

Indeed, oxidations on sulfur atoms yielded the corresponding sulfoxide and sulfone whether in the aliphatic [aldicarb (44)], aralkyl [mesurol (2)], or diaryl [abate (16)] series. The similar oxidation of amines to N-oxides (185), hydroxylamines (185), azobenzenes (12), and even nitro compounds [diuron to 3,4-dichloronitrobenzene (182), doubtless via the dichloroaniline] is well recognized. The facile oxidative desulfuration of phosphorothionates to their oxygen analogs appears to be a general biochemical reaction (32, 181).

The epoxidation of olefins also is general as shown by the metabolism of aldrin, heptachlor, and isodrin (24); biological epoxidation of aromatic bonds has been suspected, too, and the isolation of naphthalene epoxide recently has been accomplished (117). Products of carbaryl metabolism, such as 5,6-dihydroxy-5,6-dihydro-1-naphthyl N-methylcarbamate (137, 226), strongly suggested an epoxide intermediate, although none has been isolated. The NIH shift observed in the ring-hydroxylation of 2,4-D—with formation of 2,5- and 2,3-dichloro-4-hydroxy-phenoxyacetic acids (81)—likewise has been suggested (96) to involve an epoxide intermediate, and epoxidation actually may account for many common metabolic hydroxylations of pesticides.

Aromatic hydroxylation is exceptionally important to the environmental fate of pesticides in that the introduction of an oxidizable, electron-releasing group prepares benzenoid compounds for further degradation. Ring oxidation is perhaps the most usual and predictable biological transformation of xenobiotics; the extensive search for its biochemical mechanism currently is inclined toward the operation of a powerfully electrophilic species reminiscent of the still hypothetical hydroxonium ion HO^+ (98, 231).

Reduction

The metabolic process of reduction is much less obvious than oxidation, perhaps because life exists in and on an oxidizing system. However, a few general examples are known in which pesticides are, in effect, hydrogenated. Most organisms appear to reduce DDE to DDD, but the reduction usually stops after replacement of a single chlorine. A few instances of further reductive dechlorination of chloroalkanes (including DDT) are known (152), but no example of a similar reaction of aromatic halides has been reported.

The biological reduction of C–C double bonds has been suggested to explain intermediates observed in the metabolism of DDT and its relatives (188), but several simple olefins (pyrethrins, rotenone, and aldrin) have not yielded the reduced products. On the other hand, double bonds between other elements often are readily hydrogenated: carbonyl groups can form the corresponding alcohol (14), and aromatic nitro compounds including parathion (107), trifluralin (90), and PCNB (35) were reduced to the related amines via intermediate nitroso compounds. The uptake of an electron by paraquat and diquat (27) and by quinones such as dichlone and chloranil (88) resulted in free-radical products.

Hydrolysis

The ability to hydrolyze is general among plants, animals, and microorganisms. A recent review (32) showed that plants, for example, hydrolyze a surprising number of pesticides. Aliphatic and aromatic carboxylic esters, amides, and nitriles; carbonic acid derivatives including carbonates, carbamates, and ureas; thiolcarbamates, dithiocarbamates, sulfites, and sulfonates; and phosphoric acid derivatives of many types all react enzymatically with water to provide the same products as are produced by acidic or alkaline hydrolysis in the laboratory. The relative rates of hydrolysis depend strongly on chemical structure—substituents on an aromatic ring, for example (86).

Other types of pesticides also react with water within living organisms. Hydrolysis of the epoxide ring of dieldrin to form the *trans*-diol provided one of its most important routes of metabolic degradation (24); halogenated aliphatic compounds such as ethylene dibromide (21) and dalapon (123) were dehalogenated; and chlorotriazines such as simazine formed the hydroxy analog (121) (the corresponding methoxy- and methylthiotriazines produced similar products). The list of examples could continue, but the point is clear: biological hydrolysis has a major influence on the environmental fate of pesticides.

Like many other types of environmental transformations, such hydrolysis is not always straightforward. The principal route for detoxication of DDT in animals is the apparent "hydrolysis" of the trichloromethyl group to a carboxyl group (DDA), a well-known laboratory procedure, but for some animals, at least (3, 188), DDA is only the end result of a multistep process which involves reduction, oxidation, and dehydrochlorination, but may not involve reaction with water at all! The "hydrolysis" of simazine to hydroxysimazine in corn operates through prior reaction with 2,4-dihydroxy-7-methoxy-1,4-benzoxazin-3-one (208), and there is good evidence that the "hydrolysis" of parathion in some instances actually is an oxidation (65). The enzymatic hydrolysis of the amide propanil by certain plants was fatally inhibited by phosphate and carbamate insecticides which had been applied for "crop protection" (153).

Other Ionic Reactions: Conjugation

Most biological transformations are degradative and provide "primary metabolism." However, organisms also can carry out certain synthetic processes in their treatment of xenobiotics, and this "secondary metabolism" usually is referred to as conjugation. The natural conjugating reagents generally confer increased polarity and water solubility on the pesticide (or its decomposition products), although the effect often appears as amelioration of a dangerous reactivity.

The powerful sulfhydryl nucleophile glutathione (GSH) characteristically displaces reactive halides and opens epoxide rings, although the stable end product usually is a thioether with cysteine (a mercapturic acid). For example, lindane was converted to S-(2,4-dichlorophenyl) glutathione by the cattle tick (40), but to 2,4-dichlorophenylmercapturic acid by the rat (92). Amines such as that resulting from metabolic reduction of dichloran (233) were acetylated, presumably by acetylcoenzyme A. Conversely, carboxylic acids such as 2,4-D formed watersoluble amides with amino acids such as glutamine via their intermediate thiol ester with CoA (145), while acids, phenols, alcohols, and amines displaced nucleotides to form glucuronides, glucosides, and sulfate esters (185).

Comparative Metabolism

Although most creatures share common metabolic pathways (Table 2), they certainly are not equally proficient. In fact, distinct metabolic peculiarities exist at the individual, species, and even the phylum level. For example, most vertebrate animals can convert acidic compounds such as phenols and carboxylic acids into esters or ethers of β-D-glucuronic acid (185), but bacteria, plants, and insects cannot do so (or do so only very selectively). Notably, the cat is almost unable to carry out this metabolically simple reaction (185).

On the other hand, bacteria, plants, and insects form β-D-glucosides (or β-D-glucose esters) with the same types of compounds, while the other animals do not; the necessary transferases appear to be missing. The ability to acetylate aromatic amines is much less uniform, and, again, a few species (such as the chicken and dog) seem unable to conform with other members of their class. Undoubtedly, as experience with comparative metabolism grows, many other peculiarities will come to light.

The extreme simplification in Table 2 also should not be considered to exclude very considerable differences in metabolic rate. For example, the relative rate of reduction of the nitro group of parathion by vertebrate livers decreased in the order rat (100) > bass (70) > chicken (40) > sparrow (below 20) (107). In the hydrolysis of parathion by blood serum (7), the mouse exhibited only 1% and the rat 2% of the rabbit's rate, although all are classed as rodents.

These differences are of the utmost significance toxicologically, and indeed they form the basis for the selective toxicity upon which application of almost all pesticides is based. Yet man's personal preferences and concerns notwithstanding, all of the metabolic abilities of mammals probably are insignificant to the long-term fate of pesticides in the environment as compared with, say, microbial oxidation. Here, especially, is revealed the importance of the other side of the metabolism picture—what Alexander (8) has aptly termed "molecular recalcitrance." From extensive examination of the (oxidative) metabolic abilities of isolated microbial cultures against phenoxy herbicides and other compounds, certain generalities emerged (9, 120, 126): (a) long, straight aliphatic side chains were more rapidly degraded than short or branched ones; (b) certain ring substituents (NO_2, Cl, CH_3) sharply reduced the rate of degradation; and (c) ring

Table 2   Comparative metabolic ability

	Bacteria	Plants	Insects	Fish	Frogs	Reptiles	Birds	Mammals
Aldrin epoxidation	+	+	+	±	+	+	+	+
Nitro reduction	+	+	+	±	−	+	+	+
Paraoxon hydrolysis	+	+	+	+	+	+	+	+
Peptide conjugation	−	Asp	Gly	Gly	Gly	Orn	Orn	GluNH$_2$
Amine acetylation	+	−	+	+	+	−	+[a]	+[b]
Glucosylation	+	+	+	−	−	−	−	−
Glucuronylation	−	−	−	±	+	+	+	+[c]

[a] Chicken

[b] Dog

[c] Cat

position was very important, an *ortho*-chlorine producing the greatest stability and a *para*-chlorine the least. Among the benzoic acids, the parent compound was rapidly degraded, the isomeric monochlorobenzoates were more stable, and the polyhalobenzoates were resistant to microbial degradation (150).

ENVIRONMENTAL FATE

Most pesticides and their breakdown products appear able to accomplish transport within and transfer between environmental compartments with rather more facility than might have been expected (111). As detailed later, firm evidence exists for the partitioning of specific pesticides between any two phases—soil and air, for example—and the concept of pesticide cycling in the environment has gained popularity (245). Cycling is logical and probably occurs, but there is no instance in which it has been demonstrated conclusively, perhaps due to several formidable difficulties. Dilution and dispersal, especially into the atmosphere, are rapid and often extreme at each stage; chemical and biochemical transformations tend to increase polarity and decrease volatility [although DDD and DDE are more volatile than DDT (223)]; and, despite the astonishing sensitivity of modern determinative methods for pesticides, adequate soil analysis is hampered by adsorption of the compounds to particles, the complex background of interferences, and the multiple pathways by which pesticides are transformed into difficulty-extracted polar products.

Biota

Whether by direct application, precipitation from the atmosphere, or uptake from soils or water, green plants usually serve as recipients for most pesticides. Extensive "market-basket" surveys by the U.S. Food and Drug Administration (69, 70) revealed that only ppb (parts per billion) levels of pesticides remained on raw food, but the detection of many thousands of times these amounts in inedible plants or other plant parts (108) suggests that both crops and wild plants may receive and retain more significant quantities.

A major part of any pesticide application often is removed by wind, rain, volatilization, and photodecomposition on leaf surfaces. For example, a graph of the residues remaining on alfalfa after treatment with methoxychlor (p. 468) shows that they "dissipate" in proportion to the time elapsed since application and approach zero asymtotically (Figure 1). Similarly, the 66 ppm leaf residue resulting from a spray application of disulfoton (1 lb/acre) to cotton was reduced to 1.8 ppm (almost all as metabolites) after 52 days (159). However, application through the soil (196) resulted in rapid uptake and translocation into the plant, and even after 35 days the lower and upper leaves still contained 83 and 36 ppm, respectively, of disulfoton and metabolites.

This "systemic" uptake is fairly common among pesticides and, although complex, depends largely on the water solubility of the neutral compound or its metabolites (72); chlorinated hydrocarbons generally exhibit little if any systemic distribution (142). Biological transformation usually starts almost at once, and, although the initial metabolic changes may be rapid, long-lived "terminal" products often represent the final result. The experiments with disulfoton on cotton (159) revealed that the parent compound was oxidized immediately to its sulfoxide, which in turn formed toxic metabolites whose levels remained almost unchanged for days (Figure 2). The same effect was observed in seedlings following seed-treatment with disulfoton (159).

The fact is that plants are seriously handicapped by lack of an adequate excretory system, and the metabolic tendency is to convert the pesticide into some neutral, water-soluble form in which it can be stored in cell vacuoles. For example, conversion of NAA to its glucose ester occurred soon after absorption and was followed by the slower formation of more stable amino acid conjugates (235). Insecticides such as carbaryl formed cholinesterase-inhibiting, glycosylated metabolites which were very persistent (137). Eventually, elements of the original pesticide structure may be bound to protein (20) or incorporated into the host's structural lignins, pectic substances, etc. (39, 154).

The complete course of plant metabolism of many—perhaps most—pesticides is obscure, often because of the diversity of products and the difficulties inherent in their isolation. Benefin was rapidly metabolized by plants (194); the initial metabolites conformed to the expected routes of N-dealkylation and nitro reduction, but eventually almost all appeared as highly polar, largely unextractable substances widely distributed in the plant. Simazine provided a similar example in which hydrolysis to hydroxysimazine was straightforward, but further degradation involving reduction, dealkylation, and ring opening obscured the precise pathways of subsequent metabolism (169). That simazine (169), thiolcarbamates (80), CDEC (115), and many other pesticides underwent essentially complete decomposition in plants followed by transpiration of the resulting amines, CO_2, and

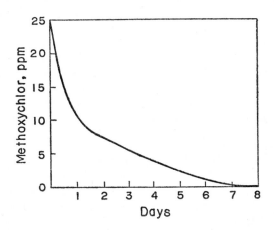

Figure 1   Dissipation of methoxychlor applied to alfalfa (83).

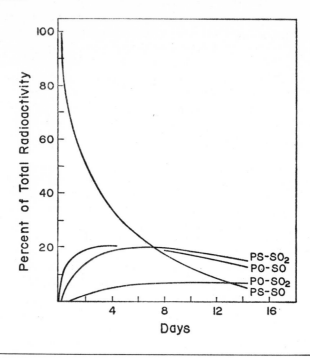

Figure 2 Metabolism of disulfoton in cotton leaves to form disulfoton sulfoxide (PS-SO) and sulfone (PS-SO₂) and the corresponding oxons (PO-SO and PO-SO₂) (1959).

other fragments or incorporation into the pools of intermediary metabolism suggests that green plants not only can destabilize pesticides for later breakdown elsewhere but contribute significantly themselves to the environmental destruction of pesticides.

One of the prime movers in the decade-old furor over pesticides in the environment has been the easily detectable occurrence of pesticide residues throughout the animal world. Wildlife, fish, other marine species, domestic animals, and even man (71, 75, 109) were shown to contain amounts of chlorinated hydrocarbon insecticides which varied from a few tenths to several hundred ppm. In most instances, food represented the only significant source, although fish are known to accumulate dieldrin primarily from water (36). The fact that a single small marine mussel filters about 60 liters of water per day and a mussel bed annually may process 22 million metric tons of water, including about 400 tons of inert matter (195), also suggests that water and silt could contribute substantially to exposure. Soil invertebrates could be expected to extract pesticides from ingested soil particles and from vapor during foraging (75, 238).

However, it is important to recognize that most animals have rather efficient systems for the metabolism and/or excretion of xenobiotics, and indeed, only the chlorinated hydrocarbon insecticides have been found to remain for more than a few days (71, 75). Most other types of compounds are cleared quite rapidly, even from doses providing many times those encountered in everyday life. At 10 mg/kg, rats metabolized and excreted 66% of the administered oral dose of parathion in 24 hr (4); trifluralin (100 mg/kg), 82% (78); and 2,4-D (10 mg/kg), 96% (131). We may assume that the minute proportions of excreted pesticides return to soil and water.

Even the now-infamous DDT was metabolized and excreted, usually as DDA (3, 188) or other oxygenated fragments including dichlorobenzophenone and *p*-chlorobenzoic acid (157), but here the clearance rate often was balanced by a continuous intake and subsequent temporary storage. A reduction of intake resulted in a reduction in both excretion rate and tissue level until a new equilibrium was reached (71). A complex of factors regulates clearance—exposure, diet, age, sex, etc—but it is apparent that some species simply are inefficient excreters and tend to accumulate certain compounds faster than they return them to the environment. Plankton (61) and fish (75), for example, may contain more than 10^4 times the environmental level of a chlorinated hydrocarbon, giving rise to the rather simplistic concept of "biological magnification" in food chains (128, 148, 170). Indeed, pesticide-transferring links in a chain have been verified in both natural (225) and model (119) ecosystems, although the expected accumulation was not observed in others (75, 148); *each transfer* depended upon the excretion efficiency of both the predator and the prey. While transfer efficiency must be less than 100% for obvious reasons, and often may be very low, Rudd's concept of a "liposphere" (210) in which certain fat-soluble substances such as DDT can persist and move with a degree of independence through and around biota deserves reflection.

Soil Environments

Pesticides can enter the soil by many routes. Of course, enormous quantities of chemicals are applied intentionally to soil, and, in view of the increasing recognition of pesticide dissipation by volatilization and photodecomposition (50), application by soil incorporation certainly will increase. However, every other form of application to crops must inadvertently lose a proportion to the soil; significant quantities are known to be transported for great distances on dust particles [pesticide-laden dust from New Mexico once fell out on Cincinnati at a rate of 4.8 tons/mi² (41)], and rainfall can add further amounts (228, 237). Of course, residues from plants and animals, living and dead, return to the soil. The soil at some time forms a vast repository for much of the world's pesticides.

Although "soil" is quite variable, it generally consists of sand (primarily silica), clay (complex aluminosilicates), metalliferous minerals, organic debris, gases, and water. Also, there generally is a sizeable component of living matter, both animal and microbial. The amount of pesticides in typical soil samples is small—often tenths of a ppm—but one must recall that the bulk of soil is so great

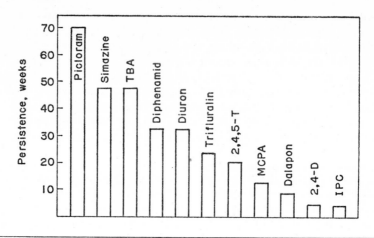

Figure 3　Persistence of common pesticides in soil (125).

that a normal pesticide application of 1 lb/acre distributed uniformly throughout the top foot provides a concentration of only 0.25 ppm. Of the organic pesticides, the chlorinated hydrocarbon insecticides are by far the most generally distributed in soils. In areas of prolonged heavy application, hundreds of pounds per acre of DDT might accumulate locally (215), representing most of the pesticide ever applied.

Many factors influence the persistence of pesticides in soil, and the term "half-life" commonly used as a measure is virtually meaningless. Kearney and co-workers (125) instead have estimated more realistic ranges of persistence for groups of structurally similar compounds (Figure 3) from large numbers of widely separated measurements, although these values are subject to large local variations. With few exceptions, pesticides seldom "persist" in soil for more than a few months, but such figures give no hint of where the chemical has gone.

All pesticides become adsorbed on soil to some degree, although the extent is regulated by temperature, the polarity of the compound, type of soil, solubility, and volatility. Adsorption usually is proportional to organic content (243), although models indicate that lignin, humic acids, or charcoal bind at least 20 times as much atrazine as does starch, pectin, or chitin. Soil minerals also bind pesticides by several mechanisms (239), but elution by subirrigation tends to move the compounds in a process analogous to chromatography to distribute them between water and the surface of particles. In fact, solvent partitioning (236) or chromatography (132) form useful predictors of pesticide mobility in soil; as expected, water-soluble herbicides such as 2,4-D exhibited near-maximum movement, phenylureas and dinitroanilines less, and most organophosphorus insecticides even less. The nonpolar chlorinated hydrocarbons and the ionic dipyridinium herbicides were essentially immobile with water, although leaching efficiency varied greatly with soil type (99, 122).

The living component of soil also sequesters pesticides, but only temporarily—"ashes to ashes, dust to dust. . . . " The strongly adsorbed compounds such as DDT are preferentially taken up by fungal mycelia (133), roots (93), and soil invertebrates (75), while the more soluble ones can move systemically into the aboveground parts of plants. These processes, too, require water; pesticides can be very persistent in a dry soil (74).

Volatility is important, and pesticide movement in soil must be considered to take place in both the dissolved and vapor state simultaneously (222). It is clear (18, 25) that pesticide movement by volatility is enhanced by water, but this is due to displacement of the compound from adsorption sites rather than "codistillation" (102). The capillary "wick" effect which draws water to the surface by evaporation brings pesticides up where they can transfer to other environmental compartments by volatility (97), blowing dust (41, 197), and—less important—surface runoff into waterways (76). An acre of moist soil can transfer many tons of water in a day, and while the consequent loss of DDT may amount to only ounces per year (102), up to 0.25 lbs/day of parathion and 2,4-D butyl ester could evaporate from the soil surface.

The soil provides a reactor as well as a repository. Its constitution offers sites for hydrolysis, oxidation, reduction, and the other pesticide reactions mentioned previously (50). For example, the hydrolysis of triazine herbicides (11, 211) and organophosphorus insecticides (171, 213) is catalyzed by clay minerals, although soil water often is sufficiently alkaline to hydrolyze many pesticides. DMTT (Mylone) and vapam underwent soil hydrolysis followed by conversion to the reactive and volatile methyl isothiocyanate (176, 230), most of which escaped as vapor, and wet soil accelerated the hydrolysis of chlorinated hydrocarbon nematicides (34). Oxidation converted aldrin to the more reactive dieldrin (143), and DDT was reduced to DDD (89) or dehydrohalogenated to DDE nonbiologically (95). The vapor above a soil containing a 2.5 ppm of technical DDT (75% p,p'-DDT, 21% o,p'-DDT, and 0.8% DDE) consisted of 41%, 43%, and 16% of these compounds, respectively (223), again indicating that soil transformations may increase loss by volatility. Such diverse substances as heptachlor (250) and trifluralin (242) were decomposed by sunlight on the soil surface.

Despite substrate inaccessibility, inadequate nutrition, and other problems which can lead to "microbial fallibility" (8), it is the large and varied soil microflora which probably plays the most important part in pesticide degradation (10). The biochemical *forte* of microbes is oxidation; the relatively short-lived 2,4-DB was found to be degraded to 2,4-dichlorophenol via 2,4-D (10); both the phenol and the phenoxy acids were hydroxylated, and the resulting chlorocatechol suffered enzymatic ring cleavage to provide a 6-carbon dicarboxylic acid (chloromuconic acid), its related lactone, and succinic acid (229) in a demonstration of the unique oxidizing ability of microorganisms. The final degradation products from 2,4-D and 2,4-DB were carbon dioxide and chloride ions (149).

Reduction and hydrolysis of parathion by pure cultures of *Bacillus subtilis* was observed (164), but not oxidation. Still, soil populations performed all these steps by apparent "co-metabolism" in which the various species carried out different degradative functions (10). DDT offered another example: in either aerobic or anaerobic (wet) soils, reduction, oxidation, and dehydrochlorination provided DDE, DDD, DDA, dichlorobenzophenone, p-chlorobenzoic acid, p-chlorophenylacetic acid (PCPA), and other products (94, 189). Recent work (189) showed that *Hydrogenomonas* degraded DDT as far as PCPA, while *Arthrobacter* then continued the degradation.

Aquatic Environments

The level and fate of pesticides in aquatic environments remain even more obscure than in biota and soil. Considerable amounts of chemicals are introduced into water intentionally to control weeds, algae, and insect larvae (principally mosquitos). Unknown but undoubtedly even larger amounts are introduced unintentionally by drift (220), presumed disposal (178), and especially by industrial and domestic sewage. For example, until recently the drainage from the San Joaquin Valley poured an estimated 1900 kg of DDT into the Pacific Ocean each year (197), and accidental flooding of an industrial dump started a wave of endrin and related compounds down the Mississippi River which resulted in the death of countless fish (172). Distinct but unquantified amounts also enter from rainfall (228, 237), soil erosion (41, 179, 197), and surface runoff (76); percolation into ground water rarely occurs (76, 174).

Monitoring data (19) provide a few general conclusions: (a) several of the most stable pesticides such as DDT and dieldrin are frequently detectable in inland waters; (b) other pesticides are detectable only in the immediate area of use; (c) pesticide levels are highly variable with both location and time; and (d) the pesticides are almost always present on suspended particles. For example, the mean DDT level in major U. S. river basins in 1957–65 was 8.2 ng/1 [maximum 149 ng/1 (19)]; filtered lake water contained an average of 0.1–0.2 ppb of DDT while suspended particulate matter contained 6000–9300 ppb (127). Bottom sediments provide a major reservoir for continuing contamination (13, 22), especially in estuaries (165), but pesticide levels in water and mud decline as the number of new applications decreases (179).

Surprisingly little information exists on pesticides in the ocean (47). From the fact that many marine animals contain distinct traces of the stable pesticides, it is apparent that the chemicals reach the sea. In addition to the direct introduction of sewage, rivers must deposit appreciable amounts in both soluble and particulate form (165), and fallout due to dust and rain could be expected to make significant additional contributions (197). The occurrence of organic surface films provides both an appropriate water-air interface and a concentrating mechanism (103, 212).

Although static tests in both the laboratory (77) and the field (22) showed dissipation of pesticides in aquatic environments, the mobility of water defied estimates of persistence under most natural conditions. Compounds often volatilize readily from the water surface (18, 25), surface slicks concentrate and move them (103, 212), and they are taken up, concentrated, and moved by aquatic plants, animals, and particulates (46, 75, 187). However, although relatively few data exist, aquatic biota appear fully capable of metabolizing pesticides (Table 2), and excretion or death principally return a compound or its metabolites to the surrounding medium.

The alkalinity of natural waters indicates that many types of chemicals will undergo nonbiological transformations in aquatic environments (p. 472). The uv transparency of water also suggests that photodecomposition must be an important force (52), but few unequivocal field examples have been reported. Perhaps this anomaly is due to the analytical obstacles posed by the concurrent operation of multiple transformation pathways. For example, phenoxy herbicides are simultaneously photooxidized, photoreduced, and hydrolyzed (62, 63), and those reactive intermediates which do not become bound to particles or other surfaces eventually precipitate as polymeric humic acids. The rapid and extensive nonbiological reduction of trifluralin and benefin in water-saturated mud (193) demonstrates the chemical complexity of the soil-water system.

Microorganisms represent a major force for pesticide transformations in water as demonstrated by the conversion of DDT to DDE and a variety of other products (163, 189) and can further confuse the interpretation of analytical results: does the detected compound represent field application, a microbial transformation, photodecomposition, or earlier environmental events at some distant location?

Air Environments

With the exception of "bugbombs" and similar devices for very limited urban use, air seldom is the target of pesticide applications. However, any aboveground application inevitably results in the air dispersal ("drift") (5, 6) responsible for a large proportion of the pesticide which disappears during spraying. Dust and other particulate matter apparently can deliver pesticides over great distances, and the detection of ppb levels of persistent compounds in rainwater (228, 237), antarctic glacial ice (186), and remote islands (197) gives convincing testimony that pesticides do occur widely in air. Pesticides and their transformation products often have remarkably high vapor pressures (97), and it is becoming recognized that significant proportions of any application inevitably join the atmosphere in gaseous as well as particulate form.

Of the environmental phases, the occurrence and significance of pesticides in air are the least understood. Local concentrations can be very high at the time of application (116)—airborne malathion and carbaryl reached 800 $\mu g/m^3$, guthion 2500 $\mu g/m^3$, and DDT 8500 $\mu g/m^3$ during orchard applications, and systox reached almost 20 mg/m^3 in greenhouse applications—but the chemicals precipitate or rapidly become diluted until nearby agricultural communities might receive maximum amounts in the order of only ng/m^3 of DDT and of malathion (227). Recent monitoring measurements with the best available techniques (224) showed that the "normal" air over an agricultural community such as Orlando, Florida, may contain as much as 2 $\mu g/m^3$ of DDT relatives and 0.5 $\mu g/m^3$ of parathion, while a city such as Baltimore, Maryland, would receive less than 1/100 of these amounts. Levels could not be correlated with rainfall, but most of each compound was associated with the particulate fraction as in the case of fenitrothion drift (249).

The total burden and "persistence" of any pesticide in air presently is imponderable but surely must represent a complex interplay of surface activity, vapor pressure, temperatures, and many other factors. Whether vapor or particulate, air-borne pesticides also must be subject to photodecomposition, but outside of a single laboratory investigation where dieldrin and trifluralin were shown to decompose (167), no specific data have appeared.

SOME CONSEQUENCES

The all-too-frequent reference to "disappearance" or "half-life" of pesticides reveals a widespread and continuing lack of information about the environmental transfer and transport of chemicals. In an attempt at quantitation, Hindin (106) applied a total of 5.6 lbs/acre of DDT by airplane to a carefully monitored corn field; of the 1.67 lbs/acre (35%) which descended to tassel level (8 ft), less than half could be accounted for after one day. No further loss was apparent after 3 weeks, when soil had accumulated 9% of application, plants 4%, and water and silt only a negligible proportion. Even then, air contained significant levels and must be considered to have provided the principal transfer medium. After a year, soil accounted for 3% of the original dose (water and silt represented a total of only 0.001%) and provided an estimated DDT halflife of 6 months—figures which do not reveal what happened to the other 97%! Similar results were observed more recently with dieldrin (30).

The inability to account for such mobile and apparently persistent compounds has led to considerable speculation concerning their whereabouts and to the logical suggestion that they may occupy the atmosphere and ocean deeps (246), not altogether coincidentally the two environmental compartments for which analytical data are the weakest. The speculation has been supported in some instances by mathematical modeling, for example in the areas of movement in soils (15), freshwater ecosystems (73, 187), distribution in vertebrates (198) including man (199), and fate in the environment as a whole (100, 246). Seemingly, a major error of many models must lie in the lack of information on, or actual failure to recognize, the rates of pesticide *transformation* by biological and nonbiological forces (e.g. atmospheric photooxidation).

The large majority of pesticides are not that persistent, let alone prone to environmental buildup or "magnification" in environmental compartments. There are no permanent organic pesticides, and degradation in the environment must be assumed to continue on to inorganic products—how rapidly depends upon application rate, climate, soil type, plant cover, and many other factors which can vary widely with geographical location.

The environmental mechanisms can transport and degrade pesticides only as long as they are not overwhelmed. To ensure satisfactory operation of the process, maximum reduction in the casual (uneconomic) introduction of chemicals seems only logical. While the design of a certain degree of intrinsic degradability into new pesticides may become possible, methods to limit environmental transfer and transport, reduce the number and level of required applications, and avoid the careless disposal of wastes are more within our reach. For example, an increase of even a few percent in application efficiency could dramatically lower the environment's pesticide burden.

Despite the staggering volume of work remaining to be done, research on the fate of pesticides already has produced many benefits: rapid advances in analytical chemistry, the biochemistry of metabolism, photochemistry, and toxicology; new approaches to agriculture, natural history, public health, and even law; and the tools and methods to help society deal with *other* kinds of environmental problems. Perhaps, too, we should hope for an added measure of humility as continuing study reveals how little any of us really knows about the chemical world in which we all must live.

Literature Cited

1. Abdel-Wahab, A. M., Casida, J. E. 1967. *J. Agr. Food Chem.* 15:479–87
2. Abdel-Wahab, A. M., Kuhr, R. J., Casida, J. E. 1966. *J. Agr. Food Chem.* 14:290–98
3. Abou-Donia, M. B., Menzel, D. B. 1968. *Biochem. Pharmacol.* 17:2143–61
4. Ahmed, M. K., Casida, J. E., Nichols, R. E. 1958. *J. Agr. Food Chem.* 6:740–45
5. Akesson, N. B., Yates, W. E. 1964. *Ann. Rev. Entomol.* 9:285–318
6. Akesson, N. B., Yates, W. E., Coutts, H. H., Burgoyne, W. E. 1964. *Agr. Aviat.* 6:72–82
7. Aldridge, W. N. 1953. *Biochem. J.* 53:117–25
8. Alexander, M. 1965. *Advan. Appl. Microbiol.* 7:35–80
9. Alexander, M. 1967. *Agriculture and the Quality of Our Environment*, ed. N. C. Brady, 331–42. Washington: AAAS. 460 pp.
10. Alexander, M. 1972. *Environmental Toxicology of Pesticides*, ed. F. Matsumura, G. M. Boush, T. Misato, 365–83. New York: Academic. 637 pp.
11. Armstrong, D. E., Chesters, G. 1968. *Environ. Sci. Technol.* 2:683–89
12. Bartha, R., Linke, H. A. B., Pramer, D. 1968. *Science* 161:582–83
13. Barthel, W. F. et al. 1969. *Pestic. Monit. J.* 3:8–66
14. Bartley, W. J., Andrawes, N. R., Chancey, E. L., Bagley, W. P., Spurr, H. W. Jr. 1970. *J. Agr. Food Chem.* 18:446–53
15. Biggar, J. W. 1971. See Ref. 161, 107–19
16. Blinn, R. C. 1968. *J. Agr. Food Chem.* 16:441–45
17. Boehme, C., Baer, F. 1967. *Food Cosmet. Toxicol.* 5:23–25
18. Bowman, M. C., Acree, F. Jr., Lofgren, C. S., Beroza, M. 1964. *Science* 146:1480–81
19. Breidenbach, A. W., Gunnerson, C. G., Kawahara, F. K., Lichtenberg, J. J., Green, R. S. 1967. *Pub. Health Rep.* 82:139–56
20. Brian, R. C. 1960. *Plant Physiol.* 35: 773–82
21. Bridges, R. G. 1956. *J. Sci. Food Agr.* 7:305–13
22. Bridges, W. R., Kallman, B. J., Andrews, A. K. 1963. *Trans. Am. Fish. Soc.* 92:421–27
23. Broadhurst, N. A., Montgomery, M. L., Freed, V. H. 1966. *J. Agr. Food Chem.* 14:585–88
24. Brooks, G. T. 1969. *Residue Rev.* 27: 81–138
25. Buescher, C. A. Jr., Bell, M. C., Berry, R. K. 1964. *J. Water Pollut. Contr. Fed.* 36:1005–14
26. Burchfield, H. P., Storrs, E. E. 1956. *Contrib. Boyce Thompson Inst.* 18: 395–418
27. Calderbank, A. 1968. *Advan. Pest Contr. Res.* 8:127–229
28. California Dep. Agriculture. 1971. *Pesticide Use Report 1970.* 107 pp.
29. Calvert, J. G., Pitts, J. N. Jr. 1966. *Photochemistry.* New York: Wiley. 830 pp.

30. Caro, J. H., Taylor, A. W. 1971. *J. Agr. Food Chem.* 19:379–84
31. Carson, R. 1962. *Silent Spring.* Boston: Houghton-Mifflin. 368 pp.
32. Casida, J. E., Lykken, L. 1969. *Ann. Rev. Plant Physiol.* 20:607–36
33. Castro, C. E. 1964. *J. Am. Chem. Soc.* 86:2310–11
34. Castro, C. E., Belser, N. O. 1966. *J. Agr. Food Chem.* 14:69–70
35. Chacko, C. I., Lockwood, J. L., Zabik, M. L. 1966. *Science* 154:893–95
36. Chadwick, G. G., Brocksen, R. W. 1969. *J. Wildl. Manage.* 33:693–700
37. Chen, Y.-L., Casida, J. E. 1969. *J. Agr. Food Chem.* 17:208–15
38. Cheng, H.-M., Yamamoto, I., Casida, J. E. 1972. *J. Agr. Food Chem.* 20: 850–56
39. Chin, W. T., Stanovick, R. P., Cullen, T. E., Holsing, G. C. 1964. *Weeds* 12:201–5
40. Clark, A. G., Hitchcock, M., Smith, J. N. 1966. *Nature* 209:103
41. Cohen, J. M., Pinkerton, C. 1966. *Advan. Chem. Ser.* 60:163–76
42. Cook, J. W. 1954. *J. Assoc. Offic. Agr. Chem.* 37:989–96
43. Cope, O. B. 1971. *Ann. Rev. Entomol.* 16:325–64
44. Coppedge, J. R., Lindquist, D. A., Bull, D. L., Dorough, H. W. 1967. *J. Agr. Food Chem.* 15:902–10
45. Courshee, R. J. 1960. *Ann. Rev. Entomol.* 5:327–52
46. Cox, J. L. 1970. *Bull. Environ. Contam. Toxicol.* 5:218–21
47. Cox, J. L. 1971. *Fish. Bull.* 69:443–50
48. Crosby, D. G. 1969. *Photooxidation of Pesticides.* Presented at 158th Nat. Meet. Am. Chem. Soc., New York
49. Crosby, D. G. 1969. *Residue Rev.* 25:1–12
50. Crosby, D. G. 1971. See Ref. 161, 86–94
51. Crosby, D. G. 1972. *Degradation of Synthetic Organic Molecules in the Biosphere,* 260–78 Washington, D.C.: NAS-NRC
52. Crosby, D. G. 1972. *Advan. Chem. Ser.* 111:173–88
53. Crosby, D. G., Hamadmad, N. 1971. *J. Agr. Food Chem.* 19:1171–74
54. Crosby, D. G., Leitis, E. 1969. *J. Agr. Food Chem.* 17:1033–35
55. Ibid, 1036–40
56. Crosby, D. G., Leitis, E., Winterlin, W. L. 1965. *J. Agr. Food Chem.* 13: 204–7
57. Crosby, D. G., Li, M.-Y. 1969. See Ref. 124, 321–63
58. Crosby, D. G., Moilanen, K. W., Nakagawa, M., Wong, A. S. 1972. See Ref. 10, 423–33
59. Crosby, D. G., Tang, C.-S. 1969. *J. Agr. Food Chem.* 17:1291–95
60. Ibid, 1041–44
61. Crosby, D. G., Tucker, R. K. 1971. *Environ. Sci. Technol.* 5:714–16
62. Crosby, D. G., Tutass, H. O. 1966. *J. Agr. Food Chem.* 14:596–99
63. Crosby, D. G., Wong, A. S. 1970. *The Effects of Light on Phenoxy Herbicides.* Presented at 160th Meet. Am. Chem. Soc., Chicago
64. Dachauer, A. C. 1962. *The Deuterium Isotope Rate Effect in Free Radical Reactions of t-Carbon Deuterated DDT and its Analogs.* PhD thesis. Fordham Univ., New York. 103 pp.
65. Dahm, P. A., Nakatsugawa, T. 1968. *The Enzymatic Oxidation of Toxicants,* ed. E. Hodgson, 89–112. Raleigh: Univ. North Carolina. 229 pp.
66. Dauterman, W. C., Viado, G. B., Casida, J. E., O'Brien, R. D. 1960. *J. Agr. Food Chem.* 8:115–19
67. Davies, A. G. 1961. *Organic Peroxides.* London: Butterworth. 215 pp.
68. Drescher, N., Otto, S. 1968. *Residue Rev.* 23:49–54
69. Duggan, R. E. 1969. *Ann. N.Y. Acad. Sci.* 160:173–82
70. Duggan, R. E., Weatherwax, J. 1967. *Science* 151:101–4
71. Durham, W. F. 1969. *Ann. N.Y. Acad. Sci.* 160:183–95
72. Ebeling, W. 1963. *Residue Rev.* 3:35–163
73. Eberhardt, L. L., Meeks, R. L., Peterle, T. J. 1971. *Nature* 230:60–62
74. Edwards, C. A. 1966. *Residue Rev.* 13:83–132
75. Edwards, C. A. 1970. *Persistent Pesticides in the Environment.* Cleveland: Chemical Rubber Co. 78 pp.
76. Edwards, C. A., Thompson, A. R., Beynon, K. I., Edwards, M. J. 1970. *Pestic. Sci.* 1:169–73
77. Eichelberger, J. W., Lichtenberg, J. J. 1971. *Environ. Sci. Technol.* 5: 541–44
78. Emmerson, J. L., Anderson, R. C. 1966. *Toxicol. Appl. Pharmacol.* 9: 84–97
79. Ernst, W., Bar, F. 1964. *Arzneimittelforsch* 14:81–84
80. Fang, S. C. 1969. See Ref. 124, 147–64
81. Faulkner, J. K., Woodcock, D. 1965. *J. Chem. Soc.* 1965:1187–91
82. Feil, V. J., Hedde, R. D., Zaylskie,

R. G., Zachrison, C. H. 1970. *J. Agr. Food Chem.* 18:120–24

83. Frear, D. E. H. 1963. *Penn. State Univ. Agr. Exp. Sta. Bull. 703.* 77 pp.

84. Frear, D. E. H. 1969. *Pesticide Index.* State College, Pa.: College Sci. Publ. 4th ed. 399 pp.

85. Freed, V. H., Witt, J. M. 1969. *Advan. Chem. Ser.* 86:70–80

86. Fukuto, T. R., Metcalf, R. L. 1956. *J. Agr. Food Chem.* 4:930–35

87. Fukuto, T. R., Metcalf, R. L., March, R. B., Maxon, M. G. 1955. *J. Econ. Entomol.* 48:347–54

88. Gause, E. M., Montalvo, D. A., Rowlands, J. R. 1967. *Biochim. Biophys. Acta* 141:217–19

89. Glass, B. L. 1972. *J. Agr. Food Chem.* 20:324–27

90. Golab, T., Herberg, J., Parka, S. J., Tepe, J. B. 1967. *J. Agr. Food. Chem.* 15:638–41

91. Gollnick, K. 1968. *Advan. Photochem.* 6:1–122

92. Grover, P. L., Sims, P. 1965. *Biochem. J.* 96:521–25

93. Grover, R., Hance, R. J. 1969. *Can. J. Plant Sci.* 49:378–80

94. Guenzi, W. D., Beard, W. E. 1967. *Science* 156:1116–17

95. Gunther, F. A. 1945. *J. Chem. Educ.* 22:238–40

96. Guroff, G. et al 1967. *Science* 157: 1524–30

97. Hamaker, J. W., Kerlinger, H. O. 1969. *Advan. Chem. Ser.* 86:39–54

98. Hamilton, G. A. 1964. *J. Am. Chem. Soc.* 86:3391–92

99. Harris, C. I. 1969. *J. Agr. Food Chem.* 17:80–82

100. Harrison, H. L. et al 1970. *Science* 170:503–8

101. Harrison, R. B., Holmes, D. G., Roburn, J., Tatton, J. O'G. 1967. *J. Sci. Food Agr.* 18:10–15

102. Hartley, G. S. 1969. *Advan. Chem. Ser.* 86:115–34

103. Hartung, R., King, G. W. 1970. *Environ. Sci. Technol.* 4:407–10

104. Heath, D. F. 1958. *J. Chem. Soc.* 1958:1643–51

105. Hilgetag, G., Teichmann, H. 1965. *Angew. Chem. Int. Ed. Engl.* 4:914–22

106. Hindin, E., May, D. S., Dunstan, G. H. 1966. *Advan. Chem. Ser.* 60: 132–45

107. Hitchcock, M., Murphy, S. D. 1967. *Biochem. Pharmacol.* 16:1801–11

108. Hoerger, F., Kenaga, E. E. 1972. *Environ. Qual. Safety* 1:9–28

109. Holden, A. V. 1970. *Pestic. Monit. J.* 4:117–35

110. Hull, H. M. 1970. *Herbicide Handbook of the Weed Society of America.* Geneva, N. Y.: Humphrey. 2nd ed. 293 pp.

111. Hurtig, H. 1972. See Ref. 10, 257–80

112. Ivie, G. W., Casida, J. E. 1970. *Science* 167:1620–22

113. Ivie, G. W., Casida, J. E. 1971. *J. Agr. Food Chem.* 19:410–16

114. Jarboe, R. H. Jr., Data, J. B., Christian, J. E. 1968. *J. Pharm. Sci.* 57:323–25

115. Jaworski, E. G. 1964. *J. Agr. Food. Chem.* 12:33–37

116. Jegier, Z. 1969. *Ann. N.Y. Acad. Sci.* 160:143–54

117. Jerina, D. M., Daly, J. W., Witkop, B., Zaltzman-Nirenberg, P., Udenfriend, S. 1968. *J. Am. Chem. Soc.* 90:6525–27

118. Joschek, H.-I., Grossweiner, L. I. 1966. *J. Am. Chem. Soc.* 66:3261–68

119. Kapoor, I. P., Metcalf, R. L., Nystrom, R. F., Sangha, G. K. 1970. *J. Agr. Food Chem.* 18:1145–52

120. Kaufman, D. D. 1966. *Pesticides and Their Effects in Soil and Water,* 85–108. Madison, Wis.: Soil Sci. Soc. Am. 150 pp.

121. Kaufman, D. D., Kearney, P. C. 1970. *Residue Rev.* 32:235–65

122. Kearney, P. C. 1970. *Residue Rev.* 32:391–410

123. Kearney, P. C., Harris, C. I., Kaufman, D. D., Sheets, T. J. 1965. *Advan. Pest Contr. Res.* 6:1–30

124. Kearney, P. C., Kaufman, D. D., Eds. 1969. *Degradation of Herbicides.* New York: Dekker. 394 pp.

125. Kearney, P. C., Nash, R. G., Isensee, A. R. 1969. *Chemical Fallout,* ed. M. W. Miller, G. G. Berg, 54–67. Springfield, Ill.: Thomas. 509 pp.

126. Kearney, P. C., Plimmer, J. R. 1970. See Ref. 161, 65–71

127. Keith, J. O. 1966. *J. Appl. Ecol.* 3(suppl.):71–85

128. Kenaga, E. E. 1972. See Ref. 10, 193–228

129. Kenaga, E. E., Allison, W. E. 1969. *Bull. Entomol. Soc. Am.* 15:85–148

130. Ketelaar, J. A. A., Gersmann, H. R., Beck, M. M. 1956. *Nature* 177:392–93

131. Khanna, S., Fang, S. C. 1966. *J. Agr. Food Chem.* 14:500–3

132. King, P. H., McCarty, P. L. 1968. *Soil Sci.* 106:248–61

133. Ko, W. H., Lockwood, J. L. 1968. *Can. J. Microbiol.* 14:1075–78

134. Kohn, G. K. 1969. *Residue Rev.* 29: 3–12

135. Koivistoinen, P., Merilainen, M. 1962. *Acta Agr. Scand.* 12:267–76
136. Koller, L. R. 1965. *Ultraviolet Radiation.* New York: Wiley. 2nd ed. 301 pp.
137. Kuhr, R. J., Casida, J. E. 1967. *J. Agr. Food Chem.* 15:814–24
138. Leffingwell, J. T. 1973. *The Photodecomposition of DDT and Related Compounds in Water.* PhD thesis. Univ. California, Davis
139. Leitis, E. 1973. *The Photodecomposition of Trifluralin.* PhD thesis. Univ. California, Davis
140. Lemin, A. J. 1966. *J. Agr. Food Chem.* 14:109–11
141. Li, C.-F., Bradley, R. L. 1969. *J. Dairy Sci.* 52:27–30
142. Lichtenstein, E. P. 1969. *Ann. N.Y. Acad. Sci.* 16:155–61
143. Lichtenstein, E. P., Schultz, K. R. 1960. *J. Econ. Entomol.* 53:192–97
144. Lightbody, H., Mathews, J. 1936. *Ind. Eng. Chem.* 28:809–11
145. Loos, M. A. 1969. See Ref. 124, 1–49
146. Ludwig, R. A., Thorn, G. D. 1960. *Advan. Pest Contr. Res.* 3:219–52
147. Lykken, L. 1972. See Ref. 10, 449–69
148. Macek, K. J. 1970. *The Biological Impact of Pesticides in the Environment,* 17–21. Corvallis: Oregon State Univ. 210 pp.
149. MacRae, I. C., Alexander, M., Rovira, A. D. 1963. *J. Gen. Microbiol.* 32:69–76
150. MacRae, I. C., Alexander, M. 1965. *J. Agr. Food Chem.* 13:72–76
151. Martin, H. 1971. *Pesticide Manual.* Droitwich, England: Brit. Crop Prot. Counc. 2nd ed. 495 pp.
152. Matsumura, F., Patil, K. C., Boush, G. M. 1971. *Nature* 230:325
153. Matsunaka, S. 1968. *Science* 160:1360–61
154. Meagher, W. R. 1966. *J. Agr. Food Chem.* 14:599–601
155. Melnikov, N. N. 1971. *Residue Rev.* 36:1–479
156. Menzie, C. M. 1969. *Metabolism of Pesticides.* US Fish Wildl. Serv. Spec. Sci. Report: Wildlife No. 127
157. Menzie, C. M. 1972. *Ann. Rev. Entomol.* 17:199–222
158. Metcalf, R. L. et al 1968. *J. Agr. Food Chem.* 16:300–11
159. Metcalf, R. L., Fukuto, T. R., March, R. B. 1957. *J. Econ. Entomol.* 50:338–45
160. Metcalf, R. L., March, R. B. 1953. *J. Econ. Entomol.* 46:288–94
161. Michigan State University. 1971. *Pesticides in the Soil: Ecology, Degradation, and Movement.* East Lansing: Michigan State Univ. 144 pp.
162. Miller, L. L., Narang, R. S. 1970. *Science* 169:368–70
163. Miskus, R. P., Blair, D. P., Casida, J. E. 1965. *J. Agr. Food Chem.* 13:481–83
164. Miyamoto, J., Kitagawa, K., Sato, Y. 1966. *Jap. J. Exp. Med.* 36:211–25
165. Modin, J. C. 1969. *Pestic. Monit. J.* 3:1–7
166. Moilanen, K. W., Crosby, D. G. 1972. *The Photodecomposition of Dexon.* Presented at West. Reg. Meet. Am. Chem. Soc., San Francisco
167. Moilanen, K. W., Crosby, D. G. 1972. *Photolysis of Pesticides in the Vapor Phase.* Presented at 163rd Meet. Am. Chem. Soc., Boston
168. Moilanen, K. W., Crosby, D. G. 1972. *J. Agr. Food Chem.* 20:950–53
169. Montgomery, M. L., Freed, V. H. 1964. *J. Agr. Food Chem.* 12:11–14
170. Moriarity, F. 1972. *New Sci.* 53:594–96
171. Mortland, M. M., Raman, K. V. 1967. *J. Agr. Food Chem.* 15:163–67
172. Mount, D. I., Putnicki, G. J. 1966. *Trans. 31st N. Am. Wildl. Nat. Res. Conf.,* 177–84
173. Mozier, A. R., Guenzi, W. D., Miller, L. L. 1969. *Science* 164:1083–85
174. Mrak, E. M. 1969. *Report of the Secretary's Commission on Pesticides and Their Relationship to Environmental Health.* Washington, D.C.: US Dep. HEW. Parts 1 and 2. 677 pp.
175. Munakata, K., Kuwahara, M. 1969. *Residue Rev.* 25:13–23
176. Munnecke, D. E., Martin, J. P. 1964. *Phytopathology* 54:941–45
177. Negherbon, W. O. 1959. *Handbook of Toxicology,* 3:467–74. Philadelphia: Saunders. 854 pp.
178. Nicholson, H. P. 1967. *Science* 158:871–76
179. Nicholson, H. P. 1969. *Proc. Wash. Acad. Sci.* 59:77–85
180. Norris, M. V., Vail, W. A., Averall, P. R. 1954. *J. Agr. Food Chem.* 2:570–73
181. O'Brien, R. D. 1965. *Ann. N.Y. Acad. Sci.* 123:156–64
182. Onley, J. H., Yip, G., Aldridge, M. H. 1968. *J. Agr. Food Chem.* 16:426–33
183. Owens, R. G., Blaak, G. 1960. *Contrib. Boyce Thompson Inst.* 20:475–97
184. Pape, B. E., Zabik, M. J. 1970. *J. Agr. Food Chem.* 18:202–7
185. Parke, D. V. 1968. *The Biochemistry*

of Foreign Compounds. Oxford: Pergamon. 269 pp.
186. Peterle, T. J. 1969. *Nature* 224:620
187. Peterle, T. J. 1970. See Ref. 148, 11–16
188. Peterson, J. E., Robison, W. H. 1964. *Toxicol. Appl. Pharmacol.* 6:321–27
189. Pfaender, F. K., Alexander, M. 1972. *J. Agr. Food Chem.* 20:842–46
190. Pimentel, D. 1971. *Ecological Effects of Pesticides on Non-Target Species.* Washington, D.C.: Off. Sci. Technol. 220 pp.
191. Pitts, J. N. Jr. 1969. *Advan. Environ. Sci.* 1:289–337
192. Plimmer, J. R. 1969. *Residue Rev.* 33:47–74
193. Probst, G. W. et al 1967. *J. Agr. Food Chem.* 15:592–99
194. Probst, G. W., Tepe, J. B. 1969. See Ref. 124, 255–82
195. Ricketts, E. F., Calvin, J. 1968. *Between Pacific Tides*, p. 396. Stanford Univ. Press. 4th ed. 614 pp.
196. Ridgway, R. L., Lindquist, D. A., Bull, D. L. 1965. *J. Econ. Entomol.* 58:-349-52
197. Riseborough, R. W., Huggett, R. J., Griffin, J. J., Goldberg, E. D. 1968. *Science* 159:1233–36
198. Robinson, J. 1967. *Nature* 215:33–35
199. Robinson, J. 1969. *Can. Med. Assoc. J.* 100:180–91
200. Rosen, J. D. 1967. *Bull. Environ. Contam. Toxicol.* 2:349–54
201. Rosen, J. D. 1967. *Chem. Commun.* 1967:189–90
202. Rosen, J. D. 1970. *J. Agr. Food Chem.* 18:494–96
203. Rosen, J. D. 1972. See Ref. 10, 435–47
204. Rosen, J. D., Siewierski, M. 1970. *J. Agr. Food Chem.* 18:943
205. Rosen, J. D., Strusz, R. F. 1968. *J. Agr. Food Chem.* 16:568–70
206. Ross, W. C. J. 1962. *Biological Alkylating Agents.* London: Butterworth. 232 pp.
207. Rosenthal, I., Frisone, G. J., Gunther, F. A. 1957. *J. Agr. Food Chem.* 5:514–17
208. Roth, W., Knüsli, E. 1961. *Experientia* 17:312–13
209. Rowlands, D. G., Lloyd, C. J. 1969. *J. Stored Prod. Res.*, 413–15
210. Rudd, R. L. 1972. See Ref. 10, 471–85
211. Russell, J. D. et al 1968. *Science* 160:1340–42
212. Seba, D. B., Corcoran, E. F. 1969. *Pestic. Monit. J.* 3:190–93

213. Sethunathan, N., Yoshida, T. 1969. *J. Agr. Food Chem.* 17:1192–95
214. Shaw, W. C., Hilton, J. L., Moreland, D. E., Jansen, L. L. 1960. *The Nature and Fate of Chemicals Applied to Soils, Plants, and Animals.* Beltsville: USDA-ARS Publ. 20-9. 221 pp.
215. Sheets, T. J. 1967. See Ref. 9, 311–30
216. Shimabukuro, R. H. 1967. *J. Agr. Food Chem.* 15:557–62
217. Slade, M., Casida, J. E. 1970. *J. Agr. Food Chem.* 18:467–74
218. Smith, A. E., Grove, J. 1969. *J. Agr. Food Chem.* 17:609–13
219. Smith, J. W., Sheets, T. J. 1967. *J. Agr. Food Chem.* 15:577–81
220. Sparr, B. I. et al 1966. *Advan. Chem. Ser.* 60:146–62
221. Spencer, E. Y. 1968. *Guide to the Chemicals Used in Crop Protection.* Ottawa: Can. Dep. Agr. Publ. 1093. 5th ed. 483 pp.
222. Spencer, W. F. 1971. See Ref. 161, 120–28
223. Spencer, W. F., Cliath, M. M. 1972. *J. Agr. Food Chem.* 20:645–48
224. Stanley, C. W., Barney, J. E., Helton, M. R., Yobs, A. R. 1971. *Environ. Sci. Technol.* 5:430–35
225. Stickel, L. F. 1968. *Organochlorine Pesticides in the Environment.* US Fish Wildl. Serv. Spec. Sci. Rep.: Wildlife No. 119
226. Sullivan, L. L., Eldridge, J. M., Knaak, J. B., Tallant, M. J. 1972. *J. Agr. Food Chem.* 20:980–84
227. Tabor, E. C. 1966. *Trans. N.Y. Acad. Sci.* 28:569–78
228. Tarrant, K. R., Tatton, J. O'G. 1968. *Nature* 219:725–27
229. Tiedje, J. M., Duxbury, J. M., Alexander, M., Dawson, J. E. 1969. *J. Agr. Food Chem.* 17:1021–26
230. Turner, N. J., Corden, M. E. 1963. *Phytopathology* 53:1388–94
231. Ullrich, V., Staudinger, H. 1969. *Microsomes and Drug Oxidations*, ed. J. R. Gillette et al, 199–225. New York: Academic. 547 pp.
232. US Geological Survey 1954. *Water Supply Papers 1299, 1300, 1460A.* Washington, D.C.: GPO
233. Van Alfen, N. 1972. *Microbial Metabolism of the Fungicide, 2,6-Dichloro-4-nitroaniline (Dichloran).* PhD thesis. Univ. California, Davis. 75 pp.
234. Van Middelem, C. H. 1966. *Advan. Chem. Ser.* 60:228–49
235. Veen, H. 1966. *Acta Bot. Neer.* 15:419–33

236. Ward, T. M., Holly, K. 1966. *J. Colloid Interface Sci.* 22:221–30
237. Wheatley, G. A., Hardman, J. A. 1965. *Nature* 207:486–87
238. Wheatley, G. A., Hardman, J. A. 1968. *J. Sci. Food Agr.* 19:219–25
239. White, J. L., Mortland, M. M. 1970. See Ref. 161, 95–100
240. Windmuller, H. G., Ackerman, C. J., Engle, R. W. 1959. *J. Biol. Chem.* 234:895–99
241. Winteringham, F. P. W., Harrison, A., Bridges, R. G., Bridges, P. M. 1955. *J. Sci. Food Agr.* 6:251–62
242. Wright, W. L., Warren, G. F. 1965. *Weeds* 13:329–35
243. Wolcott, A. R. 1971. See Ref. 161, 128–38
244. Wong, A. S., Crosby, D. G. 1972. *The Photodecomposition of Penta-chlorophenol in Water.* Presented at West. Reg. Meet. Am. Chem. Soc., San Francisco
245. Woodwell, G. M. 1967. *Sci. Am.* 216: 24–31
246. Woodwell, G. M., Craig, P. P., Johnson, H. A. 1971. *Science* 174: 1101–7
247. Yamamoto, I. 1969. *Residue Rev.* 25:161–74
248. Yamamoto, I., Kimmel, E. C., Casida, J. E. 1969. *J. Agr. Food Chem.* 17:1227–36
249. Yule, W. N., Cole, A. F. W., Hoffman, I. 1971. *Bull. Environ. Contam. Toxicol.* 6:289–96
250. Zabik, M. J., Schuetz, R. D., Burton, W. L., Pape, B. E. 1971. *J. Agr. Food Chem.* 19:308–13

Ann. Rev. Plant Physiol. 1973. 24:493–518

BIOCHEMICAL GENETICS OF HIGHER PLANTS[1]

✦ 7557

Oliver E. Nelson Jr.[2] *and Benjamin Burr*[3,4]

Laboratory of Genetics, University of Wisconsin, Madison

CONTENTS

The investigations that might logically be discussed under this inclusive title are numerous and diverse in nature. They encompass the attempts of geneticists to elucidate the biochemical basis for mutant phenotypes and the attempts of plant physiologists to utilize mutant genes as experimental tools in investigations of biosynthetic pathways. Owing to this diversity of subject matter, we intend to be selective in this review, emphasizing areas in which considerable progress has been made, in which potentialities for progress exist, in which the existence of mutant strains has markedly facilitated investigations, or simply those areas closest to our particular interests. Further, since one of us reviewed the general

[1] For supplementary bibliographic material (216 references, 17 pages) order NAPS document 02087 from ASIS, c/o Microfiche Publications, 305 East 46th St., New York, NY 10017. Remit in advance for each NAPS accession number $1.50 for microfiche or $5.00 for each photocopy (15¢ for each additional page over 30). Make checks payable to Microfiche Publications.

[2] The preparation of this review was supported in part by Grant No. GB-15104 from the National Science Foundation.

[3] Present address: Service de Biochimie Cellulaire, Institut Pasteur, Paris 15è.

[4] Acknowledgment: We thank Ming Tu Chang for her able assistance in the literature survey.

area 6 years ago (96), we intend in most instances to cover only the intervening span of years. In light of this eclectic discussion of the recent publications, we are placing on file with NAPS[1] a list of references from the last 6 years that are germane to the field but could not be discussed in a short review.

The use of strains carrying a mutant gene as probes in physiological investigations has been limited naturally to those mutants in which the biosynthetic path or process affected is known. These instances are few but have stimulated a number of fruitful investigations. See, for example, the review of Porter & Anderson (106) on carotenoid biosynthesis or that of Levine (69) on mutations affecting the photosynthetic apparatus. It has not been possible to detect induced mutations easily because of diploidy or higher levels of ploidy in the sporophytic generation of higher plants. Nor have good selective techniques been available to aid in the isolation of numbers of mutants that affect steps in a particular biosynthetic pathway. Recent advances in plant tissue culture techniques, in the derivation of haploid callus cultures, and in the induction in these cultures of haploid plantlets which can be diploidized, however, set the stage for genetic manipulations that may well relieve such constraints on induction and identification of desired mutants. These developments allow the cells of higher plants to be handled in much the same manner as microbial populations. Considering, in addition, the totipotency of plant cells so that for some species one can obtain adult plants from callus tissue, there emerges the outline of new systems of investigating plant genetics and of plant breeding that may yield striking advances in the future.

SOMATIC CELL GENETICS

Starting with the report by Guha & Maheshwari (46) in 1964 that the culture of anthers from *Datura* could yield embryo-like structures, there has been considerable activity in the field. The derivation of haploid plants from anther cultures has been recently reviewed by Nitsch (100), who reported successful attempts with various *Nicotiana* and *Datura* species, and also *Oryza* and *Brassica*. Nitsch emphasizes the ease with which diploid *Nicotiana* plants may be derived from the haploid plants resulting from anther culture.

While anther cultures may be the easiest method of isolating haploid cell cultures or plants, it is not the only route for haploid derivation. For the past 20 years maize breeders have, on a small scale, derived homozygous inbred lines by the two-stage procedure of detecting the haploid plants which occur spontaneously at a low frequency and then inducing diploid formation. This subject has been reviewed recently by Chase (20). The detection of spontaneous haploid plants in pepper (*Capsicum annuum* L.) has also been reported by Pochard & Dumas de Vaulx (105). Kermicle (62) has shown that maize plants homozygous for the *indeterminate gametophyte* mutation will produce androgenetic haploids at a relatively high frequency. It is a reasonable supposition that seeds with haploid embryos are produced at low frequencies in many plant species. Genetic marker systems often can be arranged to facilitate the detection and rescue of such plants as shown for maize by Chase (20). Haploid callus cultures can then be derived from the plants.

For mutation induction and detection, cell cultures grown as suspensions are most easily manipulated. Suspension cultures can be grown for most species that will grow in callus culture (147). For some species, cells grown in suspension culture can be induced to differentiate into plantlets again (148).

Auxotrophic mutants following mutagenization of haploid cells have been recovered by Carlson in the fern *Todea barbara* (L.) (15) and in tobacco *Nicotiana tabacum* (L.) (16). In *Todea*, haploid spores were treated with ethyl methanesulphonate as a mutagenic agent. Following germination on minimal medium to produce a three or four-celled prothallus, the sporelings were allowed to take up 5-bromodeoxyuridine (BUdR) for 48 hours in the dark, after which the sporelings were exposed to fluorescent lights in the expectation that nonmutant sporelings, having grown on the minimal medium and incorporated BUdR into the DNA of their chromatin, would be sensitive to near-visible radiation. Such sensitivity results in chromosomal aberrations, cell death, and ultimately plant death. The 1.7 percent of the sporelings that survived were presumably those that were auxotrophic owing to a mutational lesion and hence not growing on the minimal medium or growing slowly enough that little BUdR was incorporated into their DNA, thus allowing survival on ensuing light exposure. Of 200 surviving plants, 33 were apparently auxotrophic mutants. The tobacco experiment followed a similar protocol except that haploid plants were derived by culturing anthers. Haploid callus cultures were derived from the plants. Finally, suspension cultures of largely haploid cells were obtained from the callus cultures. Of 119 callus isolates following mutagenization, BUdR incorporation, and light exposure, 6 were apparent auxotrophs, but all were leaky: i.e. grew to some extent on minimal medium but more rapidly when a particular nutritional supplement was supplied. Carlson suggested that the explanation may lie in the genetic constitution of *N. tabacum*, an allopolyploid species. The haploid genome thus contains the complete genomes from two closely related species, each with a complete set of functional genes. Under this explanation, the induced mutation is assumed to affect a particular locus in one genome, leaving unimpaired the homoeologous locus in the other. The functioning of the unchanged locus allows some growth by the partial auxotroph. If this is the correct explanation, a second round of mutagenization and selection might result in the production of complete auxotrophs.

These six presumptive *Nicotiana* mutants required in different instances hypoxanthine, biotin, para-amino benzoic acid, arginine, lysine, and proline. The *Todea* mutants required one of a variety of different supplements such as an amino acid, a vitamin, or a nitrogen base. Previous attempts have been made to derive auxotrophic mutants from higher plants (notably tomato and *Arabidopsis*) by applying mutagenic agents to seeds and allowing the resulting plants to undergo meiosis and self-fertilization in order to detect the mutations induced in these diploid systems. The result of such endeavors has been a preponderance of mutants requiring thiamin as a supplement (96). It is not clear that any complete auxotrophs other than thiamin auxotrophs have been isolated. Seemingly, metabolic lesions in the developing embryos conditioned by various mutations cannot be compensated by cross-feeding from competent maternal tissue except for the mutations affecting steps in thiamin biosynthesis. Completely auxotrophic mutant plants derived from mutagenized haploid cells in cultures would allow a test of this hypothesis explaining the absence of other than thiamin auxotrophs in the usual type of mutant derivation.

Plants have been derived from four of the six partially auxotrophic *Nicotiana tabacum* calluses reported by Carlson (16), although no further studies have been reported. In most instances, mutant clones of cells would be most useful to investigators following differentiation, although this does not preclude the utilization of mutant cell cultures as investigative tools. Heimer & Filner (51) and Widholm (153) have reported tobacco cell variants that are resistant to the presence of threonine and 5-methyltryptophan, respectively, in the culture medium. The threonine-resistant mutant appears based in the insensitivity of the nitrate uptake system to threonine. The 5-methyltryptophan-resistant variant reported by Widholm has an altered anthranilate synthetase that is more resistant to inhibition by *l*-tryptophan or 5-methyltryptophan. The tryptophan pool in the mutant cells is 10 times that in the normal cells, supporting the hypothesis that feedback inhibition of anthranilate synthetase by tryptophan plays an important role in regulating the quantity of tryptophan produced. The data reported on the anthranilate synthetase of the resistant variant do not allow a decision as to whether normal enzyme is also present, as would be expected since the cell cultures were derived from diploid plants. The mutational event altering the structural gene for anthranilate synthetase on one chromosome would not affect the synthetase locus on the homologous chromosome, and these resistant cells should produce both altered and normal enzyme.

With these demonstrations that for some species of higher plants it is possible to derive haploid cell cultures by various techniques, induce mutations in these cells and select for mutant cells which can give rise to haploid plantlets, and finally from haploid plants derive diploid plants for subsequent biochemical and genetic investigations, we see the broad outline of experimental protocols that could relieve the stringent restriction on correlated biochemical and genetic investigations in higher plants. This restriction has been the difficulty of isolating desired mutations in diploid systems where one is required to screen very large populations, usually without the aid of selective techniques. Added to this impediment is the fact that most auxotrophic mutants do not seem to survive embryonic development when borne on heterozygous, phenotypically normal plants. When all the necessary steps from derivation of haploid cell cultures through induction of plantlet development from callus tissue and ultimately recovery of diploid plants from the haploid plantlets are possible for more plant species and, in particular, for those that are studied genetically most frequently, there will undoubtedly be enhanced interest in the field followed by substantial advances.

It would be appropriate here to mention other advances in the manipulation of plant cells although these are not integral with the steps described above. The first point is the facile production of plant protoplasts from different tissues in a number of species, usually by enzymatic digestion of cell wall components (24). However, protoplasts generally have not survived in culture on a defined medium, and thus an important technical advancement was made by Nagata & Takebe (90, 91), who were able to obtain large quantities of protoplasts from tobacco leaves. When these cells were plated on a solid agar medium, 60 percent regenerated cell walls and formed colonies. These colonies differentiated and produced whole plants when transferred to a suitable medium. A major expectation has been that the availability of plant protoplasts would afford the opportunity for fusion in culture of cells derived from different donors within the same species or from different species. Such a fusion of protoplasts from different species to produce hybrid cells which give rise to plants that are interspecific hybrids has recently been reported by Carlson, Smith & Dearing (18). By the methods discussed above, they produced a hybrid between two *Nicotiana* species, *langsdorffii* and *glauca*. Since it is possible to produce this hybrid by crossing, the "parasexual" hybrid could be compared to the sexual hybrids. It was shown to be similar in karyotype and morphology. The investigators were aided by the known inability of cell cultures from the parental species to grow on media lacking growth substances, whereas the hybrid cells do so readily. This attribute of the hybrid cells provided a selective screen for their recovery, a circumstance which may be more difficult to arrange in other situations. The demonstration is, nonetheless, of considerable importance for plant genetics and plant breeding.

It is important to point out that while hybrid cells have been produced in cultures of mammalian cells, usually chromosomes from one of the parental lines are rapidly lost and selective techniques are required to maintain a chromosome or a chromosome fragment from one parent in the "hybrid" nucleus. Hybrid plant cells produced in culture are more likely to be useful for the production of interspecific hybrids which have not been obtained because of barriers to normal crossing or for the exchange of genetic information that would not otherwise be possible.

MUTATIONS AFFECTING THE PHOTOSYNTHETIC APPARATUS

Mutants that affect synthesis of photosynthetic pigments, chloroplast organization, or electron transport systems are relatively common in higher plants and have been intensively investigated. Levine (69) recently has reviewed such investigations in both algae and higher plants, as have Levine & Goodenough (70) in *Chlamydomonas*. Considering these recent and comprehensive reviews of the area, we will deal only with recent investigations of the chlorophyll-deficient mutants that fix CO_2 per unit chlorophyll at a rate greatly in excess of the wild type rate. Such mutants have been reported in tobacco by Schmid & Gaffron (123), in the garden pea by Highkin et al (53), in soybean by Keck et al (60, 61) in *Lespedeza procumbens* by Clewell & Schmid (23), and in cotton by Benedict, McCree & Kohel (9). At saturating light intensities, which for all mutants except the soybean mutant are higher than for corresponding wild type plants, the mutant plants fix as much CO_2 per unit leaf area as wild type controls. Because the chlorophyll content in the various mutants is less than wild type, CO_2 fixation expressed on a chlorophyll basis is higher than in normal plants. The investigations of these mutants provide clear demonstrations that the rate limitation on CO_2 fixation is not imposed by chlorophyll content. Clewell & Schmid (23) have suggested that at high light intensities most of the light striking normal leaves is wasted because of the inability of the remainder of the photosynthetic system to cope with the flux of primary photoproducts.

The soybean mutant, for which the saturating light intensity is not higher than for normal controls, has been intensively investigated (60, 61). The viable mutant plants (light green) are heterozygous for the mutant allele. The homozygous mutants are unable to survive. [This is true also of the tobacco mutant of Schmid & Gaffron (123).] The authors refer to plastids extracted from the heterozygous plants as mutant plastids. Electron transport and phosphorylation in both cyclic and noncyclic phosphorylation systems were 3–5 times faster than normal when calculated on a chlorophyll basis and 1.5–2.5 times faster on the basis of lamellar protein.

This rate change is attributable both to a higher first order rate constant for plastoquinone oxidation and to a three to fivefold larger pool of reducible plastoquinone. The authors suggest, as has Witt (156) on other grounds, that plastoquinone oxidation is the rate-limiting step in electron transport. The basis for faster plastoquinone kinetics is not clear but may be connected with the lower lipid content of the mutant plastids. This might, in turn, facilitate the diffusion of plastoquinone from reducing to oxidizing sites. As is typical of most chlorophyll-deficient mutants, there are fewer lamellar stacks and more single lamella in the heterozygous mutant plastids. Possibly in all these mutants the lower chlorophyll content is a consequence of altered chloroplast structure.

In the previous review under this title (96), an intriguing question considered was the genetic autonomy of certain cellular organelles—to what extent does the DNA known to be present in an organelle code for the structure of the organelle and for the functions unique to the organelle. The review by Tewari (141), the symposia proceedings edited by Boardman et al (11) and Miller (87), and the book of Kirk & Tilney-Bassett (64) provide critical insights into the partial functional autonomy of the organelles—a situation in which information encoded in nuclear genes can be shown unambiguously, by the use of mutant strains, to be necessary for most steps in the formation of photosynthetic pigments, for the activity of some chloroplast and mitochondrial enzymes, and also for some components of the electron transport chain (64, 95, 96, 150).

With the postulated origin from symbiotic inclusions of cellular organelles such as chloroplasts and mitochondria (see Margulis 82 and Cohen 25), geneticists have been presented the problem of attempting to explain how and why genes encoding information for a number of organelle functions were transferred to the host nucleus in the course of evolution. The simplest answer would seem to be the integration of an entire organelle genome into the host nuclear genetic material followed by mutational loss or deletion from the organelle genome of many genes supporting functions of the organelle. However, Raff & Mahler (112) have recently challenged this prevalent view of organelle evolution. If, as they postulate, the mitochondrion arose in a protoeukaryote from the association of a prokaryotic mesosome and a plasmid which contributed the DNA component, the apportionment of control of organelle function between nuclear genes and organelle DNA is still intrinsically interesting. It is not, however, the genetic problem posed by symbiotic origin of mitochondria or chloroplasts.

ISOZYMES, ALLELIC AND NONALLELIC

Isozymes, multiple molecular forms of enzymes capable of catalyzing the same reaction, may be coded by different alleles at the same locus or by genes at different loci (nonallelic). For some multimeric enzymes, the active form may be an aggregate of monomers coded by two or more different loci. Markert (83) and Kaplan (59) have discussed the genetic and nongenetic factors underlying isozyme variability. The investigations of allelic isozymes in different plant species generally fall into one of three broad categories. Population geneticists have estimated the extent of enzyme polymorphism present in populations, and by comparison between different populations within a species have endeavored to ascertain whether different alleles at a locus confer a selective advantage under particular environmental circumstances or whether they are neutral in terms of fitness. Crow (28) has reviewed the subject of neutral mutations. The investigations of allopolyploid species and the presumptive parental species have been designed to test the proposed phylogenetic relationship. Finally, there are investigations intended to be informative concerning the subunit structure of the functional enzyme or the factors regulating enzymatic activity in different tissues and stages of development.

It should be pointed out that in all these investigations enzymatic activity is detected by specific stains following electrophoresis on polyacrylamide or starch gels and incubation in a reaction mixture containing the substrate for enzyme action. Hence enzymes such as various dehydrogenases, esterases, peroxidases, acid or alkaline phosphatases or aminopeptidases are extensively examined while many enzymes (less amenable to such techniques) receive no attention. Such investigations of alterations in electrophoretic mobility by geneticists are not often correlated with studies of possible changes in kinetics or stability of the enzyme.

The report of Lewontin & Hubby (75) showed that for natural populations of *Drosophila pseudobscura* about 39 percent of the loci in the genome were polymorphic over the whole species, and on the average, an individual was heterozygous for about 12 percent of its loci. The implications of this widespread polymorphism [which has now been demonstrated for man and numerous animal species (48, 129)] for population genetics theory nave been discussed by Lewontin & Hubby (75). Such studies have been extended also to plant populations. Kahler & Allard (58) have investigated esterase activity in the plumule of barley seedlings and found seven loci producing esterase enzymes. Extensive polymorphism was found for four loci. One locus was invariant, and the situation with regard to the two remaining was not clear. Clegg & Allard (22) have reported on the allelic frequencies at five loci in 14 populations of the slender wild oat *Avena barbata*, growing in different regions of California. They found the allelic substitutions in these populations to be nonrandom and to be correlated with the environmental conditions. Neither of these findings is in accord with the hypothesis that the various alleles at the loci scrutinized are neutral in terms of selective advantage.

The use of isozymic differences to ascertain phylogenetic relationships has been briefly reviewed by Mitra & Bhatia (88). In their investigations of alcohol dehydrogenase and esterase-1 activity in extracts of dry seeds of diploid, tetraploid, hexaploid, and octoploid wheats, they found that the number of variant enzyme bands increased with ploidy level owing to the existence of parental enzyme bands plus hybrid bands formed by heteromonomer association. An indication that such techniques may not be totally reliable in assessing relationship has come from the work of Brewer, Sing & Sears (12) with hexaploid wheat. They assayed electrophoretically 38 nullisomic-tetrasomic lines (lacking one pair of chromosomes contributed originally by one of the diploid progenitors but having four of a homoeologous chromosome from one of the other presumed parents) for 12 enzymes. Only for alkaline phosphatase were electrophoretic differences seen, and these could be localized to chromosomes 4B and 4D. For the other 11 enzymes, no differences were found. The absence of observed differences presents a problem since the presumed diploid progenitors displayed many differences in electrophoretic mobilities for these enzymes. The authors suggested that either the diploid species have undergone divergence since contributing a genome to the formation of hexaploid wheat or that there has been selection for identity of the bands possibly via infrequent recombination between homoeologous chromosomes.

The results of Hart (49, 50) with regard to alcohol dehydrogenase (ADH) activity in hexaploid wheat are more in accord with expectations. The extracts showed three bands of ADH activity following electrophoresis. Investigations of nullisomic-tetrasomic lines indicated that genes coding for ADH subunits are located in chromosomes 4A, 4B, and 4D—the homoeologous chromosomes from each of the three diploid progenitors. Results were in accord with the hypothesis that the 4A chromosome locus codes for monomer α, the 4B locus for β, and 4D for δ with random association of the monomers to produce six combinations of active dimers. The three observed zones of activity are presumably ADH-1 $\alpha\alpha$, ADH-2 $\alpha\beta$ and $\alpha\delta$, ADH-3 $\beta\beta$, $\beta\delta$, and $\delta\delta$.

Carlson (17) has shown in *Datura* that it is possible to localize to chromosomes genes coding for enzymes for which no electrophoretic variants have been detected. This is accomplished using aneuploid lines [in this case, trisomics $(2N+1)$ with three rather than two chromosomes of a given set] and making quantitive estimates of enzyme activity in extracts of callus tissue derived from the normal plant or the various primary trisomics. Such studies also could have been carried out on tissues from intact plants. The finding that activity for a given enzyme in a particular trisomic culture was approximately 150 percent of that in the diploid culture was taken as a preliminary indication that the structural gene for that enzyme was located on the chromosome in triplicate in that trisomic line. Confirmation should come from similar tests of cultures from the two secondary trisomics (in which one of the two chromosome arms is present twice in the added chromosome while the other arm is absent). The expectation would be that extracts of one secondary trisomic would have enzymatic activity equivalent to the diploid since it lacked the chromosome arm on which the gene was located. The other secondary trisomic should give enzymatic activity twice that of the diploid, since the chromosome arm carrying the gene would be present twice giving four copies of the locus. Such an expectation was realized for lactate dehydrogenase on chromosome 3, malate dehydrogenase on chromosome 5, isocitrate dehydrogenase on chromosome 2, and 6-phosphogluconate dehydrogenase on chromosome 3. Carlson points out that the successful application here of this technique, which requires that enzyme activity be linearly proportional to the number of structural genes present, implies that gene activity is regulated on the transcriptional level and that the rate of transcription is the same for each copy of the gene present.

The genetic control of plant isozyme activity has received considerable attention. The subject has been reviewed recently (120). Many instances exist of straightforward control of electrophoretic variants of a particular enzyme by different alleles at a given locus, with the added possibility that if the active enzyme is polymeric, subunits of different electrophoretic mobility may combine at random to form a hybrid enzyme (if a dimer) or hybrid enzymes (if of a higher order of aggregation) that have different mobilities than the homopolymeric associations. The number of hybrid bands observed in plants heterozygous for alleles producing subunits of different electrophoretic mobility allow inferences as to the degree of aggregation that produces an active enzyme (119). In general, however, our interest lies in what examination of such systems may disclose about the regulation of gene activity.

Maize has been a favored organism for such studies owing to the genetic variability present and the ease of genetic manipulation. For this species, alcohol dehydrogenase (ADH) activity has been intensively investigated. Nevertheless, disagreement still exists between investigators as to the genetic basis for the two zones of ADH activity extracted from the scutellar tissue of developing maize seeds. Schwartz (125) has concluded that slow-migrating, intensely staining band or bands (Set I) are formed by dimers of monomers produced by alleles of the Adh_1 gene. The less active, faster-migrating Set II bands result from the dimers formed between AdH_1 monomers and the monomer produced by an unlinked Adh_2 gene. The Adh_2 gene product is inactive as a monomer or homodimer. In homozygous plants, the migration rate of the Set II band is always positively correlated with the migration rate of the Set I band. In support of his hypothesis, Schwartz reported Adh_1 mutants produced by treating $Adh_1{}^S/Adh_1{}^S$ homozygotes (where $Adh_1{}^S$ indicates the allele at Adh_1 locus that produces an enzyme that is slowly anodal-migrating) with ethyl methanesulphonate and then pollinating the plants with pollen from Adh_1^F/Adh_1^F plants (Adh_1^F being an allele producing a fast-migrating ADH). He found four mutants with altered mutation rates in the Set I region, and in each mutant there was a corresponding change in the migration of the Set II band. Scandalios (118) has postulated that the genetic basis for the ADH activities is two closely linked genes with the product of one gene active in the dimer form (Schwartz's Set I activity) and the product of the other active as a monomer (Schwartz's Set II activity). The concomitant variation in migration rates of Set I and Set II enzymes is explained in this hypothesis as a consequence of the close linkage of the structural genes for the slow or the fast-migrating activities. In support of this hypothesis, Scandalios (121) reported presumed recombinants between the closely linked Adh genes. The improbability of obtaining such an array of recombinant zygotes has been noted by Lewontin (74).

Of further interest concerning the tissue-specific control of ADH activity are the following observations. Efron & Schwartz (38) have shown that ADH activity in embryo extracts is rapidly lost when root extracts are added. The basis of this irreversible reaction is the presence of two easily dialyzable factors. One of these is present in the root extract and the other in the embryo extract. Neither alone will inhibit enzymatic activity. The authors suggested that this two-factor system may account for the rapid loss in ADH activity in the germinating seeds where it might be detrimental. It is not known if these factors specifically inhibit ADH activity or are capable also of affecting other enzymes. Schwartz (127, 128) has reported on the structure of monomers as it relates to ability to dimerize or to the stability of the dimers. In one instance, the monomers of two Adh_1 mutants with null activity but producing a product that cross-reacts with antiserum against the wild type enzyme were shown to be either incapable of forming dimers with monomers produced by active alleles or else preferentially forming homodimers. Also, a temperature-sensitive Adh_1 allele with low homodimer activity forms heterodimers with monomers produced by the $Adh_1^{c(t)}$ allele that are as active as $C^{(t)}$ homodimers. Further, the heterodimers are resistant to temperatures or dialysis conditions that completely inactivate the homodimers produced by the temperature-sensitive allele. It is suggested that the basis of temperature sensitivity lies in fewer or weaker intradimer bands in the homodimer with this condition corrected, partially at least, in the heterodimer. Schwartz (126) has advanced a model designed to explain activity differences between alleles and to allow for apparent changes in activity between tissues or between developmental stages. The hypothesis invokes a limiting factor that is necessary for activity (transcription) and for which the alleles compete with different degrees of effectiveness. It is then assumed that the competitive ability of an allele changes with time during development or from tissue to tissue. This model offers a formal explanation of the changes in activity of different Adh_1 alleles but relies on a change in the competitive ability of alleles for which a molecular explanation is difficult. Efron (36) has identified a second locus, Adh_r, linked with the Adh_1 (ca 17 crossover units) at which a particular allele Adh_r^L reduces the activity produced by the

$Adh_1{}^S$ allele but not by the $Adh_1{}^F$ allele. In a second paper (37), he has attempted to explain the effect in terms of Schwartz's limiting factor hypothesis.

In an investigation of genetic control of α-amylase activity in developing maize endosperms, Chao & Scandalios (19) have reported the effects both of gene dosage and differential allelic expression of alleles at the Amy_1 locus. At this locus, two electrophoretic variants are produced by two codominant alleles A and B. From the cross $B/B \times A/A$, where B/B is the female parent, both enzymes are found in the developing endosperm but only A in the scutellum for 8 days with both A and B expressed thereafter. In the reciprocal cross, only the isozyme typical of A is expressed either in the endosperm or in the scutellum as long as it can be checked. In either F_1 progeny, both isozymes were present in the root and shoot tissues or in the young leaves. This appears to be formally similar to the situation Kermicle (63) described for the R locus (one of the complementary series of genes affecting anthocyanin synthesis in the maize seed) where it can be shown that certain alleles are temporarily modified in their pigment producing capabilities in the developing seed by passage through the male gametophyte.

Efron (35) has reported a tissue-specific effect on the electrophoretic mobility of the acid phosphatase produced by alleles at the Ap_1 locus in maize. In plants homozygous for any of the alleles known at the locus, there is one electrophoretic species in the pollen, a second in the scutellum, while both forms are found in the leaf. As yet it is not possible to discriminate between several explanations—that the enzyme exists in different conformers in the pollen and in the scutellum with both in the leaf, that a changed group is conjugated with the enzymatic protein in the pollen or the scutellum or both, or that the enzyme is complexed with different proteins in the pollen and scutellum.

The majority of the foregoing discussion has been concerned with allelic isozymes. The instances of isozymes specifically produced in particular tissues, organs, or generations and coded by separate loci are particularly interesting, both to geneticists and plant physiologists, since they represent the end product of evolutionary specialization that has produced enzyme systems specifically adapted for particular milieus. Examples have been reported in higher plants. Akatsuka & Nelson (2) concluded on the basis of genetic and biochemical studies that the starch granule-bound nucleoside diphosphate sugar-starch glucosyl transferases of the maize embryo and maize endosperm were coded by different genes. Dickinson & Preiss (33) noted that in seeds homozygous for the *shrunken-2* mutation (which greatly reduces adenosine diphosphate glucose pyrophosphorylase activity in developing seeds) the activity associated with the embryo was about 16 percent of the total activity but only 1 percent in the case of normal seeds. Since the homozygous mutant seeds have only 8–10 percent as much activity in the endosperm as do normal seeds, the proportion of enzyme activity attributable to the embryonic tissue in the mutant could be explained if the *shrunken-2* mutation affected the endosperm activity but not the embryo activity. This would be possible if different structural genes coded for the activities found in the embryo and in the endosperm. Subsequently, Preiss, Lammel & Sabraw

(108) demonstrated that the embryo enzyme could be distinguished from the endosperm enzyme on biochemical grounds. Salamini, Tsai & Nelson (117) demonstrated the presence of a unique glucose phosphate isomerase in maize embryos that was distinguished from an endosperm enzyme by a difference in electrophoretic mobility and a different response to the presence of ATP in the reaction mixture. This evidence, however, is not as strong as that in the two instances above (2, 33) where it was possible to demonstrate that mutations that condition null or little activity of an endosperm enzyme do not affect activity of an embryo enzyme.

Chourey (21) has reported interallelic complementation between various mutant alleles at the *shrunken-1* locus in maize with restoration of normal phenotypes in heterozygous seeds. Unfortunately, the metabolic lesion associated with this locus is not known although extracts of nonmutant seeds and some mutants show on electrophoresis a densely staining protein band attributable to this locus. This is the first report in higher plants of intragenic complementation sufficient to restore a normal phenotype. Li & Rédei (76) have shown for *Arabidopsis* that three temperature-sensitive alleles at the *pyrimidine* (*py*) locus are capable of complementing in two combinations py^{4ts}/py^{3ts} and py^{4ts}/py^{5ts}, but the complementation was still incomplete when compared to wild type.

Warner et al (151) found that two maize inbreds, B14 and Oh 43, produce an F_1 hybrid with a higher level of nitrate reductase activity than either parent. Advanced generations revealed that the inbreds differ at two loci in genes affecting nitrate reductase activity with one inbred possessing the dominant alleles at one locus and the recessive alleles at the other. This relationship was reversed in the other inbred. The Oh 43 inbred had a higher rate of enzyme synthesis when induced and a more rapid rate of enzyme decay in vitro or in vivo. The higher level of enzyme activity in the hybrid was attributed to the intermediate rate of enzyme synthesis and decay which was found in the hybrid.

MUTATIONS THAT AFFECT THE SYNTHESIS OF STORAGE PROTEINS IN MAIZE

The maize mutants *floury-2* and *opaque-2*, which characteristically increase the content of protein-bound lysine and tryptophan (the limiting essential amino acids in maize seeds for human or monogastric animal nutrition), do so by decreasing the amount of the alcohol-soluble zein proteins synthesized and permitting a secondary increase in the amount of the water- and salt-soluble proteins (97). Thus the alteration in the amino acid profiles of the mutants as compared to normal maize is based on the change in the relative proportions of proteins with disparate amino acid compositions that are synthesized in the mutants and normal maize. The reasons for the assumption that such changes will be the usual if not the only basis for significant changes in the amino acid composition of the collective proteins of plant seeds have been detailed (97).

Although the changes in the solubility fractions of the endosperm proteins resulting from these mutations have been extensively investigated, the primary mutational lesion remains unidentified. Dalby & Davies (30) and Wilson & Alexander (155) found independently that *opaque-2* kernels had much enhanced ribonuclease activity compared to nonopaque controls. Dalby & Davies showed that at 21 days postpollination the ribonuclease activity of W64 $o_2/o_2/o_2$ endosperms was ca sevenfold higher than W64A $+/+/+$. Since the alcohol-soluble zein proteins comprise more than 50 percent of the total protein synthesized in the normal maize endosperm, there are either numerous copies of the RNA messages for zeins present or the messages are more stable than those for other proteins or both. There is no evidence that allows a choice between these alternatives, but if the second were correct, the enhanced ribonuclease activity in the mutant would be expected to be more disruptive of zein synthesis than if the first pertained. Under this view the synthesis of greater quantities of albumins and globulins in the mutant is then a secondary response following the partial repression of zein synthesis.

Mehta et al (86) have also studied development of ribonuclease activity in *opaque-2* kernels and substantiate the previous observations (30, 155) of much enhanced activity. Dalby & Cagampang (29), in a further study of ribonuclease activity in developing seeds of *opaque-2* and normal maize, have reported that in *opaque-2* endosperms the rate of ribonuclease synthesis from 10–15 days postpollination was more rapid than in normal endosperms. After 15 days the rates of synthesis were the same for both genotypes and were exponential for a period of 11 days. The authors concluded that the activity of the *opaque-2* gene ceased at 15 days. Such an interpretation assues the mutant allele is active rather than conditioning a loss of normal activity. We think this unlikely in view of the completely recessive nature of the mutant allele since the opaque phenotype is seen only in $o_2/o_2/o_2$ endosperms and the changes in amino acid profile are seen only in homozygous mutant endosperms (7). Dalby & Cagampang (29) report only minor differences in ribonuclease activity in heterozygous endosperms when compared to normal endosperms.

The same authors make an interesting point regarding ribonuclease activity, namely, the lack of increased activity in *floury-2* endosperms. The *floury-2* mutant conditions a similar change in amino acid profile to that conditioned by *opaque-2*, and the change is based also on a similar change in proportions of proteins synthesized (56). And although the albumin fraction is increased as it is in *opaque-2*, ribonuclease activity is not increased. This suggests that a rise in ribonuclease activity is not an inevitable consequence of partial derepression of synthesis of the albumin fraction.

Sodek & Wilson (132) injected leucine ^{14}C or lysine ^{14}C into *opaque-2* and normal plants just below the developing ear and, by monitoring the incorporation of the label into proteins formed in the seed, found a striking difference betwen mutant and normal. When lysine was injected, most of the label incorporated into protein in *opaque-2* was present in lysine, but in normal seeds much of the label was found in glutamate and proline. When labeled leucine was injected, most of the label recovered was in leucine for both genotypes. For the *opaque-2* mutant, Denic (32) found that in vitro protein synthesizing systems incorporated more lysine and less leucine relative to a microsomal fraction from normal maize. These differences are in accord with the composition of the proteins synthesized by these genotypes in vivo. The incorporation of leucine or lysine by microsomal fractions from either genotype was independent of the source of the transfer RNAs. This finding would seem to exclude any role of the transfer RNAs in producing the mutant effect.

Wolf et al (157) were unable to detect protein granules (rich in zein) in cross-sections of *opaque-2* endosperms in the light microscope while such granules were conspicuous in normal endosperms. Electron micrographs, however, revealed that the protein granules were present but reduced to approximately 1/20 the diameter of those present in normal endosperms. Goodsell (44) found that normal and *opaque-2* kernels taken from segregating ears contained significantly different quantities of potassium with the mean normal content being 0.37 percent and the *opaque-2* 0.52 percent.

It has been suggested that both *opaque-2* and *floury-2* are regulatory genes for zein synthesis (97). There are other instances in which it may be possible to manipulate regulation of protein synthesis advantageously. Riley & Ewart (114) have produced chromosome addition lines in which a pair of rye chromosomes has been added to the genome of hexaploid wheat thus constituting seven addition lines. The protein content and amino acid profile of the addition lines were then measured. They found approximately a 9 percent increase in lysine in the chromosome I addition line and a striking increase in crude protein (3.0 to 3.9 percent units) in the addition lines for rye chromosomes II, IV, and V. The addition line for rye chromosome II gave the largest increase. Jagannath & Bhatia (55) also observed that the addition of rye chromosome II to the genome of spring wheat results in increases in protein content of 2 to 5 percent units with no new protein species being formed.

MUTANTS THAT AFFECT STARCH SYNTHESIS IN THE MAIZE ENDOSPERM

Mutants of maize that affect starch synthesis in the maize endosperm have been studied intensively (27), and in several instances the metabolic lesion associated with a mutation has been reported (98, 144). The mutants *shrunken-2* and *brittle-2*, which were associated with loss of adenosine diphosphate glucose (ADPG) pyrophosphorylase activity by Tsai & Nelson (144), have been shown by Dickinson & Preiss (33), using a more sensitive assay, to have 8 to 10 percent of normal activity. The greatly reduced starch synthesis capabilities of these mutants with low ADPG pyrophosphorylase activity is evidence of the key role of nucleoside diphosphate sugars in starch synthesis, as is the evidence that *waxy* mutants lack the starch granule-bound nucleoside diphosphate sugar-starch glucosyl transferase present in all other stocks (98). At the same time, there is renewed interest in the possibility that a phosphorylase(s) might play a role in starch synthesis. Tsai & Nelson (145) have reported two phosphorylases in developing maize endosperm that increase during the period of starch synthesis but do not appear to be present in the endosperm during germination and mobilization of the starch reserves. De Fekete (31) and Vieweg & De Fekete (149) have concluded on the basis of studies with labeled glucose-1-PO$_4$(^{14}C) that phosphorylase(s) are responsible for starch synthesis in the cotyledon of bean (*Vicia faba*) and the bundle sheath cells of maize leaves. Such studies must be interpreted cautiously, however, since glucose-1-PO$_4$ is also a substrate for ADPG pyrophosphorylase and could produce labeled ADPG. Thus if nucleoside diphosphate sugar-starch glucosyl transferases are also present, labeled glucose would be incorporated in starch without the intermediation of phosphorylase.

The much-reduced phosphorylase activity in endosperms produced by the *shrunken-4* mutant of maize, together with reduced starch synthesis, was taken by Tsai & Nelson (146) to be evidence for a synthetic role of phosphorylase. A re-examination of enzymatic activities in the *shrunken-4* mutant by Burr & Nelson (14) has led to the conclusion that the primary effect of the mutation is on a step in pyridoxal phosphate synthesis with the effect on the phosphorylases being secondary. The activities of all enzymes tested which require pyridoxal phosphate as a cofactor were drastically reduced. Thus the reduced starch synthesis in *shrunken-4* kernels cannot be taken as evidence of the essential involvement of phosphorylase(s) in starch synthesis in the maize endosperm. Total protein content is lower in the mutant as has been noted by Preiss et al (109), who also concluded that the primary effect of the mutation was not on phosphorylase production. The resolution of this question awaits the identification of mutants that affect a structural gene for phosphorylase.

It has been demonstrated that the *waxy* mutants of maize have only very low activity for the starch granule-bound nucleoside diphosphate sugar-starch glucosyl transferase (2). The question of whether the starch granule-bound nucleoside diphosphate sugar-starch glucosyl transferase activity is an artifact reflecting the amylose content of the starch granule has been raised by Akazawa & Murata (4), who reported that amylose but not amylopectin would bind firmly the soluble adenosine diphosphate glucose-starch glucosyl transferase present in developing rice endosperms [such soluble glucosyl transferases are present also in developing maize endosperms (3, 40, 102)]. The implication is that *waxy* starch granules lack the glucosyl transferase because the mutant cannot synthesize amylose for an unknown reason. Therefore, soluble glucosyl transferase cannot be bound to amylose and deposited in the starch granule. Akatsuka & Nelson (3) have argued against this conclusion, showing that for a *Wx* diploid gene dosage series there is a linear proportionality between bound glucosyl transferase activity and the number of *Wx* alleles present (from one *Wx* allele through three *Wx* alleles). This fact suggests that either the *Wx* allele is the structural gene for the functional starch-granule bound glucosyl transferase or the *Wx* allele acts to specify the number of sites on the starch granule to which soluble glucosyl transferase can be bound. More recently, Tanaka & Akazawa (140) have shown that the substrate specificity of a soluble adenosine diphosphate glucose-starch glucosyl transferase was not altered by binding to amylose. The enzyme still utilized as a substrate only ADPG, whereas the bound glucosyl transferase utilized both ADPG and uridine diphosphate glucose. They concluded that in all probability the soluble and bound transferases were different enzymes. The only evidence contrary to such a conclusion is the report of Frydman & Cardini (41) that the bound glucosyl transferase of the starch grains of several species lost the ability to use uridine diphosphate glucose as a substrate upon mechanical disruption without losing the ability to utilize ADPG.

The polysaccharide branching and debranching enzymes in the *sugary-1* mutant of maize have received attention in several laboratories. This mutant (the sweet corn of commerce) is interesting since the principal carbohydrate storage product is a much-branched compound of high molecular weight that is water soluble. It is usually referred to as phytoglycogen. Manners, Rowe & Rowe (80) and Lavintman & Krisman (66) have reported in *sugary-1* kernels the presence of an enzyme that can branch amylopectin in addition to the previously reported Q enzyme that branches amylose. Black et al (10) have shown the amylopectin-branching enzyme to be absent from nonmutant maize and the *amylose-extender* mutant but present in the *dull*, *sugary-2*, and *waxy* mutants in addition to the *sugary-1* mutant. Manners & Rowe (81) have reported three debranching enzymes in *sugary-1* kernels, namely, an R-enzyme that acts only on amylopectin, an isoamylase that can debranch both amylopectin and phytoglycogen, and an $\alpha(1–6)$ glucosidase that can hydrolyze pullulan. Lee, Marshall & Whelen (67) have separated the pullulanase activity on an hydroxyapatite column into two fractions with identical specficity. It is not clear whether these debranching enzymes are unique to the *sugary-1* mutant since comparable studies on non-mutant maize are not reported.

Cox & Dickinson (26) have screened a series of mutants for the ability of leaf disks to accumulate starch when glucose is supplied in the medium. Five chlorophyll-deficient mutants had greatly reduced capacity to accumulate starch when compared with normal siblings. One suspects that in these mutants the ATP required for the conversion of glucose to substrates for starch-synthesizing enzymes is absent in the mutant leaves owing to reduced photophosphorylation.

MUTATIONS AFFECTING CAROTENE SYNTHESIS

Mutants affecting carotenoid synthesis in tomato fruits have been investigated intensively and have been invaluable tools in elucidation of the biosynthetic pathways leading to carotenoid formation in this species (96, 106). The original biosynthetic sequence involving successive dehydrogenation steps leading to lycopene followed by cyclization to give the cyclic carotenoid pigments was proposed by Porter & Lincoln (107) on the basis of the effects of gene substitutions at the *R*, *T*, and *B* loci. Porter & Anderson (106) have modified the original proposal in some respects, and the proposed pathway has received considerable experimental support with regard to the early biosynthetic steps (45). Recent evidence has resolved the uncertainty regarding the cyclization process—whether the cyclic carotenoids, δ-carotene and α-carotene which contain an α-ionone ring are formed by a separate pathway or synthesized from the cyclic carotenoids that contain only β-ionone rings (γ and β-carotene) by a double bond shift within the ring. The results of Tomes (142, 143), who studied carotene production in *B* (conditioning the formation of large quantities of β-carotene) and *Del* (enhanced formation of δ-carotene) double mutants, suggested that the two carotenes are formed by separate paths and compete for lycopene as a substrate. Further, the *Del* or *B* genes are effective only in genotypes where the formation of lycopene is not blocked. Williams, Britton & Goodwin (154) arrived at the same conclusion following labeling experiments with the δ-carotene mutant. Kushwaha et al (65) have demonstrated the conversion of phytoene-^{14}C by soluble enzyme systems derived from the fruit plastids of four different genotypes of tomatoes to various acyclic and cyclic carotenes typical of the geotype from which the enzyme systems were obtained. The authors further showed the conversion of lycopene-^{3}H to α, β, γ and δ-carotene by soluble enzymes from the plastids of red tomato fruits which do not form cyclic carotenes in vivo, suggesting that the red fruits either lack an essential cofactor for the synthesis of the cyclic derivatives or that solubilization frees the enzymes from an inhibitor present in vivo. The synthesis of cyclic carotenoids from lycopene in vitro by enzyme systems from red, high δ-carotene, and high β-carotene genotypes is strong evidence that this is the major pathway for the synthesis of these compounds although it does not exclude some synthesis via a possible alternative route from neurosporene via β-zeacarotene and γ-carotene to β-carotene. Raymundo et al (113) assert that the effect of dimethyl sulfoxide on the synthesis of β-carotene in the *lutescent* and high-β mutants of tomato indicates two operative paths of synthesis.

There are intriguing investigations of the genetic control of synthesis of a variety of other plant products. Although space does not suffice to discuss these studies, we should like to direct the reader's attention to the investigations of Lein (68) and Josefsson (57) on the control of glucosinolate production in rape, *Brassica napus*, and to those of Wettstein-Knowles (152), Macey & Barber (78), and Netting et al (99) on the control of wax formation on the leaves of *Hordeum vulgare* and *Brassica oleracea*, respectively.

GENETIC EFFECTS ON GROWTH FACTOR SYNTHESIS AND RESPONSE TO GROWTH FACTORS

It has been known for a number of years that some types of genetic dwarfism in peas (13) and maize (104) result from blocks in the synthesis of gibberellic acid (GA). This is also true for other species as shown by Proano & Greene (111) for a radiation-induced mutant in bean (*Phaseolus vulgaris*), and by Suge & Murakami (136) for a rice mutant. In other instances where the growth of a variant can be much influenced by GA application, the endogenous levels of GA are comparable with those of normal plants. Halevy et al (47) found that although runner strains of peanut could be converted to the erect form by the application of GA, the endogenous GA levels of runner and erect varieties were similar. The runner plants, however, had three different GA inhibitors in their leaves while erect plants had only two. The third compound present only in runner plants had a chromatographic behavior similar to abscisic acid in three solvent systems. Stoddart (135) reported that a nonflowering red clover mutant could be induced by GA to undergo stem elongation and initiation of floral primordia only under long day conditions. The mutant plants degraded GA$_3$ more rapidly than did normal plants under long day conditions. Stoddart has suggested that the low endogenous supply of GA in the mutant results from this rapid degradation and that exogenous GA is needed to keep the level above a threshold required for floral initiation.

In an inquiry into the basis of GA action, Spiker & Chalkley (133) have found that with a dwarf pea mutant responsive to GA there was no discernible difference in histones from treated and untreated plants. Russell & Galston (116) demonstrated that in a dwarf pea mutant where dwarfism is light dependent and phytochrome mediated, the phytochrome derivative Pfr is stable and remains active, but in normal plants it decays rapidly.

High endogenous auxin concentrations have been implicated as the probable basis of spontaneous tumor formation in the interspecific hybrid *Nicotiana làngsdorffii* x *N. glauca*. Bayer & Ahuja (8) detected significantly higher auxin (probably IAA) concentrations in a second spontaneously tumor-producing interspecific hybrid *N. longiflora* x *N. debneyi-tabacum* and in the tumor-producing segregants in advanced generations. The genetic basis of tumor production is a segment of a *longiflora* chromosome interacting with the *debneyi-tabacum* genome, and since both tumor production and enhanced auxin levels are associated with this segment, the authors suggested that the physiological basis of tumor production is the enhanced level of growth hormone in the tissues. Doering & Ahuja (34) have demonstrated in tomato that tumor production results from the insertion of a dominant gene from *Lycopersicon chilense* into the *L. esculentum* genome. While growth of tissue from these tumors is not independent of exogenous growth factors (as is the case with the tobacco tumor tissue), the tomato tumor tissue does have different growth requirements than normal tissue, being inhibited by levels of IAA or 2-4 D that are beneficial to normal tissue suggesting a higher endogenous auxin level in the mutant. The production of tumors is observed only in the greenhouse, so environmental conditions obviously have an over-riding role. Under greenhouse conditions, GA application results in rapid stem elongation and a cessation of tumor production, but after the induced elongation has run its course, the leaves develop tumors and continue to do so until death (1).

Gelinas & Postlethwait (43) have found that in the dominant maize mutant *Knotted* (where spontaneous outgrowths arise on the leaves) there is a higher level of inhibitors for IAA oxidase activity compared to normal plants and that the levels of these inhibitors increase with the number of *Knotted* alleles present. The major inhibitor is an unidentified ester of ferulic acid.

Tal & Imber (54, 137–139) have investigated extensively the physiological basis of wilting in tomato conditioned by the *flacca* mutation. The mutant plants wilt because their stomata resist closure under conditions that close those of normal plants. The mutant plants, which have a tenfold lower concentration of abscisic acid-like substances in their leaves, undergo a phenotypic shift towards normality with daily applications of *dl*-abscisic acid. If applications cease, mutant plants resume their wilting propensity. With aging, mutant plants are less prone to wilt and concomitantly have increased abscisic acid-like activity. The activity of other growth-regulating hormones is also altered in mutant plants. Auxin-like activity is enhanced in mutant plants as is kinetin-like activity. It is assumed, however, that the higher kinetin-like activity resulting in resistance to stomatal closing is a consequence of the lower abscisic acid-like activity. Additionally, the hormonal imbalance in the mutant plants appears to extend to an increased resistance to water absorption by the roots. Simpson & Saunders (131) have demonstrated that for either tall or dwarf peas, water stress leading to wilting induces a four to ninefold increase in abscisic acid-like activity. This increase results in stomatal closing.

Hertel et al (52) have utilized the maize mutant *amylose extender* to support the statolith theory of geotropic response. In this mutant where the coleoptile starch granules are considerably smaller than in normal maize, the lateral redistribution of *ae/ae* starch granules is 30 to 40 percent less in response to horizontal exposure. Also, with a geotropic stimulus, the lateral transport of auxin is 40 to 80 percent less and response curvature by mutant coleoptiles significantly less than with normal.

INCOMPATIBILITY REACTIONS IN PLANTS

The molecular basis of incompatibility reactions following pollination and leading to self-sterility has been intriguing to geneticists because of the specificity of the reaction and the existence of numerous alleles at the loci determining sterility.

In a gametophytic self-incompatibility system, a pollen grain carrying a particular allele (S_1) will be unable to effect fertilization if the style is produced by a plant also bearing that allele ($S_1/-$). In sporophytic incompatibility systems, the reaction of a pollen grain is determined by the genotype of the plant that produces the grain and not its own genotype. Arasu (5) has recently reviewed incompatibility systems in plants.

Immunological and biochemical techniques have demonstrated that there are specific antigens or proteins in pollen or styles of different self-incompatibility genotypes of several species. Lewis (71) found that rabbits individually immunized with pollen of one of several genotypes produced antisera specific for that genotype of *Oenothera organensis*, a species with gametophytic incompatibility. After these antisera were absorbed with heterologous pollen extracts, they reacted only with homologous pollen extracts. There was no cross-reaction with extracts from a third pollen genotype. Lewis employed the sensitive ring test in which one layers the extract on top of the antiserum in a capillary tube. In the assay, the immune precipitate develops in a matter of minutes. In a later study, Mäkinen & Lewis (79) were able to make antisera to pollen samples of homozygous plants. They employed the less sensitive Ouchterlony technique to observe immune reactions and found that antisera against pollen containing allele S_6 also reacted with a self-compatible mutant of this allele, $S_6 1$. They observed that the antigens diffused from the pollen, and it was not necessary to extract for their determination. In neither study was there an effect on pollen of either genotype by placing compatible and incompatible pollen on the same style or by a prior pollination with incompatible pollen. Thus they ruled out an immune-type response of the style against incompatible pollen (i.e. the response must be preformed). The data also suggest that the reaction is not general but probably occurs at the surface of the pollen tube. A recent study of Ockendon (101) by scanning electron microscopy of compatible and incompatible pollinations of *Brassica oleracea* (sporophytic incompatibility) tends to bear out this hypothesis. Ockendon observed that compatible pollen tubes penetrated the papillae covering the surface of the style but that incompatible pollen tubes generally spread over the surface of the papillae. However, it should be noted that in gametophytic systems the interaction takes place within the style.

More recently, Nasrallah & Wallace (93, 94) found genotype-specific antigens in the style of *Brassica oleracea*. Antisera prepared in rabbits against stigmatic homogenates of three homozygous incompatibility genotypes, S_1/S_1, S_2/S_2, and S_3/S_3, gave evidence that each genotype contained a unique antigen correlated with the presence of a particular S allele. Such homozygous lines can be prepared by making self-pollinations before the flowers open, at which time the incompatibility system is not operative. The stigmas of the heterozygous plants S_1/S_2, S_1/S_3, S_2/S_3 contained the two antigens typical of the S alleles present. In F_2 populations, the genotypes as determined by antigens present in each plant were the same as the genotypes determined by reciprocal crosses with the parental lines. The S antigens are not detectable in the pollen or in any other plant part. Nor are they present in the young flower buds until just before anthesis. Nasrallah, Barber & Wallace (92) demonstrated electrophoretic separation of proteins corresponding to the previously detected antigens. Discrete bands were detected for 5 out of 6 tested genotypes. Although there must be an allele-specific product in the pollen, this has not yet been detected, and the authors suggest the stigmatic S proteins are repressors of enzymes needed for pollen germination. Pandey (103) found that banding patterns of peroxidase activity in *Nicotiana alata* and in hybrids of *N. alata* with *N. langsdorffii* were correlated with incompatibility genotypes. He postulated, as did Lewis (72) earlier, that in incompatible pollinations, two identical dimers combine with the aid of an allosteric molecule to form an active tetramer which in this case is an IAA oxidase. The destruction of IAA results in the repression of genes producing growth substances for pollen tube growth. Nasrallah et al (92) found no electrophoretic variation in peroxidase activity in *Brassica* which could be correlated with incompatibility genotypes.

Ascher (6) has presented a similar model to explain the basis of gametophytic incompatibility. The model invokes the existence of two operons—one a "low velocity growth" operon for pollen tube growth and the second a "high velocity growth" operon. The low velocity operon initiates growth with pollen reserves. If the style and pollen grain contain the same allele, they produce monomers of the same type that combine to form a repressor of the high velocity operon, thus preventing its activation when stylar metabolites reach the pollen tube. The model assumes, of course, a mechanism allowing dimerization only of a stylar monomer with a pollen monomer but not pollen monomers with each other or stylar monomers with each other. The physical basis of such a differentiation may be found in the observations of Lewis & Crowe (73) that in *Prunus avium* the sterility locus behaves as though bipartite, with one segment producing a pollen product and the second a stylar product. The evidence for such complexity is derived from an investigation of mutations (spontaneous or X-ray induced) that affect only the pollen or the stylar component of the incompatibility reaction.

Although the reports cited above and the hypotheses developed to explain the observations have discounted an immune type of response, not all geneticists have discarded such explanations. Linskens (77) has developed a model that is similar to the immune response theory earlier favored by geneticists concerned with the basis of self-incompatibility. The model involves repression of synthesis of a protein (the antibody) by the product of an enzyme that is similar structurally to the antibody. The antigen, by complexing with the enzyme, derepresses synthesis of the so-called antibody which is then available to complex with the antigen. Such a model explains his observations on flowering branches of *Petunia* placed in solutions of ^{14}C-leucine or ^{14}C-glucose prior to pollination with compatible or incompatible pollen. After pollination, stylar extracts were subjected to electrophoresis. Styles which were treated with ^{14}C-leucine and then self-pollinated had a large peak of radioactivity which migrated toward the anode while the peak moved toward the cathode following outcrossing. There was no corresponding peak from unpollinated extracts, and the peak of radioactivity moved independently of the bulk of the protein. Styles exposed to ^{14}C-glucose showed no difference in their radioactive components following compatible or incompatible pollinations. This study suggests that a reaction product formed in both compatible and incompatible pollinations but that the compatible differed from the incompatible product.

It is interesting to speculate whether the various incompatibility alleles are really alternative expressions of a number of closely linked loci which are under unified control allowing the expression of one locus per chromosome. Two reasons prompt this speculation. One is the existence of the complex self-sterility loci found in *Oenothera organensis* and *Prunus avium* by Lewis & Crowe (73). Secondly, the studies just reviewed indicate a lack of cross-reaction between products of incompatibility alleles, as would be expected if alleles differed from each other by a single base change giving rise to a single amino acid substitution. There might be multiple incompatibility loci tightly linked, as postulated for some plant disease resistance factors (Shepherd & Mayo 130) or the major histocompatibility system found in a number of mammals.

BIOCHEMICAL CORRELATES OF GENES AFFECTING DEVELOPMENTAL PROCESSES

An interesting recent investigation into the biochemical basis of a profound morphological change is that of Scholz & Rudolph (115, 124) on the tomato mutant *chlornerva*. This mutant completely lacks the ninhydrin-positive "normalizing factor" present in normal tomato plants. The factor is apparently ubiquitous in flowering plants. The application of μg quantities of "normalizing factor" to the leaves of mutant plants results in normal development of plants that would otherwise be dwarf, partially chlorophyll deficient, and without flowers. The "normalizing factor" has a mol wt of 350 to 500, is not enzymatically degraded, but can be degraded by 6N HCL. The resultant fragments apparently do not correspond to any of the 20 amino acids usually found in proteins. The factor is eluted from an amino acid analyzer column by pH 3.2 citrate buffer (0.2 M) at a position between serine and glutamate. The "normalizing factor" is apparently synthesized in all parts of the plants because in grafts of mutant plants with normal either as stock or scion the factor is found throughout the plant.

The activities of oxidative enzymes, particularly peroxidases, have been found to be altered in various mutants in several species. It should be noted that zymograms of plant extracts may contain ca 12 bands with peroxidase activity, and most investigations offer no clue to the in vivo role of a particular isozyme. Nor do they exclude some of the possible roles. However, Price & Stebbins (110) demonstrated that the enhanced peroxidase activity found in leaves and spikes of *calcaroides* barley and attributable to some of the cathodically migrating peroxidases does not result in increased IAA oxidase activity.

In peas (*Pisum sativum*) Müller (89) has found two short internode mutants to have peroxidase (as well as catalase, laccase, and tyrosinase) activities greater than normal controls. Schertz et al (122) detected much enhanced peroxidase activity in leaves of a dwarf (*dwarf-3*) sorghum mutant as compared to a normal control. The seven zymogram bands with peroxidase activity were all enhanced in mutant extracts.

Stebbins & Gupta (134) have demonstrated that the expression of the barley spike character conditioned by the *hooded* mutation could be modified towards normality by injecting sodium azide, hydrazine, hydroxylamine, or phenylboric acid into the reproductive meristem of mutant plants. The phenotypic shift was proportional to the extent to which the peroxidase activity was lowered. The injection of IAA gave a similar response. The authors suggested that the mutant phenotype results from the modification of the IAA oxidase system in developing meristems. It is interesting that phenylboric acid, which decreased peroxidase activity in the barley spike, increased peroxidase (and other oxidative enzyme) activity in tomato and garden peas as reported by Mathan (85), who also noted the production of phenocopies of a pea flower mutant by phenylboric acid application. Previously, Mathan (84) had produced phenocopies of *lanceolate* plants in tomato by use of the same compound. Although he has interpreted these results to show a primary effect of phenylboric acid on the oxidative enzymes and a subsequent effect of elevated levels of enzyme activity on morphology, the true situation is unlikely to be simple since the phenocopy affects leaf form in tomato and flower form in the pea.

As noted (110), there is no evidence in these investigations that links the elevated oxidative enzyme activity directly with the mutant gene, and it is possible that in some or all of these instances the enhanced peroxidase activity is a secondary consequence of an unidentified primary effect of the gene.

CONCLUSIONS

In this review we have emphasized recent developments in tissue culture techniques and somatic cell genetics in higher plants because we believe that extensions of these techniques to species much used for genetic investigations will remove major impediments to coordinated biochemical and genetic investigations of higher plants. This accounts for our assessment that the most significant advances in the biochemical genetics of higher plants over the last 6 years lie in this area of supportive techniques. The availability of these techniques should prove an effective stimulus to research with higher plants, particularly in the area of biochemical genetics.

Literature Cited

1. Ahuja, M. R., Doering, G. R. 1967. *Nature* 216:800–1
2. Akatsuka, T., Nelson, O. E. 1966. *J. Biol. Chem.* 241:2280–86
3. Akatsuka, T., Nelson, O. E. 1969. *J. Jap. Starch Soc.* 17:99–114
4. Akazawa, T., Murata, T. 1965. *Biochem. Biophys. Res. Commun.* 19:21–26
5. Arasu, N. T. 1968. *Genetica* 39:1–24
6. Ascher, P. D. 1966. *Euphytica* 15:179–83
7. Bates, L. S. 1966. *Proc. High Lysine Corn Conf. Purdue Univ.*, ed. E. T. Mertz, O. E. Nelson, 61–66. Washington, D. C.: Corn Ind. Res. Found.
8. Bayer, M. H., Ahuja, M. R. 1968. *Planta* 79:292–98
9. Benedict, C. R., McCree, K. J., Kohel, R. J. 1972. *Plant Physiol.* 49:968–71
10. Black, R. C., Loerch, J. D., McArdle, L. J., Creech, R. G. 1966. *Genetics* 53:661–68
11. Boardman, N. K., Linnane, A. W., Smillie, R. M., Eds. 1971. *Autonomy and Biogenesis of Mitochondria and Chloroplasts.* Amsterdam: North Holland
12. Brewer, G. J., Sing, C. F., Sears, E. R. 1969. *Proc. Nat. Acad. Sci. USA* 64:1224–34
13. Brian, P. W., Hemming, H. G. 1955. *Physiol. Plant.* 8:669–81
14. Burr, B., Nelson, O. E. 1972. *Ann. N. Y. Acad. Sci.* 210:129–38
15. Carlson, P. S. 1969. *Genet. Res.* 14:337–39
16. Carlson, P. S. 1970. *Science* 168:487–89
17. Carlson, P. S. 1972. *Mol. Gen. Genet.* 114:273–80
18. Carlson, P. S., Smith, H. H., Dearing, R. D. 1972. *Proc. Nat. Acad. Sci. USA* 69:2292–94
19. Chao, S. E., Scandalios, J. G. 1971. *Genetics* 69:47–61
20. Chase, S. S. 1969. *Bot. Rev.* 35:117–67
21. Chourey, R. S. 1971. *Genetics* 68:435–42
22. Clegg, M. T., Allard, R. W. 1972. *Proc. Nat. Acad. Sci. USA* 69:1820–24
23. Clewell, A. F., Schmid, G. H. 1969. *Planta* 84:166–73
24. Cocking, E. C. 1969. *Ann. Rev. Plant Physiol.* 23:29–50
25. Cohen, S. S. 1970. *Am. Sci.* 58:281–89
26. Cox, E. L., Dickinson, D. B. 1971. *Biochem. Genet.* 5:15–25
27. Creech, R. G. 1968. *Advan. Agron.* 20:275–322
28. Crow, J. F. 1969. *Proc. 12th Int. Congr. Genet.* 3:105–13
29. Dalby, A., Cagampang, G. B. 1970. *Plant Physiol.* 46:142–44
30. Dalby, A., Davies, A. I. ab 1967. *Science* 155:1573–75
31. De Fekete, M. A. R. 1968. *Planta* 79:208–21
32. Denic, M. 1970. In *Improving Plant Protein by Nuclear Techniques.* 381–89. Vienna: Int. At. Energy Ag. 458 pp.

33. Dickinson, D. B., Preiss, J. 1969. *Plant Physiol.* 44:1058–62
34. Doering, G. R., Ahuja, M. R. 1967. *Planta* 75:85–93
35. Efron, Y. 1970. *Genetics* 65:575–83
36. Efron, Y. 1970. *Science* 170: 751–53
37. Efron, Y. 1971. *Mol. Gen. Genet.* 111:97–102
38. Efron, Y., Schwartz, D. 1968. *Proc. Nat. Acad. Sci. USA* 61:568–91
39. Evans, J. J. 1968. *Plant Physiol.* 43: 1037–41
40. Frydman, R. B., Cardini, C. E. 1964. *Biochem. Biophys. Res. Commun.* 14:353–57
41. Frydman, R. B., Cardini, C. E. 1967. *J. Biol. Chem.* 242:312–17
42. Galston, A. W. 1967. *Am. Sci.* 55: 144–60
43. Gelinas, D. A., Postlethwait, S. N. 1969. *Plant Physiol.* 44:1553–59
44. Goodsell, S. F. 1968. *Crop Sci.* 8:281–82
45. Goodwin, T. W. 1963. In *The Biosynthesis of Vitamins and Related Compounds*, 14:270–319. London, New York: Academic. 366 pp.
46. Guha, S., Maheshwari, S. C. 1964. *Nature* 204:497
47. Halevy, A. H., Ashri, A., Ben-Tal, Y. 1969. *Science* 164:1397–98
48. Harris, H., Hopkinson, D. A. 1972. *Ann. Hum. Genet.* 36:9–19
49. Hart, G. E. 1970. *Proc. Nat. Acad. Sci. USA* 66:1136–41
50. Hart, G. E. 1971. *Mol. Gen. Genet.* 111:61–65
51. Heimer, Y. M., Filner, P. 1970. *Biochim. Biophys. Acta* 215:152–65
52. Hertel, R., de la Fuente, R. K., Leopold, A. C. 1969. *Planta* 88: 204–14
53. Highkin, H. R., Boardman, N. K., Goodchild, D. J. 1969. *Plant Physiol.* 44:1310–20
54. Imber, D., Tal, M. 1970. *Science* 169:592–93
55. Jagannath, D. R., Bhatia, C. R. 1972. *Theor. Appl. Genet.* 42:89–92
56. Jiménez, J. R. 1966. See Ref. 7, 74–79
57. Josefsson, E. 1971. *Physiol. Plant.* 24:161–75
58. Kahler, A. L., Allard, R. W. 1970. *Crop Sci.* 10:444–48
59. Kaplan, N. O. 1968. *Ann. N. Y. Acad Sci.* 151:382–99
60. Keck, R. W., Dilley, R. A., Allen, C. F., Biggs, S. 1970. *Plant Physiol.* 46:692–98
61. Keck, R. W., Dilley, R. A., Ke, B. 1970. *Plant Physiol.* 46:699–704
62. Kermicle, J. L., 1969. *Science* 166: 1422–24
63. Kermicle, J. L., 1970. *Genetics* 66: 69–85
64. Kirk, J. T. O., Tilney-Bassett, R. A. E., Eds. 1967. *The Plastids.* London/San Francisco: Freeman. 608 pp.
65. Kushwaha, S. C., Suzue, G., Subbarayan, C., Porter, J. W. 1970. *J. Biol. Chem.* 245:4708–17
66. Lavintman, N., Krisman, C. R. 1964. *Biochim. Biophys.* Acta 89:193–96
67. Lee, E. Y. C., Marshall, J. J., Whelan, W. J. 1971. *Arch. Biochem. Biophys.* 143:365–74
68. Lein, K. A. 1972. *Z. Pflanzenzücht.* 67:243–56
69. Levine, R. P. 1969. *Ann. Rev. Plant Physiol.* 20:523–40
70. Levine, R. P., Goodenough, U. W. 1970. *Ann. Rev. Genet.* 4:397–408
71. Lewis, D. 1952. *Proc. Roy. Soc. London, Ser. B* 140:127–35
72. Lewis, D. 1964. *Proc. 11th Int. Congr. Genet.* 2:657–63
73. Lewis, D., Crowe, L. K. 1954. *Heredity* 8:357–63
74. Lewontin, R. C. 1970. *Science* 167: 1519
75. Lewontin, R. C., Hubby, J. L. 1966. *Genetics* 54:595–609
76. Li, S. L., Rédei, G. P. 1969. *Genetics* 62:281–88
77. Linskens, H. F. 1965. *Proc. 11th Int. Congr. Genet.* 3:629–36
78. Macey, M. J. K., Barber, H. N. 1970. *Phytochemistry* 9:13–23
79. Mäkinen, Y. L. A., Lewis, D. 1962. *Genet. Res.* 3:352–63
80. Manners, D. J., Rowe, J. J. M., Rowe, K. L. 1969. *Carbohyd. Res.* 8:72–81
81. Manners, D. L., Rowe, K. L. 1969. *Carbohyd. Res.* 9:107–21
82. Margulis, L. 1970. *Origin of Eukaryotic Cells.* New Haven: Yale Univ. Press. 348 pp.
83. Markert, C. L. 1968. *Ann. N. Y. Acad. Sci.* 151:14–40
84. Mathan, D. S. 1965. *Am. J. Bot.* 52: 185–92
85. Mathan, D. S. 1968. *Nature* 219: 1065–66
86. Mehta, S. L., Srivastava, K. N., Mali, P. C., Naik, M. S. 1972. *Phytochemistry* 11:937–41
87. Miller, P. L., Ed. 1970. *The Develop-

ment and Interrelationship of Cell Organelles. Symp. Soc. Exp. Biol., Vol. 24

88. Mitra, R., Bhatia, C. R. 1971. *Genet. Res.* 18:57–69

89. Müller, H. P. 1969. *Phytochemistry* 8:1867–71

90. Nagata, T., Takebe, I. 1970. *Planta* 92:301–8

91. Ibid. 99:12–20

92. Nasrallah, M. E., Barber, J. T., Wallace, D. H. 1970. *Heredity* 25: 23–27

93. Nasrallah, M. E., Wallace, D. H. 1967. *Heredity* 22:519–27

94. Nasrallah, M. E., Wallace, D. H. 1967. *Nature* 213:700–1

95. Nass, M. M. K. 1969. *Science* 165: 25–36

96. Nelson, O. E. 1967. *Ann. Rev. Genet.* 1:245–68

97. Nelson, O. E. 1969. *Advan. Agron.* 21:171–94

98. Nelson, O. E., Rines, H. W. 1962. *Biochem. Biophys. Res. Commun.* 9:297–300

99. Netting, A. G., Macey, M. J. K., Barber, H. N. 1972. *Phytochemistry* 11:579–85

100. Nitsch, J. P. 1972. *Z. Pflanzenzücht.* 67:3–18

101. Ockendon, D. J. 1972. *New Phytol.* 71:519–22

102. Ozbun, J. L., Hawker, J. S., Preiss, J. 1971. *Plant Physiol.* 48:765–69

103. Pandey, K. K. 1967. *Nature* 213: 669–72

104. Phinney, B. O. 1956. *Proc. Nat. Acad. Sci. USA* 42:185–89

105. Pochard, E., Dumas de Vaulx, R. 1971. *Z. Pflanzenzücht.* 65:23–46

106. Porter, J. W., Anderson, D. G. 1967. *Ann. Rev. Plant Physiol.* 18:197–228

107. Porter, J. W., Lincoln, R. E. 1950. *Arch. Biochem.* 27:380–403

108. Preiss, J., Lammel, C., Sabraw, A. 1971. *Plant Physiol.* 47:104–8

109. Preiss, J., Ozbun, J., Hawker, J. S., Greenberg, E., Lammel, C. 1972. *Ann. N. Y. Acad. Sci.* 210:265–78

110. Price, H. J., Stebbins, G. L. 1971. *Genetics* 68:539–46

111. Proano, V. A., Greene, G. L. 1968. *Plant Physiol.* 43:613–18

112. Raff, R. A., Mahler, H. R. 1972. *Science* 177:575–82

113. Raymundo, L. C., Griffiths, A. E., Simpson, K. L. 1970. *Phytochemistry* 9:1239–45

114. Riley, R., Ewart, J. A. D. 1970.

115. Rudolph, A., Scholz, G. 1972. *Biochem. Physiol. Pflanz.* 163:156–68

116. Russell, D. W., Galston, A. W. 1968. *Planta* 78:1–10

117. Salamini, F., Tsai, C. Y., Nelson, O. E. 1972. *Plant. Physiol.* 50:256–61

118. Scandalios, J. G. 1967. *Biochem. Genet.* 1:1–9

119. Scandalios, J. G. 1968. *Ann. N. Y. Acad. Sci.* 151:274–93

120. Scandalios, J. G. 1969. *Biochem. Genet.* 3:37–79

121. Scandalios, J. G. 1969. *Science* 166: 632–24

122. Schertz, K. F., Sumpter, N. A., Sarkissian, I. V., Hart, G. E. 1971. *J. Hered.* 62:235–38

123. Schmid, G. H., Gaffron, H. 1967. *J. Gen. Physiol.* 50:2131–44

124. Scholz, G., Rudolph, A. 1968. *Phytochemistry* 7:1759–64

125. Schwartz, D. 1969. *Science* 164: 585–86

126. Schwartz, D. 1971. *Genetics* 67: 411–25

127. Ibid, 515–19

128. Schwartz, D. 1971. *Proc. Nat. Acad. Sci. USA* 68:145–46

129. Selander, R. K., Johnson, W. E. 1972. *Proc. 17th Int. Congr. Zool.* In press

130. Shepherd, K. W., Mayo, G. M. E. 1972. *Science* 175:375–80

131. Simpson, G. M., Saunders, P. F. 1972. *Planta* 102:272–76

132. Sodek, L., Wilson, C. M. 1970. *Arch. Biochem. Biophys.* 140:29–38

133. Spiker, S., Chalkley, R. 1971. *Plant Physiol.* 47:342–45

134. Stebbins, G. L., Gupta, V. K. 1969. *Proc. Nat. Acad. Sci. USA* 64:50–56

135. Stoddart, J. L. 1966. *J. Exp. Bot.* 17:96–107

136. Suge, H., Murakami, Y. 1968. *Plant Cell Physiol.* 9:411–14

137. Tal, M., Imber, D. 1970. *Plant Physiol.* 46:373–77

138. Ibid 1971. 47:849–50

139. Tal, M., Imber, D., Stai, C. 1970. *Plant Physiol.* 46:367–73

140. Tanaka, Y., Akazawa, T. 1968. *Plant Cell Physiol.* 9:405–10

141. Tewari, K. K. 1971. *Ann. Rev. Plant Physiol.* 22:141–68

142. Tomes, M. L. 1967. *Genetics* 56 227–32

143. Ibid 1969. 62:769–80

144. Tsai, C. Y., Nelson, O. E. 1966. *Science* 151:341–43

Genet. Res. 15:209–19

145. Tsai, C. Y., Nelson, O. E. 1968. *Plant Physiol.* 43:103–12
146. Tsai, C. Y., Nelson, O. E. 1969. *Genetics* 61:813–21
147. Tulecke, W., Taggart, R., Colavito, L. 1965. *Contrib. Boyce Thompson Inst.* 23:33–46
148. Vasil, V., Hildebrandt, A. C. 1965. *Science* 150:889–92
149. Vieweg, G. H., De Fekete, M. A. R. 1972. *Planta* 104:257–66
150. Wagner, R. P. 1969. *Science* 163:1026–31
151. Warner, R. L., Hageman, R. H., Dudley, J. W., Lambert, R. J. 1969. *Proc. Nat. Acad. Sci. USA* 62:785–92
152. Wettstein-Knowles, P. von 1972. *Planta* 106:113–30
153. Widholm, J. M. 1972. *Biochim. Biophys. Acta* 261:52–58
154. Williams, R. J. H., Britton, G., Goodwin, T. W. 1967. *Biochem. J.* 105:99–105
155. Wilson, C. M., Alexander, D. E. 1967. *Science* 155:1575–76
156. Witt, H. T. 1968. In *Proc. Fifth Nobel Symp.*, ed. S. Claesson. New York: Interscience
157. Wolf, M. J., Khoo, U., Seckinger, H. L. 1967. *Science* 157:556–57

Ann Rev. Plant Physiol. 1973. 24:519–70

PLANT RESPONSES TO
WATER STRESS

♦ 7558

Theodore C. Hsiao

Laboratory of Plant-Water Relations, Department of Water Science and
Engineering, University of California, Davis

CONTENTS

INTRODUCTION

A decade has passed since plant responses to water stress were reviewed in this series (299), although writing of a review nature has been voluminous in symposium proceedings and books (60, 105, 156–158, 162, 230, 264, 268, 298). The rising demand for knowledge in this field seems to be stimulated mainly by awareness of the importance of water in food production in developing areas, by wakened concern for water as a critical resource in the industrialized nations, and, on a very different level, by progress in understanding the physical aspects of plant-water relations. Regrettably, the proliferation in words has at times not been matched by significant new findings and progress in analysis and understanding. Many studies on plant responses to water deficits (stress) were carried out by investigators concerned with agricultural production, environment and resources, and macroscopic physics of soil, plant, and atmospheric water. As expected, the physiological and metabolic aspects of these studies were often weak and, on the other hand, studies carried out by metabolism-oriented biologists frequently slighted important physical facets. Nevertheless, laudable investigations, especially during the last few years, have been sufficient to warrant optimism about substantial progress in the near future.

This review is a highly personal analysis of what we currently know (or rather do not know) of plant responses to water deficits, of how water deficits may theoretically affect plant processes, of some pitfalls in data interpretation, and of needs in future research. It is hoped that the strongly advocated views, many of which will likely be proven wrong, and the little-tested working hypotheses or speculations will serve as an impetus to progress. No pretense is made here as to completeness in covering the literature. Listing of more recent papers is reasonably comprehensive in the three volumes on water deficits edited by Kozlowski (156–158), and earlier papers are covered in the fine 1961 review by Vaadia et al (299). There are recent reviews covering the following aspects: water in relation to biological macromolecules and membranes (142, 143, 280); effects of water stress on metabolism and detailed physiology (31, 48, 162, 175, 206, 264, 265, 267, 293, 300); effects of water stress on growth, morphology, and ontogeny (84, 163, 248, 265, 325); pathogen-induced water stress and consequences (57);

yields as related to water stress and use (70, 140, 248); and drought resistance (226, 227). Also relevant to water stress and desiccation is the penetrating review of Mazur on freezing injury (185).

To keep within reasonable bounds, attention here is focused mainly on responses in the vegetative tissue of higher plants to water stress. The intriguing (though very different) process of water-initiated seed germination is discussed only as it bears on stress effects in the vegetative stage.

Life evolved in the medium of water. The number of places in the plant complex where water has a crucial role must be astronomical, ranging from photolysis of water in photosynthesis to hydrophobic bonding of macromolecules and to the maintenance of form in nonwoody tissue. Moreover, the plant, being a highly integrated organism with numerous controls, should exhibit extensive secondary and tertiary alterations in addition to the primary effects if stress is imposed for any appreciable duration. Thus, it is no great surprise that the literature leaves one with the impression that almost any parameter one cares to look at is changed by water stress, provided that the stress is strong and long enough. This is reminiscent of the situation delineated in the earlier literature for the deficiency of various mineral nutrients. Moreover, water stress in many cases actually causes specific changes similar to changes induced by nutrient deficiencies. For example, free amino acids and sugars accumulate during water stress (13, 275) as in potassium deficiency (119, 274) and ribonuclease increases when either water (293) or potassium (118) is deficient. Interestingly enough, deficiency in various mineral nutrients can result in similar changes in the plant even though the nutrients may be totally different in metabolic functions (274). Thus, many of the changes observed under nutrient or water deficiencies seem to represent general patterns of modulation in plants under adversity, being of little value in determining the underlying causes or mechanisms. Therefore, it is essential to examine water stress effects within the reference frame of stress severity and time courses. Only then can we hope to unravel the causal relations and the sequence of complex events which constitute plant responses to water stress.

From the foregoing it is clear that it would be necessary to discuss first in this review the parameters commonly used to indicate the degree of water stress so that results of different studies can be compared on a semiquantitative basis of stress severity. This section is followed by the main body, on observed plant responses and changes elicited by water stress. The last section considers how the primarily physical effects of a shortage of water might be transduced into alterations in metabolism.

PARAMETERS INDICATING PLANT WATER STATUS

During the past half decade, water potential (Ψ) has gained wide acceptance as a fundamental measure of plant water status. Reliance on the absolute values of Ψ as an indicator of physiological water stress, however, needs to be tempered with caution since evolutionary and physiological adaptation to environment could markedly influence the level of Ψ at which water stress sets in (p. 557). The chemical potential of water is Ψ, expressed as energy per unit volume and with the chemical potential of pure water at atmospheric pressure and the same temperature as the datum (reference) point (264). Since diffusion pressure deficit (DPD, equivalent to negative Ψ) has traditionally been expressed in atmospheres, Ψ is usually given also in pressure units (1 bar $= 10^6$ dynes $cm^{-2} = 10^6$ ergs $cm^{-3} = 0.987$ atm). The chemical potential of water in the plant is affected by hydrostatic pressure or tension, colligative effects of solutes, and interaction with matrices of solids (cell wall) and of macromolecules. Hence Ψ is the algebraic sum of the component potentials arising from the effect of pressure (ψ_p), of solutes (ψ_s), and of matrix (ψ_m).

Leaves and roots of herbaceous plants commonly consist of more than 80% water when turgid. As tissue water content decreases, changes in Ψ, ψ_p, and ψ_s are described by the classical Höfler diagram, which has been confirmed experimentally with up-to-date measurements (e.g. 82). In the case of fully turgid tissue, the initial decreases in tissue water content cause large decreases in Ψ (becoming more negative). Decreases in ψ_p are usually much more marked than decreases in ψ_s and account for the major part of the diminution in Ψ. Decreases in ψ_s follows the simple osmotic relationship with solution volume. After more water is lost and ψ_p falls to a negligible level [at $\Psi \simeq -12$ to -16 bars for some crop plants (15, 82)], decreases in ψ_s alone account for most of the further decrease in Ψ. At this point the change in Ψ per unit change in tissue water is small compared with that in the turgid state because ψ_p ceases to be a factor. The other component potential ψ_m is very close to zero in well-watered leaves and fleshy tissue. In many species, ψ_m does not become significant numerically until much of the tissue water (e.g. 50%) is lost (24, 315, 321). So unless the tissue is badly dehydrated, the component potentials of concern in most cases are ψ_p and ψ_s.

Another commonly used indicator of plant water status is relative water content, or RWC (311), which at one time had been less accurately termed as relative turgidity. RWC is the water content (on a percentage basis) relative to the water content of the same tissue at full turgor (after floating on water to "constant" weight). Clearly, RWC is related to Ψ of the same tissue, though the relationship is dependent on species and stages of growth (46, 82, 155, 194, 249), on long-term alterations induced by environment (129), and possibly even on the short-term water history of the plant (137). A major shortcoming is that RWC is a rather insensitive indicator of water status when water deficit is not severe. In nearly saturated tissue, where a small change in water content could markedly affect ψ_p, a change of a few bars in Ψ may correspond to a change of only several percentage points in RWC. This is about the size of the random error in RWC measurements in many studies (15). Perhaps for this reason, in some studies RWC showed no significant change although physiological processes were affected by the mild stress employed (e.g. 115, 120).

Tissue water content (percent of fresh weight) and fresh weight have also been used as indicators of water status. Unfortunately, water content or fresh weight of tissue at full turgor is normally not given as a reference. Water content can be very misleading because of its superficial resemblance to RWC. Small changes in the water content of vegetative tissue usually correspond to much larger changes in RWC (15, 184). For example, for a decrease in water content from 85% at full turgor to 80% under stress, the corresponding values of RWC are, respectively, 100% and 71%. Water content is used most frequently in studies of seeds and lower plants. With one notable exception (318, 319), virtually nothing is known about the relationship between water content and Ψ of such tissue. Hence, quantitative comparisons of stress severity among different materials are not possible.

Still less direct but sometimes useful indicators of plant water status are leaf thickness and stem diameter. An especially attractive method is that of β-ray gauging of relative leaf thickness (15). Properly calibrated against leaf Ψ or RWC, it allows a virtually continuous and nondestructive estimation of water status while carrying out other physiological measurements, such as CO_2 assimilation, on the same leaf (150, 295). Methods for measuring plant water status were reviewed in detail (15, 26). A recent booklet (316) gives practical instruction and information on techniques.

Regrettably, visual wilting is still used in some instances as the sole indicator of water status. In addition to being dependent on turgor pressure (ψ_p), wilting is also a function of the mechanical properties of cell wall and tissue. It is well known that different species may wilt at very different Ψ or RWC. An extreme example is an oil palm which does not wilt visibly even when fatally desiccated (159, p. 32). Furthermore, as water stress develops, physiological processes are often affected before wilting becomes apparent (1, 34, 63, 204, 245).

To compare the severity of stress among studies that have used different indicators of plant water status, it is convenient to block out, very loosely, some general ranges of stress levels. For the purpose of this review, mild stress is considered to entail a lowering of plant Ψ by several bars or of RWC by as much as 8 or 10 percentage points below corresponding values in well-watered plants under mild evaporative demand. Moderate stress refers to a lowering of Ψ by more than a few bars but less than 12 or 15 bars, or a lowering of RWC by more than 10 but less than 20 percentage points. Stress would be severe if Ψ is lowered more than 15 bars or RWC more than 20 percentage points. The term desiccation is reserved here for cases where more than half of the tissue water is removed.

OBSERVED RESPONSES TO WATER STRESS

Transpiration and Stomata

Overall stomatal closure and transpiration reduction in response to water deficiency have been long established. Emphasis herein will be on attempts to quantify these effects and delineate the underlying processes.

TRANSPIRATION   It is well documented that stomatal closure is the main cause for transpiration decline as water stress develops. Much of the early confusion on the cause of this transpiration decline was clarified by the introduction (165, 240, 241) and use of resistance network analysis. Quantitative treatment of transpiration (and CO_2 assimilation) in terms of resistances to gas transport has been discussed and reviewed extensively (47, 80, 128, 264). Transpiration is directly proportional to the gradient of water vapor concentration from the internal evaporation surface to the bulk air outside the leaf, and inversely proportional to the total resistance to water vapor transport of the air boundary layer and of the leaf. Since stomata control only one part of the total resistance, their closure will vary in effect with magnitude of stomatal resistance (r_s) relative to that of boundary layer resistance (r_a) and cuticular resistance (r_c). For example, with r_a and r_s connected in series, when r_a is high (nearly still air) and r_s quite low (wide open stomata), partial closure of stomata would have only a small effect on transpiration. Similarly, with r_c and r_s connected in parallel, when r_c is abnormally low due to cuticle damage by rust sporulation, closure of stomata also would not reduce transpiration markedly (58). In addition, increased stomatal resistance may not cause proportional decreases in transpiration rate because diminished dissipation of heat by vaporization and the consequent rise in leaf temperature increase the water vapor concentration inside the leaf (239).

LEAF TEMPERATURE   The question often arises as to the possible effects on metabolism of elevated leaf temperature during water stress. Many processes in the plant [e.g. CO_2 assimilation (63, 243)] have rather broad temperature optimums. The magnitude of the temperature elevation must be considered along with the sensitivity of the process in question to temperature change in the given temperature ranges. For a given reduction in transpiration due to stomatal closure, the increase in leaf temperature would depend strongly on environmental factors, particularly the radiation load on the leaf and the heat transfer coefficient of the air. The interactions, although complex, have been analyzed successfully by using the energy-balance approach (239, 241). In most situations, the rise in leaf temperature accompanying substantial reduction in transpiration has been calculated or measured to be only a few degrees (80, 235). Therefore, it would be reasonable to assume that elevation in leaf temperature does not play a general role in water stress effects. Under a specific combination of circumstances, however, temperature elevation might have a dominant impact, especially in thick leaves or massive structures such as flower heads of onion (90) and fruits.

"WALL" RESISTANCE TO TRANSPIRATION   From time to time, the possibility has been raised (e.g. 48, 299) that nonstomatal factors in the leaf, often referred to as "mesophyll" or "wall" resistance, cause significant reductions in transpiration as water stress develops. The more recent literature has been summarized and discussed (67, 264). Mechanisms proposed to explain an increase in "wall" resistance with high transpiration rate and water stress involve either a localized lowering of water activity at the surface of the evaporating cell wall or an increase in diffusive resistance to water vapor at the wall. Earlier experimental assessment of changes in "wall" resistance was inconclusive because of failure to estimate accurately changes in vapor concentration gradient or in other resistances affecting transpiration. A study not confounded by those complications was conducted by Fischer (67) on leek leaves with the epidermis removed. His data show that "wall" resistance to water loss in leek rose with water deficit but was insignificantly small (0.15 sec cm^{-1}) even after the mesophyll lost 50% of its water. Jarvis & Slatyer (130) confirmed with intact cotton that the "wall" resistance is small in turgid leaves, but reported that it may rise with moderate water deficits to a significant level which is nevertheless still minor compared with the expected stomatal resistance. It is not certain whether there is a true difference between the two species used in those two investigations, because the method used by Jarvis & Slatyer was much less direct than that of Fischer. In any event, it can be concluded that stomatal closure is generally the dominant mechanism in restricting transpiration rates in mesophytes during development of water stress.

SENSITIVITY OF STOMATA TO STRESS   Recent data have given a more quantitative basis to relationships between stomatal opening and leaf water status. Most data clearly demonstrate a threshold level of Ψ or RWC above which leaf resistance [usually determined with a diffusion porometer (189, pp. 46–48)], and therefore stomatal opening, remained constant. Overall, the threshold value of Ψ turned out to be rather low (negative), being about -7 to -9 bars for tomato (56) and for the adaxial stomata of beans (139), -10 to -12 bars for soybean (28) and the abaxial stomata of beans (139), and -12 to -16 bars for grape (164) and for greenhouse-grown cotton leaves (137). Adaxial and abaxial stomata have been observed to differ in response to water stress in some cases (139, 242) but apparently not in others (249). There is also a report (106) of a threshold Ψ of -18 bars for stomata of beet leaves, but the estimate of Ψ was most likely too low since the tissue was immersed in osmotica of KCl for the isopiestic Ψ determination. The Ψ of a well-watered control was found to be -10 bars by the same technique. In studies when RWC was used as an indicator of water status, and leaf chamber data were used to calculate leaf resistance to water vapor diffusion (mainly stomatal), the threshold values were about 80 to 85% RWC for cotton (295) and beans (58). The above results indicate that stomata are rather insensitive to mild water stress. However, this conclusion probably cannot be generalized, since there are direct or indirect indications that stomata of other species may be sensitive to small water deficits (28, 63, 72, 195, 242, 272, 273). The finding of a threshold water status for stomatal response is not new. Earlier results on this point have been summarized by Ståfelt (272). Detailed microscopic observations by Ståfelt (272) also showed that the optimum water content for stomatal opening can be actually something less than the tissue water content at full turgor. Full turgor can cause some stomatal closure, presumably because of excessive back pressure from the epidermal cells surrounding the guard cells.

Once the threshold water status for stomatal closure was reached, leaf resistance increased sharply, rising 20- or 30-fold with a further drop in Ψ of less than 5 bars (56, 137, 139) or a further decrease of 15 to 20% in RWC (58, 295). Such large increases in leaf resistance may be taken as indicative of almost complete stomatal closure. Curiously, the resistance seemed to continue to rise with further increases in water deficit, showing no signs of leveling off within the range of water status tested (56, 58, 137, 139, 164, 195, 295). Aside from leaf water status, there is some evidence that water vapor content of the air may be very important in determining stomatal opening. Raschke (242) reported that the diffusive resistance of maize leaves at the same water deficit was up to several times as great in dry air (nearly zero humidity) as in moist air.

A very interesting exception to the above generalization on stomata being sensitive to moderate stress is found in a xerophytic *Acacia*, brigalow. Stomata in brigalow phyllodes apparently remain partly open even at a Ψ as low as -50 bars, thus permitting substantial transpiration and CO_2 assimilation to continue at a Ψ value too low for mesophytes to remain viable (301). Stomata in plants possessing crassulacean acid metabolism and "inverted" stomatal rhythm also appear to close with increasing water stress in a manner similar to that in mesophytes (154).

The growing environment may possibly influence stomatal response to water stress (137, 139). Jordan & Ritchie (137) reported that stomata of cotton grown in the field did not close even at a leaf Ψ of -27 bars, in contrast to greenhouse-grown cotton, which exhibited marked closure at -16 bars. A difference in sensitivity between field and growth-chamber grown onion was also reported (195). It is not known in these cases if CO_2 concentrations were different between the field and artificial environment when stomatal measurements were taken; but if these results are further substantiated, they suggest remarkable adaptation of the stomatal apparatus to the growing condition. Some early data (231, and as cited in 272) suggest that light may modify stomatal response to water deficit. At higher light levels, more deficit seemed to be required to induce closure. Limited recent data appear to support this notion, showing that the light required to saturate stomatal opening possibly increased with increasing leaf water deficit (106). Moreover, stomata in *Pelargonium* leaves that are slightly deficient in water opened more slowly in light and closed more quickly upon darkening than did controls (317). It has also been mentioned that stomatal response to water stress was attenuated by oxygen-free air (296).

MECHANISMS OF STOMATAL RESPONSE   Stomatal opening and closing result from turgor differences between guard cells and the surrounding subsidiary or epidermal cells (189). The major solute accumulated in light by guard cells leading to turgor buildup and opening was shown in the past few years to be potassium (71, 78, 121, 122, 254). Conversely, closure in the dark is caused by a loss of potassium from guard cells. Stomatal interactions with environmental factors such as light and CO_2 are complex and appear to be mediated by several underlying processes (117; 189, p. 101; 272), the sum of which is a net gain or loss of guard cell potassium and turgor with the consequent stomatal movement.

A widely held notion is that water deficits, by reducing leaf turgor, would directly reduce stomatal opening since opening is turgor dependent. The situation may be more complex, however. Painstaking early experiments of Ståfelt (272) with three species showed that as mild water deficit develops, there is a marked loss of solutes from guard cells (as indicated by plasmolytically determined ψ_s) concurrent with stomatal closure. Thus, a part of the stress effect may not be direct but is linked to the regulation of osmotic solutes in guard cells. An intriguing question is how the moderate stress used in Ståfelt's experiments is transduced so rapidly into a physiological change. Some evidence suggests that moderate to severe water stress causes an elevation of internal CO_2 concentration sufficient to account for a part of the stomatal closure (189, p. 95). However, mild water stress has been shown to reverse the opening elicited by CO_2-free air (273), indicating that at least a part of the closing process is independent of changes in CO_2 concentration.

Along with recent discoveries that abscissic acid (ABA) rises markedly in leaves subjected to water stress (see below) and that exogenous ABA is a potent (113, 134, 198) and fast-acting (52, 163, 198) inhibitor of stomatal opening, there are suggestions that stress affects stomata via its effect on ABA levels (e.g. 52, 163) or on plant hormonal balance, specifically the balance between ABA and cytokinins (175, 200, 285). However, no critical experiments on the postulated role of hormones have yet been done. ABA that accumulated during 4 hr of moderate to severe stress (191, 324) appeared to be within the range of exogenous ABA concentrations effective in causing stomatal closure (52, 198) or in inhibiting opening (113). Cummins, Kende & Raschke (52) pointed out that the rapidity and ready reversibility of the action of ABA on stomata would make it a good modulator of stomatal behavior. However, although tissue ABA increases rapidly when water stress sets in (323), stomata possibly close even faster (10, 92, 272, 273, 317), thus raising the question of whether the accumulation of ABA from stress is fast enough for it to be the modulator of stomatal response.

As for the postulate that stomatal opening is reduced during stress by a concerted effect of depressed cytokinin level and rise in ABA (175), there are serious arguments against it. The idea is based on the reported substantial reduction in cytokinin activity of extracts of stressed leaves (see p. 542) and the finding that kinetin can promote stomatal opening within a few hours of application. Unfortunately, the promotive effect of kinetin is quite small if treatment is limited to short duration (hours) so that leaf senescence is not a factor (175, 188, 198). Further, stomata of many species and apparently of younger leaves do not respond to kinetin (175). Finally, the inhibition of stomata by applied ABA (113, 198, 200) or prior water stress (5, 68) is not substantially reversed by kinetin. It is worth pointing out that stomata are sensitive to numerous chemicals of diverse nature (e.g. 68, 78, 327). Small responses to an exogenous substance such as kinetin may be incidental and should be interpreted in functional terms only with extreme caution.

Transitory stomatal opening in leaves upon sudden deprivation of water, and closing upon sudden release of water stress, have been noted for decades (109; 264, pp. 254–55). Recently, Raschke (242, 244) elegantly confirmed the earlier conclusion that the transient changes are primarily mechanical, resulting from a time lag between pressure changes in epidermal cells and guard cells. With a sudden cutback in water supply, turgor falls faster in epidermal cells than in guard cells, with stomata opening transitorily as a result. The reverse occurs when plants are suddenly watered, with turgor rising first in the epidermal cells.

Literature on the associated phenomenon of stomatal oscillation with short periods has been reviewed (16, 112).

AFTEREFFECT ON STOMATA   For many years, workers have noted that the occurrence of leaf water deficits partially inhibited stomatal opening for some time subsequent to the apparent recovery in plant water status (89, 197, 272). In a recent detailed study, Fischer, Hsiao & Hagan (72), by assaying poststress stomatal opening in leaf disks floated on water to ensure full turgor, clearly showed that the aftereffect was not due to a persistent water deficit. Allaway & Mansfield (5) reached the same conclusion. The aftereffect was more or less proportional to the maximum water deficit (minimal RWC) reached before rewatering, and was less in *Vicia faba* than in tobacco (72). Stomata regained most of their normal opening potential within 1 day of rewatering (5, 72), though full recovery sometimes required as long as 5 days (72). In a wide-ranging study, Fischer (68), though unable to affix the cause, concluded from experiments involving a delicate exchange of epidermal strips between stressed and control *Vicia* leaves that a major part of aftereffect resides in guard cells, with only a minor part in the mesophyll. Poststress and control leaves responded fairly similarly to CO_2-free air and had similar or identical CO_2 compensation points, thus ruling out changes in internal CO_2 concentration as a major factor (5, 68). Potassium supply to guard cells was apparently not involved, since the aftereffect remained evident when opening was assayed with epidermal strips taken from stressed *Vicia* and floated on dilute KCl solution (68). It has been speculated (5, 175, 179) that the basis for the aftereffect may be an accumulation of an inhibitor of stomatal opening, possibly ABA. In view of the rapid metabolism or degradation of ABA (4, 52, 163) and the prolonged period required for recovery from the aftereffect, it remains to be shown that prior ABA accumulation is the principal cause of the aftereffect. The aftereffect may act as a mechanism preventing excessive transpiration during periods when water supply is short (5) but would also similarly suppress CO_2 assimilation in light (295).

CO_2 Assimilation in Light

AT THE LEAF LEVEL   There is almost unanimity that much of the reduction in CO_2 assimilation in light during water stress is due to stomatal closure, impeding the inward passage of CO_2. Again, this is analyzed best in terms of a network of resistances to gas transport, and the literature should be consulted (79, 80, 128, 183). Logically, as the first approximation, responses of CO_2 assimilation to stress should be similar to stomatal responses to stress. Numerous studies have supported this view, showing a close parallel between CO_2 assimilation and either transpiration or stomatal resistance in the time course of development and release of water stress (29, 33, 106, 317). Also pertinent is a hand-in-hand relation between CO_2 assimilation and stomatal opening when the latter is undergoing rapid changes. Short-term stomatal oscillation causes strikingly similar and inphase oscillation in CO_2 assimilation (14, 112, 295), while leaf water content or Ψ oscillates 180° out of phase (14, 112, 169). Transitory stomatal opening upon a sudden cutoff of the water supply to leaves is associated with a transitory increase in CO_2 assimilation in spite of the increase in leaf water deficit (10). Overall, CO_2 assimilation in many species is sensitive to moderate but not to mild stress (33, 63, 164, 295), as are the stomata of these species, and exhibits the threshold effect (33, 106, 170, 309) as do stomata. Some species appear to be sensitive even to mild stress, but their stomata are apparently equally sensitive (32, 33, 170, 245). Brigalow photosynthesized substantially even at a tissue Ψ of -50 bars; but again, the rate was closely correlated with stomatal conductivity for the whole range of tissue Ψ (301).

Still, none of the above facts rules out possible nonstomatal effects of water stress on CO_2 assimilation in addition to the dominant stomatal effect. In fact, substantial evidence of nonstomatal effects has been published in the past few years. Also reported, however, have been equally persuasive data indicating that moderate stress affects only stomata and not other factors in net CO_2 uptake. The apparent conflict in findings is probably related to differences in species and possibly in experimental techniques. Severe stress or desiccation may of course be expected to have effects other than just stomatal closure, as is discussed below.

The most convincing data demonstrating the absence of nonstomatal effects come from work on so-called mesophyll resistance in cotton. CO_2 assimilation shares the gaseous portion of the transport pathway between the air and cell surface in the leaf interior with water vapor and is affected by the vapor phase resistances controlling transpiration. It is also affected, however, by an additional resistance, the mesophyll resistance r_m' (the prime designates resistance to CO_2, in contrast to water). All resistances to CO_2 assimilation other than that offered by air boundary layer, stomata, and cuticle are represented by r_m'. These consist of liquid phase transport and biochemical resistances which include terms for CO_2 dissolution at the cell wall, transport in solution to the chloroplast, and carboxylation. Unfortunately, when used in the most straightforward way, accumulated systematic errors in measurements and assumptions are included because r_m' is a "remainder" after all other resistances have been accounted for. Troughton (295) determined r_m' for cotton leaves at varied water status in air of normal O_2 content and found it to remain nearly constant until RWC dropped to below 75%, then r_m increased markedly. In a second study, Troughton & Slatyer (296) improved the techniques by passing air through the leaves instead of over them, so that inaccuracies involved in deducting r_a' and r_s' were eliminated, and by deriving r_m' from the slope of the line of CO_2 assimilation vs intercellular CO_2 concentration (266). Their data showed r_m' to be essentially constant down to 56% RWC. In that study, O_2-free air was used to ensure that photorespiration was not a factor. The implication is that the sensitivity of r_m' to less severe water stress observed in the first study with air of normal O_2 content (295) is perhaps only apparent and came about through the erroneous assumption that the CO_2 compensation point remained unchanged with stress when in fact it may have been raised by stress-caused photorespiration rise (265). Regardless, the overall conclusion from those two studies is the same in that net photosynthesis under normal conditions would decline under severe stress partly as a result of nonstomatal factors. In contrast, moderate stress of one day or less in cotton appears to affect CO_2 assimilation virtually entirely by closing stomata. When stomatal influences were eliminated by removing the epidermis from tobacco leaves, photosynthesis did not decline until more than 40% of tissue fresh weight was lost through rapid (0.5 hr) drying (92). Photosynthesis was measured by incorporating $^{14}CO_2$ for only 20 sec, however.

Limited data on soybean showed r_m' to increase only very slightly when leaf Ψ was decreased to -41 bars (28). Interestingly, as mentioned above, soybean stomata also seem to remain open at moderate stress (28). In tomato, r_m' in normal air increased when leaf RWC fell below 80% (59). The behavior of beet leaves is perhaps similar (106). The r_m' of bean leaves was reported in 1966 (81) to increase readily with water deficit, but plant water status was not measured and the leaf-chamber technique of that time was less exact. Some early results of Boyer (23), often mentioned as ascribing a nonstomatal factor for depressing CO_2 assimilation in stressed plants, were confounded by long-term exposure to NaCl. Still earlier results (e.g. 255, 260) along the same line are questionable on the ground of techniques. For example, Scarth & Shaw (255) found in a time-course study that CO_2 assimilation in *Pelargonium* declined continuously with leaf water loss whereas stomatal opening, determined with a viscous flow porometer, apparently first increased slightly and then decreased. It is now known from time-course studies with *Vicia faba* leaves that upon some water loss a viscous flow porometer can indicate an apparent slight opening of stomata, probably because intercellular space in the leaf is enlarged by the water loss, although the stomata are actually closing as measured microscopically (Hsiao, unpublished).

It has been reported that the CO_2 compensation point of maize leaves (ordinarily about zero) became measurable when leaf RWC was decreased to 85%, and that it rose steeply to ambient CO_2 concentration, indicative of no net photosynthesis, when RWC was about 73%, or an estimated Ψ of -12 bars. This work (88) has several puzzling aspects. For example, in contrast with drying in air, floating the leaves on mannitol of -18 bars had little effect on the CO_2 compensation point. In addition, the data are in direct conflict with other work on maize demonstrating a substantial net CO_2 assimilation and only slight increases in r_m' (calculated assuming CO_2 compensation point to be zero) at a leaf Ψ of -16 bars (27, 28). Evolution of O_2 in light by leaflets of *Elodea*, an aquatic plant, was strongly inhibited when cells were plasmolyzed in mannitol (66). The effect appeared to be attributable mainly to plasmolysis and not directly to water stress.

More persuasive evidence for nonstomatal effects of water stress on CO_2 assimilation comes from recent work with tobacco and sunflower. Forcing air through tobacco leaves to eliminate r_a' and r_s' and assuming the CO_2 compensation point to stay constant, Redshaw & Meidner (245) found calculated r_m' to increase linearly with decreases in RWC and to double or triple as RWC decreased from 95 to 85% in value. Their r_m' seemed inordinately high for an annual crop, being 6 or 7 sec cm^{-1} for well watered controls in contrast to values of 2 to 3 sec cm^{-1} found in the literature. Redshaw & Meidner speculated that the increase with water stress could be real or could be only apparent, reflecting an elevation of photorespiration, but they did not test the latter possibility with O_2-free air. Limited data of Fischer (68) showed the CO_2 compensation point of tobacco to be increased from 60 to 120 $\mu l/1$ by severe water stress (65% RWC), an increase perhaps too small to account for the large apparent increase in r_m' reported by Redshaw & Meidner. Another plant which exhibits nonstomatal effects is sunflower. Among the several lines of evidence reported by Boyer, perhaps the most convincing is the effect of water stress of sunflower leaves on chloroplast function in vitro (32), as discussed in the subsequent section. With increasing stress, reduction in CO_2 assimilation by the intact plant appeared to parallel the loss in Hill activity in vitro and also the increase in stomatal resistance (32). Also reported (30) was a lack of response of CO_2 assimilation by stressed plants to increases in external CO_2, presumably evincing effects of stress not related to gas-transport resistances such as r_s'. The control leaves, however, did not respond markedly nor linearly to increasing CO_2 concentrations, as might be expected on the basis of other studies (266). Boyer (30, 32) also argued for nonstomatal effects in sunflower on the basis of responses of control and stressed leaves to light intensity. Data are needed on r_m' of sunflower to facilitate comparisons with cotton (296) and tobacco (245). A report that CO_2 compensation point of sunflower increased to about 300 $\mu l/1$ when leaf Ψ dropped to -12 bars (88) must be viewed with skepticism since Boyer clearly observed substantial CO_2 assimilation at a leaf Ψ of -16 or -17 bars (27), and detected only declines, not increases, in light and dark respiration down to -19 bars (30).

In summary, nonstomatal effects on net CO_2 assimilation brought about by mild or moderate water stress seem to be established. Differences among species appear obvious, particularly between cotton and tobacco or sunflower. Oxygen evolution in vivo should be checked along with CO_2 assimilation, especially in species other than sunflower. More attention should also be given to photorespiration and to response of stressed plants to elevated external CO_2 concentration. However, the finding of nonstomatal effects in some species should not be allowed to detract from the dominant influence that stomata normally have on CO_2 assimilation during water stress. What is possible is that these species have a concerted response to stress of several components in the CO_2-assimilation complex. Seeming to suggest this are the striking parallels in behavior between stomata and r_m' (245) and between stomata and chloroplast O_2 evolution (32) with varying degrees of stress.

When stressed plants are rewatered, CO_2 assimilation recovers readily but not necessarily fully. In some cases when prior stress was not severe or prolonged, full recovery was achieved in a fraction of a day (295); but with prolonged or severe stress, recovery after rewatering may require one to several days (7, 33, 106, 295). Accounting for part or all of the slowness in recovery may be the aftereffect on stomata of stress and, in some cases, persistent tissue water deficit (29).

Water stress reduced CO_2 assimilation by plants possessing crassulacean acid metabolism in darkness or light. The reductions were correlated with reductions in transpiration (154).

AT THE SUBCELLULAR LEVEL   An early study reported that Hill activity of isolated chloroplasts was unaffected or even increased by water stress of wheat
leaves (294). A loss of Hill activity in chloroplasts isolated from leaves of Swiss
chard (*Beta vulgaris*) subjected to stress was first reported by Nir & Poljakoff-
Mayber, but only if stress had been very severe (211). The capacity for cyclic
photophosphorylation appeared to be more resistant than Hill activity but was
also reduced when more than 50% of the leaf water was lost (211). In later
studies, Boyer & Bowen (32) presented convincing evidence of an inhibition of
Hill activity of isolated chloroplasts by mild to moderate stress in sunflower, and
by moderate to severe stress in pea, and Fry (77) showed such inhibition by severe
stress (the only level used) of cotton leaves. It should be noted that the assays of
chloroplasts from stressed and control leaves were carried out in the same
medium, so that the stress effect was apparently residual. On the other hand,
Santarius (250) found that light-dependent ATP formation in fodder beet (*Beta
vulgaris*) leaves in vivo was not decreased until 40 or 50% of the leaf water was
lost, and that light-dependent NADP reduction and phosphoglyceric acid
(PGA) utilization were not affected until water loss was more than 50 or 60%.
The reason for the apparently conflicting results is not known. Santarius (250)
used a light exposure of only 20 sec for his assays, and there is a possibility that
ATP formation, NADP reduction, and PGA turnover may be affected by water
stress more readily in the steady-state conditions than in the initial periods of
exposure to light. Leaf contents of phosphorylated compounds, particularly
hexose phosphate and PGA, have been shown to be markedly and reversibly
reduced by severe stress lasting several days (320).

    As is the case with stressing the leaves, results obtained by stressing isolated
chloroplasts also are variable. Santarius & Ernst (251) found that Hill activity
was not reduced in "broken" (osmotically shocked) spinach or sugar beet
chloroplasts until the external osmoticum of sucrose or lutrol was increased to
3 M, corresponding to a loss of 90% of chloroplast water. With sorbitol as the
osmoticum, no reduction was observed at concentrations up to 3 M. Cyclic
photophosphorylation was more sensitive, being reduced noticeably when the
osmoticum reached 1 to 2 M. In contrast, CO_2 fixation by apparently intact
spinach chloroplasts was found by Plaut (232) to be depressed by increasing the
osmoticum (sorbitol) from 0.33 M to only 0.5 M, corresponding to a reduction in
Ψ from -8 to -12 bars. CO_2 fixation by maize chloroplasts also declined when
mannitol was increased beyond the optimal concentrations of 0.1 to 0.3 M
(220). Effects of stress in vitro were quickly reversed when chloroplasts were
returned to the optimum concentration of osmotica (232, 251).

    Plaut's finding that a 4-bar decrease in Ψ of the assay medium was sufficient
to reduce CO_2 fixation by chloroplasts superficially resembles findings of Boyer
& Bowen (32) that a decrease in leaf Ψ of similar magnitude resulted in a residual
inhibition of Hill activity in subsequently isolated chloroplasts. The two cases
are dissimilar, however. Chloroplasts are essentially osmotic sacks with high differential permeability to water and extremely low moduli of elasticity (215).
Hence, when isolated, their ψ_p is close to zero and their ψ_s is virtually the same as
Ψ of the medium. When Ψ of the medium was lowered from -8 bars to -12

bars, the volume of isolated chloroplasts should have diminished by about one third and the concentration of chloroplast solutes increased by more than one third. In contrast, a comparable decrease in Ψ of intact cells at close to full turgor would result mainly in a decrease in ψ_p and only small or insignificant changes in cell and chloroplast volume and solute concentration. The finding that the effects of stress applied in vitro are readily reversible in vitro (232, 251), whereas the effects of stress in vivo are not easily reversed in vitro (32, 77, 211), supports the contention that a moderate lowering of leaf Ψ is different in effect from a lowering of Ψ of the same magnitude by the addition of osmoticum to a chloroplast suspension.

Mild to moderate stress of barley seedlings hardly affected the activities of ribulose-1, 5-diphosphate carboxylase and phosphoribulokinase extracted from leaves (120). Activities of these enzymes in spinach chloroplasts seemed to be reduced by stress in vitro with excessive osmoticum, but only when the chloroplasts were assayed intact (232). This suggests a possible importance of the integrity of organelles in stress effects.

Among the processes quite sensitive to mild stress is light-induced chlorophyll formation in etiolated leaves, first noted by Virgin (304). Reductions in RWC of a few percentage points (304) or in Ψ of 3 bars (21) caused marked reduction in chlorophyll accumulation. The inhibition of the formation of chlorophyll is apparently caused by a lessened ability to form protochlorophyll (304). This ability was restored, however, within 2 hr after the tissue regained turgidity (304). Whether this effect of water stress found in etiolated tissue is significant in green leaves is not clear. Chlorophyll content of leaves declined only slightly with 1 or 2 days of mild stress (120). Chloroplasts isolated from severely stressed leaves contained as much or more chlorophyll as those from control (294).

Summarizing, good evidence indicates that mild to severe stress of leaves of some species results in inhibition of Hill activity of subsequently isolated chloroplasts, but it is not yet totally conclusive that this effect is meaningful in terms of in vivo photosynthesis. Some enzymic components of the photosynthetic complex seem to be resistant to very severe stress. Results obtained with in vitro stress on chloroplasts are difficult to extrapolate to behavior in vivo, because the changes in organelle volumes in vitro and in vivo may well be very different for the same reduction in Ψ.

LICHENS, BRYOPHYTES, AND FERNS   Many nonvascular plants, lacking stomata, and some ferns with stomata thrive in environments where they undergo frequent desiccation and rehydration (226). CO_2 assimilation in light in these organisms has been found to correlate closely with tissue water content, at least within a particular water content interval. Only a few examples are given here. Progressive drying of water-saturated lichen thalli causes the rate of net CO_2 assimilation to first increase, reaching a broad maximum, then decline with further drying (146). The initial increase may be related to a reduction in length of the liquid path for CO_2 transport in the thallus. Considerable assimilation seemed to persist even when water saturation (roughly equivalent to RWC) dropped to about 10% or lower. Interestingly, the RWC range for maximum CO_2 assimilation tended to be lower for lichen species and races (?) found at dry sites than for those at moist sites (146). Assimilation in some mosses and liverworts appears to be related more simply to water content, showing a continuous decline with each decrement of thallus water (174, 269). Again, some assimilation persists after most of the water is lost (174, 269). Unfortunately, none of those studies determined Ψ, nor was it clear how the light-receiving tissue area changes with water content.

The CO_2 assimilation of one fern which tolerates desiccation has been studied in relation to its leaf RWC (281). Assimilation was virtually zero in leaves at 3 to 20% RWC. Between RWC of 30 and 90%, assimilation increased linearly with water content. Respiration followed a similar pattern. Upon the rehydration of desiccated fern leaves, increases in assimilation followed increases in RWC closely, without appreciable lag. Thus, extensive repair or reorganization of chloroplasts was probably not involved in the recovery process. Stomata were not given adequate consideration in this study (281).

The above results with nonvascular plants and a fern might tempt one to infer a central role for nonstomatal components in mediating the effects of water stress in Spermatophytes. Such extrapolation (48, 269) would not be justified, however, since these two groups of plants, because of adaptive selection, are almost certain to differ greatly in responses to desiccation.

Respiration

Effects of water stress on respiration were recently reviewed briefly (48, 265, 293). Early results were often conflicting, showing either no change, an increase, or a decline in dark respiration with water stress (114). Some of the contradiction was most probably due to differences in stress severity and to prolonged stress (several weeks) giving rise to much less direct effects. Recent data demonstrate that dark respiration is generally suppressed, more or less proportionately but not very markedly, by moderate to severe stress (27, 30, 33, 74, 129). Some results also showed a similar depression in apparent light respiration (30). A notable exception is the much reproduced data of Brix (33) on loblolly pine, which showed that with increasing water deficit, dark respiration decreased first at moderate stress, then increased at severe stress to levels above the control, and finally declined again at extremely severe stress. Respiration does not appear to be as sensitive to stress in *Chlorella* as in higher plants, being reduced only when external osmotica were at about −20 bars (97).

Rather complex responses were observed when excised maize roots were stressed in plasmolyzing concentrations of mannitol or glycerol of −11 or −21 bars. It should be kept in mind here that the effects could be due to some specific effect of plasmolysis rather than water stress. When transferred to the osmoticum, consumption of O_2 first increased to as much as twice that of the nonstressed control, and then decreased after 15 min to below the control level (96). In contrast, CO_2 evolution exhibited no such transient increase. Little or no inhibition of steady-state respiration was observed if the osmoticum used was easily taken up by the tissue (102). Deplasmolysis of vacuolated root segments, subsequent to plasmolysis in mannitol of −16 or −21 bars, caused pronounced reductions in respiration. Nonvacuolated root tips, in contrast, exhibited no such reductions upon deplasmolysis (96).

Respiration and the swelling and contraction of mitochondria isolated from water-stressed, etiolated maize shoots have been investigated (17, 196). A lowering of tissue Ψ by more than 4 or 5 bars resulted in reductions in state III and state IV mitochondrial O_2 uptake in a standard assay medium (17). Mitochondria from tissue at -35 bars did not respire to all in the same medium. Phosphorylation was not uncoupled by tissue water stress, as inferred from respiratory control and ADP/O ratios. Passive swelling of isolated mitochondria was promoted by a decrease of tissue Ψ of more than 8 to 10 bars (196). This stress effect, however, was observed only when the swelling medium did not contain phosphate. Severe tissue stress also suppressed the mitochondrial contraction that normally accompanies NADH oxidation in the presence of calcium (196). These results suggest possible alterations, not readily reversible in vitro, in the membrane structure of the organelle by moderate to severe tissue water stress. Reversibility in vivo has not been examined. The observed changes in isolated mitochondria appear to be the likely cause for the reduction in respiration in tissue under stress, though conclusive data are lacking.

Osmotically stressing isolated plant mitochondria also reduced state III respiration and phosphorylation, though not state IV respiration (74). Stress effects were observed when Ψ of the medium was lowered by about 10 bars with sucrose. However, since mitochondria, like chloroplasts, are essentially osmotic sacks (178), this should correspond to a loss of about one-half of the organelle water and volume and may not be comparable to a decrease in Ψ of 10 bars in vivo, as pointed out earlier in connection with stress effects on chloroplasts.

Upon severe desiccation of maize roots (about 70% loss of fresh weight), structural alterations in mitochondria were observed (210) and an increase in mitochondrial cytochrome oxidase was measured biochemically but apparently not histochemically (213). Supposedly a loss of 35% of fresh weight also caused changes in mitochondria, though much less markedly (210). Since these findings pertain to severe stress or desiccation, it is questionable that they would be relevant in interpreting effects of moderate stress.

Cell Growth and Cell Wall Synthesis

Growth in this case is the irreversible enlargement of cells. A minute irreversible increase in cell dimensions may occur with changes limited to the cell wall. Sustained enlargement, however, is accompanied by differentiation, a proliferation of membranes and organelles, and increases in protein and cell-wall material per cell (e.g. 41).

ROLE OF TURGOR AND SENSITIVITY TO STRESS   Turgor pressure has long been considered crucial in cell expansion, supplying the necessary push or pressure from inside (44, 176, 299; but contrast 39[1]). Much of the original evidence was obtained

[1] Points raised by Burström (39) were answered in detail by Ray, Green & Cleland 1972. *Nature* 239:163–64.

in connection with auxin effects on *Avena* coleoptiles, although restriction of growth by water stress in general had earlier been considered in terms of turgor reduction (161, 299). Regrettably, with the shift of attention to metabolic and molecular aspects of stress physiology in the mid-1960s, the importance of water uptake and the resulting turgor as a physical force needed for cell growth has at times been almost overlooked or ignored (48, 84). It is now abundantly clear that in many species cell expansion is one of the plant processes most sensitive to water stress, if not the most sensitive of all. Reduction in cell size has been well correlated with reductions in the Ψ of media in which growing tissue is immersed (55, 91, 148). Steady-state enlargement of leaves of maize (1, 27), sunflower (25), and soybean (27) was slowed by any reduction in leaf Ψ to values below approximately -2 bars. Growth was completely halted by a drop of leaf Ψ to about -4 bars in sunflower (25), -7 bars in maize (1), and -12 bars in soybean (27).

Diurnal variation in shoot or leaf Ψ in plants growing under daily cycles of light and dark has been well documented (83, 137, 151). In many species leaf enlargement is so sensitive to water stress that it may be largely confined to the night (25), a fact observed in the field many years ago (e.g. 177). Some data relating growth to tissue Ψ were based on growth through one or more diurnal cycles (34, 135, 172, 195) and hence are probably complicated by different growth rates during dark and light. Regardless, these results also indicate that growth has a pronounced sensitivity to stress. Other data on the time course of growth of many species during cycles of depletion and replenishment of water, though not providing adequate assessment of plant Ψ or RWC, strongly suggest a very close dependence of growth on water status (127, 150, 259, 306, 309, 312).

In addition to the pronounced sensitivity of growth to water stress and the accepted central role of turgor in cell enlargement, several lines of evidence should be cited as specific indications that water stress directly and physically reduces growth by reducing cell turgor. To begin with, enlargement responses to changes in water status are simply too rapid to be mediated by metabolism. In an elegant study, Green and co-workers (94, 95) continuously monitored the elongation of *Nitella* internode cells and cell ψ_p in vivo and found that a small variation in ψ_p caused apparently immediate acceleration or deceleration of elongation. In that study, ψ_p changes were the result of changes in polyethylene glycol (PEG) or mannitol concentration of the external medium. In the case of higher plants, Acevedo, Hsiao & Henderson (1, 116) found that rewatering the soil of very mildly stressed maize permitted virtually instant (within seconds) resumption of rapid elongation of young leaves. Rapid leaf response to the addition of water to the soil is explainable since the xylem transmits rapidly changes in tension (ψ_p) between roots and leaves (116, 244). Placing maize roots in PEG, -2 bars or lower in Ψ, caused immediate growth stoppage followed by recovery to a new and much reduced steady-state rate (1). Another pertinent observation is that the reduced elongation during a very mild and short (ca 1 hr) stress of maize leaves could be made up completely by a rapid transitory phase of growth ("stored" growth) following the release of stress so that there was no net reduction through stress in total elongation for the period (1). This was taken as an indication that metabolic events necessary for growth perhaps continued unabated during the brief stress and that only a lack of turgor prevented expansion. Other evidence is found in work of Greacen & Oh (93) on root growth. Roots growing in a soil encounter back pressure exerted by the soil which can slow root growth. If the role of turgor is to provide the necessary cell pressure for growth, increasing back pressure from the soil should be equivalent to reducing root ψ_p by the same amount in causing growth reduction. Apparently this is true (93).

GROWTH ADJUSTMENTS DURING AND AFTER STRESS   Green (94, 95) analyzed *Nitella* growth in terms of an equation relating growth rate to ψ_p:

$$\text{Rate} = E_g(\psi_p - \psi_{p,th})$$

where E_g is a coefficient termed gross extensibility of the cell, and $\psi_{p,th}$ is the threshold turgor (also known as critical pressure) below which extension would not occur. A similar though more complex equation which takes into account water transport into the cell is given by Lockhart (176). The existence of $\psi_{p,th}$ was first appreciated in work with *Avena* coleoptile (42) and has been demonstrated or indicated in *Nitella* (94), plant leaves (25, 27, 148, 172), stem (172), and roots (93). The consequence of a finite $\psi_{p,th}$ is that growth stops before ψ_p falls to zero, in agreement with the aforementioned high sensitivity of growth to stress. It should be noted that $\psi_{p,th}$ for *Nitella* is apparently altered within a fraction of an hour following turgor changes and that E_g is not necessarily constant (94, 95). These alterations in $\psi_{p,th}$ and E_g are probably closely linked to metabolism.

According to the equation, when water stress develops and turgor is lowered, growth rate must decrease. However, if E_g, $\psi_{p,th}$, and ψ_p are modified in the right direction in response to stress, growth will recover, at least partially, while tissue Ψ remains at the reduced value. The *Nitella* data (94) showed that with a small step-down in turgor (≤ 1 bar), $\psi_{p,th}$ and possibly E_g adjusted quickly in such a way as to permit resumption of growth at the original rate with the reduced turgor. With larger step-downs in turgor, adjustments were too small to compensate for the lowered ψ_p, so that growth resumed but at a rate slower than the original. The adjustment period ranged from a fraction of an hour to hours, depending on the severity of stress. Data on maize (1) showed that the time course of growth adjustment to step-wise changes in Ψ of the root medium strikingly resembled that in *Nitella;* however, ψ_p was not estimated, hence the results are only suggestive of adjustment in E_g and possibly $\psi_{p,th}$. Interestingly, step-wise changes in evaporative demand elicited growth adjustment similar to those obtained with changes in root osmoticum (116). Growth adjustments appeared to be about as fast in maize as in *Nitella*, but were of lesser magnitude.

The ability to adjust $\psi_{p,th}$ and E_g substantially and rapidly in response to stress would presumably enable a plant under mild stress to grow at reduced turgor. Growth of soybean leaves over a 24-hr period was found to persist to some extent even with an 8-bar reduction in leaf Ψ (27). The growth of *Lolium* also seemed to be relatively resistant to water stress (172, 309). It should be fruitful to compare growth adjustments of these plants with those of maize, which ceased to grow with a 4-bar reduction in leaf Ψ (1).

Provided that stress had been very mild and short, the release of stress permitted resumption of growth at a transitory fast rate followed by a steady-state rate within about 1 hr. The steady-state rates were similar before and after stress (1, 94, 95). Thus, presumably, there was ready readjustment in E_g and $\psi_{p,th}$ to the pre-stress levels. When mild stress was prolonged (6 hr or days) in maize (1) and sunflower (27), recovery after stress release was much more gradual.

Aside from extensibility and the threshold turgor, growth adjustment during stress can take place through an accumulation of solutes (osmoregulation), thus maintaining ψ_p while Ψ is reduced. In *Nitella* (94) subjected to water stress this accumulation was very slow and small (e.g. ψ_p, after being lowered by 3 bars, increased less than 1 bar in 10 hr at constant external Ψ). The situation seems to be similar in *Avena* coleoptiles (44) unless exogenous sucrose or KCl is provided (224). Is osmoregulation or osmotic adjustment more significant in other higher plant systems subjected to water stress? It is well known that many plants adjust osmotically under saline conditions, but this adjustment normally requires moderate or high concentrations of absorbable solutes in the root medium (23; 217; 264, pp. 301–6). Osmotic adjustment during water stress has been mentioned often in the early literature (51, p. 101), but much of the early data are equivocal because ψ_p was not estimated and the lower ψ_s could have resulted from water loss and not a net solute increase. A more recent investigation (247) took the water loss into account and suggested that there was some osmotic adjustment in shoots of plants after roots were placed in PEG for 1 or 2 days. Partial osmotic adjustment in shoots, based on estimates of ψ_p and ψ_s, was demonstrated for several species subjected to 2 weeks of stress with roots in PEG-4000 (172), but possible complications due to PEG side effects (141, 173, 190, 247) cannot be ruled out during such long-term exposure. Growth resumption under mild stress after initial stoppage (1, 27) may reflect alterations in $\psi_{p,th}$ and E_g, not necessarily osmotic adjustment. Kleinendorst & Brouwer (150) noted that when maize was water-stressed by cooling the roots, leaf water content and growth recovered gradually after a few hours. They attributed the recovery to osmotic adjustment, and reported increases in soluble carbohydrates preceding growth recovery. By my estimates, however, the observed increase in carbohydrate represented only a lowering of ψ_s of perhaps 1 bar or less. An equally plausible explanation for much of the growth resumption may be that the roots adapted to the cold, their permeability to water increasing (162, p. 184) to permit recovery in leaf water content. Furthermore, resumption of growth was much less in maize maintained under stress for hours by placing the roots in PEG (1). A tentative conclusion from the scanty data is that osmotic adjustments to water stress probably occur only slowly and to a limited extent in shoots of many species.

Reasonably conclusive evidence of osmotic adjustment came from a study with roots of germinating pea (93). Roots grown for 2 days in a soil ranging in Ψ from -2.8 to -8.3 bars were able to maintain their ψ_p at about 5 bars through changes in their ψ_s. As might be expected, their growth was also maintained at roughly the same rate within this range of soil Ψ. Whether leaves are truly less able than roots to adjust osmotically remains to be determined. It is nevertheless tempting to speculate that a difference in the ability to adjust may be a part of the reason for the often observed increase in root-to-shoot ratio during stress brought about by soil drying (62, 228) or low atmospheric humidity (110). Curiously, in spite of the obvious importance of osmoregulation as a possible mechanism in plant adaptation to water stress, almost no definitive study on this point has been made. Such studies are well within our present technical capabilities.

To conclude with regard to growth adjustments, growth can be expected to slow markedly with mild water stress unless $\psi_{p,th}$ and E_g shift substantially. Even then, such shifts can only compensate for a few bars of reduction in ψ_p and would not be beneficial if ψ_p falls to zero. Sustained growth under moderate stress, even at a much diminished rate, would necessitate substantial solute buildup and osmotic adjustment, which, in the shoots of some species, seems to proceed only slowly if at all.

ROOT GROWTH AND SOIL MECHANICAL IMPEDANCE   One special aspect of root growth deserves mentioning—interaction with the mechanical impedance offered by the soil and with soil moisture. It is well known that compaction of soil can greatly restrict root proliferation, particularly when the soil is low in water (228). That is expectable since, as mentioned, soil impedance can be considered as a back pressure against root enlargement. The mechanics of root growth in soils have been reviewed by Barley & Greacen (12).

Taylor & Ratliff (290) found that in a soil compacted to various degrees, root growth of nontranspiring seedlings was essentially not affected by soil Ψ ranging from about -0.2 bar down to -7 bars for cotton and -13 bars for peanuts, provided comparisons were made at the same soil strength as indicated by needle penetrometer resistance. On the other hand, root growth was sharply curtailed by increases in soil mechanical resistance, regardless of soil Ψ. Apparently, effects of soil water on the growth of roots (228, 229) in many cases might be only indirect in that a higher soil water level often reduces soil strength (288). How can this be reconciled with our understanding of the role of turgor in growth, which might lead one to conclude that root growth should be sensitive directly to both mild water stress and soil compaction? Some insight is provided by work of Greacen & Oh (93). As mentioned, those workers demonstrated, for root growth, an apparent equivalence between soil back pressure and reductions in root ψ_p; and they showed that roots under mild to moderate water stress seem able to adjust osmotically within limits to maintain a constant differential between ψ_p and soil back pressure. However, ψ_s and ψ_p of roots did not adjust to increases in soil strength to the same degree as to water stress. They proposed that the difference in adjustment accounts for root growth's relative insensitivity to soil Ψ and marked sensitivity to soil strength. The idea is rational, but much more data are needed.

It is of interest to mention that improved measurements of the pressure that a growing root can exert axially have generally confirmed the limited data Pfeffer obtained at the end of the last century (12, 87). The pressure generally falls in the range of 9 to 15 bars but varies widely from root to root (61, 289) and was correlated with ψ_s of stem segments from the same plants (289).

CELL WALL SYNTHESIS   Closely tied in with growth is cell wall metabolism (44). Indeed it is difficult to see how expansion can go on for any length of time without wall synthesis. Apparent cell wall synthesis, as measured by incorporation of labeled glucose into wall material, has been known for some time to be substantially suppressed by water stress in *Avena* coleoptiles (43, 222) and leaves of other species (233, 234). Different wall fractions seemed to be affected to different extents (222, 233, 234). Wall synthesis appears to be quite sensitive to a drop in Ψ of a few bars (43). There is some indication that the effect of Ψ reduction is due mainly to decreased ψ_p and not to lowered ψ_s (222). The question arises whether curtailed wall synthesis is a cause or result of reduced growth during stress. Unfortunately the time-course of glucose incorporation and growth rates following the onset of water stress has apparently not been determined. Such studies would indicate the sequence of events and help elucidate the causal relation.

For stress and incorporation periods of several hours, growth is correlated with wall synthesis within limits (43, 222). Nevertheless, data of Cleland (43) showed that while glucose incorporation was substantially decreased by a decrease in coleoptile ψ_p of 3 or 4 bars, growth was almost stopped by the same decrease, thus pointing to growth as the more sensitive process. *Avena* cell wall softens to some extent when growth is prevented by a lack of turgor, as indicated by a short period of growth more rapid than the steady-state rate ("stored" growth) after turgor is restored (e.g. 45; but see 44, p. 212). The detailed data on "stored" growth in *Nitella* (94, 95) and intact maize (1) have already been discussed. These results are strongly suggestive of some wall synthesis under stress in the absence of growth. A tentative conclusion is that growth is more sensitive to stress than is cell-wall synthesis, and that reduced cell enlargement results from causes other than retarded wall synthesis, i.e. it is due to a lack of turgor.

In view of the extreme sensitivity of growth to water stress and the likelihood that the plant may have many feedback controls linking metabolism to cell expansion, there is a real possibility that many of the observed alterations in metabolism during water stress, including suppressed wall synthesis, are the indirect result of reduced growth (1, 115).

Cell Division

It has often been stated that cell division appears less sensitive to water stress than does cell enlargement (264, pp. 289–90; 267; 299). Much of the previous data bearing on this point, relating to cell size and cell number in plants grown under different water regimes, have been tabulated by Ordin (223). Actually, the situation is not clear cut, as can be seen from some more recent data. Gardner & Nieman (83) incubated growing radish cotyledons on mannitol of various Ψ and subsequently determined the DNA content which they used as an estimate of cell number. Increments in DNA per cotyledon during 28 hr were reduced by more than half in the presence of -1 to -2 bars of mannitol, but only little further reduction occurred with further lowering of the medium Ψ to -16 bars. This finding was extended in a later study (148), which determined that a decrease in estimated tissue ψ_p of about 1 bar inhibited the DNA increment by some 30% while reducing mean cell length only slightly. A further decrease in ψ_p of 2 bars, however, stopped cell expansion but still permitted over 50% of the DNA increment (148). Although DNA increment may at times not be closely correlated with cell division (37), it nevertheless is basic to the process. From a study using days to weeks of very mild stress, Terry, Waldron & Ulrich (291) reported that cell multiplication in sugar beet leaves was inhibited while cell volume was affected very little. It is regrettable that they did not measure leaf Ψ, since their stress appeared to be so mild (root media Ψ lowered by as little as 0.03 bar) as to raise the question of whether the changes were really due to water deficit. In another study, increments in both cell size and number in the secondary xylem of ash twigs in 18 days were found to be markedly reduced by 1 bar of PEG in the medium (55). In contrast, in a study of long-term onion root growth with mannitol as the osmoticum ranging from -2 to -12 bars, cell size was found to be much more sensitive to stress than was the duration of the cell-division cycle (91). One drawback of all these investigations, perhaps minor, was the protracted exposure of tissue to PEG or mannitol. Possible complication with PEG has already been mentioned. Mannitol is also known to have side effects (126, 173) and may be taken up slowly by the tissue (102).

Overall, the situation seems to be that cell division, like cell expansion, can be inhibited by rather long exposure to mild water stress. It is not at all certain, however, that the stress effect on cell division is direct. One may expect that the meristematic cell must attain a minimal size before division can take place; that is, some enlargement must follow each division before the next mitosis. Doley & Leyton (55) deduced that ash cambial initials need to expand to a diameter of 6μ before division commences. Therefore, it may be speculated that the effect of water stress is possibly indirect via suppressed cell expansion, a hypothesis that would explain the susceptibility of division to very mild stress observed in some cases (84, 148). There is also a possibility of alterations in growth regulators taking place during prolonged stress and thus affecting cell division. What is needed to resolve some of these uncertainties are time-course studies of stress effects on cell expansion and division, with detailed monitoring of Ψ, mitotic index, as well as the number and size of cells in the meristematic and enlarging zones (not of the whole tissue as often done before).

Hormones and Ethylene

The research on hormonal changes from water stress was covered in a recent review that also included some speculative hypotheses on interactions between hormones and controls of plant water balance (175). The discussion here is mainly on cytokinins, ABA, and ethylene and abscission. It has been speculated that water stress affects auxin and gibberellin levels (175, 285) but direct data are lacking. Extractable indoleacetic acid oxidase activity was reported to be increased by a brief stress with 10-bar mannitol (53) and by moderate to severe stress in the field (54). Darbyshire (54) hypothesized that auxin level may be reduced by enzymic degradation in stressed plants and that retardation of growth during stress may result from lack of auxin-induced wall loosening as well as from reduced turgor. Since cell expansion appears much more sensitive to water stress than does the level of the oxidase, it is hard to see how that hypothesis would be applicable in a straightforward way.

CYTOKININ ACTIVITY   Following the finding of Kende (145) that root xylem exudate contains cytokinins, Itai and co-workers observed that days of water stress resulting in wilting (123, 124) or of salinity stress (123) lowered cytokinin activity in root xylem exudate obtained from sunflower after the plants were rewatered. The reduction in cytokinin activity appeared not very pronounced and the degree of stress was not quantitatively assessed. With salinity stress, the subsequent depression in cytokinins was related linearly to NaCl concentration of the medium in which roots had been stressed (123). It is not clear in this case how the observed effects should be apportioned between water deficit and toxicity from NaCl.

A more recent study with tobacco (125) may be of more potential significance. Less than 30 min of wilting substantially reduced the cytokinin activity in root exudate following rewatering. Also, the extractable cytokinin activity in excised leaves was reduced by drying for 30 min to 75% of initial fresh weight, and a slight reduction was detected after only 10 min of drying. Extractable cytokinin activity recovered partially when stressed leaves were rehydrated for 18 hr in water-saturated air. Since the stress effect was so rapid, Itai & Vaadia (125) suggested that the activity loss was the result of inactivation. There are, however, some contradicting data. Work by others (199) at the same institution on the same *Nicotiana* species showed that 1 day of wilting (slight?) from low humidity plus salinity did not affect extractable cytokinin activity in the leaf although transpiration was depressed and leaf ABA content increased. Stress might have been more severe in the former study. In any event, this apparently important finding of very rapid stress effects on cytokinins (125) needs substantiation, particularly with more data on the degree of water stress.

ABSCISSIC ACID   One of the more exciting recent findings in research on water stress is the dramatic accumulation of ABA in plants during stress. Unfortunately, as is the case for cytokinins, most of the studies on ABA have not assessed plant water status adequately. Wright (323) first observed an increase in a growth inhibitor, deduced to be ABA, when excised wheat leaves were kept in a wilted state. With a loss of 9% of fresh weight of the leaves, the increase was estimated to be severalfold in 2 hr (323) and as much as 40-fold in 4 hr (324). Wright & Hiron (324) confirmed that the inhibitor is ABA by optical rotary dispersion analysis. This observation of phenomenal ABA accumulation during water stress has been extended by others to intact plants of several species (191,

200, 204, 326). Mizrahi and co-workers (199–201) also demonstrated that ABA increased pronouncedly in tobacco subjected to salinity.

Excised wheat leaves dried for 10 min did not show an increase in ABA although 6% of their initial fresh weight was lost. When leaves were held at this water content for another 40 min, however, ABA increased significantly (323). After water was withheld from the soil, ABA increased in sugar cane leaves before wilting appeared (204). Hence, only mild to moderate stress appeared to be necessary to induce the ABA increase (199, 200). In fact, limited data indicate that ABA seemed to accumulate most readily in wheat leaves if the loss in fresh weight did not exceed 9% (323). In briefly salinated plants the ABA level returned to the control level within 2 days of transfer to nonsaline medium (201).

The increase in ABA during water stress has been shown by Milborrow & Noddle (191) to arise from de novo synthesis and not through release of a bound form. A supposed decline in bound ABA and rise in free ABA upon salination of tobacco have been mentioned (175, p. 259), but Zeevaart (326) found that wilting actually caused some increase in hydrolyzable (presumably "bound") ABA along with a marked increase in free ABA.

The postulate that accumulated ABA modulates stomatal behavior during water stress has already been discussed (p. 527). A possible role of elevated ABA level in retarding cell growth under stress has not been investigated. Since growth depends directly on turgor, however, one might expect the physiological significance of rise in ABA in plants under water stress to lie elsewhere, although slow recovery in growth after a stress of more than a few hours (1, 27) may possibly be related to stress-induced buildup of ABA.

ETHYLENE AND ABSCISSION   Ethylene has long been known for its ability to induce abscission (236) and abscission is a known response to water stress in some plants. Abscission of developing flowers and young fruits after water-stressed plants are irrigated is an ever-present problem in managing irrigation of cotton (103, 278). A recent study on cotton (187) determined that abscission of bolls or leaves was more or less proportional to the daily minimal water deficit as indicated by predawn leaf Ψ. When the rooting medium dried enough that predawn leaf Ψ dropped to around -8 bars, abscission of leaves and bolls of 2-month-old plants was potentiated and took place after rewatering. In young cotton seedlings, the threshold predawn Ψ for subsequent abscission of cotyledonary leaves was found to be lower, being about -17 bars (136). Usually leaf Ψ would of course be considerably lower during the day than at predawn (135).

Some evidence suggests that abscission induced by water stress may be mediated through internal ethylene production. Ethylene production by petioles on intact cotton plants tended to increase within hours when water deficits developed, and declined quickly on rewatering in some cases though not all (186). Variations from plant to plant were large, and quantitative relations remain to be established. Other evidence comes from results obtained with exogenous ethylene and CO_2 (136). Treatment with ethylene apparently did not induce abscission in cotton plants at high Ψ but greatly enhanced abscission in stressed plants. The threshold water deficit required to induce abscission was reduced by exogenous ethylene. Thus water stress seemed to predispose the leaves to ethylene action. CO_2, an inhibitor of ethylene action in many cases (236), was found to counteract, within limits, the promotive effect of exogenous ethylene on stressed seedlings. Unfortunately, the effect of CO_2 on abscission in stressed plants in the absence of exogenous ethylene was not determined. Possible interactions between ABA and ethylene in stress-induced abscission are yet to be examined.

Nitrogen Metabolism

PROTEIN SYNTHESIS IN VEGETATIVE TISSUE   For some time water stress had been supposed to reduce the ratio of protein to amino acids or the protein content in the plant (299), though some results showed an apparent increase (284, 305). More recently Shah & Loomis (258) found that both soluble and total protein contents of sugar beet leaves (per gram of dry matter) declined progressively in a matter of days when water was withheld. The contents apparently decreased measurably even before first wilting, and at severe wilting became as low as half the protein of the well-watered control. Comparison with earlier (and less firm) results on other species indicates that sugar beet is possibly more sensitive in this regard. Soluble protein content in wheat leaves was reported by Todd and co-workers to be either hardly affected (294) or markedly reduced (283) when RWC was lowered to about 60%. A decrease in protein content may reflect a retardation of protein synthesis or an acceleration of degradation.

Several studies have examined effects of water stress on the ability of tissue to incorporate labeled amino acids into proteins. Those studies, however, were really on the aftereffect of stress on incorporation, for the tissue was floated on a solution of the label for uptake, and the plants had been rewatered hours before to obtain turgid tissue. Ben-Zioni, Itai & Vaadia (18) first reported a substantial aftereffect on incorporation of leucine by leaf disks from tobacco which had been stressed mildly to moderately for 2 days. Incorporation into protein of the label taken up by tissue was about halved by prior stress. Apparently the effect was not due to stress altering the size of endogenous leucine pools (18). Complete recovery in incorporation required 3 or 4 days. Inhibition of amino acid incorporation was observed (212) also in root apices which had been dried in air of controlled humidity, but only if water loss was more than 30% of the original fresh weight. If root apices were not rehydrated first as in the preceding case but only incubated for 1 hr in labeled amino acids directly after stress, a prior loss of 11% of fresh weight in water was sufficient to reduce incorporation (as percent of uptake) slightly, and the uptake of amino acids markedly (212). These results (212) suggest that protein synthesis is susceptible to a brief water stress of mild or moderate intensity, but that an aftereffect requires more intense or prolonged stress.

Exposing roots to saline medium of several bars for days similarly depressed subsequent incorporation of amino acids into protein in leaf disks (18, 123). The depressive effects on incorporation in root apices were much more severe with Na_2SO_4 than with NaCl at the same ψ_s (138), attesting that salinity has effects other than those of water stress.

Problems involved in incorporation studies—the time required for sufficient incorporation, the necessity of floating tissue on aqueous solutions, the difference in label uptake induced by stress, and possible changes in pool sizes—are avoided by studying polyribosomes and inferring therefrom protein synthesis activities. Supposedly the more active protein synthesis is in a cell, the larger will be the proportion of ribosomes in the polymeric form and the larger will be the polymers. Hence, polysome profiles provide a means of examining protein synthesis activity in the tissue at any given instant. There have been cursory indications that polysomes may be reduced by water stress (86, 314). A detailed study was carried out by Hsiao (115). In that study, water stress of etiolated maize seedlings caused a shift from polymeric to monomeric form of the ribosomes in rapidly growing meristematic tissue. The shift started about 30 min after the initiation of stress, when Ψ of the tissue began to decline measurably. A decline in tissue Ψ of a few bars within 3 hr caused a large portion of ribosomes to change from the polymeric to monomeric form (115). When stress was induced much more rapidly, causing a loss of 10% of fresh weight in 15 min, a pronounced decline in polysomes in young green shoots of maize was also observed (237). Others have reported a loss of dimers and trimers and a gain of monomers when root apices lost more than 50% of their water (212).

Cycloheximide, an inhibitor of peptide chain elongation and termination, blocked the changes in proportion of polysomes induced by stress. The data have been interpreted to indicate that the stress effect depends on peptide chain completion and is not the result of random fragmentation of polysomes (115). Rewatering maize seedlings subjected to a brief and mild stress caused the ribosomes to revert to the polymeric form (115, 237). Recovery can be complete within a few hours of rewatering (115, 237), but there was a lag period that apparently depended on the duration and degree of prior stress (115). The rapidity of response to stress and quick reversibility by rewatering suggest that stress effects on protein synthesis are mainly at the translation level. Previously, however, modulations at the transcription level have been hypothesized to occur in stressed tissue because of claimed alterations in RNA (212; 265, pp. 63–64; 293).

Also investigated has been the ability of ribosomes prepared from stressed tissue to incorporate amino acids (237). Stressing the plants mildly to moderately reduced in vitro amino acid incorporation. There was a good correlation between in vitro incorporation and the proportion of polysomes as affected by water stress. Stress apparently affected the ribosomes and polysomes, not the supernatant factors (237). One study (253) has reported that water stress reduced the activity of supernatant factors, but in that study tissue was severely desiccated and stress was confounded by drying at high temperature. Percent stimulation by poly(U) of phenylalanine incorporation was about the same in ribosome preparations from either control or stressed tissue, thus suggesting that mRNA may not be the limiting factor in the stressed preparation (237). However, the stimulation by poly(U) was small in all preparations and further study on this point is needed.

It is well known that RNase increases in tissue subjected to rather long and severe stress (293), and increased RNase activity has been suggested as a cause of the shift from polysomes to monosomes (20, 86). Our data do not support that view. To start with, during the early part of stress, polysome level declined substantially while tissue RNase level showed no change (115, 237). Moreover, if RNase were a key factor, the quick recovery in polysomes after rewatering would require that mRNA be synthesized extremely rapidly and, even less plausibly, that the accumulated RNase be inactivated just as quickly. Further, the shape of polysome profile from stressed tissue was not consistent with the random cleavage of mRNA linking ribosomes (237); random cleavage would be expected if RNase activity were the cause of low proportion of polysomes. Thus, the early stress effect seems not to be mediated by RNase, although that does not rule out a possible role of the enzyme during prolonged stress.

Itai and co-workers (18, 123) have advocated a key role for cytokinins in modulating protein synthesis during water and salinity stress. This view is based on the reduction in cytokinins in leaves by stress (see previous section) and on the reported ability of applied kinetin or benzyladenine to alleviate a part of stress-effected reduction in amino acid incorporation (18, 123) or in leaf protein content (258). Actually, data on alleviation of stress effects by kinetin are conflicting. Sometimes kinetin did not reverse the effect of stress on incorporation at all (125), or actually depressed incorporation by control roots and roots salinized in Na_2SO_4 but stimulated incorporation by roots salinized in NaCl (138). Also the reversible changes in polysomes with stress and rewatering appear too rapid (115, 237) to be consistent with the view that stress affects protein synthesis via changes in cytokinins. Cytokinins in root exudate remained depressed for hours after rewatering of plants stressed for 30 min (125). Also leaf cytokinin level did not recover readily after a brief stress (125).

In summary, protein synthesis in rapidly growing tissue appears to be readily and reversibly reduced by very mild water stress. The dynamic responses to stress and stress release may be controlled at the translation level. The basis for the response is still obscure; however, indirect data argue against the involvement of RNase or cytokinins, at least during the early period of stress.

PROTEIN SYNTHESIS IN SEEDS AND MOSSES   Protein synthesis in relation to water status has also been studied in systems very different from the vegetative tissues of higher plants. Work on water uptake and polysome and protein synthesis in germinating seeds (180, 181) is well known. Ribosomes from dry wheat embryos are inactive in incorporating amino acids, but upon imbibition, incorporating capacity rises rapidly, lagging about 10 min behind water uptake (181). The enhancement in incorporating capacity is accompanied closely by increases in polysomes. When embryos were allowed to imbibe for 30 min and then desiccated to their original weight, they yielded ribosomes essentially as active as those from the undesiccated control (181). This behavior is in sharp contrast to the marked loss of in vitro activity of ribosomes from maize leaves subjected briefly to moderate or severe stress (237). With embryos at a more advanced stage of germination (1 day of imbibition), however, desiccation caused a marked loss in the incorporating activity of ribosomes (40).

In its response to desiccation, the moss *Tortula ruralis* seems to have a protein synthesis machinery similar to that of the wheat embryo at very early stage of imbibition (181). After complete and rapid (90 min) desiccation, some polysomes were still evident in ribosomes prepared from dried *Tortula* tissue (19). On complete rehydration (within 2 min), protein synthesis appeared to resume almost immediately and the proportion of polysomes increased substantially within a fraction of an hour (20). Bewley (20) pointed out that since *Tortula* undergoes frequent desiccation and rehydration in its native habitat, the ability to conserve components of the protein synthesis complex when desiccated so that protein synthesis can resume rapidly after rehydration is an important aspect of adaptation of the moss to the environment.

NUCLEIC ACIDS   Except for the work on germinating seeds discussed below, little or no progress has been made recently on water stress effects on nucleic acids. Results from a decade ago, though repeatedly covered in recent reviews (48, 206, 265, 293), merit reevaluation. Much of the reported dramatic changes in base composition caused by stress (147, 313) seem implausible now and might possibly be attributed to poor analytical techniques. For instance, a marked change in RNA base composition claimed in one study (313) was absent in a second study (314). The data of Gates & Bonner (85), often cited for increased RNA degradation but unimpaired RNA synthesis during stress, were probably complicated by problems such as differential uptake of radioactive phosphate, changes in the size of precursor pools, and leaf senescence induced by prolonged stress. In light of current knowledge on the amazing complexity of the controls operating in the cell at both the transcription and translation levels, information on total or bulk fractions of nucleic acids as related to water deficits is of minimal value. What may be deduced from previous results (86, 282, 314) is that much of the RNA is probably not readily altered by brief stress that is mild or moderate. Perhaps that is the reason for the current dearth of research activities in this area. An exception may possibly be sugar beet: its leaf RNA and DNA, expressed either per unit dry weight or per cell, declined significantly with apparently only moderate stress (258). A recent study (21) invoked water stress to explain reduced uracil incorporation into RNA of greening leaves, but water status was not purposefully varied between treatments and there was no direct evidence of water being deficient.

Results were more interesting with germinating seeds subjected to severe desiccation (40). Ribosomes isolated from embryos of seeds desiccated for 2 days, after either 1 or 3 days of imbibition, fully retained their ability to incorporate phenylalanine when the synthetic messenger poly-U was provided. DNA-RNA competition hybridization tests showed that most of the mRNA was apparently preserved during 2 days of desiccation if prior imbibition had been for only 1 day. With 3 days of prior imbibition, the same desiccation treatment appeared to alter or destroy the majority of the mRNA. Therefore, examination of the differences between embryos that imbibed for 1 day and 3 days may yield clues as to the basis for the marked change in sensitivity to dehydration.

PROLINE AND OTHER AMINO ACIDS   Total free amino acids in leaves are often increased if rather severe water stress lasts several days (13). Amides frequently increase (13, 292), but proline has the most pronounced rise. The increase in proline, reported first by Kemble & Macpherson (144), can amount to as much as 1% of the leaf dry matter in many species (13, 246, 262, 277). Although the rise during severe stress is dramatic when initial proline content is very low, sketchy indications are that the level of proline may be insensitive to mild stress (246, 277). The change takes place only if there are adequate carbohydrates in the tissue (277; but see 276), and it is readily reversed by rewatering (246). In several species, roots did not accumulate proline when desiccated (277) although accumulation in leaves proceeded readily in darkness (277).

The accumulated proline apparently came from de novo synthesis (13, 292), with glutamate as a precursor (13, 203). A pronounced increase in the incorporation of labeled precursors was observed even after only 2 hr of severe stress (26% fresh weight loss) (203). Some evidence indicates that carbohydrates are the ultimate source of the carbon skeleton (277). Several workers (13, 246, 277) suggested that proline may serve as a storage compound for reduced carbon and nitrogen during stress. This is consistent with the observed decline of previously accumulated proline when water-deficient leaves are kept for more than 1 day in darkness (277) and with the finding that proline is a source of respiratory CO_2 under carbohydrate-deficient conditions (276). Along the line that proline accumulation is possibly beneficial to the plant under stress, it is interesting that the ability of 10 barley varieties to accumulate proline under severe stress has been positively correlated with their drought resistance ratings (262).

NITROGEN FIXATION   A recent study (271) of stress effects on detached soybean nodules (ca 70% water content when turgid) may have significant implications for nitrogen nutrition of legumes under water stress. A loss of 10% of original nodule fresh weight slowed acetylene reduction activity, while a loss of more than 20% of fresh weight almost stopped it. Nitrogen reduction appeared to be similarly affected.

Enzyme Levels

Changes in enzyme activities caused by water stress were reviewed recently (293). Only a few salient points are discussed here. Todd (293) concluded from a list of some 25 enzymes affected by water deficits that severe stress or desiccation generally lowers enzyme levels although moderate to severe stress often raises the levels of enzymes involved in hydrolysis and degradation. Among the enzymes examined, those that appear to be reduced most readily by stress are nitrate reductase and phenylalanine ammonia-lyase. The first observation made on nitrate reductase by Mattas & Pauli (184), though often cited (206, 293) as water stress effect, was confounded in that stress was brought about not only by withholding water but also by raising the temperature to 38°. Such high temperature per se could cause sharp declines in tissue nitrate reductase (221). Subsequent studies did establish, however, that water stress alone lowers the level of the enzyme (11, 120). Only 1 day of mild stress caused a 20% decrease in extracted activity and also stopped growth (120); longer or more severe stress reduced the activity to 50% or less of that of the control (6, 11, 120). The activity recovered to as much as the control level within 24 hr of rewatering (11, 120).

In view of the inhibition of protein synthesis by water stress, it may be suspected that levels of enzymes with short half-lives would be reduced because of suppressed protein synthesis (11). Nitrate reductase is turned over quickly; some data do indicate that the decrease in the level of this enzyme during stress is through suppressed synthesis (6). Nitrate, the inducer of the enzyme, was present in sufficient amounts in stressed tissue and was apparently not the limiting factor (6, 120).

The level of phenylalanine ammonia-lyase was also found to decrease with mild to moderate water stress and to recover readily with rewatering (11). Phenylalanine ammonia-lyase too has a short half-life and may therefore be susceptible to stress through suppressed synthesis (11).

Among the enzymes whose activity is raised by moderate to severe water stress are α-amylase (270, 293) and ribonuclease (237, 293). These enzymes have been repeatedly observed to increase, though the functional significance of the increase is obscure. Amylase may presumably catalyze starch hydrolysis in vivo, but the evidence is circumstantial, i.e. observed increases in sugars and decreases in starch during stress (270, 275, 277). The observed decreases in starch, however, may, be due to reduced photosynthesis. In one study (275), the sugar accumulated in the dark was sucrose, not glucose as would be expected had hydrolysis catalyzed by α-amylase been the underlying process. Indirect evidence suggesting that increased RNase is not responsible for the reduction in polysome levels at the beginning of stress has already been discussed. Whether RNase accelerates RNA hydrolysis in prolonged stress remains to be determined. This problem is of course common to all attempts to relate enzyme activity in vitro to that in vivo. The likelihood of hydrolytic enzymes being separated from their substrates by compartmentation in the cell must always be considered.

In contrast to the increase in amylase induced in leaves by stress, α-amylase formation in germinating seeds is depressed by stress. Amylase production in barley aleurone layer was substantially inhibited by lowering the Ψ of the incubation medium with 0.2 M or more concentrated mannitol (132) and other sugars (133), corresponding to a lowering of Ψ of a few bars. α-Amylase was not synthesized in crested wheatgrass seeds kept at -60 bars and was synthesized more slowly in seeds at -20 bars than in control seeds on moistened paper towels (318). Once formed, the enzyme is stable in situ upon repeated air drying and rehydration, at least during the early period of germination (318). It was proposed (133) that the inhibition of amylase production by sugars constitutes a negative feedback loop, placing the synthesis of the hydrolytic enzymes under the osmotic modulation of hydrolytic products from the seed endosperm.

The level of phosphoenol pyruvate carboxylase in green barley leaves was reduced noticeably, though not markedly, by days of mild to moderate stress (120). It would be interesting to examine effects of such stress on this enzyme in plants possessing the C_4-dicarboxylic acid pathway of photosynthesis.

Transport Processes in the Liquid Phase

The coverage of transport processes here is confined to the effects of water stress on ion uptake and transport, on translocation of photosynthetic products, and on xylem resistance to water flow.

ION UPTAKE AND TRANSPORT   The well-known hypothesis of Crafts & Boyer (49) depicts ions as being taken up actively at the plasmalemma of epidermal and cortical cells of the root, then transported passively down their activity gradients through the symplasm into xylem vessels in the stele. Accumulated evidence, however, indicates that there is probably a second active step in the stele, with xylem parenchyma cells secreting ions into the vessels (171). Two possible kinds of stress effects may be visualized: (a) an influence on ion transport from reduced transpiration and hence reduced water flow; and (b) stress effects on the active transport mechanism and on membrane permeability.

Various experiments have demonstrated that high transpiration can increase ion uptake by roots. The enhancement in uptake is more pronounced if roots already contain high levels of nutrients (36) and if concentration of the ion in question in the medium is high (22, 35). Broyer & Hoagland (36) originally proposed that high concentration of the ion in root xylem may limit transport and that high transpiration facilitates uptake because it dilutes the xylem fluid. After some controversy which has been reviewed (35), this explanation seems to have received at least partial acceptance (22, 35, 99, 111). How high ion concentration or activity in the xylem reduces active uptake by the cortical cells which are some distance away from the xylem is not established. One plausible explanation is that both active ion "pumping" steps, one at the plasmalemma of the epidermal and cortical cells and one in the xylem parenchyma, are somehow reduced by high activity of the ion "downstream" (111). However, several other explanations (99; 162, pp. 244–47), including increased passive efflux from the xylem caused by high concentration of the ion in the vessels (22), have also been proposed.

The above discussion implies that low transpiration, in addition to slowing root uptake under some conditions, may reduce ion transport from the root to the shoot. Such reductions have been observed frequently (22, 80, 111), though they may not be marked because of the compensating increase in xylem concentration at low transpiration rates (e.g. 64).

Aside from curtailing water flow through the plant, stress may also directly affect the active transport mechanisms and the passive permeability of the plasmalemma. Before discussing the data obtained with vascular plants bearing on the latter points, it is pertinent to outline a very interesting finding on the marine alga *Valonia*. Gutknecht (104) has confirmed and extended early results indicating an effect of turgor change on salt uptake by this giant-celled alga. He varied cell ψ_p

through a mercury manometer which was connected to the cell through a fine-tipped micropipet inserted into the vacuole. Reducing ψ_p from 1 to 0 bar tripled the active influx of potassium, the predominant intracellular cation, into the vacuole and also reduced potassium efflux. The effect was readily reversed by raising ψ_p back to 1 bar. Fluxes of urea, in contrast, were not affected by changes in ψ_p. The mechanism by which the small change in ψ_p affects active transport is intriguing. Gutknecht pointed out that the function of this effect of pressure may be to maintain a rather constant ψ_p in growing cells subjected to varying external salinity to facilitate cell expansion.

Responses of ion transport mechanisms to changes in water status seem to be quite different in higher plants. With higher plants there is no indication of a general high sensitivity to small turgor changes, and water stress or turgor reduction seems to reduce ion transport, not accelerate it. Greenway and co-workers (98–100) lowered the nutrient solution Ψ by 5 bars with mannitol and found that transport of ions (phosphate, bromide, and sodium) from the root to the shoot of tomato was strongly suppressed while total uptake from the solution was less affected. When roots were allowed to take up labeled phosphate first and then stressed, movement of the prior accumulated phosphate to the shoot was markedly reduced by stress (99). Thus, the effect on transport to the shoot is independent of the effect on uptake into the roots. Movement to the shoot remained retarded for several hours after stress was alleviated (98). Strongly plasmolyzing concentrations of mannitol (e.g. −10 bars) increased the passive permeability of root membranes to ions since the efflux of ions [potassium, bromide, phosphate (101), and sulfate (66)] from roots into the medium was enhanced. Sulfate uptake into excised maize roots was inhibited by isotonic (0.3 M) or hypertonic mannitol but stimulated by hypotonic mannitol (66). In contrast, net influx of potassium into pea leaf pieces in light was maximum and net efflux in dark was minimum when the sucrose concentration of the medium was isotonic (216).

Unfortunately, it is not possible to deduce from these results exactly what happens to ion transport during water stress in general. One shortcoming of all these studies is that mannitol or other small molecules were used as osmotica at concentrations high enough to cause plasmolysis of at least some cells. Media at −5 bars used in the study of Greenway and colleagues (98–100) were considered to be hypotonic since ψ_s of tomato roots was assumed to be −7 bars (101). It is well known, however, that ψ_s of individual cells in tissues varies widely and may span a range of several to many bars (51, p. 86; 122). Therefore, it is likely that a medium 2 bars higher than the average cell ψ_s, as used by Greenway et al, would plasmolyze a portion of the cells. Plasmolysis, separating plasmalemma from the wall and presumably tearing some plasmodesmata, may well have its own unique effect on ion transport. Therefore, it will be necessary to confirm the results with an osmoticum such as PEG-6000, which apparently does not penetrate the wall to cause plasmolysis (190).

It is probable that nutrient uptake and transport are slowed by moderate to severe water stress via one or more of the effects outlined above. For plants growing in the field, however, changes in the soil brought about by water depletion may be so pronounced as to overshadow the effects of stress on uptake and transport in the plant. Drying of the soil can reduce rates of nutrient transport to the roots, and in the case of nutrients such as phosphorus, also reduce the amounts in the soil solution. In addition, stress that is sufficient to slow root growth curtails the exploration of new soil volumes for nutrients. The influence of water on the availability of soil nutrients to plants has been reviewed (303).

Slatyer pointed out that while nutrient accumulation is frequently reduced by water stress, the demand for nutrients is also reduced because of suppressed growth (265). It is probable that with certain nutrients, if they are already slightly deficient, water stress would aggravate the deficiencies. On the other hand, it is not likely that inorganic nutrition would generally be a key factor in plant function under stress, especially if the stress lasted only a few days.

PHOTOSYNTHATE TRANSLOCATION   Understanding of stress effects on photosynthate translocation appears to have progressed little since it was reviewed (48; 265, pp. 69–70; 308) several years ago. Only a few remarks are made here.

There is ample evidence for frequent reduction in photosynthate translocation in plants under moderate to severe water stress (50). Since phloem transport is thought to depend on gradients of hydrostatic pressure, suppressed translocation has at times been considered to result from a direct disturbance of phloem pressure gradients through reductions in tissue Ψ. Actually stress effects on the gradients should be complex, as is made obvious by the following consideration. Under water stress, xylem flow is reduced so that xylem Ψ gradients may become less steep even though absolute values of Ψ are lower. The phloem, being adjacent to the xylem, should be influenced by this alteration so that its Ψ gradient (opposite in direction to that of xylem) may actually become more steep if other things remain unchanged. Hence, the effects of stress on the gradients may be hard to predict.

Several reviewers have emphasized the importance of source and sink strengths for photosynthate in considering stress effects on translocation (50, 265, 308). Stress reduces source strength by reducing photosynthesis and reduces sink strength by inhibiting growth, thus lessening translocation. This is well illustrated in results of Wardlaw (309). Moderate stress markedly reduced the velocity of photosynthate movement out of a mature leaf; but the reduction was virtually offset if other mature leaves were removed to eliminate sources competing for the diminished sink brought about by reduced growth. Available evidence suggests that the phloem conducting system remains functional under moderate or severe stress (265, 308) and that the main reason for the often-observed reduction is the change in photosynthesis and in utilization of assimilates. The possibility has not been examined that the influence of stress on translocation may be partly mediated by metabolic modulation of phloem loading and unloading, that is, the transport of sugars into and out of the long distance conducting system.

XYLEM RESISTANCE TO WATER FLOW   Remarks here are confined to those increases in resistance that apparently are caused by cavitation of xylem vessel columns under water stress. Not considered are variations in plant resistances with alteration in water flow rates (e.g. 279) or possible effects of stress on tissue (92) or on membrane (e.g. 149) permeability to water. Water stress increases tension (decreases ψ_p) in the xylem. Should some water columns in the vessels break or cavitate, stress can be aggravated by the increased resistance to water flow. Milburn & Johnson (193) used a sound amplifier to detect vibrations apparently generated by cavitations in petioles and leaves of a number of species. With rapid changes in water supply and demand, the number of vibrations per minute in *Ricinus* leaves was closely related to the expected sudden tension increases in the xylem. These vibrations are suggestive that cavitation can take place under water stress (but see 219). Milburn (192) studied the kinetics of water uptake by *Ricinus* leaves that were detached and stressed (to as low as 80% RWC). Instead of following the simple decay kinetics first described by Weatherley for leaves with very small water deficits, water uptake through petioles by stressed leaves exhibited a complex pattern of two or more phases, in accordance with what might be expected if xylem resistance suddenly lessened after some water were taken up by the stressed leaves. Milburn (192) proposed that when water is resupplied to the leaf containing cavitated xylem channels, xylem tension falls with water uptake and reaches a level at which the water vapor space (virtually vacuums) created in vessels refills, at least partly. Hence, xylem resistance declines and approaches the prestress values. Further support of this idea comes from detailed measurements of leaf Ψ during the recovery of soybean from water stress (29). Again the data suggest a sudden decrease in resistance to water flow in stressed plants (to $\Psi = -15$ bars) after the initial few bars of increase in Ψ upon rewatering. Boyer (29) showed that after stressed plants were rewatered, leaf Ψ could remain below the prestress level and the level of the control for several days if stress had been rather severe, presumably because of the elevated resistance to water flow. Thus, detrimental aftereffects of severe stress could be physical as well as metabolic.

Comparative Sensitivities and Overall Responses

The preceding sections elaborated individually on the multitude of physiological and metabolic changes brought about by water stress. Here we compare the various processes as to sensitivity to water stress and attempt to deduce the sequence of the changes in the plant as stress develops. The levels of water stress at which various plant processes have generally been found to be first affected are compared in Table 1. Much of the assigned sensitivity to stress is tentative, for several reasons. To begin with, since some data provide only tenuous indications of plant water status, compiling the table involved considerable guesswork. In addition, some processes may merely appear to be relatively insensitive because detecting techniques are inadequate. This problem is particularly acute if, cumulative quantities rather than rates are measured for detecting changes in a process. Moreover, sensitivity to stress may differ among species, as mentioned in previous sections on stomatal opening and CO_2 assimilation. It is probable that as more plant species and processes are studied, more species differences in behavior under stress will become apparent. Species differences may not complicate the

Table 1 Generalized Sensitivity to Water Stress of Plant Processes or Parameters[a]

Process or Parameter Affected	SENSITIVITY TO STRESS →			Remarks
	Very sensitive		Relatively insensitive	
	REDUCTION IN TISSUE Ψ REQUIRED TO AFFECT PROCESS[b] →			
	0 bar	10 bars	20 bars	
Cell growth	▬▬▬ ▬ ▬			
Wall synthesis	▬▬▬▬			Fast growing tissue
Protein synthesis	▬▬▬▬			Fast growing tissue
Protochlorophyll formation	▬▬▬			Etiolated leaves
Nitrate reductase level	▬▬▬▬			
ABA Accumulation	▬ ▬ ▬▬▬▬			
Cytokinin level	▬▬▬▬			
Stomatal opening	▬ ▬ ▬▬▬▬▬▬▬▬▬ ▬▬ ▬▬▬			Depends on species
CO₂ Assimilation	▬ ▬ ▬ ▬▬▬▬▬▬▬▬ ▬▬ ▬▬▬			Depends on species
Respiration	▬ ▬ ▬▬▬▬			
Proline accumulation	▬ ▬ ▬ ▬▬▬▬			
Sugar accumulation		▬▬▬▬▬▬		

[a] Length of the horizontal lines represents the range of stress levels within which a process becomes first affected. Dashed lines signify deductions based on more tenuous data.

[b] With Ψ of well-watered plants under mild evaporative demand as the reference point.

picture unduly, however, for one gathers from the literature that, broadly speaking, if species A is less sensitive to water stress than species B with regard to one process, other processes in A also tend to be less sensitive than those in B. Growth conditions could also alter sensitivity to stress, although data on this point are scanty. Further, sensitivity may presumably also be dependent on the duration of stress. Table 1 is compiled on the basis that changes caused by stress become apparent within a matter of minutes or hours after tissue water status is reduced quickly (within hours) to the specified stress level.

When water stress develops rather gradually in the plant, as when water is withheld from plants growing in a substantial volume of soil, what physiological and metabolic changes would be expected, and in what sequence? Unfortunately, only a few studies followed two or more distinctly different processes in the same plant as stress developed (1, 120, 309) or directly compared sensitivity to stress of different processes in the same plant (27). Therefore the answer to this question, mainly deduced from Table 1 but enforced in certain cases by other results and results of these comparison studies, is necessarily very tentative. The first change is most likely a slowing down of shoot and leaf growth brought about by reduced ψ_p (1, 27, 120). This is probably followed closely by a reduction in cell wall and protein synthesis in tissue with high growth potential. As tissue Ψ decreases further, cell division may slow and levels of some enzymes, such as nitrate reductase, start to decline. Stomata may begin to close, with a consequent reduction in transpiration and CO_2 assimilation, and ABA probably begins to accumulate. By now, secondary and tertiary changes in the plant are likely to be numerous and possibly include many of the changes listed here. As stress continues and tissue Ψ decreases still further, declines in respiration, translocation of photosynthate, and leaf cytokinins may become substantial. Levels of some hydrolytic enzymes are likely to increase, and ion transport can be slowed. Finally, water deficits become severe enough to cause marked proline accumulation, and CO_2 assimilation becomes very low or nil. Parts of the xylem may cavitate and be blocked by vapor space. Senescence induced by stress probably becomes visible in older leaves. If the plant is now rewatered, cell growth and CO_2 assimilation of young and mature leaves would resume readily but may not reach the original rates for several days. Old leaves may shed. Other aforementioned changes would also be reversed, at least in younger tissue. Effects on subsequent plant growth and development, however, may be longlasting, as discussed below.

Relations to Long-Term Growth and Yield

Relations between long-term growth or crop yield and water stress are often studied only empirically (248). Several reviewers, however, have analyzed those relations in light of known physiological effects of stress (70, 84, 265). Further, a beginning appears to have been made experimentally at placing those relations on a sound physiological footing (69, 310). Current knowledge is still much too limited, however, for many of the physiological and metabolic alterations brought about by stress to be accurately linked to long-term growth and yield. Nevertheless, one finding, elaborated on above (that cell growth is generally more sensitive to water stress than are stomatal opening and CO_2 assimilation), is directly applicable to the analysis of total yield of dry matter in relation to stress and is discussed here in that context. Also touched on briefly are the complexity introduced when yield consists of only one part of the plant, and the consequent need for better information on the relations between water stress and plant morphogenesis.

In view of the finding that cell growth is usually much more susceptible to water stress than is CO_2 assimilation, it should no longer be assumed, as often done before, that dry matter production is not affected if plant water status does not fall to a level that reduces stomatal opening and photosynthesis. Stress that is mild enough not to affect photosynthesis can reduce the development of leaf surface area. Whether such reduction will affect dry matter yield depends on whether leaf area, i.e. leaf area index (leaf area per unit land area), is limiting the crop's assimilation of CO_2. Fischer & Hagan (70) elaborated on this aspect of stress-yield relations. One implication of the above consideration is that the sensitivity of dry matter yield to stress should be greater in a growing crop with a low leaf area index than in a crop with a high leaf area index.

As pointed out earlier, growth in the field in many cases may be largely restricted to the night, the period when water status is most favorable. Suppose, though, that daytime tissue water deficit could be reduced or eliminated (possibly by certain irrigation techniques under development). How would that affect long-term dry matter production? At a first glance it may appear that there would be little effect on dry matter yield if mid-day stress is not sufficient to reduce photosynthesis. Actually, the answer is not so simple and is likely to depend on the following factors: (*a*) whether accumulation of assimilates in the leaf inhibits photosynthesis, as various indirect evidence suggesting such an inhibition in some plants has been reviewed (207); (*b*) whether leaf area is low enough to limit total assimilation by the crop; (*c*) whether assimilates accumulated during the day exceed the requirement for growth at night—that is, whether the diurnal favorable period for growth is long enough to utilize all the assimilates. These considerations point to some obvious areas where detailed physiological studies are needed.

In considering yields in relation to water stress, the simplest to analyze is production of total dry matter. The situation is still more complex when the yield considered is only a part of the total plant material, such as grain or storage organs. Then the yield will usually depend more on the developmental stages at which stress is applied and on sensitivity to stress in the different developmental stages. The problem is too complex for adequate treatment here, and some recent studies (69, 103, 187, 309, 310), a recent review (265), and a notable review of earlier work (70) should be consulted. What is done here is to point out that information is very meager on physiological events underlying the morphogenetic effects of stress at different stages of plant development. Detailed observations on changes in morphogenesis under stress are not usually accompanied by accurate knowledge of water status of the tissue. In addition, the stress period is often ill defined and too long (weeks), making a complex problem still more complex by introducing too many secondary and tertiary changes. It should also be noted that studies of stress effects on harvestable yield often give inadequate attention to an important factor, namely, which leaves on the plant supply most of the assimilates to the harvestable organ. The interaction between water stress and leaf area development at different plant developmental stages in determining yield should be influenced by whether assimilates come from many leaves or mainly from only specific leaves, such as the flag leaf in wheat (65, 308).

MECHANISMS UNDERLYING RESPONSES TO WATER STRESS

Having discussed the complex plant responses to water stress, we now consider possible physical and chemical links between changes in water status and metabolic or physiological effects. Some aspects elaborated on here have been briefly discussed by others (31, 218).

Examination of Possibilities

The loss of tissue water may be expected to have the following physical and chemical effects: 1. The chemical potential (Ψ) or activity of cellular water is reduced. 2. Turgor pressure decreases in the cell. 3. Small molecules and macromolecules become more concentrated as cell volume is reduced by water loss. 4. Spatial relations in plasmalemma, tonoplast, and membranes of organelles are altered by volume changes. 5. Macromolecules may be affected, through removal of water of hydration or through modification of the structure of adjacent water. The possibilities 1 through 5 will now be discussed individually.

REDUCED WATER ACTIVITY   At first glance, changes in water activity with plant water stress (Possibility 1) appear to be of physiological consequence since water participates as either a reactant or product in numerous metabolic reactions. Upon closer examination, however, it becomes apparent that the changes in activity of water that are associated with the water stress normally encountered in mesophytes are extremely small. Using the thermodynamic relation of

$$\Psi = (RT/\bar{v}_w) \ln a_w$$

where R, T, $\bar{v}_w$, and a_w are respectively the gas constant, absolute temperature, partial molal volume of water, and activity of water, it is easily calculated that a_w is 0.993 for a Ψ of -10 bars, and 0.978 for a Ψ of -30 bars, at 20°C. Hence the activity of water would be reduced, at most, only 3 or 4% in mesophytes suffering very severe water stress. Therefore, explanations for metabolic effects of moderate or even severe stress that are based on reduced water activities limiting important hydrolytic reactions (3, p. 220; 132) are apparently not tenable. Of course, very severe desiccation, corresponding to reductions in Ψ of hundreds of bars (e.g. drying to constant weight in atmosphere of 50% relative humidity), would reduce water activity substantially.

In spite of the fact that tissue Ψ is a very useful indicator of plant water status and that gradients of Ψ govern water transport, the above discussion suggests that Ψ values per se may not be crucial in determining plant behavior. This conclusion, stemming from theoretical consideration, is consistent with certain experimental findings. Effects of reductions in Ψ on growth (222, 224) and on some metabolic processes (102) were minimized or eliminated if turgor was maintained by providing the tissue with an easily absorbable solute although tissue Ψ remained at the reduced level. Some observations on plant adaptation to the natural habitat are also pertinent. There is no question that halophytes such as mangroves grow well under conditions where their tissue Ψ remains below -30 bars (257). An interesting case is also a comparison of tissue at the top and bottom of tall trees. At the top, Ψ may be 10 bars lower than at the bottom (257), yet growth and function at the top are apparently unimpaired. The maintenance of turgor, rather than a high value of Ψ, is probably the critical factor in all these cases.

REDUCED TURGOR PRESSURE   Decreases in ψ_p (Possibility 2) can be substantial even with mild stress. Since ψ_p in well-watered leaves may be of the order of 5 to 9 bars (15, 25, 43, 58, 82, 139, 148), stresses entailing a reduction in Ψ of 10 bars or more would be expected to lower ψ_p to nearly zero. The critical role of turgor as a physical force in cell growth and stomatal movement has already been elaborated on. In these cases it is not so much the absolute ψ_p as the pressure differential across the cell wall boundary that is important. Changes in pressure differential across the wall may also affect spatial relations of enzymic reaction systems located at the plasmalemma or between the cell wall strands. Compression of a macromolecular complex against the cell wall or between strands of the wall could conceivably be prerequisite for its function. Water outside the plasmalemma, filling the cell wall interstices and being open to the air, should be at nearly atmospheric pressure. Reduced ψ_p within the protoplasm due to water stress could markedly alter the pressure differential across the plasmalemma-cell wall boundary, thus affecting the juxtaposition of components in reaction complex correctly located at the boundary. It can be speculated that changes in potassium influx in *Valonia* caused by small changes in ψ_p (p. 551) may be mediated by such a mechanism.

How would changes in ψ_p affect reactions or cellular components located in the bulk liquid phase, where the pressure is isotropic instead of anisotropic? Considerable data are now available dealing with effects of high hydrostatic pressure on biological reactions and processes both in vivo and in vitro, and these were recently reviewed and summarized (328). Unfortunately, almost all such studies started with a first pressure increment of at least 69 bars above atmospheric pressure. Nevertheless, it is possible to determine by interpolation that the rates of most of the diverse reactions and physiological processes examined are changed merely a few percent, if at all, by a pressure of 5 or 10 bars (328). For example, according to data of Landau on *Escherichia coli* (168), a pressure increase of 10 bars on a cell suspension may be expected to depress the induction of β-galactosidase by perhaps 2%, and the synthesis of the enzyme after induction not at all. The latter process was affected only by pressures in excess of 275 bars. One study (330) suggests that jack bean urease may lose no more than a few percent of its activity under a pressure of 10 bars. Flowers & Hanson (74) subjected soybean mitochondria to pressure of 9.2 and 11.5 bars and observed no effect on oxidative phosphorylation. Examples of other processes studied in detail are bacterial luminescence (131), synthesis of protein and RNA in Hela cells (168) and of DNA in *Tetrahymena* (329), and cell division (182). None of these processes appeared to be significantly altered by pressure of a few bars. The only exception that came to my attention is a limited study (166) claiming that the in vitro activity of a cold sensitive form of potato ATPase was increased more than threefold by a pressure of 5 bars. Further characterization and delineation of this ATPase system are needed.

Configuration of macromolecules may be of concern in examining pressure effects because of the significant volume change which can result from hydrophobic bond formation (202). However, from the work already cited on pressure effects, it can be inferred that structures of many biological macromolecules are not altered noticeably by a few bars of pressure. This conclusion is supported by direct data from structural studies of some pure proteins (202). Theoretically, for a reaction in liquid to be affected substantially by a pressure variation of several bars, either the volume change of activation ($\Delta V^{\ddagger}$) or volume change of reaction (ΔV) should be very large (131) and should be greater than those volume changes already reported for various biological reactions (209, 328). Therefore, changes in ψ_p caused by water stress should have minimal or no direct effect on reactions inside the plasmalemma.

CONCENTRATION OF MOLECULES   The concentration of molecules as water is lost from cells (Possibility 3) is certain to affect reactions if the water loss is substantial. Not only do increased activities of reactants or products alter reaction rates directly, but increased inhibitors, activators, and ionic strength of the solution could affect enzyme activities. Todd (293) also suggested a possibility for a substantial decrease in cellular pH with large losses of tissue water. With a mild to moderate water stress, however, the loss of water is small and the concentrating effect is minimal. For example, a reduction in RWC from 95 to 80% corresponds to a 16% reduction in cell volume or an increase of roughly 19% in concentration. For effects to be substantial with this minor change in concentration the enzymes involved would probably have to be allosteric (8). If a reaction is accelerated by water stress due to an increase in substrate concentration, the enzymes should be sensitive to modulation by positive modifiers (effectors) since the concentration of the modifier will also increase. Modulation by negative modifiers would tend to cancel the effect of the increase in substrate concentration (9). For a reaction to be slowed by the concentrating effect of water stress, however, the negative modifier involved must be highly effective to counter the increases in substrate concentration. Such highly sensitive negative modulation is known for some plant enzymes. For example, under the right set of conditions, ADPglucose formation catalyzed by ADPglucose pyrophosphorylase can be reduced to a fraction of the original rate by an increase of a few percent in the concentration of inorganic phosphate (252).

SPATIAL EFFECTS   One aspect of spatial effects—that induced by pressure differences across the boundary at the cell wall—has been mentioned. In addition, spatial relations of enzyme complexes located on membranes of organelles inside the cell and on the plasmalemma may change with changes in cell volume (Possibility 4). A loss of water from a cell, in addition to causing a decrease in cell volume and hence decreases in the dimensions of the plasmalemma, will also affect the dimensions of tonoplast and organelles since the latter behave as osmotic sacks. As Ψ of the cytoplasm diminishes, water moves rapidly out of the vacuole and organelles to reestablish equilibrium in Ψ, causing a contraction of their volumes. How large may the changes in membrane dimensions be with moderate to severe water stress? Assuming a water loss of 20% from the protoplasm of a turgid spherical cell, which roughly corresponds to a 20% reduction in RWC, the decrease in the surface area of the plasmalemma would be about 14%, and the decrease in the surface linear dimension, 7%. It is difficult to find data bearing directly on the significance of dimensional changes of this magnitude to the functionality of membranes and associated enzyme systems. It may be reasoned, however, that such changes occur routinely and frequently during the life of the cell, independent of water stress. Nobel (214) demonstrated a light-induced chloroplast volume change of similar magnitude in vivo. Contraction and swelling of mitochondria occur readily in the test tube and have been studied extensively (e.g. 107). Presumably, under the right conditions, these changes would take place also in vivo. Therefore, if the effects of mild or moderate water stress are mediated by spatial relations on organelle membranes, there must exist some mechanism which distinguishes volume changes due to water stress from volume changes that are a part of normal physiology. Some experimental results also argue against the contention that general dimensional changes be considered a key factor in mediating stress effects. When effects of stressing the tissue are compared with effects of stressing isolated organelles, much larger changes in organelle volumes appear to be necessary to obtain stress effects in the latter case (pp. 532-33, 535). Again, the possible importance of spatial effects are more easy to visualize for very severe stress or desiccation.

MACROMOLECULAR STRUCTURES   Macromolecular structure may conceivably be altered by water stress (Possibility 5). This possibility has been loosely postulated many times (31, 48, 206, 227, 299) but has not been rigorously examined. Usually it is reasoned that conformation of the macromolecules should be dependent on adequate hydration and perhaps also on some special and hypothetical water structure which pervades most of the cytoplasm. Water stress supposedly affects hydration and the special water structure and would thus modify macromolecular conformations and metabolism. In considering Possibility 5, the more hypothetical aspect, that of stress affecting macromolecules through alterations in the supposed special cell water structure is discussed first. Macromolecular hydration as related to function is taken up later.

Some speculative opinions hold that one way or another, much of the water in rather concentrated solutions of macromolecules or in the cell is strongly immobilized and icelike. The reference given in the water relations literature for the postulated extensive icelike water around macromolecules and its importance in macromolecular functionality is usually that of Klotz. The proposal of Klotz (152, 153) and variations thereof have been controversial and often criticized (142, discussion of 153, 208). Basically, the ordering of water by macromolecules is due to the apolar groups on the surface of the latter insulating, in a sense, the water in their vicinity against disruptive electrostatic influences from polar or ionic groups (142). Hence, water molecules in this situation can remain essentially quadruply hydrogen-bonded to their neighbors, forming a layer of "ice" adjacent to the macromolecule. There is agreement that some water is thusly immobilized or "frozen" right next to the macromolecular surface. However, the long distance (in extreme cases, supposedly of the order of microns) immobilization effect of Klotz is in doubt. After detailed and careful review of the literature, Kavanau (152, p. 47) concluded in 1964 that the weight of the more conclusive evidence argues against the "ice cage" extending beyond one to a few molecular layers, although water molecules may be subject to slight ordering for some additional distance. Since that time a number of speculative papers have been published on long distance immobilization, but Kavanau's conclusion seems to remain valid (167). The confusing nature of the information on this subject is not surprising, since even the structural model of pure water is still in a state of flux and "subject to change without notice" (75). In view of this unsatisfactory state of affairs, it may be more profitable to leave the controversy to the physical chemists and focus our attention on what is known of the state of water in the plant.

Is there any indication of extensive and strong binding and immobilization of water in plant cells? There is no question that plant tissue or constituents "bind" some water, although the energy with which the water is held was not clearly defined for much of the early data. These data have been reviewed and summarized (51, pp. 24–31; 160). "Bound" water is given recognition in the current conception of matrix potential ψ_m, representing alteration of energy of water by other than osmotic and pressure forces. Relationships between ψ_m and the water content of tissue have been estimated. Curves for fleshy tissue of potato, asparagus, and mangel (315) showed that about 20 to 30% of tissue water was held with matrix forces greater than 0.3 bar, and that about 15 to 20% was held with forces greater than 1.0 bar. The curve for sunflower leaves was similar (24). Since the chemical activity of water at -1.0 bar is greater than 0.999, it is seen that the large majority of the tissue water in this case is hardly bound at all. A similar conclusion can be inferred less directly from studies of component potentials of leaf water with several species (82, 321). In rhododendron and yew, matrix forces were reported to be stronger but were attributed mainly to the cell wall, which occupied more volume in these species (24). Also, it is likely that the technique used overestimated matrix forces in these species (297).

The extent of immobilization of water can also be glimpsed from rates of isotopic exchange of liquid water. In tissues devoid of layers impervious to water, such as roots, the half-time of equilibration (the time required for tissue water to reach 50% of the isotope content of the external solution) is very short, about 0.5 to 1 min (225, 238, 322). Woolley (322) estimated the apparent self-diffusion coefficient of water to be about one-fifth as great in maize root as in free water. The retardation of diffusion can be largely accounted for by the fact that membranes in the tissue, though very permeable to water, nevertheless constitute a barrier to its movement. It follows that the bulk of water molecules in roots do not differ substantially in mobility from the molecules in free water. There is an indication in these data, however, that a small amount of tissue water may be strongly immobilized. Most results suggest a slow exchange phase involving only a very minor fraction of the tissue water. After the initial rapid exchange, at least several hours seem required before equilibration in specific radioactivity between water of the tissue and the medium becomes complete (225, 238, 302). The slow phase has not been investigated systematically, but possibly corresponds to "bound" water or to water held with substantial matrix forces.

The information available on plants then does not indicate pronounced immobilization or even extensive weak structuring of tissue water. This is in agreement with the conclusion reached earlier by Kavanau (142) regarding macromolecular solutions. Since the postulated pervasion of special icelike water in the cell is improbable, the idea that water stress (even if severe) affects macromolecular functionality and metabolism via effects on this special water structure becomes moot.

Although extensive "ice cages" do not seem to exist in plant tissue, it is still possible that reduced water activity in the tissue affects macromolecules by reducing their hydration to beyond a critical level, or by other means. It is well known that virtually all protein and nucleic acids contain some water of hydration or crystallization (167, 286). Removal of this water, held mainly in the polar and ionic regions of the molecules (73; 286, pp. 131–32 & 337), would be expected to cause denaturation, possibly by permitting excessive interaction among the charge groups of the macromolecules. How much water is actually so involved? The energy with which the water of hydration is held is usually not well defined and the line between water of hydration and "free" water of the solution depends on the method of determination. Nonetheless, various determinations on macromolecules in solution or as solids by a variety of physicochemical techniques have delineated a general range of 0.1 to 1.0 g water of hydration per gram dry weight, with the majority of the macromolecules falling within the span of 0.2 to 0.6 g water per gram dry weight. These measurements were mainly made on animal proteins (38; 73; 167; 205; 286, pp. 58, 359 & 395), though some were made on DNA (108, 205, 307), urease, and two plant viruses (286, pp. 58 & 359). Denatured tRNA seems to be an exception, being reported to contain 1.7 g water of hydration per gram dry weight (167). Superficially the general range is similar to the amounts of "bound" water measured in various plant tissues (160). Indications are that when the water of hydration is retained, the conformations of macromolecules are essentially native. In fact, this is the implicit assumption underlying all structure studies by X-ray diffraction of macromolecules with small amounts of water of crystallization. An outstanding example is the work on DNA (76) which helped to establish the double helix structure. The Watson-Crick helix, commonly thought to be essentially the conformation in vivo, was easily maintained under high relative humidity. Transformation to the less "native" A form occurred at 75% humidity, that is, at a Ψ in the negative hundred-bar range.

Since measured water of hydration for macromolecules generally does not exceed 1 g per gram dry weight and is usually considerably less, it would be reasonable to anticipate that, even with very severe water stress corresponding to losses of up to 50% of water in the vegetative tissue, the water left is likely to be enough to keep the macromolecules well hydrated. Certainly it is not warranted to speak of changes in general protein hydration except in the case of severe desiccation. This conclusion seems to be consistent with the sketchy data which came to my attention on the activity or conformation of proteins at different water activities. Some powdered enzyme preparations (e.g. 263) and enzymes in dried food or matrix (2) have been shown to be quite active in catalysis when kept under humidity levels equivalent to water activities of 0.9, 0.7, or less. In these cases, of course, Ψ is in the negative hundred-bar range. Substantial phosphorylation of sugars and nucleotides was observed in crested wheatgrass seeds down to -130 bars (319). In an optical rotatory dispersion study of β-lactoglobulin in solution (287), changes indicative of denaturation did not occur until more than 20% of the solvent water volume was replaced with dimethylformamide. On a mole fraction basis, this should represent a reduction in Ψ by more than 50 bars. Another pertinent fact is that in the common biochemical practice of using urea to denature many proteins, a very high concentration (6 to 8 M) has to be employed. This attests to the resistance of the proteins to small changes in water activity and water structure.

Therefore, general indications are that the maintenance of macromolecular structure and function, though definitely requiring the presence of some water, is probably not hindered by stress levels pertinent to living mesophytic plants. The possibility cannot be ruled out, however, that certain proteins are particularly sensitive. The foregoing conclusion is made with full realization of the central role of water in sustaining hydrophobic bonding or interactions in proteins, nucleic acids, and membranes (256, 261, 286). The point is that in considering water stress it is not a question of whether water is important; rather, it is a question of how much water can be removed before hydrophobic bonds are weakened or altered. The evidence discussed above suggests that variation in water would be inconsequential until it is reduced to a thickness of a few molecular layers around the macromolecules.

Concluding Discussion on Possible Mechanisms

One of the most intriguing aspects of water stress physiology is how mild or moderate stress is transduced into alterations in metabolism. The foregoing considerations make it seem unlikely that mild stress could, by any of the mechanisms mentioned, damage biochemical components or organelles of the cell; yet mild stress does have pronounced effects (Table 1). It is more probable that changes in metabolism elicited by mild stress represent plant regulatory responses rather than damage. This in turn implies that many of the changes in plant processes brought about by stress arise indirectly. Still, the question remains on what are the direct effects of water stress and how do they lead to secondary effects. Among the five possibilities examined above, the direct effect of turgor change can be taken as proven in the case of reduced cell growth. Turgor change is also a likely candidate for causing other direct effects of mild stress because changes in ψ_p are pronounced when a cell that is close to full turgor loses water. Since ion transport in *Valonia* is immediately altered by small changes in pressure differentials across the cell wall (p. 551), it is possible that some specialized ion transport systems in higher plants, such as stomatal potassium transport, may also respond directly to turgor changes. Unfortunately, progress in this area is hindered because ion transport by stomatal guard cells is difficult to separate experimentally from transport by other cells in the leaf. It has already been mentioned that for ion uptake by higher plant cells in general, there is no indication of a highly sensitive dependence on cell turgor.

When growth is slowed or stopped by a reduction in ψ_p, what secondary effects would this have? It is hard to visualize that protein synthesis, cell wall synthesis, membrane proliferation, etc. will continue unabated if cell expansion is stopped for an appreciable length of time. For maintaining balance of metabolites, the plant has probably evolved controls which slow down synthesis of cell building blocks when low turgor prevents expansion. This may be a likely explanation for the susceptibility of cell wall synthesis and polyribosomes (hence protein synthesis) in growing tissue to very mild stress. It should also be interesting to examine the effect of mild stress on lipid synthesis in such tissue. There are proposals on how cell wall synthesis may be coupled to growth (44); but possible coupling of protein synthesis to growth has not been explored.

Various other changes may also be expected in growing cells under mild to moderate water stress, on the basis of suppressed enlargement of cell volume, for the alteration in volume can be marked. Root apex of maize (5 mm long) increases its volume by about 6% every 10 min under favorable growing conditions (Hsiao, unpublished). Cell enlargement in its most rapid expanding zone must be still faster. Therefore, a slowing or cessation of growth should result in a quick accumulation of many metabolites, which in turn could affect various processes. On this basis, I like to make a strong plea that in studying stress effects on metabolic processes, a clear differentiation be made between growing and nongrowing tissue and in the case of young tissue, growth should always be monitored along with other changes.

There seems to be little doubt that in nongrowing or slowly growing tissue some metabolic parameters, such as protochlorophyll formation (304) and nitrate reductase levels (11, 120), are also susceptible to mild stress. The underlying mechanisms here are still harder to hypothesize. Regarding possibilities other than turgor changes, the lowering of water activity (Possibility 1) is the least likely to be a mechanism underlying stress effects. Changes in molecular and ionic concentrations (Possibility 3) and spatial relations (Possibility 4) may mediate stress effects within the limitations outlined earlier. Critical assessment of two latter possibilities must await much ingenuity and advancement in the understanding of plant metabolic controls and organelle behavior.

ACKNOWLEDGMENTS

I am grateful to Dr. John M. Duniway for reading the manuscript and for many helpful criticisms and suggestions. I thank Dr. John Dillé and Miss Karen Hildestad for assistance in the literature search, and Mr. Kel Deming, Mrs. Barbara Grady, and Mrs. Peggy McClay for help in the preparation of the manuscript.

Literature Cited

1. Acevedo, E., Hsiao, T. C., Henderson, D. W. 1971. *Plant Physiol.* 48: 631–36
2. Acker, L. W. 1969. *Food Technol.* 32: 27–40
3. Addicott, F. T., Lynch, R. S. 1955. *Ann. Rev. Plant Physiol.* 6:211–38
4. Addicott, F. T., Lyon, J. L. 1969. *Ann. Rev. Plant Physiol.* 20:139–64
5. Allaway, W. G., Mansfield, T. A. 1970. *Can. J. Bot.* 48:513–21
6. Arriaga, C., Boyer, J. S., Hageman, R. H. 1972. *Plant Physiol.* 49 (suppl.): 49
7. Ashton, F. M. 1956. *Plant Physiol.* 31:266–74
8. Atkinson, D. E. 1966. *Ann. Rev. Biochem.* 35:85–124, Part I
9. Atkinson, D. E. 1969. *Current Topics in Cellular Regulation*, ed. B. L. Horecker, E. R. Stadtman, 1:29–43. New York: Academic. 314 pp.
10. Balasubramaniam, S., Willis, A. J. 1969. *New Phytol.* 68:663–74

11. Bardzik, J. M., Marsh, H. V. Jr., Havis, J. R. 1971. *Plant Physiol.* 47: 828–31
12. Barley, K. P., Greacen, E. L. 1967. *Advan. Agron.* 19:1–43
13. Barnett, N. M., Naylor, A. W. 1966. *Plant Physiol.* 41:1222–30
14. Barrs, H. D. 1968. *Physiol. Plant.* 21: 918–29
15. Barrs, H. D. 1968. See Ref. 156, 1:235–368
16. Barrs, H. D. 1971. *Ann. Rev. Plant Physiol.* 22:223–36
17. Bell, D. T., Koeppe, D. E., Miller, R. J. 1971. *Plant Physiol.* 48:413–15
18. Ben-Zioni, A., Itai, C., Vaadia, Y. 1967. *Plant Physiol.* 42:361–65
19. Bewley, J. D. 1972. *J. Exp. Bot.* 23: 692–98
20. Bewley, J. D. 1973. *Can. J. Bot.* 51: 203–6
21. Bourque, D. P., Naylor, A. W. 1971. *Plant Physiol.* 47:591–94
22. Bowling, D. J. F. 1968. *Planta* 83: 53–59
23. Boyer, J. S. 1965. *Plant Physiol.* 40: 229–34
24. Ibid 1967. 42:213–17
25. Ibid 1968. 43:1056–62
26. Boyer, J. S. 1969. *Ann. Rev. Plant Physiol.* 20:351–64
27. Boyer, J. S. 1970. *Plant Physiol.* 46: 233–35
28. Ibid, 236–39
29. Ibid 1971. 47:816–20
30. Ibid 1971. 48:532–36
31. Boyer, J. S. 1973. *Phytopathology* 63. In press
32. Boyer, J. S., Bowen, B. L. 1970. *Plant Physiol.* 45:612–15
33. Brix, H. 1962. *Physiol. Plant.* 15:10–20
34. Brouwer, R. 1963. *Acta Bot. Neer.* 12:248–61
35. Brouwer, R. 1965. *Ann. Rev. Plant Physiol.* 16:241–66
36. Broyer, T. C., Hoagland, D. R. 1943. *Am. J. Bot.* 30:261–73
37. Brunori, A. 1967. *Caryologia* 20: 333–38
38. Bull, H. B. 1944. *J. Am. Chem. Soc.* 66:1499–1507
39. Burström, H. G. 1971. *Nature* 234: 488
40. Chen, D., Sarid, S., Katchalski, E. 1968. *Proc. Nat. Acad. Sci. USA* 61: 1378–83
41. Chrispeels, M. J., Vatter, A. E., Hanson, J. B. 1966. *J. Roy. Microsc. Soc.* 85 (Part 1):29–44
42. Cleland, R. A. 1959. *Physiol. Plant.* 12:809–25
43. Cleland, R. 1967. *Planta* 77:182–91
44. Cleland, R. 1971. *Ann. Rev. Plant Physiol.* 22:197–222
45. Cleland, R., Bonner, J. 1956. *Plant Physiol.* 31:350–54
46. Connor, D. J., Tunstall, B. R. 1968. *Aust. J. Bot.* 16:487–90
47. Cowan, I. R., Milthorpe, F. L. 1968. See Ref. 156, 1:137–93
48. Crafts, A. S. 1968. See Ref. 157, 2:85–133
49. Crafts, A. S., Broyer, T. C. 1938. *Am. J. Bot.* 25:529–35
50. Crafts, A. S., Crisp, C. E. 1971. *Phloem Transport in Plants.* San Francisco: Freeman. 481 pp.
51. Crafts, A. S., Currier, H. B., Stocking, C. R. 1949. *Water in the Physiology of Plants*, 1–240. New York: Ronald
52. Cummins, W. R., Kende, H., Raschke, K. 1971. *Planta* 99:347–51
53. Darbyshire, B. 1971. *Plant Physiol.* 47:65–67
54. Darbyshire, B. 1971. *Physiol. Plant.* 25:80–84
55. Doley, D., Leyton, L. 1968. *New Phytol.* 67:579–94
56. Duniway, J. M. 1971. *Physiol. Plant Pathol.* 1:537–46
57. Duniway, J. M. 1973. *Phytopathology* 63. In press
58. Duniway, J. M., Durbin, R. D. 1971. *Plant Physiol.* 48:69–72
59. Duniway, J. M., Slatyer, R. O. 1971. *Phytopathology* 61:1377–81
60. Eastin, J. D., Haskins, F. A., Sullivan, C. Y., van Bavel, C. H. M., Eds. 1969. *Physiological Aspects of Crop Yield.* Madison: Am. Soc. Agron., Crop Sci. Soc. Am. 396 pp.
61. Eavis, B. W., Ratliff, L. F., Taylor, H. M. 1969. *Agron. J.* 61:640–43
62. El Nadi, A. H., Brouwer, R., Locher, J. Th. 1969. *Neth. J. Agr. Sci.* 17: 133–42
63. El-Sharkawy, M. A., Hesketh, J. D. 1964. *Crop Sci.* 4:514–18
64. Emmert, F. H. 1972. *Plant Physiol.* 50:332–35
65. Evans, L. T., Rawson, H. M. 1970. *Aust. J. Biol. Sci.* 23:245–54
66. Falk, H., Lüttge, U., Weigl, J. 1966. *Z. Pflanzenphysiol.* 54:446–62
67. Fischer, R. A. 1968. *J. Exp. Bot.* 19: 135–45
68. Ibid 1970. 21:386–404
69. Fischer, R. A. 1973. *Proc. UNESCO Symp. 1970: Plant Responses to Climatic Factors.* In press
70. Fischer, R. A., Hagan, R. M. 1965. *Exp. Agr.* 1:161–77

71. Fischer, R. A., Hsiao, T. C. 1968. *Plant Physiol.* 43:1953–58
72. Fischer, R. A., Hsiao, T. C., Hagan, R. M. 1970. *J. Exp. Bot.* 21:371–85
73. Fisher, H. F. 1965. *Biochim. Biophys. Acta* 109:544–50
74. Flowers, T. J., Hanson, J. B. 1969. *Plant Physiol.* 44:939–45
75. Frank, H. S. 1970. *Science* 169:635–41
76. Franklin, R. E., Gosling, R. G. 1953. *Nature* 171:740–41
77. Fry, K. E. 1970. *Plant Physiol.* 45:465–69
78. Fujino, M. 1967. *Sci. Bull. Fac. Educ.*, Nagasaki Univ., No. 18:1–47
79. Gaastra, P. 1959. *Meded. Landbouwhogesch. Wageningen* 59:1–68
80. Gale, J., Hagan, R. M. 1966. *Ann. Rev. Plant Physiol.* 17:269–82
81. Gale, J., Kohl, H. C., Hagan, R. M. 1966. *Isr. J. Bot.* 15:64–71
82. Gardner, W. R., Ehlig, C. F. 1965. *Plant Physiol.* 40:705–10
83. Gardner, W. R., Nieman, R. H. 1964. *Science* 143:1460–62
84. Gates, C. T. 1968. See Ref. 157, 2:135–90
85. Gates, C. T., Bonner, J. 1959. *Plant Physiol.* 34:49–55
86. Genkel, P. A., Satarova, N. A., Tvorus, E. K. 1967. *Sov. Plant Physiol.* 14:754–62
87. Gill, W. R., Bolt, G. H. 1955. *Agron. J.* 47:166–68
88. Glinka, Z., Katchansky, M. Y. 1970. *Isr. J. Bot.* 19:533–41
89. Glover, J. 1959. *J. Agr. Sci.* 53:412–16
90. Goltz, S. M., Tanner, C. B., Millar, A. A., Lang, A. R. G. 1971. *Agron. J.* 63:762–65
91. González-Bernáldez, F., López-Sáez, J. F., Garcia-Ferrero, G. 1968. *Protoplasma* 65:255–62
92. Graziani, Y., Livne, A. 1971. *Plant Physiol.* 48:575–79
93. Greacen, E. L., Oh, J. S. 1972. *Nature New Biol.* 235:24–25
94. Green, P. B. 1968. *Plant Physiol.* 43:1169–84
95. Green, P. B., Erickson, R. O., Buggy, J. 1971. *Plant Physiol.* 47:423–30
96. Greenway, H. 1970. *Plant Physiol.* 46:254–58
97. Greenway, H., Hiller, R. G. 1967. *Planta* 75:253–74
98. Greenway, H., Hughes, P. G., Klepper, B. 1969. *Physiol. Plant.* 22:199–207
99. Greenway, H., Klepper, B. 1968. *Planta* 83:119–36
100. Greenway, H., Klepper, B. 1969. *Physiol. Plant.* 22:208–19
101. Greenway, H., Klepper, B., Hughes, P. G. 1968. *Planta* 80:129–41
102. Greenway, H., Leahy, M. 1970. *Plant Physiol.* 46:259–62
103. Grimes, D. W., Miller, R. J., Dickens, L. 1970. *Calif. Agr.* (March) 4–6
104. Gutknecht, J. 1968. *Science* 160:68–70
105. Hagan, R. M., Haise, H. R., Edminster, T. W., Eds. 1967. *Irrigation of Agricultural Lands.* Madison: Am. Soc. Agron. 1180 pp.
106. Hansen, G. K. 1971. *Acta Agr. Scand.* 21:163–71
107. Hanson, J. B., Miller, R. J., Dumford, S. W. 1968. *Plant Physiol.* 43:811–14
108. Hearst, J. E., Vinograd, J. 1961. *Proc. Nat. Acad. Sci. USA* 47:825–30
109. Heath, O. V. S. 1959. *Plant Physiology: A Treatise*, ed. F. C. Steward, 2:193–250. New York: Academic. 758 pp.
110. Hoffman, G. J., Rawlins, S. L., Garber, M. J., Cullen, E. M. 1971. *Agron. J.* 63:822–26
111. Hooymans, J. J. M. 1969. *Planta* 88:369–71
112. Hopmans, P. A. M. 1971. *Meded. Landbouwhogesch. Wageningen* 71:1–86
113. Horton, R. F. 1971. *Can. J. Bot.* 49:583–85
114. Hsiao, T. C. 1966. *Environmental Biology*, ed. P. L. Altman, D. S. Dittmer, 486–88. Bethesda: Fed. Am. Soc. Exp. Biol. 694 pp.
115. Hsiao, T. C. 1970. *Plant Physiol.* 46:281–85
116. Hsiao, T. C., Acevedo, E., Henderson, D. W. 1970. *Science* 168:590–91
117. Hsiao, T. C., Allaway, W. G., Evans, L. T. 1973. *Plant Physiol.* 51:82–88
118. Hsiao, T. C., Hageman, R. H., Tyner, E. H. 1968. *Plant Physiol.* 43:1941–46
119. Hsiao, T. C., Hageman, R. H., Tyner, E. H. 1970. *Crop Sci.* 10:78–82
120. Huffaker, R. C., Radin, T., Kleinkopf, G. E., Cox, E. L. 1970. *Crop Sci.* 10:471–74
121. Humble, G. D., Hsiao, T. C. 1969. *Plant Physiol.* 44:230–34
122. Humble, G. D., Raschke, K. 1971. *Plant Physiol.* 48:447–53
123. Itai, C., Richmond, A., Vaadia, Y. 1968. *Isr. J. Bot.* 17:187–95

124. Itai, C., Vaadia, Y. 1965. *Physiol. Plant.* 18:941–44
125. Itai, C., Vaadia, Y. 1971. *Plant Physiol.* 47:87–90
126. Jackson, W. T. 1965. *Physiol. Plant.* 18:24–30
127. Jarvis, M. S. 1963. *The Water Relations of Plants*, ed. A. J. Rutter, F. H. Whitehead, 289–312. London: Blackwell. 394 pp.
128. Jarvis, P. G., 1971. *Plant Photosynthetic Production: Manual of Methods*, ed. Z. Šesták, J. Čatský, P. G. Jarvis, 566–631. Hague:W. Junk. 818 pp.
129. Jarvis, P. G., Jarvis, M. S. 1965. See Ref. 268, 167–82
130. Jarvis, P. G., Slatyer, R. O. 1970. *Planta* 90:303–22
131. Johnson, F. H., Eyring, H. 1970. See Ref. 328, 1–44
132. Jones, R. L. 1969. *Plant Physiol.* 44:101–4
133. Jones, R. L., Armstrong, J. E. 1971. *Plant Physiol.* 48:137–42
134. Jones, R. J., Mansfield, T. A. 1970. *J. Exp. Bot.* 21:714–19
135. Jordan, W. R. 1970. *Agron. J.* 62:699–701
136. Jordan, W. R., Morgan, P. W., Davenport, T. L. 1972. *Plant Physiol.* 50:756–58
137. Jordan, W. R., Ritchie, J. T. 1971. *Plant Physiol.* 48:783–88
138. Kahane, I., Poljakoff-Mayber, A. 1968. *Plant Physiol.* 43:1115–19
139. Kanemasu, E. T., Tanner, C. B. 1969. *Plant Physiol.* 44:1547–52
140. Kaufmann, M. R. 1972. See Ref. 158, 3:91–124
141. Kaufmann, M. R., Eckard, A. N. 1971. *Plant Physiol.* 47:453–56
142. Kavanau, J. L. 1964. *Water and Solute-Water Interactions.* San Francisco: Holden-Day. 101 pp.
143. Kayushin, L. P., Ed. 1969. *Water In Biological Systems.* New York: Consultants Bur. 112 pp.
144. Kemble, A. R., Macpherson, H. T. 1954. *Biochem. J.* 58:46–49
145. Kende, H. 1965. *Proc. Nat. Acad. Sci. USA* 53:1302–7
146. Kershaw, K. A. 1972. *Can. J. Bot.* 50:543–55
147. Kessler, B., Frank-Tishel, J. 1962. *Nature* 196:542–43
148. Kirkham, M. B., Gardner, W. R., Gerloff, G. C. 1972. *Plant Physiol.* 49:961–62
149. Kiyosawa, K., Tazawa, M. 1972. *Protoplasma* 74:257–70
150. Kleinendorst, A., Brouwer, R. 1970.

Neth. J. Agr. Sci. 18:140–48
151. Klepper, B. 1968. *Plant Physiol.* 43:1931–34
152. Klotz, I. M. 1958. *Science* 128:815–22
153. Klotz, I. M. 1960. *Protein Structure and Function.* Brookhaven Symp. Biol. No. 13:25–48
154. Kluge, M., Fischer, K. 1967. *Planta* 77:212–23
155. Knipling, E. B. 1967. *Physiol. Plant.* 20:65–72
156. Kozlowski, T. T., Ed. 1968. *Water Deficits and Plant Growth*, Vol. 1. New York: Academic. 390 pp.
157. Kozlowski, T. T., Ed. 1968. *Water Deficits and Plant Growth*, Vol. 2. New York: Academic. 333 pp.
158. Kozlowski, T. T., Ed. 1972. *Water Deficits and Plant Growth*, Vol. 3. New York: Academic. 368 pp.
159. Ibid, 1–64
160. Kramer, P. J. 1955. *Handbuch der Pflanzenphysiologie*, ed. W. Ruhland, I:223–43. Berlin: Springer Verlag
161. Kramer, P. J. 1959. *Advan. Agron.* 11:51–70
162. Kramer, P. J. 1969. *Plant and Soil Water Relationships—A Modern Synthesis.* New York: McGraw-Hill. 482 pp.
163. Kriedemann, P. E., Loveys, B. R., Fuller, G. L., Leopold, A. C. 1972. *Plant Physiol.* 49:842–47
164. Kriedemann, P. E., Smart, R. E. 1971. *Photosynthetica* 5:6–15
165. Kuiper, P. J. C. 1961. *Meded. Landbouwhogesch. Wageningen* 61:1–49
166. Kuiper, P. J. C. 1971. *Biochim. Biophys. Acta* 250:443–45
167. Kuntz, I. D. Jr., Brassfield, T. S., Law, G. D., Purcell, G. V. 1969. *Science* 163:1329–31
168. Landau, J. V. 1970. See Ref. 328, 45–70
169. Lang, A. R. G., Klepper, B., Cumming, M. J. 1969. *Plant Physiol.* 44:826–30
170. Larcher, W. 1960. *Bull. Res. Counc. Isr. Sect. D* 8:213–24
171. Läuchli, A. 1972. *Ann. Rev. Plant Physiol.* 23:197–218
172. Lawlor, D. W. 1969. *J. Exp. Bot.* 20:895–911
173. Lawlor, D. W. 1970. *New Phytol.* 69:501–13
174. Lee, J. A., Stewart, G. R. 1971. *New Phytol.* 70:1061–68
175. Livne, A., Vaadia, Y. 1972. See Ref. 158, 3:255–75
176. Lockhart, J. A. 1965. *Plant Biochemistry*, ed. J. Bonner, J. E. Varner,

826-49. New York: Academic. 878 pp.
177. Loomis, W. E. 1934. *Am. J. Bot.* 21: 1-6
178. Lorimer, G. H., Miller, R. J. 1969. *Plant Physiol.* 44:839-44
179. Mansfield, T. A., Jones, R. J. 1971. *Planta* 101:147-58
180. Marcus, A., Feeley, J. 1965. *J. Biol. Chem.* 240:1675-80
181. Marcus, A., Feeley, J., Volcani, T. 1966. *Plant Physiol.* 41:1167-72
182. Marsland, D. 1970. See Ref. 328, 259-312
183. Maskell, E. J. 1928. *Proc. Roy. Soc. London* 102(B):488-533
184. Mattas, R. E., Pauli, A. W. 1965. *Crop Sci.* 5:181-84
185. Mazur, P. 1969. *Ann. Rev. Plant Physiol.* 20:419-48
186. McMichael, B. L., Jordan, W. R., Powell, R. D. 1972. *Plant Physiol.* 49: 658-60
187. McMichael, B. L., Jordan, W. R., Powell, R. D. 1973. *Agron. J.* In press
188. Meidner, H. 1967. *J. Exp. Bot.* 18: 556-61
189. Meidner, H., Mansfield, T. A. 1968. *Physiology of Stomata.* London: McGraw-Hill. 179 pp.
190. Michel, B. E. 1971. *Plant Physiol.* 48: 513-16
191. Milborrow, B. V., Noddle, R. C. 1970. *Biochem. J.* 119:727-34
192. Milburn, J. A. 1966. *Planta* 69:34-42
193. Milburn, J. A., Johnson, R. P. C. 1966. *Planta* 69:43-52
194. Millar, A. A., Duysen, M. E., Wilkinson, G. E. 1968. *Plant Physiol.* 43:968-72
195. Millar, A. A., Gardner, W. R., Goltz, S. M. 1971. *Agron. J.* 63:779-84
196. Miller, R. J., Bell, D. T., Koeppe, D. E. 1971. *Plant Physiol.* 48:229-31
197. Milthorpe, F. L., Spencer, E. Y. 1957. *J. Exp. Bot.* 8:413-37
198. Mittelheuser, C. J., Van Steveninck, R. F. M. 1969. *Nature* 221:281-82
199. Mizrahi, Y., Blumenfeld, A., Bittner, S., Richmond, A. E. 1971. *Plant Physiol.* 48:752-55
200. Mizrahi, Y., Blumenfeld, A., Richmond, A. E. 1970. *Plant Physiol.* 46: 169-71
201. Mizrahi, Y., Blumenfeld, A., Richmond, A. E. 1972. *Plant Cell Physiol.* 13:15-21
202. Morita, R. Y., Becker, R. R. 1970. See Ref. 328, 71-83
203. Morris, C. J., Thompson, J. F., Johnson, C. M. 1969. *Plant Physiol.*

44:1023-26
204. Most, B. H. 1971. *Planta* 101:67-75
205. Mrevlishvili, G. M., Privalov, P. L. 1969. See Ref. 143, 63-66
206. Naylor, A. W. 1972. See Ref. 158, 3:241-54
207. Neales, T. F., Incoll, L. D. 1968. *Bot. Rev.* 34:107-25
208. Némethy, G. 1965. *Fed. Proc.* 24: S38-41
209. Neville, W. M., Eyring, H. 1972. *Proc. Nat. Acad. Sci. USA* 69:2417-19
210. Nir, I., Klein, S., Poljakoff-Mayber, A. 1969. *Aust. J. Biol. Sci.* 22:17-33
211. Nir, I., Poljakoff-Mayber, A. 1967. *Nature* 213:418-19
212. Nir, I., Poljakoff-Mayber, A., Klein, S. 1970. *Isr. J. Bot.* 19:451-62
213. Nir, I., Poljakoff-Mayber, A., Klein, S. 1970. *Plant Physiol.* 45:173-77
214. Nobel, P. S. 1968. *Plant Physiol.* 43: 781-87
215. Nobel, P. S. 1969. *Biochim. Biophys. Acta* 172:134-43
216. Nobel, P. S. 1969. *Plant Cell Physiol.* 10:597-605
217. Oertli, J. J. 1968. *9th Int. Congr. Soil Sci. Trans.* 1:95-107
218. Oertli, J. J. 1969. See Ref. 60, 85-88
219. Oertli, J. J. 1971. *Z. Pflanzenphysiol.* 65:195-209
220. O'Neal, D., Hew, C. S., Latzko, E., Gibbs, M. 1972. *Plant Physiol.* 49: 607-14
221. Onwueme, I. C., Laude, H. M., Huffaker, R. C. 1971. *Crop Sci.* 11:195-200
222. Ordin, L. 1960. *Plant Physiol.* 35: 443-50
223. Ordin, L. 1966. See Ref. 114, 493-99
224. Ordin, L., Applewhite, T. H., Bonner, J. 1956. *Plant Physiol.* 31:44-53
225. Ordin, L., Kramer, P. J. 1956. *Plant Physiol.* 31:468-71
226. Parker, J. 1968. See Ref. 156, 1:195-234
227. Parker, J. 1972. See Ref. 158, 3:125-76
228. Pearson, R. W. 1966. See Ref. 230, 95-126
229. Peters, D. B. 1957. *Soil Sci. Soc. Am. Proc.* 21:481-84
230. Pierre, W. H., Kirkham, D., Pesek, J., Shaw, R., Eds. 1966. *Plant Environment and Efficient Water Use.* Madison: Am. Soc. Agron., Soil Sci. Soc. Am. 295 pp.
231. Pisek, A., Winkler, E. 1953. *Planta* 42:253-78 (German)
232. Plaut, Z. 1971. *Plant Physiol.* 48:591-95

233. Plaut, Z., Ordin, L. 1961. *Physiol. Plant.* 14:646–58
234. Ibid 1964. 17:279–86
235. Poljakoff-Mayber, A., Gale, J. 1972. See Ref. 158, 3:277–306
236. Pratt, H. K., Goeschl, J. D. 1969. *Ann. Rev. Plant Physiol.* 20:541–84
237. Ramagopal, S., Hsiao, T. C. Submitted
238. Raney, F., Vaadia, Y. 1965. *Physiol. Plant.* 18:8–14
239. Raschke, K. 1956. *Planta* 48:200–38 (German)
240. Raschke, K. 1958. *Flora* 146:546–78 (German)
241. Raschke, K. 1960. *Ann. Rev. Plant Physiol.* 11:111–26
242. Raschke, K. 1970. *Plant Physiol.* 45:415–23
243. Raschke, K. 1970. *Planta* 91:336–63
244. Raschke, K. 1970. *Science* 167:189–91
245. Redshaw, A. J., Meidner, H. 1972. *J. Exp. Bot.* 23:229–40
246. Routley, D. G. 1966. *Crop Sci.* 6:358–61
247. Ruf, R. H., Eckert, R. E., Gifford, R. O. 1967. *Soil Sci.* 104:159–62
248. Salter, P. J., Goode, J. E. 1967. *Crop Responses to Water at Different Stages of Growth.* Commonwealth (G.B.) Bur. Hort. Plant. Crops. 246 pp.
249. Sanchez-Diaz, M. F., Kramer, P. J. 1971. *Plant Physiol.* 48:613–16
250. Santarius, K. A. 1967. *Planta* 73:228–42
251. Santarius, K. A., Ernst, R. 1967. *Planta* 73:91–108
252. Sanwal, G. G., Greenberg, E., Hardie, J., Cameron, E. C., Preiss, J. 1968. *Plant Physiol.* 43:417–27
253. Satarova, N. A., Tvorus, E. K. 1971. *Sov. Plant. Physiol.* 18:532–38
254. Sawhney, B. L., Zelitch, I. 1969. *Plant Physiol.* 44:1350–54
255. Scarth, G. W., Shaw, M. 1951. *Plant Physiol.* 26:581–97
256. Scheraga, H. A. 1963. *The Proteins: Composition, Structure and Function,* ed. H. Neurath, 1:478–594. New York:Academic. 665 pp.
257. Scholander, P. F., Hammel, H. T., Bradstreet, E. D., Hemmingsen, E. A. 1965. *Science* 148:339–46
258. Shah, C. B., Loomis, R. S. 1965. *Physiol. Plant.* 18:240–54
259. Shevelukha, V. S. 1971. *Sov. Plant Physiol.* 18:121–29
260. Shimshi, D. 1963. *Plant Physiol.* 38:713–21
261. Sinanoğlu, O., Abdulnur, S. 1965.

Fed. Proc. 24:S12–23
262. Singh, T. N., Aspinall, D., Paleg, L. G. 1972. *Nature New Biol.* 236:188–90
263. Skujins, J. J., McLaren, A. D. 1967. *Science* 158:1569–70
264. Slatyer, R. O. 1967. *Plant-Water Relationships.* London:Academic. 366 pp.
265. Slatyer, R. O. 1969. See Ref. 60, 53–83
266. Slatyer, R. O. 1970. *Planta* 93:175–89
267. Slavík, B. 1965. *The Growth of Cereals and Grasses,* ed. F. L. Milthorpe, J. D. Ivins, 227–40. London: Butterworths
268. Slavík, B., Ed. 1965. *Water Stress in Plants.* The Hague: W. Junk. 322 pp.
269. Ibid, 195–202
270. Spoehr, H. A., Milner, M. W. 1939. *Proc. Am. Phil. Soc.* 81:37–78
271. Sprent, J. I. 1971. *New Phytol.* 70:9–17
272. Stålfelt, M. G. 1955. *Physiol. Plant.* 8:572–93
273. Ibid 1961. 14:826–43
274. Steinberg, R. A. 1951. *Mineral Nutrition of Plants,* ed. E. Truog, 359–86. Univ. Wisconsin Press
275. Stewart, C. R. 1971. *Plant Physiol.* 48:792–94
276. Stewart, C. R. 1973. *Plant Physiol.* 51:508–11
277. Stewart, C. R., Morris, C. J., Thompson, J. F. 1966. *Plant Physiol.* 41:1585–90
278. Stockton, J. R., Doneen, L. D., Walhood, V. T. 1961. *Agron. J.* 53:272–75
279. Stoker, R., Weatherley, P. E. 1971. *New Phytol.* 70:547–54
280. Stowell, R. E., Ed. 1965. *Cryobiology. Fed. Proc.* 24:Suppl. 15. 324 pp.
281. Stuart, T. S. 1968. *Planta* 83:185–206
282. Stutte, C. A., Todd, G. W. 1968. *Crop Sci.* 8:319–21
283. Ibid 1969. 9:510–12
284. Subbotina, N. V. 1962. *Sov. Plant Physiol.* 9:62–65
285. Tal, M., Imber, D. 1971. *Plant Physiol.* 47:849–50
286. Tanford, C. 1961. *Physical Chemistry of Macromolecules.* New York: Wiley. 710 pp.
287. Tanford, C., De, P. K., Taggart, V. G. 1960. *J. Am. Chem. Soc.* 82:6028–34
288. Taylor, H. M., Gardner, H. R. 1963. *Soil Sci.* 96:153–56
289. Taylor, H. M., Ratliff, L. F. 1969. *Agron. J.* 61:398–402
290. Taylor, H. M., Ratliff, L. F. 1969. *Soil Sci.* 108:113–19

291. Terry, N., Waldron, L. J., Ulrich, A. 1971. *Planta* 97:281–89
292. Thompson, J. F., Stewart, C. R., Morris, C. J. 1966. *Plant Physiol.* 41:1578–84
293. Todd, G. W. 1972. See Ref. 158, 3:177–216
294. Todd, G. W., Basler, E. 1965. *Phyton* 22:79–85
295. Troughton, J. H. 1969. *Aust. J. Biol. Sci.* 22:289–302
296. Troughton, J. H., Slatyer, R. O. 1969. *Aust. J. Biol. Sci.* 22:815–27
297. Tyree, M. T., Hammel, H. T. 1972. *J. Exp. Bot.* 23:267–82
298. UNESCO 1973. *Proc. UNESCO Symp. 1970: Plant Responses to Climatic Factors.* In press
299. Vaadia, Y., Raney, F. C., Hagan, R. M. 1961. *Ann. Rev. Plant Physiol.* 12:265–92
300. Vaadia, Y., Waisel, Y. 1967. See Ref. 105, 354–72
301. van den Driesche, R., Connor, D. J., Tunstall, B. R. 1971. *Photosynthetica* 5:210–17
302. Vartapetyan, B. B. 1965. See Ref. 268, 72–80
303. Viets, F. G. Jr. 1972. See Ref. 158, 3:217–39
304. Virgin, H. I. 1965. *Physiol. Plant.* 18:994–1000
305. Wadleigh, C. H., Ayers, A. D. 1945. *Plant Physiol.* 20:106–32
306. Wadleigh, C. H., Gauch, H. G. 1948. *Plant Physiol.* 23:485–95
307. Wang, J. H. 1955. *J. Am. Chem. Soc.* 77:258–60
308. Wardlaw, I. F. 1968. *Bot. Rev.* 34:79–105
309. Wardlaw, I. F. 1969. *Aust. J. Biol. Sci.* 22:1–16
310. Ibid 1971. 24:1047–55
311. Weatherley, P. E. 1950. *New Phytol.* 49:81–87
312. Wellensiek, S. F. 1957. *Control of the Plant Environment*, ed. J. P. Hudson, 3–15. London:Butterworths. 240 pp.
313. West, S. H. 1962. *Plant Physiol.* 37:565–71
314. West, S. H. 1966. *Proc. 10th Int. Grassland Congr.*, 91–94
315. Wiebe, H. H. 1966. *Plant Physiol.* 41:1439–42
316. Wiebe, H. H., Ed. 1971. *Utah Agr. Exp. Sta. Bull.* 484. 71 pp.
317. Willis, A. J., Balasubramaniam, S. 1968. *New Phytol.* 67:265–85
318. Wilson, A. M. 1971. *Plant Physiol.* 48:541–46
319. Wilson, A. M., Harris, G. A. 1968. *Plant Physiol.* 43:61–65
320. Wilson, A. M., Huffaker, R. C. 1964. *Plant Physiol.* 39:555–60
321. Wilson, J. W. 1967. *Aust. J. Biol. Sci.* 20:329–47
322. Woolley, J. T. 1965. *Plant Physiol.* 40:711–17
323. Wright, S. T. C. 1969. *Planta* 86:10–20
324. Wright, S. T. C., Hiron, R. W. P. 1969. *Nature* 224:719–20
325. Zahner, R. 1968. See Ref. 157, 2:191–254
326. Zeevaart, J. A. D. 1971. *Plant Physiol.* 48:86–90
327. Zelitch, I. 1965. *Biol. Rev.* 40:463–82
328. Zimmerman, A. M., Ed. 1970. *High Pressure Effects on Cellular Processes.* New York:Academic. 324 pp.
329. Zimmerman, S. B., Zimmerman, A. M. 1970. See Ref. 328, 179–210
330. ZoBell, C. E. 1970. See Ref. 328, 85–130

Ann. Rev. Plant Physiol. 1973. 24:571–98

GIBBERELLINS: THEIR PHYSIOLOGICAL ROLE[1]

Russell L. Jones

Department of Botany, University of California, Berkeley

CONTENTS

Interest in both the chemistry and physiology of the gibberellins has grown considerably since Paleg (131) reviewed the subject in 1965. Because of this Lang (105), in a more recent treatment of the gibberellins, restricted his scope to that of the structure and metabolism of these hormones. This review is also limited in its perspective and will deal primarily with the area of the physiological role of the gibberellins in plants. Some aspects of GA action have been dealt with in more recent reviews. In particular, Key (97) and Glasziou (56) provided documentation of the effects of these regulators on nucleic acid and protein metabolism respectively, while Yomo & Varner (187) reviewed recent advances in investigations of the GA response in aleurone tissue of barley.

Even though this review is restricted to physiological aspects of the role of the GAs, it cannot attempt to be a comprehensive evaluation of this topic. A

[1] Abbreviations used: ABA [abscisic acid]; AMO-1618 [2′ isopropyl-4′-(trimethyl-ammonium chloride)-5′methylphenyl piperidine carboxylate]; B-995 [*n*-dimethylamino-succinamic acid]; CCC [2-chloroethyltrimethylammonium chloride]; GA(s) [gibberel-lin(s)]; GA_1–GA_n [the chemically characterized gibberellins]; Phosfon D [tributyl-2-, 4-dichlorobenzylphosphonium chloride].

mere survey of the morphological effects induced by the application of GA would require considerably more space than that allowed for a review of this nature.

THE GIBBERELLINS AS ENDOGENOUS REGULATORS

Previous reviews of the subject of GA action have persistently argued the case that GA's are effective as endogenous regulators of plant growth and development via a stimulation of auxin metabolism (21, 38, 131). The evidence is now overwhelming that the GA's are themselves endogenous regulators, although they may interact directly or indirectly with auxins and the other plant hormones. The argument that the GAs function endogenously through an effect on auxin metabolism derives from observations that GA application stimulates IAA production in some plants (102, 103, 128, 159), and that some of the physiological effects of GA are similar to those of auxin, e.g. in fruit set and apical dominance. These arguments, however, are based on specialized responses, e.g. parthenocarpy in apple which can be stimulated by IAA or GA (42) or on responses where the role of GA is equivocal as is the case for apical dominance (132, 153, 156). When the role of auxins in responses which are considered diagnostic for the gibberellins are compared, the conclusion that GAs are a class of hormones distinct from auxins is inescapable. Thus gibberellins promote substrate mobilization in barley (129, 186), normalization of growth habit in genetic or physiological dwarfs (22, 135), germination of many photodormant seeds (1), germination of fern spores (179), and growth of dormant buds (44, 178), while auxins have no significant effect on these processes.

The responses documented above are responses to applied GAs and thus do not provide evidence that GAs function directly to regulate normal growth patterns. Evidence provided by work with inhibitors of GA biosynthesis, which was comprehensively reviewed by Lang (105), has shown an excellent correlation between inhibition of GA biosynthesis and reduced stem growth. Together with the fact that GA application can overcome growth inhibition resulting from growth retardant application, these observations provide strong support for the role of GAs as endogenous regulators of growth (see Lang 105). Still further support comes from an examination of the GA status of plants with dwarfed growth habits. In some genetic dwarfs of corn (*Zea mays*) and rice (*Oryza sativa*), dwarfism is related to the absence or reduced levels of endogenous GA-like substances providing a strong relationship between GA levels and growth (136, 168).

Perhaps the major area of ignorance in our knowledge of gibberellin physiology concerns the regulation of the GA status of the plant. Although chemists and biochemists have made great advances in the study of the chemistry and pathways of GA biosynthesis (see 105), our knowledge of the metabolism of the physiologically functional GAs is sparse. Thus limited information is available on the role of the individual GAs found in extracts—are all GA-like substances extracted from the plant functional as physiological regulators, or are some precursors or degradation products of a single functional GA? Until answers to these questions are obtained, many investigations into the physiology of the endogenous GAs will remain fruitless. Thus extraction of tissues and measurement of the endogenous GA levels will provide information which is at best equivocal if information concerning GA turnover is not known.

Several reports have now provided evidence for differences in the rate of GA turnover in various plant tissues (5, 37, 100, 142, 166, 190). Zeevaart (190) found that while the level of extractable GA-like substances in spinach plants under long and short days was the same, the rate of degradation was greater in the long day grown plants. It is inferred that since GA levels were similar under long and short days, transfer of plants to long days resulted in a marked increase in GA biosynthesis (165, 190). In a study of the GA status of gynoecious and monoecious lines of *Cucumis sativus*, Atsmon et al (5) found that differences in levels of diffusible GA-like substances were also related to increased GA turnover. When ^{3}H-GA$_1$ was applied to the two seedling types, more label was recovered by diffusion from the monoecious as opposed to the gynoecious line (5). Since no difference was apparent in rates of transport or uptake of the labeled GA, it was concluded that the difference in recovery between the two lines was caused by degradation (5). The experiments of Atsmon et al (5) must be interpreted with some caution, however, since these workers did not establish that the labeled GA$_1$ used in the study was in fact a native GA of cucumber. Thus turnover of nonendogenous GAs may indeed be quite different from that of an endogenous GA. It is apparent from the foregoing that any evidence for increased GA biosynthesis in plants after various treatments, e.g. illumination, photoperiod shift, or temperature change, must be analyzed in light of these observations concerning GA turnover. Thus, experiments designed merely to measure the level of free GAs by solvent extraction will be of questionable utility.

Using a combination of diffusion and extraction methods, it is possible to obtain evidence for changes in GA production under various experimental conditions (5, 7, 37, 83, 85, 86). Experiments with diffusion of GA-like substances from isolated plant parts also point to the conclusion that not all GAs which are extractable from a tissue are mobile in that tissue (5, 37, 83). In peas, *Silene armeria* and *Cucumis*, a discrepancy exists between the number of GA-like substances detected by extraction and those found in diffusates (5, 37, 77, 83). Extracts of pea and *Silene* shoot tips contain GA$_1$ or GA$_3$-like and GA$_5$-like substances; however, the GA$_5$-like component is absent from diffusates of such shoot tips (37, 77, 83). Since the leaves of the shoot tip are the source of GA, and the internodes below this tip are the target tissues (84, 85), measurement of GA-like substances moving in the internode might be expected to indicate the physiologically functional GA. Thus, the GA$_5$-like component found only in tissue extracts of pea and *Silene* could be the immediate precursor of the transported or diffusing GA.

In cucumber the relationship between diffusible and extractable GAs is even more complex than in pea and *Silene*. There is evidence for at least three or four GA-like substances in root bleeding sap and tissue extracts of this species, but only one is found in diffusates of the apical bud (5). This situation in cucumber demonstrates clearly the need to resolve the relationship between diffusible and extractable GAs.

In addition to obtaining more information concerning the synthesis, degradation, and transport of the endogenous GAs, the presence and role of conjugated GAs must be more rigorously examined. Evidence reviewed by Lang (105) and recent publications (9, 11, 64, 121, 158) have indicated that GA glucosides may play a role in controlling the status of free GAs within the plant. Conjugation of GAs to a sugar moiety could be important in the sequestering of GAs (e.g. in seeds and dormant trees) and in the subsequent transport of GAs over long distances as in the spring sap of many deciduous trees (10, 155). The possibility that conjugated GAs function as regulators of the level of free GA has been highlighted by the recent observation of Nadeau et al (121) that ABA enhances the conversion of ^{3}H-GA$_1$ to ^{3}H-GA$_1$-glucoside, ^{3}H-GA$_8$-glucoside, ^{3}H-GA$_8$ and an unidentified ^{3}H-GA conjugate in aleurone layers of barley. The physiological role of GA conjugation in barley aleurone is unclear (121). The stimulation of the formation of GA glucosides or other inactive forms of GA from ^{3}H-GA$_1$ in ABA-treated aleurone layers could not account for the inhibition of GA-induced α-amylase production in this tissue since inhibition of α-amylase synthesis is not stoichiometrically related to the formation of biologically inactive GA products (121).

A possible explanation for the role of ABA in GA metabolism comes from the work of Musgrave et al (120). These workers examined the fate of ^{3}H-GA$_1$, ^{3}H-GA$_8$ and ^{3}H-GA$_8$-methyl ester applied to isolated aleurone layers of barley and found that the rate of conversion of these GAs to more polar metabolites (conjugates) was inversely related to their biological activity in the system (120). Thus GA$_8$ which elicits a weak response in aleurone cells relative to GA$_3$ (88) is accumulated to a greater extent than is GA$_1$ (120). It is possible that GA conjugates function to sequester GAs, especially when the potential response to GA is blocked by the presence of an inhibitor (120, 121) or where the GA molecule itself possesses limited biological activity (120). This discussion raises the obvious question of the fate of conjugated GAs in tissues such as the aleurone layer following removal of ABA or the biologically less active GAs (120, 121). Are these conjugates hydrolyzed and, if so, what is the fate of the unconjugated GAs which are liberated? Answers to these questions would help to define more clearly the role of GA conjugates in vivo.

VARIATIONS IN ENDOGENOUS GIBBERELLIN LEVELS

Changes in the levels of GAs in response to changes in environmental conditions are now well documented. Thus light, photoperiod, vernalization, and stratification treatments have all been shown to bring about quantitative and qualitative changes in the GA levels of higher plants.

Light

Central among the photomorphogenic problems studied over the past 15 years has been the phenomenon of red light inhibition of stem growth in pea (*Pisum sativum* L.). In this species, both normal and dwarf varieties are found which differ only in their growth habit in red light (107). Normal and dwarf peas grow to similar heights in darkness or far-red light, but in red light the growth of both is reduced, the reduction of growth in the dwarf variety being more severe than that in the normal (107). Such dwarf plants have been referred to as physiological dwarfs as opposed to true genetic dwarfs which possess stunted growth habits under all conditions of illumination. In 1955, Brian et al (22) showed that dwarfism in pea could be reversed by GA_3, while Lockhart (107) showed that GA_3 could completely overcome the inhibitory effect of red light in both normal and dwarf plants. Based on kinetic analysis of the effects of red light and GA on pea stem growth, Lockhart (108) concluded that red light inhibited growth by either (a) inhibition of GA production, (b) promotion of GA breakdown, or (c) reduction of the responsiveness of the tissue to GA.

Examination of the endogenous GAs of normal and physiological dwarf varieties of pea by diffusion or extraction have, with one exception (99), consistently shown no quantitative or qualitative difference in the level of this regulator (83, 96, 141). It has been suggested that red light inhibition of stem growth in peas is related to the plant's ability to respond to the endogenous GAs. Thus in light grown normal peas the response to the endogenous GA_5- and GA_1-like components is approximately equal, while in the physiological dwarfs the response to the GA_5-like component is reduced (96). This interpretation of red light inhibition of growth is lent credence by the fact that plant extracts contain substances which interfere with the response of dwarf peas to applied GA (101). Although the chemical identification of this inhibitor has not been established, estimation of the ABA levels of normal and dwarf pea plants grown in light or darkness indicates that this inhibitor plays no role in controlling stem elongation (95).

Experiments with reciprocal grafting of normal and dwarf root systems and shoot tips of pea by McComb & McComb (117) have shown that dwarfism in pea variety Progress No. 9 is not controlled by factors which are transmittable across a graft union. This evidence suggests that physiological dwarfism of pea results from a change in the ability of stem tissue to respond to the endogenous GA rather than from the presence of a defined chemical substance whose synthesis is affected by light. Further, these workers also showed that "slender" pea varieties having the alleles la and cry^s are not affected by AMO-1618, indicating that their rate of growth is not related to decreased endogenous GA levels (117).

Similar results and conclusions have been drawn from investigations of dwarfism in peach (184), wheat (145), and rice (168), although in the latter case, as with corn (136), dwarf mutants are found which are deficient in endogenous GAs (136). The genetically defined mutants of corn provide an example of the anomalous relationship between growth habit, response to applied GA, and endogenous GA levels (136). Thus, tall varieties of corn possess at least two endogenous GAs, dwarf mutants d_1 and d_2 contain about one-half the GA levels of the tall mutant, and dwarf mutants an_1, d_3, and d_5 contain no detectable endogenous GAs (136). The experience with the response of dwarf varieties to GA indicates that great caution should be exercised in extrapolating data obtained from the application of hormones exogenously to the function of these hormones as endogenous regulators of growth and development.

Although there is a telling amount of evidence arguing against the involvement of endogenous GAs in light-mediated inhibition of stem growth, the participation of this hormone in other photomorphogenic events is more convincing. The unrolling of dark grown cereal leaves is stimulated by exposure to red light of approximately 660 nm for short time periods (15, 110, 111, 140, 146). In barley, exposure of leaf sections to 1.3 $\mu w/cm^2$ of light of 660 nm for 30 min results in a significant increase in the level of GA-like substances found in extracts (146). Incubation of leaf sections in AMO-1618 or CCC, however, results in the elimination of the red-light induced increase in GA levels, suggesting that new synthesis of the hormone is occurring (146). In wheat leaves a similar increase in the level of GA-like substances is observed after exposure to red light (15, 110). Since both red light and GA stimulate unrolling of what leaves, it is suggested that the effect of red light may be brought about by changes in the level of GA-like substances (15, 100). In both barley and wheat, the increase in the level of GA appears to be transient since in both tissues the level of GA-like substances declines, indicating that rapid turnover of the GA may be occurring (15, 146).

An interesting relationship has recently been demonstrated between GA distribution and phototropic stimulation in sunflower (*Helianthus annuus*) shoot tips (134). Unilateral illumination of sunflower shoot tips results in the appearance of eight times more diffusible GA from the shaded side of the shoot tip compared to the illuminated side (134). More significantly, this unequal distribution of diffusible GA can be detected prior to the onset of the phototropic growth curvature (134). Phillips (134) postulates that lateral GA distribution plays an important role in the establishment of the phototropic response in sunflower. Similar redistribution of diffusible GAs has been shown by Phillips (133) in geotropically stimulated shoot tips of sunflower. As with the phototropically stimulated shoots, the diffusible GAs of the geotropically stimulated apical bud are redistributed—approximately 10 times more GA appearing in diffusates from the lower half of the internode (133). Geotropically stimulated shoot tips produce considerably more diffusible GAs than those of comparable upright plants (133). This evidence suggests that the distribution of diffusible GAs between upper and lower halves of geotropically stimulated shoot tips does not result from increased GA destruction, rather, it seems to suggest that increased GA biosynthesis and transport are involved (133).

These observations of Phillips (133, 134) may be of considerable significance in the interpretation of the tropic responses of light grown shoots. Such shoots do not respond to applied auxin to any significant degree, although marked responses are found to applied GAs (85, 133). It is possible, therefore, that in such shoots redistribution of GA rather than auxin functions to regulate differential growth. Several questions remain to be answered concerning the role of the GAs in tropistic responses. Thus it must be established that the differential redistribution of GAs in the subapical region of the step tip as determined by agar diffusion is also maintained in those regions of the stem which exhibit differential growth response to tropistic stimuli. The redistribution of auxins in tropistically responding organs is maintained by the polarity of the auxin transport mechanism

(181); however, evidence for the existence of polar GA transport is equivocal (40, 66, 70, 89). It is necessary also to exclude the possibility that redistribution of GAs results indirectly from increased rates of cell division and/or cell elongation in the tropistically responding shoot (133).

Photoperiod

There are now several well-documented examples of changes in GA metabolism following a shift in photoperiod.

In spinach, Zeevaart (190) has clearly documented changes in GA metabolism which are related to petiole growth. When spinach is transferred from short to long days, increased petiole growth can be detected within one long day (190). Although no significant qualitative or quantitative difference was observed in the level of extractable GAs from long and short day grown spinach, there were marked differences in the metabolism of GAs—the turnover of GA being far more rapid in long day plants (190). Since extractable GA levels of long and short day plants were found to be equal, Zeevaart (190) concluded that the higher rate of turnover in the long day plants reflected a higher rate of synthesis in these plants. Spinach plants grown under long day conditions also possess a higher sensitivity to applied GAs when compared with similar plants grown under short days (190). This increased sensitivity to GA, combined with increased GA bio-synthesis under long days, results in increased growth of petioles (190).

In *Silene armeria* (37) and clover (166), changes observed in GA metabolism following a shift from short to long days have been correlated with both growth and flowering. In *S. armeria*, exposure of plants to 4–6 long days results in a ten-fold increase in diffusible GAs (37). Examination of extracts from plants treated similarly, however, indicates only a twofold increase in GA levels, demonstrating, as in spinach, a marked increase in GA turnover when plants are exposed to long photoperiods (37, 190). In clover, flowering is also induced by long photoperiods, and following exposure to such photoperiods, Stoddart (165, 166) was able to demonstrate a significant increase in the level of extractable GAs. Using ^{3}H-GA$_1$, Stoddart (166) also showed that metabolism of applied GA was enhanced under long day conditions.

There are several other publications which purport to show qualitative and/or quantitative changes in GA levels following shifts in photoperiod (e.g. 29, 58, 112). In these reports, however, the relationship of the changes in GA levels to the light regime are tenuous since flower primordia and buds were visible when the plants were extracted (29, 112).

Changes in photoperiod also influence other developmental events in plants, e.g. the onset and breaking of dormancy (44, 176, 178). In birch (*Betula pubescens*), short day conditions bring about a cessation of cambial activity within 2–3 weeks of the beginning of the short photoperiods (44, 176). These changes in cambial activity are preceded by a marked reduction in the level of GA-like substances (44). In *Ribes nigrum*, onset of bud dormancy induced by short days is also correlated with reduced GA-like activity in extracts (176). In this case, however, it is unclear whether reduced GA levels resulted from a photoperiodically induced change in biosynthesis or were merely a result of the reduced growth rate typically induced by short days (176).

Temperature

Many developmental changes in higher plants result from shifts in temperature regimes, but the most common are those following vernalization and stratification. Vernalization, which is defined as the promotion of flower formation after exposure to a period of low temperature, can be replaced in some cases by GAs (104). Several workers have thus investigated the possibility that cold treatment results in increased GA levels in such vernalized plants. Harada (63) found an increase in the levels of substances possessing biological activity in the oat mesocotyl assay of extracts from vernalized *Althaea rosea* and *Chrysanthemum* plants. This increase coincided with the minimum length of the cold period necessary to produce flowering in these two species (63). The relevance of this report must remain uncertain, however, since the oat mesocotyl assay does not discriminate between GA-like and auxin-like activity. Thus the increase in biological activity observed by Harada (63) could correspond to peaks of auxin activity. An increase in the level of GA-like substances has been reported in winter wheat, winter rye, and the winter annual *Brassica napus* by Chailakhyan & Lozhnikova (30), and in radish by Suge (167). However, in these reports, plants which were vernalized were then exposed to long day conditions for varying time periods which were as long as 19 days prior to extraction (30, 167). Thus, although these workers dealt with plants which had been vernalized, they did not establish that the vernalization process itself resulted in the increased production of GA-like substances. Significantly, examination of GA levels immediately after the termination of the cold treatment in *Lunaria* failed to indicate an increase in the level of GA-like substances (188). It is apparent from the diffuse information at hand that the role of endogenous GAs in vernalization is not at all clear. It should be pointed out that Lang (104), in his review of this subject, suggested that GAs play no direct role in the process of thermoinduction, but rather he suggested that they may play a role in establishing the thermoinduced state.

Stratification, which is defined as a cold requirement for overcoming the dormancy of some seeds, can also be substituted, in many cases, by GA (1, 13, 19, 52, 53, 98). This observation prompted an investigation of changes in the levels of GA-like substances in seeds both during and after periods of stratification. Hazel (*Corylus avellana*) seeds have dormant embryos if stored under dry conditions for a few weeks after harvest, but this dormancy can be broken by GA treatment (19, 52, 53). Stratification (12 weeks at 5°C) of hazel seeds results in an increase in the level of GA-like substances (52, 53, 148). This increase in GA levels with chilling is small, however, when compared with the amount of GA that is required to overcome the dormant condition (53). Because of this descrepancy between the levels of endogenous GA found after chilling and the concentration of GA required to overcome dormancy, Bradbeer (17) examined the timing of GA production in vernalized hazel seed. He found that CCC was not effective in inhibiting germination if applied during stratification but was effective when the seeds were treated with CCC at the germinating temperature of 20°C (17). Bradbeer (17) suggested that stratification (5°C) prepared the seed for GA synthesis which occurred subsequently at 20°C (17). This conclusion was confirmed by estimating the relative amount of GA-like substances formed during chilling and the subsequent 20°C period (149). During this 20°C period, sufficient GA-like substances could be isolated from the seed to account for the stimulation of germination by these substances (149). The appearance of GA after stratification of hazel is possibly a result of new synthesis since CCC, AMO-1618, B995, and Phosfon D prevent its production (150).

In tulip bulbs, a similar increase in GA levels occurs when bulbs are transferred from 13°C to 18°C (6). In this case, however, there is evidence that cold regulates the conjugation of GA-like substances, since at 13°C there was a marked increase in the level of conjugated GAs with a concomitant decrease in free GAs, while at 18°C the reverse occurred (6). This role of cold in regulating endogenous GA levels in tulips emphasizes the importance of conjugated GAs in organs which overwinter or possess an extended dormancy period (see Lang 105).

It is apparent from this discussion of the endogenous GAs of plants that attention must be given to factors other than the mere presence of GA-like substances in extracts. Before we can speculate on the role of the endogenous GAs we must have answers to questions concerning the regulation of their synthesis, degradation, transport, and the sensitivity of the tissue to the endogenous GAs.

RESPONSES OF PLANTS TO GIBBERELLIN

For convenience, the responses of plants to GAs can be divided into two groups, namely morphological responses and biochemical responses. Although this is a convenient form of categorization, it is recognized as being an unnatural division since all morphological responses to GA must be preceded by biochemical events at the subcellular level. Since these subcellular events have not been characterized, these morphological manifestations of the GA response will be discussed separately.

Morphological Responses to Applied GA

It is clear that all aspects of the growth and development of higher plants from seed germination to fruit set can be affected by GAs. Even though responses to GAs are diverse, all aspects of GA effects on growth, with the possible exception of their effects on flower formation, can be discussed as effects of these regulators on cell elongation, cell division, or both.

The evidence that GAs affect growth by stimulating cell elongation comes from work with gamma plantlets of wheat, epicotyl sections of lentil, and internode sections of *Avena* (60, 126, 147). Irradiation of wheat seeds with 500 kr from a cobalt gamma source results in the production of morphologically "normal" seedlings by the process of cell elongation alone (60, 147). Application of GA to such seedlings elicits a further growth response (60, 147). Since gamma irradiation of the embryos results in a complete inhibition of cell division, GA must be stimulating growth by affecting the process of cell elongation (60, 147). Paleg (131) indicated caution in the interpretation of such results since irradiation could affect the mobilization of substrates within the endosperm. Thus Paleg (131) argued that the effect of irradiation could be mediated indirectly via events within the endosperm. Haber et al (59) countered this argument by showing that in barley, gamma irradiation did not affect the response of aleurone layers to GA_3. Kefford & Rijven (94) also showed that the response of gamma irradiated wheat embryos to GA_3 was independent of the mobilization of stored endosperm substrates. Thus, when embryos are isolated from irradiated seeds they retain the ability to respond by cell elongation to applied GA_3 (94). Although these results with irradiated plant tissue indicate that GA_3 can promote growth by stimulating cell elongation, they do not rule out an involvement of GA_3 in the process of cell division. Indeed, since the occurrence of cell division in these tissues is eliminated by gamma irradiation, it is not possible to rule out an effect of GA on this process as Haber and his associates have suggested (59).

Further evidence that GAs can support elongation growth in the absence of cell division comes from the studies of the growth of nonirradiated wheat coleoptiles. Wright (182, 183) has shown that growth in the wheat coleoptile can be separated into three distinct phases, the earliest of which proceeds by cell elongation alone. Gibberellic acid also promotes growth of wheat coleoptiles during this phase of elongation growth (182, 183). GA_3 suppresses cell division in the intercalary meristem of *Avena* internodes but stimulates growth in this tissue via cell elongation (76, 93). Extension growth in decapitated epicotyls of lentil can also be stimulated by GA_3, and an analysis of this growth response indicates that it occurs predominantly by cell elongation (125, 126). Thus in epicotyl sections, GA_3 stimulates growth by 450% in a 48 hr period but cell number increases by only 16% (125, 126). The growth of *Phaseolus vulgaris* leaf disks in light is also stimulated by GA_3 through its effects on cell expansion alone (57, 68). The effect of GA on leaf growth, however, is complex. Thus GA_3-stimulated growth of leaf disks in darkness occurs by a combination of cell division and expansion (57, 68). A similar observation was made for the growth of *Vicia* seedlings by Butler (27), who showed that light had a considerable influence on cell division. At light intensities of 0.1 fc cell division was inhibited; however, at 10.0 fc or above, an inhibition of cell elongation was observed (27). These two observations indicate that attention should be paid to conditions of illumination in those experiments designed to demonstrate effects of hormones on growth, particularly since low levels of light have a marked effect on the regulation of cell division (27). Thus the effects observed may be governed more by the presence or absence of light rather than the applied growth regulator (27, 57, 68, 69).

GA-stimulated growth of stems, particularly those of rosette plants, involves an increase in both cell size and cell number [the reader is referred to the review by Sachs (151) for a discussion of this topic]. In *Hyoscyamus* and *Samolus*, GA_3 causes an increase in mitotic frequency in the subapical regions of the stem (152). Sachs et al (152) concluded that in these two species growth during the 72 hr period after GA$_3$ application was solely by cell division since no increase in cell size occurred during that period (152). Further evidence for the role of GA in stimulating cell division comes from the work of Wareing and others on the regulation of cambial activity (20, 43, 116, 157, 175, 177). In deciduous trees cambial divisions cease with the onset of short days, but division can be initiated in dormant twigs by GA application (43, 175, 177). GA application has also been shown to increase cell number in petioles, leaves, and stems of caulescent plants, and these observations have also led to the conclusion that GA stimulates growth via its effect on cell division (see Sachs 151).

The evidence for the role of GAs as cell division factors has received considerable criticism because of the inherent difficulty in separating any effects GA might have on division from effects it is known to have on cell expansion. Thus, it can be argued that since GA is known to effect cell elongation, its effect on cell division is secondary and a direct result of GA-induced elongation. Investigators who favor a role for GA in stimulating cell division counter these arguments by pointing to experiments showing an increase in cell number, but not in cell size, following GA treatment (152). These arguments are not entirely convincing, however, since they presuppose a direct relationship between cell size and cell division. Since both cell elongation and cell division occur in rapidly elongating tissues, it is probable that GAs function to regulate both processes.

The control of sex expression in higher plants affords an example that will illustrate the subtle control which the GAs can exert on plant form via variations in both cell division and cell elongation (see review by Nitsch 127). The development of pistillate or staminate flowers can be controlled to varying degrees by growth regulators, photoperiod, and also the genome (127). Gibberellins have been shown to be powerful modifiers of sex expression, particularly in cucurbits, and their effect is to favor the production of staminate flowers (2, 62, 127, 164). By careful manipulation of the GA level of cucurbit plants, either by application of GA or by addition of inhibitors of GA biosynthesis, the ratio of staminate to pistillate flowers can be changed (2, 62, 164). Since flower development involves the integrated transformation of the vegetative apex to a reproductive structure by cell division and cell elongation, it follows that any change in floral morphology induced by GA must result from subtle changes in both processes (127).

The role of auxin in various growth responses has received considerable attention during the last decade. Using sensitive methods for determining changes in cell wall plasticity and growth rate, many of the parameters of auxin-induced growth have been defined. Unfortunately, few investigators have applied the knowledge gained from these studies to growth responses elicited by GAs. Thus, the relationship between cell wall plasticity and auxin-induced growth is now well documented, but only three reports have appeared concerning the effect of GA on wall plasticity.

In an investigation of radiation-induced growth inhibition of pea, Lockhart (109) reasoned that reduced growth rate could be brought about by changes in cell turgor mediated by changes in wall plasticity or by changes in the osmotic concentration of the cell. Although no difference was found in the osmotic concentrations of cell sap obtained from pea plants grown in darkness or light, estimation of wall extensibility indicated that light markedly reduced both elasticity and plasticity of internode cells (109). Reduced cell wall plasticity (and growth) resulting from irradiation could be overcome by application of GA_3, suggesting that GA_3 also functions to stimulate growth by increasing cell wall plasticity (109). On the other hand, Yoda & Ashida (185) reported that GA did not affect the plasticity of Alaska pea internodes. It should be noted, however, that these workers used dark grown internode sections of pea which Giles & Myers (55) showed to be unresponsive to applied GA.

Cleland et al (39) reinvestigated the effect of GA on extensibility of cucumber hypocotyl sections. This tissue is particularly well suited for such an investigation since it responds to both GA and IAA. As expected, auxin caused an increase in wall plasticity; however, GA_3 was without effect (39). These workers suggested that GA-stimulated growth of cucumber hypocotyl tissue resulted from a change in the osmotic concentration of the cell (39). It is apparent from these above discrepancies and the paucity of other well-documented experiments that the effect of GA on cell extensibility must be more rigorously examined in a wider variety of responding tissues.

Another aspect of the auxin response which has been elegantly characterized is the timing of growth initiation (12, 45). Evans & Ray (45) showed that a linear rate of growth was established within 10 min of IAA application to *Avena* coleoptile sections, while Barkley & Evans (12) confirmed the latent period of 10 min in the response of pea internode sections to IAA. Such analyses have indicated that the auxin-induced growth response is independent of the synthesis of of new "informational RNA" or of "enzymatic protein" (45). Thus, the early response to auxin precludes an effect of this regulator on the transcriptional process although an effect on translation could not be excluded (45). Similar information would be invaluable for an evaluation of the role of GA in promoting extension growth. The experimental material of choice for these experiments would be epicotyl or hypocotyl tissue which possesses a marked response to GA as isolated section.

It is becoming increasingly evident that certain aspects of floral induction which can be influenced by the GAs are not related to any direct effects these regulators may have on either cell division or cell elongation. It has been known for some time that application of certain GAs to certain rosette long day plants stimulate flower formation under noninductive daylength conditions (104). It has been argued, however, that in such rosette plants, this effect of GA on flowering is mediated through an effect on stem elongation and not on florigen synthesis per se (104).

The evidence against a direct effect of GA on flower formation has been discussed by Lang (104) and can be summarized as follows: 1. Treatment of caulescent short day plants with GAs has no effect on flower formation (104). 2. The effect of GAs on long day plants are restricted to those having a rosette habit; application of GA to caulescent long day plants are without effect (104). 3. GAs are most effective when applied to the stem apex, suggesting an effect on stem growth, whereas processes related to photoinduction are known to occur in leaves (104, 189). 4. In the majority of cases where flower induction is mediated by changes in photoperiod, flower primordia are visible microscopically before a change in stem growth rate is evident. Similar plants treated with GA respond first by increased stem elongation and later, if at all, by flower and bud formation (104, 180). 5. Correlations between endogenous GA levels correlate more closely with stem and petiole growth rather than flowering (37, 190). 6. Treatment with inhibitors of GA biosynthesis during photoinduction does not inhibit flower formation in all long day plants (8, 37, 169, 190).

The role of GA in stimulating flower formation in *Bryophyllum*, however, is direct (189). Thus, application of GA_3 to a leaf of debudded plants results in the production of the floral stimulus in that leaf (189). Zeevaart (189) concludes that the role of GA_3 in *Bryophyllum* is directly on the production of the floral stimulus rather than on the growth of the stem.

As with flowering of long day plants in short days, the action of GA in stimulating flowering in cold-requiring plants is unclear (see Lang 104, pp. 1497–1502). In *Chrysanthemum* (63), GA causes the formation of a graft-transmissable flower-inducing factor in nonthermoinduced plants, and in this respect the effect of GA is similar to its effect in *Bryophyllum* (99). In other cases, however, the effect of GA can be seen to be primarily on stem growth and resembles the effect of GA on rosette long day plants (104). As Lang (104) points out, though the effects of GA on cold-requiring plants resemble those on long day plants, the role of GA in vernalization is "at present less clear and straightforward than in photoinduction."

Nevertheless, the role of GAs in flowering of *Chrysanthemum* (63) and *Bryophyllum* (189) suggests its involvement in processes other than cell division and elongation.

BIOCHEMICAL ASPECTS OF THE GIBBERELLIN RESPONSE

Cereal Aleurone

Following the discovery in 1960 by Paleg (129) and Yomo (186) that GA_3 could substitute for the embryo in the initiation of events leading to substrate mobilization, considerable attention has been focused on the role of GA_3 in this process. Haberlandt (61) was the first to recognize that the diastatic activity responsible for starch digestion originated in the seed coat layers of cereals, and he was also able to demonstrate that a growing embryo was necessary for the production of the diastase. The relationship between the embryo and other seed parts in barley (*Hordeum vulgare* L.) is now well understood with respect to the process of substrate mobilization. Gibberellins similar to GA_3 and GA_1 produced by the embryonic axis (61, 113–115, 143, 144) are transported to the aleurone layer presumably via the vasculature of the nodal region (113, 115) and scutellum. This GA induces the de novo synthesis of hydrolytic enzymes within cells of the aleurone layer (14, 49, 71), whereupon the enzymes are secreted into the endosperm where hydrolysis of stored substrates proceeds. These substrates are absorbed by the scutellum and are translocated to the growing embryo to maintain early development of the seedling (see review by Yomo & Varner 187).

The attractiveness of the GA₃ response in cereal aleurone as a system for the study of the biochemical basis of GA action derives from several unique features of this tissue: (*a*) gibberellins are the natural trigger of enzyme production in vivo; (*b*) its response is confined to this one hormone class; (*c*) the target tissue consists of only one cell type; (*d*) the cells of the tissue do not undergo division; (*e*) the cells can be isolated free of other tissues and still respond to GA₃; and (*f*) the response of the isolated aleurone tissue is independent of the addition of substrates.

The response of isolated aleurone tissue to GA₃ typically has a lag period which varies from 4–20 hr after addition of the hormone. The appearance of an enzyme with laminarase (β, 1–3, glucanase) activity occurs within 4 hr of GA₃ addition (81), while α-amylase and protease are secreted after a lag of 8–10 hr (35, 71). Ribonuclease, on the other hand, does not appear in the incubation medium until 18–20 hr after GA₃ addition although its synthesis begins after 8–10 hr of GA₃ (35).

Using H_2O^{18} and D_2O, the synthesis of these four enzymes has been shown to be de novo (14, 49, 71). Thus, although considerable quantities of protein are found in protein bodies (aleurone grains) within the aleurone layer (78), this protein is a storage protein only and is hydrolyzed to amino acids for de novo synthesis of the secreted hydrolases. Inhibition of α-amylase production by inhibitors of protein synthesis is consistent with the observation that this synthesis is de novo (23, 173). Production of α-amylase can also be inhibited by inhibitors of RNA synthesis, e.g. actinomycin D and 6-methylpurine (31, 36). These observations led Chrispeels & Varner (36) to conclude that α-amylase production was dependent on "enzyme specific RNA." This observations has been interpreted by others as evidence for the production of mRNA in such GA₃-treated tissues. Yomo & Varner (187) point out, however, that "at the present time we can neither rule in nor rule out the possibility that gibberellic acid induces the transcription of certain kinds of RNA specific for the synthesis of α-amylase and four other hydrolases."

An increase in the synthesis of ribosomes (and presumably rRNA) following GA₃ treatment of aleurone tissue is indicated by the work of Evins (46) and Evins & Varner (48). During the 8–10 hr lag period of α-amylase production, there is an increase in the total number of ribosomes isolated from GA₃-treated cells relative to controls, and significantly there is an increase in the number of ribosomes associated as polyribosomes (Figure 1 A–D; and 46, 48). Since the GA₃-stimulated increase in ribosome aggregation (Figure 1 B) is greater than the stimulation of ribosome synthesis (Figure 1 A), it can be concluded that GA also stimulates polysome formation (or ribosome aggregation) directly (46, 48). In addition to this stimulation of polyribosomes formation, there is an increase in membrane, particularly endoplasmic reticulum (ER), production (79, 80). Johnson & Kende (74) have shown that the activities of two enzymes of the cytidine diphosphate-choline pathway of lecithin biosynthesis are stimulated within 2 hr of GA_3 treatment. Choline kinase, the first enzyme of this pathway shown in (1) does not increase following GA_3 treatment; however, phosphorylcholine-cytidyl transferase and phosphorylcholine-glyceride transferase increase significantly following GA_3 treatment (Figure 2; and 74).

$$\text{1, 2}$$

(1) CHOLINE $\xrightarrow{\text{ATP} \quad \text{ADP}}$ PHOSPHORYL CHOLINE $\xrightarrow{\text{CTP} \quad \text{PP}_i}$ CDP-CHOLINE $\xrightarrow{\text{DIGLYCERIDE} \quad \text{CMP}}$ LECITHIN

CHOLINE KINASE	PHOSPHORYLCHOLINE	PHOSPHORYLCHOLINE
	CYTIDYL TRANSFERASE	GLYCERIDE TRANSFERASE

Evins & Varner (47) showed that the incorporation of C^{14}-choline into a semi-purified, acid insoluble ER fraction was stimulated up to tenfold by GA_3 (Figure 1 E). As with the increase in ribosome synthesis and polyribosome formation, the increase in ER synthesis precedes GA-induced hydrolase synthesis (47).

The synthesis of ribosomes and membranes (particularly ER) are concomitant events when assayed biochemically and they correlate with the appearance in the cell of "stacked" rough ER (79, 80). Are these events which occur during the lag phase related to hydrolase production? Evins (46) has shown that nascent proteins isolated from GA treated cells have a high tryptophan tyrosine ratio—a characteristic feature of α-amylase, while unpublished data by Evins (see 46) suggests that polysomes isolated from GA_3-treated cells have more α-amylase associated with them relative to control polysomes. It is tempting to speculate that these polysomes bound to the ER are functional in hydrolase synthesis. Tata (171) and others have suggested that "topographical segregation" of protein synthesis could be achieved in cells by binding of polyribosomes to membranes. Johnson & Kende (74) have hypothesized that in barley aleurone, synthesis of hydrolases occurs only on polyribosomes attached to newly synthesized ER. They suggest that the role of GA may lie in the production of membranes for the attachment of polyribosomes for translation of preexistent mRNAs (74). It is clear from the evidence presented above that GA could function to regulate protein synthesis at the translational level in a quantitative or qualitative manner. A quantitative form of control could be exerted by changes in ribosome synthesis (46, 48), while control of a qualitative nature could result from a change in the amount of ER for ribosome attachment (47, 74, 79).

Recent experiments by Carlson (28) also support the hypothesis that GA exerts its effect in aleurone cells by controlling post-transcriptional events. By examining the thermolability of α-amylase molecules produced in the presence and absence of 5-fluorouracil (Fu), Carlson (28) showed that two subpopulations of the enzyme appeared. Carlson (28) reasoned that if translational control of enzyme production was operative in aleurone cells, then treatment of tissue with Fu either before or during GA treatment should result in the production of α-amylase molecules with increasing thermolability. On the other hand, if transcriptional controls were operative, treatment with Fu prior to GA_3 application should give rise to thermostable molecules, while treatment with Fu during GA_3 treatment

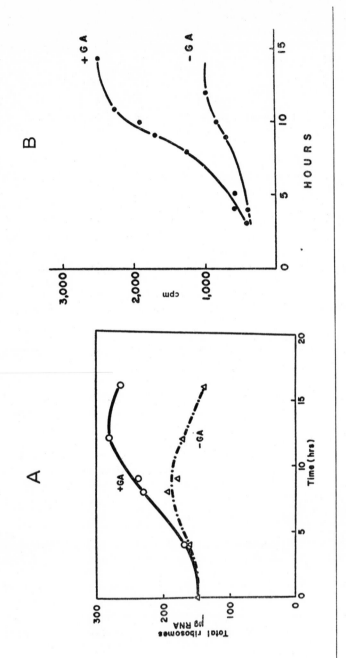

Figure 1   The effect of GA₃ on ribosome, polyribosome, and ER synthesis in barley aleurone layers.

A: GA₃ and total ribosome content.
B: GA₃ and polyribosome formation.

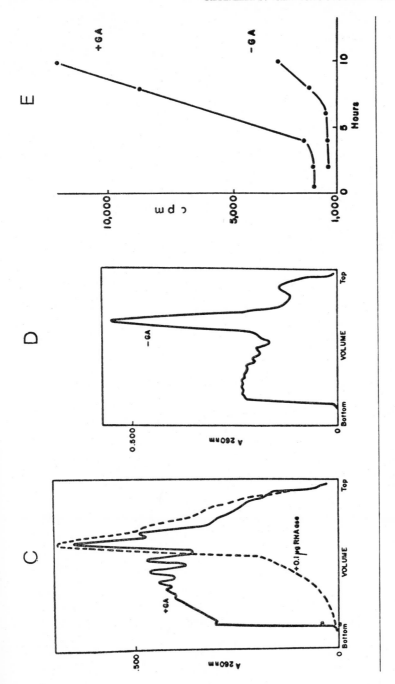

C: polyribosome profiles from GA$_3$-treated aleurone layers.
D: polyribosome profiles from control aleurone layers.
E: GA$_3$ and ER production.
A–D from Evins (46), and E from Evins & Varner (47).

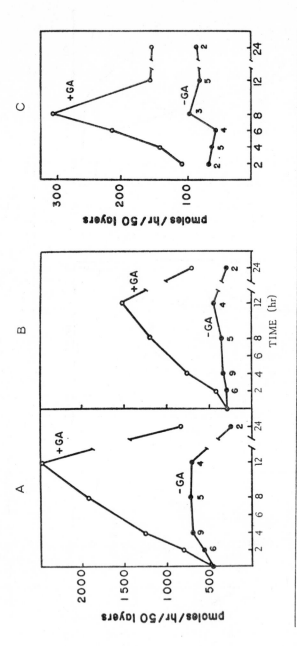

Figure 2   The effect of GA$_3$ on phosphorylcholine-cytidyl transferase activity from an 11,000×g pellet (A) and 44,000×g pellet (B), and on phosphoryl-choline-glyceride transferase activity from a composite 500–44,000×g pellet (C). From Johnson & Kende (74).

should result in the production of thermolabile α-amylase molecules (28). Since two populations of the enzyme were found when Fu was applied before and during GA_3 treatment, Carlson (28) deduced that control was probably at the post-transcriptional level.

The observations on events occurring in the lag phase of the GA_3 response referred to above concern the regulation of hydrolase synthesis. However, mobilization of the stored endosperm substrates requires the export of the newly synthesized hydrolases from the aleurone cell cytoplasm to the endosperm. A role for GA_3 in regulating the export of materials from aleurone cells has now been established (81, 119, 174). The release—without concomitant synthesis—of β,1-3, glucanase from aleurone cells is stimulated within 4 hr of GA_3 treatment (81). Similarly Melcher & Varner (119) have shown that the release of stored protein and peptides from aleurone cells is controlled by GA_3.

Although the cytological events occurring in aleurone tissue following GA_3 treatment have been well documented, little evidence has been obtained for the subcellular pathway of enzyme release (79, 80, 87). There is no good evidence for the participation of secretory vesicles in the releasing process, rather attempts to isolate putative α-amylase and β,1-3, glucanase containing vesicles have been fruitless (82). Evidence obtained from an examination of cell wall hydrolysis (170), together with the negative evidence concerning secretory vesicles, suggest that enzymes and other products may be released from the cytoplasm across the plasmalemma without the participation of a discrete secretory organelle (82). This suggests that GA_3 may selectively change the permeability characteristics of the cell membrane. Evidence for such a GA_3-induced change in aleurone plasmalemma permeability has recently been obtained in the writer's laboratory. Release of potassium and magnesium ions from aleurone cells occurs only after treatment of the tissue with GA_3. This effect of GA_3 treatment on ion release does not result from the release of these ions from a complexed form since free ions are present in the tissue prior to GA_3 treatment. Rather, the effect of GA_3 is to cause a change in the permeability characteristics of these cells allowing for the controlled release of these ions.

These effects of GA_3 on the permeability of aleurone cells to ions, and storage protein and peptides (119), suggest that this hormone could regulate cellular events by controlling the permeability of cell membranes. In this regard it should be noted that in the aleurone cell the substrates for lipid and protein synthesis and the reserve phosphate are all located in distinct membrane bound organelles (78). It is possible therefore that GA_3 could initiate the events leading to hydrolase production by causing changes in the permeability of these organelles leading to the availability of substrates for synthetic metabolism. It should be pointed out in this context that mobilization of storage proteins alone cannot limit hydrolase production. This conclusion is based on the observation that bromate inhibition of stored protein hydrolysis can be completely overcome by the addition of casein hydrolysate (119).

Although the investigations of the GA_3 response in barley aleurone have not as yet indicated the site or mode of action of this hormone, they help to pinpoint the biochemical and physiological roles in which the GAs are involved. Thus, it is clear that in this system GA₃ affects ribosome production, membrane synthesis, and membrane permeability. Can investigations on such a highly differentiated cell be extrapolated to responses in other less specialized and "growing" plant cells? This question cannot, and probably need not, be answered. An answer to the role of GA₃ in a specialized system would be sufficiently satisfying, and if such an investigation gives a clue to other GA responses, this would be even more satisfying.

Other Cereal and Dicotyledonous Seed

It is apparent that in the past decade investigators have focused their attention on the response of the barley aleurone to GA₃. Other cereal seeds, however, respond similarly to applied GA, and these species are now being exploited as systems for study of biochemical aspects of GA effects.

Paleg and co-workers have turned their attention to a study of the response of the wheat (*Triticum aestivum*) aleurone to GA₃ (41). As with barley, wheat aleurone cells respond to applied GA₃ by producing α-amylase within 6–8 hr of hormone treatment; however, the wheat aleurone offers several advantages over that of barley (41, 130). The principal asset of the wheat aleurone system rests with the fact that the tissue can be readily isolated in gram amounts by mechanical means while retaining its response to GA₃. (Mechanical isolation of barley aleurone is also possible, but its response to GA₃ is either considerably reduced or eliminated.) Collins, Jenner & Paleg (41) investigated the change in soluble nucleotides of GA₃-treated wheat aleurone and found that although nucleotide content did not change appreciably, the specific radioactivities of ATP, CTP, GTP, and UTP following ³²P labeling increased after 30 min of hormone addition (41). More specifically, these workers demonstrated that the increase in specific activity of CTP exceeded that of the other nucleotides (41). Based on this observation, Collins et al (41) suggested that increased metabolism of CTP reflected an important role for this nucleotide in phospholipid metabolism. This, they argued, implied a role for GA in the synthesis of membrane components (41). This suggestion supports the observations on aleurone tissue of barley which also imply a role for GA in the synthesis of cell membrane components (47, 74, 79).

An interesting relationship exists between GAs and the process of germination in the seed of oat (*Avena* species). In *Avena sativa*, germination proceeds immediately upon imbition of the seed, whereas in *Avena fatua* L. germination proceeds only after a period of after-ripening or GA₃ treatment (123). It is now clear that in unripened *A. fatua* seeds dormancy results from the failure of the embryo to produce a GA-like substance (124, 161, 162). The role of either exogenous or endogenous GA in promoting germination of *A. fatua* seeds has now been resolved into two distinct parts (161). As in barley, aleurone tissue of wild oat responds to GA₃ application by producing α-amylase which can function in the process of substrate mobilization (123, 161, 163). Substrates (sugars, amino acids) alone, however, cannot relieve the dormancy of wild oat embryos (124, 161). Only when dormant embryos are incubated with GA_3 does growth of the embryonic axis ensue (124, 161). It can be shown that embryo dormancy is restricted to events occurring within the embryonic axis and is not dependant on mobilization of stored substrates (32). Thus, germination as measured by radicle protusion precedes amylase production by at least 24 hr (32, 34).

Several workers have investigated the biochemical basis of embryo dormancy in wild oat (33, 34, 160). Simmonds & Simpson (160) found that after-ripening (or GA_3 treatment) of embryos resulted in an increased participation of the pentosephosphate pathway of glucose metabolism. These authors speculate that such a diversion in the pathway of glucose metabolism could result in the preferential synthesis of important intermediates such as nucleotides, NADPH, flavones, etc, which would be essential for the initiation of embryo growth (160). Chen & Varner (33) have shown that after-ripened embryos of *A. fatua* can synthesize sucrose via the enzyme sucrose synthetase more readily than their dormant counterparts. These workers argued that since sucrose is the primary transport sugar, any limitation on its synthesis from hexoses would limit the export of reserve carbohydrates and fats to the developing embryo (33).

Although the role of GA in relieving embryo dormancy is yet to be resolved, it occupies a central role in seed germination. This role of GA is overlooked by many authors (see 1) who imply that GA-stimulated germination of cereal seeds results solely from increased mobilization of stored endosperm reserves by hydrolases produced in the aleurone layer.

Although the GA response in cereal seeds has attracted the most attention from physiologists, dicotyledonous seeds are proving equally attractive as experimental systems for studying biochemical and physiological aspects of the GA response. One such system is the hazel (*Corylus avellana* L.) seed. This seed exhibits two distinct forms of dormancy: primary dormancy which results from the presence of inhibitory substances present in the testa and pericarp of the seed, and secondary dormancy which results from reduced GA production following storage under moisture-free conditions (17). Although it is not known whether the locus of inhibition lies within the embryonic axis or cotyledons, dormancy can be relieved by addition of GA_3 (52, 53). Exogenously applied GA stimulates growth of the embryonic axis and cotyledonary cell expansion within 4 days of hormone application (18, 19, 138, 139). Prior to these visible morphological effects induced by GA, changes in the metabolism of both embryo and cotyledons can be detected. Jarvis et al (72, 73) and Pinfield & Stobart (139) found that GA_3 promoted both DNA and RNA metabolism in embryonic axes of hazel within 12 hr of GA_3 treatment. Jarvis et al (72, 73) showed further that GA stimulated DNA template activity and RNA polymerase activity twofold within 24 hr of hormone treatment.

The metabolism of DNA and RNA in cotyledons of hazel is also stimulated by GA_3; however, in contrast to the embryo, this stimulation is not apparent until 48 hr after pretreatment with the hormone (139). Pinfield & Stobart (139) suggested that there were two distinct responses to GA in hazel seed: the first being in the embryo and the second an independent or partly independent response in the cotyledons.

Following the GA-stimulated changes in nucleic acid synthesis in cotyledons there are changes in protein, fat, and nucleotide metabolism. Isocitrate lyase, an enzyme of the glyoxylate bypass, is stimulated severalfold 48 hr following GA treatment in cotyledons (137). The pattern of acetate-2-^{14}C metabolism in hazel cotyledons also suggests that the glyoxylate cycle is activated after GA treatment resulting in increased sucrose production from fats (18, 183). On the basis of their studies of intermediary metabolism in nondormant cotyledons of hazel, Pinfield (137) and Bradbeer & Colman (18) suggest that germination results from a diversion of the metabolism of the tissue from one which is basically catabolic to one which is anabolic. Thus, GA could function to stimulate the activity or synthesis of a group of specific enzymes (e.g. isocitrate lyase, malate synthetase, sucrose synthetase, etc) which would function to divert the metabolism of 2-carbon fragments, resulting in the synthesis of new intermediates which could serve to support growth of the new seedling.

Growing Tissues

The number of reports showing effects of GAs on DNA, RNA, and protein metabolism in growing tissues are extensive (see reviews by Key 97 and Glasziou 56). Many of these reports deal only with gross changes in nucleic acid or protein metabolism and as such tell us little about the way in which GA might promote growth in these tissues. Indeed, it would be surprising if GA did not influence nucleic acid or protein synthesis in those tissues responding to the hormone. Most studies, however, do not discriminate between an increased rate of production of the same nucleic acids or proteins and a qualitative change in the spectrum of nucleic acid or proteins produced.

In 1965 Nitsan & Lang (125) demonstrated that GA$_3$-stimulated elongation growth was dependent on the synthesis of DNA. In lettuce hypocotyl and lentil epicotyl tissue cell elongation occurring in the presence or absence of GA$_3$ could be inhibited by 5-fluorodeoxyuridine (FuDR) (125). Significantly, the growth inhibition caused by FuDR could be completely reversed by thymidine but not by uridine (125). Further, Nitsan & Lang (126) showed that in lentil epicotyls GA$_3$-stimulated extension growth was accompanied by a marked increase in DNA synthesis. Holm & Key (67) and Broughton (24, 25) were unable to demonstrate a dependence between DNA synthesis and cell elongation in soybean and pea respectively. Treatment of sobyean epicotyls with FuDR did not cause inhibition of GA-induced growth, although in this tissue DNA synthesis was inhibited by FuDR (67). However, these workers did show that continued DNA synthesis was a requirement for growth in soybean epicotyls, although it is clear that this synthesis is related to the requirement for continued cell division rather than elongation (67). Likewise, the marked elongation growth of dwarf pea internodes elicited by GA$_3$ cannot be blocked by FuDR (24, 25). These findings are also supported by the observation of Haber and associates (59, 60) and

Rose & Adamson (147) on GA$_3$-stimulated growth of gamma irradiated tissues. Since GA stimulates growth of gamma irradiated wheat and lettuce seeds where no cell division or DNA synthesis occurs (59, 60, 147), it follows that DNA synthesis is not an absolute prerequisite for cell elongation.

Atsmon & Lang (4) and Atsmon (3) have reinvestigated the relationship between DNA synthesis and GA-induced growth. These workers have suggested that the small number of cell divisions (and consequently DNA synthesis) occurring in the lentil epicotyl may indeed be necessary for the maintenance of GA$_3$-induced cell elongation (see Key 97, p. 465; also 3, 4). Thus inhibition of cell division in epicotyl or hypocotyl tissue could result, indirectly, in an inhibition of elongation (3, 4). Atsmon (3) has provided other circumstantial evidence which speaks against the hypothesis of Nitsan & Lang (125, 126). Thus colchicine, which is an effective inhibitor of cell division but not of DNA synthesis, also inhibits GA-induced growth in some tissues (3).

The weight of evidence suggests that cell elongation, whether controlled by endogenous or exogenous GA, is not directly dependent on DNA synthesis. These investigations, however, do suggest a requirement for continued cell division in those cells adjacent to the elongating zone during GA-induced growth (4, 67, 97). Do these cell divisions merely serve to provide cells to the elongating region, or do they provide cells which produce substances that are necessary for the continued elongation growth in the tissue?

One of the most striking demonstrations of an effect of GA$_3$ on RNA metabolism comes from the work of Johri & Varner (75). When nuclei are isolated from light-grown dwarf pea internodes in the presence of GA$_3$, an enhanced rate of synthesis of an RNA species differing in nearest neighbor frequency and size distribution from control nuclei is observed (75). In an examination of the effect of GA$_3$ and light on the RNA content of dwarf peas using DNA-RNA hybridization techniques, Thompson & Cleland (172) were unable to confirm the appearance of a new RNA species in light-grown internodes treated with GA for 36 hr. The report of Thompson & Cleland (172) is not necessarily at variance with the report of Johri & Varner (75), since hybridization-competition experiments might not be expected to resolve changes in a small fraction of RNA such as that new fraction of RNA induced by GA$_3$. Support for a role of GA in the stimulation of RNA synthesis in dwarf pea internodes comes from experiments showing increased RNA polymerase activity associated with chromatin of GA$_3$-treated internodes (118). McComb et al (118) showed that the increase in RNA polymerase activity occurred in the absence of any increase in DNA template and that the increase in enzyme activity preceded the observed growth response in this tissue. It should be pointed out, however, that increased RNA polymerase activity following GA$_3$ treatment may reflect only a quantitative rather than a qualitative change in the rate of transcription.

Using cytochemical and autoradiographic methods, Nagl (122) found that GA$_3$ stimulated the formation of numerous nucleoli in endopolyploid nuclei of *Phaseolus vulgaris* endosperm. The nucleoli or nucleolar bodies are not only produced by the main nucleolus but also from other regions of the polytene chromosomes (122). Histochemical tests indicate that these bodies contain RNA and that actinomycin D prevents their appearance (122). On the basis of this evidence Nagl (122) suggests that GA_3 functions to stimulate gene activity in this tissue. Although, as in other cases, the stimulation of gene activity referred to by Nagl (122) may be of a quantitative nature, the appearance of new nucleolar bodies in *Phaseolus* nuclei suggests a qualitative effect of GA_3 on RNA synthesis in this tissue.

In his review of the GAs in 1965, Paleg (131) pointed to the interesting corollary between GA treatment of tissues and the loss of stored starch from such tissues (16, 50, 51). Paleg (131) suggested that GA-mediated hydrolysis of stored substrates within plant cells could play a significant role in the control of elongation growth. This observation of Paleg (131) stimulated an intensive search for various hydrolytic enzymes in tissues which respond to GA (see review by Glasziou 56). Kaufman et al (92) examined the correlation between invertase activity and extension growth of *Avena* internode sections. Although a four-to fivefold increase in invertase activity occurred within 6 hr of GA_3 treatment, extension growth was initiated within 40 min of hormone addition (92). Seitz & Lang (154) followed the development of invertase activity in GA_3-treated lentil (*Lens culinaris*) epicotyls and in this tissue demonstrated that the peak in the rate of increase of enzyme activity preceded the maximum rate of elongation. A correlation between invertase development and growth is apparent in sugar cane and, as in *Avena* and lentil, both growth and invertase activity are stimulated by GA_3 (54, 65).

Among the many other examples of changes in enzyme development with GA_3 treatment in elongating tissues are α-amylase in corn (90, 91), pea (26), *Pharbitis* (91), and tobacco (106); invertase in tobacco (106); β-fructo-furanosidase in pea (91), and peroxidase in sugar cane (54). Are these GA_3-stimulated changes in enzyme activity the cause of the increased growth rate in stem tissue, or are they involved in the maintenance of the higher rate of cell elongation? The same question can be asked of the changes in both DNA and RNA synthesis which have been correlated to GA-stimulated growth (97). Answers to these questions cannot be provided until refined analyses are made of the kinetics of GA-induced growth. Thus in the reports of Kaufman et al (92) and Seitz & Lang (154), growth was measured over periods of hours rather than minutes. Experience with the response of tissues to auxins has indicated that maximum rates of growth can be established within 10 min of hormone treatment (12, 45). Such kinetic analyses of the auxin-induced responses have led to the conclusion that increased rates of RNA and protein metabolism are required for the maintenance of the increased rate of growth (97).

CONCLUDING REMARKS

The last decade has seen considerable progress in our understanding of the chemistry of the Ga's. Nearly 40 GAs have been chemically characterized, while much is also known of the chemistry of the GA conjugates. In contrast, most investigations of the physiology of these compounds has been characterized by the lack of effort and imagination. For many physiologists the well-tried techniques of extraction and bioassay still suffice as the basic experimental approach. In many cases, however, it is not clear what the investigator hopes to learn by using such techniques. Physiologists can no longer design their experiments to ask merely whether the plant contains GA-like substances. Answers must be obtained to some of the more basic questions concerning the endogenous GAs of plants. Are all the extractable GAs of a plant functional as growth regulators within that plant? How are the GAs metabolized, and does this metabolism play a direct role in the regulation of endogenous GA levels as recent evidence suggests? What is the nature of the GA transport system—indeed, does a specific GA transport system exist as for auxin? How do the kinetics of the GA response compare with that of auxin? It is clear that we have made little progress in answering some basic questions. Although much of the lack of progress results from the fact that physiologists have not approached these questions directly, a contributory factor is perhaps the decline, during the last 5 years, in the study of the physiology of whole plants. Thus, physiologists have tended to study the physiology of specialized tissues, e.g. the cereal aleurone. However, this effort has been at the expense of the physiology of the whole plant.

Literature Cited

1. Amen, R. D. 1968. *Bot. Rev.* 34:1–31
2. Atsmon, D. 1968. *Ann. Bot. London* 32:877
3. Atsmon, D. 1971. *Isr. J. Bot.* 20:336
4. Atsmon, D., Lang, A. 1968. *Isr. J. Bot.* 17:225
5. Atsmon, D., Lang, A., Light, E. 1968. *Plant Physiol.* 43:806–10
6. Aung, L. H., De Hertogh, A. A., Staby, G. 1969. *Plant Physiol.* 44:403–6
7. Bailiss, K. W., Wilson, I. M. 1967. *Ann. Bot. London* 31:195–211
8. Baldev, B., Lang, A. 1965. *Am. J. Bot.* 52:408–17
9. Barendse, G. W. M. 1971. *Planta* 99:290–301
10. Barendse, G. W. M., Kende, H., Lang, A. 1968. *Planta* 43:815–22
11. Barendse, G. W. M., Kok, N. J. J. 1971. *Plant Physiol.* 48:476–79
12. Barkley, G. M., Evans, M. L. 1970. *Plant Physiol.* 45:143–47
13. Baskin, J. M., Baskin, C. C. 1970. *Planta* 94:250–58
14. Bennett, P. A., Chrispeels, M. J. 1972. *Plant Physiol.* 49:445–47
15. Beevers, L., Loveys, B. R., Pearson, J. A., Wareing, P. F. 1970. *Planta* 90:286–94
16. Boothby, D., Wright, S. T. C. 1962. *Nature* 196:389–90
17. Bradbeer, J. W. 1968. *Planta* 78:266–76
18. Bradbeer, J. W., Colman, B. 1967. *New Phytol.* 66:5–15
19. Bradbeer, J. W., Pinfield, N. J. 1967. *New Phytol.* 66:515–23
20. Bradley, M. V., Crane, J. C. 1957. *Science* 126:972
21. Brian, P. W. 1966. *Int. Rev. Cytol.* 19:229–66
22. Brian, P. W., Elson, G. W., Hemming, H. G., Radley, M. 1955. *J. Sci. Food Agr.* 5:602–12
23. Briggs, D. E. 1963. *J. Inst. Brew.* 69:13–19
24. Broughton, W. J. 1968. *Biochim. Biophys. Acta* 155:308–10
25. Broughton, W. J. 1969. *Ann. Bot. London* 33:227–43
26. Broughton, W. J., McComb, A. J. 1971. *Ann. Bot. London* 35:213–28
27. Butler, R. D. 1963. *J. Exp. Bot.* 14:142–53
28. Carlson, P. S. 1972. *Nature* 237:39–41
29. Chailakhyan, M. K., Lozhnikova, V. N. 1961. *Sov. Plant Physiol.* 16:392–99
30. Chailakhyan, M. K., Lozhnikova V. N. 1962. *Fiziol. Rast.* 9:21–31
31. Chandra, G. R., Varner, J. E. 1965. *Biochim. Biophys. Acta* 108:583–92

32. Chen, S. S. C., Chang, J. L. L. 1972. *Plant Physiol.* 49:441–42
33. Chen, S. S. C., Varner, J. E. 1969. *Plant Physiol.* 44:770–74
34. Ibid 1970. 46:108–12
35. Chrispeels, M. J., Varner, J. E. 1967. *Plant Physiol.* 42:398–406
36. Ibid, 1008–16
37. Cleland, C. F., Zeevaart, J. A. D. 1970. *Plant Physiol.* 46:392–400
38. Cleland, R. E. 1969. In *The Physiology of Plant Growth and Development*, ed. M. B. Wilkins, 49–81. London: McGraw-Hill
39. Cleland, R. E., Thompson, M. L., Rayle, D. L., Purves, W. K. 1968. *Nature* 219:510–11
40. Clor, M. A. 1967. *Nature* 214:1263–64
41. Collins, G. G., Jenner, C. F., Paleg, L. G. 1972. *Plant Physiol.* 49:404–10
42. Crane, J. C. 1964. *Ann. Rev. Plant Physiol.* 15:303–26
43. Digby, J., Wareing, P. F. 1966. *Ann. Bot. London* 30:539–48
44. Eagles, C. F., Wareing, P. F. 1964. *Physiol. Plant.* 17:697–709
45. Evans, M. L., Ray, P. M. 1969. *J. Gen. Physiol.* 53:1–20
46. Evins, W. H. 1971. *Biochemistry* 10:4295–4303
47. Evins, W. H., Varner, J. E. 1971. *Proc. Nat. Acad. Sci. USA* 68:1631–33
48. Evins, W. H., Varner, J. E. 1972. *Plant Physiol.* 49:348–52
49. Filner, P., Varner, J. E. 1967. *Proc. Nat. Acad. Sci. USA* 58:1520–26
50. Flemion, F. 1961. *Contrib. Boyce Thompson Inst.* 20:57–70
51. Flemion, F., Topping, C. 1963. *Contrib. Boyce Thompson Inst.* 22:17–22
52. Frankland, B., Wareing, P. F. 1962. *Nature* 194:313–14
53. Frankland, B., Wareing, P. F. 1966. *J. Exp. Bot.* 17:596–611
54. Gayler, K. R., Glasziou, K. T. 1969. *Planta* 84:185–95
55. Giles, K. W., Myers, A. 1966. *Phytochemistry* 5:193–96
56. Glasziou, K. T. 1969. *Ann. Rev. Plant Physiol.* 20:63–88
57. Greulach, V. A., Haesloop, J. G. 1958. *Am. J. Bot.* 45:566–70
58. Grigorieva, N., Kucherov, Ya., Lozhnikova, V. N., Chailakhyan, M. K. 1971. *Phytochemistry* 10:509–17
59. Haber, A. H., Foard, D. E., Perdue, S. W. 1969. *Plant Physiol.* 44:463–67
60. Haber, A. H., Luippold, H. J. 1960. *Am. J. Bot.* 47:140–44

61. Haberlandt, G. 1890. *Ber. Deut. Bot. Ges.* 8:40–48
62. Halevy, A. H., Rudich, Y. 1967. *Physiol. Plant* 20:1052–58
63. Harada, H. 1962. *Rev. Gen. Bot.* 69:201–97
64. Harada, H., Yokota, T. 1970. *Planta* 92:100–4
65. Hatch, M. D., Glasziou, K. T. 1963. *Plant Physiol.* 38:344–48
66. Hertel, R., Evans, M. L., Leopold, A. C., Sell, H. M. 1969. *Planta* 85:238–49
67. Holm, R. E., Key, J. L. 1969. *Plant Physiol.* 44:1295–1302
68. Humphries, E. C., Wheeler, A. W. 1960. *J. Exp. Bot.* 11:1–85
69. Humphries, E. C., Wheeler, A. W. 1963. *Ann. Rev. Plant Physiol.* 14:385–410
70. Jacobs, W. P., Kaldewey, H. 1970. *Plant Physiol.* 45:539–41
71. Jacobsen, J. V., Varner, J. E. 1967. *Plant Physiol.* 42:1596–1600
72. Jarvis, B. C., Frankland, B., Cherry, J. H. 1968. *Plant Physiol.* 43:1734
73. Jarvis, B. C., Frankland, B., Cherry, J. H. 1968. *Planta* 83:257–66
74. Johnson, K. D., Kende, H. 1971. *Proc. Nat. Acad. Sci. USA* 59:260–76
75. Johri, M. M., Varner, J. E. 1968. *Proc. Nat. Acad. Sci. USA* 59:260–76
76. Jones, R. A., Kaufman, P. B. 1971. *Plant Physiol.* 24:491–97
77. Jones, R. L. 1968. In *Biochemistry and Physiology of Plant Growth Substances*, ed. F. Wightman, G. Setterfield, 73–84. Ottawa: Runge
78. Jones, R. L. 1969. *Planta* 85:359–75
79. Ibid. 87:119–33
80. Ibid. 88:73–86
81. Jones, R. L. 1971. *Plant Physiol.* 47:412–16
82. Jones, R. L. 1972. *Planta* 103:95–109
83. Jones, R. L., Lang, A. 1968. *Plant Physiol.* 43:629–34
84. Jones, R. L., Phillips, I. D. J. 1964. *Nature* 204:497–99
85. Jones, R. L., Phillips, I. D. J. 1966. *Plant Physiol.* 41:1381–86
86. Jones, R. L., Phillips, I. D. J. 1967. *Planta* 72:53–59
87. Jones, R. L., Price, J. E. 1970. *Planta* 94:191–202
88. Jones, R. L., Varner, J. E. 1967. *Planta* 72:155–61
89. Kato, J. 1958. *Science* 128:1008–9
90. Katsumi, M. 1970. *Physiol. Plant.* 23:1077–84
91. Katsumi, M., Fukuhara, M. 1969. *Physiol. Plant.* 22:68–75
92. Kaufman, P. B., Ghosheh, N.,

Ikuma, H. 1968. *Plant Physiol.* 43: 29–34

93. Kaufman, P. B., Petering, L. B., Adams, P. B. 1969. *Am. J. Bot.* 56: 918–27
94. Kefford, N. P., Rijven, A. H. G. C. 1966. *Science* 151:104–5
95. Kende, H., Kays, S. E. 1971. *Naturwissenschaften* 58:524–25
96. Kende, H., Lang, A. 1964. *Plant Physiol.* 39:435–40
97. Key, J. L. 1969. *Ann. Rev. Plant Physiol.* 20:449–74
98. Khan, A. H., Heit, C. E. 1969. *Biochem. J.* 113:707–12
99. Köhler, D. 1965. *Planta* 65:218–24
100. Köhler, D. 1971. *Z. Pflanzenphysiol.* 65:404–9
101. Köhler, D., Lang, A. 1963. *Plant Physiol.* 38:555–60
102. Kuraishi, S., Muir, R. M. 1963. *Naturwissenschaften* 50:337–38
103. Kuraishi, S., Muir, R. M. 1964. *Plant Cell Physiol.* 5:61–69
104. Lang, A. 1965. In *Handbook of Plant Physiology*, ed. W. Ruhland, 15/1:1380–1536. Berlin: Springer-Verlag
105. Lang, A. 1970. *Ann. Rev. Plant Physiol.* 21:537–70
106. Lee, T. T., Rosa, N. 1969. *Can. J. Bot.* 47:1591–98
107. Lockhart, J. A. 1956. *Proc. Nat. Acad. Sci. USA* 42:841–48
108. Lockhart, J. A. 1959. *Plant Physiol.* 34:457–60
109. Ibid 1960. 35:129–35
110. Loveys, B. R., Wareing, P. F. 1971. *Planta* 98:109–16
111. Ibid, 117–27
112. Lozhnikova, V. N. 1967. *Dokl. Bot. Sci.* 168:92–94
113. Macleod, A. M., Palmer, G. H. 1966. *J. Inst. Brew.* 72:580–89
114. Macleod, A. M., Palmer, G. H. 1967. *Nature* 216:1342–43
115. Macleod, A. M., Palmer, G. H. 1969. *New Phytol.* 68:295–304
116. DeMaggio, A. E. 1966. *Science* 152: 370–72
117. McComb, A. J., McComb, J. A. 1970. *Planta* 91:235–45
118. McComb, A. J., McComb, J. A., Duda, C. T. 1970. *Plant Physiol.* 46: 221–23
119. Melcher, U., Varner, J. E. 1971. *J. Inst. Brew.* 77:456–61
120. Musgrave, A., Kays, S. E., Kende, H. 1972. *Planta* 102:1–10
121. Nadeau, R. L., Rappaport, L., Stolp, C. F. 1972. *Planta* 107:315–24
122. Nagl, W. 1971. *Planta* 96:145–51

123. Naylor, J. M. 1966. *Can. J. Bot.* 44: 19–32
124. Naylor, J. M., Simpson, G. M. 1961. *Can. J. Bot.* 39:281–95
125. Nitsan, J., Lang, A. 1965. *Develop. Biol.* 12:358–76
126. Nitsan, J., Lang, A. 1966. *Plant Physiol.* 41:965–70
127. Nitsch, J. P. 1965. See Ref. 104, 1537–1647
128. Nitsch, J. P., Nitsch, C. 1959. *Bull. Soc. Fr. Physiol. Veg.* 5:20–23
129. Paleg, L. G. 1960. *Plant Physiol.* 35: 293–99
130. Ibid 1962. 37:798–803
131. Paleg, L. G. 1965. *Ann. Rev. Plant Physiol.* 16:291–322
132. Phillips, I. D. J. 1969. *Planta* 86: 315–23
133. Ibid 1972. 105:234–44
134. Ibid 1972. 106:363–67
135. Phinney, B. O. 1957. *Proc. Nat. Acad. Sci. USA* 42:185–89
136. Phinney, B. O. 1960. In *Plant Growth Regulation*, ed. R. M. Klein, 489–95. Iowa State Univ. Press, Ames
137. Pinfield, N. J. 1968. *Planta* 82:337–41
138. Pinfield, N. J. 1968. *J. Exp. Bot.* 19: 452–59
139. Pinfield, N. J., Stobart, A. K. 1969. *New Phytol.* 68:993–99
140. Poulson, R., Beevers, L. 1970. *Plant Physiol.* 46:509–14
141. Radley, M. 1958. *Ann. Bot. London* 22:297–307
142. Ibid 1963. 27:373–77
143. Radley, M. 1967. *Planta* 75:164–71
144. Ibid 1969. 86:218–23
145. Ibid 1970. 92:292–300
146. Reid, D. M., Clements, J. B., Carr, D. J. 1968. *Nature* 217:580–82
147. Rose, R. J., Adamson, D. 1969. *Planta* 88:274–81
148. Ross, J. D., Bradbeer, J. W. 1968. *Nature* 220:85–86
149. Ross, J. D., Bradbeer, J. W. 1971. *Planta* 100:288–302
150. Ibid, 303–8
151. Sachs, R. M. 1965. *Ann. Rev. Plant Physiol.* 16:73–96
152. Sachs, R. M., Bretz, C., Lang, A. 1959. *J. Exp. Cell Res.* 18:230
153. Scott, T. K., Case, D. B., Jacobs, W. P. 1967. *Plant Physiol.* 42:1329–33
154. Seitz, K., Lang, A. 1968. *Plant Physiol.* 43:1075–82
155. Sembdner, G., Weiland, J., Aurich, O., Schreiber, K. 1968. In *Plant Growth Regulators*, Monogr. 31 Soc. Chem. Ind., 70–86

598   JONES

156. Shein, T., Jackson, D. I. 1971. *Ann. Bot. London* 35:555–64
157. Shininger, R. L. 1971. *Plant Physiol.* 47:417–22
158. Schreiber, K., Weiland, J., Sembdner, G. 1970. *Phytochemistry* 9:189–98
159. Skytt-Anderson, A., Muir, R. M. 1969. *Physiol. Plant.* 22:354–63
160. Simmonds, J. A., Simpson, G. M. 1971. *Can. J. Bot.* 1833–40
161. Simpson, G. M. 1965. *Can. J. Bot.* 43:793–816
162. Ibid 1966. 44:115–16
163. Simpson, G. M., Naylor, J. M. 1962. *Can. J. Bot.* 40:1660–73
164. Splittstoesser, W. E. 1970. *Physiol. Plant.* 23:762–68
165. Stoddart, J. L. 1962. *Nature* 194: 1064–65
166. Stoddart, J. L. 1966. *J. Exp. Bot.* 17: 96–107
167. Suge, H. 1970. *Plant Cell Physiol.* 11:729–35
168. Suge, H., Murakami, Y. 1968. *Plant Cell Physiol.* 9:411–14
169. Suge, H., Rappaport, L. 1968. *Plant Physiol.* 43:1208–14
170. Taiz, L., Jones, R. L. 1970. *Planta* 92:73–84
171. Tata, J. R. 1968. *Nature* 219:331–37
172. Thompson, W. F., Cleland, R. E. 1972. *Plant Physiol.* 50:289–92
173. Varner, J. E. 1964. *Plant Physiol.* 39: 413–15
174. Varner, J. E., Mense, R. 1972. *Plant Physiol.* 49:187–89

175. Wareing, P. F. 1958. *Nature* 181: 1744–45
176. Wareing, P. F. 1969. *Symp. Soc. Exp. Biol.* 23:241–62
177. Wareing, P. F., Hanney, C. E. A., Digby, J. 1964. in *The Formation of Wood in Forest Trees*, ed. M. H. Zimmermann, 323–44. New York: Academic
178. Wareing, P. F., Saunders, P. F. 1971. *Ann. Rev. Plant Physiol.* 22:261–88
179. Weinberg, E. S., Voeller, B. R. 1969. *Proc. Nat. Acad. Sci. USA* 64:835–42
180. Wellensiek, S. J. 1967. *Z. Pflanzenphysiol.* 56:33–39
181. Wilkins, M. B. 1966. *Ann. Rev. Plant Physiol.* 17:379–408
182. Wright, S. T. C. 1961. *Nature* 190: 699–700
183. Wright, S. T. C. 1966. *J. Exp. Bot.* 17:165–76
184. Wylie, A., Ryugo, K. 1971. *Plant Physiol.* 48:91–93
185. Yoda, S., Ashida, J. 1960. *Plant Cell Physiol.* 1:99–105
186. Yomo, H. 1960. *Hakko Kyokaishi* 18:600–2
187. Yomo, H., Varner, J. E. 1971. In *Current Topics in Developmental Biology*, ed. A. A. Moscona, A. Monray, 6:111–14. New York: Academic
188. Zeevaart, J. A. D. 1968. See Ref. 77, 1357–70
189. Zeevaart, J. A. D. 1969. *Planta* 84: 339–47
190. Zeevaart, J. A. D. 1971. *Plant Physiol.* 47:821–27

REPRINTS

The conspicuous number aligned in the margin with the title of each article in this volume is a key for use in ordering reprints.

Available reprints are priced at the uniform rate of $1 each postpaid. Payment must accompany orders less than $10. A discount of 20% will be given on orders of 20 or more. For orders of 100 or more, any *Annual Reviews* article will be specially printed and shipped within 6 weeks.

The sale of reprints of articles published in the *Reviews* has been expanded in the belief that reprints as individual copies, as sets covering stated topics, and in quantity for classroom use will have a special appeal to students and teachers.

AUTHOR INDEX

SUBJECT INDEX

A

Absicisic acid (ABA)
  and cytokinin, 420-21
  on gibberellin conjugation to
    sugar, 574
  growth inhibition by, 376-
    77
  mutations affecting tomato
    wilt, 510-11
  in regulation of ripening,
    210-11
  in water stress, 527, 542-
    43
Abscission
  of cotton leaves in water
    stress, 543
  indoleacetic acid oxidase on,
    364
Absorption
  of salt and water, 8
  of water during transpiration,
    20
Acetabularia
  chloroplasts of, 56
Acetabularia mediterrania
  electrogenic pump in, 34
Acetate
  in fatty acid synthesis, 293,
    303
  photoassimilation of, 92
Acetic acid esters
  of hypobaric storage,
    453
Acetyl-CoA carboxylase
  cellular function of, 289
  isolation of, 289
  localization of, 289
Acid phosphatase
  activity in ripening, 205-6
  isozymes of, 503
Acid proteinase
  properties of, 179-80
Acyl carrier protein
  on fatty acid synthetase,
    289-90
  properties of, 289
Acyl-CoA desaturase
  properties of, 291
Acyl transferase
  activity in microsomes,
    291
Adenosine 3'-phosphate 5'-
    phosphosulfate (PAPS)
  occurrence of, 390
  reductase system, 394
  as sulfate donor, 389, 393
  synthesis of, 385, 397
Adenosine 5'-phosphosulfate
  (APS)

conversion to adenosine 3'-
    phosphate 5'-phosphosul-
    fate, 397
  in dissimilatory sulfate
    reduction, 408-9
  reductase complex, 394
  as sulfate donor, 393
  synthesis of, 384-85, 397
  transferase, 394, 397
Adenosine triphosphatase
    (ATPase)
  enhancement of, 39
  in ion transport, 34
  in plasma membrane, 43
  and sodium pump, 33
Adenosine triphosphate (ATP)
  in active ion transport,
    43
  in amino acid incorporation,
    61-63
  in chemiosmotic theory,
    39
  concentration
    in cytoplasm, 246-47
    in phloem, 231
    in relation to photophos-
      phorylation, 108
  in electron transport, 104
  energy source for ion trans-
    port, 38
  in phosphate transport,
    233
  in photophosphorylation, 91-
    96, 98, 105-8
  production in C_4 plants,
    276-78
  on protein synthesis initia-
    tion, 57-58
  protochlorophyll synthesis,
    148
  synthesis of, 33
  transport as charged sub-
    stance, 26
  of water stress, 532
S-Adenosylmethionine-magne-
    sium protoporphyrin
    methyl transferase
  assay for, 147
  function of, 146-47
Adenylate system
  function in photosynthesis,
    105-7
Adiantum
  phytochrome action in devel-
    opment of, 122
i^6Ado
  in cytokinin activity, 430-
    33
  inhibition of tRNA methyla-
    tion, 439

metabolism of, 434-35
  source of, 436, 439
Adsorption
  of pesticides by soil,
    282
Aeration
  in water absorption, 20
Aesculus
  aminoacyl tRNA synthetase
    of, 55
Aging genes
  absence of, 198
Aging of leaves
  and photosynthetic rates,
    108
Agrobacterium tumefaciens
  crown gall disease, 355-
    56
  cytokinins and, 367-68
  gibberellins and, 374
  indoleacetic acid oxidase
    and, 362
Air environments
  pesticides in, 485-86
Alanine
  conversion to pyruvate in
    spore germination, 318
Alcohol
  in oxime pathway, 79-81
Alcohol dehydrogenase
  in genetic studies, 501-2
Aldoxime
  to amides and non-nitrogen-
    ous compounds, 79-81
  of amino acids to, 72-75
  to glucosinolates, 77-79
  growth regulation by, 70
  to nitriles, cyanogenic gly-
    cosides, and hydrocyanic
    acid, 75-77
  properties of, 69-88
Aleurone grains
  in protein synthesis,
    584
  proteolytic enzymes in, 175,
    176
Aleurone layer
  of gibberellin in, 583
Algae
  action potentials in, 40
  potassium transport in,
    32
Allium cepa
  sodium pump in, 33
Alpine plants
  respiration in, 282
Amino acid
  incorporating systems of
    general, 61-64
  localization of, 63-64

of cytokinins, 415-44
Moss
  gametophytes of, 115-28
  photoassimilation in, 111
  water stress on photosynthesis in, 534
Muramic acid
  presence in spore walls, 314
Mutation
  induction of, 494-95
  rate in microbial spores, 397
Mycorrhizae
  ectotrophic, 233, 235
  endotrophic, 233, 235
  on growth, 233
  on phosphate
    content, 233
    movement to roots, 235
    uptake by plant, 234
  in plant competition, 234
Myrosinase
  in oxime metabolism, 75

N

Nasturtium officinale
  formation of secondary plant products in, 70, 74, 78
Nematode
  effect of hormones on, 368
Nepenthes
  endopeptidase in, 185
Nernst equation
  in diffusion potentials, 28-31
Neurospora crassa
  electrogenic pumps in, 35, 38
  potassium transport in, 29-30
  protein synthesis in, 57, 64
Nicotiana tabacum
  anther cultures of, 495
Nicotine adenine dinucleotide (NAD)
  cellular concentration of, 247
  membrane penetration by, 243
  and protochlorophyll synthesis, 148
Nicotine adenine dinucleotide, reduced (NADH)
  role in sulfate reduction, 404
Nicotine adenine dinucleotide phosphate (NADP)
  cellular concentration of, 247
  in chlorophyll synthesis, 163
  with malic dehydrogenase,

268, 272-73
  membrane penetration by, 243
  water stress on and reduction of, 532
Nicotine adenine dinucleotide phosphate, reduced (NADPH)
  oxidase of
    in fatty acid synthesis, 290
  production in C_4 plants, 275-78
  in sulfate reduction, 404
Nitella
  action potentials in, 40
  cell appearance in, 36
  cytoplasmic phosphate ester pool in, 241
  cytoplasmic phosphate labeling patterns in, 242
  electropotential measurements in, 27-40
  hydrogen ion efflux pump in, 35
  phosphate labeling patterns in, 242
  phosphate uptake in, 96
  photoassimilation in, 94
  plasmodesmata of, 41
  potassium transport in, 29
Nitella clavata
  energy source of, 39
Nitella flexilis
  cell depolarization by light, 38
Nitella translucens
  anions in, 34
  chloride uptake in, 39
  electrogenic hydrogen ion pump, 36-37
  ouabain on, 33
  resting potential and light, 37
Nitrate reductase
  isoenyzmes of, 504
  water stress on, 549
Nitrilase
  occurrence and activity of, 80
Nitriles
  in oxime pathway, 71, 75-77
Nitrogen
  oximes in metabolism of, 69-88
  on photophosphorylation, 100-8
Nitrogen fixation
  in blue-green algae, 100-1
  oximes in, 69
  water stress on, 548
Nongreen cells
  carbon dioxide fixation in, 273
Nonpolar organic substance

transport of, 26
"Normalizing factor"
  as biochemical correlate of gene, 514
Nucleic acids
  content and base composition of
    in mosses, 116
  metabolism in bryophyte development, 124
  water stress on, 548
Nucleoli
  effect of gibberellin on, 593
Nucleoside diphosphate sugars
  in starch synthesis, 506
Nutrition
  plant
    early contributions to understanding of, 4
  mineral
    early contributions to understanding of, 4, 8

O

Oats
  anion accumulation in seedlings of, 33-34
  calcium on coleoptiles of, 34
  electrogenic pumps, 35
  in etiolated tissue, 38
  indole-3-acetic acid upon coleoptiles of, 42
  plasmodesmata of, 41
  red-far red and electrical potential, 242
  sodium efflux pump in, 32
  see also Avena
Ochromonas danica
  protein synthesis in, 63
Oleic acid
  labeling pattern, 291
  NADH and NADPH in incubation medium of, 292
  as precursor of linolenic acid, 293
Olive knot
  indoleacetic acid in, 357-58
Onoclea sensibilis
  atypical strain of, 123
  gametophytes of, 119-20
Organic acids
  as charged substance in transport, 26
  in photoassimilation, 92
Osmosis
  theory and pressure gradients in, 4
  in water absorption, 17
Ouabain
  and sodium pump, 33
Oxaloacetic acid
  in C_4 cycle, 269
α-Oxidation of fatty acids,

CUMULATIVE INDEXES

CONTRIBUTING AUTHORS, VOLUMES 15-24

A

Abeles, F. B., 23:259
Åberg, B., 16:53
Addicott, F. T., 20:139
Agurell, S., 15:143
Aleem, M. I. H., 21:67
Allsopp, A., 15:225
Altschul, A. M., 17:113
Amesz, J., 20:305
Anderson, D. G., 18:197
Anderson, L., 17:209
Anderson, W. P., 23:51
Armstrong, D. J., 21:359
Avron, M., 19:137

B

Baker, J. E., 16:343
Barber, D. A., 19:71
Barrs, H. D., 22:223
Bartnicki-Garcia, S., 15:327
Bassham, J. A., 15:101
Beevers, L., 20:495
Bendall, D. S., 19:167
Benson, A. A., 15:1
Bergersen, F. J., 22:121
Bidwell, R. G. S., 21:43
Bieleski, R. L., 24:225
Birch, A. J., 19:321
Birnstiel, M., 18:25
Bishop, N. I., 17:185
Black, C. C. Jr., 16:155; 24:253
Boardman, N. K., 21:115
Bollard, E. G., 17:77
Bond, G., 18:107
Bonner, W. D. Jr., 19:295
Bopp, M., 19:361
Boulter, D., 21:91
Boyer, J. S., 20:351
Brandes, H., 24:115
Branton, D., 20:209
Brouwer, R., 16:241
Brown, J. S., 23:73
Brown, S. A., 17:223
Burr, B., 24:493
Burris, R. H., 17:155
Butler, G. W., 17:77
Butler, W. L., 15:451; 16:383

C

Canny, M. J., 22:237

Carlile, M. J., 16:175
Carns, H. R., 17:295
Casida, J. E., 20:607
Castelfranco, P. A., 24:129
Cathey, H. M., 15:271
Chailakhyan, M. Kh., 19:1
Chance, B., 19:295
Cheniae, G. M., 21:467
Cleland, R., 22:197
Clements, H. F., 15:409
Cocking, E. C., 23:29
Cook, G. M. W., 22:97
Crane, J. C., 15:303
Criddle, R. S., 20:239
Crosby, D. G., 24:467
Cumming, B. G., 19:381

D

Douthit, H. A., 24:311
Dugger, W. M., 21:215

E

Engleman, E. M., 17:113
Epstein, E., 15:169
Eschrich, W., 21:193
Evans, H. J., 17:47
Evans, L. T., 22:365

F

Fork, D. C., 20:305
Forsyth, W. G. C., 15:443
Fowden, L., 18:85
Franke, W., 18:281
Fredga, A., 16:53

G

Gaffron, H., 20:1
Gale, J., 17:269
Galston, A. W., 19:417
Gardner, W. R., 16:323
Gates, D. M., 19:211
Glasziou, K. T., 20:63
Goedheer, J. C., 23:87
Goeschl, J. D., 20:541
Goldsmith, M. H. M., 19:347
Golueke, C. G., 15:387
Green, D. E., 18:147

Green, P. B., 20:365
Gunning, B. E. S., 23:173

H

Haber, A. H., 19:463
Hagan, R. M., 17:269
Hageman, R. H., 20:495
Hall, A. E., 22:431
Hall, R. H., 24:415
Halperin, W., 20:395
Hampton, R. O., 23:389
Hansen, E., 17:459
Hanson, K. R., 23:335
Haselkorn, R., 17:137
Hassid, W. Z., 18:253
Hatch, M. D., 21:141
Haupt, W., 16:267
Henckel, P. A., 15:363
Hendricks, S. B., 21:1
Heslop-Harrison, J., 18:325
Hiesey, W. M., 16:203
Higinbotham, N., 24:25
Hill, R., 19:167
Hillman, W. S., 18:301
Hind, G., 19:249
Hodson, R. C., 24:381
Hsiao, T. C., 24:519
Hutner, S. H., 15:37

I

Ikuma, H., 23:419

J

Jackson, W. A., 21:385
Jaffe, M. J., 19:417
Jefferies, R. L., 15:169
Jones, B. L., 24:47
Jones, R., 24:571

K

Kadyrov, Ch. Sh., 22:185
Kawashima, N., 21:325
Kefeli, V., 22:185
Kessler, E., 15:57
Key, J. L., 20:449
Kirk, J. T. O., 21:11
Kirkwood, S., 16:393
Kolattukudy, P. E., 21:163
Kosuge, T., 21:433
Kramer, P. J., 24:1
Kretovich, W. L., 16:141

643